UNIVERSE

UNIVERSE

FOURTH EDITION

William J. Kaufmann III

W. H. Freeman and Company
New York

To Connie and Michael, with love

Cover image: The greenish nebula on the front cover, an interstellar cloud located about 1300 light-years from Earth, seems about to devour a spiral galaxy (seen edge-on). Radiation from the stars is eroding and elongating the nebula by blowing its gas and dust. Because of its shape, the nebula is called a *cometary globule,* but it is no comet! Its "head" alone measures about $1\frac{1}{2}$ light-years across.

Cometary globules are normally dim—because they shine primarily by reflecting light from nearby stars—and most are bluish. This nebula, however, has a muddy greenish hue because the reflected starlight reaches us only after it has filtered through a dusty cloud. The fine dust scatters and loses blue light, which has short wavelengths, more efficiently than longer wavelengths.

This remarkable photograph was taken by David Malin using the 3.9-meter Anglo-Australian Telescope. Many of his fine pictures grace the pages of this text.

Illustration credits are listed on pages 611–613.

Library of Congress Cataloging-in-Publication Data

Kaufmann, William J.
Universe / William J. Kaufmann III.—4th ed.
    p.   cm.
Includes bibliographical references and index
ISBN 0-7167-2379-4
1. Astronomy.   2. Cosmology.   I. Title.
QB43.2.K38   1994
520—dc20   93-27842

Printed in the United States of America

Fourth printing 1996, RRD

CONTENTS OVERVIEW

CONTENTS

STARS AND STELLAR EVOLUTION

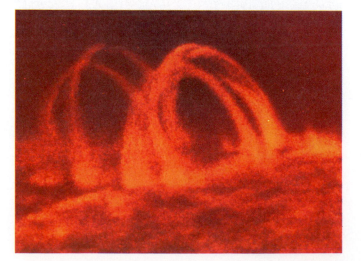

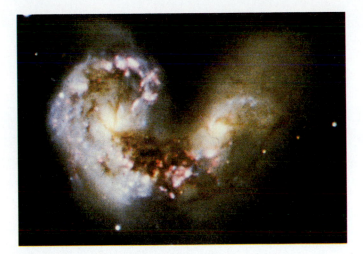

PREFACE

I wrote this book for the student about to explore the mysteries of the universe, from the Earth around us to the dim and distant reaches of the heavens. Armed with the powers of observation, the laws of physics, and the resourcefulness of the human mind, astronomers survey alien worlds and follow the life cycles of stars. With our growing understanding of the evolution of the universe, even space and time take on new meanings. From its first edition, *Universe* has therefore been designed to guide students, including students who may be less than at ease in mathematics and the physical sciences, faced with some of the most profound questions ever asked: from the creation of the cosmos to the formation of the planets and the motions of the stars.

I also wrote this book for the instructor who wants to inspire questions about the nature of any scientific inquiry—about how we ever come to know what we know. Astronomers today must investigate realms far removed from daily experience, including objects too vast, distant, or intangible ever to sample directly; many of the phenomena that we observe occurred very long ago. Showing how scientists reason is therefore central to this fourth edition.

Our developing picture of the universe guides the features of this book

Astronomy reveals the fascinating nature of our physical universe and how it is comprehended. This twin emphasis determines both the text's organization and each of its features:

- A chapter or part typically begins with questions and observations, **showing how scientists discover new physical principles.**
- A **modular part structure** allows instructors to probe these questions in the order and depth desired.
- **Optional boxes** further develop the underlying reasoning, with many *new* worked examples as a guide to the calculations.
- The **many color illustrations**—from *newly obtained* photographs to *new* supercomputer simulations—are carefully coordinated with the text to make the

observations essential to astronomy vivid and easy to understand.

- **Frequent highlighting** and an *added* range of end-of-chapter questions reinforce important principles.
- **Essays by six leading astronomers**—two *new* to this edition—let the reader share in the excitement of scientific discovery.
- Recent discoveries are interwoven throughout, bringing our picture of the universe *fully up to date.*
- In particular, this fourth edition has allowed *expanded* coverage of the stars and galaxies, while trimming some of the details of planetary astronomy.
- I have also *revised* overly intricate discussions so that **complex topics are more easily covered** or omitted at the discretion of each instructor.

As we shall now see in more detail, each of these features reflects the text's consistent themes and pedagogical emphasis. Let me begin with those features that build most closely on the strengths of previous editions.

Chapters are organized to emphasize scientific discovery

The text's traditional Earth-outward organization stresses how our understanding of the universe has developed. By moving outward from the planets to the stars and galaxies, we begin with celestial objects first observed by ancient astronomers using visible light.

The first six chapters introduce the foundations of astronomy, including naked-eye observations of eclipses and planetary motions and such basic tools as Kepler's laws, the properties of light, and the design of telescopes. Discussion of the formation of the solar system in Chapter 7 prepares the reader for the next ten chapters, which cover the solar system in outward order from the Sun.

Chapter 18 introduces stellar astronomy with the star that is most familiar to us: the Sun. The properties of stars are further described in Chapter 19, after which stellar evolution is described chronologically—from birth to death. Molecular clouds, star clusters, nebulae, neutron stars, and black holes are then presented in a sequence that mirrors the natural life of a star. These chapters thus provide a unified description of the wide variety of objects that astronomers find scattered about the heavens.

A survey of the Milky Way in Chapter 25 is followed by two chapters on galaxies and quasars. The final chapters, on cosmology, further develop our understanding of the universe and its history. All of the "standard" cosmological issues are discussed in Chapter 28, and it would be quite appropriate to end an astronomy course with that chapter. Chapter 29 was created for those who wish to explore topics in which research is today proceeding at a fevered pitch. The student will learn that physicists are at the brink of tackling questions like: Why is there only one dimension of time yet three dimensions of space?

However, *Universe* does not rely solely on its Earth-outward sequence to show how our knowledge of the universe has grown. It is designed throughout to begin with historic questions and exciting observations. These observations, in turn, raise new questions that draw the reader on to the outer limits of our universe and our understanding.

Related chapters are grouped together for greater flexibility

Because this book is designed for both one- and two-term courses, it is more comprehensive than its sister text, *Discovering the Universe*. To make that comprehensive coverage accessible to a wide range of students, *Universe* has a flexible, modular structure that permits instructors to teach the topics in the order desired.

The book's 29 chapters are divided into two nearly equal parts, the first half dealing with introductory material and planetary astronomy, and the second half treating stars, galaxies, and cosmology. Instructors may emphasize either half of the book according to preference, covering more or less of the detail as time and the preparation of the students permit. For instance, a course on stellar astronomy could begin with the introductory chapters on gravitation and light (Chapters 4 and 5) and proceed directly to Chapter 18, which introduces the Sun and stars, without any loss of continuity.

Optional boxes with worked examples permit varied depth of coverage

Universe necessarily contains a little more mathematics than *Discovering the Universe*, but boxed inserts set aside much of the technical material. These boxes help instructors tailor their course to an appropriate depth of coverage and level of difficulty, and they permit students to locate and review topics more easily. All boxes may simply be omitted without loss of continuity.

Many boxes review the formulas that led to important discoveries. Others contain information, such as orbital and physical data for each planet, that will be useful for working the exercises. Still others take an idea raised in the text a step further.

For this edition, boxes that contain useful calculations always conclude with a highlighted worked example. The examples make the formulas easier to understand and better prepare readers for the end-of-chapter questions. The bright, new design of this edition also helps students to distinguish and to locate more easily those of the boxes that contain reference data useful in working questions.

Color illustrations are an essential part of this book

One glance at a color photograph of a planet's cloudtops or of the glowing gases of a nebula reveals significant details that cannot be gleaned from a black-and-white view. X-ray, infrared, and radio views of the sky can now be displayed comprehensibly in computer-generated false-color images. *Universe* was therefore the first to introduce full color to an astronomy text, and its selection of illustrations has been substantially updated to reflect how astronomy is practiced today.

Photographs new to this edition include a high-resolution map of Venus in Chapter 11, based on altimeter measurements. The next several chapters benefit from Hubble Space Telescope images, while Chapter 21 includes new images of 47 Tuscanae from the Anglo-Australian Telescope and of Betelgeuse from the U.S. National Optical Astronomy Observatories. Sally Heap's remarkable Hubble Space Telescope image of the central star of NGC 2440 has been added to Chapter 22, and Tony Tyson's stunning image of gravitational lensing in Abell 2218 elucidates discussion of dark matter in Chapter 25.

Artists' renderings within the text have been widely praised for their pedagogical effectiveness. This edition has corrected and simplified the labeling of many illustrations while adding new ones. For example, Chapter 18 now illustrates temperature changes and convection patterns within the Sun's layers. Particularly exciting are new supercomputer simulations of events that no Earth-based observer could ever have experienced. For example, Chapter 10 adds a colorful simulation of the formation of Mercury, and Chapter 22 vividly depicts instabilities that develop immediately after core bounce during the onset of a supernova explosion.

Chapters repeatedly summarize and review the main ideas

Part of helping students gain an appreciation for astronomy is guiding them through the most important ideas. Each chapter begins with a brief, one-paragraph abstract that gives the reader a clear idea of the chapter's contents.

The chapter headings, in the form of declarative sentences, then highlight main concepts. Read consecutively, these headings also make it easier to review the chapter.

Important terms are set in bold type where they are first defined, and they are defined again in the glossary at the back of the book, which also includes useful page references to the text. Each chapter concludes with both a list of these key terms and a summary of the essential facts in outline form.

End-of-chapter questions begin with a short summary of the relevant formulas; these "Tips and Tools" give students just enough of a hint to get them started. An annotated list of further readings rounds out each chapter's presentation.

End-of-chapter questions have been significantly enlarged

For this edition, the sets of questions have been rearranged and greatly enlarged. They are now grouped more simply into review, advanced, and discussion questions for ease of use. Answers to all questions that require computation appear at the end of the book, while the *Instructor's Manual* has step-by-step solutions that can be posted at the instructor's discretion.

In response to instructors' requests, greater care has been taken to include questions whose answers require reasoning rather than just memorization. Hence the range of questions is considerably expanded. To enhance student interest in astronomy, observational activities are also included as in previous editions, and attendant star charts are located at the back of the book.

Two additional essays introduce the personal views of leading astronomers

It is again a great pleasure to include the thoughts of renowned astronomers on topics of current interest. Six essays give this edition a depth and quality that could not have otherwise been possible:

"Why Astronomy?" (following Chapter 1), Sandra M. Faber, University of California, Santa Cruz, and Lick Observatory

"Astrology and Astronomy" (following Chapter 3), Owen Gingerich, Smithsonian Astrophysical Observatory and Harvard University

"Exploring the Planets" (following Chapter 12), Marcia Neugebauer, Jet Propulsion Laboratory

"The Solar Corona" (following Chapter 18), Arthur B. C. Walker, Jr., Stanford University

"The Great Attractor" (following Chapter 26), Alan Dressler, Carnegie Institution of Washington

"The Edge of Spacetime" (following Chapter 28), Stephen W. Hawking, University of Cambridge

I am particularly grateful to Dr. Gingerich, who has revised his essay so as to develop still further the historical background to modern science, and to Dr. Neugebauer and Dr. Walker, whose essays are *new* to this edition and greatly enhance its coverage.

This edition has an increased emphasis on stellar and galactic astronomy

The good judgment and extensive classroom experience of numerous text users, reviewers, and colleagues have inspired me to expand the treatment of stellar and galactic astronomy at the expense of planetary astronomy in this edition. Some consolidation in the planetary chapters involved thorough reorganization with the student in mind. For instance, Chapter 7 has been streamlined and reorganized, with the addition of two new sections, to make this overview of the solar system easier to follow, while the Moon's history, previously scattered throughout Chapter 9, has now been collected in one location. Among many other examples, in Chapter 16 the discussion of the interiors of Uranus and Neptune has been combined, providing a better lead-in to coverage of their magnetic fields.

Beginning with the Sun in Chapter 18, I have added significant material. For example, more details of the proton–proton chain have been spelled out in a box accompanying Chapter 18, and Chapter 19, on the nature of stars, has a much improved discussion of the distance modulus and the luminosity function near the Sun. Chapter 21 has been beefed up to include overcontact systems and long-period (Mira-type) variables. Chapter 22 now describes dredge-up in red giants and AGB stars; also included are a brief discussion of carbon stars and a characterization of Type Ia, Type Ib, and Type II supernovae.

Chapter 25 on the Milky Way now includes high-velocity stars, along with a box on the Local Bubble and the Geminga Pulsar. New topics in Chapter 27 include a discussion of nonthermal emission, superluminal motion, and the Eddington limit.

Coverage has been brought thoroughly up to date

The extensive revision of the sections on both planetary and stellar astronomy has given me the chance to update this edition to within only weeks of publication—right through the summer of 1993. Some of the additions are reflected in the dramatic new color images already mentioned. Among many other important changes, Chapter 8 on the Earth now includes discussion of repeated cycles of supercontinent formation and breakup. The latest ideas about Mercury's iron core have been added to Chapter 10, and Chapter 11 is much enhanced by the recent discoveries about Venus from *Magellan*.

The latest word about such planned and ongoing missions as *Cassini* and the ill-fated *Mars Observer* has been included, and in Chapter 18 the discussion of solar neutrinos has been updated to include latest results from GALLEX and SAGE. The discussion of Sagittarius A* in Chapter 25 has also been updated; the entire section on galaxy evolution has been rewritten and is now based primarily on Hubble Space Telescope observations by Alan Dressler and others. Finally, the Afterword on the search for extraterrestrial intelligence now includes a discussion of NASA's Microwave Observing Project, as well as ongoing projects at Ohio State, Berkeley, and Harvard and other recent proposals.

This edition also makes many complex topics more manageable

An important result of this revision has been to make difficult discussions more accessible to students. I also took this opportunity to restructure several chapters so that they begin with basic observations; speculative topics are left to the end of the chapter, where they may be more easily omitted. I have in fact rechecked every section for clarity at the sentence level, eliminated several more challenging boxes, and moved some complex text discussions to new boxes, where they can be better supported by worked examples or simply omitted.

The more flexible coverage will be evident, for example, in Chapter 5, which now leaves many details of spectral lines and energy-level diagrams to a box, so as to make the discussion of the Bohr model clearer. Other boxes in that chapter include added worked examples of Wien's law, the Stefan–Boltzmann law, and Doppler shifts. A new box takes up details of the solar spectra in Chapter 19, a box in Chapter 25 applies Kepler's laws to the Sun's orbit in our Galaxy, while a box in Chapter 27 compares the redshift and cosmological parameters.

Some of the cuts, particularly in planetary astronomy, have already been mentioned, but encyclopedic aspects of the previous edition have generally been curtailed. For example, a box listing *all* the successful missions to the Moon has been deleted, although a short section on manned lunar exploration remains.

Chapter 18 has been turned inside-out, to begin with basic observation of the Sun's atmosphere rather than with the physics of the solar interior. Also at the request of many instructors, I have clarified the model of spiral-arm formation in Chapter 25.

I have completely rewritten Chapter 27 on quasars and active galaxies to follow a historical approach, from the discovery of high redshifts to the "unified model," to show more clearly how science is done. Finally, Chapter 29 now places highly speculative issues (GUTs, TOEs, and Kaluza–Klein theories) only in the second half of the chapter, and Feynman diagrams no longer appear.

Supplementary materials include a greater variety of still and moving images

The new **Instructor's Manual and Resource Guide**, prepared by George A. Carlson of Citrus College, contains chapter synopses and outlines, hints for teaching and discussion, a current list of resources for teaching that includes audiovisual materials as well as readings, and fully worked out solutions to the computational problems in the book.

Both **Overhead Transparencies** and **Slides** containing a selection of 100 color drawings from the text are also available. A **Computerized Test Bank**, available in Macintosh and IBM formats, has been revised and updated by T. Alan Clark and William J. F. Wilson of the University of Calgary to include some 2000 multiple-choice questions, indexed by chapter and topic. Newly included are chapter outlines, which are convenient for class distribution or as a basis for lectures. A printed test bank is also available.

New for this edition are **Active Sun Videos** produced by Lockheed Research Laboratory. These self-contained videos highlight recent advances in high-resolution solar astronomy that are difficult to convey in a textbook. The series features recent discoveries about the photosphere, sunspots, and the corona.

In addition, two 45-minute *chapter-specific* Video Lectures are available. These tapes of my lectures were made during a meeting of the Astronomical Society of the Pacific at the University of Wisconsin. *Black Holes and Warped Spacetime* covers most of Chapter 24. *Cosmology and the Creation of the Universe* is designed to be an entertaining introduction to Chapter 28.

Also available for the first time is *Gems of Hubble,* an **Electronic PictureBook** of images recently obtained by the Hubble Space Telescope. These exceptional images, each with a detailed caption, are part of a HyperCard stack and so may be quickly selected and presented using a Macintosh computer.

For more information, please contact your W. H. Freeman representative.

Acknowledgments

I would like to begin by thanking the many instructors who responded to questionnaires about previous editions of *Universe* and especially the dozens of people who sent in unsolicited detailed comments and suggestions. I also wish to thank the many people whose advice on the first and second editions has had an ongoing influence:

Robert Allen, University of Wisconsin, La Crosse
Alice L. Argon, Harvard–Smithsonian Center for Astrophysics
David Van Blerkom, University of Massachusetts
John M. Burns, Mt. San Antonio College
Bruce W. Carney, University of North Carolina
Bradley W. Carroll, Weber State University
Roger B. Culver, Colorado State University
James N. Douglas, University of Texas at Austin
David S. Evans, University of Texas at Austin
George W. Ficken, Jr., Cleveland State University
Andrew Fraknoi, Astronomical Society of the Pacific
Owen Gingerich, Harvard University
Paul F. Goldsmith, University of Massachusetts
J. Richard Gott III, Princeton University
Austin F. Gulliver, Brandon University
Bruce Hanna, Old Dominion University
Paul Hodge, University of Washington
Douglas P. Hube, University of Alberta
Icko Iben, Jr., Pennsylvania State University
John K. Lawrence, California State University, Northridge
Laurence A. Marschall, Gettysburg College
Dimitri Mihalas, University of Illinois
L. D. Opplinger, Western Michigan University
John R. Percy, University of Toronto
Terry Retting, University of Notre Dame
Kenneth S. Rumstay, Valdosta State College
Richard Saenz, California Polytechnic State University
Thomas F. Scanlon, Grossmont College
Richard L. Sears, University of Michigan
David B. Slavsky, Loyola University of Chicago
Joseph S. Tenn, Sonoma State University
Gordon B. Thomson, Rutgers University
Virginia Trimble, University of California, Irvine
Bruce A. Twarog, University of Kansas
Donat G. Wentzel, University of Maryland
Nicholas Wheeler, Reed College
Raymond E. White, University of Arizona

Foremost among those deserving of thanks are eight instructors who scrutinized the third edition. Their many specific suggestions profoundly affected the writing of the fourth edition:

Robert R. J. Antonucci, University of California, Santa Barbara
Robert J. Dukes, Jr., College of Charleston
James L. Regas, California State University, Chico
Tina Riedinger, University of Tennessee
James A. Roberts, University of North Texas

Isaac Shlosman, University of Kentucky
Michael L. Sitko, University of Cincinnati
Charles R. Tolbert, University of Virginia

I am especially grateful to Dr. Antonucci, whose comments inspired me to recast Chapter 27. Carlton Pryor, Rutgers University, also kindly discussed his experience with the previous edition.

I am also deeply grateful to those who carefully reviewed manuscript for the third edition:

David Burstein, Arizona State University
Gordon M. MacAlpine, University of Michigan
John Mathis, University of Wisconsin
C. R. O'Dell, Rice University
James A. Rose, University of North Carolina
Richard L. Sears, University of Michigan
George Wolf, Southwest Missouri State University
Don York, University of Chicago

The following individuals were most helpful with my requests for photographs:

David F. Malin, Anglo-Australian Observatory
Rudolph E. Schild, Center for Astrophysics
Richard Schmidt, U.S. Naval Observatory
James D. Wray, McDonald Observatory

Many others have participated in the preparation of this book, and I thank them for their efforts. Foremost among them are John Haber, development editor, and Jerry Lyons, publisher, who provided support and encouragement. It was a pleasure to work with Georgia Lee Hadler, my project editor; Alice Fernandes-Brown, who created the design; Christine McAuliffe, illustration coordinator; José Fonfrias, page makeup artist; and Sheila Anderson, production coordinator. They all deserve special thanks for their unfailing concern for quality. I also acknowledge Paul Monsour, whose copyediting significantly improved the manuscript; Patrick Shriner, who oversaw the supplemental materials; the people at Vantage Art Studio, who drafted many of the illustrations; and Tomo Narashima and George Kelvin, whose marvelous airbrush artistry makes the drawings jump off the pages of the book.

Although we have made a valiant effort to make this an error-free edition, some mistakes may have crept in. I would appreciate hearing from anyone who finds an error or who wishes to comment on the text. You may write to me in care of the publisher. I will personally respond to all correspondence.

William J. Kaufmann III
Department of Physics
San Diego State University

ASTRONOMY AND THE UNIVERSE

STARS AND INTERSTELLAR GAS New stars are forming in the clouds of interstellar gas and dust shown in this photograph. The gases glow because of the radiation emitted by newborn stars. Clouds of interstellar dust block light; they appear as dark regions silhouetted against glowing background gas. The Horsehead Nebula is in the upper left; the Orion Nebula is toward the lower right. Both nebulae are about 1500 light-years from Earth. (Royal Observatory, Edinburgh)

ASTRONOMY, the study of the universe, addresses profound questions that people have pondered since the dawn of civilization. This chapter introduces the scope of astronomy by previewing the contents of the book. We see that by exploring the planets, astronomers learn about the formation and evolution of the solar system. By examining stars and nebulae, they fathom the life cycles of stars. By observing galaxies, they uncover important clues about the creation of the universe. Such delvings into the mysteries of the universe demonstrate that the physical world obeys certain rules called the laws of physics. To appreciate the nature of the universe more fully, we equip ourselves with mathematical tools, like powers-of-ten notation and units of measure. In addition to the deep satisfaction each new discovery brings, astronomy gives us a sweeping perspective from which we can appreciate our existence on Earth.

The splendor of the star-filled night sky is one of the central experiences of life. Gazing out into the heavens, we see thousands of stars scattered from horizon to horizon. The delicate mist of the Milky Way traces a faerie path across the starry firmament, and the entire spectacle swings slowly overhead from east to west as the night progresses. No light show, no artist's brush, no poet's words can truly capture the beauty of this breathtaking panorama (Figure 1-1).

For thousands of years people have looked up at the heavens and contemplated the universe. Like our ancestors, we find our thoughts turning to profound questions as we gaze at the stars. How was the universe created? Where did the Earth, Moon, and Sun come from? What are the planets and stars made of? And how do we fit in? What is our place in the cosmic scope of space and time?

To wonder about the universe is a particularly human endeavor. Our curiosity, our desire to explore and discover, is a quality that distinguishes us from other animals. The study of the stars transcends all boundaries of culture, geography, and politics. In a literal sense, astronomy is a universal subject—its subject is the entire universe.

1-1 Astronomers use the laws of physics and construct testable theories and models to understand the universe

Astronomy has a rich heritage that dates back to the myths and legends of antiquity. The heavens were thought to be populated with demons and heroes, gods and goddesses. Astronomical phenomena were explained as the result of supernatural forces and divine intervention.

The course of civilization was greatly affected by one realization: *The universe is comprehensible.* This first awareness is one of the great gifts to come to us from ancient

FIGURE 1-1 The Starry Sky The star-filled sky is a beautiful and inspiring sight. This photograph, taken from northern Mexico, shows Halley's Comet and a portion of the Milky Way (upper right). To get a good view of the heavens, you must be far from city lights. (Courtesy of D. L. Mammana)

Greece. Greek astronomers discovered that by observing the heavens and carefully thinking about what they saw, they could learn something about how the universe operates. For example, as we shall see in Chapter 3, they measured the size of the Earth and understood and predicted eclipses.

Greek astronomers used observation and logic to explore physical reality. Their approach evolved into the **scientific method,** which has become a dominant force in science and society over the past four hundred years. In essence, the scientific method requires that our ideas about the world around us be consistent with what we actually observe. Specifically, a scientist trying to understand some phenomenon proposes a **hypothesis,** which is a collection of ideas that seems to explain the phenomenon. The hypothesis cannot be in conflict with existing observations and experiments, because a discrepancy with known facts implies that the hypothesis is wrong. The scientist then considers how the implications of the hypothesis might lead to predictions that can be tested. Only after a hypothesis has accurately forecast the results of new experiments and observations does the scientist feel confident that the hypothesis is on firm ground.

Scientists describe reality in terms of **models,** which are hypotheses that have withstood observational or experimental tests. A model tells us about the properties and behavior of some object or phenomenon. A familiar example is a model of the atom, which scientists picture as electrons orbiting a central nucleus. Another example, which we shall encounter in Chapter 18, is a model of the Sun that tells us all about physical conditions (for example, temperature, pressure, density) throughout the Sun's interior.

A body of related hypotheses can be pieced together into a self-consistent description of nature called a **theory.** An example from Chapter 24 is Einstein's general theory of relativity, which describes the effect of gravity on space and time. Without models and theories there is no understanding and no science, only collections of facts.

The scientific method requires that the scientist be open-minded, willing to discard even the most cherished ideas if they fail to agree with observation and experiment. As new discoveries are made, old models and theories must be modified or discarded in favor of more comprehensive explanations of the world around us.

Theories that accurately describe the workings of physical reality can have a significant effect on civilization. For example, the seventeenth-century scientist Isaac Newton succeeded in describing how the planets orbit the Sun. The motions of the planets, unhampered by air resistance or friction, reveal some of the most fundamental laws of nature in their simplest form. From Newton's work we obtained our first complete, coherent description of the behavior of the physical universe. As we shall see in Chapter 4, the resulting body of knowledge, called **Newtonian mechanics,** explains the concepts of force, mass, acceleration, momentum, and energy. This understanding had immediate practical application in the construction of machines, buildings, and bridges. It is no coincidence that the Industrial Revolution followed hard on the heels of these theoretical and mathematical advances inspired by astronomy.

Astronomers use Newtonian mechanics, along with other physical principles (usually called the **laws of physics**), to interpret their observations and to understand the phenomena of the universe. Light and its relationship to matter are of particular importance. By studying how objects absorb and emit radiation, astronomers acquire the skills needed to analyze and interpret the wealth of information coming to us from the stars and galaxies. As a result, astronomers obtain fundamental information about these celestial objects as well as the evolution of the universe.

The laws of physics, particularly those pertaining to light and optics, can also be used to develop new tools for examining and exploring the universe. Until recently, everything we knew about the distant universe was based on visible light. Astronomers would peer through telescopes to observe and analyze visible starlight. By the end of the nineteenth century, however, scientists had begun discovering forms of light invisible to the human eye: X rays and gamma rays, radio waves and microwaves, and ultraviolet and infrared radiation.

As we shall see in Chapter 6, astronomers have recently constructed telescopes that can detect even nonvisible forms of light (Figure 1-2). Whether located on Earth or in orbit

FIGURE 1-2 A Telescope in Space This painting shows astronauts constructing a telescope, called the Large Deployable Reflector, at an Earth-orbiting space station. Astronomers who are planning this ambitious project hope that the telescope will be launched early in the twenty-first century. (NASA)

FIGURE 1-3 Saturn and Its Satellites This composite photograph shows the planet Saturn along with several of its moons. Space probes to the planets have provided us with a wealth of information about other worlds. This new knowledge gives us important insights into the formation and evolution of the solar system, as well as an appreciation of the variety found in nature. (NASA)

above the obscuring effects of the atmosphere, these instruments give us views of the universe vastly different from anything our eyes can see. This new information is crucial not only to our understanding of familiar objects like the Sun, but also such exotic objects as neutron stars, pulsars, quasars, and black holes.

1-2 By exploring the planets, astronomers uncover clues about the formation of the solar system

In this book, we take three major steps out into the universe: from the planets to the stars to the galaxies. The star we call the Sun and all the celestial bodies (including the Earth) that orbit it make up the **solar system.** We explore the solar system in Chapters 7 through 17, beginning with the Earth and the Moon and then on to the other planets, moving outward toward the frigid depths of space where comets spend most of their time.

During the 1970s and 1980s Soviet and American spacecraft visited all the planets except Pluto (Figure 1-3). We have flown over Mercury's cratered surface, we have peered beneath Venus's poisonous cloud cover, we have discovered enormous canyons and extinct volcanoes on Mars. We have found active volcanoes on a moon of Jupiter, we have seen the rings of Saturn and Uranus close up, we have looked down on the active atmosphere of Neptune. Along with the moon rocks brought back by the Apollo astronauts, this new

information has revolutionized theories about the creation and evolution of the solar system. We have come to realize that many of the planets and their satellites were shaped by collisions with other objects. Craters on the Moon and on many other worlds stand in mute testimony to innumerable impacts by interplanetary rocks. More importantly, the Moon may itself be the result of a catastrophic collision between the Earth and a planet-sized object shortly after the solar system was formed. As we shall see, such a collision could have torn sufficient material from the primordial Earth to create the Moon.

Throughout our journey across the solar system, we shall find that our discoveries are relevant to the quality of human life here on Earth. Until recently, our knowledge of geology, weather, and climate was based on data from only the planet Earth. Since the advent of space exploration, however, we have been able to compare and contrast other worlds with our own. This new knowledge gives us valuable insight into the origins and extent of our natural resources.

1-3 By studying stars and nebulae, astronomers discover how stars are born, grow old, and die

As we begin our study of the Sun and other stars in Chapters 18 and 19, we shall again see the surprising impact of astronomy on the course of civilization. In the 1920s and 1930s, physicists figured out how the Sun shines. At its center, thermonuclear reactions convert hydrogen into helium. This violent process releases a vast amount of energy, which eventually makes its way to the Sun's surface and escapes as light (Figure 1-4). By 1950 physicists could reproduce these

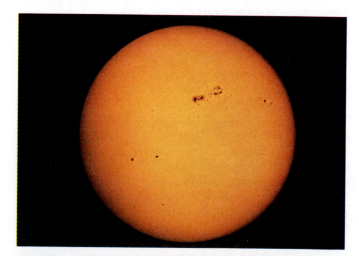

FIGURE 1-4 Our Star—the Sun The Sun is a typical star. Its diameter is about 1.39 million kilometers (roughly a million miles), and its surface temperature is about 5500°C (10,000°F). The Sun draws its energy from thermonuclear reactions occurring at its center, where the temperature is about 15 million degrees Celsius. (Celestron International)

FIGURE 1-5 A Thermonuclear Explosion Understanding the Sun's source of energy has given humanity the ability to build thermonuclear weapons. The hydrogen bomb and the thermonuclear reactions at the Sun's center both operate under the same basic physical principle: the conversion of matter into energy. This thermonuclear detonation on October 31, 1952, had an energy output, or "yield," equivalent to 10.4 million tons of TNT. (Defense Nuclear Agency)

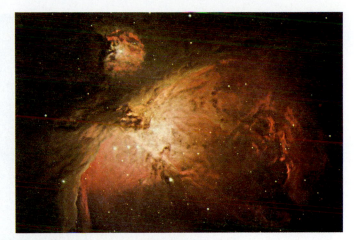

FIGURE 1-6 The Orion Nebula This beautiful nebula is a fine example of a stellar "nursery" where stars are born. Intense radiation from the newly formed stars causes the surrounding gases to glow. Many of the stars embedded in this nebula are less than a million years old. The Orion Nebula is about 1500 light-years from Earth, and the distance across the nebula is about 5 light-years. (Anglo-Australian Observatory)

thermonuclear reactions here on Earth (Figure 1-5). Hydrogen bombs operate on the same basic principles as do the thermonuclear reactions that produce energy at the Sun's center. But while these thermonuclear weapons hold the specter of destroying life on our planet, peaceful applications of this same process may provide a clean source of electrical energy early in the twenty-first century.

In Chapter 20 we look deep into space to find star clusters and clouds of glowing gas, called **nebulae** (singular **nebula**), scattered across the sky. These beautiful objects can tell us much about the lives of stars. Stars are born in huge clouds of interstellar gas and dust such as the Orion Nebula, shown in Figure 1-6. After millions or billions of years, stars die. As we shall see in Chapter 22, some end their lives with a spectacular detonation called a **supernova** (plural **supernovae**) that blows the star apart. The Crab Nebula seen in Figure 1-7 is a striking example of a supernova remnant.

During their death throes, stars return gas to interstellar space. This gas contains heavy elements created by thermonuclear reactions in the stars' interiors. Interstellar space thus becomes enriched with newly manufactured atoms and molecules. The Sun and its planets were formed from interstellar material enriched by preceding generations of stars. We therefore realize that virtually everything we touch, including the atoms in our bodies, was created deep inside ancient stars.

Dying stars can produce some of the strangest objects in the sky. In Chapter 23 we shall see that some dead stars

become **pulsars,** which emit pulses of radio waves, or **bursters,** which emit powerful bursts of X rays. In Chapter 24 we encounter massive dead stars called **black holes,** which are surrounded by gravity so powerful that nothing—not even light—can escape. Many of these bizarre stellar corpses have been discovered only recently by Earth-orbiting telescopes that detect nonvisible light, such as the X rays emitted by gases falling toward a black hole.

FIGURE 1-7 The Crab Nebula This nebula is a fine example of a supernova remnant. A dying star exploded, leaving behind this beautiful funeral shroud of glowing gases blasted violently into space. In fact, these gases are still moving outward, at about 1000 km/s (roughly 2 million miles per hour). The Crab Nebula is 6300 light-years from Earth, and the distance across the nebula is about 6 light-years. (Lick Observatory)

FIGURE 1-8 A Galaxy This spectacular galaxy contains several hundred billion stars. The galaxy's spiral arms are outlined by many nebulae, which are sites of active star formation. This galaxy, with a diameter of about 130,000 light-years, is about 65 million light-years from Earth. (Anglo-Australian Observatory)

1-4   By observing galaxies, astronomers learn about the creation and fate of the universe

Stars are not spread uniformly across the universe but are grouped together in huge assemblages called **galaxies.** Galaxies come in a wide range of shapes and sizes. A typical galaxy, like our own Milky Way, contains several hundred billion stars. As we shall see in Chapter 26, some galaxies are much smaller, containing only a few hundred million stars. Others are veritable monstrosities that devour neighboring galaxies in a process called "galactic cannibalism."

The Milky Way Galaxy has arching spiral arms like those of the galaxy shown in Figure 1-8. These arms are particularly active sites of star formation. As we explore our Galaxy in Chapter 25, we find that its center is emitting vast quantities of energy. Some astronomers suspect that this energy output is caused by gas falling into an enormous black hole at the galactic center.

Some of the most intriguing galaxies appear to be in the throes of violent convulsions, and are rapidly expelling matter. The centers of these strange galaxies, which may harbor extremely massive black holes, are often powerful sources of X rays and radio waves.

Even more dramatic sources of energy are found still deeper in space. At distances of billions of light-years from Earth, we find the mysterious **quasars.** Although quasars look like stars (Figure 1-9), they are the most distant and most luminous objects in the sky. A typical quasar shines with the brilliance of a hundred galaxies. As explained in Chapter 27, recent observations imply that quasars draw their awesome energy from enormous black holes.

The motions of clusters of galaxies reveal that we live in an expanding universe. Extrapolating into the past, we learn that the universe must have been born from an incredibly dense—perhaps infinitely dense—state roughly 15 billion years ago.

Most astronomers believe that the universe began with a cosmic explosion, known as the **Big Bang,** that occurred throughout all space at the beginning of time. Events shortly

FIGURE 1-9 A Quasar Quasars, believed to be the most distant objects in the universe, are the most luminous objects ever seen. At first glance, a quasar is easily mistaken for a faint star. This quasar is about 5 billion light-years from Earth. (Palomar Observatory)

after the Big Bang dictated the present nature of the universe. Chapters 28 and 29 describe the significant progress that astronomers are making in understanding these cosmic events. This knowledge may even reveal the origin of some of the most basic properties of physical reality. In addition, the motions of distant clusters of galaxies may tell us the ultimate fate of the universe—whether it will continue expanding forever or someday stop and collapse back in on itself.

Today's astronomy owes a great debt to astute observers and thinkers spanning many centuries. Curiosity, insight, and creativity together with observation and disciplined inquiry fostered the development of science. Given the human need to know and understand, it seems inevitable that we would begin to unravel even the deepest mysteries of the universe. The basic tools for this endeavor were among the first stirrings of human comprehension: systems of measurement.

1-5 Astronomers use angles to denote the apparent sizes and positions of objects in the sky

Astronomers use angles and a system of angular measure to denote the positions and apparent sizes of objects in the sky. An **angle** is the opening between two lines that meet at a point. Angular measure describes the shape, or size, of an angle exactly. The basic unit of angular measure is the **degree**, designated by the symbol °. A full circle is divided into 360°, and a right angle measures 90°. As shown in Figure 1-10, the angle between the two "pointer stars" in the Big Dipper is about 5°.

Astronomers use **angular measure** to describe the apparent size of a celestial object. For example, imagine looking up

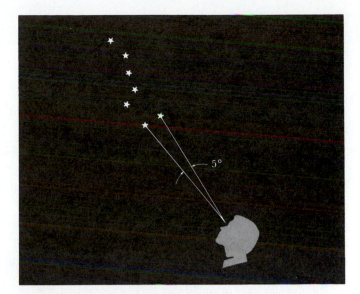

FIGURE 1-10 The Big Dipper The Big Dipper is an easily recognized grouping of seven bright stars. The angular distance between the two "pointer stars" at the front of the Big Dipper is about 5°. For comparison, the angular diameter of the Moon is about $\frac{1}{2}°$.

at the full moon. The angle covered by the Moon's diameter is nearly $\frac{1}{2}°$. We therefore say that the **angular diameter** (or angular size) of the Moon is $\frac{1}{2}°$. Alternatively, astronomers say that the Moon *subtends* an angle of $\frac{1}{2}°$. Ten full moons could fit side by side between the two pointer stars in the Big Dipper.

The adult human hand held at arm's length provides a means of estimating angles, as shown in Figure 1-11. For

FIGURE 1-11 Estimating Angles with the Human Hand Various parts of the adult human hand extended to arm's length can be used to estimate angular distances and sizes in the sky.

example, your fist covers an angle of 10°, whereas the tip of your finger is about 1° wide. Various segments of your index finger extended to arm's length can be similarly used to estimate angles a few degrees across.

To talk about smaller angles, we subdivide the degree into 60 **minutes of arc** (abbreviated 60 arc min or 60′). A minute of arc is further subdivided into 60 **seconds of arc** (abbreviated 60 arc sec or 60″). Thus,

$$1° = 60 \text{ arc min} = 60'$$

$$1' = 60 \text{ arc sec} = 60''$$

The *Astronomical Almanac* for 1993, for example, states that on February 2, Venus had an angular diameter of 28.34 seconds of arc as viewed from Earth. That is a convenient, precise statement of how big the planet appeared in Earth's sky on that date.

The angular size of an object can be converted into a linear size (measured in kilometers or miles, for example) if we know the distance to the object. A method of conversion is described in Box 1-1.

1-6 Powers-of-ten notation is a useful shorthand system of writing numbers

Astronomy is a subject of extremes. As we examine various environments, we find an astonishing range of conditions, from the incredibly hot, dense centers of stars to the frigid, near vacuum of interstellar space. To describe such divergent conditions accurately, we need a wide range of both large and small numbers. Astronomers avoid such confusing terms as "a million billion billion" by using a standard shorthand system: All the cumbersome zeros that accompany a large number are consolidated into one term consisting of 10 followed by an **exponent,** which is written as a superscript and called the **power of ten.** The exponent indicates how many zeros you would need to write out the long form of the number. Thus,

$$10^0 = 1$$

$$10^1 = 10$$

$$10^2 = 100$$

$$10^3 = 1,000$$

$$10^4 = 10,000$$

and so forth. The exponent tells you how many tens must be multiplied together to give the desired number. For example, ten thousand can be written as 10^4 ("ten to the fourth") because $10^4 = 10 \times 10 \times 10 \times 10 = 10,000$.

With this notation, numbers are written as a figure between one and ten multiplied by the appropriate power of ten. The distance between the Earth and the Sun, for exam-

---

BOX 1-1

The Small-Angle Formula

Astronomers are usually concerned with objects that subtend tiny angles in the sky. If you know the distance to an object, you can convert its angular size into a linear size (in kilometers or miles, for example) by using the **small-angle formula.** Suppose that an object subtends an angle α (measured in seconds of arc) and is at a distance d from the observer, as in the diagram below. The small-angle formula tells us that the linear size (D) of the object is given by the expression

$$D = \frac{\alpha d}{206,265}$$

where 206,265 is the number of arc seconds in 360° divided by the circumference of a circle whose radius equals 1.

EXAMPLE: On March 30, 1993, Jupiter was at a distance of 666.4 million kilometers from Earth. Jupiter's angular diameter on that date was 44.20 seconds of arc, so we can calculate the planet's diameter as follows:

$$D = \frac{44.20 \times 666,400,000 \text{ km}}{206,265} = 142,800 \text{ km}$$

This answer agrees reasonably well with an equatorial diameter of 142,984 km as determined from spacecraft flybys.

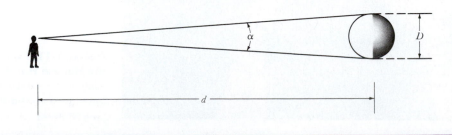

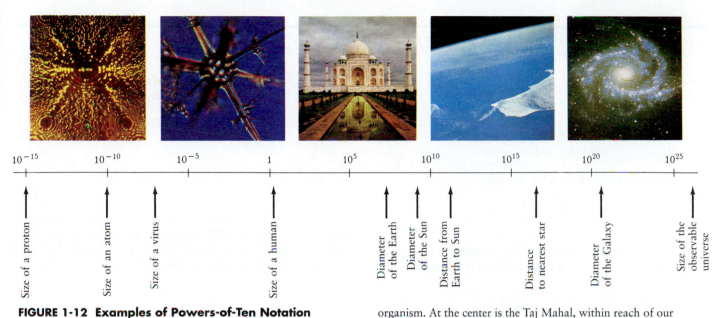

| 10^{-15} | 10^{-10} | 10^{-5} | 1 | 10^{5} | 10^{10} | 10^{15} | 10^{20} | 10^{25} |

Size of a proton — Size of an atom — Size of a virus — Size of a human — Diameter of the Earth — Diameter of the Sun — Distance from Earth to Sun — Distance to nearest star — Diameter of the Galaxy — Size of the observable universe

FIGURE 1-12 Examples of Powers-of-Ten Notation
The scale gives the sizes of objects in meters, ranging from sub-atomic particles at the left to the entire observable universe on the right. The photograph at the left shows tungsten atoms, 10^{-10} meter in diameter. Second from left is a crystalline skeleton, 10^{-4} meter (0.1 millimeter) in size, of a diatom—a single-celled organism. At the center is the Taj Mahal, within reach of our unaided senses. On the right, looking across the Indian Ocean toward the South Pole, we see the curvature of the Earth, 10^{7} meters in diameter. At the far right is a galaxy, 10^{21} meters (100,000 light-years) in diameter. (Courtesy of Scientific American Books; NASA; AAT)

ple, can be written as 1.5×10^{8} km, which, once you get used to it, is more convenient than writing "150,000,000 kilometers" or "one hundred and fifty million kilometers."

This shorthand system can be applied to numbers that are less than one by using a minus sign in front of the exponent. A negative exponent tells you to divide by the appropriate number of tens, for example, $10^{-1} = \frac{1}{10}$. The location of the decimal point is as follows:

$$10^{0} = 1$$
$$10^{-1} = 0.1$$
$$10^{-2} = 0.01$$
$$10^{-3} = 0.001$$
$$10^{-4} = 0.0001$$

and so forth. Perhaps the easiest way of understanding negative exponents is to realize that they tell you how many tenths must be multiplied together to give the desired number. For example, one ten-thousandth can be written as 10^{-4} ("ten to the minus four") because $10^{-4} = \frac{1}{10} \times \frac{1}{10} \times \frac{1}{10} \times \frac{1}{10} = 0.0001$. Expressed with this notation, the diameter of a hydrogen atom is 1.1×10^{-8} cm. That is more convenient than saying "0.000000011 centimeters" or "eleven-billionths of a centimeter."

Using the powers-of-ten notation, one can write familiar numerical terms as follows:

$$\text{one thousand} = 1000 = 10^{3}$$
$$\text{one million} = 1,000,000 = 10^{6}$$
$$\text{one billion} = 1,000,000,000 = 10^{9}$$
$$\text{one trillion} = 1,000,000,000,000 = 10^{12}$$

and also

$$\text{one-thousandth} = 0.001 = 10^{-3}$$
$$\text{one-millionth} = 0.000001 = 10^{-6}$$
$$\text{one-billionth} = 0.000000001 = 10^{-9}$$
$$\text{one-trillionth} = 0.000000000001 = 10^{-12}$$

Because powers-of-ten notation bypasses all the awkward zeros, we can conveniently describe the size of objects as small as atoms or as big as galaxies (Figure 1-12). Powers-of-ten notation also makes it easy to multiply and divide numbers, as described in Box 1-2.

1-7 Astronomical distances are often measured in astronomical units, parsecs, or light-years

As we turn toward the stars in the second half of this book, we shall find that some of our traditional units of measure become cumbersome. It is fine to use kilometers or miles to

---

BOX 1-2

Arithmetic with Exponents

Using powers-of-ten notation, it is easy to multiply numbers; simply add up the exponents.

EXAMPLE:

$$10^2 \times 10^3 = 10^{2+3} = 10^5$$

To divide numbers, remember the rule

$$10^{-n} = \frac{1}{10^n}$$

In other words, dividing by 10^6, for instance, is the same as multiplying by 10^{-6}. You again add up the exponents.

EXAMPLE:

$$\frac{10^4}{10^6} = 10^4 \times 10^{-6} = 10^{4-6} = 10^{-2}$$

Usually a computation involves numbers multiplied by powers of ten. In such cases, to perform multiplication or division, you can treat the numbers separately from the powers of ten.

EXAMPLE: The numerical example worked out in Box 1-1 can be done straightforwardly by using exponents:

$$D = \frac{44.20 \times 666,400,000}{206,265} = \frac{4.420 \times 10 \times 6.664 \times 10^8}{2.06265 \times 10^5}$$

$$= \frac{4.420 \times 6.664 \times 10^{1+8-5}}{2.06265} = 1.428 \times 10^5$$

Negative exponents are often used with units of measure. For instance, speed in kilometers per second can be written as "km/s" or as "km s^{-1}." Similarly, the speed of light in a vacuum (usually designated by the letter "c") is written as

$$c = 3 \times 10^8 \text{ m/s} = 3 \times 10^8 \text{ m s}^{-1}$$

EXAMPLE: The average density of a typical rock (that is, its mass divided by its volume) is 3 grams per cubic centimeter. This can be written as

$$3 \text{ g/cm}^3 = 3 \text{ g cm}^{-3}$$

---

give the diameters of craters on the Moon or the heights of volcanoes on Mars. But it is as awkward to use kilometers to express distances to stars or galaxies as it would be use millimeters or inches to talk about the distance from New York to San Francisco. Astronomers have therefore devised new units of measure.

When discussing distances across the solar system, astronomers use a unit of length called the **astronomical unit** (abbreviated AU), which is the average distance between the Earth and the Sun:

$$1 \text{ AU} = 1.496 \times 10^8 \text{ km} = 93 \text{ million miles}$$

Thus the distance between the Sun and Jupiter can be conveniently stated as 5.2 AU.

To talk about distances to the stars, astronomers have two different units of length. The **light-year** (abbreviated ly) is the distance that light travels in one year:

$$1 \text{ ly} = 9.46 \times 10^{12} \text{ km}$$

or, alternatively,

$$1 \text{ ly} = 63,240 \text{ AU}$$

This distance is roughly equal to 6 trillion miles. Proxima Centauri, the nearest star other than the Sun, is 4.3 light-

years from Earth, for example. Physicists often measure interstellar distances in light-years, because the speed of light is one of nature's most important numbers.

Many astronomers prefer to use another unit of length, the parsec (abbreviated pc), because its definition is closely related to a method of measuring distances to the stars (you might want to glance ahead to Figures 19-1 and 19-2). Imagine taking a journey far into space, beyond the orbits of the outer planets. As you look back toward the Sun, the Earth's orbit subtends a smaller angle in the sky the farther you are from the Sun. The distance at which 1 AU subtends an angle of 1 arc sec, as shown in Figure 1-13, is defined as one **parsec**:

$$1 \text{ pc} = 3.09 \times 10^{13} \text{ km} = 3.26 \text{ ly}$$

Thus, the distance to the nearest star can be stated as 1.3 pc as well as 4.3 ly. Whether you choose to use parsecs or light-years is a matter of personal taste.

For even greater distances, astronomers commonly use **kiloparsecs** and **megaparsecs** (abbreviated kpc and Mpc), in which the prefixes simply mean "thousand" and "million," respectively:

$$1 \text{ kpc} = 10^3 \text{ pc}$$

$$1 \text{ Mpc} = 10^6 \text{ pc}$$

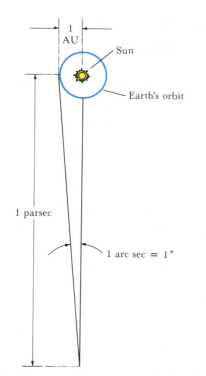

FIGURE 1-13 A Parsec The parsec, a unit of length commonly used by astronomers, is equal to 3.26 light-years. The parsec is defined as the distance at which 1 AU perpendicular to the observer's line of sight subtends an angle of 1 arc sec.

earthly experience. Thus a scientist can do experiments in a laboratory to determine the properties of light or the behavior of atoms, and then use this knowledge to investigate the structure of the universe.

The discovery of fundamental laws of nature has had a profound influence on humanity. Four centuries of scientific advances have now made their way into our lives. The world around us is filled with examples in technology, commerce, medicine, entertainment, and transportation. In the near future we can look forward to enjoying the benefits of space technology. Weightlessness and the near vacuum of space will enable us to manufacture a wide range of exceptional substances, from exotic alloys to ultrapure medicines (Figure 1-14).

The dreams of Jules Verne and H. G. Wells pale in comparison to the reality of today. Ours is an age of exploration and discovery more profound than any since Columbus and Magellan set sail. We have walked on the Moon, dug into the Martian soil. We have discovered active volcanoes and barren ice fields on the satellites of Jupiter. We have visited the shimmering rings of Saturn. Never before has so much been revealed in so short a time. As you proceed through this book, you will come to appreciate the power of the human mind to reach out, to explore, to observe, and to comprehend. In this way, modern astronomy enables us to transcend the limitations of our bodies and the brevity of human life.

For example, the distance from Earth to the center of our Milky Way Galaxy is about 8 kpc, and the rich cluster of galaxies in the direction of the constellation of Virgo is 20 Mpc away.

Some astronomers prefer to talk about thousands or millions of light-years rather than kiloparsecs and megaparsecs. Once again, the choice is a matter of personal taste.

To the possible annoyance of some people, astronomers use whatever yardsticks seem best suited for the issue at hand and do not restrict themselves to one system of measurement. For instance, an astronomer might say that the supergiant star Antares has a diameter of 860 million kilometers and is located at a distance of 150 parsecs from Earth. Units of measure are discussed further in Box 1-3.

1-8 Astronomy is an adventure of the human mind

An underlying theme of this book is that the universe is rational. It is not a hodgepodge of unrelated things behaving in unpredictable ways. Rather, we find strong evidence that fundamental laws of physics govern the nature of the universe and the behavior of everything in it. These unifying concepts enable us to explore realms far removed from our

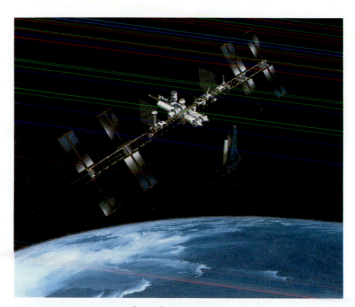

FIGURE 1-14 An Earth-Orbiting Space Station A new generation of high-technology materials could be manufactured in space. Exotic alloys, foam metals, ultrapure semiconducting crystals, and rare vaccines are among the obvious practical applications of zero-gravity industry. This artist's conception shows the first phase of the space station now being planned by NASA. (Courtesy of Lockheed)

BOX 1-3

Units of Length, Time, and Mass

To understand and appreciate the universe, we shall need to describe a wide range of phenomena. Astronomers generally use units that are best suited to the topic at hand. For instance, interstellar distances are conveniently expressed in either light years or parsecs, whereas the diameters of the planets are more comfortably presented in kilometers.

Most scientists prefer to use a version of the metric system called the International System of Units, abbreviated **SI**. In SI units, length is measured in meters (m), time is measured in seconds (s), and mass is measured in kilograms (kg). How are these basic units related to other measures?

When discussing objects on a human scale, sizes and distances are usually expressed in millimeters (mm), centimeters (cm), and kilometers (km). These units of length are related to the meter as follows:

$$1 \text{ millimeter} = 0.001 \text{ m}$$

$$1 \text{ centimeter} = 0.01 \text{ m}$$

$$1 \text{ kilometer} = 1000 \text{ m}$$

To make conversions from the English system of inches (in.), feet (ft), and miles (mi), it is necessary to know that

$$1 \text{ inch} = 2.54 \text{ cm}$$

$$1 \text{ foot} = 0.3048 \text{ m}$$

$$1 \text{ mile} = 1.609 \text{ km}$$

Each of these equalities can also be written as a fraction equal to 1. For instance, you can write

$$\frac{0.3048 \text{ m}}{1 \text{ ft}} = 1$$

These fractions are then used to cancel the unwanted units and replace them with the desired units.

EXAMPLE: Suppose a NASA publication tells you that "the *Saturn V* rocket used to send astronauts to the Moon stands about 363 feet tall." You can convert this to meters as follows:

$$363 \text{ ft} \times \frac{0.3048 \text{ m}}{1 \text{ ft}} = 111 \text{ m}$$

For the convenience of readers used to the English system, we shall often give measurements in both systems; thus we might note that "the diameter of Mars is 6794 km (4222 mi)."

When discussing very small distances, such as the size of an atom, astronomers often use the micrometer (μm) or the nanometer (nm), which are related to the meter as follows:

$$1 \text{ μm} = 10^{-6} \text{ m}$$

$$1 \text{ nm} = 10^{-9} \text{ m}$$

Thus $1 \text{ μm} = 10^3 \text{ nm}$.

BOX 1-4

Astronomy as a Profession

Many thousands of people enjoy astronomy as a hobby. Small telescopes can be purchased in department stores, and most large cities have astronomy clubs. Attending a "star party" on a cool, clear night to view planets and nebulae with friends can be a very enjoyable experience. But what does it mean to pursue astronomy as a career?

Astronomy is a physical science, and to be an astronomer you need a strong background in both physics and mathematics. It is advisable to major in physics at the undergraduate level. Formal training in astronomy occurs in graduate schools, where it typically takes six years to obtain the Ph.D. degree. A doctorate is usually the minimum requirement for employment as a professional astronomer. In addition, most newly graduated astronomers spend two to five years doing postdoctoral research to hone their talents. Finally, at an age of approximately 30, the budding astronomer is ready to enter the job market.

The American Astronomical Society (AAS) is the primary professional organization for astronomers. Its membership totals approximately 6000 persons, about 4000 of whom are full-time professional astronomers in the United States. In a recent survey of AAS members, 55% were employed by colleges and universities. As a professor, an astronomer teaches classes and conducts research, often at major observatories in Arizona, Hawaii, or Chile.

As the accompanying pie chart shows, about 16% of all astronomers are employed at federally funded research and development centers like Los Alamos National Laboratory and Lawrence Livermore National Laboratory. The government agencies, especially the National Aeronautics and Space Administration (NASA), employs an additional 14%. Industry and aerospace companies employ roughly 9% of all astronomers.

If you are thinking about a career in astronomy, you

The basic unit of time is the second (s). It is related to other units of time as follows:

$$1 \text{ minute} = 60 \text{ s}$$

$$1 \text{ hour} = 3600 \text{ s}$$

$$1 \text{ day} = 86,400 \text{ s}$$

$$1 \text{ year} = 3.16 \times 10^7 \text{ s}$$

According to SI, speed is properly measured in meters per second (m/s). Quite commonly, however, speed is also expressed in km/s and mi/hr:

$$1 \text{ km/s} = 10^3 \text{ m/s}$$

$$1 \text{ km/s} = 2237 \text{ mi/hr}$$

$$1 \text{ mi/hr} = 0.447 \text{ m/s}$$

$$1 \text{ mi/hr} = 1.47 \text{ ft/s}$$

In addition to using kilograms, astronomers sometimes express mass in grams (g) and in solar masses ($M_\odot$), where the subscript $\odot$ is the symbol denoting the Sun. As we shall see, it is especially convenient to use solar masses when discussing the masses of stars and galaxies. These units are related to each other as follows:

$$1 \text{ kg} = 1000 \text{ g}$$

$$1 \ M_\odot = 1.99 \times 10^{30} \text{ kg}$$

On the Earth, a mass of 1 kg weighs 2.20 pounds. Thus we can relate the English pound (lb) to the gram and kilogram:

$$1 \text{ kg} = 2.20 \text{ lb}$$

$$1 \text{ lb} = 453.6 \text{ g}$$

These two equalities require a word of caution: They are valid only on the Earth's surface. When we discuss gravity in Chapter 4, we shall see that weight is a measure of the pull of gravity; on Earth gravity pulls downward with a strength such that 1 kilogram weighs 2.20 pounds. But elsewhere in the universe, where the strength of gravity is different from that on the Earth, a kilogram would weigh something other than 2.20 pounds. In fact, in space far from any sources of gravity, a kilogram of matter would have no weight at all.

Finally, it is often informative to describe a substance by its **density**, which is mass divided by volume. Because the density of water is 1 g/cm³, which is an easy number to remember, some people prefer to measure density in grams per cubic centimeter. Since everyone is familiar with water, it is immediately obvious that rock is 3 times denser than water, if we note that the density of a typical rock is 3 g/cm³. According to SI, however, density should be measured in kilograms per cubic meter. It is therefore convenient to know that

$$1 \text{ g/cm}^3 = 1000 \text{ kg/m}^3$$

should know that the field is small and job opportunities are limited. As of 1990, the average income of all professional astronomers was approximately $50,400 per year. The starting salary of an assistant professor is considerably less than this. Women account for only 11% of the membership of the AAS, according to recent surveys. However, 11% is a higher percentage than other physical sciences and the AAS is actively trying to increase the involvement and status of women in astronomy.

At present, there is a surplus of astronomers seeking academic positions at colleges and universities. Recent studies suggest, however, that this employment picture may soon improve. During the late 1960s more Ph.D.'s in astronomy were awarded than in any comparable period before or since. The people who received these degrees will be reaching retirement age around the year 2000, and the present rate of incoming new Ph.D.'s will not suffice to replace them. Thus, astronomers could be in short supply early in the twenty-first century.

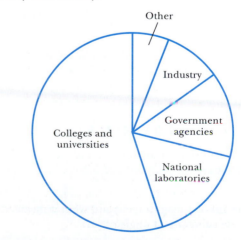

KEY WORDS

Terms preceded by an asterisk are discussed in the boxes.

angle	*density	minute of arc (′)	scientific method
angular diameter	exponent	model	second of arc (″)
angular measure	galaxy	nebula (*plural* nebulae)	*SI units
astronomical unit (AU)	hypothesis	Newtonian mechanics	*small-angle formula
Big Bang	kiloparsec (kpc)	parsec (pc)	solar system
black hole	laws of physics	power of ten	supernova (*plural* super-novae)
burster	light-year (ly)	pulsar	theory
degree (°)	megaparsec (Mpc)	quasar	

KEY IDEAS

• The universe is comprehensible.

• The scientific method is a procedure for formulating hypotheses. These are tested by observation or experimentation in order to build consistent models or theories that accurately describe phenomena in the universe.

Observations of the heavens have led to discovery of some of the fundamental laws of nature.

• Exploration of the planets provides information about the origin and evolution of the solar system, as well as the history and resources of the Earth.

• Study of the stars and nebulae provides information about the origin and history of the Sun and the solar system.

• Observations of galaxies provide information about the origin and history of the universe.

• Astronomers use angles to denote the positions and sizes of objects in the sky.

The size of an angle is measured in degrees, minutes of arc, and seconds of arc.

• The powers-of-ten notation system is a convenient shorthand for writing numbers.

• A variety of distance units, including the parsec and the light-year, are used by astronomers.

REVIEW QUESTIONS

1. What are degrees, minutes of arc, and seconds of arc used for? What are the relationships between these units of measure?

2. With the aid of a diagram, explain what it means to say that the Moon subtends an angle of $\frac{1}{2}°$.

3. How many seconds of arc equal $1°$?

4. What is an exponent? How are exponents used in powers-of-ten notation?

5. What is the advantage of using powers-of-ten notation?

6. How is an AU defined? Give an example where this unit of measure would be convenient to use.

7. What is the advantage to the astronomer of using the light-year as a unit of distance?

8. What is a parsec and how is it related to a kiloparsec and a megaparsec?

9. Write the following numbers using powers-of-ten notation: (a) ten million, (b) four hundred thousand, (c) six one-hundredths, (d) seventeen billion, (e) your age, (f) the number of pages in this book.

10. Give the word or phrase that corresponds to the following standard abbreviations: (a) km, (b) cm, (c) s, (d) km/s, (e) mph, (f) m, (g) m/s, (h) hr, (i) yr, (j) g, (k) kg. Which of these are units of speed? (*Hint:* You may have to refer to a dictionary. All of these abbreviations should be part of your working vocabulary.)

ADVANCED QUESTIONS

Tips and tools . . .

The size of an astronomical object is related to the angle it subtends by the small-angle formula given in Box 1-1. The method for converting from one unit of measure to another is illustrated in Box 1-3. The volume of a sphere is $\frac{4}{3}\pi r^3$, where r is the sphere's radius and π (pi) is a constant approximately equal to 3.142.

11. According to information listed in the *Astronomical Almanac* for 1993, on August 7 the planet Uranus was at a distance of 18.687 AU from Earth. In an appendix at the end of this book, you can find that the diameter of Uranus is 51,118 km. What was the angular size of Uranus as seen from Earth on August 7, 1993?

12. The bright star Sirius is 2.65 pc from Earth. **(a)** What is the distance to Sirius in kilometers? **(b)** How long does it take for light emanating from Sirius to reach the Earth?

13. The Sun's mass is 1.99×10^{30} kg, three-quarters of which is hydrogen. The mass of a hydrogen atom is 1.67×10^{-27} kg. How many hydrogen atoms does the Sun contain?

14. The average distance to the Moon is 384,000 km, and the Moon subtends an angle of $\frac{1}{2}°$. Calculate the diameter of the Moon.

15. At what distance would a person have to hold a quarter (which has a diameter of about 2.5 cm) in order for the quarter to subtend an angle of **(a)** $1°$? **(b)** 1 arc min? **(c)** 1 arc sec?

16. The speed of light is 3×10^8 m/s. How long does it take light to go from the Sun to the Earth? (*Hint:* An object traveling at velocity v for a time t covers a distance d given by $d = vt$.)

17. The mass of the Earth is 5.98×10^{24} kg, and its diameter is 12,756 km. Divide the Earth's mass by its volume to find the average density of the Earth in kilograms per cubic meter (kg/m^3). How does the Earth's density compare with the density of water, which is 1 g/cm^3?

18. A hydrogen atom has a radius of about 5×10^{-9} cm. The radius of the observable universe is about 15 billion light-years. How many times larger than a hydrogen atom is the universe?

19. Explain where the number 206,265 in the small-angle formula came from.

20. The diameter of the Sun is 1.4×10^{11} cm, and the distance to the nearest star, Proxima Centauri, is 4.3 ly. If the Sun were reduced to the size of a basketball (about 30 cm in diameter), at what distance would Proxima Centauri be from the Sun on this reduced scale?

21. How many Suns would it take, laid side by side, to reach to the nearest star?

22. When *Voyager 2* sent back pictures of Neptune during its historic flyby of that planet in 1989, the spacecraft's radio signals traveled for four hours to reach the Earth. How far away was the spacecraft?

23. Suppose your telescope can give you a clear view of objects and features that subtend angles of at least 2 arc sec. What is the diameter of the smallest crater you can see on the Moon?

24. What is the age of the universe in seconds?

DISCUSSION QUESTIONS

25. How do astronomical observations and experiments differ from those of other sciences?

26. Scientists assume that "reality is rational." Discuss what this means and the thinking behind it.

OBSERVING PROJECTS

27. Look up at the sky on a clear, cloud-free night. Is the Moon in the sky? If so, does it interfere with your ability to see the fainter stars? Why do you suppose astronomers prefer to schedule their observations on nights when the Moon is not in the sky?

28. Can you see the Milky Way from where you live on a dark, clear, moonless night? If so, briefly describe its appearance. If not, what seems to be interfering with your ability to see the Milky Way?

29. Look up at the sky on a clear, cloud-free night and note the positions of a few prominent stars relative to such reference markers as rooftops, telephone poles, and treetops. Also note the location from where you made your observations. A few hours later, return to that location and again note the positions of the same bright stars that you observed earlier. How have their positions changed? From these changes, can you deduce the general direction in which the stars appear to be moving?

FOR FURTHER READING

Asimov, I. *The Measure of the Universe.* Harper & Row, 1983. An extensive journey through the universe that uses powers of ten.

Chaisson, E. *Cosmic Dawn.* Little, Brown, 1981. This excellent book presents the scientist's worldview by discussing the origins of matter and life.

Dickinson, T. *The Universe and Beyond.* Camden House, 1986. This beautifully illustrated, nontechnical coffee table book takes the reader on a tour of modern astronomy. Contrary to what the title suggests, the book stays firmly within our universe.

Gore, R. "The Once and Future Universe." *National Geographic,* June 1983. A well-written, beautifully illustrated introduction to modern astronomy.

Graham-Smith, F., and Lovell, B. *Pathways to the Universe.* Cambridge University Press, 1988. An introduction to astronomy, written by two distinguished British astronomers and consisting of 21 topics, from "How We Became Astronomers" to "Cosmology."

Hartmann, W., and Miller, R. *Cycles of Fire.* Workman, 1987. An oversized, colorful album of paintings by two noted astronomical artists, accompanied by an excellent nontechnical tour of stars, nebulae, and galaxies.

Morrison, P., and others. *Powers of Ten.* Scientific American Books, 1982. A tour of the universe where each step corresponds to a power of ten.

Seielstad, G. *Cosmic Ecology.* University of California Press, 1983. This eloquent survey traces evolutionary processes from the Big Bang to the development of intelligence on Earth.

SANDRA FABER

Why Astronomy?

SANDRA M. FABER, Professor of Astronomy at the University of California, Santa Cruz, and Astronomer at the Lick Observatory, was intrigued by the origins of the universe when she entered Swarthmore College. She completed her doctorate in astronomy at Harvard University, but did her dissertation at the Carnegie Institution's Department of Terrestrial Magnetism, where she was influenced by the distinguished woman astronomer Vera Rubin. Dr. Faber chaired the now legendary group of astronomers called the "Seven Samurai," who surveyed the nearest 400 elliptical galaxies and discovered a new mass concentration, the "Great Attractor." She is the recipient of many honors and awards for her research, including the Bok Prize from Harvard University in 1978. She was elected to the National Academy of Sciences in 1985 and won the coveted Dannie Heineman Prize from the American Astronomical Society, given in recognition of a sustained body of especially influential astronomical research, in 1986. She also is on the Board of Trustees of the Carnegie Institution.

As you study astronomy, it is appropriate for you to ask, why am I studying this subject? What good is it for human beings in general and for me in particular? Astronomy, it must be admitted, does not offer the same practical benefits as other sciences. So how can astronomy be important to your life?

On the most basic level, I like to think of astronomy as providing the introductory chapters for the ultimate textbook on human history. Ordinary texts start with recorded history, going back some 3000 years. For events before that, we consult archeologists and anthropologists, who tell us about the early history of our species. For knowledge of the time before that, we consult paleontologists, biologists, and geologists about the origin and evolution of life and the evolution of our planet, altogether going back some five billion years. Astronomy tells us about the vast stretch of time before the origin of the Earth, the ten billion years or so that saw the formation of the Sun and solar system, the Milky Way Galaxy, and the origin of the universe in the Big Bang. Knowledge of astronomy is essential for a well-educated person's view of history.

Astronomy challenges our belief system and impels us to put our "philosophical house" in order. Take the origin of the world, for example. According to Genesis in the Bible, the world and all in it were created in six days by the hand of God. However, the ancient Egyptians believed that the Earth arose spontaneously from the infinite waters of the eternal universe, Nun, and Alaskan legends taught that the world was created by the conscious imaginings of a creator deity named Father Raven.

The modern astronomical story of the creation of the Earth differs from all these in that it claims to be buttressed by physics and by observational fact. The Sun, it is asserted, formed via gravitational collapse from a dense cloud of interstellar dust and gas about five billion years ago. The planets coalesced at the same time as condensates within the swirling solar nebula. This process took not weeks or days but several hundred thousand years. Confidence in this theory stems from our looking out into the Galaxy and seeing young stars actually form in this way.

At issue here, really, is the deep question of how we are to gain information about the nature of the physical world—whether by revelation and intuition or by logic and observation. Where science stops and faith begins is a thorny issue for all human beings, but particularly so for astronomers—and for astronomy students.

Astronomy cultivates our notions about cosmic time and cosmic evolution. It is all too easy, given the short span of human life, to overlook the fact that the universe is a dynamic place. When I first entered astronomy, I found myself handicapped by a conservative mind-set that assumed, subtly, that celestial objects were unchanging. This might seem strange for someone whose avowed interest was in learning how galaxies formed. Such are the vagaries of the human mind! Fortunately I lost this bias as I grew older, mainly, I think, by observing evolution all around me. Many things that had seemed immutable to me as a child—social structures, customs, even the physical environment—turned out not to be so. With that realization, the scales fell from my eyes and I was able to accept cosmic evolution as a core concept in my thinking about the universe.

Following closely on the concept of evolution is the concept of fragility—if something can change, it might even actually disappear someday. The most obvious example is the limited lifetime of our Sun: In another five billion years or so, the Sun will enter its death throes, during which it will swell up and brighten to

1000 times its present luminosity and, in the process, incinerate the Earth.

Five billion years is far enough in the future that neither you nor I feel any personal responsibility for preparing the human race to meet this challenge. However, there are many other cosmic catastrophes that will befall us in the meantime. The Earth will be hit by sizable pieces of space junk; eroded impact craters show this happens all the time—every few million years or so. Debris thrown up into the atmosphere from these events could be so extensive as to totally alter the climate for an extended period of time, possibly leading to mass extinctions. There will be enormous volcanic eruptions that will make Mount St. Helens's look like a small firecracker. Another Ice Age, which will totally disrupt modern agriculture, is virtually certain within the next 20,000 years unless we cook the Earth ourselves first by burning too much fossil fuel.

Closer to the human sphere, such common notions as the inevitability of human progress, the desirability of endless economic growth, the ability of the Earth to support its growing human population, are all based on limited experience and may not—probably *will* not—prove viable in the long run. Consequently, we must rethink who we are as a species and what our proper activity on Earth is. Contemplating cosmic time puts us into the appropriate frame of mind for grappling with these issues. These are problems that are very long range and involve the whole human race, yet are totally under human control and are vital to our long-term survival and well-being.

Astronomy is essential to developing a perspective on human existence and its relation to the cosmos. What question could be grander and at the same time more relevant to us? We can hardly contemplate human existence in general without taking at least a brief look at our own lives.

Earth is a mote in the vast cosmic sea, and our lives, in the evolutionary scheme of the universe, would seem at face value to be insignificant. Should that thought terrify us, depriving our lives of any meaning?

Should we seek to provide meaning by introducing an intelligent creator? Or should we be fully comforted merely to know that this great universe has given birth to us in an entirely natural, perhaps even inevitable, chain of events?

Not long ago I visited the southern hemisphere for the first time to observe the sky from Chile. At that latitude, the center of the Milky Way passes overhead, where it makes a grand show, not the miserable, fuzzy patch that is visible, low on the horizon, from North America. Stepping out on the observatory catwalk one morning before dawn, I saw for the first time the galactic center in all its glory soaring overhead. At that instant, my perceptions underwent a profound alteration. One moment I was standing in an earthbound "room" with the Galaxy painted on the "ceiling." The next instant, the walls of this earthly room fell away, and I was floating freely in outer space, viewing the center of the Galaxy from an outpost on its periphery. The sense of depth was breathtaking: Sparkling star clusters and dense dust clouds were etched against the blazing inner bulge of the Galaxy many thousands of light-years distant. I saw the Galaxy as an immense, three-dimensional object in space for the first time. It became "real." At that moment, my sense of place in the universe underwent a fundamental and irreversible change. I remained a citizen of the Earth, but I also became a citizen of the Galaxy.

Many astronomers believe that the ultimate, proper concept of "home" for the human race is our universe. It seems more and more likely that there are a very large number of other universes. Virtually none of them would appear to resemble ours; they may have different numbers of space and time dimensions—perhaps even no space or time—and very different physical laws. The vast majority probably are incapable of harboring intelligent life as we know it.

The parallel with Earth is striking. Among the solar system planets, only Earth can support human life. Among the great number of planets which likely exist in our Galaxy, only a small fraction are apt to be such that we could call them home. The fraction of hospitable universes is likely to be smaller still. Our universe would seem, then, to be the ultimate instance of "home": a sanctuary in a vast sea of inhospitable universes.

I began this essay by talking about history and ended with questions that border on the ethical and religious. Astronomy is like that: It offers a modern-day version of Genesis—and of the Apocalypse, too. Astronomy stimulates and challenges us on our deepest, most human levels. I hope that, during this course, you will be able to take time out to contemplate the broader implications of what you are studying. This is one of the rare opportunities in life to think about who you are and where you and the human race are going. Don't miss it.

The Milky Way as seen from Cerro Tololo Observatory. (National Optical Astronomy Observatories)

KNOWING THE HEAVENS

CIRCUMPOLAR STAR TRAILS This long exposure, aimed at the south celestial pole, shows the rotation of the sky. The foreground building is part of the Anglo-Australian Observatory, which houses one of the largest telescopes in the southern hemisphere. During the exposure, someone carrying a flashlight walked along the dome's outside catwalk. Another flashlight made the wavy trail at ground level. (Anglo-Australian Observatory)

BEFORE THE dawn of recorded history, our ancestors divided the starry sky into constellations. Modern astronomers still use many of the constellations described by ancient Babylonians and Greeks. These starry patterns, named for mythical figures, help us find our way around the sky, which is conveniently described as the celestial sphere. The annual path of the Sun across the constellations of the zodiac, together with the Earth's equator and poles, serve as guideposts. We find that the seasons are related to the tilt of the Earth's axis of rotation and that the axis itself is slowly changing its orientation. The motions of the Earth are crucial in timekeeping and the calendar, which are traditional, practical applications of astronomy.

The roots of astronomy reach back to the very beginning of civilization. This ancient heritage is most apparent in the **constellations** (from the Latin word meaning "group of stars"). Perhaps it was shepherds tending their flocks at night, or families sitting by the fire, or priests studying the heavens who first imagined pictures among groupings of stars.

You may already be familiar with some of these patterns in the sky, such as the Big Dipper, which is actually part of a large constellation called Ursa Major (the Great Bear). Many of these constellations, such as Orion in Figure 2-1, have names derived from the myths and legends of antiquity. Although some star groupings vaguely resemble the figures they are supposed to represent, most do not.

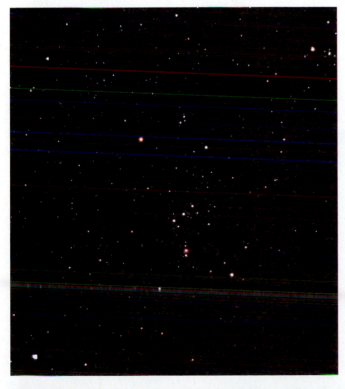

FIGURE 2-1 Orion Orion is a prominent winter constellation. From the United States, Orion is easily seen high above the southern horizon from December through March. The photograph was a four-minute exposure on Kodak film. Because of the time exposure, the colors of the stars are quite noticeable. The fanciful drawing of Orion is from an 1835 star atlas. (Courtesy of R. Mitchell and Janus Publications)

FIGURE 2-2 Diurnal Motion of the Night Sky This time lapse photograph records the motion of the stars across the night sky over the Kitt Peak National Observatory near Tucson, Arizona. The star images are trailed because of the Earth's rotation. (NOAO)

Looking at the sky on a clear, dark night, you might think that you can see millions of stars. Actually, the unaided human eye can detect only about 6000 stars. Because half of the sky is below the horizon at any one time, you can see roughly 3000 stars at most. Like constellations, many stars have ancient names; star names and catalogues are discussed in Box 2-1.

2-1 Eighty-eight constellations cover the entire sky

On modern star charts, the entire sky is divided into 88 constellations. Some constellations cover large areas (Ursa Major being one of the biggest), others very small areas (Crux being the smallest). The constellations are used to

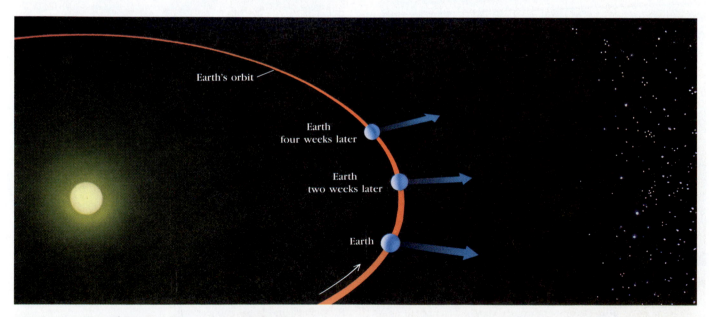

FIGURE 2-3 Our Changing View of the Night Sky As we orbit the Sun, the nighttime sky of the Earth gradually turns toward different parts of the sky. Thus, the constellations that we can see slowly change from one night to the next.

specify certain regions of the sky. For instance, we might speak of "the nebula M42 in Orion" much as we would refer to "the Ural Mountains in Russia."

The Earth rotates from west to east once every 24 hours, which is why we have day and night. Because of this rotation, stars rise in the east and set in the west, as do the Sun and Moon. You can easily observe this basic astronomical fact yourself. Go outdoors soon after dark, find a spot away from bright lights, and note the patterns of stars in the sky. A few hours later, check again. You will find that the entire pattern of stars (including the Moon, if it is visible) has shifted its position. New constellations will have risen above the eastern horizon, and some will have disappeared below the western horizon. If you can check again before dawn, you will find low in the western sky the stars that had just been rising when the night began. This **diurnal** (daily) **motion** of the stars is very apparent in time exposure photographs (Figure 2-2).

The constellations that you can see in the sky change slowly over the course of a year, because the Earth is orbiting the Sun (Figure 2-3). Because the Earth takes a full year to go once around the Sun, the darkened, nighttime side of the Earth is gradually turned toward different parts of the heavens. If you follow a particular star on successive evenings, you will find that it rises approximately four minutes earlier each night.

Constellations can help you orient yourself and find your way around the sky. For instance, if you live in the northern hemisphere, you can use the Big Dipper in Ursa Major to find the north direction by drawing a straight line through the two stars at the front of the Big Dipper's bowl (Figure 2-4). The first moderately bright star you come to is Polaris, also

called the North Star because it is located almost directly over the Earth's north pole. If you draw a line from Polaris straight down to the horizon, you will find the north direction.

By drawing a line through the two stars at the rear of the bowl of the Big Dipper, you can find Leo (the Lion). As Figure 2-4 shows, that line points toward Regulus, the brightest star in the "sickle" tracing the lion's mane. By following the handle of the Big Dipper, you can locate the bright reddish star Arcturus in Boötes (the Shepherd) and the prominent bluish star Spica in Virgo (the Virgin). The saying "Follow the arc to Arcturus and speed to Spica" may help you remember these stars, which are conspicuous in the evening sky during the spring and summer.

During the winter in the northern hemisphere, you can see some of the brightest stars in the sky. Many of them are in the vicinity of the "winter triangle," which connects bright stars in the constellations of Orion (the Hunter), Canis Major (the Large Dog), and Canis Minor (the Small Dog), as shown in Figure 2-5. The winter triangle is high above the southern horizon during the middle of winter at midnight.

A similar feature, called the "summer triangle," graces the summer sky. This triangle connects the brightest stars in Lyra (the Harp), Cygnus (the Swan), and Aquila (the Eagle) (Figure 2-6). A conspicuous portion of the Milky Way forms a beautiful background for these constellations, which are nearly overhead during the middle of summer at midnight.

A set of selected star charts for the evening hours of all twelve months of the year is included at the end of this book. You may well find stargazing an enjoyable experience, and these star charts will help you identify many well-known constellations.

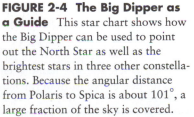

FIGURE 2-4 The Big Dipper as a Guide This star chart shows how the Big Dipper can be used to point out the North Star as well as the brightest stars in three other constellations. Because the angular distance from Polaris to Spica is about 101°, a large fraction of the sky is covered.

FIGURE 2-5 The "Winter Triangle"
This star chart shows the eastern sky as it appears during the evening in December. Three of the brightest stars in the sky make up the "winter triangle," which is about 26° on a side. In addition to the constellations involved in the triangle, Gemini (the Twins), Auriga (the Charioteer), and Taurus (the Bull) are shown.

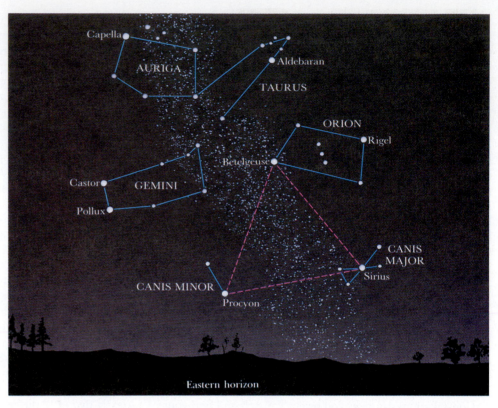

BOX 2-1

Star Names and Catalogues

The customs for naming stars have changed over the centuries. Many of the brightest stars in the sky have Arabic names assigned in medieval times, when astronomy was widely studied among Islamic nations. The diagram shows the Big Dipper with the Arabic names for its seven brightest stars. The North Star received its formal name, Polaris, later, when Latin was the language used by European astronomers.

As you might suspect, memorizing exotic star names is an unwelcome burden for many astronomers. A simpler system, invented by Johann Bayer in 1603, uses the constellation names and the 24 lowercase letters of the Greek alphabet:

α	alpha	ι	iota	ρ	rho
β	beta	κ	kappa	σ	sigma
γ	gamma	λ	lambda	τ	tau
δ	delta	μ	mu	υ	upsilon
ε	epsilon	ν	nu	φ	phi
ζ	zeta	ξ	xi	χ	chi
η	eta	o	omicron	ψ	psi
θ	theta	π	pi	ω	omega

In constructing a star name, a Greek letter is used with the Latin possessive form of the name of the constellation in which the star is located. The 12 constellations of the zodiac with their Latin possessives are as follows:

Constellation	Possessive
Aries	Arietis
Taurus	Tauri
Gemini	Geminorum
Cancer	Cancri
Leo	Leonis
Virgo	Virginis
Libra	Librae
Scorpius	Scorpii
Sagittarius	Sagittarii
Capricornus	Capricorni
Aquarius	Aquarii
Pisces	Piscium

In most cases the brightest star in the constellation is α, the second brightest is β, the third is γ, and so on. For example, the brightest star in Libra (the Scales) is called α Librae. This name is more informative than its Arabic name, Zubenelgenubi.

In Johann Bayer's system, only the brightest two dozen stars in a constellation can be named. However, astronomers are often interested in very faint stars, many of which are too dim to be seen with the naked eye. Astronomers

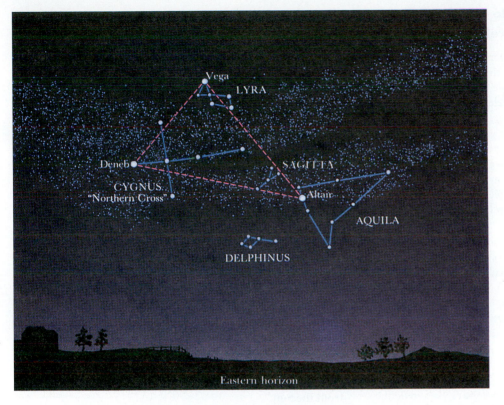

FIGURE 2-6 The "Summer Triangle" This star chart shows the northeastern sky as it appears in the evening in June. The angular distance from Deneb to Altair is about 38°. In addition to the three constellations in the triangle, the faint constellations of Sagitta (the Arrow) and Delphinus (the Dolphin) are shown.

refer to these fainter stars by using designations from standard star catalogues.

One of the first major star catalogues, called the **Bonner Durchmusterung** (Bonn Comprehensive Survey), was produced in Germany in the mid-1800s by F. W. Argelander at the Bonn Observatory. This catalogue lists the positions of 320,000 stars. Each star in the catalogue is designated by a "BD number." For example, the star BD+5° 1668 is a dim star in the constellation of Monoceros (the Unicorn).

Another commonly used catalogue is the **Henry Draper Catalogue**, which was compiled in the United States around 1920. (It is named after a physician whose widow financed the project.) Stars from this catalogue are listed by their "HD numbers." For example, HD 87901 is α Leonis (also called Regulus), the brightest star in Leo (the Lion).

Tens of thousands of faint nonstellar objects such as galaxies, nebulae, and star clusters are scattered across the sky. As with the faint stars, astronomers refer to these objects by their catalogue designations. Roughly a hundred of the brightest nonstellar objects are listed in a well-known catalogue by the eighteenth-century French astronomer Charles Messier. Astronomers often refer to these objects by their "M numbers." For example, the Orion Nebula (see Figure 1-6), the 42nd object on Messier's list, is called M42.

During the nineteenth century William Herschel and his son John Herschel extended this list to include nearly 5000 faint nonstellar objects. Their list was later enlarged by J. L. E. Dreyer and published in 1888 as the **New General Catalogue**. During the next 20 years new entries to this catalogue and its supplements expanded the list to include nearly 15,000 objects. Astronomers frequently refer to galaxies and nebulae by their "NGC numbers." For example, the Crab Nebula (see Figure 1-7) is called NGC 1952.

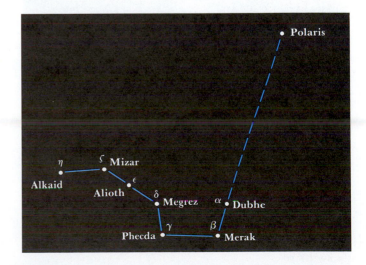

2-2 It is often convenient to imagine that the stars are located on the celestial sphere

Many ancient societies believed the Earth to be at the center of the universe. They also imagined the stars to be attached to the surface of a huge sphere whose center was the Earth. This imaginary sphere, called the **celestial sphere,** is still a useful concept.

The stars are actually scattered at various distances from Earth. Many of the brightest stars you can see without a telescope are 10 to 1000 light-years away. These distances are so immense that all the stars appear to be equally remote, as if fixed to that imaginary sphere. We can therefore use the celestial sphere as a reference for specifying the locations of objects in the sky.

The Earth is at the center of the celestial sphere, as shown in Figure 2-7. If we project the Earth's equator out into space, we obtain the **celestial equator.** The celestial equator divides the sky into northern and southern hemispheres, just as the Earth's equator divides the Earth into two hemispheres.

We can also project the Earth's north and south poles into space, thus obtaining the **north celestial pole** and the **south celestial pole.** The two celestial poles are the points of intersection between the Earth's axis of rotation extended out into space and the celestial sphere (see Figure 2-7).

Together, the celestial equator and poles are used to define a coordinate system with which you can specify the positions

of objects anywhere in the sky. The most commonly used coordinate system, which uses two angles (right ascension and declination) that are quite like longitude and latitude on Earth, is described in Box 2-2.

2-3 The seasons are caused by the tilt of the Earth's axis of rotation

In addition to rotating on its axis every 24 hours, the Earth revolves around the Sun in about $365\frac{1}{4}$ days. Earth's seasonal changes are caused by the tilt of the Earth's axis of rotation with respect to the plane of the Earth's orbit around the Sun.

The Earth's axis of rotation is not perpendicular to the Earth's orbit; instead it is tilted about $23\frac{1}{2}°$ away from the perpendicular, as shown in Figure 2-8. As it orbits the Sun, the Earth maintains this tilted orientation, with the Earth's north pole pointing toward Polaris. Thus, during part of the year we find the northern hemisphere tilted toward the Sun and the southern hemisphere tilted away. As a result, we have summer in the north and winter in the south. Half a year later the situation is reversed, with winter in the northern hemisphere (now tilted away from the Sun) and summer in the southern hemisphere. As Figure 2-8 illustrates, spring and fall occur between winter and summer, when the two hemispheres are receiving roughly equal amounts of illumination from the Sun.

As the Earth moves along its orbit, the Sun's position with respect to the background stars gradually shifts. The Sun therefore traces out a path in the sky during the year. This apparent path of the Sun on the celestial sphere is called the **ecliptic.** Because of the tilt of the Earth's axis of rotation, the ecliptic and the celestial equator are inclined to each other by about $23\frac{1}{2}°$ (Figure 2-9). Because there are $365\frac{1}{4}$ days in a year and $360°$ in a circle, the Sun appears to move along the ecliptic at a rate of approximately $1°$ per day.

As Figure 2-9 demonstrates, the ecliptic and the celestial equator intersect at only two points, which are exactly opposite each other on the celestial sphere. Both points are called **equinoxes** (from the Latin words meaning "equal night") because when the Sun appears at either of these points, day and night are each about 12 hours long at all locations on Earth.

The **vernal equinox** marks the beginning of spring in the northern hemisphere, when the Sun moves northward across the celestial equator on about March 21. The **autumnal equinox** marks the moment when fall begins in the northern hemisphere (about September 22), when the Sun moves southward across the celestial equator.

The northern and southern hemispheres experience opposite seasons at any given time. For example, March 21 marks the beginning of autumn for people in Australia. Obviously, these astronomical terms come from a time when virtually all astronomers lived north of the equator.

Between the vernal and autumnal equinoxes, there are two other significant locations along the ecliptic. The point on the ecliptic farthest north of the celestial equator is called the **summer solstice** (from the Latin for "solar standstill,"

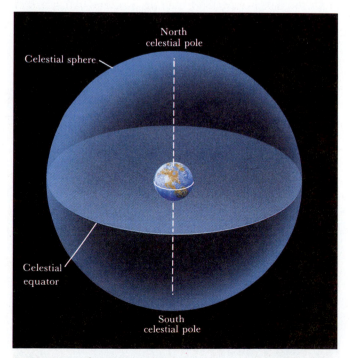

FIGURE 2-7 The Celestial Sphere The celestial sphere is the apparent sphere of the sky. The celestial equator and poles are obtained by projecting the Earth's equator and axis of rotation out into space. The celestial poles are therefore located directly over the Earth's poles.

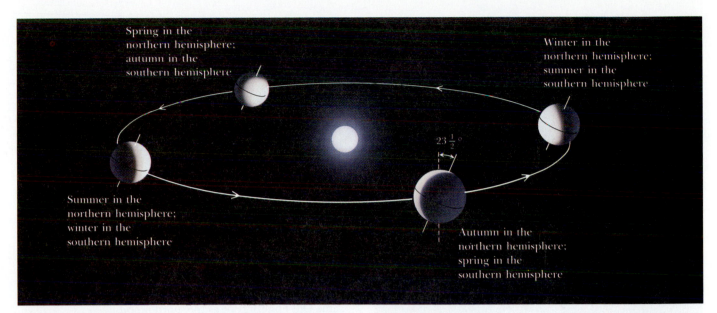

FIGURE 2-8 The Seasons The Earth's axis of rotation is inclined $23\frac{1}{2}°$ away from the perpendicular to the plane of the Earth's orbit. The Earth maintains this orientation (with its north pole aimed at the north celestial pole, near the star called Polaris) throughout the year as it orbits the Sun. Consequently, the amount of solar illumination and the number of daylight hours at any location on Earth vary in a regular fashion throughout the year.

BOX 2-2

Celestial Coordinates

To denote the positions of objects in the sky, astronomers use a system based on right ascension and declination. Declination corresponds to latitude. The **declination** of an object is its angular distance north or south of the celestial equator, measured along a circle passing through both celestial poles, as shown in the diagram.

Right ascension corresponds to longitude. Astronomers measure right ascension from a specific point, called the vernal equinox, on the celestial equator. This point is one of two locations where the Sun crosses the celestial equator during its apparent annual motion. In the Earth's northern hemisphere, spring is said to begin when the Sun reaches the vernal equinox in late March. The **right ascension** of an object in the sky is the angular distance from the vernal equinox eastward along the celestial equator to the circle used in measuring its declination (see diagram). Following traditional practice, astronomers measure this angular distance in time units (hours, minutes, and seconds) corresponding to the time required for the celestial sphere to rotate through this angle.

In catalogues of stars, galaxies, and nebulae, the positions of objects are given by their right ascension and declination. For example, the coordinates of the bright star Rigel (β Orionis) for the year 2000 are R.A. = 5^h 14^m 32.2^s, Decl. = $-8°$ $12'$ $06''$ where R.A. and Decl. are common abbreviations for right ascension and declination, respectively. Furthermore, a minus sign indicates that the star is south of the celestial equator; a plus sign (or no sign at all) indicates that an object is north of the celestial equator. These coordinates are very useful because they tell an astronomer precisely where an object is located.

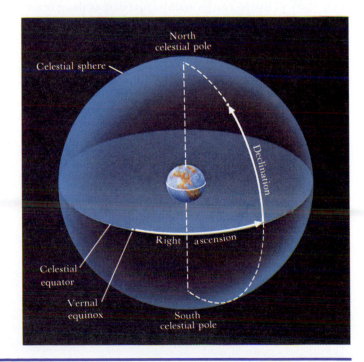

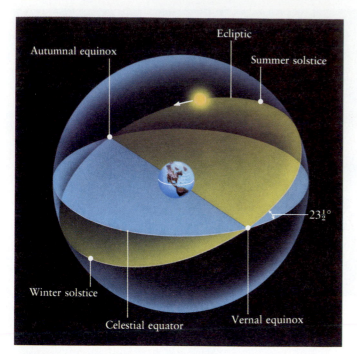

FIGURE 2-9 Equinoxes and Solstices The ecliptic is the apparent annual path of the Sun as projected onto the celestial sphere. Inclined to the celestial equator by $23\frac{1}{2}°$ because of the tilt of the Earth's axis of rotation, the ecliptic intersects the celestial equator at two points, called equinoxes. The northernmost point on the ecliptic is called the summer solstice, and the corresponding southernmost point, the winter solstice. The Sun is shown in its position for approximately September 1.

indicating that the Sun is as far from the celestial equator as it can get). It marks the location of the Sun at the moment summer begins in the northern hemisphere (about June 21). At the beginning of winter, the Sun is farthest south of the celestial equator, at a point called the **winter solstice** (about December 21).

Seasonal changes in the Sun's daily path across the sky are diagrammed in Figure 2-10. On the first day of spring or the first day of fall (when the Sun is at one of the equinoxes), the Sun rises directly in the east and sets directly in the west. Day and night are of equal duration.

During the summer months, when the northern hemisphere is tilted toward the Sun, the Sun rises in the northeast and sets in the northwest. It spends more than 12 hours above the horizon and passes high in the sky at noon. The summer is hot not only because of the extended daylight hours, but also because of the more nearly perpendicular angle at which midday sunlight strikes the ground. This nearly perpendicular impact of sunlight heats the ground more efficiently than when the Sun is lower in the sky, as it is during nonsummer months.

The point in the sky directly overhead is called the **zenith**, as shown in Figure 2-10. At the summer solstice, the Sun is as far north of the celestial equator as it gets, giving the north-

ern hemisphere the greatest number of daylight hours. There are even certain locations on Earth (north of the Arctic Circle, as discussed in Box 2-3) where the Sun does not set at all during summer nights.

During the winter months, when the northern hemisphere is tilted away from the Sun, the Sun rises in the southeast. Daylight lasts for less than 12 hours as the Sun skims low over the southern horizon and sets in the southwest. Night is longest when the Sun is at the winter solstice. In fact, north of the Arctic Circle a winter night lasts for all day, as described in Box 2-3.

2-4 Precession is a slow, conical motion of the Earth's axis of rotation

Ancient astronomers realized that the Moon orbits the Earth in roughly four weeks. The word *month* comes from the same Old English root as the word *moon*.

As seen from the Earth, the Moon is never far from the ecliptic. In other words, the Moon's path through the constellations is close to the Sun's path. The Moon's path remains within a band called the **zodiac** that extends about 8° on either side of the ecliptic. Twelve famous constellations (listed in Box 2-1) lie along the zodiac. The Moon is generally found in one of these 12 constellations. As it moves along its orbit, the Moon appears north of the celestial equator for about two weeks and then south of the celestial equator for about the next two weeks.

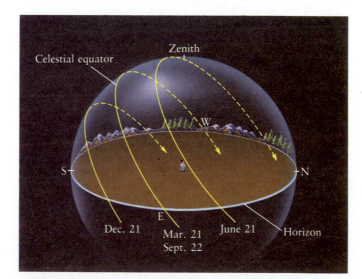

FIGURE 2-10 The Sun's Daily Path On the first day of spring and the first day of fall, the Sun rises precisely in the east and sets exactly in the west. During summer in the northern hemisphere, the Sun rises in the northeast and sets in the northwest. The Sun reaches its northernmost position at the summer solstice. In the winter, the Sun rises in the southeast and sets in the southwest, reaching its southernmost position at the winter solstice.

BOX 2-3

Tropics and Circles

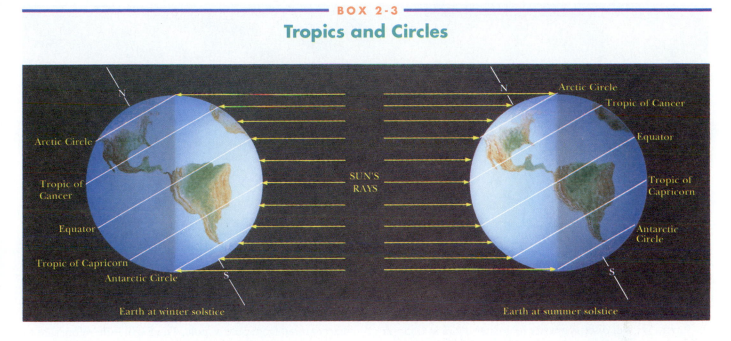

The inclination of the ecliptic to the celestial equator, called the **obliquity,** produces some noteworthy events at particular locations on Earth. For example, on any date during the year the Sun appears directly overhead at "high noon" along a band of locations encircling the Earth. The northernmost band where this happens is called the **Tropic of Cancer,** which circles the Earth at a latitude of $23\frac{1}{2}^{\circ}$ north of the equator, as shown in the illustration above. On the day of the summer solstice, the Sun is at the zenith at high noon all along the Tropic of Cancer.

The corresponding southern location is called the **Tropic of Capricorn,** $23\frac{1}{2}^{\circ}$ south of the equator. On the day of the winter solstice, when the Sun has reached its southernmost declination, the Sun passes directly overhead at high noon at all locations along this tropic.

There are also regions on Earth near the two poles where the Sun spends many consecutive days either above or below the horizon. For example, at the north pole, the Sun rises on the day of the vernal equinox, stays above the horizon for six months, and then sets on the day of the autumnal equinox. During the next six months, the north pole is subjected to six continuous months of night, because the Sun is too far south to be seen.

North of the **Arctic Circle** you can see the Sun for 24 continuous hours on at least one day of the year. The Arctic Circle lies $23\frac{1}{2}^{\circ}$ south of the north pole (that is, at a latitude of $90^{\circ} - 23\frac{1}{2}^{\circ} = 66\frac{1}{2}^{\circ}$ N).

The corresponding region around the south pole is bounded by the **Antarctic Circle,** as shown in the accompanying diagram. While the Earth's north polar regions are being subjected to continuous night near the time of the winter solstice, antarctic explorers enjoy the "midnight sun."

The Earth's inclination and your own latitude conspire to keep some constellations permanently above your horizon and others forever hidden from your view. For example, as viewed from the United States, the north celestial pole is always above the horizon. In fact, the elevation of the north celestial pole above the horizon is exactly equal to your latitude on Earth. As the Earth turns, constellations near the north celestial pole revolve about the pole, never rising or setting. Similarly, some constellations around the south celestial pole are too far south to be seen from northern latitudes.

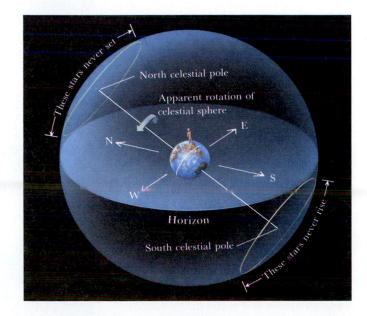

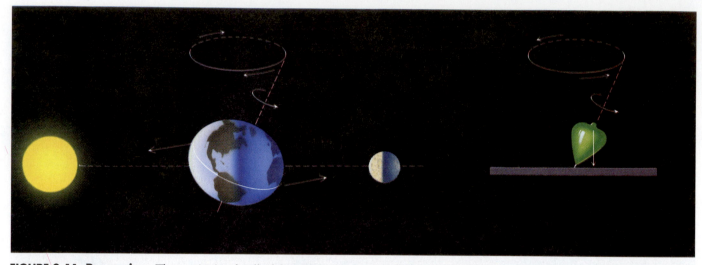

FIGURE 2-11 Precession The gravitational pull of the Moon and the Sun on the Earth's equatorial bulge causes the Earth to precess. As the Earth precesses, its axis of rotation slowly traces out a circle in the sky, like the shifting axis of a spinning top.

Both the Sun and the Moon exert a gravitational pull on the Earth. We will discuss gravity in greater detail in Chapter 4. For now, it is sufficient to realize that gravity is the universal attraction of matter for other matter.

The gravitational pull of the Sun and the Moon affects the Earth's rotation because the Earth is slightly fatter across the equator than it is from pole to pole: The equatorial diameter of the Earth is 43 kilometers (27 miles) larger than the diameter measured from pole to pole. The Earth is therefore said to have an "equatorial bulge." Because of the gravitational pull of the Moon and the Sun on this bulge, the orientation of the Earth's axis of rotation gradually changes.

The Earth behaves somewhat like a spinning top, as illustrated in Figure 2-11. If the top were not spinning, gravity

FIGURE 2-12 The Path of the North Celestial Pole As the Earth precesses, the north celestial pole slowly traces out a circle among the northern constellations. At present, the north celestial pole is near the moderately bright star Polaris, which serves as the "pole star." Twelve thousand years from now the bright star Vega will be the pole star.

would pull the top over on its side. When the top is spinning, gravity causes the top's axis of rotation to trace a circle, producing a motion called **precession.**

As the Sun and Moon move along the zodiac, each spends half its time north of the Earth's equatorial bulge and half its time south of it. The gravitational pull of the Sun and Moon tugging on the equatorial bulge tries to "straighten up" the Earth. In other words, as sketched in Figure 2-11, the gravity of the Sun and Moon tries to pull the Earth's axis of rotation perpendicular to the plane of the ecliptic. But the Earth is spinning. As with the toy top, the combined actions of gravity and rotation cause the Earth's axis to trace out a circle in the sky. As the axis precesses, it remains tilted about $23\frac{1}{2}°$ to the perpendicular.

The Earth's rate of precession is fairly slow. At present, the Earth's axis of rotation points within 1° of the star Polaris. In 3000 BC, it was pointing near the star Thuban in the constellation of Draco (the Dragon). In AD 14,000, the "pole star" will be Vega in Lyra (the Harp). It takes 26,000 years for the north celestial pole to complete one full precessional circle around the sky (Figure 2-12). Of course, the south celestial pole executes a similar circle in the southern sky.

As the Earth's axis of rotation precesses, the Earth's equatorial plane also moves. Because the Earth's equatorial plane defines the location of the celestial equator in the sky, the celestial equator precesses as well. The intersections of the celestial equator and the ecliptic define the equinoxes, as we saw in Figure 2-9, so these key locations in the sky also shift slowly from year to year. In fact, the entire phenomenon is often called the **precession of the equinoxes.** Today the vernal equinox is located in the constellation Pisces (the Fishes). Two thousand years ago, it was in Aries (the Ram). Around the year AD 2600, the vernal equinox will move into Aquarius (the Water Bearer). Incidentally, astrological terms like the "Age of Aquarius" involve boundaries in the sky that are not recognized by astronomers and are generally not even related to the positions of the constellations.

As discussed in Box 2-2, the astronomer's system of locating heavenly bodies by their right ascension and declination is tied to the positions of the celestial equator and the vernal equinox. Because of precession, these positions are changing, and so the coordinates of stars in the sky are constantly changing too. These changes are very small and gradual, but they add up over the years. To cope with this difficulty, astronomers always make note of the date (called the **epoch**) for which a particular set of coordinates is precisely correct. Consequently, star catalogues and star charts are periodically updated. Most new catalogues and star charts are prepared for the epoch 2000. The coordinates in these reference books, which are precise for January 1, 2000, will require very little correction over the next few decades.

2-5 Keeping track of time is traditionally a responsibility of astronomers

Since the dawn of civilization, people have needed accurate timekeeping. Ancient Egyptians wanted to know when the

Nile would flood. Farmers needed to know when to plant crops. Priests had to schedule religious observances. Indeed, the proper timing of seasonal religious events has always been important in most cultures. Even today we have many reasons to keep the calendar consistent with the cycle of the seasons. For example, many of our holiday traditions involve seasonal observances.

Astronomers have traditionally been responsible for telling time. In measuring short time intervals, we want our system of timekeeping to reflect the position of the Sun in the sky. The Sun's position determines whether we are awake or asleep and whether it is time for breakfast or for dinner.

Thousands of years ago, the sundial was invented to keep track of **apparent solar time.** To obtain more formal measurements, astronomers use the **meridian,** which is a north–south circle on the celestial sphere that passes through the zenith (the point directly overhead) and both celestial poles, as shown in Figure 2-13. By definition, the Sun crosses the meridian above the horizon at local noon. At local midnight, the Sun makes the opposite crossing below the horizon, although this crossing cannot be observed directly. The crossing of the meridian by any object in the sky is called a **meridian transit.** If the crossing occurs above the horizon, it is an upper meridian transit. An **apparent solar day** is formally defined as the interval between two successive upper meridian transits of the Sun as observed from any fixed spot on the Earth.

Early observers soon realized that the Sun is not a good timekeeper. Most obviously, the apparent solar day (as measured by a device such as an hourglass) varies from one season to another. More subtly, the speed of the Sun's slow eastward progress against the background stars varies over the course of a year.

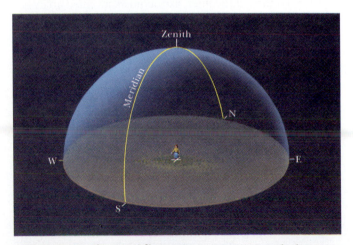

FIGURE 2-13 The Meridian The meridian is a circle that passes through the observer's zenith and the two celestial poles. The passing of celestial objects across the meridian can be used to measure time.

FIGURE 2-14 Why the Sun Is a Poor Timekeeper There are two main reasons that the Sun is a poor timekeeper. (a) The Earth's speed along its orbit varies throughout the year. Hence, the apparent speed of the Sun along the ecliptic is not constant. (b) The projection of the Sun's daily progress along the ecliptic onto the celestial equator varies throughout the year. Hence, the Sun's net daily eastward progress against the stars is not constant.

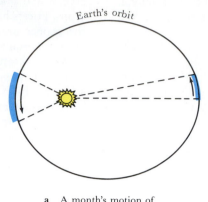

a A month's motion of the Earth along its orbit

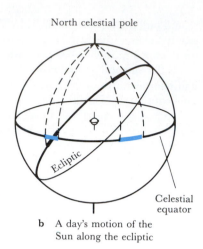

b A day's motion of the Sun along the ecliptic

There are two main reasons for the Sun's nonuniform motion. First, the Earth's orbit is not a perfect circle. The orbit is actually an ellipse, as shown in exaggerated form in Figure 2-14a. As we shall learn in Chapter 4, the Earth moves more rapidly along its orbit when it is near the Sun than when it is farther away. The speed of the Sun along the ecliptic thus varies throughout the year. The Sun appears to move more than 1° per day along the ecliptic in January, when the Earth is nearest the Sun, and less than 1° per day in July, when the Earth is farthest from the Sun.

Even if the Sun did move along the ecliptic at a uniform rate, it would still not be a good timekeeper. Because the ecliptic is inclined by $23\frac{1}{2}°$ to the celestial equator, a significant part of the Sun's motion near the equinoxes is in a north–south direction (Figure 2-14b). The net daily eastward progress in the sky is then somewhat foreshortened. In contrast, the Sun's motion is parallel to the celestial equator at the time of the solstices, so there is no comparable foreshortening around the beginning of summer or winter.

To avoid these difficulties, astronomers invented an imaginary object called the **mean sun** that moves along the celestial equator at a uniform rate. (In this context, *mean* is a synonym for "average.") The mean sun is sometimes slightly ahead of the real Sun in the sky, sometimes behind. As a result, mean solar time and apparent solar time can differ by as much as a quarter of an hour at certain times of the year.

Since the mean sun moves at a constant rate, it serves as a fine timekeeper. A **mean solar day** is the interval between successive upper meridian transits of the mean sun. It is exactly 24 hours long, the average length of an apparent solar day. Our clocks and wristwatches measure mean solar time.

Time zones were invented for convenience in commerce, transportation, and communication. In a time zone, everyone agrees to set all clocks and watches to the mean solar time for a meridian of longitude that runs approximately through the center of the zone. Time zones around the world are generally centered on meridians of longitude at 15° intervals. Going from one time zone to the next requires you to change the time on your wristwatch by exactly one hour. The time zones for North America are shown in Figure 2-15.

Although it is natural to want our clocks and method of timekeeping to be related to the Sun, astronomers often use a system that is related to the stars. **Sidereal time**, which is based on the apparent motion of the stars rather than that of the Sun, is useful when aiming a telescope. Most observatories are therefore equipped with a sidereal clock, as discussed in Box 2-4.

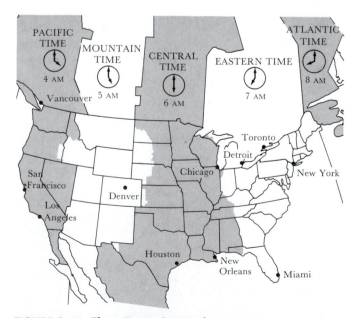

FIGURE 2-15 Time Zones in North America For convenience, the Earth is divided into 24 time zones, generally centered on 15° intervals of longitude around the globe. Consequently, there are four time zones across the continental United States, giving a three-hour time difference between New York and California.

BOX 2-4

Sidereal Time

If you want to observe a particular star or galaxy, it must obviously be above the horizon. It is best to view an astronomical object when it is high in the sky to minimize the distorting effects of the Earth's atmosphere, which increase as you view closer to the horizon. Ideally, the object should be on or near the upper meridian.

To tell which objects are crossing the upper meridian, most observatories are equipped with a **sidereal clock,** which tells sidereal time. Sidereal time is simply the right ascension of any object on the meridian. As new objects cross the meridian, sidereal time progresses.

EXAMPLE: Suppose you want to observe the bright star Regulus in the constellation of Leo (the Lion). From reference books, such as the *Astronomical Almanac,* you find that the Epoch 2000.0 coordinates of this star are

$$\text{R.A.} = 10^h\ 08^m\ 22.2^s$$

$$\text{Decl.} = +11°\ 58'\ 02''$$

Since its right ascension is $10^h\ 07^m\ 58.4^s$, Regulus will be on the upper meridian when the sidereal time is about 10:08. A sidereal clock is useful because it tells you the right ascension of those objects near the meridian and thus best suited for observation.

From this example, you can see why astronomers measure right ascension in units of time. Because the Earth rotates once a day, it takes nearly 24 hours for all the stars to pass across the meridian. By measuring right ascension in units of time, astronomers have a convenient relationship between time and star positions.

It is sometimes convenient to convert between angular and sidereal time. For example, you might want to know how many degrees correspond to one hour of right ascension. Because 24 hours of right ascension takes you all the way around the celestial equator, we have the relationship that $24^h = 360°$. Therefore, the following equivalents hold:

$$1\ h = 15°$$

$$1\ m = 15'$$

$$1\ s = 15''$$

$$1° = 4\ m$$

$$1' = 4\ s$$

$$1'' = 0.067\ s$$

Sidereal time is useful only to people such as astronomers and navigators who deal with the stars. It is different from the time on your wristwatch. In fact, a sidereal clock

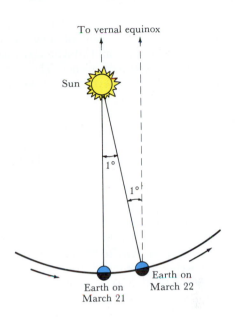

and an ordinary clock even tick at different rates because they are based on different astronomical objects. Ordinary clocks are related to the position of the Sun, while sidereal clocks are based on the position of the vernal equinox, the location from which right ascension is measured.

Regardless of where the Sun is, the sidereal day starts when the vernal equinox crosses the upper meridian. A **sidereal day** is the time between two successive upper meridian passages of the vernal equinox. In contrast, an apparent solar day is the time between two successive upper meridian crossings of the Sun. These two kinds of day are not equal for reasons illustrated in the diagram. Because the Earth moves along its orbit around the Sun, the Earth must rotate through nearly 361° to get from one local noon to the next. This extra 1° of rotation corresponds to four minutes of time, which is the amount by which a solar day exceeds a sidereal day. To be precise,

$$1\ \text{sidereal day} = 23^h\ 56^m\ 4.091^s$$

where the hours, minutes, and seconds are in mean solar time. Of course, one day according to your wristwatch is one mean solar day, which is exactly 24 hours of solar time long.

The sidereal clock measures sidereal hours, minutes, and seconds—dividing the sidereal day into exactly 24 sidereal hours. This explains why the sidereal clock ticks at a slightly different rate from your wristwatch. Measurements of right ascension are always expressed in sidereal hours, minutes, and seconds. All time measurements in this book are expressed in mean solar time unless otherwise stated.

2-6 Astronomical observations led to the development of the modern calendar

Just as the day is a natural unit of time based on the Earth's rotation, the year is a natural unit of time based on the Earth's revolution about the Sun. Unfortunately, nature has not arranged things for our convenience. The year does not divide into exactly 365 whole days. Ancient astronomers realized that the length of a year is approximately $365\frac{1}{4}$ days, and Julius Caesar established the system of "leap years" to account for this extra quarter of a day. By adding an extra day to the calendar every four years, Caesar hoped to ensure that seasonal astronomical events, such as the beginning of spring, would occur on the same date year after year. The Roman Catholic Church then became concerned because Easter kept shifting to progressively earlier dates.

Caesar's system would have been perfect if the year were exactly $365\frac{1}{4}$ days long and if there were no precession. Unfortunately, this is not the case. To be more accurate, astronomers now use several different types of years. For example, the **sidereal year** is defined to be the time required for the Sun to return to the same position with respect to the stars. It is equal to 365.2564 mean solar days, or $365^d\ 6^h\ 9^m\ 10^s$.

The sidereal year is the true orbital period of the Earth around the Sun, but it is not the year on which we base our calendar. Like Caesar, most people want annual events to fall on the same date each year. For example, we want the first day of spring to occur on March 21. But spring begins when the Sun is at the vernal equinox, and the vernal equinox moves slowly against the background stars because of precession. Therefore, to set up a calendar we make use of the **tropical year,** which is equal to the time needed for the Sun to return to the vernal equinox. This period is equal to 365.2422 mean solar days, or $365^d\ 5^h\ 48^m\ 46^s$. Because of precession, the tropical year is 20 minutes and 24 seconds shorter than the sidereal year, a difference already known to the ancient Greeks.

In the second century BC, Hipparchus calculated the length of the tropical year to within six minutes of the currently accepted value. (Hipparchus was also the first person to detect the precession of the equinoxes when he compared his own observations with those of Babylonian astronomers three centuries earlier.) Caesar's assumption that the tropical year equals $365\frac{1}{4}$ years was less precise—off by 11 minutes and 14 seconds. This tiny error adds up to about three days every four centuries. Although Caesar's astronomical advisers were aware of the discrepancy, they felt that it was too small to matter. However, by the sixteenth century the first day of spring was occurring on March 11.

To straighten things out, Pope Gregory XIII instituted a calendar reform in 1582. He began by dropping ten days (October 5, 1582, was proclaimed to be October 15, 1582), which brought the first day of spring back to March 21. Next he modified Caesar's system of leap years.

Caesar had added February 29 to every calendar year that is evenly divisible by four. Thus, for example, 1984, 1988, and 1992 were all leap years with 366 days. But we have seen that this system produces an error of about three days every four centuries. To solve the problem, Pope Gregory decreed that only the century years evenly divisible by 400 should be leap years. For example, the years 1700, 1800, and 1900 (which would have been leap years, according to Caesar) were not leap years in the improved Gregorian system. But the year 2000—which can be divided evenly by 400—is a leap year.

We use the Gregorian system today. It assumes that the year is 365.2425 mean solar days long, which is very close to the true length of the tropical year. In fact, the error is only one day in every 3300 years. That won't cause any problems for a long time.

KEY WORDS

Terms preceded by an asterisk are discussed in the boxes.

* Antarctic Circle
apparent solar day
apparent solar time
* Arctic Circle
* *Astronomical Almanac*
autumnal equinox
* *Bonner Durchmusterung*
celestial equator
celestial sphere
constellation
* declination

diurnal motion
ecliptic
epoch
equinox
* *Henry Draper Catalogue*
mean solar day
mean sun
meridian
meridian transit
* *New General Catalogue*
north celestial pole

* obliquity
precession
precession of the equinoxes
* right ascension
* sidereal clock
* sidereal day
sidereal time
sidereal year
south celestial pole
summer solstice

time zone
tropical year
* Tropic of Cancer
* Tropic of Capricorn
vernal equinox
winter solstice
zenith
zodiac

KEY IDEAS

- It is convenient to imagine the stars fixed to the celestial sphere with the Earth at its center.

  The surface of the celestial sphere is divided into 88 regions called constellations.

- The celestial sphere appears to rotate around the Earth once in each 24-hour period; of course, it is actually the Earth that is rotating.

  The poles and equator of the celestial sphere are determined by extending the axis of rotation and the equatorial plane of the Earth out to the celestial sphere.

  The positions of objects on the celestial sphere are described by specifying their right ascension (in time units) and declination (in angular measure).

- The Earth's axis of rotation is tilted at an angle of about $23\frac{1}{2}^{\circ}$ from the perpendicular to the plane of the Earth's orbit.

  The seasons are caused by the tilt of the Earth's axis.

  Equinoxes and solstices are significant points in the sky determined by the relationship between the Sun's path on the celestial sphere (the ecliptic) and the celestial equator.

- The orientation of the Earth's axis of rotation moves slowly, a phenomenon called precession.

  Precession is caused by the gravitational pull of the Sun and Moon on the Earth's equatorial bulge.

  Precession of the Earth's axis causes precession of the equinoxes along the ecliptic.

- Because the system of right ascension and declination is tied to the position of the vernal equinox, the date (or epoch) of observation must be specified when giving the position of an object in the sky.

- Timekeeping is an important concern of astronomers.

  Apparent solar time is based upon the apparent motion of the Sun across the celestial sphere, which varies with the seasons.

  Mean solar time is based upon the motion of an imaginary mean sun along the celestial equator, which produces a uniform mean solar day of 24 hours.

  Sidereal time is based upon the apparent motion of the celestial sphere; the sidereal year is the actual orbital period of the Earth.

- The tropical year is the period between two passages of the Sun across the vernal equinox; leap year corrections are needed because the tropical year is not exactly 365 days.

REVIEW QUESTIONS

1. How are constellations useful to astronomers? Are there any stars in the sky that are not part of a constellation?

2. Why can't a person in Australia use the Big Dipper to find the north direction?

3. What is the celestial sphere, and why is this ancient concept still useful today?

4. What is the celestial equator, and how is it related to the Earth's equator? How are the north and south celestial poles related to the Earth's axis of rotation?

5. From what location on Earth does the north celestial pole appear on the horizon?

6. Where do you have to be on Earth in order to see the Sun at the zenith? If you stay at that location for a full year, on how many days will the Sun pass through the zenith?

7. Where do you have to be on Earth in order to see the south celestial pole directly overhead? What is the maximum possible elevation of the Sun above the horizon at that location? On what date can this maximum elevation be observed?

8. Using a diagram, explain why the tilt of the Earth's axis relative to the Earth's orbit causes the seasons as we orbit the Sun.

9. What are the vernal and the autumnal equinoxes? What are the summer and winter solstices? How are these four points related to the ecliptic and the celestial equator?

10. At what point on the horizon does the vernal equinox rise?

11. How does the daily path of the Sun across the sky change with the seasons?

12. What causes precession of the equinoxes? How long does it take for the vernal equinox to move 1° along the ecliptic?

13. Why is the fictitious mean sun a better timekeeper than the actual Sun in the sky?

14. Why is the time given by a sundial not necessarily the same as the time on your wristwatch?

15. Why is it convenient to have the Earth divided into time zones?

16. What is the difference between the sidereal year and the tropical year? Why are calendars based on the tropical year?

17. When is the next leap year? Will the year 2000 be a leap year?

ADVANCED QUESTIONS

Tips and tools . . .

Take the time to become familiar with various types of star charts. For instance, you should compare the monthly star charts published in such magazines as *Sky & Telescope* and *Astronomy* with more detailed maps of the heavens, like those in Wil Tirion's *Sky Atlas 2000.0* (Sky Publishing Corporation and Cambridge University Press, 1981). You may also find it useful to examine a *planisphere*, a device consisting of two rotatable disks. The bottom disk shows all the stars in the sky (for a particular latitude), while the top one is opaque with a transparent, oval window through which only some of the stars can be seen. By rotating the top disk, you can immediately see which constellations are above the horizon at any time of the year.

18. On December 1 at 10:00 PM you look toward the eastern horizon and see the bright star Procyon rising. At approximately what time will Procyon rise two weeks later, on December 15?

19. How would the sidereal and solar days change (**a**) if the Earth's rate of rotation increased, (**b**) if the Earth's rate of rotation decreased, and (**c**) if the Earth's rotation were *retrograde* (i.e., if the Earth rotated about its axis opposite to the direction in which it revolves about the Sun)?

20. What is the sidereal time when the vernal equinox rises?

21. On what date is the sidereal time nearly equal to the solar time?

22. Consult a star map of the southern hemisphere and determine which, if any, bright southern stars could someday become south celestial pole stars.

23. Are there stars in the sky that never set? Are there stars that never rise? Does your answer depend on the observer's location on Earth?

24. Using a diagram, demonstrate that your latitude on Earth is equal to the altitude of the north celestial pole above your northern horizon.

25. Go to a library that has recent issues of the *Astronomical Almanac.* In the section entitled "Bright Stars," look up the coordinates of a star of your choice for two successive years. By how much did the right ascension and declination of that star change?

26. Using a star map, determine which bright stars, if any, could someday mark the location of the vernal equinox. Give the approximate years when this should happen.

27. What is the right ascension of a star that is on the meridian at midnight at the time of the autumnal equinox?

28. Suppose that you live at a latitude of 40° N. What is the elevation of the Sun above the southern horizon at noon at the time of the winter solstice?

DISCUSSION QUESTIONS

29. Examine a list of the 88 constellations. Are there any constellations whose names obviously date from modern times? Where are these constellations located? Why do you suppose they do not have archaic names?

30. Describe the seasons if the Earth's axis of rotation were tilted (**a**) 0° and (**b**) 90° to its orbital plane.

OBSERVING PROJECTS

31. On a clear, cloud-free night, use the star charts at the end of this book to see how many constellations of the zodiac you can identify. Which ones were easy to find? Which were difficult?

32. Examine the star charts that are published monthly in such popular astronomy magazines as *Sky & Telescope* and *Astronomy.* How do they differ from the star charts at the back of this book? On a clear, cloud-free night, use one of these star charts to locate the celestial equator and the ecliptic. Note the inclination of the Milky Way to the ecliptic and celestial equator. What do your observations tell you about the orientation of the Earth and its orbit relative to the rest of the Galaxy?

33. Suppose you wake up before dawn and want to see which constellations are in the sky. Explain how the star charts at the end of this book can be quite useful, even though chart times are given only for the evening hours. Which chart most closely depicts the sky at 4:00 AM tomorrow morning? Set your alarm clock for 4:00 AM to see if you are correct.

FOR FURTHER READING

Allen, R. *Star Names: Their Lore and Meaning.* Originally published 1899. Dover, 1963. A classic book detailing the origins of constellations and star names.

Berry, R. *Discover the Stars.* Harmony, 1987. An excellent manual featuring clear, beautiful star charts showing the whole sky, along with enlarged, detailed views.

Brown, R. H. *Man and the Stars.* Oxford University Press, 1978. This book includes nice sections on the history of calendars, timekeeping, and navigation, as well as a discussion of the interaction between astronomy and philosophy.

Chartrand, M. *Skyguide.* Western Publications, 1982. This pocket-sized book gives a concise introduction to observational astronomy and has superb illustrations and excellent star charts.

Gallant, R. *The Constellations: How They Came To Be.* Four Winds Press, 1979. The stories of 50 constellations are accompanied by star charts and useful information.

Gingerich, O. "The Origin of the Zodiac." *Sky & Telescope,* March 1984. A brief article on the constellations by a distinguished historian of astronomy.

Kunitzsch, P. "How We Got Our Arabic Star Names." *Sky & Telescope,* January 1983. This fascinating article gives many historical insights into star names and Islamic astronomy during the Middle Ages.

Motz, L., and Nathanson, C. *The Constellations: An Enthusiast's Guide to the Night Sky.* Doubleday, 1988. A useful guide to the constellations that gives mythology, observing information, and astronomical data.

Ridpath, I., and Tirion, W. *Universe Guide to Stars and Planets.* Universe, 1985. All 88 constellations are covered in exquisitely drawn star charts along with sky maps for different seasons.

ECLIPSES AND ANCIENT ASTRONOMY

THE CARACOL AT CHICHÉN ITZÁ This ancient Mayan observatory in the Yucatán was built around AD 1000. Its architecture is based on alignments with important celestial events. Mayan astronomers developed a very accurate calendar and measured the motions of celestial bodies with great precision. Special significance was associated with the planet Venus, which inspired sacrificial rites and other ceremonies. (Courtesy of E. C. Krupp)

ANCIENT CULTURES made important observations that set the stage for modern astronomy. This chapter retraces their discoveries, beginning with how the phases of the Moon relate to its motion about the Earth. Ancient astronomers ingeniously attempted to measure the size of the Earth and the distances from Earth to the Sun and the Moon. Eclipses of the Sun and the Moon, which are dramatic phenomena, played important roles in many early theories. Indeed, the first astronomers understood eclipses well enough to predict their occurrences fairly reliably. Another enduring legacy is the magnitude scale, handed down to us from a Greek astronomer in the second century BC and still used to denote the apparent brightness of objects in the sky.

The beauty of the star-filled night sky or the drama of an eclipse may well suffice to make astronomy fascinating. However, even the early Greeks had practical reasons for studying the universe. They knew the connection between the seasons and the relative orientations of the Sun and Earth. Like many other seafaring cultures, they were aware that the position of the Moon influences the tides.

Ancient civilizations placed great emphasis on careful astronomical observation. Hundreds of impressive monuments that dot the British Isles, such as Stonehenge (Figure 3-1), provide evidence of this preoccupation with astronomy.

Alignments of these stones show where the Sun and Moon rise and set at key times during the year, such as solstices. Similar monuments have also been found in the United States. For example, the Plains Indians constructed a circular ring of stones, called the Medicine Wheel, atop a windswept plateau in Wyoming. Certain stones in this ring are markers for the rising points of specific bright stars, including the Sun at the summer solstice.

Aztec, Mayan, and Incan architects in Central and South America also designed buildings to be astronomically oriented. For example, at the ruined city of Tiahuanaco, in

FIGURE 3-1 Stonehenge This astronomical monument was constructed nearly 4000 years ago on Salisbury Plain in southern England. The monument originally consisted of 30 blocks of gray sandstone, each 4 meters (13 feet) high, set in a circle 30 meters (98 feet) in diameter. These stones were topped with a continuous circle of smaller stones. Inside this circle there still exists a geometric arrangement of other stones, most notably a horseshoe-shaped set of larger stones that open toward the northeast. (Courtesy of the British government)

Bolivia, the walls of the Temple of the Sun were aligned north–south and east–west with an accuracy of better than one degree. The great Egyptian pyramids, built 5000 years ago, are likewise oriented north–south and east–west with remarkable precision.

In addition to temples and tombs, the ancients constructed what were apparently astronomical observatories. One of the best examples lies in the Mayan city of Chichén Itzá on the Yucatán Peninsula. Built nearly a thousand years ago, the Caracol's cylindrical tower contains windows aligned with the northernmost and southernmost rising and setting points of both the Sun and the planet Venus. A similar four-story adobe building, probably constructed during the fourteenth century, is located at the Casa Grande site in Arizona. These structures all bear witness to careful, patient astronomical observations by the peoples of many cultures.

3-1 Lunar phases are caused by the Moon's orbital motion

Ancient astronomers knew that the Moon shines by reflected sunlight and that it orbits the Earth. They deduced these facts by observing the Moon's changing appearance from night to night.

Although ancient astronomers did not have the benefit of photographs from space, like Figure 3-2, they realized that the Moon is a sphere. The hemisphere of the Moon that faces the Sun is always illuminated, while the hemisphere that faces away from the Sun is always in darkness. As the Moon orbits the Earth, we see different **lunar phases,** depending on the amount of the Moon's illuminated hemisphere exposed to our Earth-based view.

The Moon orbits the Earth in about four weeks, during which time the Moon completes a full cycle of its phases (Figure 3-3). Its phase depends upon where the Moon is relative to the Sun as seen from the Earth. For example, when the Moon is in the same part of the sky as the Sun, the dark hemisphere of the Moon faces the Earth. This phase, during which the Moon is not visible, is called **new moon.**

During the next seven days, more of the Moon's illuminated hemisphere becomes exposed to our view, resulting in a phase called **waxing crescent moon.** At **first quarter moon,** we see half of the illuminated hemisphere and half of the dark hemisphere.

During the next week, still more of the illuminated hemisphere can be seen from Earth, giving us the phase called **waxing gibbous moon.** When the Moon stands opposite the Sun in the sky, we see the fully illuminated hemisphere in the phase called **full moon.**

Over the following two weeks, we see less and less of the illuminated hemisphere as the Moon continues along its orbit. This movement produces the phases called **waning gibbous moon, last quarter moon,** and **waning crescent moon.**

Figure 3-4 shows the Moon at various positions on its orbit. Note, for instance, that the angle between the Sun

FIGURE 3-2 The Earth and the Moon The Moon orbits the Earth every 27.3 days, at an average distance of 384,400 kilometers (238,900 miles). This picture was taken in 1992 by the *Galileo* spacecraft on its way toward Jupiter. (NASA)

and Moon at both first and last quarter is 90°, when we see exactly half of the Moon's illuminated hemisphere and half of the darkened hemisphere. When the Moon is a crescent, the angular distance between the Sun and Moon is less than 90°, which exposes less than half of the illuminated hemisphere to our Earth-based view. When the Moon is gibbous, the angle between the Moon and the Sun is greater than 90°, exposing more than half of the Moon's illuminated hemisphere to our view.

Figure 3-4 also shows local time around the globe, from noon, where the Sun is overhead, to midnight on the opposite side of the Earth. These time markings can be used to correlate the phase and position of the Moon with the time of day. For example, at first quarter the Moon is 90° east of the Sun in the sky, and hence moonrise occurs approximately at noon. At full moon the Moon is opposite the Sun in the sky and thus moonrise occurs at sunset.

The four weeks that the Moon takes to complete one cycle of its phases inspired our ancestors to invent the concept of a month. For historical reasons, none of which have much to do with the heavens, the calendar we use today has months of differing lengths. In contrast, astronomers find it useful to define two other types of months, depending on whether the Moon's motion is measured relative to the stars or to the Sun. Neither corresponds exactly to the familiar months of our usual calendar.

FIGURE 3-3 The Moon's Phases As the Moon orbits the Earth, varying amounts of the Moon's illuminated hemisphere are exposed to our Earth-based view. The "age" refers to the time that has elapsed since new moon phase. It takes about $29\frac{1}{2}$ days for the Moon to go through all its phases. Although its phases change, the Moon always keeps the same side facing the Earth. Consequently, Earth-based observers always see the same craters and lunar mountains, regardless of the Moon's phase. (Lick Observatory)

Waxing crescent
(age: 4 days)

First quarter
(age: 7 days)

Waxing gibbous
(age: 10 days)

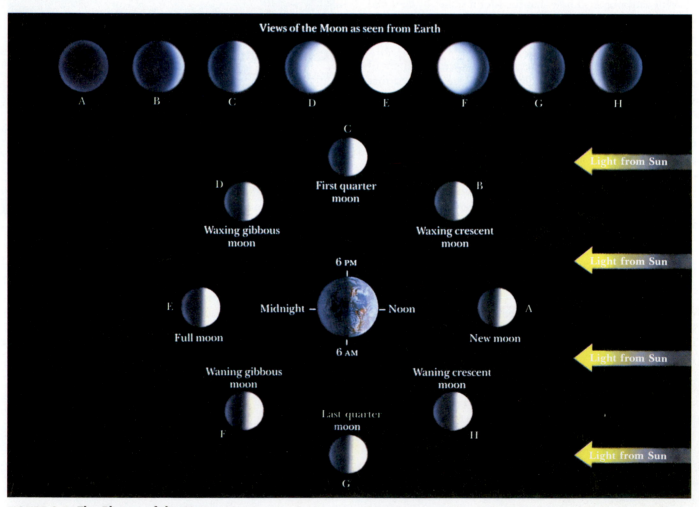

FIGURE 3-4 The Phases of the Moon This diagram shows the Moon at eight locations on its orbit (as viewed from above the Earth's north pole) along with the resulting lunar phases. Light from the Sun illuminates one half of the Moon. As the Moon orbits the Earth, we see varying amounts of the Moon's illuminated hemisphere. Local time on the Earth is also indicated.

| Full moon (age: 14 days) | Waning gibbous (age: 20 days) | Last quarter (age: 22 days) | Waning crescent (age: 26 days) |

The **sidereal month** is the time it takes the Moon to complete one full orbit of the Earth, measured with respect to the stars. This true orbital period is equal to 27.3 days. The **synodic month,** or **lunar month,** is the time it takes the Moon to complete one cycle of phases (i.e., from new moon to new moon, or full moon to full moon) and thus is measured with respect to the Sun (rather than the stars).

The synodic month is longer than the sidereal month because the Earth is orbiting the Sun while the Moon goes through its phases. As Figure 3-5 shows, the Moon must travel *more* than 360° along its orbit to complete a cycle of phases (e.g., from one new moon to the next). Because of this extra distance, the synodic month is equal to about 29.53 days—about two days longer than the sidereal month.

Both the sidereal and synodic month vary somewhat, the latter by as much as half a day. The reason is that the Sun's gravity sometimes causes the Moon to speed up or slow down slightly in its orbit, depending on the relative positions of the Sun, Moon, and Earth.

3-2 Ancient astronomers measured the size of the Earth and attempted to determine distances to the Sun and Moon

More than two thousand years ago, centuries before sailors of Columbus's era crossed the oceans, Greek astronomers were fully aware that the Earth is not "flat." During a lunar eclipse, when the Moon passes through the Earth's shadow, these astronomers noted that the edge of the shadow is always circular. Because a sphere is the only shape that always casts a circular shadow from any angle, they concluded that the Earth is spherical.

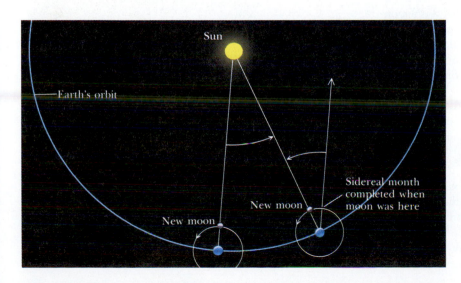

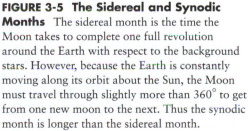

FIGURE 3-5 The Sidereal and Synodic Months The sidereal month is the time the Moon takes to complete one full revolution around the Earth with respect to the background stars. However, because the Earth is constantly moving along its orbit about the Sun, the Moon must travel through slightly more than 360° to get from one new moon to the next. Thus the synodic month is longer than the sidereal month.

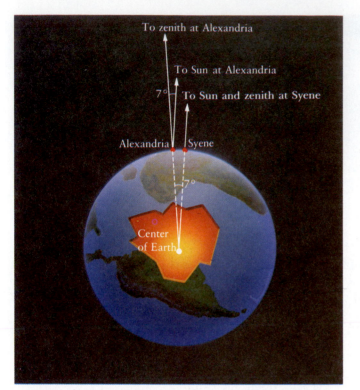

FIGURE 3-6 Eratosthenes' Method of Determining Earth's Diameter Eratosthenes noticed that the Sun is about 7° south of the zenith at Alexandria when it is directly overhead at Syene. This angle is about one-fiftieth of a circle, so the distance between Alexandria and Syene must be about one-fiftieth of the Earth's circumference.

Around 200 BC, the Greek astronomer Eratosthenes devised a way to measure the circumference of the Earth. He had been intrigued by reliable reports from the town of Syene in Egypt (near the modern Aswan) that the Sun there shone directly down vertical wells on the first day of summer. Eratosthenes knew that the Sun never appeared at the zenith at his home in the Egyptian city of Alexandria, which is on the Mediterranean Sea almost due north of Syene. Rather, on the summer solstice in Alexandria, the position of the Sun at local noon was about 7° south of the zenith (Figure 3-6). This angle is one-fiftieth of a complete circle, so he concluded that the distance from Alexandria to Syene must be one-fiftieth of the Earth's circumference.

In Eratosthenes' day, the distance from Alexandria to Syene was said to be 5000 stades. Therefore, Eratosthenes found the Earth's circumference to be

$$50 \times 5000 \text{ stades} = 250,000 \text{ stades}$$

Unfortunately, no one today is sure of the exact length of the Greek unit called the stade. One guess is that the stade was about $\frac{1}{6}$ kilometer, which would mean that Eratosthenes obtained a circumference for the Earth of about 42,000 kilometers, which is within 5% of the modern value of 40,000 kilometers.

Eratosthenes was only one of several brilliant astronomers to emerge from the so-called Alexandrian school, which by his time already had a distinguished tradition. One of the first Alexandrian astronomers, Aristarchus of Samos, had proposed a method of determining the relative distances to the Sun and Moon, perhaps as long ago as 280 BC.

Aristarchus knew that the Sun, Moon, and Earth form a right triangle at the moment of first or last quarter moon, with the right angle at the location of the Moon (Figure 3-7). He estimated that, as seen from Earth, the angle between the Moon and the Sun at first and third quarters is 3° less than a right angle (that is, 87°). Using the rules of geometry, Aristarchus concluded that the Sun is about 20 times farther from us than the Moon is. We now know that the average distance to the Sun is about 390 times larger than the average distance to the Moon. It is nevertheless impressive that people were trying to measure distances across the solar system more than 2000 years ago.

Using lunar eclipses, Aristarchus also made an equally bold attempt to determine the relative sizes of the Earth, Moon, and Sun. From his observations of how long the Moon takes to move through the Earth's shadow, Aristarchus estimated the diameter of the Earth to be about three times larger than the diameter of the Moon. To determine the diameter of the Sun, Aristarchus simply pointed out that the Sun and the Moon have the same angular size in the sky. Therefore, their diameters must be in the same proportion as their distances. In other words, because Aristarchus believed the Sun to be 20 times farther from the Earth than the Moon, he concluded that the Sun must be 20 times larger than the Moon. Once Eratosthenes had measured the Earth's circumference, astronomers of the Alexandrian school could estimate the diameters of the Sun and Moon as well as their distances from Earth.

For comparison, Table 3-1 summarizes some ancient and modern measurements of the sizes of the Earth, Moon, and Sun and the distances between them. Although some of these ancient measurements are far from the modern values, our

---

TABLE 3-1

A Comparison of Ancient and Modern Measurements

	Ancient measure (km)	Modern measure (km)
Earth's diameter	13,000	12,756
Moon's diameter	4,300	3,476
Sun's diameter	9×10^4	1.39×10^6
Earth–Moon distance	4×10^5	3.84×10^5
Earth–Sun distance	10^7	1.50×10^8

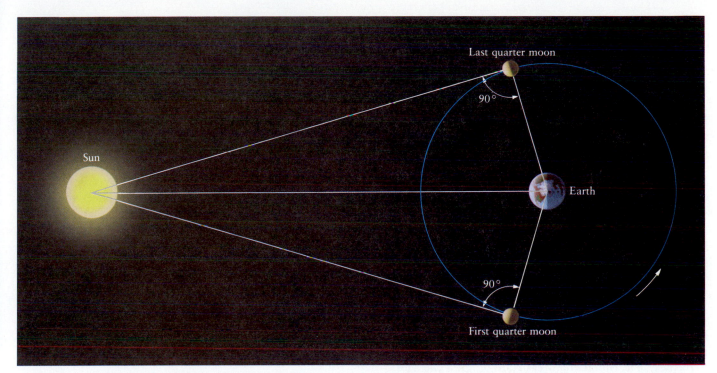

FIGURE 3-7 Aristarchus's Method of Determining Distances to the Sun and Moon Aristarchus knew that the Sun, Moon, and Earth form a right triangle at first and last quarter phases. Using geometrical arguments, he calculated the relative lengths of the sides of these triangles, thereby obtaining the distances to the Sun and Moon.

ancestors' achievements still stand as an impressive application of observation and reasoning that constituted an important step toward the development of the scientific method.

3-3 Eclipses occur only when the Sun and Moon are both on the line of nodes

A **lunar eclipse** occurs when the Moon at full phase passes through the Earth's shadow. This can happen only when the Sun, Earth, and Moon are in a straight line at full moon.

A **solar eclipse** occurs when the Earth passes through the Moon's shadow. As seen from Earth, the Moon moves in front of the Sun. Once again, this can happen only when the Sun, Moon, and Earth are aligned. However, the Moon must be between the Earth and the Sun for a solar eclipse. Therefore, a solar eclipse can occur only at new moon.

Both new moon and full moon occur at intervals of $29\frac{1}{2}$ days, but solar and lunar eclipses happen much less frequently because the Moon's orbit is tilted slightly out of the plane of the Earth's orbit, as shown in Figure 3-8. The angle

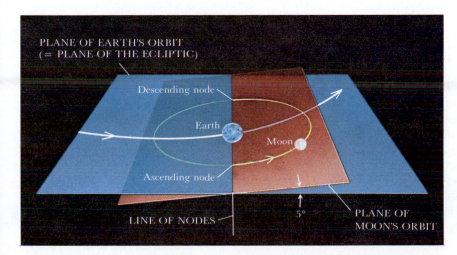

FIGURE 3-8 The Inclination of the Moon's Orbit The plane of the Moon's orbit is slightly tilted with respect to the plane of the Earth's orbit. These two planes intersect along a line called the line of nodes.

BOX 3-1

Some Details of the Moon's Orbit

The Moon's orbit about the Earth is an ellipse, as sketched in the accompanying diagram. The Moon is said to be at perigee when it is nearest the Earth and at apogee when it is farthest from Earth. The line connecting the points of perigee and apogee passes through the Earth and is called the **line of apsides** (diagram *a*).

The center-to-center distance from the Earth to the Moon can vary by more than 50,000 km, from a minimum of 356,410 km (221,510 mi) at perigee to a maximum of 406,697 km (252,763 mi) at apogee. Consequently, the apparent angular size of the Moon as seen from Earth varies over the course of a month. At perigee, the Moon has an angular diameter of 33′ 31″, whereas at apogee the Moon's angular diameter is only 29′ 22″. The average apparent size of the Moon is 31′ 5″, which corresponds to the average Earth–Moon distance of 384,400 km (238,900 mi).

The average speed of the Moon along its orbit is 1.02 km/s (2280 mi/hr). As seen from Earth, the Moon appears to move eastward through the constellations from one day to the next. The Moon's daily eastward progress averages

13.2° (which is 360° divided by the 27.3 days in the sidereal month). In one hour, the Moon moves slightly more than $\frac{1}{2}°$, which is just a bit more than its own diameter. As a result of this motion, moonrise is about 50 minutes later from one day to the next.

The mutual gravitational attraction between the Earth and the Moon keeps the Moon in orbit about the Earth. However, the Sun's gravitational pull is also constantly tugging on the Moon, causing both the line of nodes and the line of apsides to shift slowly over the years. The line of nodes gradually moves westward, as sketched in diagram *b*. This movement is called the **regression of the line of nodes.** It takes 18.61 years for the line of nodes to complete one full rotation.

While the line of nodes is moving westward among the constellations, the Sun's gravity causes the line of apsides to move eastward. It takes 8.85 years for the line of apsides to complete one full rotation. This effect, sketched in diagram *c*, is simply called the **rotation of the Moon's orbit.**

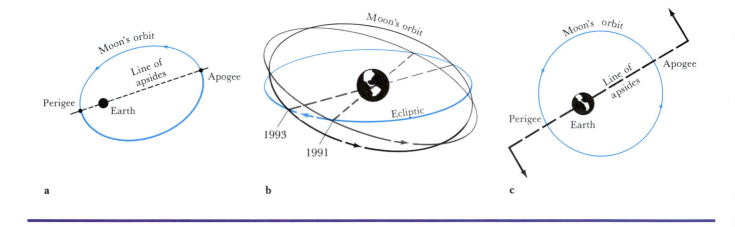

a   **b**   **c**

---

between the plane of the Earth's orbit and the plane of the Moon's orbit is about 5°. Because of this tilt, new moon and full moon usually occur when the Moon is either above or below the plane of the Earth's orbit. When the Moon is not in the plane of the Earth's orbit, the Sun, Moon, and Earth cannot align perfectly, and thus an eclipse cannot occur.

The plane of the Earth's orbit and the plane of the Moon's orbit intersect along a line called the **line of nodes.** The line of nodes passes through the Earth and is pointed in a particular direction in space. Eclipses can occur only when both the Sun and the Moon lie on or very near the line of nodes, because only then do the Sun, Earth, and Moon lie in a straight enough line, as shown in Figure 3-9.

Anyone who wants to predict eclipses must know the orientation of the line of nodes. But the line of nodes is gradually shifting because of the gravitational pull of the Sun on the Moon, as described in Box 3-1. As result, the line of nodes moves slowly westward. Astronomers calculate such details to fix the dates and times of upcoming eclipses.

There are at least two—but never more than five—solar eclipses each year. The last year in which five solar eclipses occurred was 1935. The least number of eclipses possible (two solar, zero lunar) happened in 1969. Lunar eclipses occur just about as frequently as solar eclipses, but the maximum possible number of eclipses (solar and lunar combined) in a single year is seven.

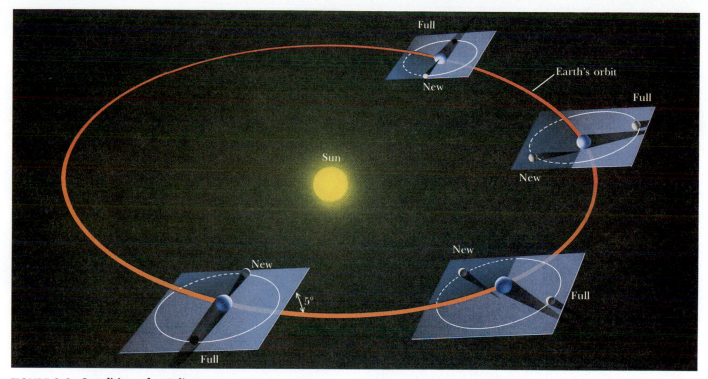

FIGURE 3-9 Conditions for Eclipses A solar eclipse occurs only if the Moon is very near the line of nodes at new moon. A lunar eclipse occurs only if the Moon is very near the line of nodes at full moon. These conditions are met in two of the circumstances (upper right and lower left) shown here. No eclipses would be possible in the other two circumstances (to the right of the Sun and at the lower right) because the Moon is not near the line of nodes.

3-4   Lunar eclipses can be either partial, total, or penumbral, depending on the alignment of the Sun, Earth, and Moon

There are three kinds of lunar eclipses, depending on exactly how the Moon travels through the Earth's shadow, which has two distinct parts, as diagrammed in Figure 3-10. The **umbra** is the darkest part of the Earth's shadow, within which no portion of the Sun's surface can be seen. The **penumbra** is not quite so dark, since only part of the Sun's surface is covered by the Earth. Most people notice a lunar eclipse only if the Moon passes into the Earth's umbra. As the umbral phase of the eclipse begins, a bite seems to be taken out of the Moon.

If only part of the Moon passes through the umbra, we have a **partial lunar eclipse.** If the Moon travels completely into the umbra, as seen in Figure 3-11, a **total lunar eclipse** occurs. Finally, when the Moon passes through only the Earth's penumbra, we see a **penumbral eclipse.** During a penumbral eclipse, none of the lunar surface is completely

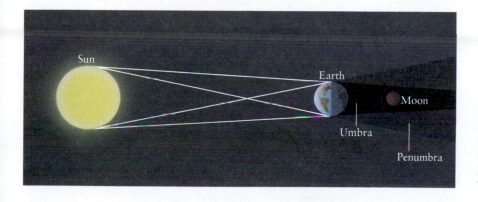

FIGURE 3-10 The Geometry of a Lunar Eclipse People on the nighttime side of the Earth see a lunar eclipse when the Moon moves through the Earth's shadow. The umbra is the darkest part of the shadow. In the penumbra, only part of the Sun is covered by the Earth.

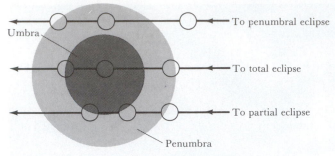

FIGURE 3-11 Various Lunar Eclipses This diagram shows the Earth's umbra and penumbra at the distance of the Moon's orbit. Different kinds of lunar eclipses can be seen, depending on the Moon's path through the Earth's shadow.

shaded by the Earth, because only part of the Sun's light is blocked by the Earth. At mid-eclipse, the Moon merely looks a little dimmer than usual. Since the Moon still looks full, penumbral eclipses are easy to miss.

As an example of the frequency of lunar eclipses, Table 3-2 lists all the eclipses, both total and partial, from 1993 through 1999. Penumbral eclipses are not included.

The Moon does not completely disappear from the sky during totality. A small amount of sunlight passing through the Earth's atmosphere is deflected into the Earth's umbra. Most of this deflected light is red, and thus the darkened Moon glows faintly in reddish hues during totality (Figure 3-12).

The maximum duration of a lunar eclipse occurs when the Moon travels directly through the center of the umbra. The Moon's speed through the Earth's shadow is roughly 1 kilometer per second (2280 miles per hour), which means that totality can last for as long as 1 hour 42 minutes.

TABLE 3-2
Lunar Eclipses, 1993–1999

Date	Percentage eclipsed (100 = total)	Duration of totality
1993 June 4	100	1^h 38^m
1993 November 29	100	0 50
1994 May 25	28	——
1995 April 15	12	——
1996 April 4	100	1 24
1996 September 27	100	1 12
1997 March 24	93	——
1997 September 16	100	1 06
1999 July 28	42	——

FIGURE 3-12 A Total Eclipse of the Moon This photograph was taken by an amateur astronomer during the lunar eclipse of September 6, 1979. The distinctly reddish color of the Moon is caused by sunlight that is deflected into the Earth's shadow by the Earth's atmosphere. (Courtesy of M. Harms)

3-5 Solar eclipses can be either partial, total, or annular, depending on the positions of the Sun, Earth, and Moon

Because of a fortunate coincidence of nature, the apparent diameter of the Moon, as seen from Earth, is almost exactly the same as the apparent diameter of the Sun. Both the Sun and the Moon have angular diameters of about $\frac{1}{2}°$. For this reason, the Moon "fits" over the Sun during a **total solar eclipse**, blocking out the dazzling solar disk and not much else. In fact, it allows the thin, hot gases (called the **solar corona**) that surround the Sun to be photographed and studied in detail during the few precious moments when the eclipse is total (Figure 3-13).

To see a total solar eclipse, you must be inside the darkest part of the Moon's shadow, also called the umbra, where the Moon completely blocks the Sun. Because the Sun and the Moon have nearly the same angular diameter, only the tip of the Moon's umbra reaches the Earth's surface, as shown in Figure 3-14. As the Earth rotates, the tip of the umbra traces an **eclipse path** across the Earth's surface. Only the people within the eclipse path are treated to the spectacle of a total solar eclipse. Figure 3-15 shows the dark spot on the Earth's surface produced by the Moon's umbra.

Immediately surrounding the Moon's umbra is the region of partial shadow called the penumbra. From this area, the Sun's surface appears only partially covered by the Moon. During a solar eclipse, the Moon's penumbra covers a large portion of the Earth's surface, and anyone standing inside the penumbra sees a **partial solar eclipse**.

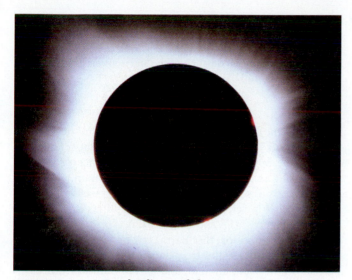

FIGURE 3-13 A Total Eclipse of the Sun During a total solar eclipse, the Moon completely covers the Sun's disk, allowing the solar corona to be seen. This halo of hot gases extends for thousands upon thousands of kilometers into space. This photograph was taken by an amateur astronomer during the solar eclipse of July 11, 1991. (Courtesy of W. Dellinges)

The width of the eclipse path depends primarily on the Earth–Moon distance during totality. The eclipse path is widest if the Moon happens to be at **perigee,** the point in its orbit nearest the Earth (see Box 3-1). Although this nearness can produce an eclipse path up to 270 kilometers (170 miles) wide, it is usually much narrower.

Sometimes the Moon's umbra does not reach down to the Earth's surface, resulting in a third type of solar eclipse, called an **annular eclipse.** When the Moon is at or near **apogee,** its farthest position from the Earth, its umbra falls short of the Earth and the Moon appears too small to cover the Sun completely. During these annular eclipses, a thin ring of light is seen around the edge of the Moon (Figure 3-16). The length of the Moon's umbra is nearly 5000 kilometers (3100 miles) shorter than the average distance between the Moon and the Earth's surface. Thus, the Moon's shadow often fails to reach the Earth, and annular eclipses are slightly more common than total eclipses.

Even during a total eclipse, most people along the eclipse path observe totality for only a few moments. The Earth's rotation, coupled with the orbital motion of the Moon, causes the umbra to race along the eclipse path at speeds in excess of 1700 kilometers per hour (1060 miles per hour).

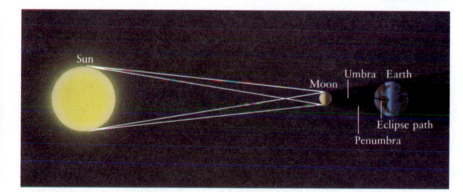

FIGURE 3-14 The Geometry of a Total Solar Eclipse During a total solar eclipse, the tip of the Moon's umbra traces an eclipse path across the Earth's surface. People within the eclipse path see a total eclipse, whereas anyone within the penumbra sees only a partial eclipse of the Sun.

FIGURE 3-15 The Moon's Shadow on the Earth This photograph was taken from an Earth-orbiting satellite during a total solar eclipse on March 7, 1970. The Moon's umbra appears as a dark spot (indicated by the arrow) on the eastern coast of the United States. (NASA)

◄ **FIGURE 3-16 An Annular Eclipse of the Sun** This composite of six exposures taken at sunrise in Costa Rica shows the progress of an annular eclipse of the Sun on December 24, 1974. Note that at mid-eclipse the limb of the Sun is visible around the Moon. (Courtesy of D. di Cicco)

Because of the high speed of the umbra, totality never lasts for more than $7\frac{1}{2}$ minutes. In a typical total solar eclipse, the angle of the Sun and the Earth–Moon distance produce a duration of totality much less than this maximum.

The details of solar eclipses are calculated well in advance and published in reference books such as the *Astronomical Almanac.* Figure 3-17 shows a typical eclipse map, which displays the areas of the Earth covered by the Moon's shadow. To illustrate the frequency of solar eclipses, Table 3-3 lists all the total, annular, and partial eclipses from 1993 to 1999.

3-6 Ancient astronomers achieved a limited ability to predict eclipses

A total solar eclipse is a dramatic event. The sky begins to darken, the air temperature falls, and winds increase as the Moon's umbra races toward you. All nature responds: Birds go to roost, flowers close their petals, and crickets begin to

FIGURE 3-17 An Eclipse Map Details of upcoming eclipses are published in astronomical reference books. Maps such as this one show the regions of the Earth that will experience either total, partial, or annular eclipse. The Moon's shadow travels generally eastward across the Earth's surface. This map shows details of an annular eclipse that will occur on May 10, 1994. (Reproduced from the *Astronomical Almanac*)

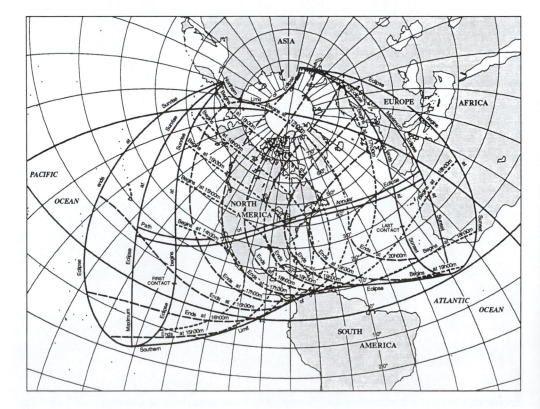

TABLE 3-3
Solar Eclipses, 1993–1999

Date	Area	Type	Notes
1993 May 21	Canada, Arctic, northern Europe	Partial	74% eclipsed
1993 November 13	Antarctic	Partial	93% eclipsed
1994 May 10	Pacific, Mexico, United States, Canada	Annular	
1994 November 3	Peru, Brazil, South Atlantic	Total	Maximum length 4 m 23 s
1995 April 29	South Pacific, Peru, Brazil, South Atlantic	Annular	
1995 October 24	Iran, India, East Indies, Pacific	Total	Maximum length 2 m 5 s
1996 April 17	South Pacific, Antarctic	Partial	88% eclipsed
1996 October 12	North Atlantic, Arctic, northern Europe	Partial	76% eclipsed
1997 March 9	China, Russia, Arctic	Total	Maximum length 2 m 50 s
1997 September 2	Australia, New Zealand, Antarctic	Partial	90% eclipsed
1998 February 26	Pacific, Central America, Atlantic	Total	Maximum length 3 m 56 s
1998 August 22	Indian Ocean, East Indies, Pacific	Annular	
1999 February 16	Indian Ocean, Australia	Annular	
1999 August 11	Atlantic, Europe, Middle East, India	Total	Maximum length 2 m 23 s

---

chirp as if evening had arrived. As totality approaches, the landscape around you is bathed in an eerie gray or, less frequently, in shimmering bands of light and dark as the last few rays of sunlight peek out from behind the edge of the Moon. Finally the corona blazes forth in a star-studded midday sky. It is an awesome sight.

In ancient times, the ability to predict eclipses must have been very desirable. The number and placement of certain holes in the ground around Stonehenge indicate some ability to predict eclipses more than 4000 years ago. One of three priceless manuscripts to survive the devastating Spanish Conquest shows that the Mayan astronomers of Mexico and Guatemala had a fairly reliable method for predicting eclipses. The great Greek astronomer Thales of Miletus is said to have predicted the famous eclipse of 585 BC, which occurred during the middle of a war. The sight was so unnerving that the soldiers put down their arms and declared peace.

In retrospect, it seems that what the ancient astronomers actually produced were eclipse "warnings" of various degrees of reliability, rather than true predictions. Working with historical records, these astronomers generally sought to discover cycles and regularities from which future eclipses could be anticipated.

To see how you might produce eclipse warnings yourself, suppose that you observe a solar eclipse in your hometown and want to figure out when you and your neighbors might see another eclipse. How would you begin?

First, remember that a solar eclipse can occur only if the line of nodes points toward the Sun at new moon (recall Figure 3-9). Second, you must know that it takes 29.53 days to go from one new moon to the next (the synodic month). Because solar eclipses occur only during the new moon, you must wait several whole lunar months for the proper alignment to occur again.

However, there is a complication. The line of nodes gradually shifts its position with respect to the background stars (see Box 3-1). It takes 346.6 days to go from one alignment of the line of nodes pointing toward the Sun to the next identical alignment. This period is called the **eclipse year.**

To predict when you will see another solar eclipse, you need to know how many whole lunar months equal some whole number of eclipse years. This information will tell you how long you will have to wait for the next virtually identical alignment of the Sun, the Moon, and the line of nodes. By trial and error, you find that the answer is 223 lunar months, or 19 eclipse years, because

$$223 \times 29.53 = 19 \times 346.6 = 6585 \text{ days}$$

This calculation is accurate to within a few hours. A more accurate calculation gives an interval, called the **saros,** of 6585.3 days. Eclipses separated by the saros interval are said to form an "eclipse series."

You might think that you and your neighbors would simply have to wait one full saros interval to go from one solar

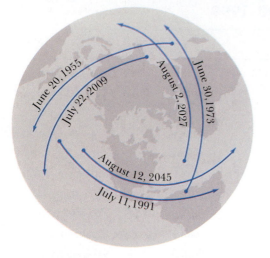

FIGURE 3-18 An Eclipse Series Eclipses separated by the saros interval (18 years 11.3 days) are said to form an "eclipse series" because their geometric details, such as the orientation of the line of nodes, are nearly identical. Even the duration of totality is nearly the same for all members of a particular eclipse series.

eclipse to the next. However, because of the extra one-third day, the Earth will have rotated by an extra 120° when the next solar eclipse of a particular series occurs. The eclipse path will thus be one-third of the way around the world from you. You must therefore wait three full saros intervals (54 years 34 days) before the eclipse path comes back around to your part of the Earth. Figure 3-18 shows a series of solar eclipse paths, each separated from the next by one saros interval.

There is evidence that some ancient astronomers knew about the saros interval. The discovery of the saros is more likely to have come from lunar eclipses than solar eclipses. If you are far from the eclipse path yet still within the Moon's penumbra, there is a good chance that you could fail to notice a solar eclipse. Even if half the Sun is covered by the Moon, the remaining solar surface provides enough sunlight for the outdoor illumination not to be greatly diminished. In contrast, everyone on the nighttime side of the Earth can see an eclipse of the Moon unless clouds block the view.

3-7   Ancient astronomers devised star catalogues and established a brightness scale that is still used

Ancient astronomers, particularly in Greece, gave humanity a new and powerful way of thinking about the world. They gave the first clear demonstration that the tools of logic and mathematics can be used to discover and understand the workings of the universe. This general approach underlies all modern science. Aristarchus even went so far as to suggest

that the motions of the planets in the sky could be explained simply if all the planets as well as the Earth orbited the Sun. As we shall see in the next chapter, this man was nearly 2000 years ahead of his time.

In addition to setting the stage for modern science, the early Greek astronomers also gave us ideas and established certain traditions that are still useful. For example, around the year 160 BC, Hipparchus built an observatory on the island of Rhodes. Over a period of several years, he compiled the first comprehensive star catalogue, which listed the coordinates and brightnesses of some 850 stars. During the course of this pioneering work, Hipparchus compared his star positions with earlier records dating back to Aristarchus's time, a century earlier. Hipparchus soon noted systematic differences that led him to conclude that the north celestial pole had shifted slightly over the previous hundred years. He had discovered precession.

While compiling his star catalogue, Hipparchus established a system to denote the brightness of stars. His system is the basis of the **magnitude scale** used by astronomers today. Hipparchus said that the brightest stars in the sky are "first magnitude." The dimmest stars visible to the unaided eye he called "sixth magnitude." To stars of intermediate brightness he assigned intervening numbers on the scale of 1 to 6.

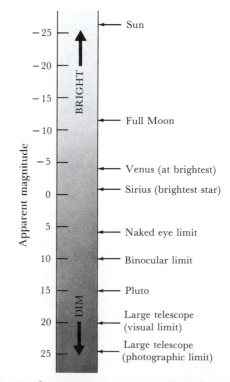

FIGURE 3-19 The Apparent Magnitude Scale Astronomers denote the brightness of an object in the sky by its "apparent magnitude." Most stars visible to the naked eye have apparent magnitudes between 1 and 6. Photography through a large telescope can reveal stars as faint as magnitude 25.

With only a few refinements, this same system is used today. As shown in Figure 3-19, the magnitude scale has been extended to include even very faint stars. For example, with a good pair of binoculars you can see stars as faint as tenth magnitude. Through some of the largest telescopes stars as dim as magnitude 20 can be seen. Photography with long exposure times reveals even dimmer stars.

Modern astronomers also use negative numbers to extend Hipparchus's scale to include very bright objects. For example, Sirius, the brightest star in the sky, has a magnitude of $-1\frac{1}{2}$. At its brightest, the planet Venus shines with a magnitude of -4. The Sun, the brightest object in the sky, has a magnitude of $-26\frac{1}{2}$.

Later in this book we shall discuss details about the brightness of stars and galaxies. We shall see how careful observation and logical thinking can extend our knowledge far beyond our limited daily experiences. Although they have been refined and elaborated, many concepts like the magnitude scale have come down to us from the people who first discovered how the human mind can unlock the secrets of the universe.

KEY WORDS

Terms preceded by an asterisk are discussed in the boxes.

annular eclipse	lunar eclipse	penumbral eclipse	solar eclipse
apogee	lunar month	perigee	synodic month
eclipse path	lunar phases	*regression of the line of nodes	total lunar eclipse
eclipse year	magnitude scale		total solar eclipse
first quarter moon	new moon	*rotation of the Moon's orbit	umbra (*plural* umbrae)
full moon	partial lunar eclipse		waning crescent moon
last quarter moon	partial solar eclipse	saros	waning gibbous moon
*line of apsides	penumbra (*plural* penumbrae)	sidereal month	waxing crescent moon
line of nodes		solar corona	waxing gibbous moon

KEY IDEAS

• Thousands of years ago, many cultures observed the apparent positions of the Sun, Moon, and planets against the background of the stars. The phases of the Moon occur because of the shifting relative positions of the Earth, the Moon, and the Sun.

With respect to the stars, the Moon completes one orbit around the Earth in a sidereal month averaging 27.3 days.

The Moon completes one cycle of phases (one orbit around the Earth with respect to the Sun) in a synodic month averaging 29.5 days, which is also called the lunar month.

The lengths of the sidereal and synodic months vary slightly because of the Sun's gravitational pull.

• Ancient astronomers made great progress in determining the sizes and relative distances of the Earth, the Moon, and the Sun.

Around 280 BC, Aristarchus attempted to measure the distances from the Earth to the Sun and the Moon by using the Moon's phases. He also tried to estimate the relative sizes of the Earth, Sun, and Moon.

Decades later, around 200 BC, Eratosthenes measured the Earth's size by comparing the position of the Sun in the sky at the same moment in two different locations.

• The line of nodes is the line where the planes of the Earth's orbit and the Moon's orbit intersect.

The gravitational pull of the Sun gradually shifts the orientation of the line of nodes with respect to the stars.

• During a lunar eclipse, the Moon moves through the Earth's shadow. This happens when the Sun and Moon are both on the line of nodes at full moon. During a solar eclipse, the Earth passes through the Moon's shadow while the Sun and Moon are both on the line of nodes at new moon.

• Depending on the relative positions of the Sun, Moon, and Earth, lunar eclipses may be penumbral, partial, or total. Solar eclipses may be partial, annular, or total.

The shadow of an object has two parts: the umbra, within which the light source is completely blocked, and the penumbra, where the light source is only partially blocked.

During a total solar eclipse, the Moon's umbra traces out an eclipse path over the Earth's surface as the Earth rotates.

During an annular eclipse, the umbra falls short of the Earth, so the limb of the Sun is visible around the Moon at mid-eclipse.

• Using an interval called the saros, it is possible to group total solar eclipses into series and to predict the time of the next eclipse in each series.

• Astronomers still use a system for denoting the apparent brightness of objects in the sky that was invented by Hipparchus around 160 BC.

Very bright objects have negative values of magnitude; very dim objects have large positive values of magnitude.

REVIEW QUESTIONS

1. Why does the Moon exhibit phases?

2. What is the phase of the Moon if it (a) rises at 3 AM? (b) is crossing the upper meridian at midnight? (c) sets at 9 PM? At what time does (d) the full moon set? (e) the first quarter moon rise? (f) the third quarter moon cross the upper meridian?

3. What is the phase of the Moon if, on the first day of spring, the Moon is located at (a) the vernal equinox, (b) the summer solstice, (c) the autumnal equinox, (d) the winter solstice?

4. How did Eratosthenes determine the size of the Earth?

5. How did Aristarchus try to estimate the distance from the Earth to the Sun and Moon?

6. What is the difference between a sidereal month and a synodic month? Which is longer? Why?

7. How many more sidereal months than synodic months are there in a year? Why?

8. Why isn't there a lunar eclipse at every full moon and a solar eclipse at every new moon?

9. What is the line of nodes and why is it important to the subject of eclipses?

10. Which type of eclipse—lunar or solar—do you think most people have seen? Why?

11. What is the difference between the umbra and the penumbra of a shadow?

12. Can a total eclipse of the Sun be followed three months later by a lunar eclipse? Why?

13. What is a penumbral eclipse of the Moon? Why do you suppose that it is easy to overlook such an eclipse?

14. How is an annular eclipse of the Sun different from a total eclipse of the Sun? What causes this difference?

15. Can one ever observe an annular eclipse of the Moon? Why?

16. What is the saros and how did ancient astronomers use it to predict eclipses?

17. Does the star Betelgeuse, whose apparent magnitude is 0.5, look brighter or dimmer than the star Pollux, whose apparent magnitude is 1.1?

ADVANCED QUESTIONS

Tips and tools . . .

It is helpful to know that the saros interval of 6585.3 days equals 18 years $11\frac{1}{3}$ days if the interval includes four leap years, but is 18 years $10\frac{1}{3}$ days if it includes five leap years. The average angular speed of the Moon along its orbit can be estimated by dividing 360° by the length of a sidereal month.

18. During a lunar eclipse, does the Moon enter the Earth's shadow from the east or from the west? Why?

19. If the Moon revolved about the Earth in the same orbit but in the opposite direction, would the synodic month be longer or shorter than the sidereal month? Why?

20. On July 11, 1991, residents of Hawaii were treated to a total solar eclipse. (a) When and over what part of the world will the next eclipse of that series occur? (b) When might the Hawaiians next expect to see an eclipse of that series?

21. During an occultation, or "covering up," of Jupiter by the Moon, an astronomer notices that it takes the Moon's edge 90 seconds to cover Jupiter's disk completely. If the Moon's motion is assumed to be uniform and the occultation was "central" (i.e., center over center), find the angular diameter of Jupiter.

DISCUSSION QUESTIONS

22. Suppose that you were an ancient astronomer given the task of building a Stonehenge-type monument to be aligned with the apparent positions and phases of the Moon. Which directions and alignments might you consider important? What sort of observations would be needed to determine these directions? How long would it take to collect the necessary data?

23. Why do you suppose that total solar eclipse paths fall more frequently on oceans than on land?

24. In his novel *King Solomon's Mines*, author H. Rider Haggard described a total solar eclipse that was reportedly seen in both South Africa and in the British Isles. Is such an eclipse possible? Why or why not?

25. Describe the cycle of lunar phases that would be observed if the Moon moved about the Earth in an orbit

perpendicular to the plane of the Earth's orbit. Is it possible for both solar and lunar eclipses to occur under these circumstances?

26. How would a lunar eclipse look if the Earth had no atmosphere?

27. Examine a listing of total solar eclipses over the next several decades. What are the chances that you might be able to travel to one of the eclipse paths? Do you think you might go through your entire life without ever seeing a total eclipse of the Sun?

OBSERVING PROJECTS

28. Observe the Moon on each clear night over the course of a month. On each night, note the Moon's location among the constellations and record that location on a star chart that also shows the ecliptic. After a few weeks, your observations will begin to trace the Moon's orbit. Identify the orientation of the line of nodes by marking the points where the Moon's orbit and the ecliptic intersect. On what dates is the Sun near the nodes marked on your star chart? Compare these dates with the dates of the next solar and lunar eclipses.

29. It is quite possible that a lunar eclipse will occur while you are taking this course. Look up the date of the next lunar eclipse in the current issue of a reference such as the *Astronomical Almanac* or *Astronomical Phenomena.* You would also be well advised to consult such magazines as *Sky & Telescope* and *Astronomy,* which generally run articles about upcoming eclipses the month before they happen. Make arrangements to observe the next lunar eclipse. If the eclipse is partial or total, note the times at which the Moon enters and exits the Earth's umbra. If the eclipse is penumbral, can you see any changes in the Moon's brightness as the eclipse progresses?

FOR FURTHER READING

Allen, D., and Allen, C. *Eclipse.* Allen & Unwin, 1987. A well-written history of our understanding of eclipses, including the mythology they engendered and the science behind them.

Cornell, J. *The First Stargazers.* Scribner, 1981. This introduction to "archeoastronomy," the archeological study of ancient astronomy, includes excellent sections about the astronomy of Stonehenge, native Americans, China, and Egypt.

Daniel, G. "Megalithic Monuments." *Scientific American,* July 1980. This article surveys stone monuments that were built by the thousands throughout prehistoric Europe, probably between 3000 and 1000 BC. Like Stonehenge, many of these structures seem to have been constructed with astronomical alignments in mind.

Gingerich, O. "Aristarchus of Samos: A Report on a Symposium." *Sky & Telescope,* November 1980. This brief article gives interesting insights into Aristarchus and the island of Samos on which he was born.

Hadingham, E. *Early Man and the Cosmos.* Walker, 1984. An introduction to prehistoric and early astronomy with nice sections on stone monuments in Britain and America. An excellent primer on archeoastronomy.

Hawkins, G. *Stonehenge Decoded.* Delta, 1965. This classic book was one of the first to decipher and popularize Stonehenge as an astronomical monument.

Hetherington, N. *Ancient Astronomy and Civilization.* Pachart, 1987. A slim volume that introduces early astronomy, much of it through translations of original writings.

Krupp, E. *Echoes of the Ancient Skies.* Harper & Row, 1983. A superb introduction to archeoastronomy. Krupp skillfully brings together the mythology, rituals, and monuments of many cultures, demonstrating both similarities and differences.

Kundu, M. "Observing the Sun during Eclipses." *Mercury,* July/August 1981. This article is a personal account of the trials and tribulations of observing the 1980 total eclipse of the Sun in India.

Menzel, D., and Pasachoff, J. "Solar Eclipse: Nature's Superspectacular." *National Geographic,* August 1970. A beautifully illustrated account of a total solar eclipse.

Sagan, C. "The Shores of the Cosmic Ocean." In Sagan, C., *Cosmos.* Random House, 1980. This chapter is an excellent recounting of Eratosthenes' experiment to measure the size of the Earth.

Stephenson, F. "Historical Eclipses." *Scientific American,* October 1982. This fascinating article shows how reliable records of solar and lunar eclipses dating back to 750 BC can be used to see if the Sun is shrinking or if the Earth's rate of rotation is changing.

Williamson, R. *Living the Sky: The Cosmos of the American Indian.* Houghton-Mifflin, 1984. This book gives a fascinating tour of the astronomical thought and monuments of native American Indian tribes.

OWEN GINGERICH

Astrology and Astronomy

OWEN GINGERICH is a senior astronomer at the Smithsonian Astrophysical Observatory in Cambridge and a Professor of Astronomy and the History of Science at Harvard University. His research interests have included modeling the solar atmosphere, interpretation of stellar spectra, and the recomputation of ancient Babylonian mathematical tables. Dr. Gingerich is a leading authority on the works of Johannes Kepler and Nicolaus Copernicus. His essays on a broad range of historical topics, from Stonehenge to Einstein, appear in *The Great Copernicus Chase and Other Adventures in Historical Astronomy*. He serves as editor of the twentieth-century section of the International Astrophysical Union's *General History of Astronomy*, and he is an active participant in such scientific organizations as the American Astronomical Society and the International Astronomical Union. Of the several hundred general interest, technical, and historical publications to his credit, many of the most accessible have appeared in *Scientific American* and *Sky & Telescope*.

In 1226 the feared Mongol conqueror Genghis Khan, about to subdue the entire known world, suddenly called off his campaign. The planet Jupiter was about to overtake Saturn, and quite possibly Genghis Khan's astrologers warned of dangerous consequences heralded by this once-in-twenty-year planetary conjunction.

In 1634 the Bohemian general Wallenstein became more and more nervous. He carried close to his chest a horoscope prepared by the astronomer Johannes Kepler, and the predictions stopped in that year with the prophecy that "according to astrological lore" the month of March would be marked with "horrible disorder." Wallenstein became increasingly troubled, brought matters to a crisis, and then, in February, was assassinated by other officers.

What are we to make of these extraordinary events? Did the distant planets have some subtle influence on these military men? Or were these self-fulfilling prophecies, events destined to happen because the belief in their inevitability was so strong that they were subconsciously determined?

A few hundred years BC, when the earliest scientists began to appreciate the relationships between the Sun's noontime height and the seasons and between the Moon and the tides, it seemed logical to suppose that the planets might also have other terrestrial influences. This astrological thinking was far more scientific than the animistic notions that had preceded it, in which capricious gods in the natural world were continually carrying out unpredictable acts, but unfortunately there was little empirical evidence behind it.

Gradually a systematic lore developed about the supposed planetary influences. Mars and Saturn were seen as generally evil, Venus and Jupiter as benign. Geometrical relationships between the Sun, Moon, and planets, such as conjunctions, oppositions (two objects 180° apart), and quadratures (90° apart), enhanced or countered these astrological effects.

The 12 zodiacal signs acquired special properties in combination with planetary movements. Furthermore, the sky was divided from horizon to horizon into a series of "houses," each of which controlled some aspect of human life such as marriage, friends, riches, or death. The configuration of the heavens and the placement of planets at the moment of birth (or conception) was believed to establish a person's personality and destiny. The part of the sky just below the eastern horizon, the ascendant, was considered the most powerful indicator in a horoscope.

The continuously changing positions of the planets within the houses, as the daily rotation of the

heavens carried the stars and planets across the sky, were thought to presage one's fortune from moment to moment. In ancient times and in the Middle Ages, many thoughtful people who passionately believed that the universe is rational succumbed to astrology. To them it gave apparent "causes" to otherwise inexplicable events, even though it lacked any systematic basis in observations. Altogether, astrology with its zodiacal signs and houses and planetary configurations was a marvelous system of folk psychology, but not a scientific inquiry.

After the Renaissance, a whole new physics arose. The ancient distinction between "terrestrial" and "celestial" matter—the Earth and the heavens—was shattered, and Isaac Newton devised gravitational law that could predict motions on Earth as well as in the sky. The new astronomy did more than displace the Earth from the center of the universe: It became a changing, continuously self-renewing enterprise—in short, a science. Astrology rapidly became merely a fossil of the old geocentric scheme.

Johannes Kepler, like many astronomers of the early Renaissance, drew up many horoscopes for his friends and employers, although he was very skeptical about the prevailing opinions on celestial influences. For the benefit of other astrologers, Kepler gave the position of the planets at his own birth, but he added: "My stars were not Mercury rising in the seventh angle in quadrature with Mars, but Copernicus and Tycho Brahe, without whose observation books everything I have brought into light would have remained in darkness; my rulers were not Saturn predominating over Mercury, but the emperors Rudolph and Matthias; not a planetary house, Capricorn with Saturn, but Upper Austria, the house of the emperor."

In some respects, Kepler was the astrologer who destroyed astrology. By insisting on an astronomy based on physical causes, he furthered the development of a logically connected scientific framework. Astrology could find neither causes nor correlations between planetary configurations and their supposed influences.

Perhaps the most extensive statistics on the effectiveness of astrological predictions were compiled in the 1960s by the French psychologist Maurice Gauquelin. He failed to find any correlations between the birth horoscopes of 25,000 celebrities and the traditional qualities associated with the various zodiacal constellations. Gauquelin also conducted an interesting experiment in which he offered free computer-generated horoscopes with the provision that the recipients evaluate how well the astrological characterization fit them. Nearly 95% of the recipients said they recognized themselves in the "psychological portrait," and 80% of their family and friends shared the opinion. What they did not know was that exactly the same computer printout had been sent to everyone!

While astrology remains an amusing game comparable to, but rather more complicated than, Chinese fortune cookies, the fact that many people continue to take it seriously can only be attributed to wishful thinking and self-fulfilling prophecies.

Of all the evaluations pro and con, I particularly enjoy the late Michael Flanders's perceptive tongue-in-cheek view of astrology:

Jupiter's passed through Orion

And coming to conjunction
  with Mars

Saturn is wheeling through
  infinite space

To its pre-ordained place
  in the stars

And I gaze at the planets in wonder

At the trouble and time they spend

All to warn me to *be careful*

In dealings involving a friend.

GRAVITATION AND THE MOTIONS OF PLANETS

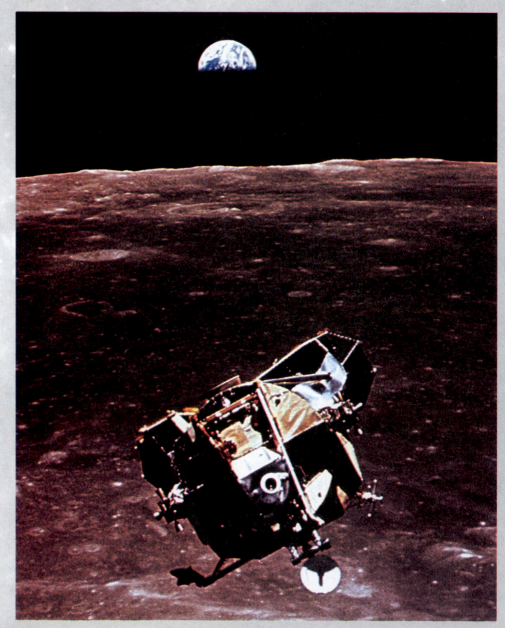

APOLLO 11 LEAVING THE MOON
The lunar module Eagle returns from the Moon after completing the first successful manned lunar landing in July 1969. This photograph was taken from the command module *Columbia,* in which the astronauts returned to Earth. All of the orbital maneuvers to dock the *Eagle* and the *Columbia* and then to set course for Earth were based on Newtonian mechanics and Newton's law of gravity. Astronomers use these same physical principles to understand a wide range of phenomena, from the motions of double stars to the rotation of an entire galaxy. (NASA)

ANCIENT ASTRONOMERS believed that the heavens rotated around a stationary Earth. During the Renaissance, several brilliant thinkers spearheaded a revolution that dethroned the Earth from its central location and laid the foundation of modern science. First Copernicus proposed a model in which the planets go around the Sun along a system of circles. Observations by Tycho Brahe and Galileo then clearly demonstrated inaccuracies in the Earth-centered view: the planets must indeed orbit the Sun. Later, Kepler's profound discovery that planetary orbits are ellipses rather than circles inspired Isaac Newton, who formulated the basic laws of physics. From his studies of planetary motion, Newton created a precise, mathematical description of the force of gravity, which holds the planets in their orbits about the Sun. In the early twentieth century, Albert Einstein proposed the radically new idea that gravity affects the curvature of space and the flow of time. This relationship between gravity, space, and time are fundamental to our present understanding of the very structure of the universe, its creation, and its ultimate fate.

It is by no means obvious that the Earth moves around the Sun. Indeed, our daily experience strongly suggests that the opposite is true. The daily rising and setting of the Sun, Moon, and stars could lead us to believe that the entire cosmos revolves about an Earth that is at the center of the universe. This is just what most people did believe for thousands of years. How, then, did inspired Renaissance thinkers finally overturn this erroneous notion and lay the groundwork for modern science?

4-1 Ancient astronomers invented geocentric cosmologies to explain planetary motions

The ancient Greeks conceived of certain principles that still guide modern scientists today. For instance, around 550 BC Pythagoras and his followers put forth the idea that nature can be described with mathematics. About 200 years later, Aristotle asserted that the universe is governed by physical laws. These two concepts would find their highest expression when the great scientists of later generations explained planetary motion using mathematics and physics.

Early Greek astronomers were among the first to leave a written record of their attempts to explain the motion of the planets. Most Greeks assumed that the Sun, the Moon, the stars, and the five planets revolve about the Earth; their view of the universe is said to be *geocentric*. A theory of the universe is called a **cosmology,** and thus Greek thinkers such as Aristotle adhered to a **geocentric cosmology.**

The Greeks and other cultures of that time knew of five planets: Mercury, Venus, Mars, Jupiter, and Saturn. These planets are quite obvious in the sky because from night to night they slowly move with respect to the stationary, or "fixed," stars in the constellations. In fact, the word *planet* comes from a Greek term meaning "wanderer." Furthermore, some of these planets are very bright in the night sky. Venus, for example, at its maximum brilliancy is 16 times brighter than the brightest star.

Observations of the positions of the planets against the stars from night to night reveal that the planets do not move at uniform rates through the sky. Explaining the nonuniform motions of the five planets was one of the main challenges facing the astronomers of antiquity. It was not easy to develop a comprehensive geocentric theory of the universe.

As seen from Earth, the planets wander primarily across the 12 constellations of the zodiac. As mentioned in Chapter 2, these constellations encircle the sky in a continuous band centered on the ecliptic. If you observe a planet as it travels across the zodiac from night to night, you will find that the planet usually moves slowly eastward against the background stars. This eastward progress is called **direct motion.** Occasionally, however, the planet will seem to stop and then back up for several weeks or months. This occasional westward movement is called **retrograde motion.** These motions are much slower than the apparent daily rotation of the sky caused by the Earth's rotation. Both direct and retrograde motions are best detected by mapping the position of a planet against the background stars from night to night over a long period. An example is the path of Mars from late 1994 through mid-1995, shown in Figure 4-1.

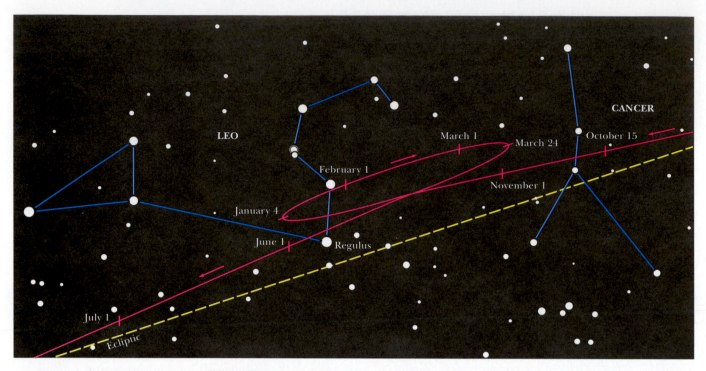

FIGURE 4-1 The Path of Mars in 1994–1995 From the fall of 1994 through the spring of 1995, Mars will move across the constellations of Cancer and Leo. From January 4 through March 25, Mars's motion will be retrograde.

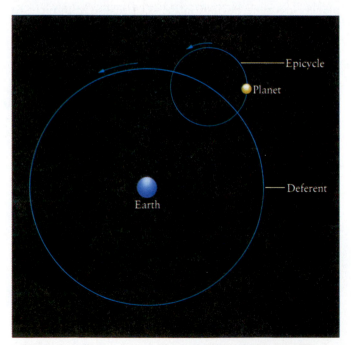

FIGURE 4-2 A Geocentric Explanation of Retrograde Motion Each planet moves along an epicycle, which in turn moves along a deferent centered approximately on the Earth. As seen from Earth, the speed of the planet on the epicycle alternately adds to or subtracts from the speed of the epicycle on the deferent, thus producing periods of either direct or retrograde motion.

The Greeks developed many theories to account for retrograde motion and the exact loops that the planets trace out against the background stars. One of the most successful and enduring ideas was expounded by the last of the great Greek astronomers, Ptolemy, who lived in Alexandria during the second century AD. The basic concept is sketched in Figure 4-2. Each planet is assumed to move in a small circle called an **epicycle**, whose center in turn moves in a larger circle, called a **deferent**, which is centered approximately on the Earth. As viewed from Earth, the epicycle moves eastward along the deferent, and both circles rotate in the same direction (counterclockwise in Figure 4-2).

Most of the time the motion of the planet on its epicycle adds to the eastward motion of the epicycle on the deferent. Thus, the planet is seen to be in direct (eastward) motion against the background stars throughout most of the year. However, when the planet is on the part of its epicycle nearest the Earth, the motion of the planet along the epicycle subtracts from the motion of the epicycle along the deferent. The planet therefore appears to slow down and halt its usual eastward movement among the constellations, ultimately going backward for a few weeks or months. Using this concept of epicycles and deferents, Greek astronomers were able to explain the retrograde loops of the planets.

Using the wealth of astronomical data in the library at Alexandria, including records of planetary positions for hundreds of years, Ptolemy deduced the sizes and rotation rates of the epicycles and deferents needed to reproduce the re-

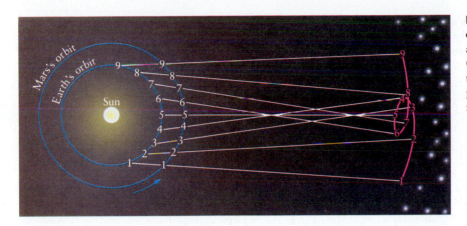

FIGURE 4-3 A Heliocentric Explanation of Planetary Motion The Earth travels around the Sun more rapidly than does Mars. Consequently, as the Earth overtakes and passes this slower-moving planet, Mars appears (from points 4 through 6) to move backward for a few months.

corded paths of the planets. After years of tedious work, Ptolemy assembled his calculations into 13 volumes, collectively called the *Almagest*. His work was used to predict the positions and paths of the Sun, Moon, and planets with unprecedented accuracy. In fact, the *Almagest* was so successful that it became the astronomer's bible, and for over 1000 years Ptolemy's cosmology endured as a useful description of the workings of the heavens.

Eventually, however, things began going awry. Tiny errors unnoticeable in Ptolemy's day accumulated over the years, especially those concerning precession. Thirteenth-century astronomers made some cosmetic adjustments to the Ptolemaic system. However, as later astronomers amended and adjusted the system to keep it consistent with the observed motions of the planets, Ptolemy's model became less and less aesthetically satisfying.

One of the guiding beliefs of many scientists is that simple, straightforward explanations of phenomena are more likely to be correct than complicated, convoluted ones. This idea is called **Occam's razor**, named after the fourteenth-century English philosopher who first expressed it. (The term *razor* refers to shaving extraneous details from an argument or explanation.) Although Occam's razor has no proof or verification, it appeals to the scientist's sense of beauty and elegance.

Half a century after the death of William of Occam, the cumbersome complexity of the Ptolemaic system led the Polish astronomer Nicolaus Copernicus to doubt its validity. His Sun-centered system allowed a simpler explanation of planetary motions and helped lay the foundations of modern physical science.

4-2 Nicolaus Copernicus devised the first comprehensive heliocentric cosmology

Imagine driving on a freeway at high speed. As you pass a slowly moving car, it appears to move backward even though it is traveling in the same direction as your car. This sort of observation inspired the ancient Greek astronomer Aristarchus to suggest a straightforward explanation of retrograde motion: All the planets, including the Earth, revolve about the Sun. The retrograde motion of Mars, for example, occurs

when the Earth overtakes and passes Mars, as shown in Figure 4-3. The occasional backward motion of a planet is merely the result of our moving viewpoint—an idea that is beautifully simple compared to an Earth-centered system with all its "circles upon circles."

Aristarchus had demonstrated that the Sun is bigger than the Earth (see Table 3-1), which also made it sensible to imagine the Earth orbiting the larger Sun. In Aristarchus's day, however, the idea of a moving Earth seemed inconceivable, given the Earth's apparent stillness and immobility. Almost 2000 years elapsed before someone had both the insight and determination to work out the details of a **heliocentric** (Sun-centered) **cosmology**. That person was a Polish lawyer, physician, economist, canon, and artist named Nicolaus Copernicus (Figure 4-4). Especially gifted in mathematics, Copernicus turned his attention to astronomy in the early 1500s.

FIGURE 4-4 Nicolaus Copernicus (1473–1543) Copernicus was the first person to work out the details of a heliocentric system in which the planets, including the Earth, orbit the Sun. (E. Lessing/Magnum)

FIGURE 4-5 Planetary Configurations The key
points along a planet's orbit are shown in this
diagram. These points identify the specific geometric
arrangements possible between the Earth, a planet,
and the Sun.

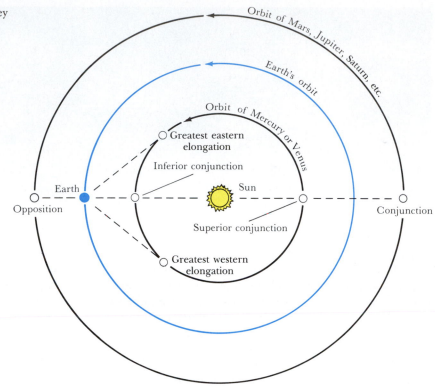

Copernicus realized that a heliocentric model enabled him to determine which planets are closer to the Sun than the Earth is, and which are farther away. Because Mercury and Venus are always observed fairly near the Sun in the sky, Copernicus concluded that their orbits must be smaller than the Earth's. Because the other visible planets—Mars, Jupiter, and Saturn—can be seen in the middle of the night, when the Sun is far below the horizon, Copernicus realized that the Earth had to be between the Sun and these planets. He therefore concluded that the orbits of Mars, Jupiter, and Saturn must be larger than the Earth's orbit. Three additional planets (Uranus, Neptune, and Pluto), discovered after the telescope was invented, also have orbits larger than the Earth's.

Various points on a planet's orbit are shown in Figure 4-5. These points indicate certain geometric relationships, or **configurations,** formed by the Earth, another planet, and the Sun. When Mercury or Venus is between the Earth and the Sun, we say that the planet is at **inferior conjunction.** When Mercury or Venus is on the opposite side of the Sun, we say that the planet is at **superior conjunction.**

The angle between the Sun and a planet, as viewed from the Earth, is called the planet's **elongation.** At **greatest eastern elongation,** Mercury or Venus is as far east of the Sun as it can be. At such times, the planet appears above the western horizon after sunset and is often called an "evening star." Similarly, at **greatest western elongation** Mercury or Venus is as far west of the Sun as it can possibly be and rises before the Sun, gracing the predawn sky as a "morning star."

Planets whose orbits are larger than Earth's have different configurations. When Mars, for example, is located behind the Sun, it is in a configuration called **conjunction.** When it is opposite the Sun in the sky, the planet is at **opposition.**

It is not difficult to determine when a planet happens to be located at one of the key positions in Figure 4-5. For example, when Mars is at opposition, it crosses the upper meridian at midnight.

Although it is easy to follow a planet as it moves from one configuration to another, these observations alone do not tell us details of a planet's actual orbit around the Sun. The Earth, from which we must make the observations, is also moving. Realizing this, Copernicus was careful to distinguish between two characteristic time intervals, or **periods,** of each planet. The **synodic period** is the time that elapses between two successive identical configurations, as seen from the Earth—from one opposition to the next, for example, or from one conjunction to the next. The **sidereal period** is the true orbital period of a planet, the time it takes the planet to complete one full orbit of the Sun relative to the stars.

The synodic period of a planet can be determined by observing the sky, but the sidereal period must be calculated. Copernicus figured out how to do this (Box 4-1), and his results are shown in Table 4-1.

With the planetary orbits now known, Copernicus devised a straightforward geometric method of determining the distances of the planets from the Sun. The mathematical details, which involve some trigonometry, are described in Box 4-2. His answers turned out to be remarkably close to the modern values (Table 4-2). From Tables 4-1 and 4-2 it is apparent that the farther a planet is from the Sun, the longer it takes to travel around its orbit.

TABLE 4-1		
Synodic and Sidereal Periods of the Planets		
Planet	Synodic period	Sidereal period
Mercury	116 days	88 days
Venus	584 days	225 days
Earth	—	1.0 year
Mars	780 days	1.9 years
Jupiter	399 days	11.9 years
Saturn	378 days	29.5 years

TABLE 4-2		
Average Distances of the Planets from the Sun (AU)		
Planet	Copernicus's value	Modern value
Mercury	0.38	0.39
Venus	0.72	0.72
Earth	1.00	1.00
Mars	1.52	1.52
Jupiter	5.22	5.20
Saturn	9.07	9.54

Copernicus compiled his ideas and calculations into a book entitled *De Revolutionibus Orbium Coelestium* (On the Revolutions of the Celestial Spheres), which was published in 1543, the year of his death. Copernicus assumed that the Earth travels around the Sun along a circular path, but he found that perfectly circular orbits could not accurately describe the paths of the other planets. He thus had to add an epicycle to each planet to account for its slight variation in

---

BOX 4-1

Synodic and Sidereal Periods

Mercury and Venus are sometimes called **inferior planets,** because their orbits lie within the Earth's. The outer planets, such as Mars, Jupiter, and Saturn, are called **superior planets,** because their orbits lie outside of the Earth's.

Consider an inferior planet (Mercury or Venus) orbiting the Sun as shown in the diagram. Let P be the sidereal period of the planet, S the synodic period of the planet, and E the sidereal period of the Earth (which Copernicus knew to be equal to nearly $365\frac{1}{4}$ days).

The rate at which the Earth moves around its orbit is $360°/E$ (nearly $1°$ per day). Similarly, the rate at which the inferior planet moves along its orbit is $360°/P$.

During the inferior planet's synodic period, the Earth covers an angular distance of $S(360°/E)$ around its orbit. In that same time, the inferior planet has covered an angular distance of $S(360°/P)$. Note, however, that the inferior planet has gained one full lap on the Earth. One lap corresponds to $360°$. Thus, $S(360°/E) + 360° = S(360°/P)$. Dividing each term of this equation by $360°\,S$ gives

$$\frac{1}{P} = \frac{1}{E} + \frac{1}{S}$$

A similar analysis for a superior planet (for example, Mars, Jupiter, Saturn) yields

$$\frac{1}{P} = \frac{1}{E} - \frac{1}{S}$$

Using these formulas, we can calculate a planet's sidereal period from its synodic period.

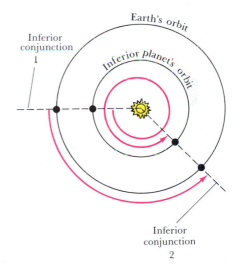

EXAMPLE: Consider the superior planet Jupiter, which has an observed synodic period of 398.9 days, or 1.092 years. (When astronomers express a time interval in years, they mean Earth years of $365\frac{1}{4}$ days.) Of course, $E = 1$ year exactly. Thus

$$\frac{1}{P} = 1 - \frac{1}{1.092} = 0.08425 = \frac{1}{11.87}$$

It takes 11.87 years for Jupiter to complete one full orbit of the Sun.

BOX 4-2

Copernicus's Method of Determining the Sizes of Orbits

Copernicus used different geometrical constructions to calculate the sizes of planetary orbits, depending on whether the planet is closer or farther from the Sun than the Earth is. As in Box 4-1, we shall therefore distinguish between inferior planets, whose orbits lie within the Earth's, and superior planets, whose orbits lie outside the Earth's.

To determine the size of an inferior planet's orbit, Copernicus measured the angle (α) between the Sun and the planet at its greatest elongation. As shown in diagram *a*, the triangle formed by the Earth, the inferior planet, and the Sun then contains a right angle. The hypotenuse of the triangle has a length of one astronomical unit (1 AU), which is the distance from the Earth to the Sun. Therefore, the radius (also measured in astronomical units) of the inferior planet's orbit is equal to sin α.

Determining the size of a superior planet's orbit is slightly more complicated. First, note the date on which the planet is at opposition. (In diagram *b*, opposition occurs when the planets are in the positions designated by 1.) After a few months, note the date on which the planet appears at an elongation of 90°. (The planets have now moved to the positions marked 2.) Using the number of days that have elapsed, determine the angle β, which is the distance the Earth has traveled, moving at roughly 1° per day. Knowing the sidereal period of the planet (see Table 4-1), you can determine the angle γ in the same way.

The triangle formed by the Sun, the Earth, and the planet contains a right angle. In addition, the short side of the triangle is exactly 1 AU long, and the angle ($\beta - \gamma$) is now known. The hypotenuse of the triangle, which is the radius of the superior planet's orbit, therefore has a length of 1/cos ($\beta - \gamma$), also measured in astronomical units.

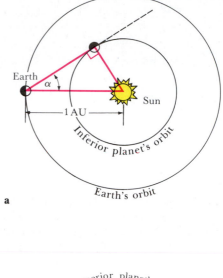

a

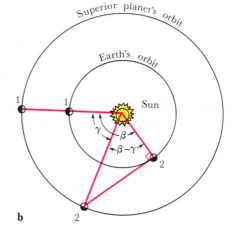

b

---

speed along its orbit. Therefore, according to Copernicus, each planet moves along a small epicycle, which in turn traces out a circular path around the Sun. The true shape of planetary orbits was still to be discovered.

4-3 Tycho Brahe made astronomical observations that disproved ancient ideas about the heavens

In November of 1572, a bright star suddenly appeared in the constellation of Cassiopeia. At first, it was even brighter than Venus, but then it began to grow dim. After 18 months, it faded from view.

Modern astronomers recognize this event as a supernova explosion, the violent death of a certain type of star (see Chapter 22). In the sixteenth century, however, the prevailing opinion was quite different. Classical teachings dating back to Aristotle and Plato argued that the heavens are permanent and unalterable. Consequently, the "new star" of 1572 could not really be a star at all, because the heavens do not change; it must instead be some sort of bright object quite near Earth, perhaps not much further away than the clouds overhead.

A 25-year-old Danish astronomer named Tycho Brahe realized that straightforward observations might reveal the distance to the new star. It is everyone's common experience that when you walk from one place to another, nearby objects appear to change position against the background of more distant objects. This phenomenon, whereby the apparent position of an object changes because of the motion of the observer, is called **parallax.** If the new star was nearby, then its position should shift against the background stars over the course of a night, because the Earth's rotation changes our viewpoint, as shown in Figure 4-6. (Actually, Tycho believed that the heavens rotate about the Earth, but the net effect is the same.) Tycho's careful observations failed to disclose any parallax, and so the new star had to be quite far away, farther from Earth than anyone had imagined. Tycho Brahe summarized his findings in a small book *De Stella Nova* (On the New Star), published in 1573.

Tycho's discovery, which flew in the face of nearly 2000 years of wisdom, attracted the attention of the Danish king, who was so impressed that he financed the construction of a magnificent observatory for Tycho on the island of Hven, just off the Danish coast. The bequest included a lavish house with servants and coastal land that provided an income. Tycho designed and supervised the construction of astronomical instruments, like the wall quadrant shown in Figure 4-7, with which he was able to measure the positions of stars and

FIGURE 4-7 Tycho Brahe at His Observatory   During the sixteenth century Tycho Brahe measured the positions of stars and planets with unprecedented accuracy. His measurements were crucial to the development of astronomy in the seventeenth century. (Photo Researchers, Inc.)

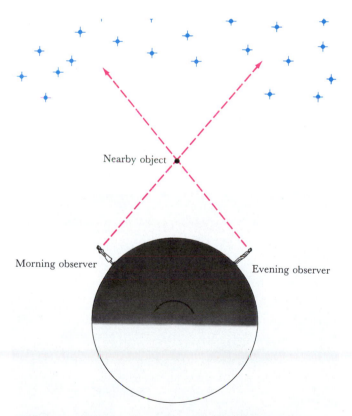

FIGURE 4-6 The Parallax of a Nearby Object   Tycho Brahe argued that if an object is near the Earth, its position relative to the background stars should change over the course of a night. Since Tycho failed to measure such changes for a supernova in 1572 and a comet in 1577, he concluded that these objects are far from the Earth.

planets with unprecedented accuracy. Tycho named his palatial observatory Uraniborg, which means "sky castle."

Using his instruments at Uraniborg, Tycho Brahe attempted to test Copernicus's ideas about the Earth going around the Sun. Tycho argued that if the Earth was in motion, then nearby stars should appear to shift their positions with respect to background stars as we orbit the Sun. Tycho failed to detect any such parallax, and so he concluded that the Earth was at rest and the Copernican system was wrong. Actually, the stars are so far away that naked-eye observations cannot possibly detect the tiny shifting of stellar positions that have since been confirmed with telescopic observations.

Tycho Brahe's astronomical records were soon to play an important role in the development of a heliocentric cosmology. From 1576 to 1597, he made comprehensive observations, measuring planetary positions with an accuracy of 1 arc min, about as well as can be done with the naked eye. Upon his death in 1601, most of these invaluable records fell into the hands of his gifted assistant, Johannes Kepler (Figure 4-8).

FIGURE 4-8 Johannes Kepler (1571–1630) Using Tycho Brahe's records of planetary positions, Kepler discovered that the planets orbit the Sun along ellipses. Kepler's three laws describe how the planets move about the Sun. (E. Lessing/Magnum)

4-4 Johannes Kepler proposed elliptical paths for the planets about the Sun

Until Kepler's time, astronomers had assumed that heavenly objects move in circles, which were considered the most perfect and harmonious of all geometric shapes. They believed that if a perfect God resided in heaven along with the stars and planets, then the motions of these bodies must be perfect too. Against this context, Johannes Kepler dared to try and explain planetary motions with noncircular curves.

Kepler first tried to use ovals for the paths of the planets around the Sun. For years he tried in vain to find ovals that would fit Tycho Brahe's observations of the planets against the background of the distant stars. Then he began working with a slightly different curve called an **ellipse.**

An ellipse can be constructed with a loop of string, two thumbtacks, and a pencil, as shown in Figure 4-9. An ellipse has two foci: Each thumbtack in the figure is at a **focus.** The longest diameter of an ellipse, called the **major axis,** passes through both foci. Half of that distance is called the **semimajor axis,** whose length is usually designated by the letter *a*.

To Kepler's delight, the ellipse turned out to be the curve he had been searching for. He published his breakthrough along with other material in 1609 in a book known today as *New Astronomy*. His important discovery is now called **Kepler's first law:**

The orbit of a planet about the Sun is an ellipse with the Sun at one focus.

Kepler also realized that planets do not move at uniform speeds along their orbits. A planet moves most rapidly when it is nearest the Sun, at a point on its orbit called **perihelion.** Conversely, a planet moves most slowly when it is farthest from the Sun, at a point called **aphelion.**

After much trial and error, Kepler found a way to describe how fast a planet moves along its orbit. This discovery, also published in *New Astronomy*, is illustrated in Figure 4-10. Suppose that it takes 30 days for a planet to go from point *A* to point *B*. During that time, a line joining the Sun and the planet sweeps out a nearly triangular area. Kepler discovered that the line joining the Sun and the planet also sweeps out an equal area during any other 30-day interval. In other words, if the planet also takes a month to go from point *C* to point *D*, then the two shaded segments in Figure 4-10 are equal in area. **Kepler's second law,** also called the **law of equal areas,** can be stated thus:

A line joining a planet and the Sun sweeps out equal areas in equal intervals of time.

Kepler was fascinated by the many harmonious relationships in the motions of the planets. His writing is filled with intriguing speculations, including musical scores intended to represent the celestial music that the planets make as they travel along their orbits. One of Kepler's later discoveries stands out because of its impact on future developments. Now called the **harmonic law** or **Kepler's third law,** it states a relationship between the size of a planet's orbit and the time the planet takes to go once around the Sun:

The squares of the sidereal periods of the planets are proportional to the cubes of their semimajor axes.

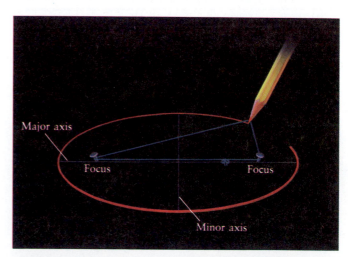

FIGURE 4-9 The Construction of an Ellipse An ellipse can be drawn with a pencil, a loop of string, and two thumbtacks. If the string is kept taut, the pencil traces out an ellipse. The thumbtacks are located at the two foci. The ellipse's major axis is the longest diameter across the ellipse; the minor axis is the perpendicular bisector of the major axis.

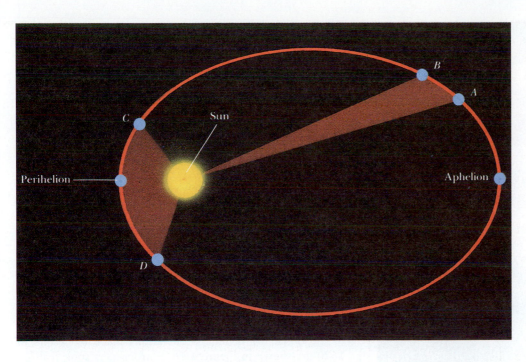

If a planet's sidereal period (P) is measured in years and the length of its semimajor axis (a) is measured in astronomical units, then Kepler's third law is simply stated as

$$P^2 = a^3$$

The length of the semimajor axis may be thought of as the average distance between a planet and the Sun. Using data from Tables 4-1 and 4-2, we can demonstrate Kepler's third law as shown in Table 4-3. This relationship can also be displayed on a graph, as in Figure 4-11.

This third law was published in 1619. It is testimony to Kepler's genius that his three laws are rigorously obeyed whenever two objects orbit each other because of their mutual gravitational attraction. Throughout this book, we shall

see that Kepler's laws have a wide range of practical applications. Kepler's laws are obeyed not only by planets circling the Sun but also by artificial satellites orbiting the Earth, by two stars revolving about each other in a double-star system, by stars in their orbits within galaxies, and even by galaxies in their orbits about each other.

TABLE 4-3

A Demonstration of Kepler's Third Law

Planet	Sidereal period P (years)	Semimajor axis a (AU)	P^2	a^3
Mercury	0.24	0.39	0.06	0.06
Venus	0.61	0.72	0.37	0.37
Earth	1.00	1.00	1.00	1.00
Mars	1.88	1.52	3.53	3.51
Jupiter	11.86	5.20	140.7	140.6
Saturn	29.46	9.54	867.9	868.3

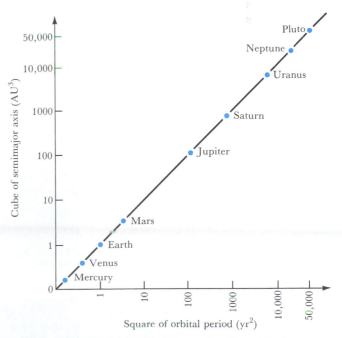

FIGURE 4-11 Kepler's Third Law On this graph, the squares of the periods of the planets (P^2) are plotted against the cubes of their semimajor axes (a^3). The fact that the points fall along such a straight line is verification of Kepler's discovery that $P^2 = a^3$.

After only a few months of observation, Galileo noticed that the apparent size of Venus as seen through his telescope was related to the planet's phase. Venus appears smallest at gibbous phase and largest at crescent phase. There is also a correlation between the phases of Venus and the planet's angular distance from the Sun. These relationships clearly support the conclusion that Venus goes around the Sun (Figure 4-14).

In 1610 Galileo discovered four moons orbiting Jupiter (Figure 4-15). He realized that they were orbiting Jupiter because they moved back and forth from one side of the planet to the other. Confirming observations made by Jesuits in 1620 are shown in Figure 4-16. Astronomers soon realized that these four moons obey Kepler's third law: The cube of a moon's distance from Jupiter is proportional to the square of its orbital period about the planet.

These telescopic observations constituted the first fresh influx of fundamentally new astronomical data in almost 2000 years. Contradicting prevailing opinion, these discoveries strongly suggested a heliocentric structure of the universe. The Roman Catholic church attacked Galileo's ideas because they could not be reconciled with certain passages in the Bible or with the writings of Aristotle and Plato. Nevertheless, there was no turning back. Although Galileo was condemned to spend his latter years under house arrest "for vehement suspicion of heresy," his revolutionary ideas were soon to inspire a sickly English boy born on Christmas Day of 1642, less than one year after Galileo died. The boy's name was Isaac Newton.

4-5 Galileo's discoveries with a telescope strongly supported a heliocentric cosmology

While Kepler was making rapid progress in central Europe, an Italian physicist was making equally dramatic observational discoveries in southern Europe. Galileo Galilei (Figure 4-12) did not invent the telescope, but he was the first to point one of these new devices toward the sky and publish his observations. What he saw, no one had ever dreamed of: mountains on the Moon and sunspots on the Sun. He also discovered that Venus exhibits phases (Figure 4-13).

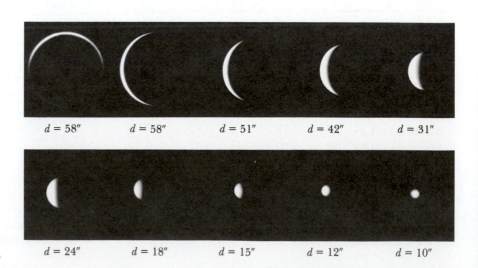

FIGURE 4-13 The Phases of Venus This series of photographs shows how the appearance of Venus changes as it moves along its orbit. The number below each view is the angular diameter of the planet in seconds of arc.

$d = 58''$ $d = 58''$ $d = 51''$ $d = 42''$ $d = 31''$

$d = 24''$ $d = 18''$ $d = 15''$ $d = 12''$ $d = 10''$

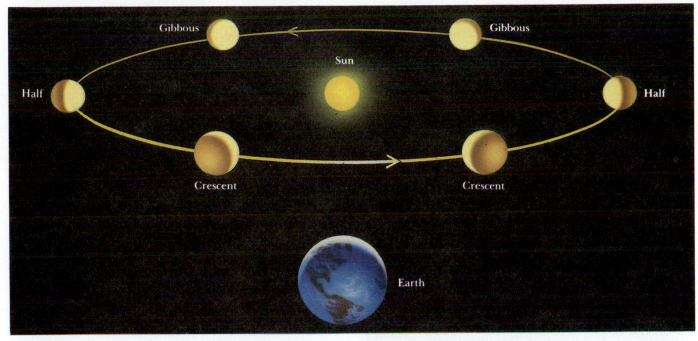

FIGURE 4-14 The Changing Appearance of Venus The phases of Venus are correlated with the planet's angular size and its angular distance from the Sun, as sketched in this diagram.

These observations clearly support the idea that Venus orbits the Sun.

FIGURE 4-15 Jupiter and Its Largest Moons This photograph, taken by an amateur astronomer with a small telescope, shows the four Galilean satellites alongside an overexposed image of Jupiter. Each satellite is bright enough to be seen with the unaided eye were it not overwhelmed by the glare of Jupiter. (Courtesy of C. Holmes)

FIGURE 4-16 Early Observations of Jupiter's Moons In 1610 Galileo discovered four "stars" that move back and forth across Jupiter from one night to the next. He concluded that these are four moons that orbit Jupiter much as our Moon orbits the Earth. This drawing shows observations made in 1620. (Yerkes Observatory)

FIGURE 4-17 Isaac Newton (1642–1727) Using mathematical techniques that he devised, Isaac Newton formulated the universal law of gravitation and demonstrated that the planets orbit the Sun according to simple mechanical rules. (National Portrait Gallery, London)

4-6 Isaac Newton formulated three laws that describe fundamental properties of physical reality

Until the mid-seventeenth century, virtually all mathematical astronomy was entirely empirical, characterized by trial and error. From Ptolemy to Kepler, essentially the same approach was used. Astronomers would work directly from data and observations, adjusting ideas and calculations until the right answers finally emerged.

Isaac Newton (Figure 4-17) introduced a new approach. He made three assumptions, now called **Newton's laws of motion.** These are quite general statements that apply to all forces and all bodies. Newton then showed that Kepler's three laws follow logically from these laws of motion and from a formula for the force of gravity that he derived from observations and his laws. He used this formula to describe the observed orbits of the planets, the Moon, and comets.

Newton's first assumption, known as **Newton's first law** or the **law of inertia,** reads as follows:

A body remains at rest, or moves in a straight line at a constant speed, unless acted upon by an outside force.

This idea tells us that a force must be acting on the planets. If there were no "outside" force acting on planets, they would leave their curved orbits and move away from the Sun along straight-line paths at constant speeds. Because this does not happen, Newton concluded that the continuous action of a force confines the planets to their elliptical orbits.

Newton's second assumption does two things: It defines the concept of **force,** and it describes how a force changes the motion of an object. To appreciate these concepts, we must first understand the quantities that describe motion: quantities such as speed, velocity, and acceleration.

Imagine an object in space. Push on the object, and it will begin to move. At any moment you can describe the object's motion by specifying both its speed and its direction. Speed and direction of motion together constitute the object's **velocity.** If you continue to push on the object, its speed will increase—in other words, it will accelerate.

Acceleration is the rate at which velocity changes. Since velocity involves both speed and direction, acceleration can result from changes in either. Also note that, contrary to popular use of the term, acceleration is not restricted to increases in speed alone. A slowing down, a speeding up, or a change in direction all involve accelerations.

An apple falling from a tree is a good example of acceleration that involves only an increase in speed. Initially, at the moment the stem breaks, the apple's speed is zero. After one second, its downward speed is 32 feet per second. After two seconds, the apple's speed is 64 feet per second. After three seconds, the speed is 96 feet per second. Because the apple's speed increases by 32 feet per second for each second of free fall, the rate of acceleration is 32 feet per second per second, or 32 ft/s^2. In other words, the Earth's gravity produces a constant acceleration of 32 ft/s^2 (9.8 m/s^2) downward, toward the center of the Earth.

A planet revolving about the Sun along a perfectly circular orbit is an example of acceleration that involves change of direction only. As the planet moves along its orbit, its speed remains constant. Nevertheless, the planet is continuously being accelerated because its direction of motion is continuously changing.

Newton's second law says that the acceleration of an object is proportional to the force acting it. In other words, the harder you push on an object, the greater is the resulting acceleration. This law can be succinctly stated as an equation. If a force (F) acts on the object, it will experience an acceleration (a) such that

$$F = ma$$

where m is the mass of the object.

The **mass** of an object is a measure of the total amount of material in the object and is usually expressed in grams or kilograms. For example, the mass of the Sun is 2×10^{30} kg, the mass of a hydrogen atom is 1.7×10^{-27} kg, and the mass of the author of this book is 81 kg. The Sun, a hydrogen atom, and the author have these masses regardless of where they happen to be in the universe.

It is important not to confuse the concepts of mass and weight. **Weight** is the force with which an object presses down on the ground (due to gravity's pull); like force, it is usually expressed in pounds or newtons. These units of measure are related to one another as follows:

$$1 \text{ newton} = 0.225 \text{ pound}$$

We can use Newton's second law to relate mass and weight. We have seen that the acceleration caused by the Earth's gravity is 32 ft/s^2, or 9.8 m/s^2. From the second law ($F = ma$), the force with which the author presses down on the ground is

$$81 \text{ kg} \times 9.8 \text{ m/s}^2 = 794 \text{ newtons} = 179 \text{ pounds}$$

Note that this answer is correct only when the author is standing on the Earth. He would weigh less on the Moon and more on Jupiter. Floating deep in space, he would have no weight at all; he would be "weightless." Nevertheless, under all these circumstances, he would always have exactly the same mass. Thus we see that mass is an inherent property of matter unaffected by details of the environment. Whenever we describe the properties of planets, stars, or galaxies, we speak of their masses, never of their weights.

Newton's final assumption, called **Newton's third law**, is the famous statement about action and reaction:

> **Whenever one body exerts a force on a second body, the second body exerts an equal and opposite force on the first body.**

For example, if you weigh 165 pounds, you are pressing down on the floor with a force of 165 pounds. Newton's third law tells us that the floor is also pushing up against your feet with an equal force of 165 pounds. (If it were not, you would fall through the floor.)

In the same way, Newton realized that, because the Sun is exerting a force on each planet to keep it in orbit, each planet must also be exerting an equal and opposite force on the Sun. However, the planets are much less massive than the Sun (for example, the Earth has only $\frac{1}{330,000}$ of the Sun's mass). So although the Sun's force on a planet is the same as the planet's force on the Sun, the planet's much smaller mass gives it a much larger acceleration, according to Newton's second law. This is why the planets circle the Sun, instead of vice versa. Newton's laws reveal the reason for our heliocentric solar system.

4-7 Newton's description of gravity accounts for Kepler's laws and explains the motions of the planets

Isaac Newton did not invent the idea of gravity. An educated seventeenth-century person had a vague sense that some force pulls things down to the ground. It was Newton, how-ever, who precisely described the action of gravity. Using his first law, Newton proved mathematically that the force acting on each of the planets is aimed directly at the Sun. This discovery led him to suspect that the force of gravity pulling a falling apple straight down to the ground is the same as the force on the planets that is always aimed straight at the Sun.

Using his own three laws and Kepler's three laws, Newton succeeded in formulating a general statement describing the nature of the force called **gravity** that keeps the planets in their orbits. Newton's **universal law of gravitation** is

> **Two bodies attract each other with a force that is directly proportional to the product of their masses and inversely proportional to the square of the distance between them.**

In other words, if two objects have masses m_1 and m_2 and are separated by a distance (r), then the gravitational force (F) between these two masses is given by the equation

$$F = G \left(\frac{m_1 m_2}{r^2} \right)$$

If the masses are measured in kilograms and the distance between them in meters, then the force is measured in newtons. In this formula, G is a number called the **universal constant of gravitation**. Laboratory experiments have yielded a value for G of:

$$G = 6.67 \times 10^{-11} \text{ newton m}^2/\text{kg}^2$$

We can use Newton's law of gravity to calculate the force with which any two bodies attract each other. For example, to compute the gravitational force between the Earth and the Sun, we substitute values for the Earth's mass ($m_1 = 5.98 \times 10^{24}$ kg), the Sun's mass ($m_2 = 1.99 \times 10^{30}$ kg), the distance between them ($r = 1 \text{ AU} = 1.5 \times 10^{11}$ m), and the value of G into Newton's equation to get

$$F_{\text{Sun–Earth}} = 6.67 \times 10^{-11} \left[\frac{5.98 \times 10^{24} \times 1.99 \times 10^{30}}{(1.50 \times 10^{11})^2} \right]$$

$$= 3.53 \times 10^{22} \text{ newtons}$$

Using his law of gravity, Newton found that he could mathematically prove Kepler's three laws. For example, whereas Kepler had to discover his third law ($P^2 = a^3$) by trial and error, Newton demonstrated mathematically that this equation follows logically from his law of gravity. Specifically, he proved that two masses (m_1 and m_2) orbiting each other obey the equation

$$P^2 = \left[\frac{4\pi^2}{G(m_1 + m_2)} \right] a^3$$

This formulation of Kepler's third law is very useful because it relates the period (P) and semimajor axis (a) of the two bodies orbiting each other to the sum of their masses ($m_1 + m_2$). This general formula is valid whenever two objects orbit each other due to their mutual gravitational attraction.

It is invaluable, for example, in the study of double stars, as we shall see in Chapter 19. If the orbital period and semimajor axis of two stars in a double star are known, an astronomer can use this formula to calculate the sum of the masses of the two stars. Additional details concerning orbits and Newtonian gravitation are presented in Box 4-3.

Newton also discovered new features of orbits around the Sun. For instance, his equations soon led him to conclude that the orbit of an object around the Sun need not be an ellipse. It could be any one of a family of curves called conic sections.

A **conic section** is any curve that you get by cutting a cone with a plane, as shown in Figure 4-18. You can get circles and ellipses by slicing all the way through the cone. You can also get two open curves called **parabolas** and **hyperbolas.** Com-

ets hurtling toward the Sun from the depths of space sometimes follow parabolic orbits.

Newton's ideas turned out to be applicable to an incredibly wide range of situations. The orbits of the planets and their satellites could now be calculated with unprecedented precision. Using Newton's laws, mathematicians proved that the Earth's axis of rotation must precess because of the gravitational pull of the Moon and the Sun on the Earth's equatorial bulge (see Figure 2-11). Similarly, all the details of the Moon's orbit (see Box 3-1) could be demonstrated mathematically with a body of knowledge built on Newton's work that is today called **Newtonian mechanics.**

A central feature of Newtonian mechanics is the idea that certain quantities—energy, momentum, and angular momentum—are conserved. In other words, in an isolated system of

---

BOX 4-3

Ellipses and Orbits

Using his universal law of gravitation and a lot of mathematics, Newton calculated many details of the planets' orbits around the Sun.

As shown in diagram a, the largest diameter across an ellipse is called the **major axis** and has a length designated as $2a$. The shortest diameter through the center of the ellipse, called the **minor axis,** has a length designated $2b$. The minor axis is the perpendicular bisector of the major axis.

The shape of an ellipse is determined by its **eccentricity** (e). The distance from the center of an ellipse to one focus is ae. Notice that if $e = 0$, we have a circle. A few examples of ellipses with different eccentricities are shown in diagram b.

Kepler's second law tells us that the velocity of a planet varies as it orbits the Sun. Newtonian mechanics can be used to calculate the speed of the planet at various points along its orbit. If P is the planet's sidereal period, then the orbital speed (v) at perihelion is given by

$$v = \frac{2\pi a}{P}\left(\frac{1+e}{1-e}\right)^{1/2}$$

and at aphelion by

$$v = \frac{2\pi a}{P}\left(\frac{1-e}{1+e}\right)^{1/2}$$

EXAMPLE: We have the following data for the Earth:

$$a = 1 \text{ AU} = 1.496 \times 10^8 \text{ km}$$

$$P = 1 \text{ yr} = 3.156 \times 10^7 \text{ s}$$

$$e = 0.0167$$

where the eccentricity of the Earth's orbit is determined from the annual variation of the Earth–Sun distance. Inserting these values into the above equations, we find that the Earth's orbital speed varies from 30.3 km/s at perihelion to 29.3 km/s at aphelion.

Although it is commonly said that the Moon goes about the Earth or that the planets revolve around the Sun, such statements are not entirely accurate. When one object seemingly orbits another, both objects are actually revolving about a common, stationary point called the **center of mass,** as shown in diagram c. To appreciate this concept, imagine placing a mass (m_1) at one end of a seesaw and a second mass (m_2) at the other end. The center of mass is the point where you would put the fulcrum to balance the seesaw.

In the case of the Moon orbiting the Earth or a planet orbiting the Sun, one object is much more massive than the other, and so the center of mass is quite near the more massive object. For instance, the center of mass of the Earth–Moon system actually lies within the Earth, whereas that for the Sun–Jupiter system lies just outside the Sun's surface.

Consider the masses m_1 and m_2 orbiting their common center of mass as shown in diagram c. The location of their center of mass is given by

$$m_1 r_1 = m_2 r_2$$

where $r_1 + r_2 = a$, the distance between the two masses. The gravitational force between these two masses is given by Newton's law:

$$F = G\left(\frac{m_1 m_2}{a^2}\right)$$

where G is the universal constant of gravitation. From these equations, it is possible to derive the following

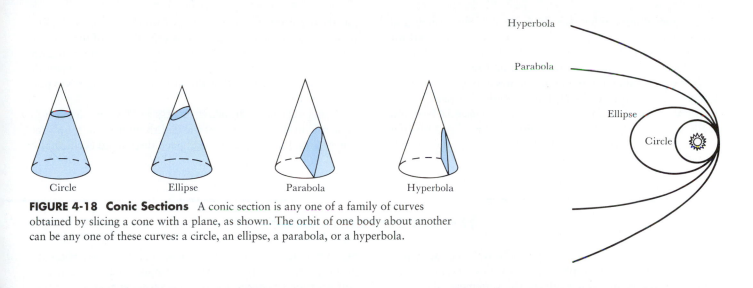

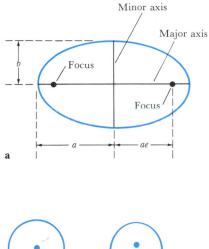

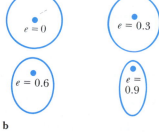

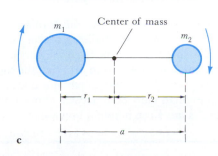

FIGURE 4-18 **Conic Sections**   A conic section is any one of a family of curves obtained by slicing a cone with a plane, as shown. The orbit of one body about another can be any one of these curves: a circle, an ellipse, a parabola, or a hyperbola.

relationship between the total separation of the two masses a and the orbital period P with which they revolve about their common center of mass:

$$P^2 = \left[\frac{4\pi^2}{G(m_1 + m_2)}\right] a^3$$

This equation is actually the most general and complete statement of Kepler's third law. It is valid in any situation where two objects revolve about each other under the influence of their mutual gravitational attraction. Astronomers find many uses for this powerful equation, from double stars, to pairs of galaxies, to satellites orbiting planets.

In the case of the solar system, the mass of the Sun is thousands of times greater than the masses of any of the planets. Consequently, for all practical purposes $m_1 + m_2$ is equal to the mass of the Sun ($M_\odot$), which is 2×10^{30} kg. Thus, Kepler's third law becomes

$$P^2 = \left(\frac{4\pi^2}{GM_\odot}\right) a^3$$

If P is measured in years and a is measured in astronomical units, then the quantity in the parentheses equals 1 and we have

$$P^2 = a^3$$

EXAMPLE: Since we know the Earth's orbital period ($P = 1$ yr $= 3.156 \times 10^7$ s) and the semimajor axis of its orbit ($a = 1$ AU $= 1.496 \times 10^{11}$ m), we can use Kepler's third law to calculate the mass of the Sun ($M_\odot$) as follows:

$$M_\odot = \frac{4\pi^2 a^3}{GP^2} = \frac{4(3.1416)^2(1.496 \times 10^{11})^3}{(6.67 \times 10^{-11})(3.156 \times 10^7)^2}$$

$$= 1.99 \times 10^{30} \text{ kg}$$

interacting objects, whatever the history of the system and however it may evolve, these three quantities measured for the entire system remain unchanged. A brief discussion of these basic conservation laws is found in Box 4-4.

Not only could Newtonian mechanics explain a variety of known phenomena in detail, but it could also predict new phenomena. For example, one of Newton's friends, Edmund Halley, was intrigued by three similar historical records of a comet that had been sighted at intervals of 76 years. Assum-

ing these records to be accounts of the same comet, Halley used Newton's methods to work out the details of the comet's orbit and predicted its return in 1758. It was first sighted on Christmas night of 1757, a fitting memorial to Newton's birthdate, and to this day the comet bears Halley's name (Figure 4-19).

Perhaps the most dramatic success of Newton's ideas was their role in the discovery of the eighth planet from the Sun. The seventh planet, Uranus, had been discovered accidentally

---

BOX 4-4

The Conservation Laws

One of the most powerful concepts in Newtonian mechanics is the idea that certain physical quantities remain unchanged as the universe evolves. The statements that express these constancies of nature are called *conservation laws,* and three of the most important conserved quantities are energy, momentum, and angular momentum.

According to the **conservation of energy,** the total energy of a physical system is constant. There may be different kinds of energy—chemical energy, thermal energy, radiation—in that system, and these different manifestations of energy may be converted one into another, but the total amount remains unchanged. In order to use this law, however, you must know what energy is and how to express it mathematically.

Energy is the ability to do work, and work is the exertion of a force over a distance. From these definitions it is possible to derive mathematical expressions for certain kinds of energy. For instance, the energy associated with motion is called **kinetic energy.** The kinetic energy (E_k) of an object of mass m moving with a velocity v is

$$E_k = \frac{1}{2}mv^2$$

Another example is the energy associated with position in a gravitational field, called **potential energy.** The potential energy (E_p) of masses m and M separated by distance r is

$$E_p = -\frac{GMm}{r}$$

where G is the universal constant of gravitation.

EXAMPLE: We can use the conservation of energy to derive a formula for the **escape speed,** the minimum speed needed by a small mass (m), such as a rocket, to escape from the gravitational attraction of a larger mass (M), such as a planet. The small mass will not fall back toward the larger one if it has just enough speed to reach $r = \infty$. At $r = \infty$ the particle has no kinetic energy because it is not

moving. It also has zero potential energy at $r = \infty$, where it no longer feels the pull of the larger mass. According to the conservation of energy, the particle's total energy, which is the sum of its kinetic and potential energies, is constant. Since the particle's total energy is equal to zero at $r = \infty$, it must be equal to zero everywhere. We can therefore write

$$E_{total} = E_k + E_p = 0$$

Substituting the expressions for kinetic and potential energy given above, we get

$$\frac{1}{2}mv^2 - \frac{GMm}{r} = 0$$

We want to know the speed ($v = V_{escape}$) needed to escape from the surface of a planet whose radius is $r = R$. Rearranging terms in the previous equation, we get

$$V_{escape} = \sqrt{\frac{2GM}{R}}$$

Note that the mass of the smaller object cancels out and does not appear in the final equation for V_{escape}. Escape speed therefore does not depend on the mass of the escaping particle. For example, knowing that the Earth's mass is 5.98×10^{24} kg and its radius is 6.38×10^6 m, we can compute the velocity needed to escape from the Earth:

$$V_{escape} = \sqrt{\frac{2 \times 6.67 \times 10^{-11} \times 5.98 \times 10^{24}}{6.38 \times 10^6}}$$

$$= 1.12 \times 10^4 \text{ m/s} = 11.2 \text{ km/s}$$

This calculation, which neglects the effects of air friction, tells us that any object—be it a bullet, a rocket, or an atom—with this speed can escape from the Earth.

The **conservation of momentum** says that the total momentum of the particles in a system remains unchanged. The momentum of an individual particle of mass m moving with velocity v is mv. Keep in mind, however, that velocity

by William Herschel in 1781 during a telescopic survey of the sky. Fifty years later, however, it was clear that Uranus was not following its predicted orbit. Two mathematicians, John Couch Adams in England and U. J. Leverrier in France, independently calculated that the gravitational pull of a yet unknown, more distant planet could explain the deviations of Uranus from its orbit. They each predicted that the planet would be found at a certain location in the constellation of Aquarius. A brief telescopic search on September 23, 1846,

has both size *and* direction, which means that momentum does also.

Angular momentum is a measure of the momentum carried by an object because of its rotation about an axis. To appreciate the definition of angular momentum, consider a mass (m) revolving about an axis, as shown in the accompanying diagram. If r is the distance of the mass from the axis of rotation and v is the linear speed of the mass, then the angular momentum (L) is

$$L = mvr$$

The **conservation of angular momentum** requires that both the size and the direction of the angular momentum never change. For a mass revolving about an axis, this means that the value of L must remain constant, and the axis of rotation must remain fixed in space.

In the coming chapters we shall often see how these conservation laws play a crucial role in determining the behavior and outcome of many features of the universe.

FIGURE 4-19 Halley's Comet Halley's Comet orbits the Sun with an average period of about 76 years. During the twentieth century, the comet passed near the Sun twice—in 1910 and again in 1986. This photograph shows how the comet looked in 1986. (Courtesy of J. Marling)

revealed the planet Neptune within $1°$ of the calculated position. (Figure 4-20 shows photographs of both Uranus and Neptune.) Before it was sighted with a telescope, Neptune was really discovered with pencil and paper.

Because it has been so successful in explaining and predicting many important phenomena, Newtonian mechanics has become the cornerstone of modern physical science. Even today, as we send astronauts to the Moon and spacecraft to the outer planets, Newton's equations are used to calculate orbits and trajectories. These applications usually employ high-speed computers, called supercomputers, that greatly facilitate the repetitive computations that are often required (Box 4-5).

In one instance, however, Newtonian mechanics was not quite in agreement with astronomical observations. During the mid-1800s, Leverrier pointed out that Mercury was not following its predicted orbit. As the planet moves along its elliptical orbit, the orbit itself rotates (or precesses), as shown in Figure 4-21. Most of Mercury's precession is caused by the gravitational pull of the other planets, and so is explained with Newtonian mechanics. There is, however, an unexplained excess rotation of Mercury's major axis amounting to only 43 arc sec per century. Although the effect is very small, all attempts to account for this phenomenon using Newtonian mechanics met with failure.

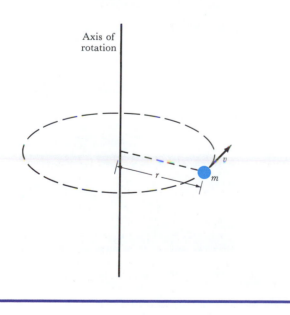

Axis of rotation

FIGURE 4-20 Uranus and Neptune The discovery of Neptune was a major triumph for Newtonian mechanics. In an effort to explain deviations in the predicted orbit of Uranus (shown on the left), astronomers predicted the existence of Neptune (shown on the right). Uranus and Neptune are nearly the same size; both have diameters about four times that of Earth. (NASA)

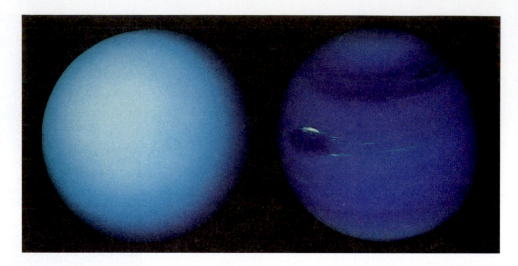

Supercomputing—A New Mode of Science

The invention of high-speed electronic computers has dramatically changed the way we do science. Because of the speed with which supercomputers can perform repetitive mathematical operations, they have revolutionized how we learn about the universe.

Previously all scientific investigation fell into two broad categories: experiment and theory. In the experimental mode, researchers learn about the world through observation, measurement, and experimentation. Galileo Galilei, who showed us the heavens with his telescope and investigated gravitational acceleration by timing balls as they rolled down an inclined plane, can be considered the founder of this technique of science. The theoretical mode was conceived by Isaac Newton, who showed us that mathematics is the fundamental language with which the regularities we observe in nature can be written as universal laws.

Although these traditional forms of science have flourished since the time of Galileo and Newton, both have distinct limitations. Much of what scientists would like to observe is too small, too far away, or too ephemeral to yield readily to experiment. Similarly, the complexity of the real world poses insurmountable problems for the theoretician, who despairs at the prospect of solving the myriad of equations that describe realistic situations, such as colliding galaxies or a thunderstorm.

In the 1940s, the Hungarian-American mathematician John von Neumann conceived of a third way of doing science: the computational mode. Von Neumann pointed out that while we know the laws of physics, we can solve them only in the very simplest cases. Realistic situations typically involve so many intertwined equations that a scientist with pencil and paper will not live long enough to solve them. Von Neumann envisioned the day when high-speed computers would be able to solve these huge sets of equations. That day has arrived.

A supercomputer is the fastest, most powerful computer available. Examples include the CM-5, shown in the accompanying photograph, which can perform hundreds of billions of mathematical operations per second. Every supercomputer enjoys its exalted status for only a few years, because continuing technological advances soon give rise to the next generation of supercomputers.

John von Neumann
(1903–1957)

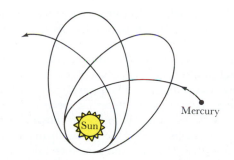

FIGURE 4-21 The Advance of Mercury's Perihelion
As Mercury moves along its orbit, the orbit slowly rotates.
Consequently, Mercury traces out a rosette figure about the Sun.
The effect is greatly exaggerated in this diagram. The excess
rotation of Mercury's orbit (beyond that predicted by Newtonian
mechanics) amounts to only 43 arc sec per century.

4-8 Albert Einstein's theory states that gravity affects the shape of space and the flow of time

Isaac Newton brought a new dimension of elegance and
sophistication to our understanding of the universe. It
remained for the insight and genius of Albert Einstein to
develop a whole new perspective on space, time, and
gravitation.

In the cosmologies of most ancient civilizations, the Earth
and its inhabitants occupied a special place at the center of
the universe. Then Copernicus's cosmology made the Earth
just one of a number of planets orbiting the Sun. We have
seen that this new approach proved quite fruitful. Within less
than a century, Kepler's accurate description of planetary
orbits led directly to Newton's universal law of gravitation.

A supercomputer is a valuable research tool because the
laws of physics can be written in a way that the supercomputer can solve them. In their basic form, the laws of
physics are valid at every point in space and at every
moment of time. But scientists seldom need to apply these
laws everywhere at all times. Instead, they use points in
space separated by distances that are small compared to the
size of the object under study. They also use time intervals
that are short compared to the duration of the process they
want to examine. By programming a supercomputer with
the laws of physics expressed only at these selected points
and time intervals, scientists can reduce a complicated
problem to a form that the machine can handle.

For example, suppose that you want to calculate the
orbit of Halley's Comet. If the Sun were the only object in
the solar system, the comet's path would be an ellipse with
the Sun at one focus. But all the planets are also pulling on
the comet, and so its path will deviate slightly from a
perfect ellipse. To calculate the true orbit, you start with the
comet and planets at their known locations in space on a
certain day. Knowing the distances between the comet, the
Sun, and the planets, you can calculate the total gravitational force acting on the comet and, using Newton's
second law, compute the acceleration that the comet experiences at that moment. This tells you where the comet is
headed in the immediate future, and you can deduce its
location on the following day. Now you can repeat the
entire calculation using the data for the comet and the
planets in their new positions. Step by step, you compute
the path of the comet across the solar system. Modern
supercomputers are ideal for performing these repetitive
computations.

Furthermore, using supercomputers, scientists can simulate phenomena that might otherwise never be observed.
For example, later in this book we shall see how the Moon
might have been created by a Mars-sized object striking the
Earth. We shall also see what happens when gas is captured
and swallowed by a black hole. We shall watch as two
galaxies, consisting of thousands of stars and gas clouds,
collide and interact with each other. These are just a few
examples of how supercomputing gives us a new, deeper
understanding of the universe.

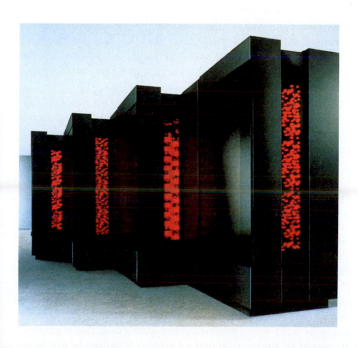

FIGURE 4-22 Albert Einstein (1879–1955) Albert Einstein demonstrated that gravity affects the curvature of space and the flow of time. These effects are conspicuous only in extremely intense gravitational fields, like that found around a black hole. (L. Jacobi, Diamond Library, University of New Hampshire)

Much later, astronomers began to explore the possibility that even the Sun may occupy no special location. Today we know that our Sun is merely one of the many stars scattered throughout the Milky Way Galaxy, which is only one of many galaxies in the universe.

Albert Einstein (Figure 4-22) introduced a powerful extension of this perspective. Einstein believed that the fundamental laws of the universe do not depend on a person's location or motion. In other words, the laws of physics should be the same whether we happen to be sitting on Earth or moving through space at a high speed.

Soon after 1900, Einstein began developing a new approach to the phenomena of electricity and magnetism. The basic properties of electricity and magnetism had been summarized earlier in four equations formulated in 1865 by the great Scottish physicist James Clerk Maxwell. Maxwell's equations are the basis of **electromagnetic theory,** which today has a wide range of practical applications, from television sets to microwave ovens. Maxwell's electromagnetic theory predicts various effects, depending upon the motion of electric charges and magnets. For example, a moving electric charge creates a magnetic "field," whereas a stationary electric charge does not. Such predictions imply some absolute or fixed spatial framework within which an object is either stationary or moving.

Einstein's goal was to eliminate this assumption of absolute space from electromagnetic theory. He wanted to rewrite Maxwell's equations so that they would depend only upon the relative motions of the observer and the electric charges or magnets. In 1905 he succeeded and published his results in a famous paper entitled "On the Electrodynamics of Moving Bodies."

Einstein's approach to electromagnetic theory introduced some revolutionary concepts to physics. To achieve his goal, Einstein had to abandon both the idea of absolute time and the idea of absolute space. This new perspective led to surprising conclusions: The length of an object depends upon its speed relative to the observer who measures it, and the rate at which a clock ticks depends upon its speed relative to the observer who monitors it. The body of knowledge that describes all these effects is called the **special theory of relativity.**

The relativity of distances and time intervals is easily noticeable only at speeds approaching the speed of light. For most phenomena, Newtonian mechanics provides accurate predictions and explanations. However, for extremely high speeds, such as those of subatomic particles in particle accelerators, physicists found that special relativity is needed to account for the results of their experiments. Using extremely accurate atomic clocks, physicists have even confirmed that a moving clock carried in jet plane ticks slightly more slowly than an identical, stationary clock. Einstein's theory has been supported by every experiment designed to test it.

Details of special relativity need not concern us here (they will be discussed in Box 24-1), but this newly discovered relativity of space and time was very important in Einstein's desire to understand gravity. Einstein found Newton's description of gravity disappointing because it assumes that space and time are absolute. However, gravity causes a falling object to accelerate and, according to the special theory of relativity, the resulting motion also affects rulers and clocks that are moving *with* the falling object. Thus Einstein argued that gravity must affect the shape of space and the flow of time.

In 1915 Einstein formulated a new theory of gravity that came to be called the **general theory of relativity.** The basic idea of general relativity is that the mass of an object alters the properties of space and time around the object. According to the theory, gravity causes space to become curved and time to slow down. Einstein eliminated the idea of a "force of gravity" from his theory.

A useful analogy is to imagine that the space near a massive object such as the Sun becomes curved like the surface in Figure 4-23. Imagine a ball rolling along this surface. Far from the "well" that represents the Sun, the ball would move in a straight line since the surface is fairly flat. If it passes near the well, however, it would curve in toward the well. If it is moving at an appropriate speed with respect to the well, it might move in an orbit around the sides of the well. In this analogy, it is the curvature of the surface that makes the ball follow a curved path near the well.

One of the first things Einstein did with his new theory was to calculate the orbits of the planets. Einstein realized that if his theory is correct, it should be able to predict accurately the well-known motions of the planets about the Sun. According to general relativity, space far from the Sun is almost flat and so objects travel along nearly straight-line paths. Near the Sun, planets and comets travel along curved paths because space itself is curved. According to the general theory of relativity, only strong gravitational fields will have easily detectable effects upon space and time in their vicinity.

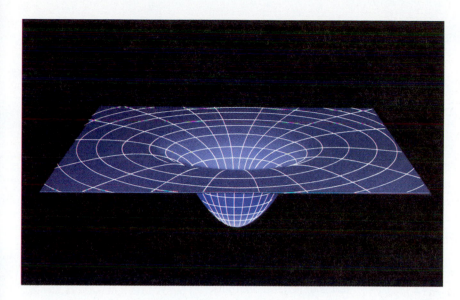

FIGURE 4-23 The Gravitational Curvature of Space According to Einstein's general theory of relativity, space becomes curved near a massive object such as the Sun. This diagram shows a two-dimensional analogy of the shape of space around a massive object.

With only one minor exception, general relativity gave almost exactly the same answers as Newtonian theory. Mercury is the only planet to pass close enough to the Sun for the curvature of space to give rise to motions that differ noticeably from those predicted by Newtonian mechanics. Indeed, throughout most of the solar system the curvature of space is so slight that Einstein's equations are essentially equivalent to Newtonian calculations. However, unlike Newtonian mechanics, Einstein's theory could account for the excess precession of Mercury's orbit, thereby explaining a phenomenon that had frustrated astronomers for half a century.

To help validate his theory, Einstein made other predictions that could be tested. With his calculations he showed that light rays passing near the surface of the Sun should appear to be deflected from their straight-line paths because the space through which they are moving is curved. In other words, gravity would bend light rays, an effect not predicted by Newtonian mechanics because light has no mass.

Figure 4-24 shows a beam of light from a star passing by the Sun and continuing on to the Earth. Because the light ray is bent, the star appears to be shifted from its actual location. The largest deflection (a mere 1.75 arc sec) occurs for light rays grazing the Sun's surface.

This prediction was first tested in 1919 during a total solar eclipse. During the precious moments of totality, when the Moon blocked out the blinding solar disk, astronomers succeeded in photographing the stars around the Sun. Careful measurements afterward revealed that the stars were shifted from their usual positions by an amount consistent with Einstein's theory. General relativity had passed another important test.

Einstein made a third prediction. He stated that because gravity causes time to slow down, clocks on the first floor of a building would tick slightly more slowly than clocks in the attic, which are farther from the Earth, as shown in Figure 4-25. This prediction was tested by analyzing light from compact stars (whose surface gravity is strong) and by using

extremely accurate clocks first developed in 1960. Again Einstein was proven correct.

The general theory of relativity now stands as our most precise and complete description of gravity. Einstein demonstrated that Newtonian mechanics is accurate only when applied to low speeds and weak gravity. If extremely high speeds or powerful gravity (such as those of neutron stars and black holes) are involved, only a relativistic calculation will give correct answers.

Einstein's general theory of relativity evoked a sensation when first proposed. Newtonian gravitation had been overthrown by a radically different approach that worked better. After the initial excitement, however, interest in general relativity rapidly waned. No one could imagine places where the curvature of space might have a major effect. It was generally believed that Newtonian mechanics would suffice in virtually all circumstances and that relativistic theory offered only a

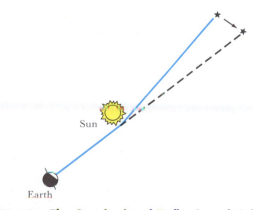

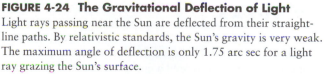

FIGURE 4-24 The Gravitational Deflection of Light Light rays passing near the Sun are deflected from their straight-line paths. By relativistic standards, the Sun's gravity is very weak. The maximum angle of deflection is only 1.75 arc sec for a light ray grazing the Sun's surface.

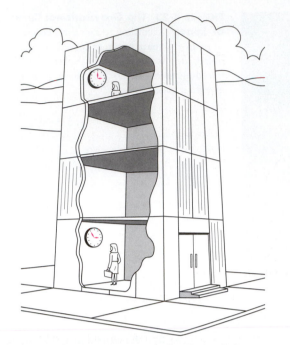

FIGURE 4-25 The Gravitational Slowing of Time A clock on the ground floor of a building is closer to the Earth than one at a higher elevation. According to general relativity, the clock on the ground floor ticks more slowly than the clock on the roof.

little more accuracy, as in calculating Mercury's orbit. The complex mathematics of relativity seemed burdensome, and Newton's simpler approach was reasonably precise in most practical circumstances.

Although Newtonian mechanics is still widely used, in recent years there has been a reawakening of interest in general relativity, primarily because of dramatic advances in our understanding of the evolution of stars. As we shall see in later chapters, we now have a reasonably complete picture of how stars are born, what happens to them as they mature, and how they die. In particular, it appears that the most massive dying stars are doomed to collapse completely in upon themselves, thereby producing some of the most bizarre objects in the universe, black holes. The gravity around one of these massive stellar corpses is so strong that it in effect punches a hole in the fabric of space. Because of this intense gravitational effect, black holes can be described only in terms of general relativity.

As we set our sights on understanding the universe as a whole, we must again turn to general relativity. All of the matter in all the stars and galaxies is responsible for the overall curvature or shape of space. From the viewpoint of general relativity, it is therefore reasonable to ask about the actual shape of the universe. We shall see that the answer suggests the ultimate fate of the cosmos.

KEY WORDS

Terms preceded by an asterisk are discussed in the boxes.

acceleration

*angular momentum

aphelion

*center of mass

configuration (of a planet)

conic section

conjunction

*conservation of angular momentum

*conservation of energy

*conservation of momentum

cosmology

deferent

direct motion

*eccentricity

electromagnetic theory

ellipse

elongation

*energy

epicycle

*escape speed

focus (of an ellipse; *plural* foci)

force

general theory of relativity

geocentric cosmology

gravity

greatest eastern elongation

greatest western elongation

harmonic law

heliocentric cosmology

hyperbola, p. 68

inferior conjunction

*inferior planet

Kepler's laws

*kinetic energy

law of equal areas

law of inertia

*major axis (of an ellipse)

mass, p. 66

*minor axis (of an ellipse)

Newtonian mechanics

Newton's laws of motion

Occam's razor

opposition

parabola

parallax

perihelion

period (of a planet)

*potential energy

retrograde motion

semimajor axis (of an ellipse)

sidereal period

special theory of relativity

superior conjunction

*superior planet

synodic period

universal constant of gravitation

universal law of gravitation

velocity

weight

KEY IDEAS

• Ancient astronomers believed the Earth to be at the center of the universe. They invented a complex system of epicycles

and deferents to explain the direct and retrograde motions of the planets.

• Copernicus's heliocentric (Sun-centered) theory simplified the general explanation of planetary motions.

In a heliocentric system, the Earth is one of the planets orbiting the Sun.

The sidereal period of a planet, its true orbital period, is measured with respect to the stars. Its synodic period is measured with respect to the Earth and the Sun (for example, from one opposition to the next).

The invention of the telescope led Galileo to new discoveries that supported a heliocentric model of the universe.

• Ellipses describe the paths of the planets around the Sun much more accurately than do circles. Kepler's three laws give important details about elliptical orbits.

• Newton based his explanation of the universe on three assumptions or laws of motion. These laws and his universal law of gravitation can be used to deduce Kepler's laws and extremely accurate descriptions of planetary motions.

The mass of an object is a measure of the amount of matter in the object; its weight is a measure of the force with which the gravity of some other object pulls on it.

In general, the path of one object about another, such as that of a comet about the Sun, is one of the curves called conic sections: circle, ellipse, parabola, or hyperbola.

• Although Newtonian mechanics adequately describes and predicts numerous phenomena, Einstein's relativistic theories are always more accurate and must be used where extremely high speeds or intense gravitational fields are involved.

The special theory of relativity eliminates the notion of absolute space and time. The motion of the observer affects measurements of distance and time.

The general theory of relativity explains that gravity causes space to be curved and time to slow down. The curvature of space and slowing of time give rise to accelerated motions.

REVIEW QUESTIONS

1. What is an epicycle, and how is it important in Ptolemy's explanation of the retrograde motions of the planets?

2. How did Copernicus explain the retrograde motion of the planets?

3. Which planets can never be seen at opposition? Which planets can never be seen at inferior conjunction?

4. At what configuration (for example, superior conjunction, greatest eastern elongation) would it be best to observe Mercury or Venus with an Earth-based telescope? At what configuration would it be best to observe Mars, Jupiter, or Saturn? Explain your answers.

5. What is the difference between the synodic and the sidereal periods of a planet?

6. What are Kepler's three laws? Why are they important?

7. A line joining the Sun and an asteroid is found to sweep out 5.2 AU^2 of space in 1990. How much area is swept out in 1991? Over a period of five years?

8. A comet with a period of 1000 years moves in a highly elongated orbit about the Sun. What is the comet's average distance from the Sun? What is the farthest it can get from the Sun?

9. The orbit of a spacecraft about the Sun has a perihelion distance of 0.5 AU and an aphelion distance of 3.5 AU. What is the spacecraft's orbital period?

10. What are Newton's three laws? Give an everyday example of each law.

11. What is your weight in pounds and in newtons? What is your mass?

12. The mass of the Moon is 7.35×10^{22} kg, while that of the Earth is 5.98×10^{24} kg. The average distance from the center of the Moon to the center of the Earth is 384,400 km. What is the size of the gravitational force that the Earth and Moon exert upon each other? How does this compare with the force between the Sun and the Earth calculated in the text?

13. Suppose that the Earth were moved to a distance of 10 AU from the Sun. How much stronger or weaker would the Sun's gravitational pull be on the Earth?

14. What is the difference between weight and mass?

15. Why was the discovery of Neptune an important confirmation of Newton's universal law of gravitation?

16. What are conic sections, and in what way are they related to the orbits of planets in the solar system?

17. Why is Einstein's general theory of relativity a better description of gravity than Newton's universal law of gravitation?

ADVANCED QUESTIONS

Tips and tools . . .

Details about sidereal and synodic periods are supplied in Box 4-1. Various useful formulas involving orbital parameters are found in Box 4-3. Data about the planets and their satellites are found in the appendixes at the back of this book.

18. The synodic period of Mercury (an inferior planet) is 115.88 days. Calculate its sidereal period.

19. Is it possible for a planet's sidereal period to equal its synodic period? Would such a planet be closer to or farther from the Sun than the Earth is? Is there a planet that nearly fits this description?

20. A satellite is said to be in a "geosynchronous" orbit if it appears always to remain over the exact same spot on Earth. (a) At what distance from the center of the Earth must such a satellite be placed into orbit? (b) Explain why the orbit must be in the plane of the Earth's equator.

21. Calculate the orbital speed of Pluto at (a) perihelion and (b) aphelion. Relevant orbital data are found in Appendix 1.

22. Suppose you have discovered an alien solar system in which a planet circles a star once every two years at an average distance of 4 AU. How does the mass of this star compare with that of our Sun?

23. What is the escape speed from the surface of (a) the Moon, (b) Jupiter, (c) the Sun? Why do you suppose that the escape velocity of the Sun is only about ten times that of Jupiter, even though the Sun is so much larger and more massive than Jupiter?

24. Suppose that you traveled to a planet whose mass is the same as Earth's but whose diameter is twice as large. Would you weigh more or less on that planet than on Earth? By what factor?

25. If you landed on Mars, would you weigh more or less than you do on the Earth? By what factor? See Appendix 2 for relevant data about Mars.

26. Look up the dates of the greatest eastern and western elongations of Mercury for any year you choose. Does it take longer to go from eastern to western elongation, or vice versa? Why do you suppose this is the case?

27. Look up orbital information for the four largest moons of Jupiter. Demonstrate that these data obey Kepler's third law.

DISCUSSION QUESTIONS

28. Which planet would you expect to exhibit the greatest variation in apparent brightness as seen from Earth? Explain your answer.

29. Use two thumbtacks, a loop of string, and a pencil to draw several ellipses. Describe how the shape of an ellipse varies as the distance between the thumbtacks changes.

OBSERVING PROJECTS

30. It is quite probable that, within a few weeks of your reading this chapter, one of the planets will be near opposition or greatest eastern elongation, making it readily visible in the evening sky. Consult a reference book such as the current issue of the *Astronomical Almanac* or the pamphlet entitled *Astronomical Phenomena* (both published by the U.S. government) to select a planet that is at or near such a configuration. At that configuration, would you expect the planet to be moving rapidly or slowly from night to night against the background stars? Verify your expectations by observing the planet once a week for a month, recording your observations on a star chart.

31. If Jupiter happens to be visible in the evening sky, observe the planet with a small telescope on five consecutive clear nights. Record the positions of the four Galilean satellites by making nightly drawings, just as the Jesuit priests did in 1620 (see Figure 4-16). From your drawings, can you tell which moon orbits closest to Jupiter and which orbits farthest? Is there a night when you could see only three of the moons? What do you suppose happened to the fourth moon on that night?

32. If Venus happens to be visible in the evening sky, observe the planet with a small telescope once a week for a month. On each night, make a drawing of the crescent that you see. From your drawings can you determine if the planet is nearer or farther from the Earth than the Sun is? Do your drawings show any changes in the shape of the crescent from one week to the next? If so, can you deduce if Venus is coming toward us or moving away from us?

FOR FURTHER READING

Christianson, G. *This Wild Abyss*. Free Press, 1978. A very readable book about the evolution of astronomy.

Cohen, I. B. "Newton's Discovery of Gravity." *Scientific American,* March 1981. An intriguing account of how Newton compared the real world with a mathematical model of it.

Drake, S. "Newton's Apple and Galileo's Dialogue." *Scientific American,* August 1980. This article shows how the work of Galileo influenced Newton's thinking.

Gardner, M. *The Relativity Explosion*. Vintage, 1976. An excellent book about Einstein's ideas and relativity theory.

Gingerich, O. "The Galileo Affair." *Scientific American,* August 1982. This article gives many insights into the life and times of Galileo, with special emphasis on his relationship with the Church.

————. "How Galileo Changed the Rules of Science." *Sky & Telescope,* March 1993. This entertaining article describes the impact of Galileo's observations on science and the Catholic church.

Kaufmann, W. *Relativity and Cosmology. 2d ed*. Harper & Row, 1977. This slim book, which emphasizes relativity, describes the evolution of our understanding of gravitation.

Kaufmann, W., and Smarr, L. *Supercomputing and the Transformation of Science*. Scientific American Library, 1993. This book describes methods and tools of supercomputing. The final chapter is devoted entirely to supercomputing in astronomy.

5

THE NATURE OF
LIGHT AND MATTER

SPECTRA AND SPECTRAL LINES When an electric spark passes through a gas, atoms of the gas emit radiation at certain wavelengths. As a result, a spectrum of that gas contains bright spectral lines. Here the spectra of various elements are displayed. Each element has a unique pattern of spectral lines. Spectra of the Sun and two common household lamps are also shown. (Courtesy of Bausch and Lomb)

ASTRONOMY IS built on an understanding of light, and every scientific discovery about light has led to important discoveries about the universe. Light is a form of energy possessing both wavelike and particle-like properties. Because the light emitted by an object depends upon the object's temperature, we can determine the surface temperature of stars. Understanding the structure of atoms lead to an understanding of why each chemical element emits and absorbs light at specific wavelengths. Astronomers use this knowledge to determine the composition of stars and galaxies. The motion of a light source also affects wavelengths, permitting us to deduce the speed at which a light source is approaching or retreating. These are but a few of the reasons why understanding light is a prerequisite to understanding the universe.

In the early 1800s, the French philosopher Auguste Comte argued that humanity would never know the nature and composition of the stars. Because they are so far beyond our earthly reach, he thought that these remote celestial worlds would forever remain unfathomable and mysterious. But only a few years after Comte's bold pronouncement, scientists began discovering the basic properties of matter and radiation. It soon became clear that by analyzing starlight we can learn many things about the stars. We now know that the stars and galaxies are made of substances found here on Earth and that all atoms, regardless of their location in the universe, obey the same physical laws. As a result, laboratory experiments conducted here on Earth can provide the insight and understanding needed to probe the distant reaches of the universe.

5-1 Light travels through empty space at a speed of 300,000 km/s

Galileo and Newton were the first to ask basic questions about light. Does light travel instantaneously from one place to another, or does it move with a measurable speed? Whatever may be the nature of light, it does seem to travel swiftly from a source to our eyes. We see a distant event before we hear the accompanying sound.

In the early 1600s, Galileo tried to measure the speed of light. He and an assistant stood at night on two hilltops a known distance apart, each holding a shuttered lantern. First Galileo opened the shutter of his lantern; as soon as his assistant saw the flash of light, he opened his own. Galileo used his pulsebeat to try and measure the time between opening his lantern and seeing the light from his assistant's lantern. From the distance and time, he hoped to compute the speed at which the light had traveled to the distant hilltop and back.

Galileo found that the measured time failed to increase noticeably, no matter how distant the assistant was stationed. Galileo therefore concluded that the speed of light is too high to be measured by slow human reaction.

The first accurate measurement of the speed of light was made in 1675 by Olaus Roemer, a Danish astronomer. Roemer had been studying the orbits of the moons of Jupiter by carefully timing the moments when the satellites passed into and out of the planet's shadow. To his surprise, their orbital motion seemed to depend on the relative positions of Jupiter and the Earth. As the Earth approached Jupiter (during the months just before an opposition), the satellites seemed to

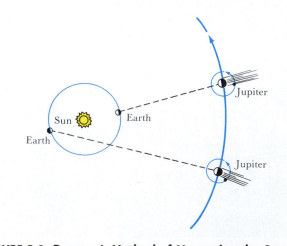

FIGURE 5-1 Roemer's Method of Measuring the Speed of Light When the Earth is near Jupiter, eclipses of Jupiter's moons occur slightly earlier than anticipated. When the Earth–Jupiter distance is large, the eclipses occur slightly later than anticipated. Roemer correctly attributed this effect to the time required for light to travel from Jupiter to the Earth.

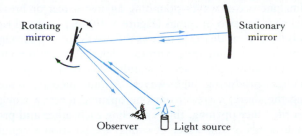

FIGURE 5-2 The Fizeau–Foucault Method of Measuring the Speed of Light Light from a light source is first reflected off a rotating mirror to a stationary mirror. The returning ray is deflected away from the path of the initial beam because the rotating mirror has moved slightly while the light was making the round trip. From the deflection angle and the dimensions of the apparatus, the speed of light can be determined.

speed up slightly in their orbits about Jupiter. As the Earth receded from Jupiter, they seemed to slow down slightly.

Roemer realized that this puzzling effect could be explained if the light needed time to travel from Jupiter to the Earth. When the Earth is closest to Jupiter, the image of a satellite disappearing into the Jovian shadow arrives at our telescopes a little sooner than it does when Jupiter and the Earth are farther apart (Figure 5-1). Roemer's measurements indicated that it takes about $16\frac{1}{2}$ minutes for light to travel across the diameter of the Earth's orbit (a distance of 2 AU). The size of the Earth's orbit was not accurately known in Roemer's day. Today, using the modern value of 150 million kilometers for the astronomical unit, Roemer's method yields roughly 300,000 km/s (186,000 mi/s) for the speed of light.

The first successful laboratory experiments to measure the speed of light were conducted in the mid-1800s in France. In 1850, Hippolyte Fizeau and Jean Foucault built the apparatus sketched in Figure 5-2. Light from a light source is reflected from a rotating mirror toward a stationary mirror

20 m away. While the light is making the round trip, the rotating mirror moves slightly. Thus the returning light ray is deflected away from the source by a small angle. By measuring this angle and knowing the dimensions of their apparatus, Fizeau and Foucault could deduce the speed of light. Once again, the answer was very nearly 300,000 km/s.

The speed of light in a vacuum is usually designated by the letter c. The modern accepted value is

$$c = 299,792.458 \text{ km/s}$$

In most calculations, you can use $c = 3 \times 10^5$ km/s $= 3 \times 10^8$ m/s. Note that this value is the speed of light *in a vacuum*. Light traveling through air, water, glass, or any other transparent substance always moves more slowly than it does in a vacuum.

The speed of light in empty space is one of the most important numbers in modern physical science. This value appears in many equations that describe atoms, gravity, electricity, and magnetism. Furthermore, according to Einstein's special theory of relativity, nothing can travel faster than the speed of light.

5-2 Light is electromagnetic radiation and is characterized by its wavelength

Light is energy. This fact is apparent to anyone who has felt the warmth of the sunshine on a summer's day. But what exactly is light? How is it produced? What is it made of? How does it move through space? Scientists have struggled with these questions for the past four centuries.

The first major breakthrough in understanding light came from a simple experiment performed by Isaac Newton in the late 1600s. Newton was familiar with what he called "the celebrated Phenomenon of Colours": A beam of sunlight passing through a glass prism is spread out into the colors of the rainbow, as shown in Figure 5-3. This rainbow, called a **spectrum** (plural **spectra**), suggested to Newton that white

FIGURE 5-3 A Prism and a Spectrum When a beam of white light passes through a glass prism, the light is broken into a rainbow-colored band called a spectrum. The numbers on the right side of the spectrum indicate wavelengths.

light must actually be a mixture of all the colors. Up to that time, it was erroneously assumed that a prism somehow added colors to the light. Newton elegantly disproved this idea by passing the spectrum through a second prism inverted with respect to the first. Only white light emerged from this second prism. So it was clear that the second prism reassembled the colors of the rainbow to give back the original beam of sunlight.

Newton assumed that light is composed of undetectably tiny particles. In fact, as we shall see later in this chapter, the modern theory of quantum mechanics, developed in the early 1900s, also takes this viewpoint. Quantum theory postulates that light is composed of tiny packets of energy called **photons**. In the mid-1600s, however, the Dutch astronomer Christian Huygens proposed a rival explanation. He suggested that light travels in the form of waves rather than particles. Scientists now realize that light possesses both particlelike and wavelike characteristics.

The wavelike aspect of light was convincingly demonstrated around 1801 by Thomas Young, an English physicist, physician, and general scholar. Young passed a beam of light through two thin, parallel slits in an opaque screen (Figure 5-4a). On a white surface some distance beyond the slits, the light formed a pattern of alternating bright and dark bands. Young reasoned that if a beam of light was a stream of particles (as Isaac Newton had argued), the two beams of light from the slits should simply form bright images of the slits on the white surface. The pattern of bright and dark bands he observed is just what would be expected, however, if light had wavelike properties. An analogy with water waves demonstrates why this is so.

Imagine ocean waves pounding against a reef or breakwater that has two openings (Figure 5-4b). A pattern of ripples is formed inside the barrier as the waves coming through the two openings interfere with each other. At certain points along the shore, crests arrive simultaneously from the two openings, producing high waves. At intermediate points along the shore, a crest from one opening meets a trough from the other opening, thus canceling each other and producing bands of still water. A similar explanation accounts for the pattern of bright and dark bands that Young observed on the white surface behind the slits in his experiment.

The discovery of the wave nature of light posed some obvious questions. Waves of what? What is it about light that goes up and down like water waves on the ocean? The answer came from a seemingly unlikely source: a comprehensive theory that described electricity and magnetism. Numerous experiments during the first half of the nineteenth century had demonstrated an intimate connection between electric and magnetic forces. A central idea to emerge from these experiments is the concept of a *field,* a region of space throughout which electric or magnetic forces are felt. Thus, an electric charge is surrounded by an electric field, and a magnet is surrounded by a magnetic field. Experiments in the early 1800s demonstrated that moving electric charges produce a magnetic field; conversely, motion in a magnetic field can give rise to an electric field.

In the 1860s, the Scottish mathematician and physicist James Clerk Maxwell succeeded in describing all the basic properties of electricity and magnetism in four equations. This mathematical achievement demonstrated that electric and magnetic forces are really two aspects of the same phe-

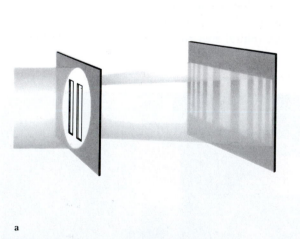

a

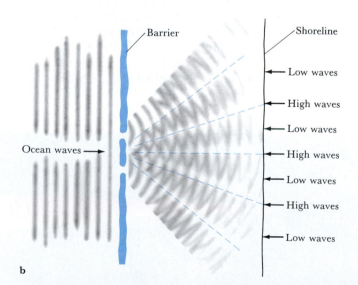

b

FIGURE 5-4 Young's Double-Slit Experiment (a) In the modern laboratory, Thomas Young's classic double-slit experiment is most easily repeated by shining light from a laser onto two closely spaced parallel slits. Alternating dark and light bands appear on a screen beyond the slits. (b) The intensity of light on the screen in the double-slit experiment is analogous to the height of water waves that strike a shore after passing through a barrier with two openings. In certain locations, ripples from both openings reinforce each other to produce extra high waves. At intermediate locations along the shoreline, ripples and troughs cancel each other, producing still water.

Barrier

Shoreline

Low waves

High waves

Low waves

Ocean waves →

High waves

Low waves

High waves

Low waves

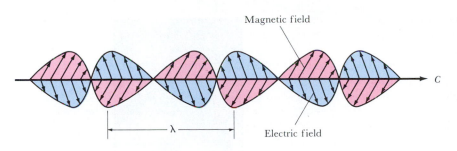

FIGURE 5-5 Electromagnetic Radiation
All forms of light consist of oscillating electric and magnetic fields that move through space at a speed of 3×10^8 m/s. The distance between two successive crests, called the wavelength of the light, is usually designated by the Greek letter λ.

nomenon, which we now call **electromagnetism.** By combining his four equations, Maxwell showed that electric and magnetic fields should travel through space in the form of waves at a speed of 3×10^8 m/s—exactly equal to the best available value for the speed of light. Maxwell's suggestion that these waves do exist and are observed as light was soon confirmed by experiments.

Because of its electric and magnetic properties, light is called **electromagnetic radiation.** It consists of perpendicular, oscillating electric and magnetic fields, as shown in Figure 5-5. The distance between two successive wave crests is called the **wavelength** of the light, usually designated by the Greek letter λ (lambda).

More than a century elapsed between Newton's experiments with a prism and confirmation of the wave nature of light. One reason for this delay is that the wavelength of **visible light** is extremely short—less than a thousandth of a millimeter—and thus is not easily detectable. To express such tiny distances conveniently, scientists use a unit of length called the *nanometer* (abbreviated nm), where 1 nm = 10^{-9} m. Experiments demonstrated that visible light has wavelengths covering the range from about 400 nm for violet light to about 700 nm for red light. Intermediate colors of the rainbow like yellow (550 nm) have intermediate wavelengths (see Figure 5-3).

Maxwell's equations place no restrictions, however, on the wavelength of electromagnetic radiation. In other words, electromagnetic waves could and should exist with wavelengths both longer and shorter than the 400- to 700-nm range of visible light. Consequently, researchers began to look for invisible forms of light: forms of electromagnetic radiation to which the cells of the human retina do not respond.

Around 1800 the British astronomer William Herschel discovered **infrared radiation** in an experiment with a prism, when he held a thermometer just beyond the red end of the visible spectrum. The thermometer registered a temperature increase, indicating that it was being exposed to an invisible form of energy. In experiments with electric sparks in 1888, the German physicist Heinrich Hertz succeeded in producing electromagnetic radiation a few centimeters long in wavelength, now known as **radio waves.** In 1895 Wilhelm Röntgen invented a machine that produces electromagnetic radiation with wavelengths shorter than 10 nm, now known as **X rays.** Modern versions of Röntgen's machine are now

found in medical and dental offices. Over the years radiation has been discovered with many other wavelengths.

Visible light occupies only a tiny fraction of the full range of possible wavelengths, collectively called the **electromagnetic spectrum.** As shown in Figure 5-6, the electromagnetic spectrum stretches from the longest-wavelength radio waves to the shortest-wavelength gamma rays.

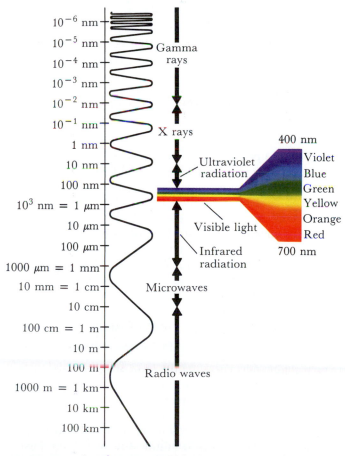

FIGURE 5-6 The Electromagnetic Spectrum The full array of all types of electromagnetic radiation is called the electromagnetic spectrum. It extends from the longest-wavelength radio waves to the shortest-wavelength gamma rays. Visible light occupies only a tiny portion of the full electromagnetic spectrum.

On the long-wavelength side of the visible spectrum, infrared radiation covers the range from about 700 nm to 1 mm. Astronomers interested in infrared radiation often express wavelength in *micrometers* or *microns* (abbreviated μm), where 1 μm = 10^3 nm = 10^{-6} m. From roughly 1 mm to 10 cm is the range of microwaves, beyond which is the domain of radio waves.

At wavelengths shorter than those of visible light, **ultraviolet radiation** extends from about 400 nm down to 10 nm. Next are X rays, which consist of wavelengths between about 10 and 0.01 nm, and beyond them are **gamma rays**. It should be noted that these rough boundaries are simply arbitrary divisions in the electromagnetic spectrum.

Astronomers who work with radio telescopes often prefer to speak of frequency rather than wavelength. The **frequency** of light is the number of wave crests passing a given point in one second. The shorter the wavelength, the higher is the frequency, because more wave crests pass by per second. Since light travels at the speed $c = 3 \times 10^8$ m/s, the frequency (ν) of light is related to its wavelength (λ) by

$$\nu = \frac{c}{\lambda}$$

For example, hydrogen atoms in space emit radio waves with a wavelength of 21.12 cm. The frequency of this radiation is therefore calculated as follows:

$$\nu = \frac{c}{\lambda} = \frac{3 \times 10^8 \text{ m/s}}{0.2112 \text{ m}} = 1.42 \times 10^9 \text{ s}^{-1}$$

$$= 1,420,000,000 \text{ cycles per second}$$

One cycle per second is also called a hertz (abbreviated Hz) in honor of Heinrich Hertz, the physicist who discovered radio waves. Often it is convenient to use the prefix *mega-* (meaning "million" and abbreviated M) or *kilo-* (meaning "thousand" and abbreviated k). Thus the frequency of hydrogen emission is 1420 MHz, or 1420 megahertz. For comparison, AM radio stations broadcast at frequencies between 535 and 1605 kHz, while FM radio stations broadcast at frequencies extending from 88 to 108 MHz.

5-3 An object emits electromagnetic radiation according to its temperature

Every object emits energy in the form of electromagnetic radiation. The amount of energy radiated by an object depends on its temperature. The hotter the object is, the more energy it emits.

The basic phenomena are familiar to anyone who has ever watched a welder or blacksmith heat a bar of iron (Figure 5-7). As the bar becomes hot, it begins to glow deep red. As the temperature rises further, the bar begins to give off a brighter, reddish-orange light. At still higher temperatures, it shines with a brilliant yellowish-white light. If the bar could

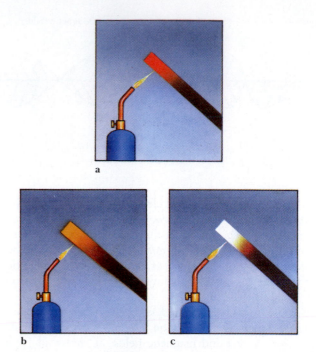

FIGURE 5-7 Heating a Bar of Iron This sequence of drawings shows how the appearance of a heated bar of iron changes with temperature. As the temperature increases, the amount of energy radiated by the bar increases. The color of the bar also changes because, as the temperature goes up, the dominant wavelength of light emitted by the bar decreases. These effects are described by the Stefan–Boltzmann law and Wien's law.

be prevented from melting and vaporizing, at extremely high temperatures it would emit a dazzling blue-white light.

As this example shows, the dominant wavelength of the emitted radiation also depends on the temperature of the object. A cool object emits most of its energy at long wavelengths, whereas a hot object emits most of its energy at short wavelengths.

To appreciate the relationship between energy and temperature that describes these phenomena, we must be able to state them quantitatively. We must therefore understand the units that scientists use to measure energy and temperature. Energy is usually measured in joules (J), named after the nineteenth-century English physicist James Joule. A **joule** is the amount of energy contained in the motion of a two-kilogram mass moving at a speed of one meter per second. The joule is a convenient unit of energy because it is closely related to the familiar watt (W): 1 W = 1 J/s. For instance, the energy consumption of a 100-watt light bulb is 100 J/s.

In discussing temperature, scientists usually prefer the Kelvin temperature scale, on which temperature is measured in **kelvin** (K) upward from **absolute zero,** the coldest possible temperature. On the familiar Celsius and Fahrenheit temperature scales, absolute zero (0 K) is −273°C and −460°F. The relationships between the Kelvin, Celsius, and Fahrenheit temperature scales are discussed in Box 5-1.

BOX 5-1

Temperatures and Temperature Scales

Three temperature scales are in common use. Throughout most of the world, temperatures are expressed in **degrees Celsius** (°C). The Celsius temperature scale is based on the behavior of water, which freezes at 0°C and boils at 100°C at sea level on Earth. This scale is named in honor of the Swedish astronomer Anders Celsius, who proposed it in 1742.

Scientists usually prefer the Kelvin scale, named after the British physicist Lord Kelvin (William Thomson), who made many important contributions to our knowledge about heat and temperature. On the Kelvin temperature scale, water freezes at 273 K and boils at 373 K. Note that we do not use *degree* (°) with the Kelvin temperature scale. For instance, room temperature is roughly 295 K, or 295 kelvin.

Because water must be heated through a change of 100 K or 100°C to go from its freezing point to its boiling point, you can see that the "size" of a kelvin is the same as the "size" of a degree Celsius. When considering temperature *changes,* measurements in kelvin and in degrees Celsius are the same.

A temperature expressed in kelvin is always equal to the temperature in degrees Celsius plus 273. Scientists prefer the Kelvin scale because it is closely related to the physical interpretation of the meaning of temperature.

All substances are made of atoms. These atoms are very tiny (typical atomic diameters are about 10^{-10} m) and are constantly in motion. The temperature of a substance is directly related to the average speed of its atoms. If something is hot, its atoms are moving at high speeds. If a substance is cold, its atoms are moving much more slowly.

The coldest possible temperature is the temperature at which atoms move as slowly as possible (they can never quite stop completely). This minimum possible temperature, called absolute zero, is the starting point for the Kelvin scale. Absolute zero is 0 K, or −273°C. Since it is impossible for anything to be colder than 0 K, there are no negative temperatures on the Kelvin scale.

In the United States, many people still use the now-archaic Fahrenheit scale, which expresses temperatures in **degrees Fahrenheit** (°F). When the German physicist Gabriel Fahrenheit introduced this scale in the early 1700s, he intended 0°F to represent the coldest temperature then achievable (with a mixture of ice and saltwater) and 100°F to represent the temperature of a healthy human body. On the Fahrenheit scale, water freezes at 32°F and boils at 212°F. Because there are 180 degrees Fahrenheit between the freezing and boiling points of water, a degree Fahrenheit is only $\frac{100}{180} = \frac{5}{9}$ as large as either a degree Celsius or a kelvin.

The following equation converts from degrees Celsius to degrees Fahrenheit:

$$T_F = \frac{9}{5} T_C + 32$$

where T_F is the temperature in degrees Fahrenheit and T_C is the temperature in degrees Celsius. To convert in the opposite direction, a simple rearrangement of terms gives the relationship

$$T_C = \frac{5}{9}(T_F - 32)$$

EXAMPLE: Consider a typical room temperature of 72°F. Using the second equation, we can convert this measurement to the Celsius scale as follows:

$$T_C = \frac{5}{9}(72 - 32) = 22°C$$

To arrive at the Kelvin scale, we simply add 273 degrees to the value in degrees Celsius. Thus,

$$72°F = 22°C = 295 \text{ K}$$

The diagram below displays the relationships between these three temperature scales.

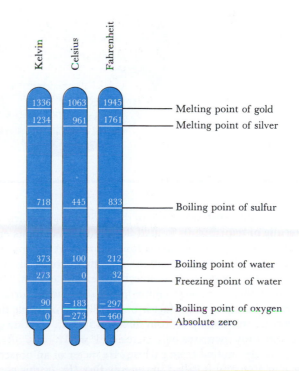

BOX 5-2

The Sun's Luminosity and Surface Temperature

Using detectors above the Earth's atmosphere, astronomers have measured the average flux of solar energy arriving at the Earth; this value, called the **solar constant,** is equal to 1370 W m^{-2}. Imagine a huge sphere of radius 1 AU with the Sun at its center, as shown in the diagram. Each square meter of that sphere receives 1370 watts of power from the Sun, so we can compute the total energy output of the Sun by multiplying the solar constant by the sphere's area. The result, called the **luminosity** of the Sun (L$_\odot$), is

$$L_\odot = 3.90 \times 10^{26} \text{ W}$$

(Astronomers use the symbol $\odot$ to denote the Sun.)

Because we know the size of the Sun, we can compute the energy flux emitted from its surface. The radius of the

Sun is R$_\odot$ = 6.96 × 10^8 m, and its surface area is $4\pi R_\odot^2$. Therefore, its energy flux is

$$F_\odot = \frac{L_\odot}{4\pi R_\odot^2} = 6.41 \times 10^7 \text{ W m}^{-2}$$

Now we can use the Stefan–Boltzmann law to find the Sun's surface temperature:

$$T_\odot^4 = \frac{F_\odot}{\sigma} = 1.13 \times 10^{15} \text{ K}^4$$

Taking the fourth root (the square root of the square root) of this value, we find the surface temperature of the Sun to be T$_\odot$ = 5800 K.

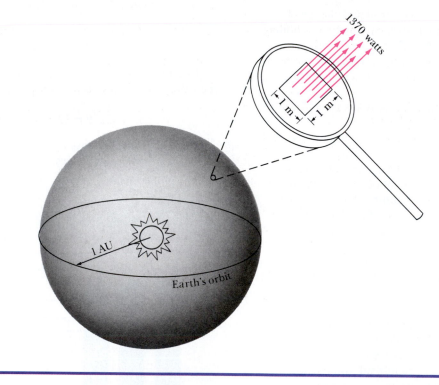

In 1879 the Austrian physicist Josef Stefan summarized the results of experiments he had done: An object emits energy at a rate proportional to the fourth power of the object's temperature (measured in kelvin). In other words, if you double the temperature of an object (for example, from 500 to 1000 K), then the energy emitted from the object's surface each second increases by a factor of $2^4 = 16$. If you triple the temperature (for example, from 500 to 1500 K), the rate of energy emission increases by a factor of $3^4 = 81$.

The energy emitted from each square meter of an object's surface in a second is called the **energy flux** (F). In this context, flux means "rate of flow," and thus F is a measure of how much energy is flowing out of the object. Using this

term, we can write Stefan's law as

$$F = \sigma T^4$$

where σ (sigma) is a constant, T is the object's temperature measured in kelvin, and F is the energy flux measured in joules per square meter per second (usually written as J m^{-2} s^{-1}). Since one watt equals one joule per second, F can also be expressed in watts per square meter.

Five years after Stefan announced this law, another Austrian physicist, Ludwig Boltzmann, showed how it could be derived mathematically from basic assumptions about atoms and molecules. The law is thus commonly known as the **Stefan–Boltzmann law.** The Stefan–Boltzmann constant σ is known from laboratory experiments to have the value

$$\sigma = 5.67 \times 10^{-8} \text{ W m}^{-2} \text{ K}^{-4}$$

Boltzmann showed that this law is best obeyed by an object called a blackbody. A **blackbody** does not reflect any light; it instead absorbs all radiation falling on it. Since it reflects no electromagnetic radiation, the radiation that it does emit is entirely the result of its temperature. Ordinary objects, like tables and textbooks, are rather poor blackbodies; they reflect light, which is why they are visible. A star, however, behaves like a blackbody because it efficiently absorbs almost all of the radiation falling upon it from the outside. Since a star is almost a perfect blackbody, astronomers can use the Stefan–Boltzmann law to relate its energy output to its surface temperature. Box 5-2 shows how the Stefan–Boltzmann law can be used to calculate the surface temperature of the Sun.

A blackbody does not necessarily look black. For instance, a star does not look black because it consists of hot, glowing gases. A blackbody is so named because at room temperature it predominantly emits radiation our eyes cannot perceive.

An object emits radiation over a wide range of wavelengths, but there is always a particular wavelength (λ_{max}) at which the emission of energy is strongest. This dominant wavelength gives a glowing hot object its characteristic color.

In 1893 the German physicist Wilhelm Wien discovered a simple relationship between the temperature of a blackbody and the dominant wavelength (λ_{max}) of the energy it emits:

$$\lambda_{max} = \frac{0.0029}{T}$$

where λ_{max} is measured in meters and T is measured in kelvin. This relation is today called **Wien's law.**

According to Wien's law, if the temperature of the blackbody increases, the dominant wavelength of its radiation must decrease. In other words, the hotter an object, the shorter the dominant wavelength of the electromagnetic radiation it emits.

From the Stefan–Boltzmann law, we see that any object with a temperature above absolute zero (0 K) emits some electromagnetic radiation. From Wien's law, we find that a very cold object with a temperature of only a few kelvin emits primarily microwaves. An object at room temperature (about 295 K) emits most of its radiation in the infrared portion of the spectrum. And an object with a temperature of a few thousand kelvin emits most of its radiation as visible light. An object with a temperature of a few million kelvin emits most of its radiation at X-ray wavelengths.

Wien's law is very useful for computing the surface temperatures of stars because the star's size or luminosity need not be known. All we need to know is the dominant wavelength of the star's electromagnetic radiation. Several examples of applying the Stefan–Boltzmann law and Wien's law to stars are given in Box 5-3.

BOX 5-3

Applications of the Radiation Laws

Stars are almost perfect blackbodies. The Stefan–Boltzmann law and Wien's law can therefore be used to relate the surface temperature, flux, and dominant wavelength of the radiation emitted by stars, as demonstrated in the following examples.

EXAMPLE: Consider energy coming from the Sun. The Sun emits energy over a wide range of wavelengths, but the maximum intensity of sunlight is at a wavelength of roughly 500 nm = 5×10^{-7} m. From Wien's law, we find the Sun's surface temperature ($T_\odot$) to be

$$T_\odot = \frac{0.0029}{\lambda_{max}} = \frac{0.0029}{5 \times 10^{-7}} = 5800 \text{ K}$$

This result agrees with the value we computed using the Stefan–Boltzmann law (see Box 5-2).

EXAMPLE: The bright star Sirius has a surface temperature of about 10,000 K. We can use Wien's law to calculate the wavelength (λ_{max}) at which Sirius emits most of its energy:

$$\lambda_{max} = \frac{0.0029}{T} = \frac{0.0029}{10,000} = 2.9 \times 10^{-7} \text{ m} = 290 \text{ nm}$$

Sirius therefore emits light most intensely in the ultraviolet.

EXAMPLE: We can use the Stefan–Boltzmann law to compare the energy flux from Sirius with that from the Sun. For the Sun, the Stefan–Boltzmann law is $F_\odot = \sigma T_\odot^4$, and similarly for Sirius we can write $F_* = \sigma T_*^4$, where the subscripts $\odot$ and $*$ refer to the Sun and Sirius, respectively. If we divide one equation by the other, the Stefan–Boltzmann constants cancel out giving us

$$\frac{F_*}{F_\odot} = \frac{T_*^4}{T_\odot^4}$$

Knowing that the temperature of Sirius is 10,000 K and the Sun is 5800 K, we get

$$\frac{F_*}{F_\odot} = \frac{(10,000)^4}{(5800)^4} = \left(\frac{10,000}{5800}\right)^4 = 8.8$$

In other words, each square meter of Sirius's surface emits 8.8 times more energy than a square meter of the Sun's surface.

5-4   In a full explanation of blackbody radiation, light is assumed to have particlelike properties

The Stefan–Boltzmann law and Wien's law describe only two basic properties of **blackbody radiation,** the electromagnetic radiation emitted by an ideal blackbody. A more complete picture is given by **blackbody curves** like those in Figure 5-8. These curves show the relationship between wavelength and the intensity of the light emitted by a blackbody at a given temperature. Blackbody curves illustrate both of the laws we have discussed: a cool object has a low curve that peaks at a long wavelength, whereas a hot object has a high curve that peaks at a short wavelength. More precisely, the total area under a blackbody curve is proportional to the energy flux (F). The wavelength of the peak of the curve is the dominant wavelength (λ_{max}).

Figure 5-9 shows how the intensity of sunlight varies with wavelength. This curve was obtained by measuring the intensity of sunlight at various wavelengths above the Earth's atmosphere. As mentioned earlier, the peak of the curve is at a wavelength of about 0.5 μm = 500 nm. The blackbody curve for a temperature of 5800 K is also plotted in Figure 5-9. Note how closely the observed intensity curve for the Sun matches the blackbody curve. This close correlation between the observed intensity curves for most stars and the

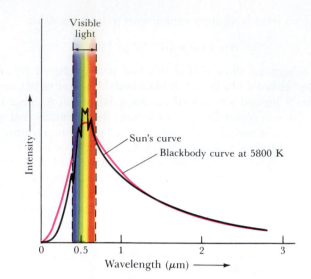

FIGURE 5-9 The Sun as a Blackbody This graph compares the intensity of sunlight over a wide range of wavelengths with the intensity of radiation coming from a blackbody at a temperature of 5800 K. Measurements of the Sun's intensity were made above the Earth's atmosphere. The Sun mimics a blackbody remarkably well.

blackbody curves is the reason astronomers are interested in the physics of blackbody radiation.

By the end of the nineteenth century, physicists realized that they had reached an impasse. All attempts to explain the characteristic shape of blackbody curves had failed.

In 1900, however, the German physicist Max Planck discovered that he could derive a formula for blackbody curves if he assumed that electromagnetic energy is emitted in discrete packets. This important assumption was verified in 1905 by Einstein, who used the idea of the particlelike nature of light to explain a phenomenon, called the **photoelectric effect,** in which a metal gives off electrons when exposed to short-wavelength radiation. Einstein won a Nobel Prize for this explanation.

The apparatus to demonstrate the photoelectric effect consists of a glass container, called a vacuum tube, from which all the air has been pumped (Figure 5-10). At either end of the vacuum tube is a metal plate called an electrode. Both electrodes are connected to a battery. In the dark or when the apparatus is illuminated by visible light, no current flows through the circuit. If a beam of ultraviolet light is focused on the negatively charged electrode, however, a current does flow through the circuit. An electric current is carried by tiny, negatively charged particles called **electrons.** (We will see below that the electron is one of the basic particles of which atoms are composed.) Somehow, the ultraviolet light liberates electrons from the negative electrode, allowing them to move across the tube to the positively charged electrode. It is surprising that a bright red light has no effect on the apparatus, whereas a dim ultraviolet light causes a current to flow.

Einstein explained the photoelectric effect by assuming that light exists in particlelike packets, or photons, of energy. The amount of energy in one photon of light is inversely

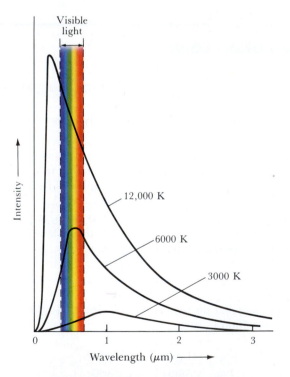

FIGURE 5-8 Blackbody Curves Three representative blackbody curves are shown here. Each curve shows the intensity of light at every wavelength that is emitted by a blackbody at a particular temperature. Note the range of wavelengths of visible light.

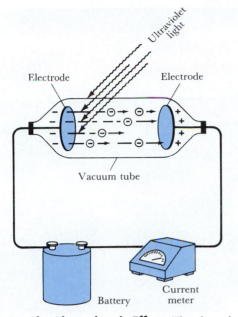

FIGURE 5-10   The Photoelectric Effect   The photoelectric effect can be demonstrated with a vacuum tube containing two electrodes attached to a battery. When no light falls on the negative electrode (at left), no electrons are liberated from its metal and no current flows. When long-wavelength red light is shone on the electrode, still no electrons are released and no current flows. But when short-wavelength ultraviolet light falls on the electrode, many electrons are liberated from the metal. These negatively charged electrons are attracted to the positively charged electrode at the opposite end of the vacuum tube and thus create an electric current.

proportional to the wavelength of the light. In other words, the longer the wavelength, the less energy there is in the photon. A single photon of red light, for instance, with its relatively long wavelength, does not have enough energy to knock an electron out of the metal electrode. So, no matter how many photons of red light strike the metal, no collision of a photon with an atom can liberate an electron. A single photon of ultraviolet radiation, however, with its much shorter wavelength, does have enough energy to knock loose an electron. Thus even a very weak beam of ultraviolet light can liberate electrons and cause a current to flow.

The relationship between the energy (E) of each photon and the wavelength (λ) of the electromagnetic radiation can be expressed in a simple equation:

$$E = \frac{hc}{\lambda}$$

where c is the speed of light and h is a constant value now called Planck's constant. Because $c/\lambda = \nu$, we can express this relationship in another form, often called **Planck's law:**

$$E = h\nu$$

where ν is the frequency of the light. Note that both equations express a relationship between a particlelike property of light (the energy E of a photon) and a wavelike property (the wavelength λ or the frequency ν).

Laboratory experiments show that Planck's constant has the value

$$h = 6.625 \times 10^{-34} \text{ J s}$$

This is a tiny value; by the standards of our everyday experience, a single photon carries a very small amount of energy (Box 5-4).

BOX 5-4

Photon Energy and the Electron Volt

A beam of light can be regarded as a stream of tiny packets of energy called photons. The energy (E) carried by a photon is directly related to its frequency (ν) by Planck's law, as discussed in the text:

$$E = h\nu$$

or, equivalently,

$$E = \frac{hc}{\lambda}$$

where h is Planck's constant, λ is the wavelength of the light, and c is the speed of light in a vacuum. Laboratory experiments reveal that

$$h = 6.625 \times 10^{-34} \text{ J s}$$

The energy carried by a single photon is very small and is conveniently expressed in a unit of energy called the **electron volt (eV)**, where

$$1 \text{ eV} = 1.602 \times 10^{-19} \text{ J}$$

Thus, Planck's constant may be written as

$$h = 4.135 \times 10^{-15} \text{ eV s}$$

EXAMPLE: Consider a photon of wavelength 620 nm, visible to the human eye as orange light. The energy of that photon is calculated as

$$E = \frac{hc}{\lambda} = \frac{(4.135 \times 10^{-15} \text{ eV s}) (3 \times 10^8 \text{ m s}^{-1})}{6.2 \times 10^{-7} \text{ m}} = 2 \text{ eV}$$

Visible light consists of photons in the energy range of approximately 2 to 3 eV. X-ray photons carry thousands of times more energy than visible photons, so their energies are conveniently expressed in thousands of electron volts (keV), where the prefix k indicates one thousand. For instance, the energy carried by an X-ray photon with a wavelength of 1 nm is 1.24 keV. Similarly, the energies of gamma rays are often expressed in millions of electron volts (MeV), where the prefix M indicates one million.

FIGURE 5-11 The Sun's Spectrum Numerous spectral lines are seen in this photograph of the Sun's spectrum. The spectrum is so long that it had to be cut into segments to fit on this page. (NOAO)

5-5   Each chemical element produces its own unique set of spectral lines

In 1814 the German master optician Joseph von Fraunhofer repeated the classic experiment of shining a beam of sunlight through a prism (see Figure 5-3), but this time he subjected the resulting rainbow-colored spectrum to intense magnification. To his surprise, Fraunhofer discovered that the solar spectrum contains hundreds of fine, dark lines, now called **spectral lines.** Fraunhofer counted over 600 such lines; today we know of about a million. A photograph of the Sun's spectrum is shown in Figure 5-11.

Half a century later, chemists discovered that they could produce spectral lines in the laboratory. Chemists had long known that many substances emit distinctive colors when sprinkled into a flame. Such experiments, called flame tests, are used to identify chemical **elements,** the fundamental substances that cannot be broken down into more basic chemicals.

About 1857 the German chemist Robert Bunsen invented a gas burner that became widely used in flame tests because it had no color of its own to be confused with the color produced by the substance sprinkled into it. A young colleague, the Prussian-born physicist Gustav Kirchhoff, suggested that light from the colored flames might best be studied by passing it through a prism (Figure 5-12). Bunsen and Kirchhoff collaborated in designing and constructing a spectroscope, a device consisting of a prism and several lenses that magnifies and focuses the spectra of flames. The two scientists promptly discovered that the spectrum from a flame consists of a pattern of thin, bright spectral lines against a dark background. Later, they found that each

chemical element produces its own unique pattern of spectral lines. Thus was born in 1859 the technique of **spectral analysis,** the identification of chemical substances by their unique patterns of spectral lines.

Bunsen and Kirchhoff also discovered new chemical elements through spectral analysis. After they had recorded the prominent spectral lines of all the then-known elements, they soon began to discover other spectral lines in the spectra of mineral samples. In 1860 they found a new line in the blue

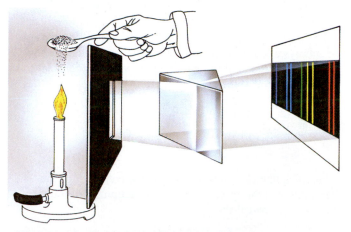

FIGURE 5-12 The Kirchhoff–Bunsen Experiment In the mid-1850s, Gustav Kirchhoff and Robert Bunsen discovered that when a chemical substance is heated and vaporized, the resulting spectrum exhibits a series of bright spectral lines. They also found that each chemical element produces its own characteristic pattern of spectral lines. (In an actual laboratory experiment, lenses would be needed to focus the image of the slit onto the screen.)

portion of the spectrum of a sample of mineral water. After isolating the previously unknown element responsible for making the line, they named it cesium (from the Latin *caesium,* "gray-blue"). The next year, a new line in the red portion of the spectrum of a mineral sample led them to discover the element rubidium (Latin *rubidium,* "red").

During the solar eclipse of 1868, astronomers found a new spectral line in light coming from the upper surface of the Sun while the main body of the Sun was hidden by the Moon. This line was attributed to a new element that was named helium (from the Greek *helios,* "sun"). Helium was not discovered on the Earth until 1895, when it was found in gases obtained from a uranium mineral.

By the early 1860s, Kirchhoff's experiments had progressed sufficiently for him to formulate three important statements about spectra that are today called **Kirchhoff's laws:**

Law 1 A hot opaque body, such as a hot, dense gas produces a **continuous spectrum**—a complete rainbow of colors without any spectral lines.

Law 2 A hot, transparent gas produces an **emission line spectrum**—a series of bright spectral lines against a dark background.

Law 3 A cool, transparent gas in front of a source of a continuous spectrum produces an **absorption line spectrum**—a series of dark spectral lines among the colors of the continuous spectrum.

Figure 5-13 illustrates how absorption and emission lines are formed. The bright lines in the emission spectrum of a particular gas occur at exactly the same wavelengths as the dark lines in its absorption spectrum. The relative temperatures of the gas cloud and its background determine which spectrum is observed: Absorption lines are seen if the background is hotter than the gas, and emission lines are seen if the background is cooler. Either can be used to determine the gas's chemical composition.

At the time of these discoveries, scientists knew that all matter is composed of atoms. An **atom** is the smallest particle

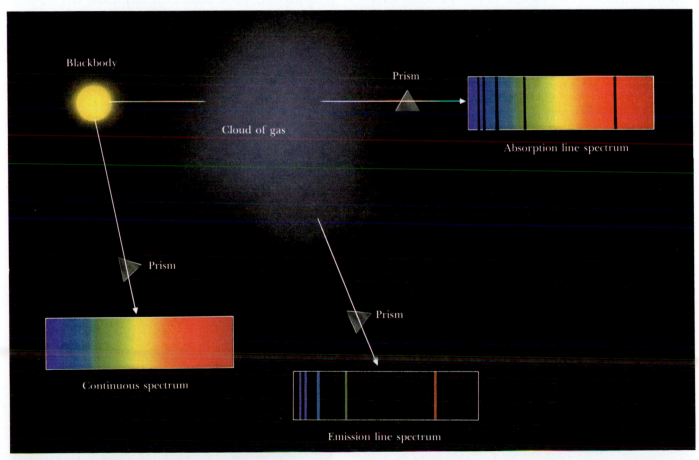

FIGURE 5-13 Continuous, Absorption Line, and Emission Line Spectra This schematic diagram summarizes how different types of spectra are produced. A hot, glowing object emits a continuous spectrum. If this source of light is viewed through a cloud of gas, dark absorption lines appear in the resulting spectrum. When the gas is viewed against a cold, dark background, the spectrum contains bright emission lines.

FIGURE 5-14 Rutherford's Experiment Alpha particles from a radioactive source are directed at a thin metal foil. Most alpha particles pass through the foil with very little deflection. Occasionally, however, an alpha particle will recoil, indicating that it has collided with the massive nucleus of an atom. This experiment provided the first evidence that the nuclei of atoms are relatively massive and compact.

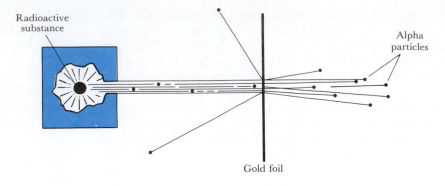

Radioactive substance

Alpha particles

Gold foil

(typical diameter = 0.1 nm) of a chemical element that still has the properties of that element. Kirchhoff's laws mean that the atoms in a gas somehow extract light of very specific wavelengths from the white light that passes through, thus creating dark absorption lines in the continuous spectrum of the white-light source. Because the atoms then radiate light of precisely these same wavelengths in all directions, an observer at an oblique angle (without the white-light source in the background) detects bright emission lines.

Why does an atom absorb light of only particular wavelengths? Why does it then emit light of only these same wavelengths? Maxwell's theory of electromagnetism could not answer these questions. The answers did not come until early in the twentieth century, when scientists began to discover the structure and properties of atoms.

5-6   An atom consists of a small, dense nucleus surrounded by electrons

The first important clue about the internal structure of atoms came from an experiment conducted in 1910 by Ernest Rutherford, a gifted chemist and physicist from New Zealand. Rutherford and his colleagues at the University of Manchester in England had been investigating the recently discovered phenomenon of radioactivity. Certain radioactive elements, such as uranium and radium, were known to emit particles. One type of particle, called an alpha particle, has about the same mass as a helium atom and is emitted from a radioactive substance with considerable speed.

In one series of experiments, Rutherford and his colleagues were using alpha particles as projectiles to probe the structure of solid matter. They directed a beam of these particles at a thin sheet of metal (Figure 5-14). Almost all the alpha particles passed through the metal sheet with little or no deflection from their straight-line paths. To the surprise of the experimenters, however, an occasional alpha particle bounced back from the metal sheet as though it had struck something quite dense. Rutherford later remarked, "It was almost as incredible as if you fired a fifteen-inch shell at a piece of tissue paper and it came back and hit you."

Rutherford concluded from this experiment that most of the mass of an atom is concentrated in a compact, massive lump of matter that occupies only a small part of the atom's volume. Most of the alpha particles pass freely through the nearly empty space that makes up most of the atom, but a few particles happen to strike the dense mass at the center of the atom and bounce back.

Rutherford proposed a new model for the structure of an atom. According to this model, a massive, positively charged **nucleus** at the center of the atom is orbited by tiny, negatively charged electrons (Figure 5-15). Rutherford concluded that at least 99.98% of the mass of an atom must be concentrated in its nucleus, whose diameter is only about one ten-thousandth the diameter of the atom.

We know today that the nucleus of an atom contains two types of particles: **protons** and **neutrons.** A proton has almost the same mass as a neutron, and each has about 2000 times as much mass as an electron. A proton has a positive electric charge, equal and opposite to that of an electron; a neutron has no electric charge. The electric force between the positively charged protons and the negatively charged electrons holds an atom together. The challenge that scientists faced in the early 1900s was to explain how these tiny particles give rise to spectral lines.

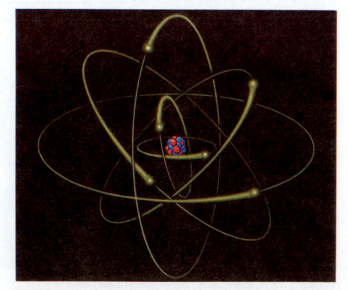

FIGURE 5-15 Rutherford's Model of the Atom Electrons orbit the atom's nucleus, which contains most of the atom's mass. The nucleus contains two types of particles: protons and neutrons.

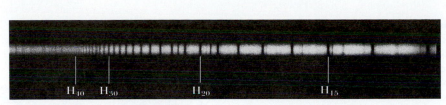

FIGURE 5-16 Balmer Lines in the Spectrum of a Star
This portion of the spectrum of the star HD 193182 shows nearly two dozen Balmer lines, from H_{13} through H_{40} (numbers are used beyond the first few Balmer lines so you don't have to memorize the entire Greek alphabet). The series converges at 364.56 nm, just to the left of H_{40}. This star's spectrum also contains the first twelve Balmer lines (H_{α} through H_{12}), but they are located beyond the right edge of this photograph. (Carnegie Observatories)

5-7 Spectral lines are produced when an electron jumps from one energy level to another within an atom

The task of reconciling Rutherford's atomic model with Kirchhoff's laws of spectral analysis was undertaken by the young Danish physicist Niels Bohr, who joined Rutherford's group at Manchester in 1911.

Bohr began by trying to understand the structure of hydrogen, the simplest and lightest of the elements. A hydrogen atom consists of a single electron and a single proton. Hydrogen has a simple spectrum consisting of a pattern of lines that begins at 656.3 nm and ends at 364.6 nm. The first spectral line is called H_{α}, the second spectral line is called H_{β}, the third is H_{γ}, and so forth. The closer you get to the short-wavelength end of the spectrum at 364.6 nm, the more spectral lines you see.

The regularity in this spectral pattern was described mathematically in 1885 by Johann Jakob Balmer, an elderly Swiss schoolteacher. The spectral lines of hydrogen at visible wavelengths are today called the **Balmer lines,** and the entire pattern from H_{α} onwards is called the **Balmer series.** Over two dozen Balmer lines are seen in the spectrum of a star shown in Figure 5-16.

Using trial and error, Balmer discovered a formula from which the wavelengths (λ) of hydrogen's spectral lines can be calculated. Balmer's formula is usually written

$$\frac{1}{\lambda} = R\left(\frac{1}{4} - \frac{1}{n^2}\right)$$

where n is an integer (whole number) greater than 2 and R is the Rydberg constant ($R = 1.097 \times 10^7$ m^{-1}), named in honor of the Swedish spectroscopist J. R. Rydberg. To get the wavelength of H_{α}, you put $n = 3$ into Balmer's formula. To get H_{β}, use $n = 4$. To get H_{γ}, use $n = 5$. To get the short-wavelength end of the hydrogen spectrum at 364.6 nm, use $n = \infty$ (the symbol ∞ stands for infinity).

Bohr realized that to fully understand the structure of the hydrogen atom, he had to be able to derive Balmer's formula using the laws of physics. He assumed first that the electron in a hydrogen atom orbits the nucleus in certain specific orbits. As shown in Figure 5-17, it is customary to label these orbits, called Bohr orbits, $n = 1$, $n = 2$, $n = 3$, and so on. Although confined to one of these allowed orbits while circling the nucleus, an electron can jump from one Bohr orbit to another.

For an electron to jump from one Bohr orbit to another, the hydrogen atom must gain or lose a specific amount of energy. For the electron to go from an inner to an outer orbit, the atom must absorb energy; for the electron to go from an outer to an inner orbit, the atom must release energy. As an example, Figure 5-18 shows an electron jumping between the $n = 2$ and $n = 3$ orbits of a hydrogen atom as it absorbs or emits an H_{α} photon.

The energy gained or released by the atom when the electron jumps from one orbit to another is the difference in energy between these two orbits. According to Planck and Einstein, the packet of energy gained or released is a photon, whose energy is given by $E = hc/\lambda$, where h is Planck's constant, c is the speed of light, and λ is the photon's wavelength. Bohr combined these ideas to prove mathematically that the

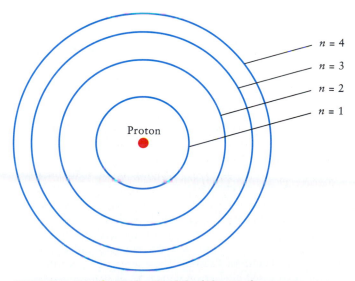

FIGURE 5-17 The Bohr Model of the Hydrogen Atom
According to the Bohr model of the hydrogen atom, an electron circles the nucleus (a proton) only in allowed orbits $n = 1, 2, 3,$ and so forth. The first four Bohr orbits are shown here.

FIGURE 5-18 The Absorption and Emission of an H$_\alpha$ Photon This schematic diagram, drawn according to the Bohr model of the atom, shows what happens when a hydrogen atom absorbs or emits a photon whose wavelength is 656.28 nm. (a) The photon is absorbed by the atom as the electron jumps from orbit $n = 2$ up to $n = 3$. (b) The photon is emitted by the atom as the electron falls from orbit $n = 3$ down to orbit $n = 2$.

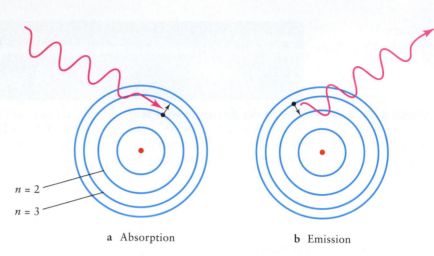

a Absorption b Emission

BOX 5-5

Atomic Structure and Energy-Level Diagrams

The Bohr model of the hydrogen atom correctly predicts series of spectral lines at visible and nonvisible wavelengths. As we have seen, Bohr's formula with $N = 2$ gives the Balmer series, the only series at visible wavelengths. The value $N = 1$ gives the **Lyman series,** which is entirely in the ultraviolet. All the spectral lines in this series involve electron transitions between the lowest Bohr orbit and all higher orbits ($n = 2, 3, 4$, and so on). This pattern of spectral lines begins with L$_\alpha$ ("Lyman alpha") at 122 nm and converges on L$_\infty$ at 91 nm. At infrared wavelengths is the **Paschen series,** for which $N = 3$. It begins with P$_\alpha$ ("Paschen alpha") at 1875 nm and converges on P$_\infty$ at 821 nm. Additional series exist at still longer wavelengths. Diagram a shows some examples of electron transitions in the Bohr atom for these series.

Today's view of the atom owes much to the Bohr model, but it is different in certain ways. The modern picture is based on **quantum mechanics,** a branch of physics dealing with photons and subatomic particles that was developed during the 1920s. As a result of this work, physicists no longer picture electrons as moving in specific orbits about the nucleus. Instead, electrons are now said to occupy certain **energy levels** in the atom.

An extremely useful way of displaying the structure of an atom is with an **energy-level diagram,** such as that for hydrogen shown in diagram b. The lowest energy level, called the ground state, corresponds to the $n = 1$ Bohr orbit. An electron can jump from the ground state up to the $n = 2$ level only if the atom absorbs a Lyman-alpha photon with a wavelength of 122 nm. The energy of a photon, usually expressed in electron volts (eV), is determined by the relationship $E = h\nu = hc/\lambda$. As explained in Box 5-4, an electron volt is a tiny amount of energy (1 eV = 1.6 ×

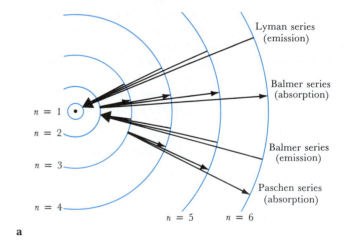

a

10^{-19} J). The Lyman-alpha photon has an energy of 10.19 eV, so the energy level $n = 2$ is shown in the diagram as having an energy 10.19 eV above the energy of the ground state (conventionally assigned a value of 0 eV). Similarly, the $n = 3$ level is 12.07 eV above the ground state, and so forth up to the $n = \infty$ level at 13.6 eV. If the atom absorbs a photon of any energy greater than 13.6 eV, an electron from the ground state will be knocked completely out of the atom. This process, in which an electron is removed from the atom, is called **ionization.**

The atoms of heavier elements have more complex energy-level diagrams. Diagram c shows the energy-level diagram for sodium, along with the wavelengths of photons absorbed or emitted in some of sodium's major electron transitions. At visible wavelengths, the sodium spectrum is dominated by two strong lines, called the

wavelength (λ) of the photon emitted or absorbed as the electron jumps from orbit N to orbit n is

$$\frac{1}{\lambda} = R \left(\frac{1}{N^2} - \frac{1}{n^2} \right)$$

In deriving this equation from the laws of physics, Bohr found that R can be expressed entirely in terms of basic physical quantities such as the speed of light, Planck's constant, and the mass and charge of the electron.

Bohr's discovery elucidated the meaning of Balmer's formula: All the Balmer lines are produced by electron transitions between the second Bohr orbit ($N = 2$) and higher orbits ($n = 3, 4, 5$, and so on). Bohr's formula also correctly predicts the wavelengths of other series of spectral lines that occur at nonvisible wavelengths. These nonvisible series of spectral lines are discussed in Box 5-5.

Bohr's ideas explain Kirchhoff's laws. Each spectral line corresponds to one particular transition between the orbits in the atoms of a particular element. An absorption line occurs when an electron jumps from an inner to an outer orbit, extracting the required photon from an outside source of energy, such as the continuous spectrum of a hot, glowing object. An emission line is produced when the electron falls back down to a lower orbit and gives up a photon. However, the wavelengths of the absorption and emission lines of a particular element are the same, since they involve the same electron orbits.

With the work of people like Planck, Einstein, Rutherford, and Bohr, the interchange between astronomy and physics came full circle. Modern physics was born when Newton set out to understand the motions of the planets. Two and a half centuries later, physicists in their laboratories had probed the properties of light and the structures of atoms. Their labors

sodium D lines, at 588.99 and 589.59 nm. These two lines are strong because they correspond to two transitions that are the primary avenue through which electrons cascade from high orbits down to the ground state. Astronomers find energy-level diagrams useful in understanding the lines they observe in the spectra of stars and nebulae.

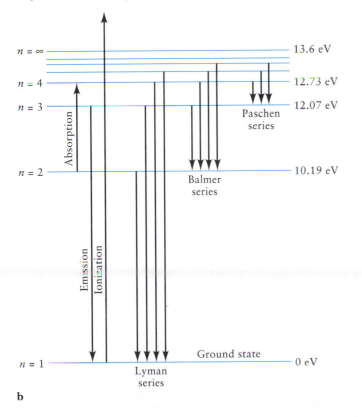

b

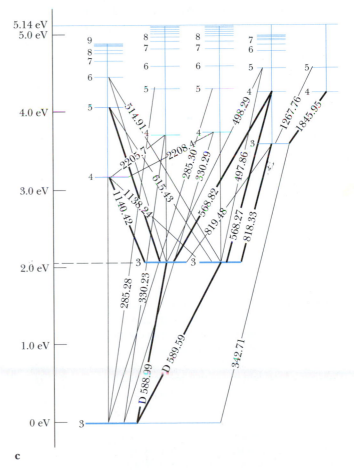

c

had immediate applications in astronomy, enabling scientists to probe the chemical and physical properties of planets, stars, and galaxies.

5-8 The wavelength of a spectral line is affected by the relative motion between the source and the observer

Christian Doppler, a professor of mathematics in Prague, pointed out in 1842 that observed wavelength is affected by motion. As shown in Figure 5-19, light waves from an approaching light source are compressed. The circles represent the peaks of waves emitted from various positions as the source moves along. Because each successive wave is emitted from a position slightly closer to you, you see a shorter wavelength than you would if the source were stationary. All the lines in the spectrum of an approaching source are shifted toward the short-wavelength (blue) end of the spectrum. This phenomenon is called a **blueshift**.

Conversely, light waves from a receding source are stretched out, so that you see a longer wavelength than you would if the source were stationary. All the lines in the spectrum of a receding source are shifted toward the longer-wavelength (red) end of the spectrum, producing a **redshift**. In general, the effect of relative motion on wavelength is called the **Doppler effect**.

Suppose that λ_0 is the wavelength of a particular spectral line from a source that is not moving. It is the wavelength that you might look up in a reference book or determine in a laboratory experiment for this spectral line. If the source is moving, this particular spectral line is shifted to a different wavelength (λ). The size of the wavelength shift is usually written as $\Delta\lambda$, where $\Delta\lambda = \lambda - \lambda_0$. Thus $\Delta\lambda$ is the difference between the wavelength listed in reference books and the wavelength that you actually observe in the spectrum of a star or galaxy.

Christian Doppler proved that the wavelength shift ($\Delta\lambda$) is governed by the following simple equation:

$$\frac{\Delta\lambda}{\lambda_0} = \frac{v}{c}$$

where v is the speed of the source measured along the line of sight between the source and the observer and c is the speed of light (3×10^5 km/s). (This equation is valid only if v is small in comparison to c.)

The Doppler shift enables astronomers to determine the line-of-sight motion of astronomical objects. As we shall see in the next chapter, astronomers use a device called a spectrograph, which they attach to a telescope, to obtain spectra of planets, stars, or galaxies. Figure 5-20 shows the spectrum of the bright star Vega obtained with such equipment. Balmer lines, like H_β in the blue-green, are quite prominent. Laboratory experiments have determined that normal wavelength λ_0 of H_β is 486.133 nm. But in Vega's spectrum, H_β is located at $\lambda = 486.111$ nm. The wavelength of H_β has thus been shortened, which means that the spectral line is blueshifted, and so Vega is coming toward us. The arithmetic of calculating Vega's speed is given in Box 5-6 along with other examples.

Speed determined from the Doppler effect is called **radial velocity**, because v is the component of the star's motion parallel to our line of sight, or along the "radius" drawn from Earth to the star. Of course, a sizable fraction of a star's motion may be perpendicular to our line of sight. The speed of this transverse movement, which is in the plane of the sky, does not affect wavelengths if the speed is small compared with c.

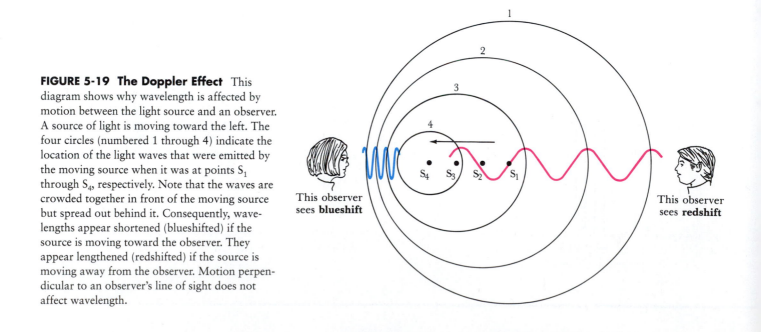

FIGURE 5-19 The Doppler Effect This diagram shows why wavelength is affected by motion between the light source and an observer. A source of light is moving toward the left. The four circles (numbered 1 through 4) indicate the location of the light waves that were emitted by the moving source when it was at points S_1 through S_4, respectively. Note that the waves are crowded together in front of the moving source but spread out behind it. Consequently, wavelengths appear shortened (blueshifted) if the source is moving toward the observer. They appear lengthened (redshifted) if the source is moving away from the observer. Motion perpendicular to an observer's line of sight does not affect wavelength.

This observer sees **blueshift**

This observer sees **redshift**

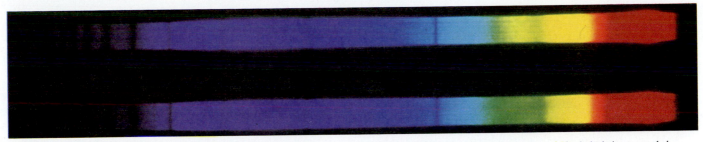

FIGURE 5-20 The Spectrum of Vega Several Balmer lines are seen in this photograph of the spectrum of Vega (also called α Lyrae). Most notable is H_β in the blue-green part of the spectrum. Since the spectral lines are shifted slightly toward the blue side of the spectrum, Vega must be approaching us. (NOAO)

The Doppler effect is an important tool in astronomy because it tells us how to calculate speeds from wavelength shifts, thereby uncovering basic information about the motions of planets, stars, and galaxies. As we shall see in coming chapters, the rotation of Venus was deduced from the Doppler shift of radar waves reflected from its surface. The Doppler shifting of spectral lines of double stars gives information about orbital speeds that can be combined with Kepler's third law to yield clues about stellar masses. Astronomers also use the Doppler effect along with Kepler's third law to measure the masses of galaxies. And finally, near the end of the book, we shall see that the redshifts of spectral lines of remote galaxies reveal that we live in an expanding universe. These are but a few examples of how Doppler's discovery has empowered astronomers in their quest to understand the universe.

BOX 5-6

Applications of the Doppler Effect

Doppler's formula relates the radial velocity of an astronomical object to a wavelength shift of its spectral lines. A few examples of how this formula is used are given below.

EXAMPLE: The prominent hydrogen line H_α has a normal wavelength of 656.285 nm, but in Vega's spectrum this line is located at 656.255 nm. The wavelength shift is calculated thus:

$$\Delta\lambda = \lambda - \lambda_0 = 656.255 \text{ nm} - 656.285 \text{ nm} = -0.030 \text{ nm}$$

Using Doppler's formula, we can calculate the star's radial velocity:

$$v = c\frac{\Delta\lambda}{\lambda_0} = 3 \times 10^5 \text{ km/s} \left(\frac{-0.030 \text{ nm}}{656.285 \text{ nm}}\right) = -13.7 \text{ km/s}$$

The minus sign indicates that the star is coming toward us.

EXAMPLE: In February 1987 a supernova now called SN 1987A exploded. About half a year later, astronomers studying the infrared spectrum of the exploded star discovered a spectral line of helium at a wavelength of 1.070 μm, which in the laboratory has a wavelength of 1.083 μm. The supernova's spectral line therefore had been subjected to a blueshift of

$$\Delta\lambda = \lambda - \lambda_0 = 1.070 \text{ μm} - 1.083 \text{ μm} = -0.013 \text{ μm}$$

and so we can compute the speed of the gases ejected by the detonation:

$$v = c\left(\frac{\Delta\lambda}{\lambda_0}\right) = c\left(\frac{-0.013}{1.083}\right) = -0.012c = -3600 \text{ km/s}$$

In other words, the star's ejected gases are approaching us at a speed of roughly one percent of the speed of light.

EXAMPLE: In the radio region of the electromagnetic spectrum, hydrogen atoms emit and absorb photons with a wavelength of 21.12 cm, giving rise to a spectral feature commonly called the "21-centimeter line." The galaxy NGC 3840 in the constellation of Leo (the Lion) is receding from us at a speed of 7370 km/s. We can predict the wavelength at which to expect the 21-cm line as follows:

$$\Delta\lambda = \lambda_0\left(\frac{v}{c}\right) = 21.12 \text{ cm} \left(\frac{7370 \text{ km/s}}{3 \times 10^5 \text{ km/s}}\right) = 0.52 \text{ cm}$$

Therefore the 21-cm line of hydrogen in this galaxy is detected at a wavelength of

$$\lambda = \lambda_0 + \Delta\lambda = 21.12 \text{ cm} + 0.52 \text{ cm} = 21.64 \text{ cm}$$

KEY WORDS

Terms preceded by an asterisk are discussed in the boxes.

absolute zero	electromagnetic spectrum	kelvin	radio waves
absorption line spectrum	electromagnetism	Kirchhoff's laws	redshift
atom	electron	*luminosity	*solar constant
Balmer line	*electron volt	*Lyman series	spectral analysis
Balmer series	element	microwaves	spectral line
blackbody	emission line spectrum	neutron	spectrum (*plural* spectra)
blackbody curves	energy flux	nucleus	Stefan–Boltzmann law
blackbody radiation	*energy level	*Paschen series	ultraviolet radiation
blueshift	*energy-level diagram	photoelectric effect	visible light
continuous spectrum	frequency	photon	wavelength
*degrees Celsius	gamma rays	Planck's law	Wien's law
*degrees Fahrenheit	infrared radiation	proton	X rays
Doppler effect	*ionization	*quantum mechanics	
electromagnetic radiation	joule	radial velocity	

KEY IDEAS

• Electromagnetic radiation (including visible light) has wavelike properties and travels through empty space at the constant speed $c = 3 \times 10^8$ m/s.

• A blackbody is a hypothetical object that is a perfect absorber of electromagnetic radiation at all wavelengths. Stars closely approximate the behavior of blackbodies.

The Stefan–Boltzmann law states that a blackbody radiates electromagnetic waves with a total energy flux (F) directly proportional to the fourth power of the temperature (T) of the object: $F = \sigma T^4$.

Wien's law states that the dominant wavelength at which a blackbody emits electromagnetic radiation is inversely proportional to the temperature of the object: λ_{max} (in meters) = 0.0029/T.

• The intensities of radiation emitted at various wavelengths by a blackbody at a given temperature are shown by a blackbody curve.

An explanation of blackbody curves led to the discovery that light has particlelike properties; the particles of light are called photons.

Planck's law relates the energy (E) of a photon to its frequency (ν) or wavelength (λ): $E = h\nu = hc/\lambda$, where h is Planck's constant.

• Kirchhoff's three laws of spectral analysis describe conditions under which different kinds of spectra are produced.

A hot object such as hot, dense gas emits a continuous spectrum covering all wavelengths.

A hot, transparent gas exposed to a source of energy produces a spectrum that contains bright (emission) lines.

A cool, transparent gas in front of a continuous source of energy produces dark (absorption) lines in the continuous spectrum.

• An atom has a small, dense nucleus composed of protons and neutrons; the nucleus is surrounded by electrons that occupy only certain orbits or energy levels.

When an electron jumps from one energy level to another, it emits or absorbs a photon of appropriate energy (and hence of a specific wavelength).

The spectral lines of a particular element correspond to the various electron transitions between energy levels in atoms of that element.

Bohr's model of the atom correctly predicts the wavelengths of hydrogen's spectral lines.

• The Doppler shift enables us to determine the radial velocity of a light source from the displacement of its spectral lines.

The spectral lines of an approaching light source are shifted toward short wavelengths (a blueshift); the spectral lines of a receding light source are shifted toward long wavelengths (a redshift).

The size of a wavelength shift is proportional to the radial velocity between the light source and the observer.

REVIEW QUESTIONS

1. Approximately how many times around the world could a beam of light travel in one second?

2. Describe an experiment in which light behaves like a wave.

3. Describe an experiment in which light behaves like a particle.

4. What is meant by the frequency of light? How is frequency related to wavelength?

5. A light source emits infrared radiation at a wavelength of 825 nm. What is the frequency of this radiation?

6. Announcers at a certain radio station say that they are at "103.3 FM on your dial," meaning that they transmit at a frequency of 103.3 MHz. What is the wavelength of the radio waves from this station?

7. Using Wien's law and the Stefan–Boltzmann law, explain the color and intensity changes that are observed as the temperature of a hot, glowing object increases.

8. The bright star Regulus in the constellation of Leo (the Lion) has a surface temperature of 12,200 K. What is the dominant wavelength (λ_{max}) of its energy emission?

9. The bright star Procyon in the constellation of Canis Minor (the Small Dog) emits the greatest intensity of radiation at a wavelength (λ_{max}) of 445 nm. What is the surface temperature of the star?

10. What wavelength of electromagnetic radiation is emitted with greatest intensity by this book? To what region of the electromagnetic spectrum does this wavelength correspond?

11. What is a blackbody? In what way is a blackbody black? If a blackbody is black, how can it emit light?

12. Explain why astronomers are interested in blackbody radiation.

13. How is the energy of a photon related to its wavelength? What kind of photons carry the most energy? What kind of photons carry the least energy?

14. How do we know that atoms have nuclei?

15. Describe the spectrum of hydrogen at visible wavelengths and show how Bohr's model of the atom explains the Balmer lines.

16. Why do different elements have different patterns of lines in their spectra?

17. What is the Doppler effect, and why is it important to astronomers?

ADVANCED QUESTIONS

Tips and tools . . .

Formulas for converting between temperature scales are found in Box 5-1. Relationships between a star's radius, its luminosity, and its surface temperature are discussed in Box 5-2. A simple application of Planck's law to calculate the energy of a photon is given in Box 5-4.

18. What is the temperature of the Sun's surface in degrees Fahrenheit?

19. Your body temperature is 98.6°F. What kind of radiation do you emit? At what wavelength do you emit the most radiation?

20. The bright star Sirius in the constellation of Canis Major (the Large Dog) has a radius of 1.67 $R_\odot$ and a luminosity of 25 $L_\odot$. What is the star's surface temperature?

21. During the 1970s and 1980s, balloon-borne instruments detected 511 keV photons coming from the direction of the center of our galaxy. What is the wavelength of these photons? To what part of the electromagnetic spectrum do these photons belong?

22. Calculate the wavelength of H_δ. Draw a schematic diagram of the hydrogen atom and indicate the electron transition that gives rise to this spectral line.

23. The bright star Procyon has a surface temperature of about 6500 K. Compared to the Sun, how much more energy is emitted each second from each square meter of Procyon's surface?

24. Imagine a star the same size as the Sun but with a surface temperature twice that of the Sun. At what wavelength would that star emit most of its radiation? How many times brighter than the Sun would that star be?

25. Imagine a star whose diameter is 10 times larger than the Sun's and whose surface temperature is 2900 K. Suppose that both the Sun and this star were located at the same distance from you. Which would be brighter? By what factor?

26. The wavelength of H_β in the spectrum of the star Megrez (Ursa Majoris) in the Big Dipper is 486.112 nm. Laboratory measurements demonstrate that the normal wavelength of this spectral line is 486.133 nm. Is the star coming toward us or moving away from us? At what speed?

27. In the spectrum of the bright star Rigel, H_α has a wavelength of 656.331 nm. Is the star coming toward us or moving away from us? How fast?

DISCUSSION QUESTIONS

28. Why do you suppose that ultraviolet light can cause skin cancer but ordinary visible light does not?

29. Compare chemical identification based on spectral line patterns with the identification of people by their fingerprints. In what ways is this a good or poor analogy?

30. Suppose you look up at the night sky and observe some of the brightest stars with your naked eye. Is there any way of telling which stars are hot and which are cool? Explain.

31. The human eye is most sensitive over the same wavelength range at which the Sun emits the greatest intensity of radiation. Suppose creatures were to evolve on a planet orbiting a star somewhat hotter than the Sun. To what wavelengths would their "eyes" most likely be sensitive?

OBSERVING PROJECTS

32. As soon as weather conditions permit, observe a rainbow. Confirm that the colors are in the same order as shown in Figure 5-3.

33. Turn on an electric stove or toaster oven and carefully observe the heating elements as they warm up. Relate your observations to Wien's law and the Stefan–Boltzmann law.

34. Obtain a glass prism (or a diffraction grating, which is probably more readily available and will be discussed in the next chapter) and look through it at various light sources, such as an ordinary incandescent light, a neon sign, and a mercury vapor street lamp. **DO NOT LOOK AT THE SUN! Looking directly at the Sun can cause blindness.** Do you have any trouble seeing spectra? What do you have to do to see a spectrum? Describe the differences in the spectra of the various light sources you observed.

FOR FURTHER READING

Bova, B. *The Beauty of Light*. Wiley, 1988. A fascinating introduction to all aspects of light. Includes sections on the eye, color, and the technology of producing and detecting light.

Cline, B. *Men Who Made a New Physics*. Signet, 1965. This book describes the history of the discoveries that led to our modern understanding of light and atoms.

Feynman, R. *QED*. Princeton University Press, 1985. Using many carefully thought-out analogies, this slim volume discusses how photons and particles interact. The author won a Nobel Prize for his work in this field.

Gingerich, O. "Unlocking the Chemical Secrets of the Cosmos." *Sky & Telescope,* July 1981. This brief article offers some fascinating insights on the work of Kirchhoff and Bunsen.

Gribbin, J. *In Search of Schroedinger's Cat*. Bantam, 1984. This excellent introduction to quantum mechanics provides both background and historical material along with a fine review of current thinking on the subject.

van Heel, A., and Velzel, C. *What Is Light?* McGraw-Hill, 1968. This book surveys our modern understanding of the phenomenon of light.

Snow, C. *The Physicists*. Macmillan, 1981. This book describes many of the people who contributed to our modern understanding of light and matter.

Sobel, M. *Light*. University of Chicago Press, 1987. An excellent nontechnical introduction to all aspects of light.

Weinberg, S. *Subatomic Particles*. Scientific American Books, 1983. This book beautifully describes the various developments, both theoretical and experimental, leading up to our modern understanding of light and matter.

OPTICS AND TELESCOPES

THE SUMMIT OF MAUNA KEA ON HAWAII Astronomers prefer to build observatories on mountaintops far from city lights, where the air is dry, stable, and cloud free. The summit of Mauna Kea, shown here in winter, offers the world's best observing site. The dome in the foreground houses the 10-m Keck reflector, the largest telescope in the world. To the right of the Keck Observatory is the NASA Infrared Telescope Facility. On the ridge in the background are (from left to right) the 3.6-m Canada–France–Hawaii telescope, the 2.2-m University of Hawaii telescope, and the U.K. Royal Observatory 3.8-m infrared telescope. (California Association for Research in Astronomy)

THE TELESCOPE is the astronomer's most important tool. Telescopes are essential because they give us bigger, brighter, sharper images of distant astronomical objects than our eyes do. Refracting telescopes, which use large lenses to collect incoming starlight, were popular in the nineteenth century, but modern astronomers prefer reflecting telescopes, which gather light with large concave mirrors. Astronomers also attach a variety of equipment to telescopes in order to record and analyze incoming starlight. Light-sensitive silicon chips, called CCDs, mounted at a telescope's focus are only the latest device for studying distant stars and galaxies. Because the Earth's atmosphere is transparent to radio waves as well as visible light, ground-based radio telescopes can view the universe at these longer wavelengths. At most other wavelengths, the Earth's atmosphere is opaque, so observations must be performed from space. Sophisticated orbiting observatories are now giving us unprecedented views of the cosmos all across the electromagnetic spectrum.

Telescopes have played a major role in astronomy ever since Galileo first looked through one and saw craters on the Moon four centuries ago. Using a telescope, we can see extremely faint objects in space far more clearly than we can with the naked eye.

Traditionally, telescopes have been used to detect visible light. By means of lenses or mirrors, light from a distant object is brought to a focus where the resulting image is viewed or photographed. Recently, astronomers have built telescopes that detect nonvisible forms of light, such as X rays and radio waves.

Although all wavelengths of electromagnetic radiation share many basic properties—for example, they all travel at the speed of light—they interact quite differently with matter. For instance, your body is transparent to X rays but not to visible light; your eyes respond to visible light but not to gamma rays; your radio detects radio waves but not ultraviolet light. Consequently, astronomers use fundamentally different kinds of telescopes for observing various wavelength ranges. For example, a radio telescope that detects radio waves from space differs from either an X-ray telescope or an ordinary optical telescope. Because optical telescopes are the most common and familiar astronomical tool, we shall discuss them in detail before turning to the more exotic instruments capable of revealing the nonvisible sky.

6-1 A refracting telescope uses a lens to concentrate incoming starlight at a focus

Although light travels at 3×10^8 m/s in a vacuum, it travels at a slower speed through a dense substance like glass. The abrupt slowing of light as it enters a piece of glass is analogous to a person's walking from a boardwalk onto a sandy beach: Her pace suddenly slows as she steps from the smooth pavement into the sand. Similarly, upon exiting a piece of glass, light resumes its original speed, just as a person stepping back onto a boardwalk easily resumes her original pace.

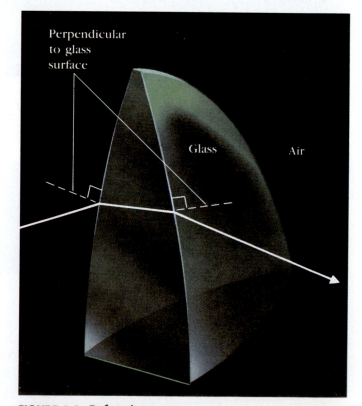

FIGURE 6-1 Refraction A light ray entering a piece of glass is bent toward the perpendicular. As the light ray leaves the piece of glass, it is bent away from the perpendicular.

As light passes from one transparent medium into another, a light ray is bent at an angle oblique to the surface between the media. This phenomenon, which is called **refraction,** is caused by the change in the speed of light. Imagine driving a car from a smooth pavement onto a sandy beach. If the car approaches the beach at an angle, one of the front wheels will be slowed by the sand before the other is, causing the car to veer from its original direction.

To describe the refraction of a light ray entering a piece of glass, imagine drawing a perpendicular to the surface of the glass at the point where the light strikes the glass, as shown in Figure 6-1. As the light ray goes from the less dense medium (such as air or a vacuum) into the denser medium (like glass), it will be bent toward the perpendicular. This happens in the same way as a car driving obliquely onto sand veers toward the direction perpendicular to the pavement–beach boundary. Upon emerging from the far side of a piece of glass, the light ray resumes its original high speed and is bent away from the perpendicular. Because the bending depends on the speed of light in the glass, different kinds of glass produce slightly different amounts of refraction.

Because of the refracting property of glass, a convex lens (one that is fatter in the middle than at the edges) causes incoming light rays to converge at a point called the **focus,** as shown in Figure 6-2. If the light source is extremely far away, the incoming light rays will be parallel and come to a focus at a specific distance from the lens known as the **focal length** of the lens.

The stars and planets are so far away that light rays from them are essentially parallel. Consequently, a lens always focuses light from an astronomical object as shown in Figure 6-2. A second lens is used to magnify and examine the image of an astronomical object formed at the focus. Such an ar-

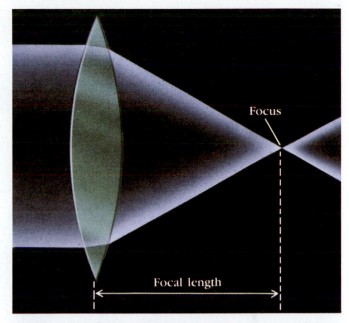

FIGURE 6-2 A Convex Lens   A convex lens causes parallel light rays to converge to a focus. The distance from the lens to the focus is the focal length of the lens.

rangement of two lenses is called a **refracting telescope,** or **refractor** (Figure 6-3). The large-diameter, long-focal-length lens at the front of the telescope, called the **objective lens,** forms the image; the smaller, shorter-focal-length lens at the rear of the telescope, called the **eyepiece lens,** magnifies the image for the observer. Galileo used a small refracting telescope for astronomical observations very soon after the device was invented in Holland for viewing distant objects on Earth.

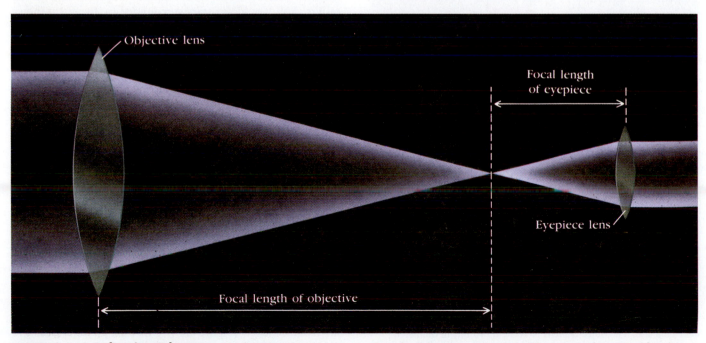

FIGURE 6-3 A Refracting Telescope   A refracting telescope consists of a large, long-focal-length objective lens and a small, short-focal-length eyepiece lens. The eyepiece lens magnifies the image formed at the focus of the objective lens.

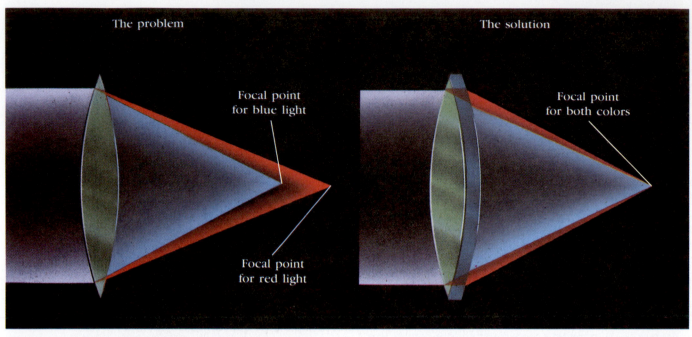

FIGURE 6-4 Chromatic Aberration A single lens suffers from a defect called chromatic aberration, in which different colors of light have slightly different focal lengths. This problem is corrected by adding a second lens made from a different kind of glass.

The **magnification,** or magnifying power, of a refracting telescope is equal to the focal length of the objective lens divided by the focal length of the eyepiece lens. For example, if the objective of a telescope has a focal length of 100 cm and the eyepiece has a focal length of $\frac{1}{2}$ cm, the magnifying power of the telescope is $100 \div \frac{1}{2} = 200$ (usually written as 200×).

If you were to build a telescope using only the instructions given so far, you would probably be disappointed with the results. You would see stars surrounded by fuzzy, rainbow-colored halos. This optical defect, called **chromatic aberration,** arises because a lens bends different colors of light through different angles, just as a prism does (recall Figure 5-3).

By adding small amounts of chemicals to a vat of molten glass, an optician can manufacture different kinds of glass. Because the speed of light varies slightly from one kind of glass to another, opticians can correct for chromatic aberration. Specifically, a thin lens can be mounted just behind the main objective lens of a telescope, as shown in Figure 6-4. By carefully choosing two different kinds of glass for these two lenses, the optician can ensure that different colors of light will come to a focus at the same point.

Chromatic aberration is only the most severe of a host of optical problems that must be solved in designing a high-quality refracting telescope. During the nineteenth century, master opticians devoted their lives to overcoming these problems, and several magnificent refractors were constructed in the late 1800s. The largest refracting telescope, completed in 1897, is located at the Yerkes Observatory, not far from Chicago (Figure 6-5). Other major refractors are listed in Box 6-1.

FIGURE 6-5 A Large Refracting Telescope This giant refractor, built in the late 1800s, is housed at Yerkes Observatory near Chicago. The objective lens is 102 cm (40 in.) in diameter, and the telescope tube is $19\frac{1}{2}$ m (64 ft) long. (Yerkes Observatory)

Few major refracting telescopes have been constructed in the twentieth century. There are many reasons for the modern astronomer's lack of interest in this type of telescope. First, because faint light must readily pass through the objective lens, the glass from which the lens is made must be totally free of defects, such as the bubbles that frequently form when molten glass is poured into a mold. Consequently, the glass for the objective lens is extremely expensive. Second, glass is opaque to certain kinds of light. Even visible light is dimmed substantially as it passes through the thick slab of glass at the front of a refractor, and ultraviolet radiation is largely absorbed by the glass lens. Third, it is impossible to produce a large lens that is free of chromatic aberration. Fourth, it is difficult to support these heavy lenses without blocking the path of light into the telescope. All these problems can be avoided by using mirrors instead of lenses.

Major Refracting Telescopes

During the late nineteenth century, master optician Alvan Clark built some of the largest refractors in the world. In 1888 he completed a refractor with an objective lens 91 cm (36 in.) in diameter for the Lick Observatory near San Jose, California. Nine years later he finished a refractor for the Yerkes Observatory that has an objective lens with a diameter of 102 cm (40 in.). The venerable astronomer shown beside the Yerkes objective in the accompanying historical photograph is George van Biesbroeck. This telescope is still the largest refractor in the world.

There are only 14 refractors with objective lenses larger than 65 cm (26 in.) in diameter, all of which are listed below. They all have extremely long focal lengths. For example, the Yerkes refractor has a focal length of 19.35 m ($63\frac{1}{2}$ ft).

Observatory	Location	Year completed	Objective diameter (cm)	Focal length (cm)
Yerkes Observatory	Williams Bay, Wisconsin	1897	102	1936
Lick Observatory	Mount Hamilton, California	1888	90	1763
Observatoire de Paris	Meudon, France	1889	83	1616
Zentralinstitut für Astrophysik	Potsdam, Germany	1899	80	1200
Allegheny Observatory	Pittsburgh, Pennsylvania	1914	76	1412
Observatoire de Nice	Mont Gros, France	1886	74	1790
Old Royal Observatory	Greenwich, England	1894	71	848
Archenhold-Sternwarte	Berlin, Germany	1896	68	2100
Institut für Astronomie	Vienna, Austria	1880	67	1050
Republic Observatory	Johannesburg, South Africa	1925	67	1092
Leander McCormick Observatory	Charlottesville, Virginia	1883	67	991
United States Naval Observatory	Washington, D.C.	1873	66	987
Royal Greenwich Observatory	Herstmonceux, England	1899	66	686
Mount Stromlo Observatory	Canberra, Australia	1956*	66	1080

*First used at Johannesburg, South Africa, in 1925.

6-2 A reflecting telescope uses a mirror to concentrate incoming starlight at a focus

To understand **reflection**, imagine drawing a perpendicular to a mirror's surface at the point where a light ray strikes the mirror, as shown in Figure 6-6. The angle between the arriving (incident) light ray and the perpendicular is always equal to the angle between the reflected ray and the perpendicular. Knowing this, Isaac Newton realized that a concave mirror would cause parallel light rays to converge to a focus, as shown in Figure 6-7. The distance between the reflecting surface and the focus is the focal length of the mirror.

In a **reflecting telescope**, or **reflector**, an image of a distant object is formed at the focus of a concave mirror. In order to view the image, Newton simply placed a small, flat mirror at a 45° angle in front of the focal point, as sketched in Figure 6-8a. This secondary mirror deflects the light rays to one side, where Newton placed an eyepiece lens to magnify the image. A reflecting telescope having this optical design is appropriately called a **Newtonian reflector.**

The magnifying power of a Newtonian reflector is calculated in the same way as for a refractor: The focal length of the primary mirror is divided by the focal length of the eyepiece. For example, if a reflector has a focal length of 2 m and the eyepiece has a focal length of 25 mm, the magnifying power is 200 cm ÷ 2.5 cm = 80×.

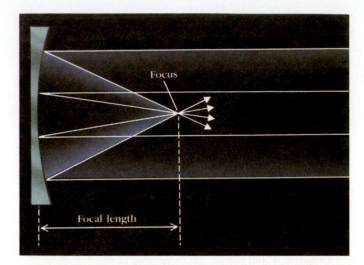

FIGURE 6-7 A Concave Mirror A concave mirror causes parallel light rays to converge to a focus. The distance between the mirror and the focus is the focal length of the mirror.

Newton's original design was modified by later astronomers. The primary mirrors of many major reflectors are so large that the astronomer can actually sit at the undeflected focal point, directly in front of the primary mirror. The "observing cage" in which the astronomer sits blocks only a small fraction of the incoming starlight. This arrangement is called a **prime focus** (Figure 6-8b).

Another popular optical design, called a **Cassegrain focus**, also has a convenient, accessible focal point. A hole is drilled directly through the center of the primary mirror and a convex secondary mirror placed in front of the original focal point reflects the light rays back through the hole (Figure 6-8c).

In a fourth design, a series of mirrors channels the light rays away from the telescope to a remote focal point. Heavy optical equipment that could not be mounted directly on the telescope is then located at the resulting **coudé focus**, named after a French word meaning "bent like an elbow" (Figure 6-8d).

To make a reflector, an optician grinds and polishes a large slab of glass into the appropriate concave shape. The glass is then coated with silver, aluminum, or a similar highly reflective substance. Defects inside the glass of a reflector, such as bubbles or flecks of dirt, do not detract from such a telescope's effectiveness, because light reflects off the surface of the glass rather than passing through it, as it would with the objective lens of a refracting telescope.

Furthermore, reflection is not affected by the wavelengths of the incoming light. Because all wavelengths are reflected to the same focus, a mirror does not have chromatic aberration. Although some chromatic aberration may arise with the smaller lens that magnifies the image, a reflector does not suffer from the major difficulties associated with a large objective lens. Finally, the mirror can be fully supported by braces on its back, so that a large, heavy mirror can be mounted without much danger of breakage or surface distortion.

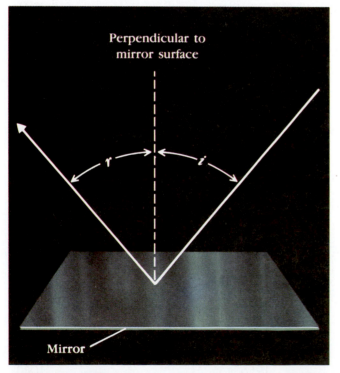

FIGURE 6-6 Reflection The angle at which a beam of light approaches a mirror, called the angle of incidence (i), is always equal to the angle at which the beam is reflected from the mirror, called the angle of reflection (r). Reflection is accurately described by the equation $i = r$.

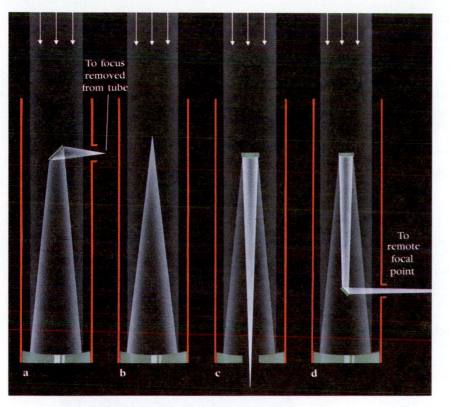

To focus removed from tube

To remote focal point

a   b   c   d

FIGURE 6-8 Reflecting Telescopes Four of the most popular optical designs for reflecting telescopes: (**a**) Newtonian focus, (**b**) prime focus, (**c**) Cassegrain focus, and (**d**) coudé focus.

Sixteen reflectors around the world have primary mirrors at least 3 meters (9.8 feet) in diameter (Box 6-2). One of the largest, a 6-meter telescope, is located in the Caucasus Mountains in southern Russia, and the famous 200-inch (5-meter) telescope is at the Palomar Observatory in southern California. In the early 1970s, a matching pair of telescopes was built in Arizona and Chile. Both have mirrors 4 meters (13.1 feet) in diameter. These twins (Figure 6-9) allow astronomers to observe the entire sky with essentially the same instrument.

Astronomers prefer large telescopes, because a large mirror intercepts and focuses more starlight than does a small mirror (Figure 6-10). A large mirror therefore produces brighter images and detects fainter stars than does a small one. The **light-gathering power** of a telescope is directly related to the area of the telescope's primary mirror, which is proportional to the square of the mirror's diameter. For example, because the 200-inch mirror at Palomar Observatory has four times the area of the 100-inch mirror at Mount Wilson Observatory, the Palomar telescope has four times the light-gathering power of the Mount Wilson telescope.

◀ **FIGURE 6-9 The 4-m Telescope at Cerro Tololo** This telescope is located on a mountaintop near Santiago, Chile. Its twin is at the Kitt Peak Observatory in Arizona. Both telescopes have been in operation since the early 1970s. (Cerro Tololo Inter-American Observatory)

A large telescope also helps achieve a second major goal: It produces star images that are sharp and crisp. A theoretical quantity called **angular resolution** gauges the extent to which fine details can be seen. Poor angular resolution causes star images to be fuzzy and blurred together.

The angular resolution of a telescope is the angle between two adjacent stars whose separate images are just barely dis-cernable, neglecting the effects of Earth's atmosphere. The smaller the angle, the sharper the image. Astronomers use empirical formulas to evaluate the angular resolution of a telescope from the diameter of its objective lens or mirror and the wavelengths being observed. Large modern instru-ments—such as those at Palomar, Kitt Peak, and Cerro Tololo—are calculated to have angular resolutions better

---

BOX 6-2

Major Reflecting Telescopes

There are 16 reflectors in the world with primary mirrors equal to or larger than 3 m (9.8 ft) in diameter. The obser-vatories where these telescopes reside are listed in the table below.

Two of the telescopes in this table have segmented primary mirrors. The Keck Telescope in Hawaii has 36 hex-agonal mirrors whose total reflective area equals one 10-m mirror (see Figure 6-12). The accompanying photograph, taken during construction, shows one of the Keck mirror segments being lowered into place. The Multiple Mirror Telescope in Arizona originally had six 1.8-m mirrors whose total reflective area equaled that of one 4.5-m mirror (see Figure 6-11). In the early 1990s, work began on converting the Multiple Mirror Telescope to a 6.5-m telescope.

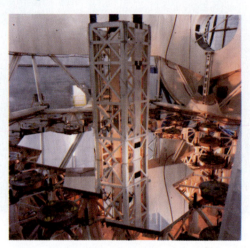

Observatory	Location	Year completed	Mirror diameter (m)
Keck Observatory	Mauna Kea, Hawaii	1993	10.0
Special Astrophysical Observatory	Zelenchukskaya, Russia	1976	6.0
Palomar Observatory	Palomar Mountain, California	1948	5.1
Whipple Observatory	Mount Hopkins, Arizona	1979	4.5
La Palma Observatory	Canary Islands	1987	4.2
Cerro Tololo Inter-American Observatory	Cerro Tololo, Chile	1974	4.0
Kitt Peak National Observatory	Kitt Peak, Arizona	1973	4.0
Anglo-Australian Observatory	Siding Spring, Australia	1975	3.9
Royal Observatory, Edinburgh*	Mauna Kea, Hawaii	1979	3.8
Canada–France–Hawaii Observatory	Mauna Kea, Hawaii	1979	3.6
European Southern Observatory	Cerro La Silla, Chile	1976	3.6
European Southern Observatory	Cerro La Silla, Chile	1989	3.6
German-Spanish Astronomical Center	Calar Alto, Spain	1983	3.5
Astrophysical Research Consortium	Apache Point, New Mexico	1993	3.5
Lick Observatory	Mount Hamilton, California	1959	3.0
Mauna Kea Observatory*	Mauna Kea, Hawaii	1979	3.0

*These two telescopes are used primarily for infrared observations.

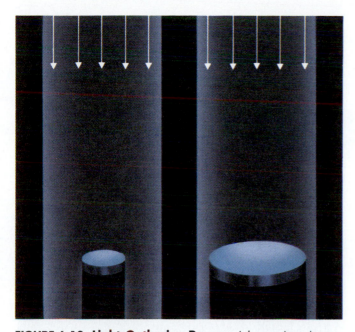

FIGURE 6-10 Light-Gathering Power A large mirror intercepts more starlight than does a small one. Large mirrors therefore produce brighter images than do small mirrors.

than 0.1 arc sec. In practice, however, these telescopes cannot achieve such exceptional sharpness. Turbulence and impurities in the air cause star images to jiggle around and twinkle. Even through the largest telescopes, a star still looks like a tiny circular blob rather than a pinpoint of light.

A more realistic measure of a telescope's best possible resolution is called the **seeing disk.** This disk is the angular diameter of a star's image broadened by turbulence. The size of the seeing disk varies from one observatory site to another and from one night to another. At Palomar and Kitt Peak, the seeing disk is typically 1 arc sec. The best conditions in the world, where the seeing disk is often as small as $\frac{1}{4}$ arc sec, have been reported at the observatories on the top of Mauna Kea, a 14,000-ft volcano on the island of Hawaii.

Building large reflectors poses significant engineering problems. Very large mirrors (more than 4 m in diameter) are so heavy that the glass actually sags under its own weight. The mirror's shape therefore changes slightly as the telescope is turned toward different parts of the sky, thereby degrading the quality of the image. Recently developed new techniques can produce thin, lightweight mirrors. A new generation of telescopes, called **new technology telescopes,** promises extremely high resolution by mounting these mirrors on computer-controlled, motorized supports that push on the glass to maintain a precise focus (Box 6-3). One such telescope has taken photographs with a resolution of 0.15 arc sec.

Another approach is to aim several smaller mirrors at the same focal point. The first instrument with this design was the Multiple Mirror Telescope, which had six mirrors, each measuring 1.8 m (5.9 ft) in diameter, mounted together as shown in Figure 6-11. Their total light-gathering power was equivalent to that of one 4.5-m mirror. This design proved so

successful that astronomers around the world are now building much larger multiple-mirror telescopes.

The first of these giant multiple-mirror instruments is the recently completed 10-m Keck telescope, on the summit of Mauna Kea in Hawaii. Thirty-six hexagonal mirrors are mounted side by side to give a primary that is 10 m (400 in.) in diameter, as shown in Figure 6-12. Each individual mirror is 5.9 ft across, 3 in. thick, and weighs 1400 lb. In 1993, engineers began installing mirrors on an identical 10-m telescope, only a short walk from the first Keck telescope.

Just as chromatic aberration plagues refracting telescopes, a defect called **spherical aberration** must be minimized when building reflecting telescopes. At issue is the precise shape of a mirror's concave surface. A spherical surface is easy to grind and polish, but different parts of a spherical mirror have slightly different focal lengths (Figure 6-13a), which results in a fuzzy image.

One common way to eliminate spherical aberration in reflecting telescopes is to polish the mirror's surface to a parabolic shape, because a parabola reflects parallel light rays to a common focus (Figure 6-13b). The only problem with this solution is that the astronomer no longer has a wide-angle view. Unlike spherical mirrors, parabolic mirrors suffer from a defect called **coma,** wherein star images far from the center of the field of view are elongated to look like tiny teardrops.

FIGURE 6-11 The Multiple Mirror Telescope This aerial photograph shows the six 1.8-m (5.9-ft) mirrors that together constitute the first multiple-mirror telescope. The total reflective area of the six mirrors is equal to that of one 4.5-m mirror. This telescope, located on the summit of Mount Hopkins in Arizona, is being refurbished to accommodate a 6.5-m mirror. (Multiple Mirror Telescope)

FIGURE 6-12 The Keck Telescope This model shows the design of a giant 10-m telescope on the summit of Mauna Kea in Hawaii. Thirty-six hexagonal mirrors, each measuring 1.8 m (5.9 ft) across, together have the same effect as one mirror 10 m (32.8 ft) in diameter. (California Institute of Technology)

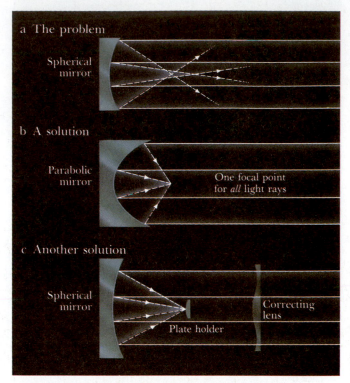

a The problem

Spherical mirror

b A solution

Parabolic mirror

One focal point for *all* light rays

c Another solution

Spherical mirror

Plate holder

Correcting lens

FIGURE 6-13 Spherical Aberration (a) Different parts of a spherically concave mirror reflect light to slightly different focal points. This difficulty can be corrected by either (b) using a parabolic mirror or (c) using a "correcting lens" in front of the mirror.

BOX 6-3

New Technology Telescopes

During the 1990s, astronomers began to build a new generation of reflecting telescopes that will greatly exceed the ability of earlier instruments. In the past, large reflectors had thick, massive mirrors. As an astronomer turned the telescope toward different parts of the sky, the mirror would sag slightly, thereby degrading the focus and producing fuzzy images. The new technology telescopes (NTTs) have thin mirrors whose precise shape is maintained by computer-activated, motorized supports called actuators.

Furthermore, the actuators are coupled to optical sensors that measure the distortion of incoming starlight caused by atmospheric turbulence. A powerful computer rapidly calculates the mirror shape needed to compensate for the distortion, and the actuators promptly deform the mirror accordingly. For example, a 3.6-m prototype NTT built by the European Southern Observatory has 19 actuators that correct the mirror's shape 100 times per second. This technique, called **adaptive optics**, effectively eliminates atmospheric distortion and produces remarkably sharp images, almost as good as if the telescope were in the vacuum of space.

At least eight new technology telescopes are at various stages of design or construction. These include the Columbus Project, an effort by the University of Arizona, which will have two 8-m mirrors. Two 8-m Advanced Technology Telescopes are also being designed by the National Optical Astronomy Observatories in Arizona. The Japan National Large Telescope will have a 7.5-m mirror, while the Magellan Project will build a single 8-m mirror to be located on a mountaintop in Chile.

One of the most ambitious new technology projects is the Very Large Telescope (VLT), being built by the European Southern Observatory. When completed, it will consist of four 8.2-m reflectors on the summit of an isolated mountain in Chile, shown in the accompanying artist's rendition. The four reflectors can be used individually, so that they can be pointed toward separate targets, or they can observe the same object simultaneously. When used together, the light of the four reflectors can be combined to give a light-gathering power equal to a single 16.4-m (53.8-ft) mirror! The first 8.2-m telescope should be finished in 1995, and all four should be completed by 1998.

An alternative design has both a high-quality image and a wide-angle field of view. Its spherical mirror minimizes coma, while a thin correcting lens at the front of the telescope eliminates spherical aberration (Figure 6-13c). The special design of this lens ensures that all light rays have the same focal point. This optical arrangement is the basic idea of the **Schmidt telescope**, named after its inventor, Bernhard Schmidt, who built the first prototype in the 1930s. Today there are more than a dozen large Schmidt telescopes at major observatories around the world. One of the largest (Figure 6-14) is located on Palomar Mountain, a short walk from the giant 5-m reflector. The reflector has a field of view only 2 arc min across, but the Schmidt telescope produces photographs covering a field 7° in diameter.

Schmidt telescopes work like cameras: The astronomer does not see the view until the photographs are developed. The resulting photographs constitute a permanent record of the sky (Box 6-4). Many of them reveal large-scale structures that had previously been overlooked by astronomers using bigger telescopes with smaller fields of view.

6-3   An electronic device is often used to record the image at a telescope's focus

The invention of photography in the nineteenth century was a boon for astronomy. By taking long exposures with a camera mounted at the focus of a telescope, an astronomer can record features too faint to be seen by simply looking through the telescope, thereby revealing details in galaxies, star clusters, and nebulae.

FIGURE 6-14 The Schmidt Telescope at Palomar This is one of the largest Schmidt telescopes in the world. Its mirror has a diameter of 1.8 m (5.9 ft), and the correcting lens at the front of the telescope measures 1.2 m (3.9 ft) across. An astronomer is shown guiding the telescope as it takes a wide-angle photograph of the sky. (Palomar Observatory)

Astronomers have long realized, however, that a photographic plate is not a very efficient light detector. Only 1 out of every 50 photons striking a photographic film triggers the chemical reaction needed to produce an image. Thus, roughly 98% of the light falling onto a photographic plate is wasted.

The most sensitive light detector currently available to astronomers is the newly invented **charge-coupled device (CCD)**. A CCD is a thin piece of silicon (Figure 6-15), not unlike the silicon chips used in pocket calculators and programmable appliances. The silicon wafer of a CCD is divided into an array of small light-sensitive squares called picture elements or, more commonly, **pixels.** For example, one of the latest CCDs has over four million pixels arranged in 2048 rows by 2048 columns. When an image from a telescope is focused on the CCD, an electric charge builds up in each pixel in proportion to the number of photons falling on that pixel. When the exposure is finished, the amount of charge on each pixel is read into a computer, where the resulting image can be stored in digital form. Over certain wavelength ranges, more than 75% of the photons falling on a CCD can be recorded.

Figure 6-16 shows an image taken with a CCD. Because of their extraordinary sensitivity and their ability to be used in conjunction with computers, CCDs are playing an increasingly important role in astronomy.

BOX 6-4

The Palomar Sky Survey

During the early 1950s, a team of astronomers spent several years photographing the sky with the Palomar Schmidt telescope in southern California. This work culminated in the famous National Geographic Society—Palomar Observatory Sky Survey. The entire sky visible from Palomar (all the northern hemisphere and the southern hemisphere down to a declination of $-33°$) was covered in 879 pairs of photographs. Each $6° \times 6°$ segment of the sky was photographed on two sets of photographic plates, one set recording primarily blue light, and the other recording primarily red light. Views of the same celestial objects in these two wavelength ranges were often strikingly different.

The accompanying photographs, taken from the Palomar sky survey, show a cluster of stars called the Pleiades that is visible to the naked eye in the constellation of Taurus (the Bull). The blue-sensitive plate is on the left, the red-sensitive plate in the center. Both pictures are printed exactly as they appear on the original plates, with black stars and white sky. Notice that the two contrasting views reveal very different features and details. For comparison, a full-color photograph is included on the right. The area covered in each of these views is about $1° \times 1\frac{1}{2}°$.

The Sky Survey was repeated in the 1980s so that astronomers could search for changes in the sky over the preceding 30 years. Schmidt telescopes in Chile and Australia have extended this wide-angle coverage to those portions of the southern sky not accessible from the Palomar Observatory.

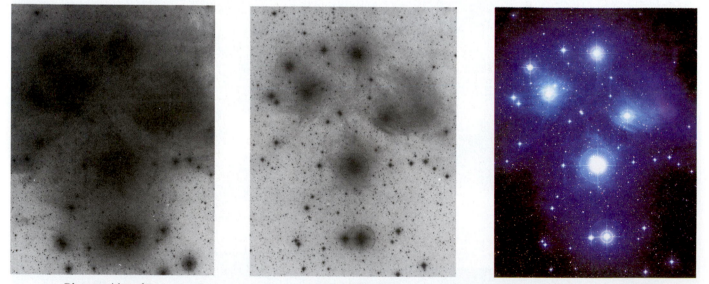

Blue-sensitive plate   Red-sensitive plate   Full-color photograph

6-4 Spectrographs record the spectra of astronomical objects

One of the astronomer's most important tools, perhaps second only to the telescope itself, is the **spectrograph**, a device that records spectra on photograph plates or a CCD. This optical device is mounted at the focus of a telescope.

In its most basic form, a spectrograph consists of a slit, two lenses, and a prism, arranged so that the spectrum of an astronomical object is focused onto a small photographic plate, as shown in Figure 6-17. First the spectrum of a star or galaxy is photographed. Then the exposed portion of the photographic plate is covered, and light from a known source (usually the emission spectrum of an element, like iron or argon) is focused on the spectrograph slit. This process produces a "comparison spectrum" above and below the spectrum of the star or galaxy, as shown in Figure 6-18. The wavelengths of the bright spectral lines of the comparison spectrum are already known from laboratory experiments and can therefore serve as reference markers.

There are drawbacks to this old-fashioned spectrograph. A prism does not disperse the colors of the rainbow evenly: Blue and violet portions of the spectrum are spread out more than the red portion. In addition, because the blue and violet wavelengths must pass through more glass than the red wavelengths (examine Figure 5-3), light is absorbed unevenly across the spectrum. Indeed, a glass prism is opaque to the near-ultraviolet part of the spectrum.

A better device for breaking starlight into the colors of the rainbow is a **diffraction grating**—or **grating** for short—a piece of glass on which thousands of closely spaced parallel lines have been cut. Some of the finest diffraction gratings

FIGURE 6-15 A Charge-Coupled Device (CCD) This tiny silicon rectangle contains 163,840 light-sensitive electric circuits that store images. At the end of each exposure, additional circuits in the silicon chip control the transfer and readout of the data to a waiting computer. (Smithsonian Astrophysical Observatory)

FIGURE 6-16 A CCD Image This color photograph was produced by combining a series of CCD images taken through colored filters at the focus of the 4-m Cerro Tololo telescope. The total exposure time was 6 hours. Extremely faint stars and galaxies with magnitude as dim as $26\frac{1}{2}$ can be seen. (Courtesy of P. Seitzer, NOAO)

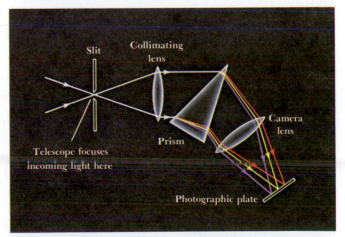

FIGURE 6-17 A Prism Spectrograph This optical device uses a prism to break up the light from a source into a spectrum. A lens, called the collimating lens, directs incoming light rays so that they enter the prism parallel to each other. A second lens, called a camera lens, then focuses the spectrum onto a photographic plate.

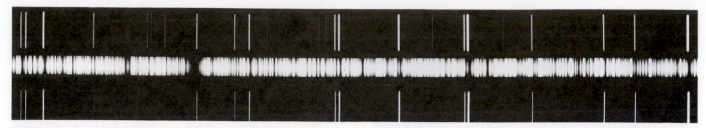

FIGURE 6-18 A Spectrogram The photographic record of a spectrum is called a spectrogram. This spectrogram shows the spectrum of a star. Numerous dark absorption lines can be seen against the brighter background of the spectrum. Above and below the star's spectrum are emission lines produced by an iron arc. These bright lines are the comparison spectrum. (Palomar Observatory)

have more than 10,000 lines per centimeter, which are usually cut by drawing a diamond back and forth across the glass. The spacing of the lines must be very regular, and the best results are obtained when the grooves are beveled. When light is shone on a diffraction grating, a spectrum is produced by the way in which various wavelengths leaving the different parts of the grating interfere with each other. Figure 6-19 shows the design of a modern grating spectrograph.

In recent years, the electronics industry has produced a variety of light-sensitive devices that are significantly better than photographic film for recording spectra. For instance, most observatories now record spectra with charge-coupled devices. A CCD, instead of a photographic plate, is placed at the focus of the spectrograph. When the exposure is finished, electronic equipment measures the charge that has accumu-

lated in each pixel. This data is used to graph light intensity against wavelength. Dark lines (absorption lines) appear as depressions or valleys on the graph, while bright lines (emission lines) in the spectrum appear as peaks. Figure 6-20 compares two ways of exhibiting a spectrum in which several absorption lines appear.

6-5 A radio telescope uses a large concave dish to reflect radio waves to a focus

For thousands of years, all the information that astronomers gathered about the universe was based on ordinary visible light. But with the discovery of nonvisible electromagnetic radiation in the nineteenth century, it seems reasonable to suppose that astronomical objects might also emit radio waves, X rays, and infrared and ultraviolet radiation. Surely views of the universe at these nonvisible wavelengths would enhance our understanding of the cosmos.

The first evidence of radio radiation coming from outer space was provided by the work of a young engineer, Karl Jansky, of Bell Telephone Laboratories. Using long antennas, Jansky began investigating the sources of the radio static that affects short-wavelength radiotelephone communication. By 1932 he realized that one particular kind of radio noise is strongest when the constellation of Sagittarius is high in the sky. The center of our Galaxy is located in the direction of Sagittarius, so Jansky concluded that he was detecting radio waves from an astronomical source.

At first only Grote Reber, an electronics engineer living in Illinois, took up Jansky's research. In 1936 Reber built the first **radio telescope** in his backyard to map radio emission from the Milky Way. Reber modeled his design after an ordinary reflecting telescope, with a parabolic "dish" (reflecting antenna) measuring 9.1 m in diameter. The radio receiver at the focal point of the metal dish was tuned to a wavelength of 1.85 m.

By 1944, when Reber completed his map of the Milky Way, astronomers had begun to take notice. Shortly after World War II, radio telescopes began to spring up around the world. Radio observatories are as common today as major optical observatories.

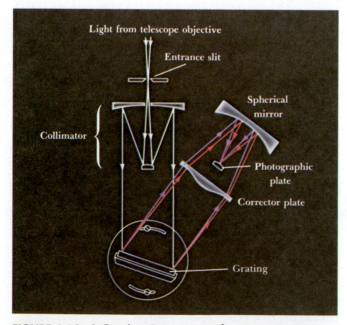

FIGURE 6-19 A Grating Spectrograph This optical device uses a reflection grating to break up the light from a source into a spectrum. The collimator ensures that light rays striking the grating are parallel. A corrector lens and mirror then focus the spectrum onto a photographic plate or a CCD.

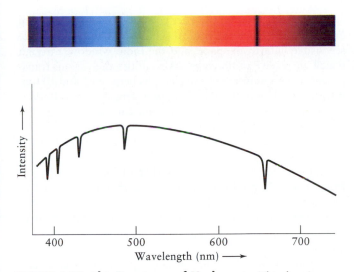

FIGURE 6-20 The Spectrum of Hydrogen This drawing compares two ways of displaying a spectrum. When a photographic plate is placed at the focus of a spectrograph, a rainbow-colored spectrum is obtained. When a CCD is placed at the focus of a spectrograph, a graph of intensity versus wavelength is produced by computer-controlled equipment attached to the CCD. Note that absorption lines appear as dips in the intensity-versus-wavelength curve.

Like Reber's prototype, the standard radio telescope has a large parabolic dish. A small antenna tuned to the desired frequency is located at the focus, and the incoming signal is relayed to amplifiers and recording instruments, typically located in a room at the base of the telescope's pier. Figure 6-21 shows a modern radio telescope.

One reason that astronomers were so slow to pursue radio emission from space was because of the poor resolution of early radio telescopes. Recall that angular resolution is the smallest angular separation between two stars that can just barely be distinguished as separate objects. That angle is proportional to the wavelength being observed. Thus angular resolution becomes worse with increasing wavelength; the longer the wavelength, the fuzzier the picture. Because radio radiation has very long wavelengths, astronomers thought that radio telescopes could produce only blurry, indistinct views.

A very large radio telescope can produce a somewhat sharper radio image, because angular resolution is inversely proportional to the diameter of the telescope. In other words, the bigger the dish, the better the resolution. For this reason, most modern radio telescopes have dishes more than 100 ft in diameter. Nevertheless, even the largest radio dish in existence cannot come close to the resolution of the best optical instruments.

A very clever technique was devised to circumvent this problem and produce high-resolution radio images. Unlike ordinary light, radio signals can be carried over electrical wires; consequently, two radio telescopes separated by many kilometers can be hooked together. This technique is called **interferometry,** because the incoming radio signals are made

to "interfere," or blend together, making the combined signal sharp and clear. The result is most impressive: The effective resolving power of two such radio telescopes is equivalent to that of one gigantic dish with a diameter equal to the distance between the two telescopes.

Interferometry was exploited for the first time in the late 1940s, when astronomers began receiving their first detailed views of radio objects in the sky. In the 1950s, telescopes separated by thousands of kilometers were linked together using radio signals to give resolving power much higher than that of optical telescopes. This technique is called **very-long-baseline interferometry** (VLBI). The best possible resolution under such a system would be obtained by two telescopes on opposite sides of the Earth. In that case, features as small as 0.0001 arc sec could be distinguished at radio wavelengths—a resolution 10,000 times better than the sharpest pictures from ordinary optical telescopes.

One of the finest arrangements of radio telescopes began operating in 1980 in the desert near Socorro, New Mexico. This system, called the Very Large Array (VLA), consists of 27 parabolic dishes, each 25 m (82 ft) in diameter. The 27 telescopes are arranged along the arms of a gigantic Y that covers an area 27 km (17 mi) in diameter. Only a portion of the VLA is shown in Figure 6-22. This system produces radio views of the sky with a resolution comparable to that of the very best optical telescopes.

Radio astronomers often use "false color" to display their radio views of astronomical objects. An example is shown in Figure 6-23. The most intense radio emission is shown in red, the least intense in blue. Intermediate colors of the rainbow represent intermediate levels of radio intensity. Black indicates no detectable radio emission. Astronomers working at

FIGURE 6-21 A Radio Telescope The dish of this radio telescope is 43 m (140 ft) in diameter. It is one of several large instruments at the National Radio Astronomy Observatory near Green Bank, West Virginia. (NRAO)

FIGURE 6-22 The Very Large Array (VLA) The 27 radio telescopes of the VLA system are arranged along the arms of a Y in central New Mexico. The north arm of the array is 19 km long; the southwest and southeast arms are each 21 km long. (NRAO)

other nonvisible wavelength ranges also frequently use false-color techniques to display images obtained from their instruments.

6-6 Telescopes in orbit around the Earth detect radiation that does not penetrate the atmosphere

The success of radio astronomy showed the value of observations at nonvisible wavelengths. But the Earth's atmosphere is opaque to many wavelengths. Very little radiation, other than visible light and radio waves, manages to penetrate the air we breathe.

The transparency of the Earth's atmosphere is graphed in Figure 6-24. Notice the **optical window,** through which we see visible light from space, and the **radio window,** through which we receive radio waves. Also notice the various transparent regions at infrared wavelengths between 1 and 10 μm. Infrared radiation within these wavelength intervals does penetrate the Earth's atmosphere and can be detected with ground-based equipment. This wavelength range is called the near-infrared, because it lies just beyond the red end of the visible spectrum.

Water vapor is the main absorber of infrared radiation from space. Infrared observatories are therefore located at sites with low humidity. For example, the 14,000-ft summit of Mauna Kea on Hawaii is exceptionally dry. Making infrared observations is the primary function of several telescopes there.

The best way of avoiding water vapor is to place a telescope in Earth orbit. In 1983 NASA launched the Infrared Astronomical Satellite (IRAS) into a 900-km-high polar orbit (Figure 6-25). During its ten-month mission, this 60-cm reflector revealed the richness of the infrared sky over wavelengths from about 10 to 100 μm. For the first time, astronomers saw dust bands in our solar system, dust disks around nearby stars, and distant galaxies that emit most of their radiation at infrared wavelengths.

In 1994 the European Space Agency (ESA) will launch an infrared telescope that will spend several years making observations at infrared wavelengths that never penetrate the Earth's atmosphere. Budget cuts have frustrated NASA plans to develop and launch SIRTF (the Space Infrared Telescope Facility), which would survey the infrared sky with unprecedented resolution.

The Earth's atmosphere is also transparent to the longest-wavelength ultraviolet light. This wavelength range, extend-

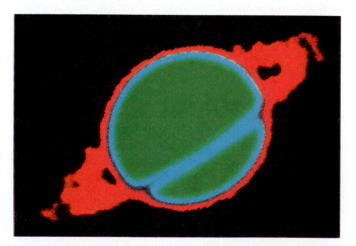

FIGURE 6-23 Optical and Radio Views of Saturn (a) This picture was taken by a spacecraft 18 million kilometers from Saturn. Sunlight reflecting from the planet's cloudtops and rings is responsible for this view. (b) This picture, taken by the VLA, shows radio emission from Saturn at a wavelength of 2 cm. Color represents the intensity of radio emission, ranging from blue for the weakest to red for the strongest. (NASA; NRAO)

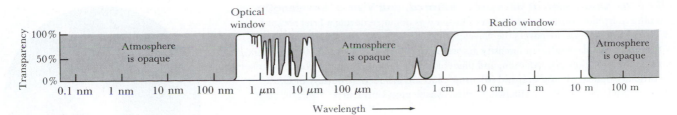

FIGURE 6-24 The Transparency of the Earth's Atmosphere This graph shows the percentage of radiation that can penetrate the Earth's atmosphere as a function of wavelength. Oxygen and nitrogen completely absorb all radiation with wavelengths shorter than 290 nm. Water vapor and carbon dioxide effectively block out all radiation from about 10 μm to 1 cm.

ing from about 400 nm down to 300 nm, is called the near-ultraviolet, because it lies just beyond the violet end of the visible spectrum. Astronomers can make ground-based observations in this wavelength range—if they do not use glass lenses in their telescopes. Because glass is opaque to the near-ultraviolet, all lenses must be made of quartz or some other nonabsorbing substance.

To see the far-ultraviolet, astronomers must make their observations from space. During the early 1970s, Apollo and Skylab astronauts carried small ultraviolet telescopes above the Earth's atmosphere to give us our first views of the ultraviolet sky. Small rockets have also lifted ultraviolet cameras

FIGURE 6-25 The Infrared Astronomical Satellite (IRAS) This satellite contained a small reflecting telescope that gave astronomers their first in-depth look at the infrared sky. Launched in 1983, IRAS contributed to research topics running the gamut from asteroids to the large-scale structure of the universe. (NASA)

briefly above the Earth's atmosphere. A typical view is shown in Figure 6-26, along with infrared and visible views.

Some of the finest ultraviolet astronomy has been accomplished by the International Ultraviolet Explorer (IUE), launched in 1978. This satellite (Figure 6-27) is built around a Cassegrain telescope with a 45-cm (18-in.) mirror and a total focal length of 6.74 m (22 ft). Observations cover the range from 116 to 320 nm.

To observe the sky at the very shortest ultraviolet wavelengths, NASA launched the Extreme Ultraviolet Explorer (EUVE) in 1992. This two-ton satellite carries four telescopes that focus high-energy ultraviolet photons by glancing them off highly polished surfaces of tapered cylinders. Three of the telescopes work together to survey the entire sky over a wavelength range of 7 to 76 nm. The fourth telescope is used to make observations of selected faint sources. Small, hot stars called white dwarfs and exploding stars called cataclysmic variables are primary targets.

Although these satellites give excellent views of the heavens at selected wavelengths, astronomers dreamed for decades of having one large telescope that could be operated at any wavelength—from the infrared through the visible range and out into the far-ultraviolet. This is the mission of the Hubble Space Telescope (HST), which was carried aloft by the Space Shuttle in 1990 (Figure 6-28).

Soon after HST was placed in orbit, astronomers discovered that the telescope's 2.4-m primary mirror suffers from spherical aberration. The mirror should have been able to concentrate 70% of a star's light into an image 0.1 arc sec in diameter. Instead, only 20% of a star's light is focused into the desired 0.1-arc-sec spot; the remaining 80 percent is smeared out over an area 1 arc sec in diameter. A star image therefore consists of a central spot of modest brightness surrounded by a hazy glow. If pictures are taken using 100% of the incoming starlight, the resulting star images are about 1 arc sec in diameter, which is no better that the images achieved at major ground-based observatories.

One way astronomers can cope with this problem is to use only the 20% of incoming starlight that is properly focused and, with computer processing, discard the remaining poorly focused 80%. An example of this procedure is shown in Figure 6-29. The ground-based image (Figure 6-29c) shows

FIGURE 6-26 Orion Seen at Ultraviolet, Infrared, and Visible Wavelengths
(**a**) An ultraviolet view of the constellation of Orion was obtained during a brief rocket flight in 1975. The 100-s exposure covers the wavelength range 125–200 nm. (**b**) The false-color view from IRAS displays infrared intensity according to color: red for strong 100-µm radiation, green for strong 60-µm radiation, and blue for strong 12-µm radiation. For comparison, (**c**) an ordinary optical photograph and (**d**) a star chart are included. (Courtesy of G. R. Carruthers, NRL; NASA; R. C. Mitchell, Central Washington University)

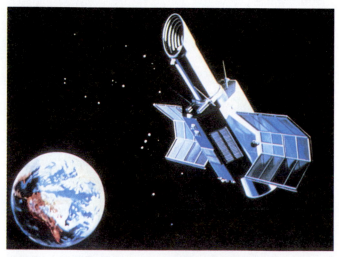

FIGURE 6-27 The International Ultraviolet Explorer (IUE) Since its launch in 1978, this 671-kg satellite has produced superb observations in the far-ultraviolet. The blue panels extending from either side of the midsection of the satellite are solar cell arrays that provide electrical power for the radio transmitters and other electronic equipment. (NASA)

FIGURE 6-28 The Hubble Space Telescope (HST) This photograph of the HST hovering above the Space Shuttle's bay was taken as the telescope was being deployed. During its anticipated 15-year lifetime, the HST will study the heavens at wavelengths from the infrared through the ultraviolet. During a return mission to the HST in 1994, astronauts will correct optical and mechanical problems with the telescope.

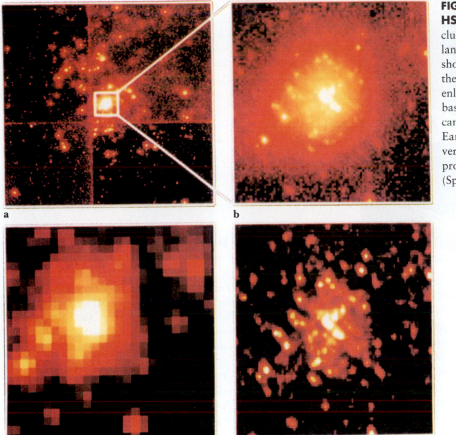

FIGURE 6-29 Computer Enhancement of HST Images These four images show a star cluster in a nearby galaxy called the Large Magellanic Cloud. (a) The raw picture from the HST shows the cluster and its surroundings. (b) Note the hazy glow that envelops the stars in this enlargement of the HST image. (c) This ground-based photograph is typical of the best image that can be obtained with a large telescope on the Earth's surface. (d) This computer-enhanced version of panel (b) demonstrates that computer processing can overcome HST's optical defects. (Space Telescope Science Institute, NASA)

details as small as 0.6 arc sec in size, whereas the computer-enhanced HST image (Figure 6-29d) achieves a resolution of about 0.1 arc sec.

Although computer enhancement produces sharp, crisp images, it is practical only for brighter objects, on which astronomers can afford to waste light. Unfortunately, many of the observing projects scheduled for the HST involve extremely dim galaxies and nebulae. These observations will be postponed until 1994, after a Shuttle mission installs new cameras equipped to compensate for the spherical aberration in the primary mirror. Meanwhile, many observing projects—particularly those that can succeed with only 20% of the incoming starlight or, like spectroscopy, do not require a pinpoint focus—will go forward as planned.

Because neither X rays nor gamma rays penetrate the Earth's atmosphere, observations at these extremely short wavelengths must also be made from space. Since these high-energy photons would simply bury themselves in the mirror of an ordinary reflecting telescope, astronomers use electronic detectors similar to Geiger counters. Astronomers got their first quick look at the X-ray sky from brief rocket flights during the late 1940s. By the early 1970s small satellites had revealed hundreds of previously unknown X-ray sources, including several good black hole candidates.

Although heroic in their day, these preliminary efforts pale in comparison to the detailed views from three huge satellites launched between 1977 and 1979, called High Energy Astrophysical Observatories (HEAO). Their X-ray and gamma-ray detectors discovered thousands of sources all across the sky. The second satellite in this series, launched near the hundredth anniversary of Albert Einstein's birth, was especially successful. Views from the Einstein Observatory appear throughout this book to illustrate the extraordinary hot objects that produce these high-energy photons.

The successor to the Einstein Observatory is ROSAT, launched into a nearly circular Earth orbit in 1990. Its instruments include an X-ray telescope and a wide-field X-ray camera. Scientists are hopeful that ROSAT, which grew from a German venture into an international project, will continue to provide a wealth of information about the X-ray sky during the next few years.

ROSAT is an important evolutionary step on the way to building AXAF, the Advanced X-Ray Astrophysics Facility, which is in the planning stages at NASA. AXAF will consist of two spacecraft, one for imaging and one for spectroscopy, that will be controlled by astronomers on the ground. If this project does not fall prey to budget cuts, AXAF could be launched in 1998.

This photograph of the Compton Observatory hovering above the Space Shuttle's cargo bay was taken as the spacecraft was being deployed in 1991. The best views of the high-energy gamma-ray sky come from this satellite, which carries four gamma-ray detectors. (NASA)

FIGURE 6-31 The Entire Sky at Five Wavelength ▶ Ranges   These five views show the entire sky at radio, infrared, visible, ultraviolet, and X-ray wavelengths. The Milky Way stretches horizontally across each picture. (a) The radio view shows the sky at a wavelength of 73 cm (411 MHz), with the brightest regions in red and the dimmest in blue. (b) The infrared view from IRAS shows several infrared wavelengths by color: 100 μm is yellow, 60 μm is red, and 12 μm is blue. (c) In the visible view Orion is at the right, Sagittarius in the middle, and Cygnus toward the left. (d) The ultraviolet view covers 135 to 255 nm and shows numerous stars around Orion and Cygnus. (e) The X-ray view from HEAO-1 covers 0.2 to 6 nm, corresponding to a range of photon energies from 6 to 0.2 keV. (Max Planck Institut für Radioastronomie; Jet Propulsion Laboratory; Griffith Observatory; Royal Observatory, Edinburgh; E. Boldt, NASA)

In 1991, the Compton Gamma Ray Observatory was carried aloft by the Space Shuttle (Figure 6-30). Named in honor of Arthur Holly Compton, an American scientist who made important discoveries about gamma rays, this satellite carries four instruments that are performing a variety of observations, giving us tantalizing views of the gamma-ray sky.

The advantages and benefits of Earth-orbiting observatories cannot be overemphasized. We are no longer limited to the narrow ranges of whatever wavelengths manage to leak through our shimmering, hazy atmosphere (Figure 6-31). For the first time we are really seeing the universe.

KEY WORDS

Terms preceded by an asterisk are discussed in the boxes.

* adaptive optics

angular resolution

Cassegrain focus

charge-coupled device (CCD)

chromatic aberration

coma

coudé focus

diffraction grating

eyepiece lens

focal length

focus (of a lens or mirror)

grating

interferometry

light-gathering power

magnification

new technology telescope (NTT)

Newtonian reflector

objective lens

optical window (in Earth's atmosphere)

pixel

prime focus

radio telescope

radio window (in Earth's atmosphere)

reflecting telescope

reflection

reflector

refracting telescope

refraction

refractor

Schmidt telescope

seeing disk

spectrograph

spherical aberration

very-long-baseline interferometry (VLBI)

KEY IDEAS

• Refracting telescopes, or refractors, produce images by bending light rays as they pass through glass lenses.

Chromatic aberration is an optical defect whereby light of different wavelengths is bent in different amounts by a lens.

Glass impurities, chromatic aberration, opacity to certain wavelengths, and structural difficulties make it inadvisable to build extremely large refractors.

• Reflecting telescopes, or reflectors, produce images by reflecting light rays to a focus point from curved mirrors.

Reflectors are not subject to most of the problems that limit the useful size of refractors, although they are

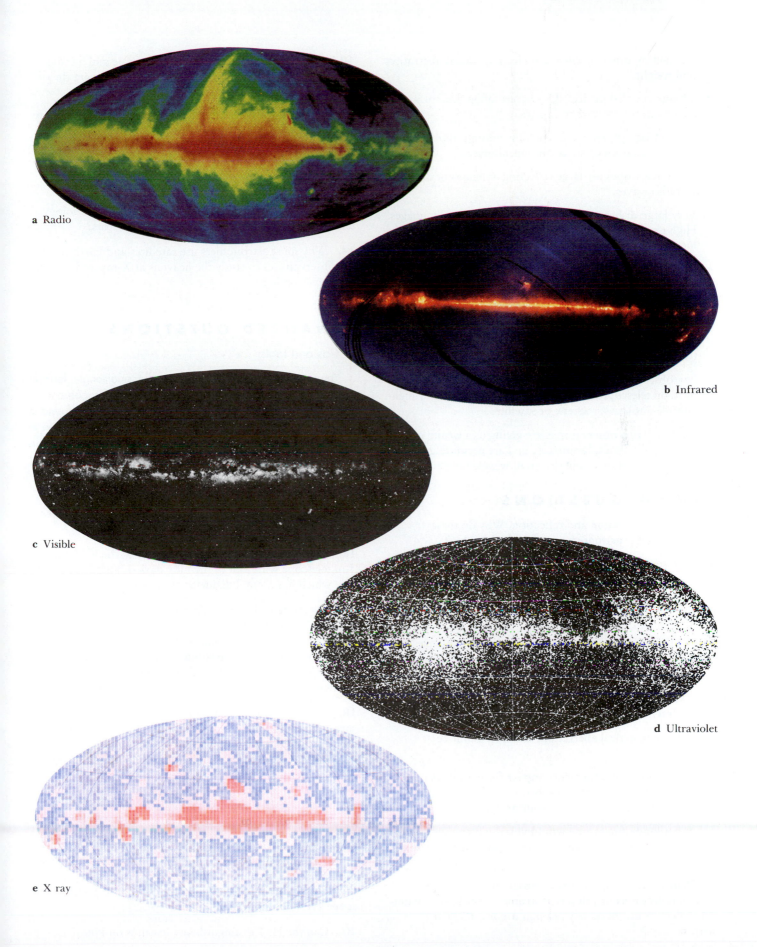

a Radio

b Infrared

c Visible

d Ultraviolet

e X ray

plagued by other problems, such as spherical aberration and weight.

- Charge-coupled devices (CCDs) are often used at a telescope's focus to record faint images.

- A spectrograph uses a diffraction grating and lenses to record the spectrum of an astronomical object.

- Radio telescopes use large reflecting antennas or dishes to focus radio waves.

    Very large dishes provide reasonably sharp radio images. High resolution is also achieved with interferometry techniques that link smaller dishes together.

- The Earth's atmosphere absorbs much of the radiation that arrives from space.

    The atmosphere is transparent chiefly in two wavelength ranges known as the optical window and the radio window. A few wavelengths in the near-infrared also reach the ground.

    For observations at other wavelengths, astronomers depend on telescopes carried above the atmosphere by high-altitude airplanes, rockets, or satellites.

    Satellite-based observatories are giving us a wealth of new information about the universe and are permitting coordinated observation of the sky at all wavelengths.

REVIEW QUESTIONS

1. Describe refraction and reflection. Why do these processes enable astronomers to build telescopes?

2. With the aid of a diagram, describe a refracting telescope.

3. What is chromatic aberration, and how can it be corrected?

4. With the aid of a diagram, describe a reflecting telescope. Describe four different ways in which an astronomer can access the focal point.

5. Explain some of the advantages of reflecting telescopes over refracting telescopes.

6. What is spherical aberration, and how can it be corrected?

7. Quite often advertisements appear for telescopes that extol their magnifying abilities. Is this a good criterion for evaluating telescopes? Explain your answer.

8. What kind of telescope would you use if you wanted to take a color photograph entirely free of chromatic aberration? Why?

9. Why do you think no major observatory has a Newtonian reflector as its primary instrument, whereas Newtonian reflectors are extremely popular among amateur astronomers?

10. Compare an optical reflecting telescope and a radio telescope. What do they have in common? How are they different?

11. Why can radio astronomers make observations at any time during the day, whereas optical astronomers are mostly limited to observing at night?

12. What are the optical window and the radio window? Why isn't there an X-ray window or an ultraviolet window?

13. What are some of the advantages of new technology telescopes over ordinary ground-based reflectors?

14. Why must astronomers use satellites and Earth-orbiting observatories to study the heavens at X-ray and gamma-ray wavelengths?

ADVANCED QUESTIONS

Tips and tools . . .

You may find it useful to review the small-angle formula discussed in Box 1-1. The area of a circle is proportional to the square of its diameter. Data on the planets are found in the appendixes at the end of this book. Examples of how to calculate magnifying power were given in the body of this chapter.

15. The observing cage in which an astronomer sits at the prime focus of the 5-m telescope on Palomar Mountain is about 1 m in diameter. What fraction of the incoming starlight is blocked by the cage?

16. Compare the light-gathering power of the Palomar 5-m telescope with that of the fully dark-adapted human eye, which has a pupil diameter of about 5 mm.

17. Several telescope manufacturers build telescopes having a design they call a "Schmidt-Cassegrain." Consult advertisements in such magazines as *Sky & Telescope* and *Astronomy* to see the appearance and cost of these telescopes. Why do you suppose they are very popular among amateur astronomers?

18. The four largest moons of Jupiter are roughly the same size as our Moon and are about 628 million kilometers from Earth at opposition. What is the size of the smallest surface features that the Hubble Space Telescope (resolution of 0.1 arc sec) can detect? How does this compare with the smallest features that can be seen on the Moon with the unaided human eye (resolution of 1 arc min)?

19. Suppose your Newtonian reflector has a mirror 20 cm (8 in.) in diameter and a focal length of 2 m. What magnification do you get with eyepieces whose focal lengths are (a) 9 mm, (b) 20 mm, and (c) 55 mm? What is the telescope's resolving power?

20. Can the HST distinguish any features on Pluto?

21. Show by means of a diagram why the image formed by a simple refracting telescope is upside down.

22. As this book went to press, several teams of astronomers around the world were designing or building new technology telescopes. Consult such magazines as *Sky & Telescope* and *Science News* to find out how some of the projects mentioned in the text are going. Are any of these telescopes in operation?

23. What are the latest plans for repairing and refurbishing the Hubble Space Telescope? Consult such magazines as *Sky & Telescope* and *Astronomy* to find out about the Space Shuttle mission to HST. When is the launch date?

DISCUSSION QUESTIONS

24. Discuss the advantages and disadvantages of using a small telescope in Earth orbit versus a large telescope on a mountaintop.

25. If you were in charge of selecting a site for a new observatory, what factors would you consider important?

OBSERVING PROJECTS

26. Obtain a telescope during the daytime along with several eyepieces of various focal lengths. If you can determine the telescope's focal length, calculate the magnifying powers of the eyepieces. Focus the telescope on some familiar object, such as a distant lamppost or tree. **DO NOT FOCUS ON THE SUN! Looking directly at the Sun can cause blindness.** Describe the image you see through the telescope. Is it upside down? How does the image move as you slowly and gently shift the telescope left and right, up and down? Examine the eyepieces, noting their focal lengths. By changing the eyepieces, examine the distant object under different magnifications. How does the field of view and the quality of the image change as you go from low power to high power?

27. On a clear night, view the Moon, a planet, and a star through a telescope using eyepieces of various focal lengths and known magnifying powers. (You may have to consult such magazines as *Sky & Telescope* or *Astronomy* to determine the phase of the Moon and the locations of the planets.) In what way does the image seem to degrade as you view with increasingly higher magnification? Do you see any chromatic aberration? If so, with which object and which eyepiece is it most noticeable?

28. Many towns and cities have amateur astronomy clubs. If you are so inclined, attend a "star party" hosted by your local club. People who bring their telescopes to such gatherings are delighted to show you their instruments and take you on a telescopic tour of the heavens. Such an experience can lead to a very enjoyable, lifelong hobby.

FOR FURTHER READING

Burns, J. O., Taylor, G. J., and Johnson, S. W. "Observatories on the Moon." *Scientific American,* March 1990. This futuristic article explains how lunar observatories would yield extraordinarily detailed views of the heavens and open new windows through which to study the universe.

Chaisson, E. J. "Early Results from the Hubble Space Telescope." *Scientific American,* June 1992. This article describes some of the eye-opening images produced by HST, in spite of its optical and mechanical problems.

Cole, S. "Astronomy of the Edge: Using the Hubble Space Telescope." *Sky & Telescope,* October 1992. Loudly proclaiming that "fainthearted astronomers need not apply," this article describes the trials and tribulations of observing with the Hubble Space Telescope.

Dyer, A. "ROSAT's Penetrating X-Ray Vision." *Astronomy,* June 1991. Written shortly after ROSAT had completed a survey of the entire X-ray sky, this article includes some remarkable pictures from this orbiting X-ray observatory.

Janesick, J., and Blouke, M. "Sky on a Chip: The Fabulous CCD." *Sky & Telescope,* September 1987. This article describes how CCDs work and why they have become an indispensable asset at well-equipped observatories.

Keel, W. C. "Galaxies Through a Red Giant." *Sky & Telescope,* June 1992. This article describes an astronomer's visit to the 6-m reflecting telescope in southern Russia.

Kellermann, K. I. "Radio Astronomy: The Next Decade." *Sky & Telescope,* September 1991. This article by a noted radio astronomer discusses hopes and plans for the future, with special emphasis on VLBI.

Kniffen, D. A. "The Gamma Ray Observatory." *Sky & Telescope,* May 1991. This article describes the Compton Gamma Ray Observatory and some features of the gamma-ray sky that it was designed to observe.

Maran, S. P. "Astro: Science in the Fast Lane." *Sky & Telescope,* June 1991. This exciting articles describes the flight of *Astro-1*—a battery of X-ray and ultraviolet telescopes carried aloft by the Space Shuttle in late 1990.

Powell, C. S. "Mirroring the Universe." *Scientific American,* November 1991. This article describes a new generation of telescopes that promise a tremendous leap in astronomers' ability to explore the universe.

Robinson, L. "Spinning a Giant Success." *Sky & Telescope,* July 1992. This article describes a new technique for making large, high-quality telescope mirrors by spinning a vat of molten glass.

Ressmeyer, R. H. "Keck's Giant Eye." *Sky & Telescope,* December 1992. Beautiful photographs of the Keck Observatory illustrate this article on the world's largest optical telescope.

OUR SOLAR SYSTEM

NEBULAE IN SAGITTARIUS

Planets are probably forming along with new stars in these clouds of gas and dust in Sagittarius. The type of planet formed depends on the temperature and available substances (rock fragments, ice crystals, and gases). In our solar system, planets composed primarily of rock formed near the Sun, whose heat drove off ices and gases. Far from the Sun, where temperatures were low, planets formed mainly from gases and ices. (Royal Observatory, Edinburgh)

PERHAPS THE most obvious fact about our solar system is that the planets fall into two distinct classes. Small, rocky planets like the Earth are found near the Sun, while huge, gaseous planets, like Jupiter, are located far from the Sun. These two classes are a direct result of conditions under which the planets formed. Seven large moons as well as numerous smaller satellites, asteroids, and comets also populate the solar system. Spectroscopy provides crucial information about the chemical composition of the planets and clues about the material from which they formed. The solar system is made of debris left over from generations of stars that lived out their lives and shed their matter into space long before the Sun was born. Our solar system was probably created from a huge cloud of interstellar gas and dust that contracted under the force of its own gravity. Most of the matter fell toward the center of that contracting cloud to form the Sun; outlying material accumulated in clumps to form the planets. Although many details of this process remain to be worked out, we seem to have a basic understanding of the creation of the solar system.

For as long as people have looked up at the heavens, they have wondered about the nature and origin of the Sun, Moon, and planets. Within the past few decades, telescopes and space probes have placed the answers to many age-old questions within our grasp. We now have a wealth of information on the Sun, planets, moons, asteroids, comets, and meteoroids that make up our niche in the universe. Many of these objects are exceedingly ancient, old enough to contain records of the cosmic events that created our solar system.

In addition to studying nearby worlds, we can observe stars and perhaps planetary systems actually forming in nebulae scattered across our Galaxy. The general process of star creation, coupled with our observations of the Sun and its satellites, gives a comprehensive picture of the birth of the solar system. For the first time, we can truly appreciate what is unique and what is commonplace about our world. We have begun to fathom our connection with the rest of the cosmos and our place in the universe.

7-1 The planets are classified as either terrestrial or Jovian by their physical attributes

A brief overview of the solar system distinguishes two classes of planets. As Figure 7-1 shows, the orbits of the four inner planets (Mercury, Venus, Earth, and Mars) are crowded in close to the Sun. In contrast, the orbits of the next four planets (Jupiter, Saturn, Uranus, and Neptune) are widely spaced at great distances from the Sun.

Most of the planets' orbits are nearly circular. As discussed in Chapter 4, Kepler discovered that these orbits are actually ellipses. Astronomers denote the elongation of an ellipse by its **eccentricity** (review Box 4-3 for details). The eccentricity of a circle is zero. Most planets have orbital eccentricities that are very close to zero. The exceptions are Mercury and Pluto. In fact, Pluto's noncircular orbit sometimes takes it nearer the Sun than its neighbor, Neptune.

The planetary orbits all lie in nearly the same plane. In other words, the orbits of the planets are inclined at only slight angles to the plane of the ecliptic. Again, however, Pluto is an exception. The plane of Pluto's orbit is tilted at about $17°$ to the plane of the Earth's orbit. Table 7-1 lists orbital characteristics of the nine planets.

When we compare the physical properties of the planets, we again find that they fall naturally into two classes—the four inner planets and the four outer ones—with Pluto as an exception. The four inner planets are small and composed primarily of rock. They are called **terrestrial planets** because they resemble the Earth (in Latin, *terra*). Mountains, craters, canyons, and volcanoes are common on their hard, rocky surfaces. The outer four planets are called **Jovian planets** because they resemble Jupiter (Jove was another name for the Roman god Jupiter). Vast, swirling cloud formations dominate the appearance of these enormous gaseous spheres. The montage of photographs in Figure 7-2 shows the distinctive appearances of these two types of planets.

The most obvious difference between terrestrial and Jovian planets is their size. The diameter of a planet can be computed from its apparent angular diameter and distance from Earth. For example, at its greatest western elongation in 1990, Venus was 1.0158×10^8 km from Earth and had an angular diameter of 24.58 arc sec. Using the small-angle formula from Box 1-1, we can calculate the diameter of Venus to be 12,100 km (7520 mi). Similar calculations demonstrate that the Earth, with its diameter of about 12,756 km

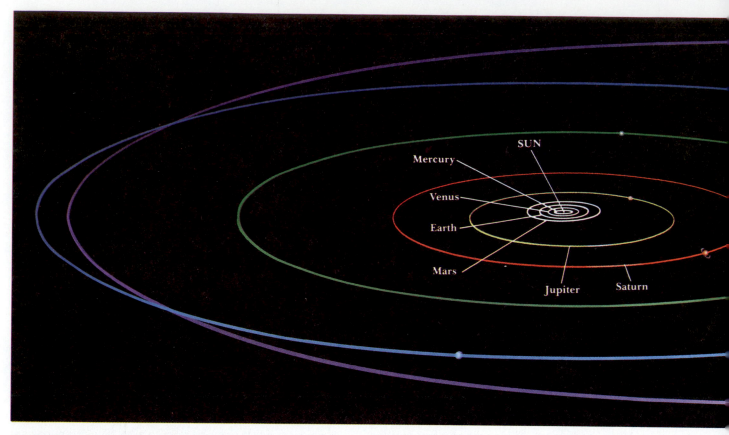

FIGURE 7-1 The Solar System This scale drawing shows the distribution of planetary orbits around the Sun. The four inner planets are crowded in close to the Sun, while the five outer planets orbit the Sun at much greater distances.

(7926 mi), is the largest of the four inner planets. In sharp contrast, the four outer planets are much larger. First place goes to Jupiter, whose equatorial diameter is about 142,984 km (88,846 mi). Pluto, however, is even smaller than the inner planets, despite being the outermost planet. Its diameter is only about 2300 km (1400 mi). Figure 7-3 shows the Sun and the planets drawn to the same scale. The diameters of the planets are given in Table 7-2.

Mass is another distinguishing characteristic between the terrestrial and Jovian planets. The four inner planets have low masses, but the next four planets have substantially greater masses. The mass of a planet is most easily deter-

---
TABLE 7-1
Orbital Characteristics of the Planets

| | Mean distance from Sun | | Orbital period (yr) | Eccentricity | Inclination to the ecliptic (°) |
	(AU)	(10⁶ km)			
Mercury	0.39	58	0.24	0.206	7.0
Venus	0.72	108	0.62	0.007	3.4
Earth	1.00	150	1.00	0.017	0.0
Mars	1.52	228	1.88	0.093	1.8
Jupiter	5.20	778	11.86	0.048	1.3
Saturn	9.54	1427	29.46	0.056	2.5
Uranus	19.19	2871	84.01	0.046	0.8
Neptune	30.06	4497	164.8	0.010	1.8
Pluto	39.53	5914	248.5	0.248	17.1

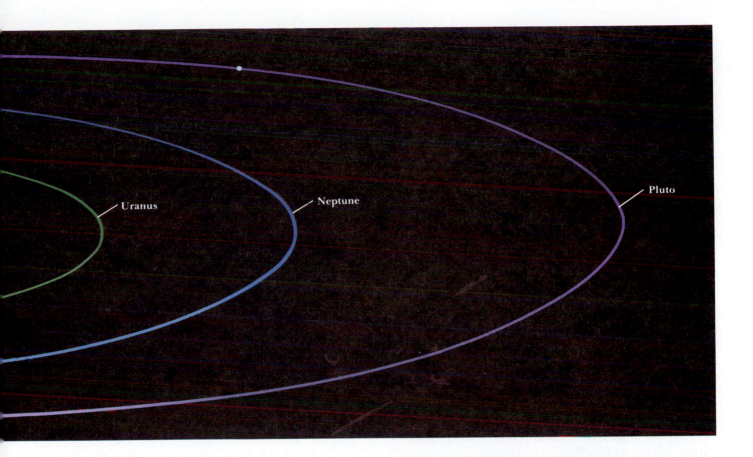

Uranus   Neptune   Pluto

mined if the planet has a satellite. Kepler's third law (see Box 4-3) relates the planet's mass to a satellite's period and semi-major axis. If a planet does not have a satellite, astronomers can track a spacecraft as it passes near the planet. The planet's gravity (which is directly related to the planet's mass) will alter the path of the body. By measuring the size of this deflection and using Newtonian mechanics, astronomers can determine the planet's mass. Again, first place goes to Jupiter, whose mass is 318 times greater than Earth's.

Average density (mass divided by volume) is a physical property that often provides important clues to the composition of an object. Scientists measure average density in kilograms per cubic meter. The four inner planets have very high average densities (see Table 7-2); the average density of the Earth, for example, is 5520 kg/m³. For comparison, the average density of a typical rock is about 3000 kg/m³, and the average density of water is 1000 kg/m³. The Earth must therefore contain a large amount of material that is denser than rock. This information provides our first clue that Earthlike planets have iron cores.

FIGURE 7-2 The Planets This montage of photographs taken ▶ by various spacecraft shows eight planets. The images are *not* reproduced to the same scale. At the top (from left to right) are Mercury, Venus, Earth, and Mars along with our Moon. At the bottom (from right to left) are Jupiter, Saturn, Uranus, and Neptune. (NASA)

FIGURE 7-3 The Sun and the Planets This drawing shows the nine planets in front of the disk of the Sun, with all ten bodies drawn to the same scale. The four planets that have orbits nearest the Sun (Mercury, Venus, Earth, and Mars) are small and are made of rock. The next four planets from the Sun (Jupiter, Saturn, Uranus, and Neptune) are large and are composed primarily of gas.

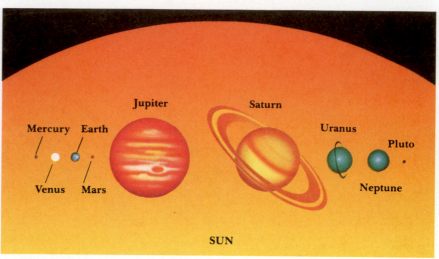

In sharp contrast, the outer planets have quite low densities. Saturn has an average density less than that of water. This information strongly suggests that the giant outer planets are composed primarily of such light elements as hydrogen and helium. All four Jovian planets probably have Earth-sized rocky cores buried beneath low-density atmospheres tens of thousands of kilometers thick.

Pluto is again an oddity. Although it is even smaller than the dense inner planets, its average density is closer to that of the giant outer planets. Pluto is probably composed of a mixture of rock and ice because its average density of 2030 kg/m^3 is between that for ice and rock.

7-2 Seven large satellites are as big as terrestrial planets

All the planets except Mercury and Venus have satellites. More than 50 satellites are known (Jupiter, Saturn, and Uranus each have at least 15), and dozens of others remain to be discovered. The known satellites fall into two distinct categories. Seven giant satellites are each roughly as big as Mercury (Table 7-3). All the other satellites are much smaller, with diameters less than 2000 km.

Interplanetary spacecraft have revealed many fascinating facts about these giant satellites. We now know that Jupiter's satellite Io is one of the most geologically active worlds in the solar system, with numerous volcanoes belching forth sulfur-rich compounds. Saturn's largest satellite, Titan, has an atmosphere nearly twice as dense as Earth's atmosphere. Figure 7-4 shows the seven giant satellites along with Mercury, all to the same scale. Like the terrestrial planets, all seven giant satellites have hard surfaces of either rock or ice.

7-3 Spectroscopy reveals the chemical composition of the Sun, planets, and stars

The most accurate determinations of the composition of the planets have come from spacecraft that have actually made

---

TABLE 7-2

Physical Characteristics of the Planets

	Equatorial diameter		Mass		Average density
	(km)	(Earth = 1)	(kg)	(Earth = 1)	(kg/m³)
Mercury	4,878	0.38	3.30×10^{23}	0.06	5430
Venus	12,100	0.95	4.87×10^{24}	0.81	5250
Earth	12,756	1.00	5.98×10^{24}	1.00	5520
Mars	6,786	0.53	6.42×10^{23}	0.11	3950
Jupiter	142,984	11.21	1.90×10^{27}	317.94	1330
Saturn	120,536	9.45	5.69×10^{26}	95.18	690
Uranus	51,118	4.01	8.68×10^{25}	14.53	1290
Neptune	49,528	3.88	1.02×10^{26}	17.14	1640
Pluto	2,300	0.18	1.29×10^{22}	0.002	2030

FIGURE 7-4 The Smaller Terrestrial Worlds Mercury, our Moon, and the largest satellites of Jupiter, Saturn, and Neptune are shown here to the same scale. Each of these worlds has its own unique characteristics, and all are comparable in size to the terrestrial planets. Only Titan possesses a substantial atmosphere. (NASA)

TABLE 7-3
The Seven Giant Satellites

Satellite	Parent planet	Diameter (km)	Average density (kg/m³)
Moon	Earth	3476	3340
Io	Jupiter	3630	3570
Europa	Jupiter	3138	2970
Ganymede	Jupiter	5262	1940
Callisto	Jupiter	4800	1860
Titan	Saturn	5150	1880
Triton	Neptune	2700	2070

direct chemical analyses of samples from a planet's atmosphere and soil. Unfortunately, we have such direct information for only four worlds: Venus, the Moon, Mars, and the Earth. In all other cases, astronomers must analyze sunlight reflected from the distant planets and their satellites. To do that, astronomers bring to bear one of their most powerful tools, spectroscopy.

In Chapter 5, we saw that white light can be separated into a spectrum of colors (review Figure 5-3). We also learned that each chemical element produces its own unique pattern of spectral lines (see Figure 5-12). **Spectroscopy** is the systematic study of spectra and spectral lines.

Spectral lines provide extremely reliable evidence about the chemical composition of distant objects. If the spectral lines of a certain chemical are found in a star's spectrum, it is safe to conclude that this chemical is present in that star's atmosphere. For example, a portion of the Sun's spectrum is shown in Figure 7-5. Also shown is the spectrum of iron over the same wavelength range. This pattern of spectral lines is iron's own distinctive "fingerprint," which no other chemical can imitate. Because some absorption lines in the Sun's spectrum coincide with the iron lines, some vaporized iron must exist in the Sun's atmosphere.

As astronomers have examined the spectra of stars and galaxies over the past century, they have always found the same chemical elements. No matter where we look—no matter how far we peer into space—the naturally occurring elements of which the Earth is made are the same building blocks for all matter everywhere in the universe. The elements are fundamental because they cannot be broken into more basic chemicals.

A listing of the chemical elements is most conveniently displayed in the form of a **periodic table** (Figure 7-6). Each element is assigned a unique **atomic number.** Elements are arranged in the periodic table in order of increasing atomic numbers, and with only a few exceptions, this sequence also corresponds to increasing average mass of the atoms of the elements. Thus hydrogen (symbol H), with atomic number 1, is the lightest element. As another example, iron (symbol Fe)

FIGURE 7-5 Iron in the Sun The upper spectrum is a portion of the Sun's spectrum from 420 to 430 nm. Numerous dark spectral lines are visible. The lower spectrum is a corresponding portion of the spectrum of vaporized iron. Several bright spectral lines are seen against a black background. The fact that the iron lines coincide with some of the solar lines proves that there is some iron (albeit a very tiny amount) in the Sun's atmosphere. (Carnegie Observatories)

1 H																	2 He
3 Li	4 Be											5 B	6 C	7 N	8 O	9 F	10 Ne
11 Na	12 Mg											13 Al	14 Si	15 P	16 S	17 Cl	18 A
19 K	20 Ca	21 Sc	22 Ti	23 V	24 Cr	25 Mn	26 Fe	27 Co	28 Ni	29 Cu	30 Zn	31 Ga	32 Ge	33 As	34 Se	35 Br	36 Kr
37 Rb	38 Sr	39 Y	40 Zr	41 Nb	42 Mo	43 Tc	44 Ru	45 Rh	46 Pd	47 Ag	48 Cd	49 In	50 Sn	51 Sb	52 Te	53 I	54 Xe
55 Cs	56 Ba	57 La	72 Hf	73 Ta	74 W	75 Re	76 Os	77 Ir	78 Pt	79 Au	80 Hg	81 Tl	82 Pb	83 Bi	84 Po	85 At	86 Rn
87 Fr	88 Ra	89 Ac	104	105	106												

58 Ce	59 Pr	60 Nd	61 Pm	62 Sm	63 Eu	64 Gd	65 Tb	66 Dy	67 Ho	68 Er	69 Tm	70 Yb	71 Lu
90 Th	91 Pa	92 U	93 Np	94 Pu	95 Am	96 Cm	97 Bk	98 Cf	99 Es	100 Fm	101 Md	102 No	103 Lr

FIGURE 7-6 The Periodic Table of the Elements The periodic table is a convenient listing of the elements, arranged according to their atomic weights and chemical properties.

has atomic number 26 and is a relatively heavy element. All the elements that appear in a single vertical column of the periodic table have similar chemical properties. For example, the elements listed in the far right column are all gases under Earth-surface conditions of temperature and pressure, and they are all very reluctant to react chemically with other elements.

In addition to nearly 100 naturally occurring elements, Figure 7-6 also includes several artificially produced elements. All these elements are heavier than uranium (symbol U) and are highly radioactive, which means that they decay into lighter elements within a short time of being created in laboratory experiments.

As discussed in Chapter 5, an atom is the smallest possible piece of an element. Typical atomic diameters are about 0.1 nm. Atoms can be broken down into three basic types of subatomic particles: *protons*, *neutrons*, and *electrons*. Box 7-1 describes atomic structure and explains why elements have different masses.

Although the number of elements is limited, a wide variety of substances exist in the universe, because atoms can combine to form various **molecules.** For example, two hydrogen atoms can combine with an atom of oxygen to form a molecule of water. Water's chemical formula, H_2O, describes the atoms in its molecule. In a similar way, two oxygen atoms can bond with a carbon atom to produce a molecule of carbon dioxide, whose formula is CO_2. The chemical formula for common table salt is NaCl, which means that a salt molecule consists of one sodium atom combined with one chlorine atom.

Molecules, like atoms, also produce unique patterns of lines in the spectra of astronomical objects. For example, Figure 7-7, a portion of the spectrum of Venus, shows a series of spectral lines caused by carbon dioxide. It is therefore reasonable to conclude that Venus's atmosphere contains this gas. Indeed, direct measurements by Soviet and American spacecraft have confirmed that carbon dioxide gas forms 96% of the Venusian atmosphere.

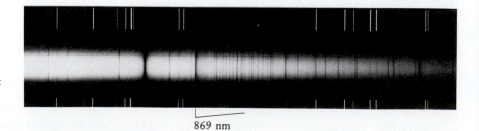

FIGURE 7-7 Carbon Dioxide on Venus This near-infrared spectrum of Venus shows a series of spectral lines beginning at 869 nm that is caused by carbon dioxide, the primary constituent of Venus's atmosphere. (Lick Observatory)

869 nm

BOX 7-1

Atoms and Isotopes

Laboratory experiments during the first part of the twentieth century revealed that the structure of an atom superficially resembles a tiny solar system. Most of the mass of an atom is concentrated in a dense **nucleus** that is composed of **protons** and **neutrons.** Orbiting this massive nucleus like small planets are **electrons.** Whereas the solar system is held together by gravitational forces, atoms are held together by electrical forces. Each proton carries a positive charge, and each electron carries an equal but opposite negative charge. The electric forces attracting the positively charged protons and the negatively charged electrons keep the atom from coming apart.

Although electrons and protons have equal but opposite charges, they have unequal masses. The electron is one of the lightest subatomic particles, with a mass of only 9.1×10^{-31} kg. The proton is nearly 2000 times heavier: Its mass is 1.7×10^{-27} kg.

A neutron has almost exactly the same mass as a proton. As its name suggests, a neutron has no electric charge—it is electrically neutral. Neutrons serve as buffers between the positively charged protons that are crowded together inside the nucleus. Typically, there are more neutrons than protons in a nucleus, especially in the case of the heaviest elements.

In the elemental state, the number of electrons orbiting an atom is equal to the number of protons in the nucleus, making the atom electrically neutral. The number of protons in an atom's nucleus equals the atomic number for that particular element. A hydrogen nucleus has 1 proton, a helium nucleus has 2 protons, and so forth—up to uranium, with 92 protons in its nucleus.

The number of protons in the nucleus of an atom determines what element that atom is. Nevertheless, the same element may have different numbers of neutrons in its nuclei. For example, consider oxygen, the eighth element

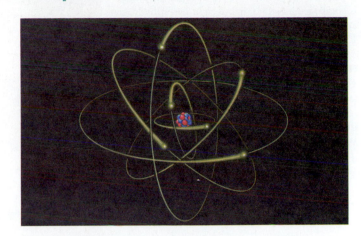

on the periodic table. Its atomic number is 8, and every oxygen nucleus has exactly 8 protons, but it can have 8, 9, or 10 neutrons. Thus, three slightly different kinds of oxygen, called **isotopes,** exist. The isotope with 8 neutrons is by far the most abundant variety. It is written as ^{16}O, or oxygen-16. The rarer isotopes with 9 and 10 neutrons are designated as ^{17}O and ^{18}O, respectively.

The superscript that precedes the chemical symbol for an element is equal to the total number of protons and neutrons in a nucleus of that particular isotope. For example, the common isotope of iron is ^{56}Fe, or iron-56, which means that its nucleus contains a total of 56 protons and neutrons. From the periodic table, however, we see that the atomic number of iron is 26. This means that every iron atom has 26 protons in its nucleus. Therefore, the number of neutrons in an iron-56 nucleus is $56 - 26 = 30$.

It is extremely difficult to distinguish chemically between the various isotopes of a particular element. Ordinary chemical reactions involve only the electrons that orbit the atom, never the neutrons buried in its nucleus.

7-4 Hydrogen and helium are abundant on the Jovian planets, whereas the terrestrial planets are composed mostly of heavy elements

Spectroscopic observations from Earth and spacecraft demonstrate that the Jovian planets are composed primarily of the lightest gases, hydrogen and helium. In contrast, chemical analysis of soil samples from Venus, Earth, and Mars demonstrate that the terrestrial planets are made mostly of heavy elements, such as iron, silicon, magnesium, sulfur, and nickel. Spacecraft images such as Figures 7-8 and 7-9 only hint at these striking differences in the chemical composition.

Temperature plays a major role in determining whether various substances exist as solids, liquids, or gases, thereby profoundly affecting the appearance of the planets. Hydro-

gen and helium are gaseous except at extremely low temperatures and extraordinarily high pressures. In contrast, the rock-forming compounds such as iron and silicon are solids except at temperatures exceeding 1000 K. Between these two extremes are substances such as water (H_2O), carbon dioxide (CO_2), methane (CH_4), and ammonia (NH_3). At low temperatures (typically below 200 to 300 K), these common chemicals solidify into the solids called ices. At somewhat higher temperatures, they can exist as liquids or gases.

Virtually no hydrogen or helium is found in the atmospheres of the terrestrial planets. The atmospheres of Venus, Earth, and Mars are instead composed of heavier gases, like carbon dioxide, oxygen, and nitrogen, all of which are made of atoms more massive than hydrogen or helium. To understand why the terrestrial planets do not have any hydrogen or helium in their atmospheres, we must look at the effects of their surface temperatures.

FIGURE 7-8 [*left*] **A Jovian Planet** This close-up view of Jupiter's turbulent cloudtops was taken by a spacecraft in 1979. Because the planet is composed mostly of lightweight gases (hydrogen and helium), its average density is only 1330 kg/m³. (NASA)

FIGURE 7-9 [*right*] **A terrestrial planet** This close-up view of Mars's rocky surface was taken by a spacecraft in 1976. Because the planet is composed mostly of heavy elements (e.g., iron, silicon, magnesium, sulfur), its average density is 3950 kg/m³. (NASA)

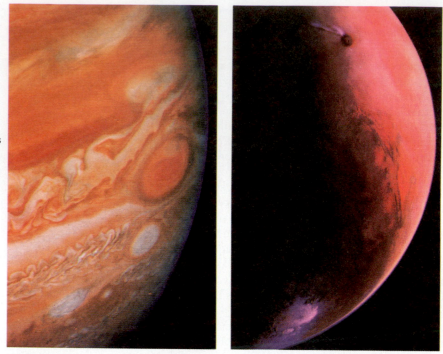

As you might expect, a planet's surface temperature is related to its distance from the Sun. (Review Box 5-1 with its discussion of temperature measurements.) The four inner planets are quite warm. For example, noontime temperatures on Mercury climb to 600 K (= 327°C = 621°F), and at noon in the middle of summer on Mars, it is sometimes as warm as 300 K (= 27°C = 81°F). The outer planets, which receive much less solar radiation, are cooler. Typical temperatures range from about 150 K (= −123°C = −189°F) in Jupiter's cloudtops to 63 K (= −210°C = −346°F) on Neptune.

Temperature is directly correlated to the speeds with which atoms or molecules of a gas move: The higher the temperature of a substance, the greater the speed of its atoms. Because of warmth from the Sun on the four inner planets, lightweight hydrogen and helium atoms move swiftly enough to escape from the relatively weak gravity of the terrestrial planets. The sparse atmospheres that do surround the terrestrial planets are primarily heavy gases such as carbon dioxide, nitrogen, oxygen, and water vapor. Far from the Sun, on the four Jovian planets, temperatures are low, and so the atoms of hydrogen and helium move slowly. The relatively strong gravity of the massive Jovian planets easily prevents even the lightest gases from escaping into space. (Box 7-2 discusses the thermal motion of atoms and the ability of a planet's gravity to retain gases.)

7-5 Small chunks of rock and ice also orbit the Sun

In addition to the nine planets, many smaller objects orbit the Sun. Between the orbits of Mars and Jupiter are thousands of rocks called **asteroids**. The largest asteroid, Ceres, has a diameter of about 900 km. The next largest, Pallas and Vesta, are each about 500 km in diameter. Still smaller ones are

increasingly numerous. There are thousands of kilometer-sized asteroids and millions that are boulder-sized or smaller. A close-up picture of an asteroid, taken by a spacecraft on its way to Jupiter, is shown in Figure 7-10. Since most asteroids orbit the Sun at distances of 2 to 3½ AU, this region of the solar system between the orbits of Mars and Jupiter is called the **asteroid belt**.

Quite far from the Sun, well beyond the orbit of Pluto, are chunks of ice called **comets**. Many comets have highly elongated orbits that occasionally bring them close to the Sun.

FIGURE 7-10 **An Asteroid** This image of the asteroid Gaspra was taken in 1991 by the *Galileo* spacecraft on its way toward Jupiter. The asteroid measures 12 × 20 × 11 km. Several thousand similar chunks of rock orbit the Sun between the orbits of Mars and Jupiter. (NASA)

even chunks of ice have survived for billions of years. Thus debris in the solar system naturally divides into two families (asteroids and comets), which can be arranged according to distance from the Sun just like the two categories of planets (terrestrial and Jovian). To further understand the chemical composition of the planets we must look to the stars, where the chemical elements are created.

7-6 The relative abundances of the elements are the result of cosmic processes

Some elements are very common, but others are very rare. Hydrogen is by far the most abundant substance in the universe—it makes up nearly three-quarters of the mass of all the stars and galaxies. Helium is the second most abundant element. Together, hydrogen and helium account for about 98% of the mass of all the material in the universe. That leaves only 2% for all the other elements combined.

There is a good reason for this overwhelming abundance of hydrogen and helium. As we shall see in Chapter 28, most astronomers think that the universe began roughly 15 billion years ago with a violent event called the Big Bang. Only the lightest elements—hydrogen, helium, and a tiny amount of lithium—emerged from the enormously high temperatures following this cosmic event. All the heavier elements were manufactured by stars later, either by nuclear fusion reactions deep in their interiors or by the violent explosions that mark the end of massive stars' lives. Were it not for stars, there would be no heavy elements in the universe today.

Near the ends of their lives, some stars cast much of their matter back out into space. This process can be a comparatively gentle one, in which a star's outer layers are gradually expelled. Figure 7-12 shows the star HD 65750, which is losing material in this fashion. Alternatively, a star may end its life with a spectacular detonation called a **supernova explosion**, which blows the star apart. Either way, the interstellar gases in the galaxy become enriched with heavy elements dredged up from the dying star's interior, where they were

FIGURE 7-11 A Comet The solid part of a comet is a chunk of ice roughly 10 km in diameter. When a comet passes near the Sun, solar radiation vaporizes some of the comet's ices, and the resulting gases form a tail millions of kilometers long. This photograph shows a comet that was seen in January 1974. (NASA)

When this happens, the Sun's radiation vaporizes some of the comet's ices, thereby producing a long flowing tail (Figure 7-11).

Astronomers believe that asteroids and comets are debris left over from the formation of the solar system. In the inner regions of the solar system, rocky fragments have been able to endure continuous exposure to the Sun's heat, but any ice originally present would have evaporated. Far from the Sun,

FIGURE 7-12 A Mass-Loss Star This star (called HD 65750) is shedding material rapidly and is surrounded by nebulosity (called IC 2220) made visible by light reflecting off dust grains. These dust grains may have condensed from material shed by the star. (Anglo-Australian Observatory)

TABLE 7-4

Abundances of the Most Common Elements

Atomic number	Element	Symbol	Relative abundance
1	Hydrogen	H	1×10^{12}
2	Helium	He	7×10^{10}
6	Carbon	C	4×10^{8}
7	Nitrogen	N	9×10^{7}
8	Oxygen	O	7×10^{8}
10	Neon	Ne	1×10^{8}
12	Magnesium	Mg	4×10^{7}
11	Silicon	Si	4×10^{7}
16	Sulfur	S	2×10^{7}
26	Iron	Fe	3×10^{7}

created. New stars that form from this enriched material thus have an ample supply of heavy elements from which to develop a system of planets, satellites, comets, and asteroids.

Stars create different elements in different amounts. For example, the elements carbon, oxygen, silicon, and iron are readily produced in the interiors of massive stars, whereas gold is created only under special circumstances. Consequently, gold is rare in our solar system but carbon is not.

A convenient way to express the relative abundances of the various elements is to say how many atoms of a particular element are found for every trillion (that is, 10^{12}) hydrogen atoms. For example, for every trillion hydrogen atoms in space, there are about 70 billion helium atoms. From spectral analysis of stars and chemical analysis of Earth rocks, Moon rocks, and meteorites, scientists have determined the relative abundances of the elements in our part of the Galaxy today. Table 7-4 lists the most abundant elements.

In addition to these ten very common elements, five elements are moderately abundant: sodium, aluminum, argon, calcium, and nickel. These elements have abundances in the range of 10^6 to 10^7 relative to the standard trillion hydrogen atoms (Figure 7-13). Most of the other elements are much

---

BOX 7-2

Thermal Motion and the Retention of an Atmosphere

A moving object possesses energy. The faster it moves, the more energy it has. As we saw in Box 4-4, energy of this type is called **kinetic energy.** If an object of mass m is moving with a speed v, its kinetic energy is given by

$$\tfrac{1}{2}mv^2$$

This expression for kinetic energy is valid for all objects, both big and small, from atoms and molecules to planets and stars, as long as their speed is slow in relation to the speed of light. If the mass is expressed in kilograms and the speed in meters per second, the energy is expressed in joules (J).

EXAMPLE: A 2-kg object moving at 300 m/s has a kinetic energy of

$$\tfrac{1}{2}(2 \times 300^2) = 90,000 \text{ J}$$

Any substance with a temperature above absolute zero possesses thermal energy. **Thermal energy** is the kinetic energy of the moving atoms or molecules in a substance. In a cold substance, the atoms move slowly; in a hot substance, they move much more rapidly.

Consider a gas, such as the atmosphere of a star or planet. If the gas is hot, it probably consists of individual atoms moving at high speeds. For example, the Sun's atmosphere consists primarily of hydrogen and helium atoms. If the gas is cool, the atoms typically combine to form molecules. For example, the Earth's atmosphere consists primarily of nitrogen (N_2) and oxygen (O_2) molecules, and the Martian atmosphere is mostly carbon dioxide (CO_2) molecules.

According to the kinetic theory of gases developed during the nineteenth century, the average amount of thermal energy per atom or molecule in a gas is given by

$$\tfrac{3}{2}kT$$

where T is the temperature of the gas in kelvin and k is the Boltzmann constant, which has the value $k = 1.38 \times 10^{-23}$ J/K.

The thermal energy of an atom or molecule is just kinetic energy, so we can write the equality

$$\tfrac{1}{2}mv^2 = \tfrac{3}{2}kT$$

where v represents the average speed of an atom or molecule in a gas with temperature T. Rearranging this equation, we obtain

$$v = \sqrt{\frac{3kT}{m}}$$

(This value is actually slightly higher than the average speed of the atoms or molecules in the gas, but it is close enough for our purposes here. If you are studying physics, you may know that v is actually the root-mean-square average speed.)

EXAMPLE: Suppose you want to know the average speed of the oxygen molecules that you breathe at a room temperature of $72°F$ ($= 22°C = 295$ K). From a reference book, you can find that the mass of an oxygen atom is 2.66×10^{-26} kg. The mass of an oxygen molecule (O_2) is twice the mass of an oxygen atom, or $2(2.66 \times 10^{-26}$ kg$) = 5.3 \times 10^{-26}$ kg. Thus the average speed is

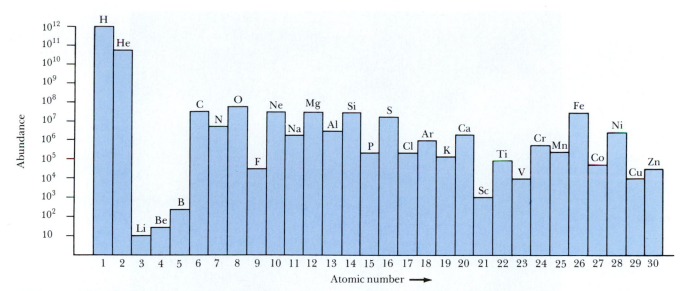

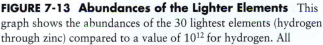

FIGURE 7-13 Abundances of the Lighter Elements This graph shows the abundances of the 30 lightest elements (hydrogen through zinc) compared to a value of 10^{12} for hydrogen. All elements heavier than zinc have abundances less than 1000 atoms per trillion atoms of hydrogen.

$$v = \left(\frac{3(1.38 \times 10^{-23})(295)}{5.3 \times 10^{-26}}\right)^{1/2} = 4.8 \times 10^2 \text{ m/s} = 0.48 \text{ km/s}$$

or almost exactly 1000 miles per hour.

Obviously, the atoms and molecules in a gas are moving rapidly, even at moderate temperatures. The speeds may even be so great that a planet's gravity is not strong enough to retain an atmosphere.

Astronomers find it useful to speak of the **escape speed** from a planet in trying to decide if an object can permanently leave the planet and escape into interplanetary space. As we saw in Box 4-4, the escape speed from a planet of mass M and radius R is given by

$$V_{escape} = \sqrt{\frac{2GM}{R}}$$

where G is the universal constant of gravitation ($G = 6.67 \times 10^{-11}$ N m^2/kg^2).

The table gives the escape speed for various objects in the solar system. For example, to get to Mars, a spacecraft must leave the Earth with a speed greater than 11.2 km/s (25,100 mi/h).

In a planet's atmosphere, some molecules move slowly, others more swiftly. Nevertheless, the average speed of a particular kind of molecule can be calculated if you know the planet's atmospheric temperature. A good rule of thumb is that a planet can retain a gas if the escape speed is at least six times greater than the average speed of the molecules in the gas. In such a case, very few molecules will be moving fast enough to escape from the planet's gravity.

EXAMPLE: Consider the Earth's atmosphere. We saw that the average speed of oxygen molecules is 0.48 km/s at room temperature. The escape speed from the Earth (11.2 km/s) is much more than six times the average speed of the oxygen molecules, so the Earth has no trouble keeping oxygen in its atmosphere.

A similar calculation for hydrogen molecules (H$_2$) gives a different result, however. At 295 K, the average speed of a hydrogen molecule is 1.9 km/s. Six times this speed is 11.4 km/s, which is slightly higher than the escape speed from the Earth. Thus, the Earth does not retain hydrogen in its atmosphere. Any hydrogen released into the air we breathe slowly leaks away into space.

	Escape speed	
	(km/s)	(mi/h)
Mercury	4.3	9,600
Venus	10.3	23,000
Earth	11.2	25,100
Moon	2.4	5,400
Mars	5.0	11,000
Jupiter	59.5	133,000
Saturn	35.6	79,600
Uranus	21.2	47,400
Neptune	23.6	52,800

FIGURE 7-14 A Dusty Region of Star Formation These young stars in the constellation of Orion are still surrounded by much of the gas and dust from which they formed. The bluish, wispy nebulosity is caused by starlight reflecting off abundant interstellar dust grains. These grains are made of heavy elements produced by earlier generations of stars. (Anglo-Australian Observatory)

rarer. For example, for every trillion hydrogen atoms in the solar system, there are only six atoms of gold.

Our solar system is made of matter created in stars that existed billions of years ago. The Sun is a fairly young star, only 5 billion years old. All the heavy elements in our solar system were created and cast off by ancient stars during the first 10 billion years of our Galaxy's existence. We are literally made of star dust (Figure 7-14).

7-7 The solar system formed in the solar nebula during the birth of the Sun

Observations of star formation elsewhere in our Galaxy suggest that our solar system formed from a vast cloud of gas and dust called the **solar nebula.** Temperatures inside this cloud determined what the types of planets formed at different distances from the Sun. To see why, we must examine how the abundant elements and their common compounds behave at various temperatures.

Just before the birth of the Sun, atoms in the solar nebula were so widely spaced that no substance could exist as a liquid. Matter in this vast cloud existed either as a gas or as tiny grains of dust and ice. At such low pressures, a substance's **condensation temperature** indicates whether it is a solid or a gas. Above its condensation temperature, a substance is a gas; below its condensation temperature, the substance solidifies into tiny specks of dust or snowflakes.

Rock-forming substances have high condensation temperatures, typically in the range of 1300 to 1600 K. Ices have much lower condensation temperatures, ranging from 100 to 300 K. The condensation temperatures of hydrogen and helium are so near absolute zero that these common elements always existed as gases during the creation of the solar system.

Initially the solar nebula was quite cold. Temperatures throughout the cloud were probably less than 50 K, which is below the condensation temperature of every common substance except hydrogen and helium. Snowflakes and ice-coated dust grains must have been scattered abundantly across the solar nebula, which had a diameter of at least 100 AU and a mass roughly two to three times that of the Sun.

The gravitational pull of the particles and gas caused them to drift toward the center of the solar nebula, which thus began to contract. Density and pressure began to increase at

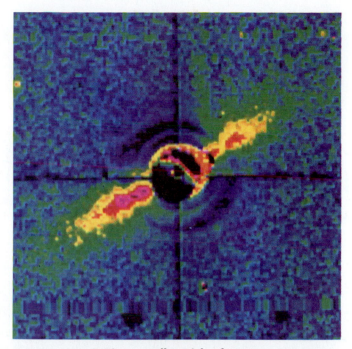

FIGURE 7-16 A Circumstellar Disk of Matter This computer-enhanced photograph made with a CCD shows a disk of matter orbiting the star ß Pictoris. The star is located at the center of the picture, behind a small, circular mask that was needed to block the star's light, which otherwise would have overwhelmed the light from the disk. The disk, seen nearly edge-on, is believed to be very young, possibly no more than a few hundred million years old. (University of Arizona and JPL)

FIGURE 7-15 The Birth of the Solar System This sequence of drawings shows three stages in the formation of the solar system. In (a) a slowly rotating cloud of interstellar gas and dust begins to contract because of its own gravity. A central condensation forms in (b) as the cloud flattens and rotates faster. In (c) a disk of gas and dust surrounds the young Sun, which has begun to shine.

the center of the solar nebula, in a concentration of matter called the **protosun**. As the solar nebula continued to contract, atoms in the cloud collided with one another with increasing speed and frequency, thereby causing temperatures deep inside the solar nebula to climb. This process, whereby the gravitational energy of a contracting gas cloud is converted into thermal energy, is called **Helmholtz contraction,** after the nineteenth-century German physicist who first described it.

The solar nebula also must have had an overall slight amount of rotation, or **angular momentum,** as it is properly called (recall Box 4-4). Otherwise, everything would have fallen straight into the protosun, leaving nothing behind to form the planets. Mathematical studies show that the combined effects of gravity and angular momentum can transform a shapeless cloud into a rotating disk that is warm at the center and cold at its edges, as sketched in Figure 7-15. The transformation of the solar nebula into a disk explains why the orbits of the planets nearly all lie in the same plane. Furthermore, astronomers find disks of material surrounding other stars. In the disk seen nearly edge-on in Figure 7-16, planets may still be forming.

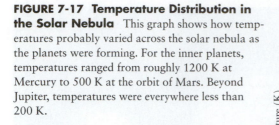

FIGURE 7-17 Temperature Distribution in the Solar Nebula This graph shows how temperatures probably varied across the solar nebula as the planets were forming. For the inner planets, temperatures ranged from roughly 1200 K at Mercury to 500 K at the orbit of Mars. Beyond Jupiter, temperatures were everywhere less than 200 K.

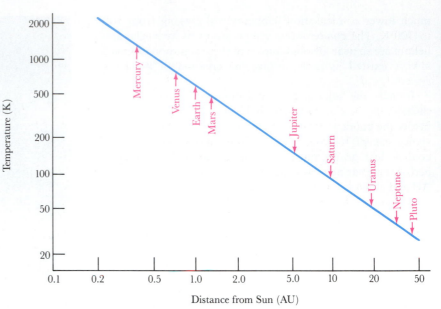

Temperatures around the newly created protosun soon climbed to 2000 K. Meanwhile, temperatures in the outermost regions of the solar nebula remained at less than 50 K. Figure 7-17 shows the probable temperature distribution throughout the solar nebula at the end of this preliminary stage in the formation of our solar system. All the common icy substances in the inner regions of the solar nebula were vaporized by the high temperatures. Only the rocky substances remained solid. That is why the four inner planets are today composed primarily of dense, rocky material. In contrast, snowflakes and ice-coated dust grains were able to survive in the cooler, outer portions of the solar nebula. As a result, the Jovian planets possess abundant ice-forming substances, as evidenced by the spectral lines of methane and ammonia that have been identified in their spectra. Indeed, many of the satellites of the Jovian planets are either partially or almost entirely composed of ices.

7-8 The inner planets formed by the accretion of planetesimals, whereas the outer planets formed by an accumulation of gases in the solar nebula

In recent years, astronomers have used computer simulations to learn more about how the inner planets formed from solid, rocky particles in the solar nebula. A computer is typically programmed to simulate a large number of particles circling a hypothetical newborn sun along orbits dictated by Newtonian mechanics. As the simulation proceeds, the particles coalesce to form larger objects, which in turn collide to form planets. From such studies, astronomers have con-

structed a reasonable scenario for the creation of the solar system.

Initially, when neighboring dust grains and pebbles in the solar nebula collided, electric and gravitational forces held them together. Over a few million years, these accumulations of dust and pebbles coalesced into roughly a billion objects called **planetesimals**, with diameters of about 10 km. During the next stage, the gravitational attraction between the planetesimals caused them to collide and coalesce into still-larger objects called **protoplanets**, which were roughly the size and mass of our Moon. This accumulation of material through the action of gravity is called **accretion**. During the final stage, these Moon-sized protoplanets collided to form the terrestrial planets. This final episode must have involved some truly spectacular, world-shattering collisions.

Computer simulations give us insights about the uniqueness of our solar system, because the astronomer can choose the number of planetesimals and their orbits at the start of the simulation. By performing a variety of simulations, each beginning with somewhat different starting conditions, it is possible to see what kinds of planetary systems are created. Such studies demonstrate that a wide range of initial conditions ultimately lead to the same result: Accretion continues for roughly 100 million years and typically forms four terrestrial planets with orbits between 0.3 and 1.6 AU from the Sun.

Figure 7-18 shows one particular computer simulation. The calculations began with 100 planetesimals, each having a mass of 1.2×10^{23} kg. This choice ensures that the total mass (1.2×10^{25} kg) equals the mass of the four terrestrial planets (Mercury through Mars) plus their satellites. The initial orbits of these planetesimals are inclined to each other by angles of less than 5° to simulate a thin layer of asteroid-like objects orbiting the protosun.

After an elapsed time simulating 30 million years, the 100 original planetesimals have coalesced into 22 protoplanets. After 79 million years, 11 larger protoplanets remain. Nearly another 100 million years elapse before the total number of growing protoplanets is reduced to six. Figure 7-18c shows four planets following nearly circular orbits after a total elapsed time of 441 million years. In this particular simulation, the fourth planet from the Sun ends up being the most massive.

The material from which the inner protoplanets accreted was rich in elements with high condensation temperatures. Iron, silicon, magnesium, and sulfur were particularly abundant, followed closely by aluminum, calcium, and nickel. The violent impacts of the planetesimals on the growing protoplanets, as well as the decay of radioactive elements, melted much of this rocky material. The terrestrial planets therefore may have begun their existence partially as spheres of molten rock.

Many details of the formation of the planets are topics of current research and debate among scientists. Two competing theories have evolved to explain why planets like the Earth have dense, iron-rich cores surrounded by mantles of less dense rock. According to the **homogeneous accretion theory,** the inner planets grew by the accretion of planetesimals of the same general composition—a mixture of rock and iron. Initially, therefore, the inner planets were quite homogeneous. As the decay of short-lived radioactive elements melted this matter, gravity caused the denser iron-rich minerals to sink to the centers of the planets, while the less dense silicon-rich minerals floated to their surfaces. This process is called **chemical differentiation.**

An alternative explanation, called the **heterogeneous accretion theory,** argues that planets formed from material condensing out of the solar nebula as it cooled. Iron and iron oxides were among the first substances to condense, and so the iron-rich cores of the terrestrial planets formed quite early. Then, as the solar nebula continued to cool, silicon-rich minerals condensed and were accreted onto the iron-rich planetary cores. An accurate account of the birth of the inner planets may involve ideas from both theories.

Like the inner planets, the outer planets probably began to form from the accretion of planetesimals. In this case, the process resulted in a rocky core for each of the Jovian planets that served as a "seed" around which the rest of the planet eventually grew. According to calculations by Peter Bodenheimer of the Lick Observatory and James Pollack of NASA Ames, the core of a Jovian protoplanet began to capture an envelope of gas as it continued to grow by accretion. Both rock and gas slowly accumulated for about a million years, until the masses of the core and the envelope became equal. From that critical moment on, the envelope pulled in all the gas it could get, dramatically increasing the protoplanet's mass and size. This runaway growth of the protoplanet continued until all the available gas was used up. The result was a huge planet with an enormously thick, hydrogen-rich atmosphere surrounding an Earth-sized core of rocky material. This scenario occurred at four different distances from the Sun, thus creating the four Jovian planets. Figure 7-19 summarizes this story of the formation of the solar system.

During the millions of years while the planets were forming, temperatures and pressures at the center of the contracting protosun continued to climb. Finally, temperatures at the center of the protosun reached 8 million kelvin, hot enough to ignite thermonuclear reactions, and the Sun was born. Sunlike stars take approximately 100 million years to form from a prestellar nebula, which means that the Sun must have become a full-fledged star at roughly the same time the accretion of the inner protoplanets was complete.

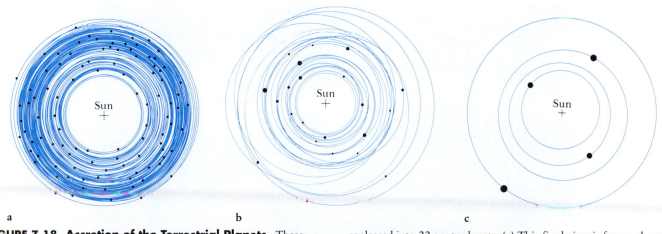

a   b   c

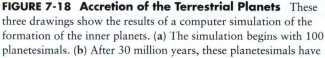

FIGURE 7-18 Accretion of the Terrestrial Planets These three drawings show the results of a computer simulation of the formation of the inner planets. (a) The simulation begins with 100 planetesimals. (b) After 30 million years, these planetesimals have coalesced into 22 protoplanets. (c) This final view is for an elapsed time of 441 million years, but the inner planets were essentially formed after 150 million years. (Adapted from G. W. Wetherill)

FIGURE 7-19 Formation of the Planets This series of sketches, spanning 100 million years, shows the major stages in the birth of the solar system. Terrestrial planets accrete from rocky material in the warm inner regions of the solar nebula. Mean-while, the huge, gaseous Jovian planets form in the cold outer regions. (a) The solar nebula in its initial stages. (b) The early solar system after 50 million years. (c) Planetary formation is nearly complete after 100 million years.

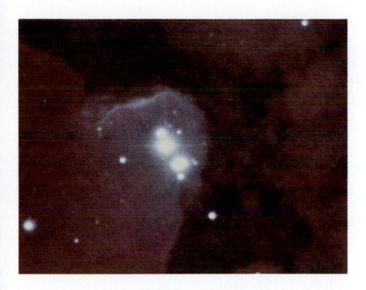

◀ **FIGURE 7-20 Newly Formed Stars** A full-fledged star is born when thermonuclear reactions ignite at its center. This ignition is often accompanied by an outpouring of particles and radiation from the star's surface that sweeps the surrounding space clean of dust and gas from which the star had formed. This photograph shows dusty material being blown away from newly created stars at the center of the so-called Trifid Nebula in the constellation of Sagittarius. (Anglo-Australian Observatory)

A newborn star adjusts fairly violently to the onset of thermonuclear reactions in its core, and often the star's tenuous outermost layers are vigorously expelled into space (Figure 7-20). This brief burst of mass loss, observed in many young stars across the sky, is called a **T Tauri wind,** after the star in Taurus (the Bull) where it was first identified. The T Tauri wind that heralded the birth of the Sun swept the solar system clean of excess gases, thereby preventing further accretion. Many small rocks were left behind to pelt the planets over the next half billion years. For all practical purposes, however, the formation of the solar system had been completed by the time the T Tauri wind began to blow.

The Sun today continues to lose matter gradually in a mild fashion. This ongoing, gentle mass loss, called the **solar wind,** consists of high-speed protons and electrons leaking away from the Sun's outer layers. In later chapters, we shall see how each planet carves out its own distinctive cavity in this solar wind.

KEY WORDS

Terms preceded by an asterisk are discussed in the boxes.

accretion	eccentricity (of an ellipse)	Jovian planet	protosun
angular momentum	*electron	*kinetic energy	solar nebula
asteroid	escape speed	molecule	solar wind
asteroid belt	Helmholtz contraction	*neutron	spectroscopy
atomic number	heterogeneous accretion theory	*nucleus	supernova explosion
average density		periodic table	T Tauri wind
chemical differentiation	homogeneous accretion theory	planetesimal	terrestrial planet
comet		*proton	*thermal energy
condensation temperature	*isotope	protoplanet	

KEY IDEAS

• The four inner planets of the solar system share many characteristics and are distinctly different from the four giant outer planets.

The four inner (terrestrial) planets are relatively small (with diameters of 5000 to 13,000 km), have high average densities (4000 to 5500 kg/m^3), and are composed primarily of rock.

The giant outer (Jovian) planets have large diameters (50,000 to 143,000 km), low average densities (700 to 1700 kg/m^3), and are composed primarily of hydrogen and helium.

• Spectroscopy, the study of spectra, provides information about the chemical composition of distant objects. Spectral lines can be used to identify the elements and chemical compounds.

• Hydrogen and helium, the two lightest elements, were formed shortly after the universe was created. The heavier elements were produced much later, in the centers of stars, and were cast into space when the stars died.

By mass, 98% of the matter in the universe is hydrogen and helium.

- The atoms of various elements can combine to form many different kinds of molecules.

- The basic planet-forming substances can be classified as gases, ices, or rock, depending on their condensation temperatures. The terrestrial planets are composed primarily of rock, whereas the Jovian planets are composed largely of gas.

- The solar system formed from a disk-shaped cloud of hydrogen and helium that also contained ice and dust particles.

  The four inner planets formed through the accretion of dust particles into planetesimals, then into larger protoplanets.

  The four outer planets probably formed through the runaway accretion of gas onto rocky protoplanetary cores.

  The Sun formed by accretion at the center of the nebula. After about 100 million years, temperatures at the protosun's center became high enough to ignite thermonuclear reactions.

  When the protosun became a star, excess gas was vigorously blown away from the Sun, thereby ending the process of planet formation.

REVIEW QUESTIONS

1. What are the characteristics of a terrestrial planet?

2. What are the characteristics of a Jovian planet?

3. In what ways does Pluto not fit the usual classification of either terrestrial or Jovian planets?

4. In what ways are the largest satellites similar to the terrestrial planets? Which satellites are these?

5. What is an asteroid? What is a comet? In what ways are these minor members of the solar system like or unlike the planets?

6. Why do astronomers almost always use the Kelvin temperature scale in their work rather than the Celsius or Fahrenheit scales?

7. What is meant by the "average density" of a planet? What does the average density of a planet tell us?

8. If hydrogen and helium account for 98% of the mass of all the material in the universe, why aren't the Earth and Moon composed primarily of these two gases?

9. What is the meant by a substance's condensation temperature? What role did condensation temperatures play in the formation of the planets?

10. What is a planetesimal? How did planetesimals give rise to the planets?

11. What is meant by accretion? What is the difference between homogeneous accretion and heterogenous accretion?

12. What occurred during the formation of our solar system to ensure that the terrestrial planets formed close to the Sun and the Jovian planets far from the Sun?

13. Explain how our current understanding of the formation of the solar system can account for the following characteristics of the solar system: (a) All planetary orbits lie in nearly the same plane. (b) All planetary orbits are nearly circular. (c) The planets orbit the Sun in the same direction that the Sun itself rotates.

14. Why is it reasonable to suppose that our solar system was formed over a relatively short period of less than half a billion years?

ADVANCED QUESTIONS

Tips and tools . . .

The volume of a sphere of radius r is $\frac{4}{3}\pi r^3$. The density of an object is its mass divided by its volume. To compute the mass of Mars, you may have to review Box 4-3 for the appropriate form of Kepler's third law. Be sure to use the same system of units (e.g., meters, seconds, kilograms) in all your calculations. Formulas relating various temperature scales are found in Box 5-1.

15. Would you expect very old stars to possess planetary systems? If so, what types of planets would they have? Explain.

16. Why do you suppose water (H_2O), methane (CH_4), and ammonia (NH_3) are comparatively abundant substances?

17. A spherical asteroid 1 km in diameter composed of rock and iron with an average density of 5000 kg/m^3 strikes the Earth with a speed of 25 km/s. What is the kinetic energy of the asteroid at the moment of impact? How does this energy compare with that released by a 20-kiloton nuclear weapon, like the device that destroyed Hiroshima? (*Hint*: 1 kiloton of TNT equals 4.2×10^{12} J.)

18. A hydrogen atom has a mass of 1.673×10^{-27} kg, and the temperature of the Sun's surface is 5800 K. What is the average speed of hydrogen atoms at the Sun's surface?

19. The Sun's mass is 1.989×10^{30} kg, and its radius is 6.96×10^8 m. What is the escape speed from the Sun's surface? Using your answer to the previous question concerning the average speed of hydrogen atoms at the Sun's surface, explain why hydrogen does not escape from the Sun.

20. Suppose a spacecraft landed on Jupiter's moon Ganymede, whose diameter is 5262 km and whose mass is 1.48×10^{23} kg. After collecting samples from the satellite's

surface, the spacecraft prepares to return to Earth. What is the escape speed from Ganymede? What is the escape speed from Jupiter at the distance of Ganymede's orbit? With what speed must the spacecraft leave Ganymede in order to begin its homeward journey?

21. Suppose you are trying to determine the chemical composition of the atmosphere of a planet by observing its spectrum. Chemicals in the Earth's atmosphere also produce spectral lines (often called *telluric lines*) in the spectra you observe. Can you think of a way of distinguishing between telluric lines and the spectral lines of the distant planet?

22. Mars has two small satellites, Phobos and Deimos. Phobos circles Mars once every 0.31891 day at an average altitude of 5980 km above the planet's surface. The diameter of Mars is 6794 km. Using this information, calculate the mass and average density of Mars.

23. At what temperature will the reading on a Celsius thermometer equal that on a Fahrenheit thermometer? At what temperature will the reading on a Kelvin thermometer equal that on a Fahrenheit thermometer?

24. What would the mass of the Earth be if it had retained hydrogen and helium in the same proportion to the heavier elements that exist elsewhere in the universe? How does your answer compare with the masses of the Jovian planets? What does your answer imply about the cores of the Jovian planets?

25. Suppose there were a planet having roughly the same mass as the Earth but located 50 AU from the Sun. What do you think this planet would be made of? On the basis of this speculation, assume a reasonable density for this planet and calculate its diameter. How many times bigger or smaller than the Earth would it be?

DISCUSSION QUESTIONS

26. Propose an explanation for the fact that the Jovian planets are orbited by terrestrial-like satellites.

27. Suppose that a planetary system is now forming around some protostar in the sky. In what ways might this planetary system turn out to be similar to or different from our own solar system?

28. Suppose astronomers discovered a planetary system in which the planets orbit a star along randomly inclined orbits. How might a theory for the formation of that planetary system differ from that for our own?

OBSERVING PROJECTS

29. There are many young stars still embedded in the clouds of gas and dust from which they formed. In the winter evening sky, for instance, is the famous Orion Nebula. In the summer night sky, the Lagoon, Omega, and Trifid nebulae are found in the Milky Way. Examine some of these nebulae with a telescope. Describe their appearance. Can you guess which stars in your field of view are actually associated with the nebulosity? To help you find some of these nebulae, the table below gives their coordinates (right ascension and declination) for the year 2000.

Nebula	Right ascension	Declination
Lagoon	$18^h 03.8^m$	$-24° 23'$
Omega	18 20.8	-16 11
Trifid	18 02.3	-23 02
Orion	5 35.4	-5 27

30. Consult such magazines as *Sky & Telescope* and *Astronomy* to determine which planets are visible in your evening sky. Examine these planets through a telescope. Describe their appearance. From what you observe, is there any way of knowing whether you are looking at a planet's surface or cloud cover?

FOR FURTHER READING

Beatty, J., and Chaikin, A., eds. *The New Solar System.* 3rd ed. Sky Publishing and Cambridge University Press, 1990. This excellent introduction to the solar system consists of twenty-three lavishly illustrated chapters, all written by noted scientists.

Briggs, G., and Taylor, F. *The Cambridge Photographic Atlas of the Planets.* Cambridge University Press, 1982. A nontechnical survey of the solar system that includes a superb collection of photographs from various space missions.

Chapman, C. *Planets of Rock and Ice.* Scribner's, 1982. The author eloquently conveys the excitement of planetary exploration while taking the reader on a nontechnical tour of the solar system.

Frazier, K. *Solar System.* Time-Life Books, 1985. This lavishly illustrated book gives an overview of our understanding of the solar system.

Moore, P., and Hunt, G. *Atlas of the Solar System.* Rand McNally, 1983. This excellent reference includes maps, photographs, and tables summarizing information about the Sun and planets.

Morrison, D. *Exploring Planetary Worlds.* Scientific American Library, 1993. This beautifully illustrated book by a noted planetary scientist summarizes our modern understanding of the solar system.

Murray, B., ed. *The Planets.* W. H. Freeman and Company, 1983. This book is a collection of ten articles about the solar system published in *Scientific American* magazine.

Wetherill, G. "The Formation of the Earth from Planetesimals." *Scientific American,* June 1981. This article describes the accretion of the terrestrial planets from the colliding planetesimals that originally orbited the Sun.

OUR LIVING EARTH

THE EARTH Astronauts often report that the Earth is the most invitingly beautiful object visible from space. Our blue-and-white world is the largest of the terrestrial planets. This photograph was taken in 1969 by *Apollo 11* astronauts on the way to the first manned landing on the Moon. Most of Africa and portions of Europe can be seen. (NASA)

THE EARTH is a geologically active world. Its surface is a rocky crust divided into huge plates that rub against one another, creating mountain ranges, volcanoes, earthquakes, and oceanic trenches. Neither Venus nor Mars, our neighboring planets, exhibits any features indicative of widespread plate motions. From seismic waves produced by earthquakes, we know that the Earth's interior consists of an iron-rich core surrounded by a thick mantle of partly molten rock. The transport of heat outward from the Earth's interior produces currents in the mantle that propel the plate motions that shape the Earth's rigid crust. Electric currents within the iron-rich core generate a planetwide magnetic field, which extends far into space and shields us from the solar wind. Earth's atmosphere is uniquely rich in nitrogen and in oxygen, a result of the activity of living organisms over billions of years. But today, by rapidly altering the chemical composition and thermal balance of Earth's atmosphere, humans are threatening the survival of many species, perhaps even our own.

Imagine an alien spacecraft approaching the inner solar system. Its occupants pass Mars with its thin atmosphere and barren desert landscapes. Closer to the Sun, they could see Venus, its shroud of corrosive clouds hiding a forbiddingly hot surface. Between these two relatively unpromising planets lies the Earth, where an ever-changing ballet of delicate white clouds plays against the darker browns and blues of its continents and oceans (Figure 8-1). Would these aliens find Earth the most inviting of the terrestrial planets? Or are we just biased by our intimate familiarity: Because we walk on its surface, drink its water, and breathe its air, we know more about the Earth than any other object in the universe. (A summary of Earth data appears in Box 8-1.)

FIGURE 8-1 The Three Largest Terrestrial Planets The Earth and Venus have nearly the same mass, size, and surface gravity. However, Venus's surface is perpetually shrouded by a dense cloud cover containing poisonous, corrosive gases. Mars is only about half the size of Earth and has only one-tenth its mass. The thin, dry Martian atmosphere does not protect the planet from the Sun's ultraviolet radiation. Both the Venusian and the Martian environments are hostile to life forms that thrive on Earth. (NASA, ESA)

BOX 8-1 START

BOX 8-1

Earth Data

Mean distance from the Sun:	$1.000 \text{ AU} = 1.496 \times 10^8 \text{ km}$
Maximum distance from the Sun:	$1.017 \text{ AU} = 1.521 \times 10^8 \text{ km}$
Minimum distance from the Sun:	$0.983 \text{ AU} = 1.471 \times 10^8 \text{ km}$
Mean orbital velocity:	29.8 km/s
Sidereal period:	365.256 days
Rotation period:	23.9345 hours
Inclination of equator to orbit:	23° 26′
Diameter (equatorial):	12,756 km
Mass:	$5.976 \times 10^{24} \text{ kg}$
Mean density:	5520 kg/m³
Escape speed:	11.2 km/s
Surface temperature range:	Maximum: 60°C = 140°F = 333 K Mean: 20°C = 70°F = 293 K Minimum: –90°C = –130°F = 183 K

8-1 Earth rocks contain clues about the history of our planet's surface

Were the aliens to land on Earth, they would find it radically different from either of its neighbors: The Earth is very wet. Nearly 71% of the Earth's surface is covered with water. An alien space probe to Earth might send back thousands of photographs such as Figure 8-2, a typical close-up view of the Earth's surface. Compared to the arid surfaces of Venus and Mars, our Sahara Desert is a veritable swamp.

Where land protrudes above the oceans, you find rocks, which are typical specimens of the Earth's outermost layer, or **crust**. By studying the kinds of rocks that are found at a particular location, geologists can deduce the history of that site. For instance, whether an area was once covered by an ancient sea or was flooded by lava from volcanoes is readily apparent from the kinds of rocks that are present.

Of course, rocks are composed of chemical elements (recall the periodic table in Figure 7-6), but individual chemical elements are rarely found in a pure state. The exceptions include gold nuggets, carbon crystals called diamonds, and a few less valuable specimens, like the sample of native sulfur shown in Figure 8-3. Such a naturally occurring solid composed of a single element or a particular chemical combination of elements is a **mineral**. Silica, for example, a common mineral composed entirely of silicon and oxygen atoms, is usually arranged in orderly rows to form crystals. A **rock** is a solid part of the Earth's crust that is composed of one or more minerals.

To identify various minerals, geologists use a number of criteria, such as the shape, hardness, and color of the crystals. Three common rock-forming minerals are shown in Figure 8-4. Each has its own unique, easily identifiable characteristics. These particular minerals are found mixed together in common rock called granite.

FIGURE 8-2 A Typical Close-Up View of Earth's Surface
Nearly three-quarters of the Earth's surface is covered with water. In contrast, there is no liquid water at all on Mercury, Venus, Mars, or the Moon.

FIGURE 8-3 Native Sulfur Sulfur is one of the very few chemical elements found in a pure state. Carbon crystals, called diamonds, and gold nuggets are also pure elements found in nature. Most minerals are chemical combinations of various elements.

FIGURE 8-4 Three Common Minerals A mineral is a specific chemical combination of elements whose atoms are typically arranged in an orderly fashion to produce a crystal. A quartz crystal is at the left. Feldspar (center) is another silicate. Mica (right) is also a common rock-forming mineral.

Geologists cannot classify rocks the same way they classify minerals, because rocks contain a mixture of mineral characteristics. In a piece of granite, for example, the tiny quartz crystals have a different color and hardness from the shiny flecks of mica. Therefore, geologists classify rocks according to how they were created. The three major categories of rocks correspond to three ways rocks form.

Igneous rocks result when minerals cool from a molten state. The formation of igneous rock can be directly observed during a volcanic eruption. Molten rock is called **magma** when it is buried below the surface and **lava** when it flows out upon the surface. Two common igneous rocks, basalt and granite, are shown in Figure 8-5.

Sedimentary rocks are those produced by the action of wind, water, or ice. For example, as winds pile up layer after layer of sand, the grains gradually become cemented together to produce sandstone. Minerals that precipitate out of the oceans can cover the ocean floor with layers of rock such as limestone. These two common sedimentary rocks are shown in Figure 8-6.

Sometimes igneous or sedimentary rocks become buried deep beneath the Earth's surface, where they are subjected to enormous pressure and high temperatures. These severe conditions change the rocks' structure, producing the third variety, called **metamorphic rock**. For example, when fine-grained igneous rock is subjected to pressure and heat, it

FIGURE 8-5 Igneous Rocks (Basalt and Granite) Igneous rocks are created when molten minerals solidify. The sample on the left is basalt, which is a fine-grained mixture of feldspar with iron-rich minerals that give the rock its dark color. A granite specimen is on the right. The fine-grained basalt typically forms when molten minerals cool near the Earth's surface; the coarser-grained granite typically forms from slower cooling deeper within the Earth's crust.

FIGURE 8-6 Sedimentary Rocks (Limestone and Sandstone) Sedimentary rocks are produced, often layer by layer, by the action of wind, water, or ice. The sample on the left is limestone, which is primarily calcite with some impurities that give it a dark color. Sandstone is on the right. A sedimentary rock is typically formed when loose particles of soil or sand are buried a short distance below the Earth's surface and are cemented into rock by chemical changes.

FIGURE 8-7 Metamorphic Rocks (Marble and Schist)
When igneous or sedimentary rocks are subjected to high temperatures and pressures deep in the Earth's crust, they are changed into metamorphic rock. Marble (left) is produced from limestone; schist (right) is formed from fine-grained igneous rock.

becomes schist. When limestone is metamorphosed, it becomes marble. These two common metamorphic rocks are shown in Figure 8-7.

8-2 Studies of earthquakes reveal the Earth's layered interior structure

Although the specimens shown in Figures 8-5, 8-6, and 8-7 are typical samples of the Earth's crust, they are not representative of what makes up our planet's interior. The densities of crustal rocks are typically 3000 to 4000 kg/m^3, but the average density of the Earth as a whole is 5500 kg/m^3. The Earth's interior must therefore be composed of a substance much denser than the crust.

Iron is a good candidate for this substance because it is the most abundant of the heavier elements (recall Table 7-4). Because of its magnetic properties, iron is also indicated by

the existence of the Earth's magnetic field. Furthermore, iron is common in meteoroids that strike the Earth, suggesting that it was abundant in the planetesimals from which the Earth formed.

Geologists strongly suspect that the Earth was entirely molten soon after its formation, about 4.5 billion years ago. Energy released by the violent impacts of numerous meteoroids and asteroids and by the decay of radioactive isotopes likely melted the solid material collected from the earlier planetesimals. Gravity caused abundant, dense iron to sink toward the Earth's center, forcing less dense material to the surface. This process of chemical differentiation then produced a layered structure within the Earth: a central core composed of almost pure iron, surrounded by a mantle of dense, iron-rich minerals. This mantle, in turn, is surrounded by a thin crust of relatively light silicon-rich minerals.

How have geologists determined this layered structure? The Earth's interior is as inaccessible as the most distant galaxies in space. The deepest wells go down only a few kilometers, barely penetrating the surface of our planet. However, geologists have learned basic properties of the Earth's interior by studying earthquakes.

Over the centuries, stresses build up in the Earth's crust. Occasionally, these stresses are relieved with a sudden vibratory motion called an **earthquake.** Earthquakes occur deep within the Earth's crust. The point on the Earth's surface directly over an earthquake's location is called the **epicenter.**

Earthquakes produce three different kinds of **seismic waves,** which travel around or through the Earth in different ways and at different speeds. Geologists use sensitive instruments called **seismographs** to detect and record these vibratory motions. The rolling motion that people feel near an epicenter travels only over the Earth's surface, like water waves on the surface of the ocean. However, two remaining kinds of waves, called primary or **P waves** and secondary or **S waves,** travel through the Earth. P waves are said to be

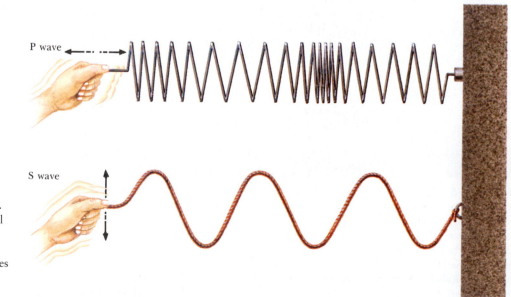

FIGURE 8-8 Seismic Waves
Earthquakes produce two kinds of waves that travel through our planet. P (or primary) waves are longitudinal waves analogous to those produced by pushing a spring in and out. S (or secondary) waves are transverse waves analogous to waves produced by shaking a rope up and down.

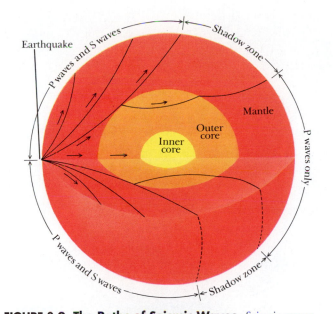

FIGURE 8-9 **The Paths of Seismic Waves** Seismic waves are refracted as they pass through the Earth. Furthermore, only the P waves can pass through the Earth's liquid outer core. From tracing the paths followed by seismic waves, geologists have deduced the dimensions of the Earth's mantle and core.

longitudinal waves because their oscillations are parallel to the direction of wave motion, like a spring that is alternately pushed and pulled. In contrast, S waves are said to be transverse waves because their vibrations are perpendicular to the direction in which the waves move. S waves are analogous to waves produced by a person shaking a rope up and down (Figure 8-8).

As seismic waves travel through the Earth, they are bent because of the varying density and composition of the Earth's interior. By studying how these waves bend, geologists can map out the general structure of the Earth's interior.

When an earthquake occurs, seismographs on the opposite side of the Earth record only P waves. The absence of S waves was first explained in 1906 by the geologist R. D. Oldham, who noted that transverse vibrations such as S waves cannot travel far through liquids. Oldham therefore concluded that our planet has a molten core. Furthermore, there is a region in which neither S waves nor P waves from an earthquake can be detected (Figure 8-9). This "shadow zone" results from the specific way in which P waves are refracted at the boundary between the solid mantle and the molten core. By measuring the size of this region, geologists have concluded that the core's diameter is about 7000 km (4300 mi). For comparison, the overall diameter of our planet is 12,700 km (7900 mi).

As the quality and sensitivity of seismographs improved, geologists discovered faint traces of P waves in an earthquake's shadow zone. In 1939, seismologist Inge Lehmann explained that some of the P waves passing though the Earth are deflected into the shadow zone by a small, solid **inner**

core at the center of our planet. The diameter of this inner core is about 2500 km (1550 mi).

The interior of our planet therefore has a curious structure: a liquid **outer core** sandwiched between a solid inner core and a solid mantle. The explanation for this arrangement lies in how temperature and pressure inside the Earth affect the melting point of rock.

Both temperature and pressure increase with increasing depth below the Earth's surface. The temperature of the Earth's interior rises steadily from about 20°C on the surface to nearly 5000°C at our planet's center (Figure 8-10).

The Earth's crust is only about 30 km thick. It is composed of rocks whose melting points are far higher than typical temperatures in the crust. Hence the crust is solid.

The Earth's **mantle**, which extends to a depth of about 2900 km (1800 mi), is largely composed of minerals rich in iron (abbreviated Fe from the Latin word *ferrum*) and magnesium. On the Earth's surface, specimens of these ferromagnesian minerals have melting points slightly over 1000°C.

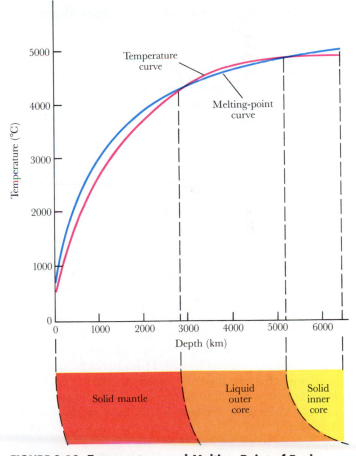

FIGURE 8-10 **Temperature and Melting Point of Rock Inside the Earth** The temperature (red curve) rises steadily from the Earth's surface to its center. By plotting the melting point of rock (blue curve) on this graph, we can deduce which portions of the Earth's interior are solid or liquid. (Adapted from F. Press and R. Siever)

However, the melting point of a substance depends on the pressure to which it is subjected: the higher the pressure, the higher the melting point. As shown in Figure 8-10, the melting point of all the mantle's minerals is everywhere higher than the actual temperature, so the mantle is primarily solid.

At the boundary between the mantle and the outer core, there is an abrupt change in chemical composition, from iron-rich minerals to almost-pure iron with a small admixture of nickel. Because this iron–nickel material has a lower melting point than the iron-rich minerals, the melting-point curve in Figure 8-10 dips below the temperature curve as it crosses from the mantle to the outer core. The melting-point curve remains below the temperature curve down to a depth of about 5100 km (3200 mi). Hence, from depths of about 2900 to 5100 km, the core is liquid.

At depths greater than about 5100 km, the pressure is more than 10^{11} N/m^2 (about 10^4 tons/in.^{2}). Because the melting point of the iron–nickel mixture under this pressure exceeds the actual temperature (see Figure 8-10), the Earth's inner core is solid.

8-3   Plate tectonics produce earthquakes, mountain ranges, and volcanoes that shape the Earth's surface

One of the most important geological discoveries of the twentieth century is the realization that our planet has an active, constantly changing crust. We have learned that the Earth's crust is divided into huge **plates** that constantly jostle each other, producing earthquakes, volcanoes, and oceanic trenches.

Anyone who carefully examines a map of the Earth might come up with the idea of moving continents. South America, for example, would fit snugly against Africa were it not for the Atlantic Ocean. As Figure 8-11 shows, the fit between landmasses on either side of the Atlantic Ocean is quite remarkable. This observation inspired the German meteorologist Alfred Wegener to advocate "continental drift"—the idea that the continents on either side of the Atlantic Ocean have simply drifted apart. After much research, Wegener published in 1924 the theory that there had originally been a single gigantic supercontinent he called Pangaea (meaning "all lands"), which began to break up and drift apart some 200 million years ago. Some geologists refined this theory, arguing that Pangaea must have first split into two smaller supercontinents, which they called Laurasia and Gondwanaland, separated by what they called the Tethys Sea. Gondwanaland later split into Africa and South America, with Laurasia dividing to become North America and Eurasia. According to this theory, the Mediterranean Sea is a surviving remnant of the ancient Tethys Sea.

Initially, most geologists ridiculed Wegener's ideas. Although it was generally accepted that the continents do "float" on the denser, somewhat plastic mantle beneath them, few geologists could accept the idea that entire continents could move around the Earth at speeds as great as several centimeters per year. The "continental drifters" could

FIGURE 8-11 Comparing the Continents   Africa, Europe, Greenland, and North and South America fit together as though they had once been joined. The fit is especially convincing if the edges of the continental shelves (shown in blue) are used, rather than today's shoreline. (Adapted from P. M. Hurley)

not explain what forces could be shoving the massive continents around.

Then, in the mid-1950s, Bruce C. Heezen of Columbia University and his colleagues began discovering long mountain ranges on the ocean floors, such as the Mid-Atlantic Ridge (Figure 8-12), which stretches all the way from Iceland to Antarctica. During the 1960s, careful examination of the ocean floor revealed that rock from the Earth's mantle is being melted and then forced upward along this Mid-Atlantic Ridge, which is therefore a long chain of underwater volcanoes. This upwelling of new material is forcing the ocean floor to spread. The eastern floor of the Atlantic Ocean is moving eastward, the western floor is moving westward. This **seafloor spreading** is in fact pushing South America and Africa apart at a speed of roughly 3 cm per year.

Seafloor spreading provided just the physical mechanism that had been missing from the theories of continental drift. With some embarrassment, geologists in the early 1960s began to find new evidence supporting the existence of large, moving plates (not simply the continents), and so was born the modern theory of crustal motion that came to be known as **plate tectonics.** Geologists today realize that earthquakes tend to occur at the boundaries of the Earth's crustal plates, where the plates are either colliding or separating. The plates' boundaries therefore stand out clearly when the epicenters of earthquakes are plotted on a map (Figure 8-13).

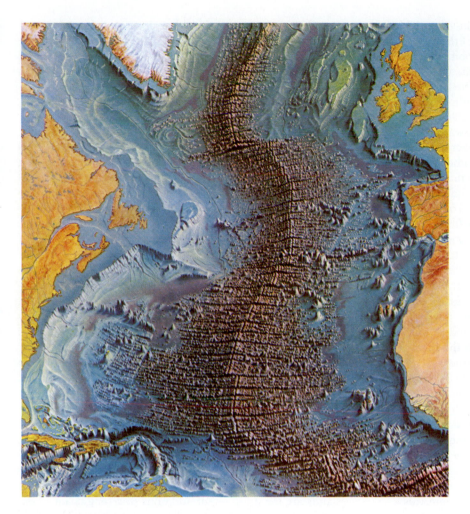

FIGURE 8-12 The Mid-Atlantic Ridge
This artist's rendition shows the floor of the North Atlantic Ocean. The unusual mountain range in the middle of the ocean floor is called the Mid-Atlantic Ridge. It is caused by lava seeping up from the Earth's interior along a rift that extends from Iceland to Antarctica. (Courtesy of M. Tharp and B. C. Heezen)

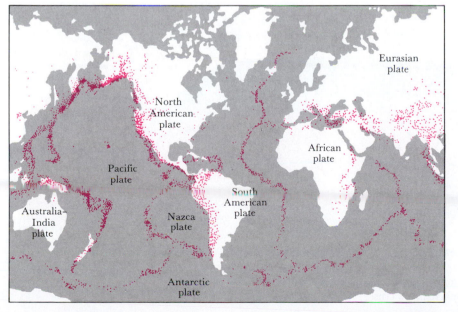

FIGURE 8-13 The Earth's Major Plates
The boundaries of Earth's major plates are the scenes of violent seismic and geologic activity. Most earthquakes occur where plates separate or collide. Plate boundaries are therefore easily identified simply by plotting earthquake epicenters on a map.

The flow of heat from the Earth's hot core outward to its cool crust is responsible for the movement of the plates. This heat transport is accomplished by **convection,** a process in which lighter, hotter material rises upward, while heavier, cooler material sinks downward. The mantle is hot enough to permit an oozing, plastic flow throughout a soft region of the upper mantle called the **asthenosphere.** As sketched in Figure 8-14, hot material (magma) seeps upward along **oceanic rifts,** where plates are separating. Cool crustal material sinks back down into the magma along **subduction zones** where plates are colliding. On top of the asthenosphere is a rigid layer, called the **lithosphere,** that is divided into plates that simply ride along the convection currents of the asthenosphere. (The boundary between crust and mantle—a boundary between different chemical compositions—is actually within the lithosphere.)

The boundaries between plates are the sites of some of the most impressive geological activity on our planet. Great mountain ranges, such as those along the western coasts of North and South America, are thrust up by ongoing collisions with the plates of the ocean floor. Subduction zones, where old crust is pushed back down into the mantle, are typically the locations of deep oceanic trenches, such as the ones off the coasts of Japan and Chile. Figures 8-15 and 8-16 show two well-known geographic features that resulted from tectonic activity at plate boundaries.

In recent years, geologists have uncovered evidence that points to a whole succession of supercontinents that once broke apart and then reassembled. Pangaea is only the most recent supercontinent in this cycle, which repeats every 500 million years. As a result, intense episodes of mountain building have occurred at roughly 500-million-year intervals.

Apparently a supercontinent sows the seeds of its own destruction because it blocks the flow of heat from the Earth's interior. As soon as a supercontinent forms, temperatures beneath it rise, much as they do under a book lying on an electric blanket. As heat accumulates, the lithosphere domes upward and cracks. Molten rock from the overheated asthenosphere wells up to fill the resulting fractures, which continue to widen as pieces of the fragmenting supercontinent move apart.

Convection currents associated with seafloor spreading carry heat outward from the Earth's interior to the crust. Here our planet differs radically from its neighbors. Venus and Mars do not have long mountain chains resembling Earth's Mid-Atlantic Ridge. Instead, both have huge volcanoes. Olympus Mons, the largest volcano on Mars, rises to an altitude of 24 km (15 mi) above the surrounding plains—nearly three times taller than Mount Everest. Similarly, the summit of Maxwell Montes on Venus attains an altitude of 11 km (7 mi).

Venus and Mars apparently transport heat outward to their surfaces largely by a mechanism called **hot-spot volcanism.** Hot spots deep in the mantles of Venus and Mars squirt molten lava up through the crust. In the absence of any tectonic activity to move the crust around, millions of years of eruptions eventually built enormous volcanoes.

Hot-spot volcanism also occurs on Earth. The Hawaiian Islands, which are in the middle of the Pacific plate, are a fine example. One hot spot in the Earth's mantle, beneath Hawaii, continuously pumps lava up through the crust. However, the Pacific plate is moving northwest at the rate of several centimeters per year. New volcanoes are thus being created over the hot spot, while older volcanoes move away from the magma source, become extinct, and eventually erode and disappear beneath the ocean. The Hawaiian Islands are the most recent additions to a long chain of extinct volcanoes that stretches all the way to Japan. This chain is a permanent record of the movements of the Pacific plate over the past 70 million years.

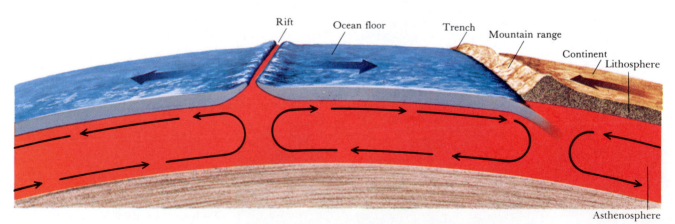

FIGURE 8-14 The Mechanism of Plate Tectonics
Convection currents in the soft upper layer of the mantle (called the asthenosphere) are responsible for pushing around rigid, low-density plates on the crust (called the lithosphere). New crust forms in oceanic rifts, where lava oozes upward between separating plates. Mountain ranges and deep oceanic trenches are formed where plates collide.

FIGURE 8-15 The Separation of Two Plates The plates that carry Egypt and Saudi Arabia are moving apart, leaving the trench that contains the Red Sea. This view, taken by astronauts in 1966, shows the northern Red Sea. Egypt is on the left, Saudi Arabia is on the right, and the Sinai Peninsula dominates the center of this view. (NASA)

FIGURE 8-16 The Collision of Two Plates The plates that carry India and China are colliding. As a result, the Himalayas have been thrust upward. In this photograph taken by astronauts in 1968, India is on the left and Tibet on the right; Mount Everest is one of the snow-covered peaks near the center. (NASA)

8-4   Absorption of sunlight and the Earth's rotation govern the behavior of our nitrogen–oxygen atmosphere

The Earth differs from its neighbors in yet another way: the chemical composition of its atmosphere. The atmospheres of both Venus and Mars consist almost entirely of carbon dioxide, whereas carbon dioxide accounts for only 0.03% of Earth's atmosphere. The Earth's atmosphere is predominantly a 4-to-1 mixture of nitrogen and oxygen, two gases that are found in only small amounts on Venus and Mars (Table 8-1). Carbon dioxide is also abundant on Earth, but it is dissolved in the oceans and chemically bound into carbonate rocks, such as limestone and marble.

As far as we know, large amounts of oxygen in a planetary atmosphere can arise only as a direct result of biological activity. Plants, for example, get energy from sunlight in a chemical process called photosynthesis that converts carbon dioxide into oxygen. From study of geological records, we know that living organisms have been active on Earth for at least the past 3 billion years, long enough to account for the chemical composition of the Earth's atmosphere today.

Because of gravity, the air weighs down upon the Earth, producing **atmospheric pressure**. The average atmospheric pressure at sea level is 14.7 pounds per square inch, or 1 atm (Box 8-2). The pressure decreases smoothly with increasing altitude, falling by roughly half with every $5\frac{1}{2}$ km.

Seventy-five percent of the mass of the Earth's atmosphere lies below an altitude of 11 km (roughly 7 mi, or 36,000 ft) in a region called the **troposphere**. All Earth's weather—clouds, rain, sleet, and snow—occurs in this lowest layer. Commercial jets generally fly at the top of the troposphere to minimize buffeting and jostling.

Over certain ranges of elevation, the air temperature decreases as we move upward. Over other ranges of elevation, the temperature rises as the altitude increases. At any elevation, the air temperature is largely determined by the interaction of sunlight and atmospheric gases. The temperature profile of the Earth's atmosphere, graphed in Figure 8-17, distinguishes layers that have different characteristics.

TABLE 8-1

Chemical Composition of Three Planetary Atmospheres

	Venus	Earth	Mars
Nitrogen	4%	77%	3%
Oxygen	Almost zero	21%	Almost zero
Carbon dioxide	96%	Almost zero	95%
Other gases	Almost zero	2%	2%

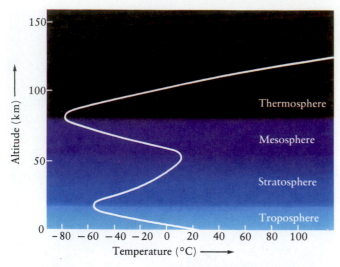

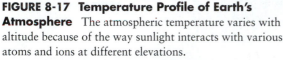

FIGURE 8-17 Temperature Profile of Earth's Atmosphere The atmospheric temperature varies with altitude because of the way sunlight interacts with various atoms and ions at different elevations.

Atmospheric temperature decreases upward, reaching a minimum of −55°C (= −67°F = 218 K) at the top of the troposphere. Above this level is the region called the **stratosphere,** which extends from 11 to 50 km (about 7 to 31 mi) above the Earth's surface. Ozone (O_3) molecules in the stratosphere efficiently absorb solar ultraviolet rays, thereby

heating the air in this layer. The temperature increases upward through the stratosphere to about 0°C (= 32°F = 273 K) at its top.

Certain chemicals, such as refrigerants and solvents, released into the air are destroying the ozone layer. If it were not absorbed by the ozone in the stratosphere, the ultraviolet radiation from the Sun would beat down on the Earth's surface. Because ultraviolet radiation breaks apart most of the delicate molecules that form living tissue, loss of the ozone layer could lead to literal sterilization of the planet.

Above the stratosphere lies the **mesosphere.** Atmospheric temperature again declines as one moves up through the mesosphere, reaching a minimum of about −75°C (= −103°F = 198 K) at an altitude of about 80 km (50 mi). This minimum marks the bottom of the **thermosphere,** above which the Sun's ultraviolet light strips atoms of one or more electrons. The stripped electrons reflect radio waves over a wide range of frequencies. The reason you can tune in a distant AM radio station is that its transmissions bounce off this radio-reflective region that surrounds our planet.

Earth's atmosphere has intricate circulation patterns. If the Earth did not rotate, heated gases would rise in the equatorial regions, flow toward the polar regions where they would cool and sink to lower altitudes, and then flow back toward the equator. Indeed, as we shall see in Chapter 11, Venus's atmosphere exhibits this simple convection pattern because Venus rotates very slowly. However, Earth's comparatively fast rotation breaks up the convection pattern as shown in Figure 8-18.

FIGURE 8-18 Circulation Patterns in Earth's Atmosphere The dominant circulation in the Earth's atmosphere consists of three convection cells in the northern hemisphere and three cells in the southern hemisphere. The Earth's rotation deflects winds in the convection cells from a purely north–south flow. Thus, at ground level in the northern temperate region (e.g., the continental United States), the prevailing winds are from the southwest toward the northeast.

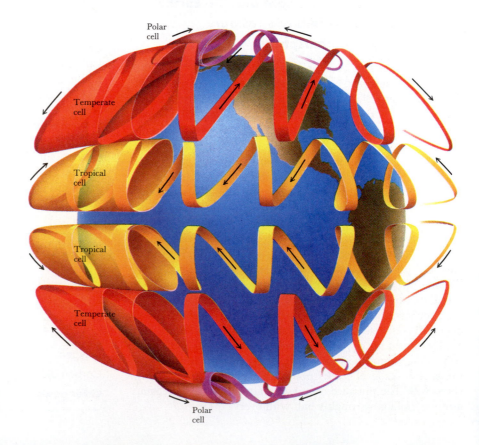

BOX 8-2

Atmospheric Pressure

At sea level on Earth, the air weighs down with a pressure of 14.7 pounds per square inch. By definition, this pressure is called one atmosphere (1 atm). Because we are familiar with this unit of measure for pressure, it is useful to express the atmospheric pressure on other planets in atmospheres also. For example, the atmospheric pressure on Venus is 90 atm, while that on Mars is only 0.01 atm.

There are, however, other ways of expressing pressure. In SI, for example, pressure is measured in newtons per square meter (N/m^2) rather than pounds per square inch. The various units are related to each other as follows:

$$1 \text{ atm} = 14.7 \text{ lb/in.}^2 = 1.01 \times 10^5 \text{ N/m}^2$$

The fact that 1 atm is equal to almost exactly 100,000 N/m^2 has prompted the definition of another unit of pressure called a **bar,** which is precisely 10^5 N/m^2. One bar is just slightly less than one atmosphere:

$$1 \text{ bar} = 0.987 \text{ atm}$$

In expressing low atmospheric pressures, it is often convenient to use the **millibar,** which is simply 0.001 bar. Thus, we say that the atmospheric pressure on Mars is about 10 millibars, approximately the same as the atmospheric pressure 30 km above the Earth's surface.

8-5 The Earth's magnetic field produces a magnetosphere that traps particles from the solar wind

As almost anyone with a compass knows, the Earth has a magnetic field. Magnetism arises whenever electrically charged particles are in motion. For example, a loop of wire carrying an electric current generates a magnetic field. Many geologists therefore suspect that the Earth's magnetic field is caused by electric currents flowing in the liquid portions of the Earth's iron core. As our planet rotates, these currents produce a magnetic field that dominates space for tens of thousands of kilometers in all directions and interacts dramatically with the solar wind.

The solar wind is a constant flow of charged particles (mostly protons and electrons) away from the outer layers of the Sun's upper atmosphere. Near the Earth, the particles in the solar wind move at speeds of roughly 400 km/s, or nearly a million miles per hour. Because this is considerably faster than the speed of sound, the solar wind is said to be "supersonic." If a planet possesses a magnetic field, this field deflects the charged particles, thereby forming an elongated "cavity" in the solar wind. This cavity is called the planet's **magnetosphere** (Figure 8-19).

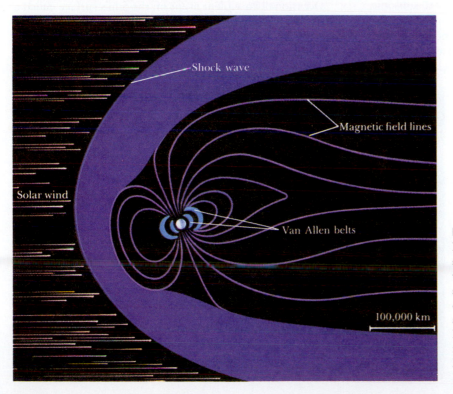

FIGURE 8-19 Earth's Magnetosphere
Earth's magnetic field carves out a cavity in the solar wind. A shock wave marks the boundary where the supersonic solar wind is abruptly slowed to subsonic speeds. Most of the particles of the solar wind are deflected around the Earth in a turbulent region colored purple in this drawing. Because of the strength of the Earth's magnetic field, our planet can trap charged particles in two huge, doughnut-shaped rings called the Van Allen belts.

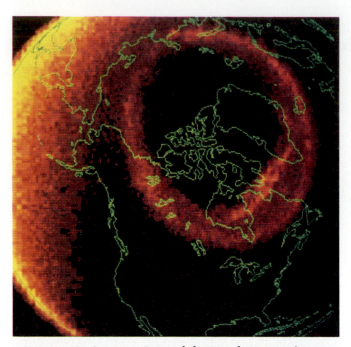

FIGURE 8-20 Aurorae Around the North Magnetic Pole This false-color image was taken at ultraviolet wavelengths by the *Dynamics Explorer 1* spacecraft. Coastlines are overlaid in green. The glowing oval of auroral emission, about 4500 km in diameter, is predominantly from oxygen atoms and molecular nitrogen. These emissions rival those from the sunlit hemisphere of the Earth. (Courtesy of L. A. Frank and J. D. Craven, The University of Iowa)

Figure 8-19 is a scale drawing of Earth's magnetosphere. When the supersonic particles in the solar wind first encounter the Earth's magnetic field, they abruptly slow to subsonic speeds. The boundary, characterized by a sudden decrease in velocity, is called a **shock wave.** Still closer to the Earth there is another boundary, called the **magnetopause,** where the outward magnetic pressure of the Earth's field is exactly counterbalanced by the impinging pressure of the solar wind. Most of the particles of the solar wind are deflected around the magnetopause just as water is deflected to either side of the bow of a ship.

Our planet's magnetic field is strong enough to trap charged particles that manage to leak through the magnetopause. These particles are trapped in two huge, doughnut-shaped rings called the **Van Allen radiation belts.** These belts were discovered in 1958 during the flight of the United States' first successful Earth-orbiting satellite. They are named after the physicist James Van Allen, who insisted that the satellite carry a Geiger counter to detect charged particles. The inner Van Allen belt, which extends over altitudes of about 2000 to 5000 km, contains mostly protons. The outer Van Allen belt, about 6000 km thick, is centered at an altitude of about 16,000 km above the Earth's surface and contains mostly electrons. The source of these charged particles is the solar wind.

The solar wind is an excellent electrical conductor. As it sweeps past the Earth's magnetic field, it causes the entire magnetosphere to act like a gigantic natural generator that produces enormous electrical currents. These currents flow along magnetic field lines down into the upper atmosphere surrounding the north and south magnetic poles. As high-speed electrons and protons collide with gases in the upper atmosphere, the gases fluoresce like the gases in a fluorescent tube. The result is a beautiful, shimmering display called the **northern lights (aurora borealis)** or **southern lights (aurora australis),** depending on the hemisphere in which the phenomenon is observed. Figure 8-20 shows a glowing ring around the Earth's north magnetic pole where charged particles streaming down onto the atmosphere continually produce aurorae.

Occasionally a violent event on the Sun's surface called a **solar flare** sends a burst of protons and electrons toward the Earth. The resulting auroral display can be exceptionally bright (Figure 8-21) and can often be seen over a wide range of latitudes.

FIGURE 8-21 The Northern Lights (Aurora Borealis) A deluge of protons and electrons from a solar flare can produce aurorae that can be seen over a wide range of latitudes. Aurorae typically occur 100 to 400 km above the Earth's surface. (Courtesy of S.-I. Akasofu, Geophysical Institute, University of Alaska)

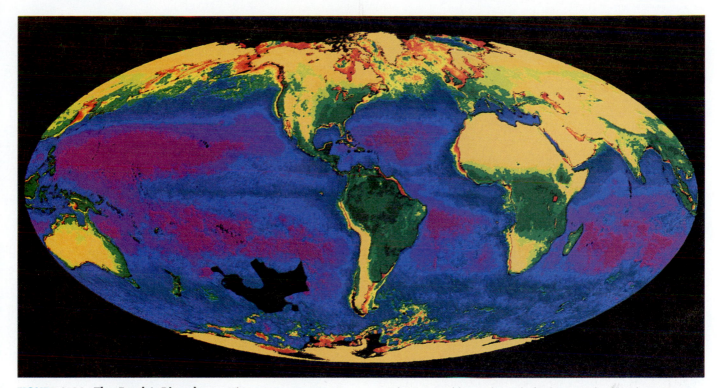

FIGURE 8-22 The Earth's Biosphere This computer-generated picture shows the plant distribution over the Earth's surface. Ocean colors in the order of the rainbow correspond to phytoplankton concentrations, with red and orange for high productivity to blue and purple for low. Land colors designate vegetation: dark green for the rain forests, light green and gold for savannas and farmland, and yellow for the deserts. (Courtesy of G. Feldman, NASA)

It is remarkable that the Earth's magnetosphere, dominated as it is by vast radiation belts completely encircling the Earth, was unknown until a few decades ago. Such discoveries remind us of how little we truly understand and how much remains to be learned even about our own planet.

8-6 A burgeoning human population is profoundly altering Earth's biosphere

One of the extraordinary characteristics of Earth is that it is covered with life. Living organisms thrive in a wide range of environments, from ocean floors to mountaintops, from frigid polar caps to blistering deserts.

Living organisms subsist in a relatively thin layer enveloping the Earth called the **biosphere,** which includes the oceans, a meter of topsoil covering the Earth's crust, and the lowest few kilometers of the troposphere. A portrait of Earth's biosphere, based on NASA satellite data, is shown in Figure 8-22. Ocean colors correspond to concentrations of microscopic, free-floating plants called phytoplankton. Land colors portray vegetation. The biosphere, which has taken hundreds of millions of years to evolve to its present state, is a delicate, highly complex system in which plants and animals depend on each other for their mutual survival.

In recent years a skyrocketing human population has begun to alter the biosphere dramatically and irreversibly. Figure 8-23 shows the sharp rise in the human population that began in the late 1700s with the onset of the Industrial

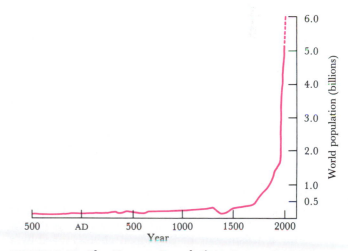

FIGURE 8-23 The Human Population Data and estimates from the U.S. Bureau of the Census, the Population Reference Bureau, and the United Nations Population Fund were combined to produce this graph showing the human population over the past 2500 years. Note how the population began to rise in the eighteenth century, and the astonishing population increase in the present century.

FIGURE 8-24 The Deforestation of Amazonia This picture from the *NOAA 9* satellite looks down onto the Brazilian state of Rondônia (outlined in yellow), located in the western Amazon rain forests where Brazil borders on Bolivia. The red dots are fires and the bluish streaks are smoke plumes. Rondônia is about the same size as the state of Oregon. (Courtesy of J. Tucker, NASA)

Revolution. In the mid-twentieth century, medical and technological advances ranging from antibiotics to high-yield grains (the so-called green revolution) fueled an unprecedented increase in the population. In 1960 there were three billion people on the Earth; in 1975, four billion. The five-billion mark was passed in 1986, and according to the United Nations Population Fund, there will be six billion people on Earth by the year 2000, and ten billion by 2025.

Every human being has basic requirements: food, clothing, and housing. We all need fuel for cooking and heating. To meet these demands, we cut down forests, cultivate grasslands, and build sprawling cities, replete with roads and factories. A striking example of this activity is occurring in the Amazon rain forest. Tropical rain forests are vital to our planet's ecology because they absorb significant amounts of carbon dioxide and release oxygen through photosynthesis. Although rain forests occupy only 7% of the world's land areas, they contain at least 50% of all species of plants and animals on our planet. Nevertheless, to make way for farms and grazing land, people simply cut down the trees and set them on fire—a process called slash-and-burn. Figure 8-24 shows the Brazilian state of Rondônia, photographed by a satellite. More than 2500 fires and numerous smoke plumes can be identified. Similar deforestation, along with extensive lumbering operations, is occurring in Malaysia, Indonesia, Papua, and New Guinea. The rain forests that once thrived in Central America, India, and the western coast of Africa are almost gone.

All this activity is adding carbon dioxide to the Earth's atmosphere faster than plants can extract it. Each year we put about 20 billion tons of CO_2 into the air we breathe. Figure 8-25 shows how the carbon dioxide content of the atmosphere has increased over the past 30 years.

Carbon dioxide is transparent to visible light but not to infrared radiation. Consequently, sunlight has no trouble entering our atmosphere and warming the ground, which then radiates at infrared wavelengths. The infrared radiation cannot escape back into space because carbon dioxide in the air absorbs it, thus warming up the Earth's atmosphere. The trapping of energy from sunlight in the atmosphere is called the **greenhouse effect.**

A mild, naturally occurring greenhouse effect has warmed the Earth's surface for billions of years. A variety of gases, most importantly water vapor and carbon dioxide, trap life-giving heat from the Sun. Without these infrared-absorbing molecules in the air, our planet would be about 30°C (54°F) cooler than it is today. So humans did not create the greenhouse effect, but we are responsible for strengthening it. Figure 8-26 shows how the global air temperature has changed during the twentieth century. If the average temperature of the Earth's surface continues to increase, the polar ice caps will melt, the sea level will rise and low-lying terrain will flood.

As we continue to populate our planet, it is clear that we face many challenges. Age-old modes of behavior that helped ensure the survival of our ancestors may not be the wisest courses of action in the future. Contemplating our behavior and our relationship to the Earth brings to mind the Chinese proverb: If we don't change the direction in which we are going, we are likely to end up where we are headed.

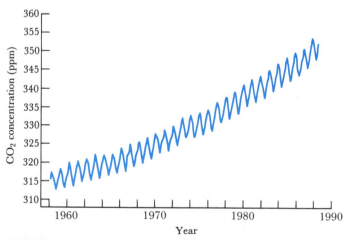

FIGURE 8-25 The Carbon Dioxide Content of the Atmosphere Measurements of carbon dioxide in parts per million (ppm) taken at the National Oceanic and Atmospheric Administration's station on Hawaii show that concentrations of this gas have increased by 10% since 1958, when observations started. The sawtooth pattern results from plants absorbing more carbon dioxide during the spring and summer.

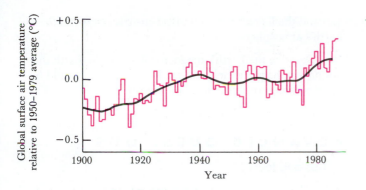

◀ **FIGURE 8-26 Change in Global Temperature** Both land and sea surface temperatures were combined to produce this graph, which seems to show a gradual warming trend over the past 90 years. The vertical scale is measured relative to the 1950–1979 average temperature.

KEY WORDS

Terms preceded by an asterisk are discussed in the boxes.

asthenosphere	hot-spot volcanism	*millibar	sedimentary rock
atmospheric pressure	igneous rock	mineral	seismic wave
aurora australis (*plural* aurorae)	inner core (of the Earth)	northern lights	seismograph
	lava	oceanic rift	shock wave
aurora borealis	lithosphere	outer core (of the Earth)	solar flare
*bar	magma	P wave	southern lights
biosphere	magnetopause	plate (lithospheric)	stratosphere
convection	magnetosphere	plate tectonics	subduction zone
crust (of the Earth)	mantle	rock	thermosphere
earthquake	mesosphere	S wave	troposphere
epicenter	metamorphic rock	seafloor spreading	Van Allen radiation belt
greenhouse effect			

KEY IDEAS

• The outermost layer, or crust, of the Earth offers clues to the history of our planet.

Its rocks are composed of minerals, naturally occurring elements, and chemical combinations of elements.

There are three major categories of rocks: igneous rocks (cooled from molten material), sedimentary rocks (formed by the action of wind, water, and ice), and metamorphic rocks (altered in the solid state by extreme heat and pressure).

• Study of seismic waves (vibrations produced by earthquakes) shows that the Earth has a small, solid inner core surrounded by a liquid outer core; the outer core is surrounded by the dense mantle, which in turn is surrounded by the thin low-density crust.

Seismologists deduce the Earth's interior structure by studying how longitudinal P waves and transverse S waves travel through the Earth's interior.

The Earth's inner and outer cores are composed of almost-pure iron with some nickel mixed in. The mantle is composed of iron-rich minerals.

Both temperature and pressure steadily increase with depth inside the Earth.

• The Earth's crust and a small part of its upper mantle form a rigid layer called the lithosphere; this layer is divided into huge plates that move about over the plastic layer called the asthenosphere in the upper mantle.

Movements of these plates, a process called plate tectonics, are caused by the upwelling of molten material at oceanic rifts, which produces seafloor spreading.

Plate tectonics is responsible for most of the major features of the Earth's surface, including mountain ranges, volcanoes, and the shapes of the continents and oceans.

• The Earth's atmosphere differs from those of the other terrestrial planets in its chemical composition, circulation pattern, and temperature profile.

The Earth's atmosphere is four-fifths nitrogen and one-fifth oxygen, with only a small amount of carbon dioxide. This abundance of oxygen is due to the biological activities of life forms on the planet.

The Earth's atmosphere is divided into layers called the troposphere, stratosphere, mesosphere, and thermosphere; ozone molecules in the stratosphere absorb ultra-violet light.

Because of the Earth's rapid rotation, the circulation in its atmosphere is complex, with three circulation cells in each hemisphere.

• The Earth's magnetic field produces a magnetosphere that surrounds the planet and blocks the solar wind from hitting the atmosphere.

A bow-shaped shock wave, where the supersonic solar wind is abruptly slowed to subsonic speeds, marks the outer boundary of the magnetosphere.

Most of the particles of the solar wind are deflected around the Earth by the magnetosphere.

Some charged particles from the solar wind are trapped in two huge, doughnut-shaped rings called the Van Allen radiation belts. A deluge of particles from a solar flare can initiate an auroral display.

• Human activity is changing the Earth's biosphere, on which all living organisms depend.

Deforestation and the burning of fossil fuels is increasing the carbon dioxide content of our atmosphere, thereby stimulating the greenhouse effect and perhaps already warming the planet.

REVIEW QUESTIONS

1. Describe the various ways in which the Earth is unique among the planets of our solar system.

2. What is the difference between a rock and a mineral?

3. What is the difference between igneous, sedimentary, and metamorphic rocks? What do these rocks tell us about the sites at which they are found?

4. How do we know about the Earth's interior, given that the deepest wells and mines go down only a few kilometers?

5. Describe the interior structure of the Earth.

6. Why is the center of the Earth not molten?

7. Describe the process of plate tectonics. Give specific examples of geographic features created by plate tectonics.

8. Explain how the outward flow of energy from the Earth's interior drives the process of plate tectonics.

9. Why do some geologists believe that Pangaea was the most recent in a succession of supercontinents?

10. Why do you suppose that active volcanoes, such as Mount St. Helens, are usually located in mountain ranges that border on subduction zones? In what way is Hawaii an exception?

11. Describe the structure of the Earth's atmosphere. Explain how biological activity has affected the chemical composition of our atmosphere. Explain how heat from the Sun and the Earth's rotation affect the circulation of the Earth's atmosphere.

12. Describe the Earth's magnetosphere. If the Earth did not have a magnetic field, do you think aurorae would be more common or less common than they are today?

13. What are the Van Allen belts?

ADVANCED QUESTIONS

Tips and tools . . .

You'll need to know that the volume of a sphere is $\frac{4}{3}\pi r^3$, where r is the sphere's radius. You may have to consult an atlas to examine the geography of the South Atlantic. Also, remember that the average density of an object is its mass divided by its volume.

14. In what way are the Earth's oceans like an atmosphere? In what way could they be considered part of the Earth's crust? How would your answers be different if the Earth were as close to the Sun as Venus is or as far from the Sun as Jupiter is?

15. Why do you suppose the Earth's surface is not riddled with craters like the surfaces of Mercury, the Moon, and Mars?

16. The Earth's primordial atmosphere probably contained an abundance of methane (CH_4) and ammonia (NH_3), whose molecules were broken apart by ultraviolet light from the Sun and particles in the solar wind. What happened to the atoms of carbon, hydrogen, and nitrogen that were liberated by this dissociation?

17. Africa and South America are separating at a rate of about 3 centimeters per year, as explained in the text. Assuming that this rate has been constant, calculate when these two continents must have been in contact.

18. Why do you suppose that worldwide television transmissions require the use of relay satellites, yet radio programs can be transmitted around the world without the use of such satellites?

19. What fractions of the Earth's total volume are occupied by the core, the mantle, and the crust?

20. Using data for the mass and size of the Earth listed in Box 8-1, verify that the average density of the Earth is 5500 kg/m^3. Assuming that the average density of material in the Earth's mantle is about 3500 kg/m^3, what must the average density of the core be?

21. The Earth's atmospheric pressure decreases by a factor of one-half for every $5\frac{1}{2}$-km increase in altitude above sea level. Construct a plot of pressure versus altitude, assuming the pressure at sea level is 1 bar. Discuss the characteristics of your graph. At what altitude is the atmospheric pressure 1 millibar?

22. On the average, temperature beneath the Earth's crust increases at a rate of 20°C per kilometer. At what depth

would water boil? (Assume the surface temperature is 20°C and neglect the effect of the pressure of overlying rock on the boiling point of water.)

DISCUSSION QUESTIONS

23. The human population on Earth is currently doubling about every 30 years. Describe the various pressures placed on the Earth by uncontrolled human population growth. Can such growth continue indefinitely? If not, what natural and human controls might arise to curb this growth? It has been suggested that overpopulation problems could be solved by colonizing the Moon or Mars. Do you think this is a reasonable solution? Explain your answer.

24. A recent scientific study suggests that the continued burning of fossil and organic fuels by humans is releasing enough CO_2 to stimulate the greenhouse effect that will melt the polar icecaps. Antarctica has an area of 13 million square kilometers and is covered by an icecap that varies in thickness from 300 meters near the coast to 1800 meters in the interior. Estimate the volume of this icecap. Assuming that water and ice have roughly the same density, estimate the amount by which the water level of the world's oceans would rise if Antarctica's ice were to melt completely. What portions of the Earth's surface would be inundated by such a deluge?

25. In 1986 scientists discovered a large hole in the Earth's ozone layer over Antarctica. The accompanying satellite image taken in 1989 shows the ozone hole in shades of purple and dark blue. Go to a library and investigate the current status of this ominous feature. Is the situation getting better or worse?

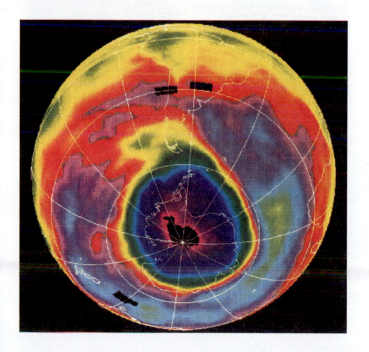

26. Cycles of supercontinents imply profound environmental changes that repeat every 500 million years. What sort of global changes might accompany the formation and breakup of a supercontinent? How might these cycles affect the evolution of life? Go to a library to research the life forms that dominated the Earth when Pangaea existed 250 million years ago and when fragments of the preceding supercontinent were as dispersed as today's continents are.

OBSERVING PROJECTS

27. When was the last earthquake near your hometown? How far is your hometown from a plate boundary? What kinds of topography (e.g., mountains, plains, seashore) dominate the geography of your hometown area? Do the topography and the frequency of earthquakes seem to be consistent with your hometown's proximity to a plate boundary?

FOR FURTHER READING

Akasofu, S.-I. "The Dynamic Aurorae." *Scientific American,* May 1989. This detailed article by a renowned expert on aurorae describes how interactions of the Earth's magnetic field and the solar wind give rise to a vast generator that powers the aurorae.

Cattermole, P., and Moore, P. *The Story of the Earth.* Cambridge University Press, 1985. This well-illustrated book gives a fine introduction to geology and the chronology of our planet.

Clark, W. C. "Managing Planet Earth." *Scientific American,* September 1989. This leadoff article to a single-topic issue of *Scientific American* explores such questions as: What kind of a planet do we want? What kind of a planet can we get?

Frohlich, C. "Deep Earthquakes." *Scientific American,* January 1989. This article explores the mysteries of deep earthquakes, which occur so far beneath the Earth's surface that high pressures and temperatures should prevent rock from fracturing suddenly.

Kelley, K., ed. *The Home Planet.* Addison Wesley and Mir, 1988. This beautiful album of photographs of Earth from space includes moving quotations from U.S. astronauts and Soviet cosmonauts.

Murphy, J. B., and Nance, R. D. "Mountain Building and the Supercontinent Cycle." *Scientific American,* April 1992. This fascinating article describes the evidence for a 500-million-year cycle of the formation and breakup of supercontinents.

Ruckelshaus, W. D. "Toward a Sustainable World." *Scientific American,* September 1989. This visionary article argues that moving people and nations toward a sustainable world will involve changes in society comparable in magnitude to the agricultural and industrial revolutions.

OUR BARREN MOON

AN ASTRONAUT ON THE MOON The Moon is a desolate, lifeless world. This view of the lunar surface shows an astronaut near a large rock. Because the Moon has no atmosphere, lunar rocks have not been subjected to weathering and thus preserve information about the early history of our solar system. American astronauts brought back a total of 382 kg (842 lb) of moon rocks from six manned lunar landings, which occurred between 1969 and 1972. (NASA)

UNLIKE THE EARTH, the Moon has remained essentially unaltered for billions of years. Long ago, debris left over from the formation of the planets rained down on the Moon, extensively cratering its rocky surface. Broad plains, called maria, bear silent witness to vast lava flows that flooded the largest impact basins. This barren world is geologically dead and lacks both an atmosphere and a magnetic field. Nevertheless, gravitational interaction between the Earth and Moon produces tides that flex the Moon, causing moonquakes. Lunar exploration gave many other insights into the Moon's history. For example, radioactive age-dating of lunar rocks brought back to Earth demonstrates that the Moon is 4.6 billion years old. The Moon was probably formed when a huge asteroidlike object struck the Earth. That impact ejected a vast amount of molten rock into space, where it cooled and coalesced to form the Moon. This plausible scenario is our first indication that catastrophic collisions between protoplanets and large planetesimals played an important role in shaping the solar system.

The Earth is a dynamic, living planet; its moon is biologically and geologically dead. The only noteworthy events to occur on the lunar surface over the past 3 billion years are occasional impacts by wayward meteoroids—and, of course, the arrival of humanity. Why did these two worlds turn out so vastly different? Lunar rocks brought back by the astronauts have provided important clues about the origin of the Moon and its relationship to the Earth. By understanding what the moon rocks tell us, we come to appreciate the scope and violence of collisions between planetesimals and protoplanets during the first billion years of our solar system.

9-1 The Moon's surface is covered with plains and craters

The Moon is so large and so near the Earth that some of its surface features are readily visible to the naked eye. Without a telescope, you can easily see dark gray and light gray areas that cover vast expanses of the Moon's rocky surface. Box 9-1 contains basic data about the Moon.

If you observe the Moon over many nights, you always see the same pattern of dark and light areas. Even as the Moon goes through its phases, you always see the same surface features because the Moon keeps the same side turned toward Earth; you never see the "back side" of the Moon. This feature of the Moon's motion, whereby the same side always faces the Earth, is called **synchronous rotation.** With patience, however, you can see slightly more than half the lunar surface, because the Moon wobbles slightly as it moves along its orbit. This periodic wobbling, called **libration,** permits us to view 59% of our satellite's surface.

With a small telescope or binoculars, you can see several different types of lunar terrain (Figure 9-1). Most prominent are the large, dark areas called **maria** (pronounced MAR-ee-uh). The singular form of this term, **mare** (pronounced MAR-ay), means "sea" in Latin and was introduced in the seventeenth century when observers using early telescopes thought these features were large bodies of water on the Moon. In fact, as we shall see later in this chapter, the maria are the remains of huge lava flows. A body of liquid water could not possibly exist on our airless satellite. Because there is no atmospheric pressure, a lake or ocean would boil furiously and evaporate rapidly into the vacuum of space. Nevertheless, we have kept their fanciful names, such as Mare Tranquillitatis (Sea of Tranquillity), Mare Nubium (Sea of Clouds), Mare Nectaris (Sea of Nectar), and Mare Serenitatis (Sea of Serenity).

The largest of the maria is Mare Imbrium (Sea of Showers). It is roughly circular and measures 1100 km (700 mi) in diameter (Figure 9-2). Although the maria seem quite smooth in telescopic views from the Earth, close-up photographs from lunar orbit reveal small craters and occasional cracks (Figure 9-3).

Perhaps the most familiar and characteristic features on the Moon are its **craters.** With an Earth-based telescope, some 30,000 craters are visible, with diameters ranging from 1 km to more than 100 km (Figure 9-4). Close-up photographs from lunar orbit have revealed millions of craters too small to be seen with Earth-based telescopes. Following a tradition established in the seventeenth century, the most prominent craters are named after famous philosophers, mathematicians, and scientists, such as Plato, Aristotle, Pythagoras, Copernicus, and Kepler.

BOX 9-1
Moon Data

Distance from Earth (center to center):	Mean: 384,400 km = 238,900 mi Minimum (perigee): 363,300 km Maximum (apogee): 405,500 km
Mean orbital velocity:	3680 km/h
Sidereal period (fixed stars):	27.322 days
Synodic period (new moon to new moon):	29.531 days
Inclination of lunar equator to orbit:	6° 41′
Diameter:	3476 km = 2160 mi
Diameter (Earth = 1):	0.272
Mass:	7.349×10^{22} kg
Mass (Earth = 1):	0.01230
Mean density:	3340 kg/m^3
Escape speed:	2.38 km/s
Surface gravity (Earth = 1):	0.166
Mean surface temperatures:	Day: 130°C = 266°F = 403 K Night: –180°C = –292°F = 93 K

FIGURE 9-1 The Moon Our Moon is one of seven large satellites in our solar system. The Moon's diameter (3476 km = 2160 mi) is slightly less than the distance from New York to San Francisco. This photograph is a composite of first quarter and last quarter views, in which long shadows enhance the surface features. (Lick Observatory)

FIGURE 9-2 Mare Imbrium from the Earth Mare Imbrium (Sea of Showers) is the largest of 14 dark plains that dominate the Earth-facing side of the Moon. Because few craters are seen on the maria, they must have been formed by lava flows late in the Moon's geologic history. (Carnegie Observatories)

FIGURE 9-3 Details of Mare Tranquillitatis Close-up views of the lunar surface reveal numerous tiny craters and cracks on the maria. Astronauts in lunar orbit took this photograph in 1969 while searching for potential landing sites for the first manned landing. (NASA)

FIGURE 9-4 The Crater Clavius This photograph, taken through the 200-in. Palomar telescope, shows one of the largest craters on the Moon. Clavius has a diameter of 232 km (144 mi) and a depth of 4.9 km (16,000 ft), measured from the crater's floor to the top of the surrounding rim. (Palomar Observatory)

Virtually all craters, both large and small, are the result of bombardment by meteoritic material. Nearly all of these craters are circular, indicating that they were not merely gouged out by high-speed rocks. Instead, the violent collisions with the Moon's surface vaporized the rapidly moving meteoroids, and the resulting explosions of hot gas produced the round craters observed today. Many large craters have a pronounced "central peak," a feature that is characteristic of a high-speed impact (Figure 9-5).

One of the surprises stemming from the early days of lunar exploration is that there is only one small mare on the Moon's far side. Except for that lone mare, craters cover the entire side of the Moon that faces away from the Earth. Cratered lunar terrain is called **terrae** (Figure 9-6). The singular form of this term, *terra*, means "land" in Latin; in this fanciful terminology, the entire lunar surface is thus covered by either "land" or "sea."

Detailed measurements by astronauts in lunar orbit demonstrated that the maria on the Moon's Earth-facing side are 2 to 5 km below the average lunar elevation. In contrast, the cratered terrain on the lunar far side are typically 4 to 5 km above the average lunar elevation. Terrae are therefore often called **highlands**. The flat, low-lying, dark gray maria cover only 15% of the lunar surface. The remaining 85% is light gray, heavily cratered highlands.

Lunar mountain ranges that rim the vast maria basins suggest ancient impacts far more violent than those that produced typical craters. One such mountain range can be seen

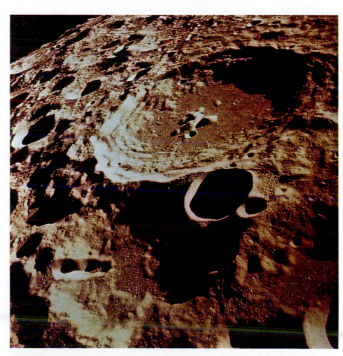

FIGURE 9-5 Details of a Lunar Crater This photograph, taken from lunar orbit by astronauts in 1969, shows a typical view of the Moon's heavily cratered far side. The large crater near the middle of the picture is approximately 80 km (50 mi) in diameter. Note the crater's central peak and the numerous tiny craters that pockmark the lunar surface. (NASA)

FIGURE 9-6 Terrae and Maria This photograph by astronauts in 1972 shows the main difference between the near and far sides of the Moon. The Earth-facing side of the Moon (toward the upper left) is characterized by dark, flat maria. The far side of the Moon (toward the lower right) has heavily cratered terrain and virtually no maria. (NASA)

around the edges of Mare Imbrium in Figure 9-2. Although they take their names from famous terrestrial ranges such as the Alps, the Apennines, and the Carpathians, the lunar mountains were not formed in the same way as mountains on Earth. The Earth's major mountain ranges were created by collisions between lithospheric plates. Analyses of moon rocks and observations from lunar orbit imply that the lunar mountain ranges were thrust up by violent impacts of the huge meteoroids that carved out the maria basins.

9-2 Manned exploration of the lunar surface was one of the greatest adventures in human history

The 1960s will long be remembered as the decade when humanity reached out and touched another world. During a few historic moments, people walked on the Moon. An incredible dream became a reality.

Lunar samples helped answer many questions about the Moon and also shed light on the origin of the Earth. Because Earth is a geologically active planet, all traces of its origins have been erased. Typical terrestrial surface rocks are only a few hundred million years old, which is only a small fraction of the Earth's age. Moon rocks, however, have been undis-

turbed for billions of years. Lunar exploration provided a valuable perspective on the creation of the Earth and the birth of the entire solar system.

Lunar missions began in 1959 when the Soviet Union sent three small spacecraft toward the Moon. American attempts to reach the Moon began in the early 1960s with Project Ranger. Equipped with six television cameras, the Ranger spacecraft transmitted close-up views of the lunar surface taken in the final few minutes before it crashed onto the Moon. Figure 9-7 shows a view from *Ranger 9* just before it hit the floor of the crater Alphonsus. A photograph of the same area taken through an Earth-based telescope demonstrates the dramatic increase in resolution that was achieved by the spacecraft.

A complete high-resolution photographic survey of the lunar surface was the mission of the Orbiter program, under which five spacecraft were launched between August 1966 and August 1967. In all, the five Orbiters sent back a total of 1950 high-resolution pictures covering $99\frac{1}{2}$% of the Moon. In some of these photographs, features as small as one meter across can be seen. Because these pictures revealed the lunar surface in close detail, they were essential in helping NASA scientists choose landing sites for the astronauts.

An important step in preparing for a manned landing on the Moon was the soft-landing of a spacecraft on the lunar

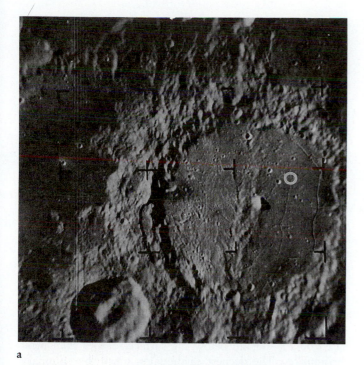

a

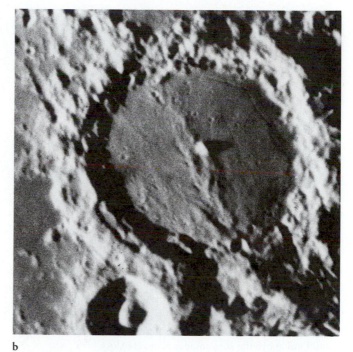

b

FIGURE 9-7 The Moon from *Ranger 9* and the Earth
(**a**) Seconds before it crashed onto the Moon, *Ranger 9* sent back this picture of the crater Alphonsus (the white circle indicates the

impact target). Note that the resolution is far better than that of the Earth-based telescopic view shown in (**b**). The diameter of Alphonsus is about 129 km (80 mi). (NASA)

surface. This was the purpose of the Surveyor program. Between June 1966 and January 1968, five Surveyors successfully touched down on the Moon, sending back pictures and data directly from the lunar surface. These missions dispelled a variety of fears, such as the worry that a spacecraft or an astronaut stepping onto the lunar surface might simply sink out of sight into a deep layer of dust. Figure 9-8 shows an astronaut visiting *Surveyor 3*, $2\frac{1}{2}$ years after it had landed on the Moon.

The six manned lunar landings between 1969 and 1972 were made in progressively more challenging terrain. The first two, *Apollo 11* and *Apollo 12,* set down in maria. The remaining four landings (*Apollo 14* through *Apollo 17*) culminated in rugged mountains east of Mare Serenitatis. These sites gave the astronauts the best opportunity to land safely and to explore a wide variety of geologically interesting features. Figure 9-9 shows one of the Apollo bases.

Between 1966 and 1976, a series of Soviet spacecraft also orbited the Moon and soft-landed on its surface. The first soft-landing, by *Luna 9,* came four months before the United States' *Surveyor 1* arrived. The first lunar satellite, *Luna 10,* orbited the Moon four months before *Orbiter 1*. In the 1970s, robotic spacecraft named *Luna 16, 20,* and *24* brought samples of rock back to Earth, and *Luna 17* and *21* landed vehicles that roamed the lunar surface. Since the successful mission of *Luna 24* in August 1976, no spacecraft has returned to the Moon.

FIGURE 9-8 An Astronaut Visits *Surveyor 3* *Surveyor 3* successfully soft-landed in Oceanus Procellarum (Ocean of Storms) in 1967. Like the four other Surveyors, it sampled the lunar soil and sent back pictures to Earth. *Surveyor 3* was visited $2\frac{1}{2}$ years later by an *Apollo 12* astronaut who took pieces of the spacecraft back to Earth for analysis. (NASA)

FIGURE 9-9 The *Apollo 15* Base The last three Apollo missions used surface vehicles called Lunar Rovers. This photograph shows the *Apollo 15* landing site at the foot of the Apennine Mountains near the eastern edge of Mare Imbrium. The hill in the background is about 5 km behind the lunar module and rises about 3 km above the surrounding plain. (NASA)

9-3   The Moon has no magnetic field but may have a small, solid core beneath a thick mantle

The Moon landings gave scientists an unprecedented opportunity to explore an alien world. The Apollo landing modules were therefore packed with an assortment of apparatus and equipment that the astronauts deployed around the landing sites (Figure 9-10). For example, all the missions carried magnetometers, which confirmed that the Moon has no magnetic field. This suggests that the Moon does not have a molten core. Careful magnetic measurements of lunar rocks, however, indicated that the Moon did have a weak magnetic field when the rocks solidified billions of years ago. The implication is that the Moon originally had a small, molten, iron-rich core, but that the core solidified as the Moon cooled, and the lunar magnetic field disappeared. This solid core is probably less than 700 km (435 mi) in diameter.

Seismometers set up by the astronauts indicated that the Moon exhibits far less seismic activity than does the Earth. Roughly 3000 **moonquakes** were detected per year, whereas a similar seismometer on Earth would record hundreds of thousands of earthquakes per year. Furthermore, typical moonquakes are far weaker than typical earthquakes. A major moonquake, which typically measures 0.5 to 1.5 on the Richter scale, would go unnoticed by a person standing near the epicenter. For comparison, a major earthquake is in the range of 6 to 8 on that scale.

Analysis of the feeble moonquakes reveals that most originate 600 to 800 km below the surface, deeper than most earthquakes. The deepest earthquakes mark the boundary between the solid lithosphere and the plastic asthenosphere. This is because the lithosphere is brittle enough to fracture and produce seismic waves, whereas the deeper rock of the asthenosphere oozes and flows rather than cracks. We may conclude that the Moon's lithosphere is about 800 km thick (Figure 9-11).

The lunar asthenosphere probably extends down to the iron-rich core, which is more than 1400 km below the lunar surface. The asthenosphere and the lower part of the lithosphere are presumably composed of relatively dense iron-rich

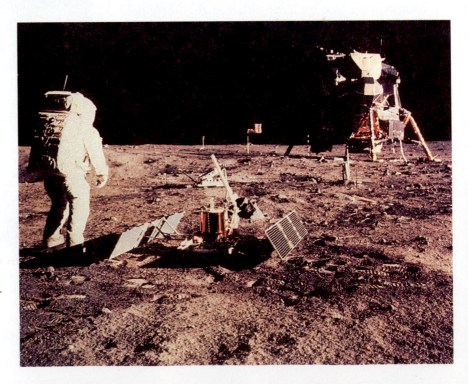

FIGURE 9-10 Seismic Equipment on the Moon This view of the *Apollo 11* base shows an astronaut standing alongside a seismometer that detected moonquakes and transmitted data back to Earth. By combining data from seismometers left on the Moon by Apollo astronauts, scientists have been able to deduce the structure of the Moon's interior. The *Apollo 11* lunar module and some additional equipment are in the background. (NASA)

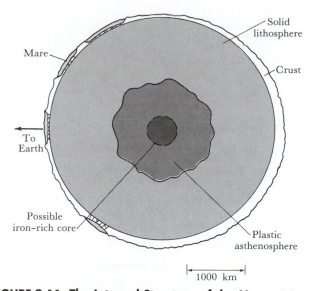

FIGURE 9-11   The Internal Structure of the Moon   Like
the Earth, the Moon probably has a crust, a mantle, and a core.
The lunar crust has an average thickness of about 60 km on the
Earth-facing side but about 100 km on the far side. The crust and
solid upper mantle form a lithosphere about 800 km thick. The
plastic (nonrigid) asthenosphere extends all the way to the base of
the mantle. If the Moon has an iron-rich core, it is probably less
than 700 km in diameter.

rocks. The upper part of the lithosphere is a less dense crust,
with an average thickness of about 60 km on the Earth-
facing side and up to 100 km on the far side. For comparison,
the thickness of the Earth's crust ranges from 5 km under the
oceans to about 30 km under major mountain ranges on the
continents.

Seismometers left on the Moon continued to transmit data
for many years, so geologists were able to monitor meteoritic
impacts. These Apollo seismometers were sensitive enough to
detect a hit by a grapefruit-sized meteoroid anywhere on the
lunar surface. Every year the Moon is struck by 80 to 150
meteoroids having masses between 100 g and 1000 kg
(roughly $\frac{1}{4}$ lb to 1 ton).

9-4   Gravitational forces between the Earth and the Moon produce tides

The seismometers left at the Apollo landing sites revealed
that moonquakes are more frequent at new moon and at full
moon. Geologists concluded that moonquakes are influenced
by tidal forces.

Tidal forces are differences in the gravitational pull at
different points in an object. The most familiar example of a
tidal force is the gravitational pull of the Moon on the Earth.
Red arrows in Figure 9-12a indicate the strength and direction of the gravitational force of the Moon at several locations on the Earth. Notice that the side of the Earth closest to
the Moon feels a greater gravitational pull than the Earth's
center does. Because of this difference, the side of the Earth

facing the Moon feels an overall pull, or tidal force, toward
the Moon. Similarly, the Earth's center is subjected to a
greater gravitational pull toward the Moon than is any place
on the Earth's hemisphere facing away from the Moon. Over
this hemisphere, the tidal forces are directed away from the
Moon. The net effect, sketched in Figure 9-12b, is a deform-
ing of the Earth, stretching it into the shape of a football.

Actually, the tides in the oceans result because *both* the
Moon and the Sun act to distort the shape of the Earth. This
tidal distortion is greatest when the Sun, Moon, and Earth
are aligned (at either new moon or full moon), producing
large shifts in water level that are called **spring tides** (Figure
9-13). At first quarter and last quarter, when the Sun and
Moon form a right angle with the Earth, the tidal distortion
of the oceans is the least pronounced, producing smaller tidal
shifts called **neap tides.**

It is easy to notice ocean tides, because the water level goes
up and down. Harder to detect are similar but much smaller
motions of the solid Earth, which is also stretched by the
gravitational forces of the Sun and Moon.

We can now see how tidal forces might trigger moonquakes. Just as the gravitational forces of the Sun and Moon
deform the Earth, so also the Sun and Earth deform the
Moon. The Moon is alternately squeezed and stretched as it
orbits our planet. Because the Earth is 81 times more massive
than the Moon, the Earth produces large tidal deformations

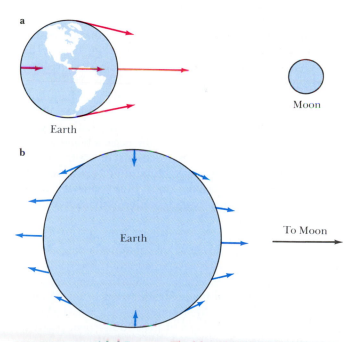

FIGURE 9-12   Tidal Forces   The Moon induces tidal forces on
the Earth because the strength of the Moon's gravity varies over
the Earth. (a) Red arrows indicate the strength and direction of the
Moon's gravitational pull at selected points on the Earth. (b) Blue
arrows indicate the strength and direction of the tidal forces acting
on the Earth. At any location, the tidal force equals the Moon's
gravitational pull at that point minus the gravitational pull of the
Moon at the center of the Earth.

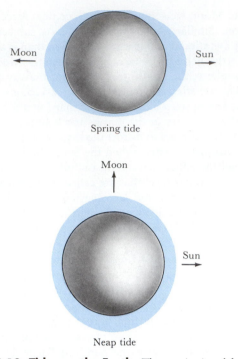

FIGURE 9-13 Tides on the Earth The gravitational forces of the Moon and Sun deform the oceans. The greatest deformation (spring tides) occurs when the Sun, Earth, and Moon are aligned. The least deformation (neap tides) occurs when the Sun, Earth, and Moon form a right angle.

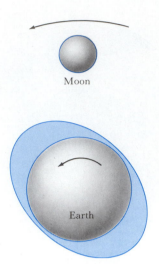

FIGURE 9-14 The Moon's Tidal Recession Because of the Earth's rapid rotation, the tidal bulge on the oceans is dragged ahead of the Moon's position in the sky. This bulge on the leading side of the Earth produces a small forward force on the Moon that causes it to spiral slowly away from the Earth.

of the lunar surface. The average tidal deformation on the Earth is about 1 foot, but about 60 feet on the Moon. The strongest gravitational stresses occur when the Moon is nearest the Earth (at perigee)—the time when the Apollo seismometers reported the highest frequency of moonquakes.

Tidal effects also explain why the same side of the Moon always faces the Earth. Under the Earth's gravitational pull, the Moon bulges slightly—much as the oceans bulge out on opposite sides of the Earth. Tidal forces that create the bulges act to keep the Moon's elongated shape aimed toward Earth. As we shall see in later chapters, most of the satellites in the solar system exhibit synchronous rotation, keeping the same side facing their planet.

The tidal interactions between the Earth and the Moon also gradually enlarge the Moon's orbit and slow the Earth's rotation. The Earth rotates about its axis faster than the Moon revolves around the Earth. The Earth's rapid rotation carries the tidal bulge of the oceans forward of the Moon's orbit. As viewed from the Moon, the tidal bulge on the Earth is always aimed slightly ahead of the Moon's own position (Figure 9-14). This bulge produces a small but constant gravitational force, which tugs the Moon forward and lifts it into a more distant orbit around the Earth. In other words, the Moon is very slowly spiraling away from the Earth, at a rate of about 4 cm per year.

As the Moon moves away from the Earth, its sidereal period becomes longer (recall Kepler's third law). Meanwhile, friction between the oceans and the Earth's surface is gradually slowing the Earth's rotation. The Earth's day is therefore becoming longer and longer, by approximately 0.002 second per century.

At some point in the distant future, the Earth will be rotating so slowly that a solar day will equal a lunar month. (Both will then equal 47 of our present solar days.) At that time, the Earth's tidal bulge will be aimed directly at the Moon, and the Moon will stop spiraling away from the Earth. The Earth will then keep its same side facing the Moon, just as the Moon presently keeps the same side facing the Earth. As we shall see in Chapter 16, Pluto and its moon have already achieved this stable configuration.

9-5 Lunar rocks were formed 3 to 4.5 billion years ago

The Apollo astronauts brought back 382 kg (842 lb) of lunar rocks, which have provided important information about the early history of the solar system. All of the lunar samples are igneous rocks, which tells us that the Moon's surface was once molten. Indeed, Moon rocks are composed mostly of the same minerals that are found in terrestrial volcanic rocks.

Astronauts who visited the maria discovered that these dark regions of the Moon are covered with basaltic rock quite similar to the dark-colored rocks formed by lava from volcanoes on Hawaii and Iceland. The rock of these low-lying lunar plains is called **mare basalt** (Figure 9-15).

In contrast to the dark maria, the lunar highlands are covered with a light-colored rock called **anorthosite** (Figure 9-16). On Earth, anorthositic rock is found only in such very old mountain ranges as the Adirondacks in the eastern United States. In comparison to the mare basalts, which have more of the heavier elements like iron, manganese, and tita-

FIGURE 9-15 Mare Basalt This 1.531-kg (3½-lb) specimen of mare basalt was brought back by *Apollo 15*. It is a vesicular basalt, so-called because of the holes, or vesicles, covering 30% of its surface. Gas must have dissolved under pressure in the lava from which this rock solidified. When the lava reached the airless lunar surface, bubbles formed as the pressure dropped. Some of the bubbles were frozen in place as the rock cooled. (NASA)

FIGURE 9-16 Anorthosite The light-colored lunar terrae are covered with an ancient type of rock called anorthosite, which is believed to be the material of the original lunar crust. This sample, called the Genesis rock by the *Apollo 15* astronauts who picked it up at the base of the Apennine Mountains, has an age of approximately 4.1 billion years. (NASA)

nium, anorthosite is rich in calcium and aluminum. Anorthosite is therefore less dense than basalt.

By carefully measuring the abundances of trace amounts of radioactive elements in lunar samples, geologists have determined that lunar rocks formed more than 3 billion years ago (Box 9-2). However, anorthosite is much older than the mare basalts. Typical anorthositic specimens from the highlands are between 4.0 and 4.3 billion years old, and one rock brought back by the *Apollo 17* astronauts is nearly 4.6 billion years old. All these extremely ancient specimens are probably samples of the Moon's original crust. In sharp contrast, all the mare basalts are between 3.1 and 3.8 billion years old.

Although lunar rocks bear a strong resemblance to terrestrial rocks, there are some important differences. For one, every terrestrial rock contains some water, but lunar rocks are totally dry. There is absolutely no evidence that water has ever existed on the Moon. Because the Moon lacks both an atmosphere and water, it is not surprising that the astronauts found no traces of life.

Meteoritic bombardment is the only source of "weathering" to which rocks on the lunar surface are subjected. In fact, the entire lunar surface is covered with a layer of fine powder and rock fragments produced by 4.5 billion years of relentless meteoritic bombardment. This layer, which ranges in thickness from 1 to 20 m, is called the **regolith** (from the Greek meaning "blanket of stone"); because the term *soil* suggests the presence of decayed biological matter, it does not properly apply to the Moon's surface.

In addition to churning up the regolith, meteoritic impacts also melt rocks to produce glass. Many lunar samples are coated with a thin layer of smooth, dark glass created when the surface of the rock suddenly melted and then rapidly solidified. Moreover, small black glass beads are common in

the lunar regolith. Presumably, these glass spheres were formed from droplets of molten rock hurled skyward by the impact of a meteoroid. Finally, many lunar samples bear the scars of high-speed meteoritic dust. Dust grains traveling at thousands of kilometers per hour produce tiny craters on the exposed surfaces of moon rocks (Figure 9-17). Because of

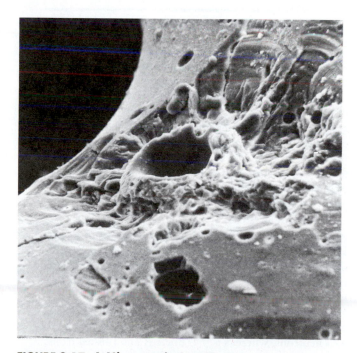

FIGURE 9-17 A Microscopic Crater The upper surfaces of many moon rocks are covered with tiny craters produced by the impact of high-speed meteoritic dust grains. These glass-lined craters are typically less than 1 mm in diameter. (NASA)

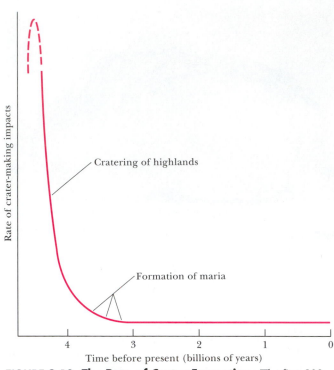

FIGURE 9-18 The Rate of Crater Formation The first 800 million years of the Moon's history was dominated by frequent crater-making impacts. Near the end of this intense bombardment, several large impacts gouged out the mare basins. For the past $3\frac{1}{2}$ billion years, the impact rate has been quite low.

these tiny, glass-lined pits, Moon rocks often seem to sparkle when held in the sunshine.

By correlating the ages of the Moon rocks with the density of craters at the sites where they were collected, geologists have found that the rate of impacts on the Moon has changed over the ages. The ancient, heavily cratered lunar highlands are evidence of intense bombardment that dominated the Moon's early history. For nearly a billion years, rocky debris left over from the formation of the planets rained down upon the Moon's young surface. As Figure 9-18 shows, this barrage extended from 4.6 billion years ago, when the Moon's surface solidified, until about 3.8 billion years ago.

The frequency of impacts gradually tapered off, however, as meteoroids and planetesimals were swept up by the newly formed planets. This crater-making era ended about the time when several large planetesimals gouged out the mare basins. A series of lava flows that occurred between 3.8 and 3.1 billion years ago flooded the mare basins with iron-rich magma that oozed up from the Moon's still-molten mantle. The relatively crater-free maria tell us that the impact rate over the past 3 billion years has been quite low.

9-6 The Moon probably formed from debris cast into space when a huge planetesimal struck the proto-Earth

Prior to the Apollo program, there were three different theories about the origin of the Moon. First, the **fission theory**

BOX 9-2

Radioactive Age-Dating

The Apollo program gave geologists the opportunity to get their hands on extremely ancient rocks. By examining these lunar samples, scientists began to piece together a history of important events that happened shortly after our solar system was created. To get the story right, however, geologists had to determine the ages of the lunar rocks. Fortunately, most rocks contain trace amounts of radioactive elements such as uranium. The relative abundances of various radioactive isotopes and their decay products provided the key that geologists needed.

As we saw in Box 7-1, every atom of a particular element has the same number of protons in its nucleus. However, different isotopes of the same element have different numbers of neutrons in their nuclei. For example, the common isotopes of uranium are ^{235}U and ^{238}U. Each isotope of uranium has 92 protons in its nucleus (correspondingly, uranium is the element 92 on the periodic chart; see Figure 7-6). However, a ^{235}U nucleus contains 143 neutrons, whereas a ^{238}U nucleus has 146 neutrons.

A radioactive nucleus is unstable because it contains too many protons or too many neutrons. It therefore ejects

particles until it becomes stable. In doing so, a nucleus may change from one element to another. Physicists refer to this transmutation as "decay."

Some radioactive isotopes decay rapidly, others decay slowly. Physicists find it convenient to talk about the decay rate in terms of an isotope's half-life. The **half-life** of an isotope is the time interval in which one-half of the nuclei decay. For example, the half-life of ^{235}U is 710 million years. This means that if you start out with 1 kg of ^{235}U, after 710 million years you will have only $\frac{1}{2}$ kg of ^{235}U left; the other $\frac{1}{2}$ kg will have turned into other elements. Several isotopes useful for age-dating rocks are listed in the table.

To see how geologists date the Moon rocks, consider the slow conversion of rubidium (^{87}Rb) into strontium (^{87}Sr). Over the years, the amount of ^{87}Rb in a rock decreases, while the amount of ^{87}Sr increases. Dating the rock is not simply a matter of measuring its ratio of rubidium to strontium, however, because the rock already had some strontium in it when it was formed. Geologists must therefore determine how much fresh strontium came from the decay of rubidium *after* the rock's formation.

holds that the Moon was pulled out from a rapidly rotating proto-Earth. Second, the **capture theory** posits that the Moon was formed elsewhere in the solar system and then drawn into orbit about the Earth by gravitational forces. Third, the **cocreation theory** proposes that the Earth and the Moon were formed at the same time but separately.

One fact used to support the fission theory is that the Moon's average density ($3340 \ kg/m^3$) is comparable to that of the Earth's outer layers, as would be expected if the Moon had been formed from material flung out by a proto-Earth's rapid rotation. However, most geologists counter this argument by pointing to the significant differences between lunar and terrestrial rocks—particularly the differences in water content and in the relative abundances of volatile and refractory elements.

Volatile elements (such as potassium and sodium) melt and boil at relatively low temperatures, whereas **refractory elements** (such as titanium, calcium, and aluminum) melt and boil at much higher temperatures. Compared to terrestrial rocks, the lunar rocks have slightly greater proportions of the refractory elements (and slightly lesser proportions of the volatile elements). This composition suggests that the Moon formed from material that was heated to a higher temperature than the rock from which Earth formed. Some of the volatile elements boiled away, leaving the young moon relatively enriched in the refractory elements.

Proponents of the capture theory use these differences to argue that the Moon formed elsewhere and was later captured by the Earth. They also note that the plane of the Moon's orbit is close to the plane of the ecliptic, as would be expected if the Moon had originally been in orbit about the Sun but was later captured by the Earth.

There are difficulties with the capture theory, however. If the Earth did capture the Moon intact, then a host of conditions must have been met entirely by chance. The Moon must have coasted to within 50,000 km of the Earth at exactly the right speed to leave a solar orbit and adopt an Earth orbit without ever actually hitting our planet.

Because it is easier for a planet to capture swarms of tiny rocks, proponents of the cocreation theory argue that the Moon formed from just such rocky debris. Great numbers of rock fragments orbited the newborn Sun in the plane of the ecliptic along with the protoplanets. Heat from the protosun could have baked the water and volatile elements out of these smaller rock fragments before their capture into orbit about the proto-Earth. Then, just as planetesimals accreted to form the proto-Earth in orbit about the Sun, the fragments in orbit about the Earth accreted to form the Moon. However, this theory fails to explain why our nearest neighbors—Venus and Mars—were not similarly bestowed with large moons.

A recently proposed fourth theory, called the **collisional ejection theory**, suggests that the Earth was struck by an object perhaps as large as Mars; this collision ejected debris from which the Moon formed. Clues provided by the lunar rocks along with what we know about the formation of the planets favor this most recent hypothesis. This theory arose in the mid-1980s, when several teams of scientists realized that the early history of the solar system was dominated by collisions between objects. As we have seen (recall Figure 7-18), smaller objects collided and fused together to form the inner planets. Some of these collisions must have been quite spectacular, especially near the end of planet forming, when most of the mass of the inner solar system was contained in the protoplanets and a few dozen large, asteroidlike objects. According to the collisional ejection theory, one such object struck the proto-Earth 4.6 billion years ago. A supercomputer simulation of this cataclysm is shown in Figure 9-19. In the simulation, energy released during the collision produces a huge plume of vaporized rock that squirts out from the point of impact. As this ejected material cools, it coalesces to form the Moon, as shown in Figure 9-20.

The collisional ejection theory is in agreement with many of the known facts about the Moon. For example, rock vaporized by the impact would have been depleted of volatile elements and water, leaving the moon rocks we now know. If the collision took place after chemical differentiation had occurred on Earth, when our planet's iron sunk to its center, then relatively little iron would have been ejected, which would account for the Moon's small iron-rich core. Furthermore, most of the debris from the collision would lie near the plane of the ecliptic, as long as the orbit of the impacting asteroid had also been in that plane. Indeed, the impact of an object large enough to create the Moon could have also tipped the Earth's axis of rotation and so inaugurated the seasons.

To do this, geologists use as a reference another isotope of strontium whose concentration has remained constant. In this case, they use ^{86}Sr, which is stable and is not created by radioactive decay; it is said to be "nonradiogenic." Dating a rock thus entails comparing the ratio of radiogenic and nonradiogenic strontium ($^{87}Sr/^{86}Sr$) in it to the ratio of radioactive rubidium to nonradiogenic strontium ($^{87}Rb/^{86}Sr$). Because the half-life for converting ^{87}Rb into ^{87}Sr is known, the rock's age can then be calculated from these ratios.

Original radio-active isotope	Half-life (billions of years)	Final stable isotope
Potassium (^{40}K)	1.3	Argon (^{40}Ar)
Rubidium (^{87}Rb)	47.0	Strontium (^{87}Sr)
Uranium (^{235}U)	0.7	Lead (^{207}Pb)
Uranium (^{238}U)	4.5	Lead (^{206}Pb)

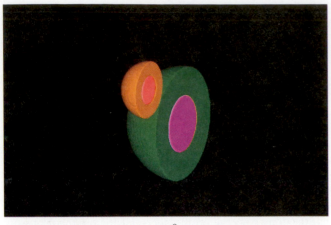

t = 0

t = 400 s

t = 800 s

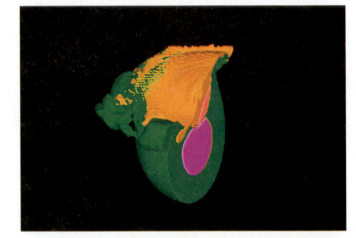

t = 1200 s

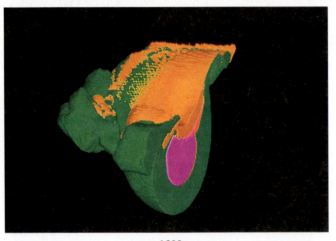

t = 1600 s

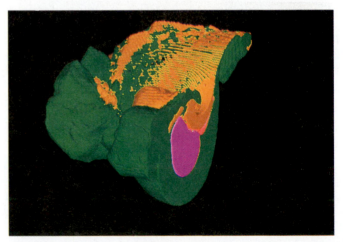

t = 2000 s ≃ ½ hr

FIGURE 9-19 An Asteroid Hitting the Earth This supercomputer simulation shows the effects of a Mars-sized asteroid hitting the Earth. The mass of the asteroid is one-tenth of the Earth's mass, and its speed at impact is 8 km/s (18,000 mi/hr). Colors identify core and mantle material of the Earth and the asteroid. Cutaway views reveal the location and movement of this material at six intervals during the half hour after the impact. According to the collisional ejection theory, the Moon formed from debris ejected by such a collision. (Courtesy of M. Kipp and J. Melosh; Sandia National Laboratories)

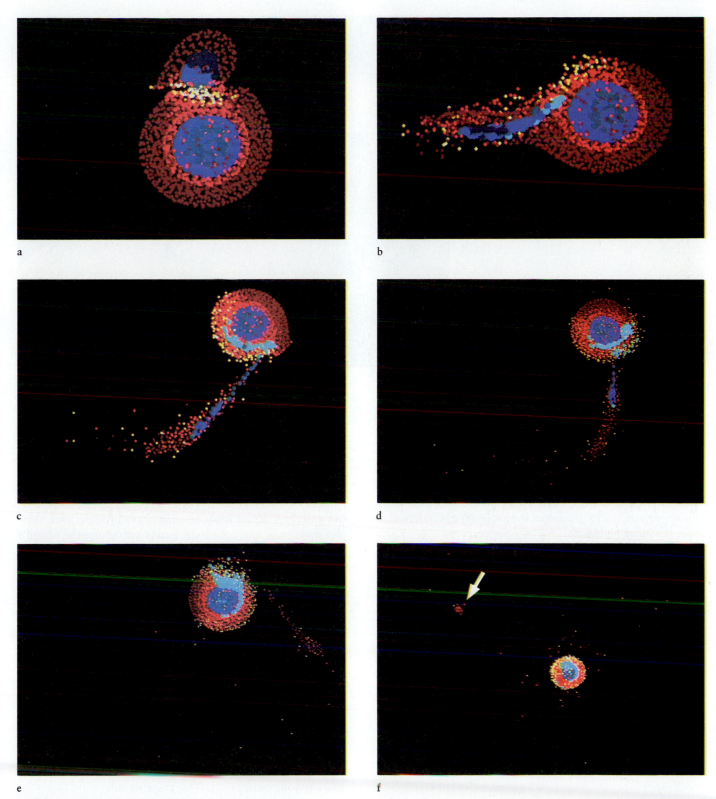

a

b

c

d

e

f

FIGURE 9-20 The Moon's Creation This supercomputer simulation shows the creation of the Moon from material ejected by the impact of a large asteroid on the Earth. To follow the ejected material as it moves away from the Earth, successive views show increasingly larger volumes of space. Blue and green indicate iron from the cores of the Earth and the asteroid; red, orange, and yellow indicate rocky mantle material. In this simulation, the impact ejects both mantle and core material, but most of the iron falls back onto the Earth. The surviving ejected rocky matter coalesces to form the Moon (indicated by arrow) during its first orbit of the Earth. (Courtesy of W. Benz)

FIGURE 9-21 Eratosthenes Eratosthenes is a young crater 61 km in diameter on the southern edge of Mare Imbrium. Another young crater, Copernicus, is near the horizon in this photograph, taken by astronauts in 1972. Eratosthenes also appears near the lower edge of the Earth-based view in Figure 9-2. (NASA)

The surface of the newborn Moon was probably molten for many years, both from heat released during the impact of rock fragments falling into the young satellite and from the decay of short-lived radioactive isotopes. As the Moon gradually cooled, low-density lava floating on the Moon's surface began to solidify into the anorthositic crust that exists today. The barrage of rock fragments ended about 3.8 billion years ago, with the impacts producing the craters that cover the lunar highlands.

Recorded among the final scars at the end of this crater-making era are the impacts of more than a dozen asteroid-size objects, each measuring at least 100 km across. As these huge rocks rained down on the young Moon, they blasted out the vast mare basins. Meanwhile, heat from the decay of long-lived radioactive elements like uranium and thorium began to melt the inside of the Moon. Then, from 3.8 to 3.1 billion years ago, great floods of molten rock gushed up from the lunar interior, filling the impact basins and creating the maria we see today.

Very little has happened on the Moon since those ancient times. A few fresh craters have been formed (Figure 9-21), but the astronauts visited a world that has remained largely unchanged for over 3 billion years. The history of our own planet was just beginning 3 billion years ago, when the first organisms began populating the new oceans.

Many questions and mysteries still remain. The six American and three Soviet lunar landings have brought back samples from only nine locations, barely scratching the Moon's surface. We still know very little about the Moon's far side and nothing at all about the Moon's poles. Is the Moon's interior molten? Does the Moon really have an iron core? Could there be ice at the Moon's poles? How old are the youngest lunar rocks? Did lava flows occur over western Oceanus Procellarum only 2 billion years ago, as crater densities there suggest? Is the Moon really geologically dead, or does it just look dead, simply because our examination of the lunar surface has been so cursory? Such questions can only be answered by returning to the Moon.

KEY WORDS

Terms preceded by an asterisk are discussed in the boxes.

anorthosite	crater	mare (*plural* maria)	spring tide
capture theory (of Moon's formation)	fission theory (of Moon's formation)	mare basalt	synchronous rotation
cocreation theory (of Moon's formation)	*half-life	moonquake	terra (*plural* terrae)
collisional ejection theory (of Moon's formation)	highlands	neap tide	tidal force
	libration	refractory element	volatile element
		regolith	

KEY IDEAS

• The Earth-facing side of the Moon displays light-colored, heavily cratered highlands and dark-colored, smooth-surfaced maria; the Moon's far side has only one small mare. Much of our knowledge about the Moon came from lunar exploration in the 1960s and early 1970s.

Analysis of seismic waves from moonquakes and meteoroid impacts indicates that the Moon has a crust thicker than the Earth's (and thickest on the far side of the Moon), a mantle with a thickness equal to about 75% of the Moon's radius, and perhaps a small, solid iron core.

The Moon's lithosphere is far thicker than that of the Earth, and the Moon's asthenosphere probably extends from the base of the lithosphere to the core.

• The Moon has no magnetic field today, although it may have had a weak magnetic field billions of years ago.

• Gravitational interactions between the Earth and the Moon produce tides in the oceans of the Earth and in the solid bodies of both worlds.

Tidal interactions have locked the Moon into synchronous rotation with the Earth and have given rise to a small constant acceleration of the Moon in its orbit, thereby causing it to spiral slowly away from the Earth.

• The anorthositic crust exposed in the highlands was formed between 4.0 and 4.3 billion years ago, whereas the mare basalts solidified between 3.1 and 3.8 billion years ago.

The Moon's surface has undergone very little change over the past 3 billion years.

Meteoroid impacts have been the only significant "weathering" agent on the Moon; the Moon's regolith ("soil" layer) was formed by meteoritic action.

All of the lunar rock samples are igneous rocks formed largely of minerals found in terrestrial rocks; the lunar rocks contain no water and also differ from terrestrial rocks in being relatively enriched in the refractory elements and depleted in the volatile elements.

• The collisional ejection theory of the Moon's origin holds that the proto-Earth was struck by a huge asteroid, and debris from this collision coalesced to form the Moon.

• The Moon was molten in its early stages, and the anorthositic crust solidified from low-density magma that floated to the lunar surface; the mare basins were created later, by the impact of planetesimals, and filled with lava from the lunar interior.

REVIEW QUESTIONS

1. What kind of features can you see on the Moon with a small telescope?

2. Why do you suppose more lunar detail is visible through a telescope when the Moon is near quarter phase than when it is at full phase?

3. Why do you suppose temperature variations between day and night on the Moon are much more severe than on the Earth?

4. Why are Moon rocks so much older than Earth rocks, even though both worlds formed at nearly the same time?

5. On the basis of moon rocks brought back by the astronauts, explain why the maria are dark colored but the lunar highlands are light colored.

6. Why are there so few craters on the maria?

7. Briefly describe the main differences and similarities between Moon rocks and Earth rocks.

8. How do we know that the maria were formed *after* the lunar highlands?

9. Why do you suppose that no Apollo mission landed on the far side of the Moon?

10. What is a tidal force? How do tidal forces produce tides in the Earth's oceans?

11. What is the difference between spring tides and neap tides?

12. Explain why moonquakes occur more frequently when the Moon is at perigee than at other locations along its orbit.

13. Why do most scientists favor the collisional ejection theory for the Moon's formation?

14. Some people who supported the fission theory proposed that the Pacific Ocean basin is the scar left when the Moon pulled away from the Earth. Explain why this idea is probably wrong.

15. Why are nearly all of the lunar maria located on the side of the Moon facing the Earth? Why are there more craters on the far side of the Moon than on the near side?

ADVANCED QUESTIONS

Tips and tools . . .

Recall that the average density of an object is its mass divided by its volume, and that the volume of a sphere is $\frac{4}{3}\pi r^3$, where r is the sphere's radius. Recall also that the acceleration of gravity on the Earth's surface is 9.8 m/s^2. You may find it useful to know that a 1-pound (1 lb) weight presses down on the Earth's surface with a force of 4.448 newtons. You might want to review Newton's law of gravitation in Chapter 4 and consult Box 9-1 and the appendixes for any additional data.

16. Using the diameter and mass of the Moon given in Box 9-1, verify that the Moon's average density is 3340 kg/m^3. Explain why this average density implies that the Moon's interior contains much less iron than the Earth's.

17. How much would an 80-kg person weigh on the Moon? How much does that person weigh on the Earth?

18. Estimate the rate at which the Moon is gaining mass from meteoritic impacts. On the basis of your calculation, how long will it be before the Moon doubles its mass? Compare your answer with the age of the universe (about 15 billion years) and explain why your extrapolation into the distant future is probably unreasonable.

19. Find the average distance between the Earth and the Moon when the length of the day and the lunar month will both be equal to 47 of our present days.

20. Can you think of any other weathering processes that might occur on the Moon besides those related to meteoritic impacts?

DISCUSSION QUESTIONS

21. Comment on the idea that without the presence of the Moon in our sky, astronomy would have developed far more slowly.

22. Compare the advantages and disadvantages of exploring the Moon with astronauts as opposed to using mobile, unmanned instrument packages.

23. Describe how you would empirically test the idea that human behavior is related to the phases of the Moon. What problems are inherent in such testing?

24. How would our theories of the Moon's history have been affected if astronauts had discovered sedimentary rock on the Moon?

25. Imagine that you are planning a lunar landing mission. What type of landing site would you select? Where might you land to search for evidence of recent volcanic activity?

OBSERVING PROJECTS

26. Use a telescope to observe the Moon. Compare the texture of the lunar surface you see on the maria with that of the lunar highlands. How does the visibility of details vary with distance from the terminator (the boundary between day and night on the Moon)?

27. Observe the Moon through a telescope every few nights over a period of two weeks between new moon and full moon. Make sketches of various surface features such as craters, mountain ranges, and maria. How does the appearance of these features change with the Moon's phase? Which features are most easily seen at a low angle of illumination? Which features show up best with the Sun nearly overhead?

28. If you live near the ocean, observe the tides to see how the times of high and low tides are correlated with the position of the Moon in the sky.

FOR FURTHER READING

Beatty, J. K. "The Making of a Better Moon." *Sky & Telescope,* December 1986. This brief but excellent article summarizes the collisional ejection hypothesis and includes a "report card" that evaluates the various theories of the Moon's origin.

Cadogan, P. *The Moon—Our Sister Planet.* Cambridge University Press, 1981. This thoughtful and thorough introduction to lunar geology was written by a scientist who participated in the analyses of Moon rocks brought back by both Apollo astronauts and Soviet automated spacecraft.

French, B. *The Moon Book.* Penguin, 1977. This layperson's introduction to the Moon was written by a noted lunar scientist.

———. "What's New On the Moon?" *Sky & Telescope,* March and April, 1977. These two articles give an excellent summary of what we have learned from the Moon rocks and the Apollo program.

Goldreich, P. "Tides and the Earth–Moon System." *Scientific American,* April 1972. This article explores the consequences of the gravitational interaction between the Earth and the Moon.

Hartmann, W. "The Moon's Early History." *Astronomy,* September 1976. This article, written by a noted planetary scientist, describes the events that shaped the Moon.

Lewis, R. *The Voyages of Apollo: The Exploration of the Moon.* Quadrangle, 1975. This book tells the story of the Apollo program.

Moore, P. *The Moon.* Rand McNally, 1981. This comprehensive yet brief atlas covers the gamut with many attractive illustrations and an excellent collection of maps and photographs.

Schmitt, H. "Exploring Taurus-Littrow: *Apollo 17.*" *National Geographic,* September 1973. This superb description of the final manned lunar landing near crater Littrow in the Taurus mountains was written by the astronaut-geologist who was there.

Weaver, K. "First Explorers on the Moon: The Incredible Story of *Apollo 11.*" *National Geographic,* December 1969. This beautifully illustrated article tells the story of the first manned lunar landing.

Wood, J. "The Moon." *Scientific American,* September 1975. This brief article gives an excellent summary of what we learned from the Moon rocks.

SUN-SCORCHED MERCURY

MERCURY AND THE MOON Mercury (left), like our Moon (right), has a heavily cratered surface and virtually no atmosphere. Mercury's diameter is 4878 km; the Moon's is 3476 km (both worlds are shown here to the same scale). For comparison, the distance from New York to Los Angeles is 3944 km (2451 mi). Daytime temperatures at the equator on Mercury reach 430°C (800°F), hot enough to melt lead and tin. One-half of Mercury was photographed at close range by the *Mariner 10* spacecraft in the mid-1970s. (NASA)

LIKE THE MOON, Mercury is heavily cratered, still bearing the scars of countless impacts that occurred soon after the planets were formed. Unlike the Moon, Mercury has a large iron core and a magnetic field. Mercury's interior is therefore similar to Earth's. Because Mercury is quite small and orbits near the Sun, Earth-based visual observations fail to reveal any surface details, although radio and radar observations have provided information about the planet's temperature and rotation. Only in 1974 did the *Mariner 10* spacecraft fly by Mercury, sending us dramatic pictures of its surface and crucial information about its magnetic field.

Until 1974 we knew little about the smallest planet to form in the warm inner regions of the solar nebula. Information about Mercury was difficult to obtain for two simple reasons: Mercury's small size and its nearness to the Sun. In fact, Mercury is so close to the Sun that even many astronomers have never seen it. Finally, in 1974, an unmanned mission to the inner planets revealed that Mercury has a dual personality: a Moonlike surface but an Earthlike interior.

10-1 Earth-based optical observations of Mercury are difficult to make and often prove disappointing

Mercury is often one of the brightest objects in the sky. Sometimes it shines with a magnitude of −1.9 at greatest brilliance, which is brighter than the brightest stars.

Like all the planets, Mercury shines by reflected sunlight. Mercury reflects about 10% of the sunlight that falls on its rocky surface. The fraction of incoming sunlight that a planet reflects is called its **albedo** (from the Latin for "whiteness"); Mercury's albedo is about 0.1, roughly comparable to that of weathered asphalt.

Although it may not be dim, Mercury is so close to the Sun that it is quite difficult to observe. Mercury circles the Sun at an average distance of only 0.387 AU (57.9 million kilometers, or 36 million miles) along an orbit that is more eccentric than that of any other planet except Pluto. Figure 10-1 is a scale drawing of the orbits of Mercury and the Earth. Box 10-1 summarizes data about the planet.

Mercury can best be seen when it is as far from the Sun in the sky as it can be, at its greatest eastern or western elongation. For a few days near the time of greatest eastern elongation, Mercury appears as an "evening star," hovering low over the western horizon for a short time after sunset.

Alternatively, near the time of greatest western elongation, Mercury can be glimpsed as a "morning star," heralding the rising Sun in the brightening eastern sky.

Because its orbit is so close to the Sun, Mercury's maximum elongation is only 28°. The celestial sphere rotates at 15° per hour (360° divided by 24 hours), so Mercury never rises more than two hours before sunrise nor sets more than two hours after sunset. Unfortunately, Mercury's elliptical orbit and its inclination to the ecliptic often place Mercury much less than 28° from the horizon at sunset or sunrise. Some elongations are thus favorable for viewing Mercury and others are not, as sketched in Figure 10-2. A total of six or seven greatest elongations (both eastern and western) occur each year (Table 10-1), but usually only two of these will be favorable for viewing the planet.

Naked-eye observations of Mercury are best made at dusk or dawn, but the best telescopic views are obtained at midday when the planet is high in the sky, far above the degrading atmospheric effects near the horizon. A yellow filter can eliminate much of the scattered blue light from the sky. The photographs in Figure 10-3, among the finest Earth-based views of Mercury, were taken at midday.

Mercury travels around the Sun faster than any other object in the solar system, taking only 88 days to complete a full orbit. Because its synodic period is 116 days (about $\frac{1}{3}$ year), Mercury passes through inferior conjunction at least three times a year. You might therefore expect to see Mercury occasionally silhouetted against the Sun. Such a passage in front of the Sun is called a **solar transit.** In fact, transits of Mercury across the Sun are not very common, because Mercury's orbit is tilted 7° to the plane of the ecliptic. Mercury therefore usually lies well above or below the solar disk at the moment of inferior conjunction.

In order for a solar transit to take place, the Sun, Mercury, and Earth must be in nearly perfect alignment. This arrangement occurs only in May and November when the Earth is

Mercury Data

Mean distance from Sun:	$0.387 \text{ AU} = 5.79 \times 10^7 \text{ km}$
Maximum distance from Sun:	$0.467 \text{ AU} = 6.98 \times 10^7 \text{ km}$
Minimum distance from Sun:	$0.307 \text{ AU} = 4.60 \times 10^7 \text{ km}$
Mean orbital velocity:	47.9 km/s
Sidereal period:	87.969 days
Rotation period:	58.65 days
Inclination of equator to orbit:	2°(?)
Inclination of orbit to ecliptic:	7° 00′ 16″
Diameter (equatorial):	4878 km
Diameter (Earth = 1):	0.382
Mass:	$3.303 \times 10^{23} \text{ kg}$
Mass (Earth = 1):	0.0553
Mean density:	5430 kg/m^3
Surface gravity (Earth = 1):	0.39
Escape speed:	4.3 km/s
Mean surface temperatures:	Day: 350°C = 662°F = 623 K
	Night: −170°C = −274°F = 103 K

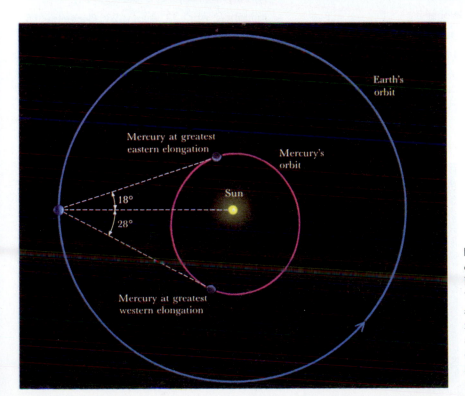

FIGURE 10-1 Mercury's Orbit Mercury orbits the Sun every 88 days. The distance between Mercury and the Sun varies from 70 million kilometers (44 million miles) at aphelion to 46 million kilometers (29 million miles) at perihelion. As seen from the Earth, the angle between Mercury and the Sun at greatest eastern or western elongation can be as large as 28° when Mercury is near aphelion or as small as 18° near perihelion.

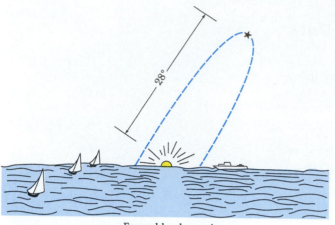
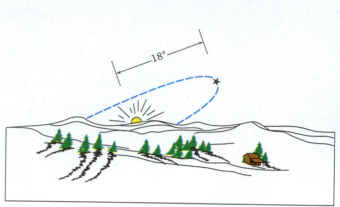
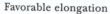

Favorable elongation

Unfavorable elongation

FIGURE 10-2 Favorable Versus Unfavorable Elongations Geometric factors such as the tilt of the Earth's axis, the inclination and ellipticity of Mercury's orbit, and the latitude of the observer on the Earth combine to make an elongation either favorable or unfavorable for viewing Mercury.

TABLE 10-1

Elongations of Mercury, 1993–1999

Year	Western elongations	Eastern elongations
1993	April 5, August 4, November 22	February 21, June 17, October 14
1994	March 19, July 17, November 6	February 4, May 30, September 26
1995	March 1, June 29, October 20	January 19, May 12, September 9
1996	February 11, June 10, October 3	January 2, April 23, August 21, December 15
1997	January 24, May 22, September 16	April 6, August 4, November 28
1998	January 6, May 4, August 31, December 20	March 20, July 17, November 11
1999	April 16, August 14, December 2	March 3, June 28, October 24

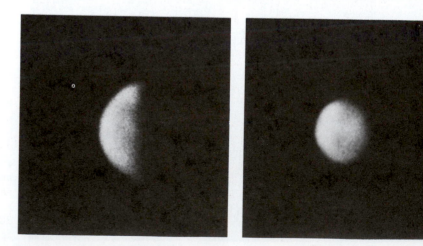

FIGURE 10-3 Earth-Based Views of Mercury These two views are among the finest photographs of Mercury ever produced with an Earth-based telescope. Hazy markings are faintly visible on the tiny planet. (New Mexico State University Observatory)

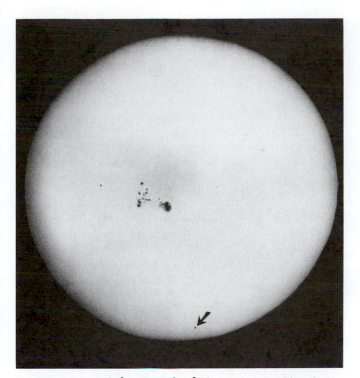

FIGURE 10-4 A Solar Transit of Mercury   Roughly a dozen solar transits of Mercury occur in each century. This photograph shows the tiny planet (arrow) silhouetted against the Sun during the transit of November 14, 1907. (Yerkes Observatory)

located near the line along which Mercury's orbit intersects the plane of the ecliptic. Only two transits will occur during the remainder of this century (on November 6, 1993, and November 15, 1999). The longest transits occur in May, when Mercury is near aphelion and therefore is traveling comparatively slowly along its orbit. The maximum duration of a solar transit is nine hours. Figure 10-4 is a photograph of one such transit. Note how tiny the planet seems in comparison to the Sun.

10-2   Mercury rotates slowly and has an unusual spin–orbit coupling

During the 1880s, the Italian astronomer Giovanni Schiaparelli attempted to make the first map of Mercury. Unfortunately, even today's best telescopes reveal only a few faint, hazy markings. Schiaparelli's views of Mercury were so vague and indistinct that he made a major error, which went uncorrected for over half a century. He erroneously believed that Mercury always kept the same side facing the Sun.

Synchronous rotation, in which the rotation period of an object equals its period of revolution, is a common phenomenon in our solar system. In the previous chapter, for example, we saw that the Moon exhibits synchronous rotation, keeping the same side exposed to our Earth-based view. The two moons of Mars and many of the satellites of Jupiter and

Saturn also keep the same side facing their parent planet. Newtonian mechanics demonstrates that this situation, also called a 1-to-1 **spin–orbit coupling,** is a stable one. But this is not how Mercury rotates. The first clue about Mercury's true rotation period came in 1962, when astronomers detected radio radiation coming from the innermost planet.

As we learned in Chapter 5, every object emits blackbody radiation. The dominant wavelength of the radiation emitted by an object depends on its temperature. Only at absolute zero does an object emit no radiation at all.

If Mercury exhibited synchronous rotation, one side of the planet would remain in perpetual, frigid darkness, quite near absolute zero. However, in 1962, radio astronomers at the University of Michigan monitoring radiation from Mercury discovered that the temperature on the planet's nighttime side is about 100 K (= −173°C = −280°F), not nearly as cold as expected. For comparison, at high noon on Mercury's equator, the temperature gets up to about 700 K (= 427°C = 800°F).

The old belief in synchronous rotation was so deeply ingrained that some astronomers refused to abandon it. They speculated that Mercury had an atmosphere whose winds carried warmth from the daytime side of the planet around to the nighttime side. In fact, there are no winds on Mercury. Its daytime temperature is too high and its gravity too weak to retain any substantial atmosphere (recall Box 7-2).

A breakthrough came in 1965, when Rolf B. Dyce and Gordon H. Pettengill used the giant 1000-ft radio telescope at the Arecibo Observatory in Puerto Rico (Figure 10-5) to bounce powerful radar pulses off Mercury. The outgoing radiation consisted of microwaves of a very specific wavelength. In the reflected signal echoed back from the planet, the wavelength had shifted. This wavelength shift, an example of the Doppler effect (see Figure 5-19), provided detailed information about Mercury's rotation. Microwaves reflected from the planet's approaching side were shortened, whereas those from its receding side were lengthened. The radar pulse went out at one specific wavelength, but it came back spread over a small wavelength range. From the width of the spread, Dyce and Pettengill deduced that Mercury's rotation period is approximately 59 days.

Giuseppe Colombo, an Italian physicist with a longstanding interest in Mercury, found this number intriguing. Colombo noted that Mercury's sidereal period is 87.969 days and that

$$\tfrac{2}{3}(87.969 \text{ days}) = 58.65 \text{ days}$$

Colombo therefore boldly speculated that Mercury's true rotation period is exactly 58 days and $15\tfrac{1}{2}$ hours. He realized that this figure would mean that Mercury is locked into a 3-to-2 spin–orbit coupling: The planet makes three complete rotations on its axis for every two complete orbits around the Sun. This is a much rarer form of dynamic stability than the 1-to-1 spin–orbit coupling that our Moon and many other satellites exhibit. Figure 10-6 shows both types of spin–orbit coupling.

FIGURE 10-5 The Arecibo Radio Telescope This is the largest telescope on Earth. The dish, which is mounted permanently in a bowl-shaped valley in Puerto Rico, measures 305 m (1000 ft) in diameter. Because the dish is not movable, this telescope can be used only to examine objects that are near the zenith. (Arecibo Observatory)

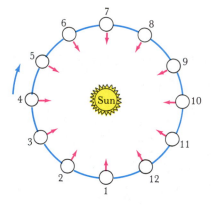

a 1-to-1 spin–orbit coupling

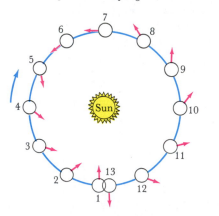

b 3-to-2 spin–orbit coupling

FIGURE 10-6 Spin–Orbit Coupling (a) The simplest kind of spin–orbit coupling is 1-to-1 coupling, where the rotation period equals the orbital period. The planet always keeps the same side facing the Sun. (b) Mercury exhibits a 3-to-2 coupling: during each revolution around the Sun, the planet rotates $1\frac{1}{2}$ times on its axis. Consequently, Mercury rotates three times about its axis during two complete orbits of the Sun.

Colombo's enlightened guess was dramatically confirmed in the mid-1970s by the *Mariner 10* spacecraft. After its first encounter with Mercury, in the spring of 1974, *Mariner 10* was placed in a 176-day orbit about the Sun. This orbital period brings the spacecraft back to the planet every two Mercurian years, just enough time for the planet to complete three rotations. During the second and third flybys, the spacecraft's cameras showed that the same side of Mercury was facing the Sun as on the first flyby. Thus it was proved that Mercury does indeed rotate three times about its axis for every two orbits it makes of the Sun.

Mercury's very slow (58.65-day) rotation period is responsible for a remarkable pause in the apparent motion of the Sun itself. Mercury's speed along its orbit varies in accordance with Kepler's second law (see Figure 4-10). Its orbital velocity is greatest (59 km/s) at perihelion and least (39 km/s) at aphelion. As seen from Mercury's surface, the Sun rises in the east and sets in the west, just as it does on Earth. When Mercury is near perihelion, however, the planet's rapid motion along its orbit outpaces its leisurely rotation about its axis. The usual east-west movement of the Sun across Mercury's sky is interrupted. The Sun actually stops and moves backward (from west to east) for a few Earth days. If you were standing on Mercury watching a sunset occurring at perihelion, the Sun would not simply set. It would dip below the western horizon and then come back up, only to set a second time a day or two later.

10-3 Photographs from *Mariner 10* reveal Mercury's heavily cratered, lunarlike surface

We acquired our first detailed knowledge about Mercury's surface in the spring of 1974, when *Mariner 10* coasted over the planet's surface. At its closest approach on March 29,

1974, *Mariner 10* passed over Mercury's darkened, nighttime side. The spacecraft's photography was therefore divided into two parts: "incoming views," taken prior to this closest approach, and "outgoing views," taken as *Mariner 10* receded from the planet. The best incoming and outgoing photomosaics of the entire planet can be seen in Figures 10-7 and 10-8. Box 10-2 summarizes the *Mariner 10* mission.

As *Mariner 10* closed in on Mercury, scientists were surprised by the Moonlike pictures appearing on their television monitors. It was obvious that Mercury is a barren, heavily cratered world. Figure 10-9 shows a typical close-up view sent back from *Mariner 10*. Astronomers believe that most of the craters on both Mercury and the Moon were produced during the 700 million years after the planets formed. As we saw in the previous chapter, the strongest evidence comes from analysis and dating of Moon rocks. Debris remaining after planet formation rained down on these young worlds, gouging out most of the craters we see today.

Although first impressions of Mercury evoke a lunar landscape, closer scrutiny of Mercury's surface reveals some significant nonlunar characteristics, such as gently rolling plains and long, meandering cliffs. For comparison, Figure 10-10 shows the Moon's southern hemisphere. Notice how the lunar craters are densely packed, with one overlapping the next. In sharp contrast, Mercury's surface has extensive plains (examine the upper half of Figure 10-9). These large, smooth areas are about 2 km lower than the cratered terrain.

As we saw in Chapter 9, the lunar maria were produced by extensive lava flows between 3.1 and 3.8 billion years ago. Primordial lava flows probably explain Mercurian plains also: As large meteoroids punctured the planet's thin, newly formed crust, lava welled up from the molten interior to flood low-lying areas. Based on the number of craters that pit them, Mercury's plains appear to have been formed near the end of the era of heavy bombardment, just over 3.8 billion years ago. Mercury's plains are therefore older than most of the lunar maria.

Mariner 10 also revealed numerous long cliffs, called **scarps,** meandering across Mercury's surface (Figure 10-11). Some scarps rise as much as 3 km (2 mi) above the surrounding plains. These cliffs probably formed as the planet cooled and contracted, causing its crust to wrinkle. Scarps are there-

FIGURE 10-7 *Mariner 10's* **Incoming View** This view of Mercury was sent back from *Mariner 10* as the spacecraft sped toward the planet. Eighteen pictures taken at 42-second intervals were assembled into this photomosaic. The spacecraft was 200,000 km (124,000 mi) from the planet at the time these pictures were taken. (NASA)

FIGURE 10-8 *Mariner 10's* **Outgoing View** This view of Mercury was sent back from *Mariner 10* as the spacecraft coasted away from the planet. Eighteen pictures taken at 42-second intervals were assembled into this photomosaic. These pictures were taken from a distance of 210,000 km (130,000 mi). (NASA)

BOX 10-2

The Exploration of Mercury

Mariner 10 is typical of the spacecraft that were launched on historic missions to the planets during the 1970s. *Mariner 10* weighed 550 kg (1100 lb) and drew its power from two large solar panels that extended from the spacecraft like wings (see first illustration). Two television cameras mounted on a steerable platform took pictures of Mercury, while a device called an infrared radiometer measured the planet's surface temperature. An ultraviolet spectrometer searched for a Mercurian atmosphere, and magnetometers measured Mercury's magnetic field. Throughout the mission, charged-particle detectors examined the solar wind.

After leaving the Earth in 1973, *Mariner 10* first flew past Venus. The trajectory of this flyby was designed so that Venus's gravitational pull would redirect the spacecraft toward Mercury. This flight was the first of several gravity-assisted missions in which a near encounter with one planet was used to catapult a spacecraft gravitationally on toward another planet. The flight path of *Mariner 10* is seen in the second illustration. After coasting past Mercury on March 29, 1974, *Mariner 10* went into an orbit whose period is exactly two Mercurian years. Consequently, Mercury and the spacecraft pass close to each other once every 176 days. However, fuel to stabilize the spacecraft lasted for only the first three encounters.

Mariner 10's pictures of Mercury covered only 45% of

the planet with a resolution of about 1 km, which is comparable to earth-based telescopic coverage of the Moon before the advent of spaceflight. Scientists therefore suspect that much remains to be discovered about this small planet.

In the late-1980s, Chen-Wan Yen at the Jet Propulsion Laboratory discovered new trajectories that could be used to place large payloads in orbit about Mercury. A spacecraft would have to make several flybys of both Venus and Mercury, taking three to five years before it could settle into orbit about Mercury. NASA is currently studying the feasibility of such a mission.

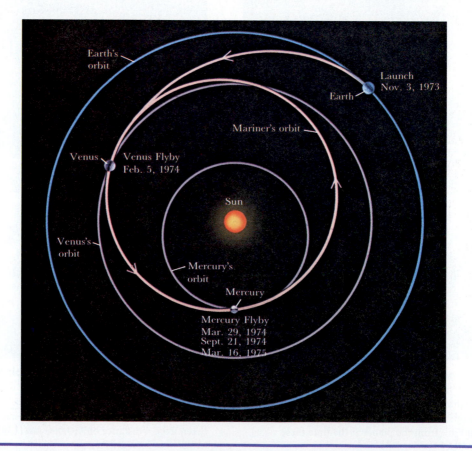

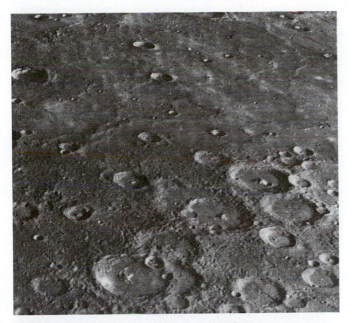

FIGURE 10-9 Mercurian Craters and Plains This view of Mercury's northern hemisphere was taken by *Mariner 10* at a range of 55,000 km (34,000 mi) from the planet's surface. Numerous craters and extensive intercrater plains appear in this photograph, which covers an area 480 km (300 mi) wide. (NASA)

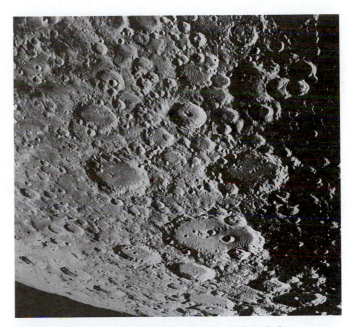

FIGURE 10-10 Lunar Craters This Earth-based photograph shows a portion of the Moon's southern hemisphere during last quarter moon. Densely packed craters fill the view, which covers an area approximately 600 km (370 mi) wide. (Carnegie Observatories)

fore ancient tectonic features unique to Mercury: Ridges and wrinkles thrust up during a period of global shrinkage. Because they seem to have formed after the lava flows that produced Mercury's plains, the episode of planetwide contraction began rather late in Mercurian history.

The most impressive feature discovered by *Mariner 10* was a huge impact basin called the Caloris Basin (from the

FIGURE 10-11 A Scarp A long, meandering cliff is seen near the horizon in this view of Mercury's northern hemisphere. The cliff, called a scarp by geologists, extends southward for several hundred kilometers. This photograph covers an area measuring roughly 550 km (340 mi) across. (NASA)

Latin word for "hot"). The Sun is directly over the Caloris Basin during alternating perihelion passages, and so it is the hottest place on the planet once every 176 days.

Mariner 10's cameras could only reveal about half of the Caloris Basin because it happens to lie on the line dividing day from night. (This dividing line is called the **terminator**.) Slightly more than half of the impact basin is hidden on the night side of the planet in Figure 10-12.

The Caloris Basin, which measures 1300 km (810 mi) in diameter, is filled and surrounded by smooth plains that resemble lunar maria. Like the lunar maria, the Caloris Basin was probably gouged out by the impact of a large meteoroid that penetrated the planet's crust. Because relatively few craters pockmark the lava flows that filled the basin, the Caloris impact must have occurred toward the end of the crater-making period that dominated the first 700 million years of our solar system.

The Caloris impact must have been a violent event that shook the planet. On the side of Mercury opposite the Caloris Basin, *Mariner 10* discovered a consequence of this impact: A jumbled, hilly region covering about 500,000 km², about twice the size of the state of Wyoming. Figure 10-13 is a wide-angle view of this area; the hills appear as tiny wrinkles that cover most of the photograph. The hills are about 5 to 10 km wide and have elevations between 100 and 1800 m (300 and 5900 ft). Geologists theorize that seismic waves from the Caloris impact became focused as they passed through Mercury. As this concentrated seismic energy reached the far surface of the planet, jumbled hills were pushed up.

FIGURE 10-12 The Caloris Basin *Mariner 10* sent back this view of a huge impact basin on the terminator. Although the center of the impact basin is hidden in the shadows (just beyond the left side of the picture), several semicircular rings of mountains reveal its extent. The outer rim of the basin is defined by a ring of mountains up to 2 km (6500 ft) high. The diameter of the basin is 1300 km (810 mi). (NASA)

FIGURE 10-13 Unusual, Hilly Terrain The tiny, fine-grained wrinkles on this picture are actually closely spaced hills, part of a jumbled, hilly terrain that covers nearly half a million square kilometers of Mercury, opposite the Caloris Basin. The large smooth-floored crater near the center of this photograph has a diameter of 170 km (106 mi). (NASA)

Features similar to those of the Caloris Basin exist on our Moon. The best example is the Orientale Basin, shown in Figure 10-14. Its diameter is 900 km (560 mi). The scarcity of fresh craters in this basin attests to its late formation. Furthermore, the theory about the seismic upthrusting of jumbled hills on Mercury is supported by the finding of similar (though less extensive) chaotic hills on the side of the Moon opposite the Orientale Basin.

10-4 Mercury has an iron core and a magnetic field, like Earth

Mercury's average density of 5430 kg/m^3 is quite similar to Earth's, which is 5520 kg/m^3. As we saw in Chapter 8, typical rocks from Earth's surface have a density of only about 3000 kg/m^3 because they are composed primarily of lightweight, mineral-forming elements. The higher average density of our planet is caused by the Earth's iron core. By studying how the Earth vibrates during earthquakes, geologists have deduced that our planet's iron core occupies about 17% of the Earth's volume.

Pressures inside a planet increase the density of rock by squeezing its atoms into smaller volumes. Because the Earth is 18 times more massive than Mercury, the weight of this larger mass compresses the Earth's core much more than Mercury's core is compressed. If the Earth were uncompressed, its density would be 4400 kg/m^3, whereas Mercury's uncompressed density would still be a hefty 5300 kg/m^3. This higher uncompressed density means that Mercury has a larger proportion of iron than the Earth does—an iron core must occupy about 42% of Mercury's volume. Mercury is therefore the most iron-rich planet in the solar system. Figure 10-15 is a scale drawing of the interior structures of Mercury and Earth.

Several theories have been proposed to account for Mercury's high iron content. According to one theory, the inner regions of the primordial solar nebula were so warm that only those substances with high condensation temperatures—like iron-rich minerals—could have condensed into solids. According to another theory, a brief episode of very powerful solar wind could have stripped Mercury of its low-density mantle shortly after the Sun formed. A third possibil-

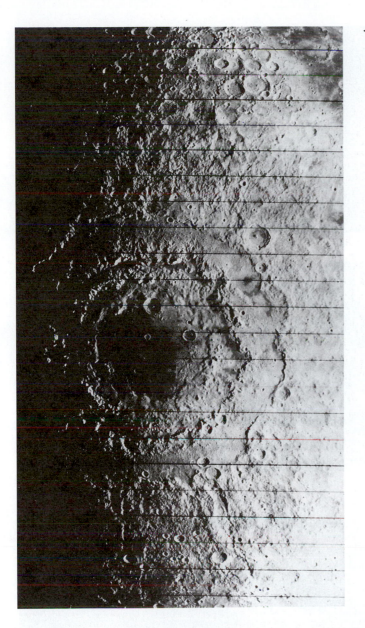

◀ FIGURE 10-14 The Orientale Basin on the Moon The Orientale Basin is one of several large impact basins on the Moon that resemble the Caloris Basin on Mercury. The outermost ring of peaks forms a circle 900 km (560 mi) in diameter. This photograph was sent back in 1967 from the *Lunar Orbiter 4* spacecraft. Black lines interrupt the picture because the photograph was transmitted to Earth in thin strips. (NASA)

cury's magnetic field therefore suggests that its core has the same structure as Earth's core (recall Figure 8-10).

Mercury's magnetic field produces an interaction with the solar winds similar to Earth's but on a much smaller scale. The solar wind is a constant flow of charged particles (mostly protons and electrons) away from the outer layers of the Sun's upper atmosphere. A planet's magnetic field can repel and deflect the impinging particles, thereby forming an elongated "cavity" in the solar wind called a **magnetosphere.**

Charged-particle detectors on *Mariner 10* mapped the structure of Mercury's magnetosphere (Figure 10-17). When the particles in the solar wind first encounter Mercury's magnetic field, they are abruptly slowed, producing a shock wave that marks the boundary where this sudden decrease in velocity occurs. Most of the particles from the solar wind are deflected around the planet, just as water is deflected to either side of the bow of a ship. Mercury's magnetosphere therefore prevents the solar wind from reaching the planet's surface. Because Mercury's magnetic field is quite weak, it is not able to capture particles from the solar wind and produce structures like the Van Allen belts that surround the Earth (recall Figure 8-19).

Mariner 10 still coasts over Mercury's sun-scorched surface every 176 days, its eyes blind and its radio voice silent. There are currently no plans to return to Mercury, and the side of the planet hidden from *Mariner 10* will remain unexplored for the rest of this century.

ity is that, during the final stages of planet formation, Mercury was struck by a large planetesimal. Supercomputer simulations show that this cataclysmic collision would have ejected much of the lighter mantle, leaving a disproportionate amount of iron to reaccumulate to form the planet we see today (Figure 10-16).

An important clue to the structure of Mercury's iron core came from *Mariner 10*'s magnetometers, which discovered that Mercury has a magnetic field similar to the Earth's but much weaker. A planetary magnetic field is similar to the magnetism that surrounds an electromagnet (a coil of wire in which electricity is flowing). According to a process called the **dynamo effect,** electric currents flowing in the liquid portions of a planet's iron core create a planetwide magnetic field. The latest theories about this process require the planet to have *both* a solid inner core and a liquid outer core. Mer-

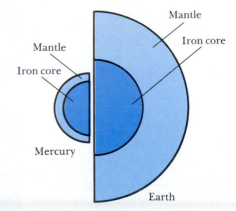

FIGURE 10-15 The Internal Structures of Mercury and Earth The diameter of Mercury's iron core is about 75% of Mercury's diameter, corresponding to 42% of its volume. Surrounding the core is a 600-km-thick rocky mantle. For comparison, the diameter of Earth's iron core is only 55% of Earth's diameter, or about 17% of its volume.

FIGURE 10-16 The Collisional Stripping of Mercury's Mantle To account for Mercury's high iron content, one theory proposes that a collision with a planet-sized object stripped Mercury of most of its rocky mantle. These four images show a supercomputer simulation of a head-on collision between proto-Mercury and a planet one-sixth its mass. Both worlds are shattered by the impact, which vaporizes much of their rocky mantles. Mercury eventually reforms from the iron-rich debris left behind. (Courtesy of W. Benz, A. G. W. Cameron, and W. Slattery)

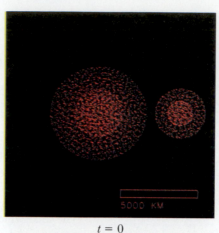
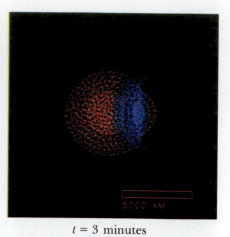

$t = 0$ $t = 3$ minutes

$t = 6$ minutes $t = \frac{1}{2}$ hour

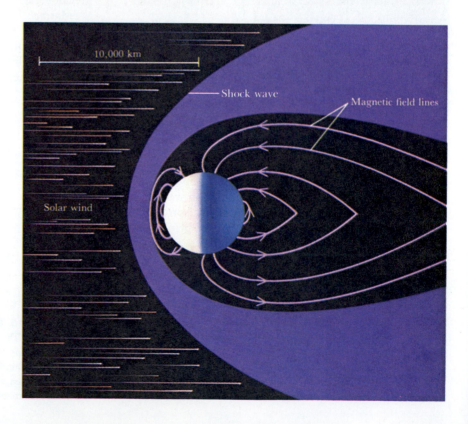

FIGURE 10-17 Mercury's Magnetosphere Mercury's weak magnetic field is just strong enough to carve out a cavity in the solar wind, preventing the impinging particles from striking the planet's surface directly. Most of the particles of the solar wind are deflected around the planet in a turbulent region colored purple in this scale drawing.

KEY WORDS

albedo

dynamo effect

magnetosphere

scarp

solar transit

spin–orbit coupling

synchronous rotation

terminator

KEY IDEAS

- At its greatest eastern and western elongations, Mercury is only 28° from the Sun, so it can be seen only briefly after sunset or before sunrise.

  Solar transits of Mercury occur about a dozen times per century.

  Poor telescopic views of Mercury's surface led to the mistaken impression that the planet always keeps the same side toward the Sun, a configuration called synchronous rotation, or 1-to-1 spin–orbit coupling.

  Radio observations and radar observations in the 1960s revealed that Mercury in fact has 3-to-2 spin–orbit coupling. For an observer on Mercury, one solar day would last for two Mercurian years.

- The *Mariner 10* spacecraft made several passes near Mercury in the mid-1970s, providing pictures of its surface.

  The Mercurian surface is pocked with craters like those of the Moon, but extensive, smooth plains exist between these craters.

  Long cliffs called scarps meander across the surface of Mercury. These scarps probably formed as the planet cooled, solidified, and shrank.

  The impact of a large object long ago formed the huge Caloris Basin and shoved up jumbled hills on the opposite side of the planet.

- Mercury has an iron core much like that of the Earth.

  The iron core of Mercury has a diameter equal to three-fourths of the planet's diameter, whereas the diameter of the Earth's core is only slightly more than one-half of the Earth's diameter.

  Mercury's magnetic field produces a magnetosphere surrounding the planet that blocks the solar wind from the surface of the planet.

REVIEW QUESTIONS

1. Why are naked-eye observations of Mercury best made at dusk or dawn, whereas telescopic observations are best made around noon?

2. Why can't you see any surface features on Mercury when it is closest to the Earth?

3. Explain why Mercury does not have a substantial atmosphere.

4. What is synchronous rotation? How is the rotation period of an object exhibiting synchronous rotation related to its orbital period?

5. How is Mercury's rotation rate related to its orbital period?

6. Explain why *Mariner 10* was able to photograph only one side of Mercury even though the spacecraft returned to the planet many times.

7. Compare the surfaces of Mercury and our Moon. How are they similar? How are they different?

8. Explain why the Sun is directly over the Caloris Basin on Mercury only during every other perihelion passage.

9. Why do astronomers believe that Mercury has a very large iron core?

10. With the aid of a drawing, describe Mercury's magnetosphere. Why do you suppose Mercury does not have Van Allen belts?

11. How can you tell an old crater from a new one?

12. What kind of tectonic features are found on Mercury? Why are they probably much older than tectonic features on the Earth?

13. Briefly describe at least one theory that would explain why Mercury has a large iron core.

ADVANCED QUESTIONS

Tips and tools . . .

You may need to refresh your memory about the small-angle formula, found in Box 1-1, and about Wien's law and the Doppler effect, which are both discussed in Chapter 5. The criteria for a planet to be able to retain an atmosphere are discussed in Box 7-2.

14. Suppose you have a superb telescope that can resolve features as small as 1 arc sec across. What is the size of the smallest surface features you should be able to see on Mercury? How does your answer compare with the size of the Caloris Basin? (*Hint:* Assume that you choose to observe Mercury when it is at greatest elongation, about 25° from the Sun.)

15. Explain why November solar transits of Mercury, which occur near the time of perihelion passage, are more common than May transits.

16. Find the value of λ_{max} for radiation coming from the sunlit side of Mercury.

17. If the albedo of Mercury were increased, would the planet's surface temperature go up or down? Explain your answer.

18. In view of Mercury's 58.65-day rotation period, what difference in wavelength is observed for a spectral line at 500 nm reflected from either the approaching or receding edge of the planet?

19. Calculate the minimum molecular weight of a gas that could in theory be retained as an atmosphere by Mercury if the average daytime temperature were 620 K. Are there any abundant gases that meet this minimum criterion? Why doesn't Mercury have an atmosphere of these gases? (*Hint:* The most convenient way to express the mass of a molecule is by its molecular weight, μ, times the mass of a hydrogen atom, $m_H = 1.67 \times 10^{-27}$ kg. The molecular weight of a molecule equals the sum of the atomic weights of its atoms, which can be looked up in a periodic table of the elements. Thus, for instance, the molecular weight of CO_2 is 12.0 + 16.0 + 16.0 = 44.0, and so the molecule's mass is $44 \times m_H$.)

20. How much would an 80-kg person weigh on Mercury? How does that compare with that person's weight on the Moon? How much does that person weigh on Earth?

21. Using the fact that the orbital period of *Mariner 10* is twice that of Mercury, calculate the length of the semimajor axis of the spacecraft's orbit.

DISCUSSION QUESTIONS

22. If you were planning a return mission to Mercury, what features and observations would be of particular interest to you?

23. What evidence do we have that the surface features on Mercury were not formed during recent geological history?

OBSERVING PROJECTS

24. Refer to Table 10-1 to determine the dates of the next two or three greatest elongations of Mercury. Consult such magazines as *Sky & Telescope* and *Astronomy* to see if any of these greatest elongations is going to be a favorable one. If so, make plans to be one of those rare individuals who has actually seen the innermost planet of the solar system. Set aside several evenings (or mornings) around the date of the favorable elongation to reduce the chances of being "clouded out." Select an observing site that has a clear, unobstructed view of the horizon where the Sun sets (or rises). Make arrangements to have a telescope at your disposal. Search for the planet on the dates you have selected and make a drawing of its appearance through your telescope.

25. **This observing project should be performed only under the direct supervision of an astronomer who knows how to point a telescope safely at Mercury.** Make arrangements to view Mercury during broad daylight. This is best done by visiting an observatory where the coordinates (right ascension and declination) of Mercury's position can be used to point the telescope. **DO NOT LOOK AT THE SUN! Looking directly at the Sun can cause blindness.**

FOR FURTHER READING

Davies, M., and others, eds. *Atlas of Mercury.* NASA SP-423, 1978. This oversized book from NASA includes many high-quality enlargements of the *Mariner 10* pictures of Mercury.

Dunne, J., and Burgess, E. *The Voyage of* Mariner 10: *Mission to Venus and Mercury.* NASA SP-424, 1978. This NASA publication gives a fine overview of the entire *Mariner 10* mission.

Hartmann, W. "The Significance of the Planet Mercury." *Sky & Telescope,* May 1976. This excellent article by a noted planetary scientist discusses the importance of what we have learned from the *Mariner 10* mission to Mercury.

Murray, B. "Mercury." *Scientific American,* May 1976. This article is one of the best summaries of what we now know about the innermost planet.

Murray, B., and Burgess, E. *Flight to Mercury.* Columbia University Press, 1977. This exciting book, which includes an excellent selection of photographs, interweaves the history of the *Mariner 10* mission with newsworthy events that often overshadowed the mission.

Strom, R. *Mercury: The Elusive Planet.* Smithsonian Institution Press, 1987. This clear, well-written overview of the *Mariner 10* mission to Mercury is a superb nontechnical introduction to the innermost planet.

———. "Mercury: The Forgotten Planet." *Sky & Telescope,* September 1990. This fascinating article explains some of the latest ideas about Mercury, including the theory that Mercury's large iron core resulted from a devastating impact 4.5 billion years ago.

CLOUD-COVERED VENUS

VENUS Venus and Earth have nearly the same size, mass, and surface gravity. However, Venus's dense carbon dioxide atmosphere efficiently traps energy from sunlight, resulting in a surface temperature on Venus of 750 K (= 480°C = 900°F), even hotter than Mercury's. Unlike Earth's clouds, which are made of water droplets, Venus's clouds are made of droplets of concentrated sulfuric acid along with sulfur dust. Active volcanoes may be responsible for maintaining these sulfur-rich clouds. This photograph was obtained by the *Pioneer Venus Orbiter* in 1979. (NASA)

VENUS AND the Earth have many similar properties, including size, mass, and average density. A thick covering of clouds had long kept astronomers from learning much more about Venus until they used radar to penetrate the clouds. Radar observations not only demonstrated that Venus rotates backward but also gave us pictures of the planet's surface, which has two large "continents" and is mostly covered with gently rolling hills. High-resolution images of the Venusian surface and other important data came later from space probes. These Soviet and American missions discovered that Venus has an extremely hot, dense atmosphere of carbon dioxide with clouds of sulfuric acid droplets. Several active volcanoes are probably responsible for the high sulfur content of the Venusian clouds. Because the greenhouse effect is responsible for Venus's high temperature, an understanding of our sister planet may give important insights into the history and future of our own world.

At first glance, Venus looks like Earth's twin. The two planets have almost the same mass, the same diameter, the same average density, and the same surface gravity. (Box 11-1 lists basic data about Venus.) However, Venus is closer to the Sun than the Earth is, so it is exposed to more intense sunlight, transforming this potentially Earthlike planet into a world that is extremely hostile to living organisms. Venus is an inferno whose crushing, poisonous atmosphere is drenched in sulfuric acid.

11-1 The surface of Venus is hidden beneath a thick, highly reflective cloud cover

Venus's orbit is almost twice as large as Mercury's. Consequently, at its greatest elongation, Venus appears about 47° away from the Sun (Figure 11-1). This distance is a comfortable one for viewing the planet without interference from the Sun's glare. At its greatest eastern elongation, Venus is seen high above the western horizon after sunset, when it is called an "evening star." At greatest western elongation the planet rises nearly three hours before the Sun. With the arrival of dawn, Venus is positioned high in the eastern sky and is often called the "morning star." Table 11-1 lists the dates of greatest eastern and western elongation from 1993 to 2000.

Venus is easy to identify because it is often one of the brightest objects in the night sky. Venus's cloud cover reflects 76% of the sunlight that falls on it (its albedo is 0.76), and Venus sometimes reaches a magnitude of −4.4 as seen from the Earth, which is 16 times brighter than the brightest star. Only the Sun and the Moon outshine Venus at its greatest brilliance.

Earth-based telescopic views of Venus reveal that it is enveloped by a thick, nearly featureless, unbroken layer of clouds. Evidence for Venus's thick atmosphere comes from observations near the time of inferior conjunction, when clouds illuminated by scattered sunlight produce a luminescent ring encircling the planet (Figure 11-2). Although this

---

TABLE 11-1

Configurations of Venus, 1993–1999

Greatest eastern elongation		Inferior conjunction		Greatest western elongation		Superior conjunction	
1993	January 19	1993	April 1	1993	June 10	1994	January 17
1994	August 25	1994	November 2	1995	January 13	1995	August 20
1996	April 1	1996	June 10	1996	August 19	1997	April 2
1997	November 6	1998	January 16	1998	March 27	1998	October 30
1999	June 11	1999	August 20	1999	October 30	2000	June 11

BOX 11-1

Venus Data

Mean distance from Sun:	$0.723 \text{ AU} = 1.082 \times 10^8$ km
Maximum distance from Sun:	$0.728 \text{ AU} = 1.089 \times 10^8$ km
Minimum distance from Sun:	$0.718 \text{ AU} = 1.075 \times 10^8$ km
Mean orbital velocity:	35.0 km/s
Sidereal period:	224.70 days
Rotation period:	243.01 days (retrograde)
Inclination of equator to orbit:	177.3°
Diameter (equatorial):	12,102 km
Diameter (Earth = 1):	0.949
Mass:	4.870×10^{24} kg
Mass (Earth = 1):	0.8149
Mean density:	5250 kg/m^3
Surface gravity (Earth = 1):	0.879
Escape speed:	10.4 km/s
Mean surface temperature:	480°C = 900°F = 750 K

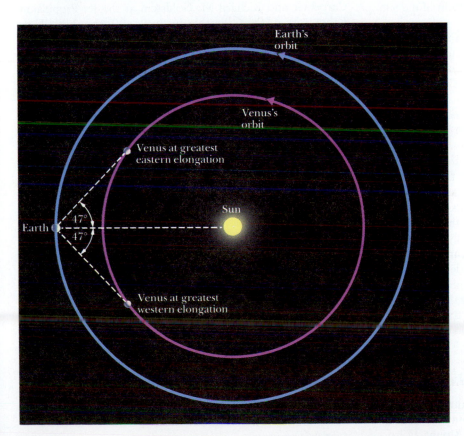

FIGURE 11-1 Venus's Orbit Venus travels around the Sun along a nearly circular orbit with a period of 224.7 days. The average distance between Venus and the Sun is 108 million kilometers (the average distance between the Earth and Sun is 150 million kilometers). At its greatest eastern elongation, Venus appears as a prominent evening star; at its greatest western elongation, it is a prominent morning star.

FIGURE 11-2 Venus at Inferior Conjunction This photograph of Venus at inferior conjunction shows an illuminated atmosphere surrounding the planet. If Venus did not have an atmosphere to scatter sunlight, a thin crescent rather than a ring would be seen. (Lowell Observatory)

highly reflective cloud cover makes Venus dazzlingly bright, it is also responsible for our longstanding ignorance of the planet. Because of this perpetual shroud, we did not even know until recently how fast Venus rotates.

Until the twentieth century, very little was gleaned from telescopic observations of Venus. Nineteenth-century observers did realize that it has an atmosphere. Otherwise, the only noteworthy viewing of Venus was largely confined to solar transits, when Venus passes directly in front of the Sun at inferior conjunction.

Solar transits of Venus are much rarer than those of Mercury. Venus is farther from the Sun than Mercury is, and the required Sun–Venus–Earth alignment occurs much less frequently. The last Venusian transits were in 1874 and 1882. Not one transit of Venus will occur during the entire twentieth century. Venusian transits always come in pairs separated by eight years. The next pair will be visible on June 8, 2004, and June 6, 2012.

The transits of 1761 and 1769 allowed the size of the solar system to be measured accurately for the first time, using the common surveying method of parallax (recall Figure 4-6). Prior to these transits, astronomers knew only the *relative* sizes of the planetary orbits. During the eighteenth-century transits, observers could simultaneously measure Venus's position from such widely separated locations as Great Britain and Tahiti (where an expedition was led by Captain James Cook). These measurements, when combined with the distances between the observation sites, revealed the actual distance between the Earth and Venus at the time of inferior conjunction. The true dimensions of the solar system had finally been determined.

11-2 Venus's rotation is slow and retrograde

In the mid-1950s, astronomers began to realize that Venus rotates backwards. Clues about Venus's rotation came in 1956 when Robert S. Richardson of the Mount Wilson

Observatory measured the Doppler shift of spectral lines in sunlight reflected from the planet. As discussed in Chapter 5 (review Figure 5-19), the wavelength of a spectral line is affected by relative motion between the source and the observer. As Venus rotates, one side of the planet approaches us and its spectral lines are thus blueshifted; the other side recedes from us and its spectral lines are redshifted (Figure 11-3). Richardson's observations indicated that Venus's rotation is **retrograde**, or backward. In other words, sunrise on Venus occurs in the west.

To accurately determine Venus's rotation rate, astronomers had to measure the rotation of the planet's surface, rather than its cloudtops. Fortunately, Venus's clouds are transparent to radio waves and microwaves. In the early 1960s, improved equipment made it possible to send microwave radiation to Venus and detect the waves reflected from its surface. The process of bouncing microwave radiation from an object is known as **radar**, an acronym for "radio detection and ranging."

The first radar observations of Venus were made in 1961 by a team of scientists from M.I.T. who sent powerful bursts of microwaves toward the planet. The microwaves left Earth at one precise wavelength, but their echoes came back spread over a small range of wavelengths: Radar waves reflected from the approaching side of the planet were shortened, whereas those from the receding side were lengthened. By measuring the wavelength spread, the M.I.T. scientists concluded that Venus's rotation is slow and probably retrograde.

These observations were repeated in 1962 by Roland L. Carpenter and Richard M. Goldstein at the Jet Propulsion Laboratory of the California Institute of Technology. They confirmed that Venus has a retrograde rotation with a sidereal period of approximately 240 days. These observations were repeated with increasing accuracy over the next several years using NASA's deep-space tracking antenna in California (Figure 11-4). Today we know that the sidereal period of Venus's retrograde rotation is 243.01 days.

Venus's retrograde rotation is surprising because the other planets (except for Uranus and Pluto) rotate about their axes

Light reflected from
receding side of
Venus is redshifted

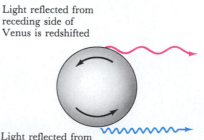

To Earth ⟶

Light reflected from
approaching side of
Venus is blueshifted

FIGURE 11-3 The Doppler Effect and Venus's Rotation Radiation—including light—reflected from Venus's approaching hemisphere is blueshifted, whereas radiation reflected from the planet's receding hemisphere is redshifted. By measuring the wavelength shift across the planet, astronomers determined that Venus's rotation is retrograde.

FIGURE 11-4 A Deep-Space Tracking Antenna Although this antenna was built by NASA to track interplanetary spacecraft, it can also be used to beam powerful pulses of microwaves toward Venus. The antenna's dish measures 64 m (210 ft) in diameter. (Jet Propulsion Laboratory)

in the same direction that they revolve about the Sun. If you could view our solar system from a great distance above the Earth's north pole, you would see all the planets orbiting the Sun counterclockwise. Closer examination would reveal that most of the planets also rotate counterclockwise on their axes. Furthermore, with very few exceptions, even the satellites of the planets move counterclockwise along their orbits.

Most of the planets and their satellites rotate in the same direction as did the primordial solar nebula. As we saw in Chapter 7, the solar nebula must have possessed angular momentum, otherwise everything would have fallen inward toward the protosun. Much of this angular momentum was tied up in the gas and planetesimals that orbited the protosun. Because the planets that formed from this material inherited this angular momentum, they tend to rotate in that same direction. It is difficult to imagine how a planet could have bucked this trend. One theory suggests that the impact of a huge planetesimal could have reversed Venus's direction of rotation. There is no direct evidence, however, that such an impact ever took place.

Venus's retrograde rotation has a curious but probably coincidental connection with the Earth's orbital motion. To understand this relationship, we must clearly distinguish between a sidereal rotation period, a solar day, and a synodic period.

Sidereal period is measured with respect to the stars, not the Sun. In other words, if you were standing on Venus and could see through the clouds, you would have to wait 243.01 Earth days to see the same star pass again overhead. Meanwhile, of course, Venus is moving around the Sun, with an

orbital period of 224.70 Earth days. Thus, Venus's sidereal rotation period is not the same as a solar day. The length of a solar day on Venus is 116.8 days, which is the time from one local noon to the next.

Remember that a planet's synodic period is the time it takes to return to the same orbital configuration, such as the interval from one inferior conjunction to the next. From the data in Table 4-1, we see that Venus's synodic period is 584 days, which is 5×116.8 days. Thus, Venus's synodic period is equal to five Venusian solar days.

Earth-based astronomers must wait 584 days between successive inferior conjunctions of Venus. Meanwhile, five solar days elapse on Venus. Consequently, at each inferior conjunction the same side of Venus is turned toward the Earth. No one has been able to explain how and why this relationship was established. It may simply be a coincidence.

11-3 The surface of Venus is very warm because of the greenhouse effect

Astronomers in the 1950s tried to deduce the surface temperature of Venus by measuring the intensity of radio waves emitted from the planet's surface. As we saw in Chapter 5, an object at any temperature emits radiation at all wavelengths (review the blackbody curves shown in Figure 5-8). By measuring the relative strengths of radiation at different wavelengths and comparing the results with blackbody curves like those in Figure 5-8, astronomers can estimate the temperature of the object emitting the radiation. Because Venus's clouds are transparent to radio waves, the relative strengths of radio waves from Venus at different wavelengths reveal the planet's surface temperature.

At first, no one could believe the results indicating that the temperature at Venus's surface is above the melting point of lead. Spacecraft that later landed on the planet also measured a high surface temperature: 750 K (= 480°C = 900°F). After their initial skepticism, astronomers quickly found a straightforward explanation—the greenhouse effect.

Perhaps you have had the experience of parking your car in the sunshine on a warm summer day. You roll up the windows, lock the car, and go on an errand. After a few hours, you return to discover that the interior of your automobile has become stiflingly hot, typically at least 20°C warmer than the outside air temperature.

What happened to make your car so warm? First, sunlight entered your car through the windows. The radiation was absorbed by the dashboard and the upholstery, raising their temperatures. Every object emits blackbody radiation appropriate to its temperature. Wien's law relates the Kelvin scale temperature (T) to the dominant wavelength (λ_{max}) at which the most intense radiation is emitted. For example, the Sun's surface temperature is about 5800 K. Thus sunlight has its greatest intensity at wavelength λ_{max} of 500 nm, in the middle of the visible spectrum. This sunlight typically raises the temperature of a car's upholstery to around 330 K (= 57°C = 135°F). At that temperature, the seats in your car reradiate the energy primarily with a λ_{max} of 8.8 μm (= 8800 nm). This radiation is in the infrared portion of the electromagnetic

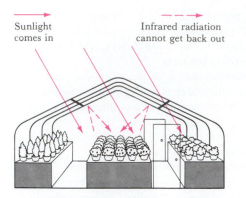

FIGURE 11-5 The Greenhouse Effect Incoming sunlight easily penetrates the windows of a greenhouse and is absorbed by objects inside. These objects reradiate energy at infrared wavelengths, to which the glass windows are opaque. The trapped infrared radiation causes the temperature inside the greenhouse to rise.

spectrum, to which your car windows are opaque. This energy is therefore trapped inside your car and absorbed by the air and interior surfaces. As more sunlight comes through the windows and is trapped, the temperature continues to rise. This phenomenon is the **greenhouse effect** (Figure 11-5).

Carbon dioxide is responsible for similar warming beneath Venus's atmosphere and, to a lesser extent, the Earth's. Like your car windows, carbon dioxide is transparent to visible light but opaque to infrared radiation. In Chapter 8, we found that the small amount of carbon dioxide in the Earth's atmosphere produces a comparatively gentle greenhouse effect that warms the Earth. Warmed by sunlight, the ground emits infrared radiation, some of which is blocked by the carbon dioxide, causing the air temperature to rise.

In contrast to Earth, 96% of Venus's thick atmosphere is carbon dioxide; the remaining 4% is mostly nitrogen (recall Table 8-1). Although most of the visible sunlight striking the Venusian clouds is reflected back into space, enough reaches the Venusian surface to heat it. The warmed surface in turn emits infrared radiation, which cannot penetrate Venus's CO_2-rich atmosphere. This trapped radiation produces the high temperatures found on Venus.

11-4 Spacecraft descending into Venus's dense, corrosive atmosphere found several sulfur-rich cloud layers

In the 1960s, both the United States and the Soviet Union began sending probes to Venus. The Americans sent fragile, lightweight spacecraft past the planet and studied the Venusian environment with remote sensing devices. The first successful American mission also revealed that Venus does not have a magnetic field (Box 11-2). The Soviets, who had

---

BOX 11-2

Venus's Ionosphere

The first successful mission to Venus was the flight of *Mariner 2* in 1962. During its three-month interplanetary voyage, this spacecraft discovered the solar wind. The solar wind, you will recall, consists of charged particles escaping from the Sun at supersonic speeds. Our first measurement of the density (1 particle/cm³) and speed (roughly 400 km/s) of the solar wind came from *Mariner 2*.

As the spacecraft passed Venus, its magnetometer failed to detect any magnetic field whatsoever. Subsequent missions confirmed this lack of a magnetic field. With no magnetic field, Venus is incapable of producing a magnetosphere to protect itself from the solar wind. Therefore the solar wind impinges directly on Venus's upper atmosphere, where it strips many of the atoms of one or more electrons.

An atom that carries an excess charge because it has gained or lost one or more electrons is called an **ion.** The ions in Venus's atmosphere have a positive electric charge, because they have lost negatively charged electrons. The electromagnetic interaction of these ions and the supersonic charged particles in the solar wind produces a shock wave where the supersonic flow of particles abruptly becomes subsonic (see illustration). Along a boundary called the **ionopause,** inside the shock wave, the pressure of the ions just counterbalances the pressure from the solar wind. The ionopause is analogous to the magnetopause that surrounds a planet with a magnetic field.

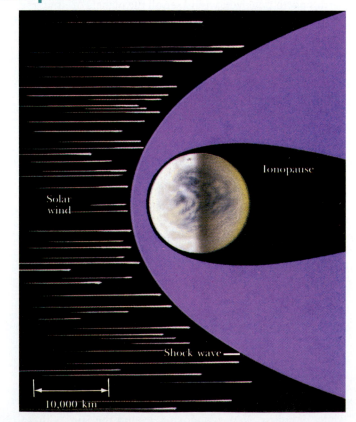

Solar wind

Ionopause

Shock wave

10,000 km

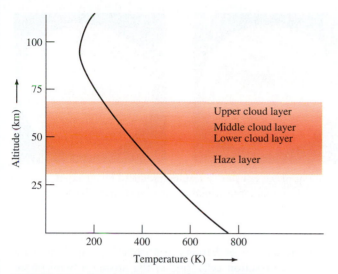

FIGURE 11-6 Temperature in the Venusian Atmosphere   The temperature in Venus's atmosphere increases smoothly from a minimum of about 170 K (= −100°C = −150°F) at an altitude of 100 km to a maximum of nearly 750 K (about 480°C, or 900°F) on the ground.

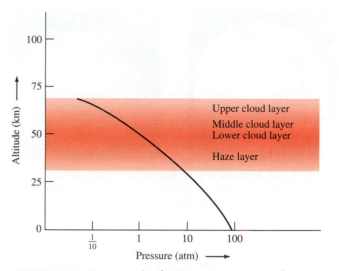

FIGURE 11-7 Pressure in the Venusian Atmosphere The pressure at the Venusian surface is a crushing 90 atm (1300 pounds per square inch). Above the surface, atmospheric pressure decreases smoothly with increasing altitude.

more powerful rockets, sent massive vehicles that plunged directly into the Venusian clouds.

Building spacecraft that could survive a descent into the Venusian cloud cover proved to be more frustrating than anyone had expected. Finally, in 1970, a Soviet probe called *Venera 7* managed to transmit data for a few seconds directly from the Venusian surface. Soviet missions during the early 1970s measured a surface temperature of 750 K (= 480°C = 900°F) and a pressure of 90 atmospheres. Recall from Box 8-2 that one atmosphere (1 atm) is the average air pressure at sea level on Earth (14.7 pounds per square inch). Thus, the Venusian atmosphere weighs down on the planet with a crushing pressure of $\frac{2}{3}$ ton per square inch, about the same as the pressure at a depth of 1 kilometer in Earth's oceans.

During the 1970s, the Soviets and Americans sent probes into Venus's atmosphere, all of which carried instruments that measured temperature and pressure as they descended to the planet's surface. The results are summarized in Figures 11-6 and 11-7. Both pressure and temperature decrease smoothly with increasing altitude. Graphs like these are important, because they show how temperature and pressure are related to the altitude above a planet's surface—fundamental information for deducing the detailed structure of a planet's atmosphere. As we saw in Chapter 8 (recall Figure 8-17), Earth's atmosphere has a much more complicated relationship between temperature and altitude. When we come to the chapters on Jupiter and Saturn, we shall see how pressure and temperature variation with altitude can profoundly affect a planet's appearance.

Soviet spacecraft discovered that Venus's clouds are confined to a 20-km-thick layer located 48 to 68 km above the planet's surface. Below the clouds is a 20-km-thick layer of haze. Beneath this haze the Venusian atmosphere is remarkably clear all the way down to the surface.

American probes also discovered distinct layers within Venus's cloud cover. An upper cloud layer extends from altitudes of 68 km down to 58 km. A denser and more opaque middle cloud layer extends from 58 km down to 52 km. Finally, the lower cloud layer, from 52 to 48 km, contains the densest and most opaque of the Venusian clouds, even though it is only 4 km thick. These layers are indicated on Figures 11-6 and 11-7.

Venus appears yellowish or yellow-orange to the human eye, and data from the spacecraft indicated why: The upper clouds contain substantial amounts of sulfur dust. Over the temperature range of these upper clouds, sulfur is distinctly yellow or yellow-orange. At lower elevations, large concentrations of sulfur compounds such as sulfur dioxide and hydrogen sulfide were found, along with droplets of sulfuric acid. Just as Earth's clouds are composed of water droplets, Venusian clouds are composed of droplets of concentrated sulfuric acid. Because of the tremendous atmospheric pressure on Venus, the droplets do not fall as a rain; instead, they are suspended in the clouds like an aerosol.

The sulfuric acid in Venus's atmosphere causes a number of chemical reactions. For example, reactions with fluorides and chlorides in surface rocks give rise to hydrofluoric acid (HF) and hydrochloric acid (HCl). Further reactions produce fluorosulfuric acid (HSO_3F), which is one of the most corrosive substances known to chemists, capable of dissolving lead, tin, and most rocks. The Venusian clouds are a cauldron of chemical reactions hostile to metals and other solid materials.

Spacecraft orbiting Venus gave astronomers the opportunity to study the global circulation of the Venusian atmosphere. For example, in the early 1980s, the American *Pioneer Venus Orbiter* sent back numerous images of Venus taken at ultraviolet wavelengths, where its atmospheric markings

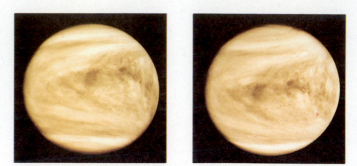

FIGURE 11-8 Venus's Cloud Patterns These four ultraviolet views of Venus were taken by the *Pioneer Venus Orbiter* in May 1980 at a distance of roughly 50,000 km. All the views show variations of the so-called V feature, produced by the rapid retrograde motion of the clouds around the planet. (NASA)

stand out best (Figure 11-8). By following individual cloud markings, scientists determined that Venus's atmosphere rotates around the planet in a retrograde direction in only four days. This rapid atmospheric motion is in sharp contrast to the slow rotation of the solid planet itself.

The American *Pioneer Venus Multiprobe* determined the dominant circulation patterns in the Venusian atmosphere after plunging into it in 1978. Warmed by the Sun, hot gases in the equatorial regions rise upward and travel in the upper cloud layer toward the cooler polar regions. At the polar latitudes, the cooled gases sink to the lower cloud layer, in which they are transported back toward the equator. Recall from Chapter 8 that this process of heat transfer, whereby hot gases rise while cooler gases sink, is called **convection**.

The circulation of Venus's atmosphere is dominated by two huge **convection cells**, one in the northern hemisphere and another in the southern hemisphere, which circulate gases between the equatorial and polar regions of the planet (Figure 11-9). These convection cells, which are almost entirely contained within main cloud layers, are called "driving cells," because they propel similar circulation cells above and below the main cloud deck somewhat like a meshed set of gears.

Because Venus's entire atmosphere rotates around the planet in only four days, strong prevailing winds blow from east to west. These winds stretch out the driving cells and produce the characteristic V-shaped, chevronlike patterns that dominate the planet's appearance.

11-5 Active volcanoes are probably responsible for Venus's clouds

There is compelling evidence for significant volcanic activity on Venus. All of the major volcanic chemicals spewed by Earth's volcanoes have been detected in Venus's atmosphere. Because many of these substances are highly reactive and very short-lived, they must be constantly replenished by new eruptions.

Antennas on *Pioneer Venus Orbiter* and the Jupiter-bound *Galileo* spacecraft have picked up further evidence of active volcanoes—radio bursts thought to be strokes of lightning. Lightning discharges have often been seen in the plumes of erupting volcanoes on Earth.

Long-term observations by *Pioneer Venus Orbiter* also suggest that sulfurous gases are being injected into the Venusian atmosphere by processes that could only be volcanic. When the spacecraft arrived at Venus in 1978, its ultraviolet spectrometer recorded unexpectedly high levels of sulfur dioxide and sulfuric acid, which steadily declined over the next several years. A similar anomalously high abundance of haze particles may have occurred in the late 1950s. Larry W. Esposito of the University of Colorado has proposed that in both the late 1950s and the late 1970s, energetic volcanic eruptions injected sulfur dioxide into the upper atmosphere. It therefore seems that Venus's sulfur-rich clouds are regularly replenished by active volcanoes.

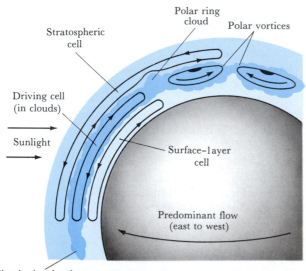

FIGURE 11-9 The Circulation in Venus's Atmosphere The main convection mechanism in Venus's atmosphere occurs in the clouds. This convection cell drives similar circulation patterns above and below the cloud layer. Circular wind patterns (vortices) in the polar regions force the cloud layer to bulge upward at polar latitudes, producing a polar ring cloud that is clearly visible in many Orbiter photographs. (Adapted from A. Seiff)

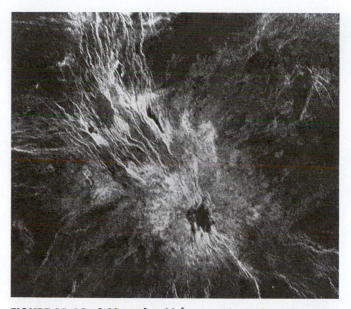

FIGURE 11-10 A Venusian Volcano This high-resolution image shows details of the lava flows and rifts that surround a large volcano. The picture, produced by Earth-based astronomers using radar, covers an area 1700 km by 1500 km—roughly three times the size of the state of Texas. The volcano's summit is the darkish spot toward the bottom center. (Courtesy of D. B. Campbell, Arecibo Observatory)

The first views of Venus's volcanoes came from Earth-based radar observations of the Venusian surface. Astronomers have often used the giant 305-m (1000-ft) dish at the Arecibo Observatory in Puerto Rico (see Figure 10-5) to map Venus because microwaves easily penetrate the Venusian clouds and reflect off its surface. The Arecibo dish is used to transmit a powerful, brief burst of microwaves toward Venus. Because these microwaves strike mountaintops on Venus slightly sooner than deep valleys, echoes from high elevations arrive back at the Arecibo dish slightly ahead of those reflected from lower elevations. In addition, Venus's rotation affects the wavelength of the microwave burst according to the Doppler effect (recall Figure 11-3). By analyzing the radar echo, which is spread out in both arrival time and wavelength, astronomers are able to construct a map of the Venusian surface. This method is successful only when Venus is near inferior conjunction. At all other times the Venus–Earth distance is too great and the radar echo too weak to produce good data.

Figure 11-10 is a radar view of Theia Mons, one of several huge volcanoes located roughly 1300 kilometers north of Venus's equator. Theia Mons rises to an altitude of 6 km and has gently sloping sides that extend over an area 1000 km in diameter. A volcano having this characteristic shape is called a **shield volcano,** because in profile it resembles an ancient Greek warrior's shield lying on the ground. Shield volcanoes exist on both Earth and Mars, the best-known example being the Hawaiian Islands. Shield volcanoes on Venus and Mars, as well as the Hawaiian Islands on Earth, are thought to be the result of **hot-spot volcanism,** whereby a hot region beneath the planet's surface extrudes molten rock over a long period of time.

By far the best images of the Venusian volcanoes come from the highly successful *Magellan* spacecraft that arrived at Venus in 1990 (Figure 11-11). Now in orbit about the planet, *Magellan* is using sophisticated radar techniques in a three-year mission to map the Venusian surface with unprecedented resolution. *Magellan*'s radar images are combined with radar altitude measurements to construct a three-dimensional map of the Venusian surface. From the resulting

FIGURE 11-11 The *Magellan* Spacecraft Since its arrival at Venus in 1990, this spacecraft has been engaged in a three-year mission to map the surface of Venus at high resolution. (NASA)

FIGURE 11-12 A Young Volcano This volcano, called Maat Mons, is the second tallest on Venus. This computer-generated view has a vertical exaggeration of 23 to 1, equivalent to stretching your house into the shape of the Washington Monument. Maat Mons actually has gentle slopes, less steep than terrestrial shield volcanoes like Hawaii's. Numerous cracks cross the gently rolling foothills at the base of the volcano. (NASA)

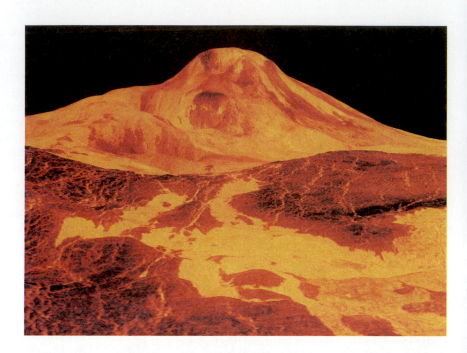

database, a supercomputer creates three-dimensional perspective views, like the one in Figure 11-12.

Some of the youngest crustal material found by *Magellan* caps the volcano called Maat Mons, shown in Figure 11-12. The radar reflectivity of lava flows descending from the summit suggest that the rock has not yet been extensively weathered by the harsh Venusian environment. Geologists estimate that the topmost material is no more than 10 million years old and could be much younger.

11-6 The absence of water on Venus suggests a hellish past

Volcanoes on Earth provide clues about the gases that were probably spewed into the original atmospheres of both Venus and Earth. For example, during the Mount St. Helens eruption of 1980 (Figure 11-13), geologists monitored the emission of substantial amounts of sulfuric acid and other sulfur compounds that are common in the Venusian clouds and atmosphere.

Water vapor and carbon dioxide are also common in volcanic vapors, but they are not equally plentiful in the atmospheres of Venus and Earth. The Earth has abundant water in

FIGURE 11-13 Mount St. Helens Eruption of May 1980 ▶
A preponderance of data suggests that the terrestrial planets obtained their atmospheres from volcanic outgassing. Geologists study this outgassing process in eruptions of volcanoes on Earth. (USGS)

its oceans, but very little carbon dioxide in its atmosphere. In contrast, Venus is very dry, but its atmosphere is mostly carbon dioxide. It is possible that the atmospheres of both planets were derived from volcanic gases. If so, what became of Earth's carbon dioxide and what happened to the water on Venus?

Carbon dioxide is indeed found on the Earth, but dissolved in the oceans and chemically bound into carbonate rocks, such as limestone and marble that formed in those oceans. Were the Earth to become as hot as Venus, so much carbon dioxide would be boiled out of the oceans and baked out of the crust that our planet would soon develop a thick, oppressive carbon dioxide atmosphere much like that of Venus.

A recently proposed theory of Venus's early history may explain its lack of water. Venus and Earth are so similar in size and mass that it is reasonable to suppose that Venusian volcanoes "outgassed" an amount of water vapor roughly comparable to the total content of Earth's oceans. Although some of this water on Venus might have originally collected in oceans, heat from the Sun would have soon vaporized the liquid to create a thick cover of water vapor clouds. Calculations demonstrate that this water vapor would have added 300 atm of pressure to the existing 90 atm of carbon dioxide. Thus the early Venusian atmosphere would have weighed down on the planet's surface with a pressure of three tons per square inch.

This thick, humid atmosphere would have efficiently trapped heat from the Sun, creating a greenhouse effect far more extreme than even that on Venus today. Calculations demonstrate that the ground temperature would have increased to 1800 K (2700°F), which is hot enough to melt rock. Indeed, the Venusian surface was probably molten down to a depth of 450 km (280 mi). Soon, however, the Sun's ultraviolet radiation striking water molecules in Venus's atmosphere may have broken them into hydrogen and oxygen atoms. The light hydrogen atoms would have then escaped into space. Oxygen, which is one of the most chemically active elements, would have readily combined with other substances in Venus's atmosphere, thus stripping Venus of its water and leaving the carbon dioxide atmosphere we find today. Hostile as it may seem, the modern environment on Venus is probably quite mild compared to that of earlier times.

Photographs from Soviet spacecraft that landed on the planet have shown us Venus's arid surface. A panoramic view taken in 1981 is shown in Figure 11-14. Russian scientists believe that this region was covered with a thin layer of lava that fractured upon cooling to create the rounded, interlocking shapes seen in the photograph. This hypothesis agrees with information obtained from the spacecraft's instruments, which indicates that the soil composition is similar to lava rocks called basalt, which are common on Earth and the Moon.

a

b

FIGURE 11-14 A Venusian Landscape (a) This color photograph from *Venera 13* shows that the rocks on the Venusian surface appear orange because the thick, cloudy atmosphere absorbs the blue component of sunlight. (b) When computer processing removes the effects of orange illumination, the true grayish color of the rocks is seen. In this panorama, the rocky plates covering the ground may be fractured segments of a thin layer of lava, or they may be crusty layers of sediment that have been cemented together by chemical and wind erosion. (Courtesy of C. M. Pieters and the Russian Academy of Sciences)

11-7 Venus is covered with gently rolling hills and two "continents"

Venus is remarkably flat. Over 80% of the planet's surface is covered with volcanic plains that are the result of numerous lava flows. Indeed, the Venusian surface is dominated by volcanic features on a scale that is unparalleled on Earth.

Magellan mapped Venus using a radar altimeter that bounced microwaves off the ground directly below the spacecraft. By measuring the time delay of the radar echo, scientists could determine the heights and depths of Venus's hills and valleys. The results are shown in Figure 11-15, which uses a color code to denote elevation. Note that two large highlands, or "continents," rise well above the generally level surface of the planet.

The continent in the northern hemisphere is Ishtar Terra, named after the Babylonian goddess of love. Ishtar Terra, approximately the same size as Australia, is dominated by a high plateau ringed by high mountains. The highest mountain is Maxwell Montes, whose summit rises to an altitude of 11 km above the average surface. For comparison, Mount Everest on Earth rises 9 km above sea level. The roughness of Maxwell's surface suggests that it is very young. Older features are quickly worn smooth by the harsh Venusian envi-

ronment, where a 1-mph wind has the impact of a 90-mph wind on Earth.

The largest Venusian continent, Aphrodite Terra (named after the Greek goddess called Venus by the Romans) is a vast belt of highlands just south of the equator. Aphrodite is 16,000 km in length and 2,000 km wide, giving it an area about one-half that of Africa. The global view of Venus in Figure 11-16 shows that most of Aphrodite is covered by vast networks of faults and fractures.

Before *Magellan,* geologists wondered if plate tectonics, the primary remolding force of Earth, had also shaped the surface of Venus. But images beamed back from the spacecraft show no evidence of Earthlike tectonics. As we saw in Chapter 8, Earth's hard outer shell, or lithosphere, is broken into about a dozen large plates that slowly shuffle across the globe. Long mountain ranges, like the Mid-Atlantic Ridge in Figure 8-12, are created where fresh magma wells up from the Earth's interior to push the plates apart. Similar features are not seen on Venus. Specifically, *Magellan* images failed to show a type of faulting that always occurs with seafloor spreading on Earth. These faults, which are perpendicular to rifts on the ocean floor, give Earth's ocean ridges a distinctly staircaselike appearance (examine Figure 8-12).

Although Venus lacks basic elements of Earthlike plate tectonics, roughly a fifth of its surface is covered by folded

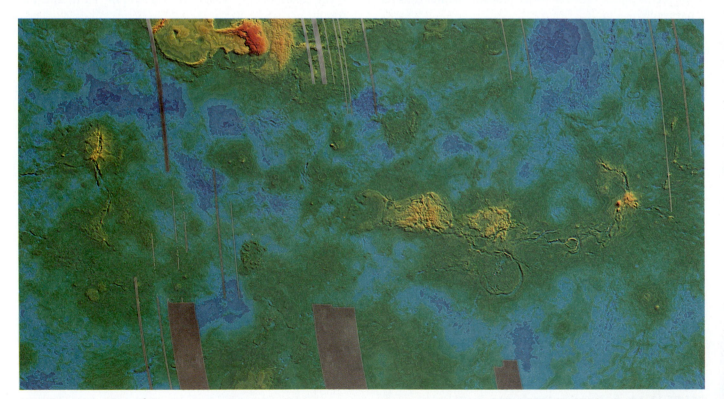

FIGURE 11-15 Map of Venus Radar altimeter measurements by *Magellan* were used to produce this map of Venus, which covers latitudes from 69° north to 69° south. The Venusian equator extends horizontally across the middle of the map. Color indicates elevation: Red corresponds to the highest, blue to the lowest. Grey areas were not mapped by *Magellan.* The elevated region in the north is Ishtar Terra, dominated by Maxwell Montes, the planet's highest mountains, which are 11 km (36,000 ft) above the planet's mean elevation. Southwest of Ishtar are the highlands of Beta Regio and Phoebe Regio. The large, scorpion-shaped feature extending along the equator is Aphrodite Terra, a continentlike highland that contains several spectacular volcanoes. (JPL; NASA)

FIGURE 11-16 A Global View of Venus In this radar image of Venus, Aphrodite Terra is the elongated, light-colored, wispy feature that wraps one-third of the way around the planet. Aphrodite is roughly parallel to Venus's equator, which extends horizontally across the middle of the picture. North is at the top. (NASA)

and faulted ridges, clear signs of more-localized tectonic activity. Deformations of Venus's surface show limited horizontal displacement, which suggests that its lithosphere is thinner and weaker than Earth's. The absence of Earthlike plate tectonics may be the result of Venus's high surface temperature, which slowed the rate at which Venus's crust could

cool, thereby causing its lithosphere to be thin and weak. It is therefore possible that Venus offers us a glimpse into the Earth's ancient history—perhaps three to four billion years ago—when Earth's newly solidified crust was neither thick enough nor strong enough to support the plate tectonic activity we see today.

Magellan has produced superb pictures of Venusian craters, which are relatively few (Figure 11-17). All told, Venus probably has only a thousand craters larger than a few kilometers in diameter. That is many more than have been found on Earth, but only a small fraction of the number on the Moon or Mercury. As we saw in Chapter 9, cratering rates were quite high during the early history of the solar system, when considerable interplanetary debris still orbited the Sun. Since that time, impact rates have declined as shown in Figure 9-18, so that the age of a planet's surface can be inferred from the number of craters it sports. Old craters are obliterated as volcanism or tectonics renews a planet's surface. Consequently, the more craters a planet has, the older is its surface. The average age of the Venusian surface seems to be about 400 million years. This is about twice the age of the Earth's surface, but much younger than the surface of the Moon or Mercury, each of which is billions of years old. Some astronomers speculate that an intense period of eruptions about 400 million years ago may have resurfaced Venus with fresh lava.

In a certain sense, Venus is the Earth's twin. Its surface tells us of our geologic past, and its atmosphere suggests that Earth may have an awesome future. Five billion years from now, as our Sun swells to become a red giant star, the oceans will boil and shroud the Earth in a thick cloud cover. As the greenhouse effect drives temperatures toward 1000 K, vast amounts of carbon dioxide will be baked out of the Earth's rocks. In looking at Venus's atmosphere, perhaps we see the distant fate of our own world.

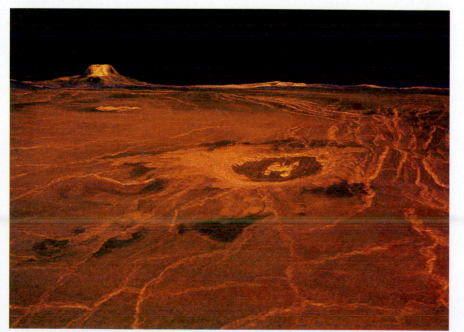

FIGURE 11-17 A Venusian Crater The crater in the foreground is 48 km (30 mi) in diameter. The volcano near the horizon is 3 km (1.9 mi) high. Note the cracks and folds in the Venusian surface toward the right side of the image. (NASA)

KEY WORDS

Terms preceded by an asterisk are discussed in the boxes.

convection   hot-spot volcanism   *ionopause   retrograde rotation

convection cell   *ion   radar   shield volcano

greenhouse effect

KEY IDEAS

• Venus is similar to the Earth in its size, mass, average density, and surface gravity, but it is covered by unbroken, highly reflective clouds that conceal its other features from Earth-based observers.

• Venus rotates slowly in a retrograde direction with a solar day of 117 Earth days and a rotation period of 243 Earth days. There are approximately two Venusian solar days in a Venusian year.

The Venusian atmosphere rotates rapidly in a retrograde direction around the planet with a period of only about four Earth days.

• Space probes reveal that 96% of the Venusian atmosphere is carbon dioxide; most of the balance of the atmosphere is nitrogen.

Venus's clouds consist of droplets of concentrated sulfuric acid, along with substantial amounts of yellowish sulfur dust. Active volcanoes on Venus may be a continual source of this sulfurous material.

Venus's clouds are confined to altitudes between 48 and 68 km above the planet's surface. A haze layer extends down to an elevation of 30 km, beneath which the atmosphere is clear.

The surface pressure on Venus is 90 atm. The surface temperature is 750 K. Both temperature and pressure decrease as altitude increases.

The circulation of the Venusian atmosphere is dominated by two huge convection currents in the cloud layers, one in the northern hemisphere and one in the southern hemisphere.

• Venus's high temperature is caused by the greenhouse effect, as the dense carbon dioxide atmosphere traps and retains energy from sunlight.

Water was lost from Venus by the action of ultraviolet radiation on the upper atmosphere. Carbon dioxide on Earth has been dissolved in the oceans and chemically bound into rocks.

• Venus has no detectable magnetic field or magnetosphere.

• The surface of Venus is surprisingly flat, mostly covered with gently rolling hills. There are two major "continents" and several large volcanoes.

The surface of Venus shows no evidence of the motion of large crustal plates, which plays a major role in shaping the Earth's surface.

REVIEW QUESTIONS

1. In earlier astronomy books, Venus is often referred to as the Earth's twin. What physical properties do the two planets have in common? In what ways are the two planets dissimilar?

2. Venus takes 440 days to move from greatest western elongation to greatest eastern elongation, but it needs only 144 days to go from greatest eastern elongation to greatest western elongation. With the aid of a diagram like Figure 11-1, explain why.

3. As seen from Earth, the magnitude of Venus changes as it moves along its orbit. Describe the main factors that determine Venus's variations in brightness as seen from Earth.

4. What is the greenhouse effect, and what role does it play in the atmospheres of Venus and the Earth?

5. Why is it hotter on Venus than on Mercury?

6. Why are there no oceans on Venus? Where has all of Venus's water gone?

7. Why is there so much carbon dioxide in Venus's atmosphere while very little of this gas is present in Earth's atmosphere?

8. Compare the ways in which Mercury, Venus, and Earth interact with the solar wind.

9. What evidence exists for active volcanoes on Venus?

10. How might Venus's cloud cover change if all of Venus's volcanic activity suddenly stopped? How might these changes affect the overall Venusian environment?

11. Describe the Venusian surface. What kinds of features would you see if you could travel around on the planet?

12. Compare Venus's continents with Earth's. What do they have in common? How are they different?

13. What kinds of tectonic features are found on Venus? Why do astronomers believe that Venus's surface was *not* molded by the kind of tectonic activity that shaped the Earth's surface?

ADVANCED QUESTIONS

> Tips and tools . . .
>
> You should recall that Wien's law relates the temperature of a blackbody to λ_{max}. The circumference of a circle of radius r is $2\pi r$. The linear speed of a point on a planet's equator is the planet's circumference divided by its rotation period.

14. At what wavelength does Venus's surface emit the most radiation? Do astronomers have telescopes that can detect this radiation? Why can't we use such telescopes to view the planet's surface?

15. What is the maximum wavelength shift of yellow light ($\lambda = 550$ nm) across the face of Venus because of its rotation?

16. Given that Venus's sidereal rotation period is 243.01 days and its orbital period is 224.70 days, prove that a solar day on Venus lasts 116.8 days. (*Hint:* Develop a formula relating Venus's solar day to its sidereal rotation period and orbital period similar to the first formula in Box 4-1.)

17. Suppose that a planet's atmosphere was opaque to visible light but transparent to infrared radiation. How would this affect the planet's surface temperature? Contrast and compare this hypothetical planet's atmosphere with the greenhouse effect in Venus's atmosphere.

18. Suppose that Venus had no atmosphere at all. How would the albedo of Venus then compare with that of Mercury or the Moon? Explain your answer.

19. Water has a density of 1000 kg/m^3, so a column of water n meters tall produces a pressure equal to $n \times 1000$ kg/m^2. On either Earth or Venus, which have nearly the same surface gravity, a mass of 1 kg weighs about 2.2 lb. Calculate how deep you would have to descend into the ocean to be subject to a pressure equal to the atmospheric pressure on Venus's surface.

DISCUSSION QUESTIONS

20. Describe the apparent motion of the Sun during a "day" on Venus relative to (**a**) the horizon and (**b**) the background stars.

21. If you were designing a space vehicle to land on Venus, what special features would you think necessary? In what ways would this mission and landing craft differ from a spacecraft designed for a similar mission to Mercury?

22. Why do you suppose that Venus does not have a magnetic field but Mercury and Earth do?

OBSERVING PROJECTS

23. Refer to Table 11-1 to see if Venus is near a greatest elongation. If so, view the planet through a telescope. Make a sketch of the planet's appearance. From your sketch, can you determine if Venus is closer to us or further from us than the Sun is?

24. Observe Venus once a week for a month and make a sketch of the planet's appearance on each occasion. From your sketches, can you determine if Venus is approaching us or moving away from us?

25. This observing project should be performed only under the direct supervision of an astronomer who knows how to point a telescope safely at Venus. Make arrangements to view Venus during broad daylight. This is best done by visiting an observatory where the coordinates (right ascension and declination) of Venus's position can be used to point the telescope. **DO NOT LOOK AT THE SUN!** Looking directly at the Sun can cause blindness.

FOR FURTHER READING

Bazilevskiy, A. T. "The Planet Next Door." Sky & Telescope, April 1989. In this article, a noted Soviet scientist describes the results and implications of the most recent Venera missions to Venus.

Beatty, J. "A Soviet Space Odyssey." *Sky & Telescope,* October 1985. This article describes the Soviet Vega missions which landed two spacecraft on Venus and floated balloon-borne instrument packages in its atmosphere.

Burgess, E. *Venus: An Arrant Twin.* Columbia University Press, 1985. This brief, nontechnical survey of our modern understanding of Venus includes results from the Pioneer Venus and Venera missions.

Burnham, R. "What Makes Venus Go?" *Astronomy,* January 1993. This article contains the latest ideas about Venus's surface features along with a nice collection of *Magellan* images.

Dyer, A. "Rendezvous with Venus." *Astronomy,* April 1991. This article describes results from the February 1990 Venus flyby of *Galileo,* which is now on the way to Jupiter.

Eicher, D. J. "*Magellan* Scores at Venus." *Astronomy,* January 1991. This article boasts an excellent collection of *Magellan* images.

Goldman, S. J. "Venus Unveiled." *Sky & Telescope,* March 1992. This brief article includes some spectacular recent images from Magellan.

Pettengill, G., and others. "The Surface of Venus." *Scientific American,* August 1980. This article gives an excellent overview of how Earth-based radar observations are used to map the surface of Venus.

Prinn, R. C. "The Volcanoes and Clouds of Venus." *Scientific American,* March 1985. This excellent article discusses evidence for active volcanoes on Venus and describes how they affect the planet's atmosphere.

Schubert, G., and Covey, C. "The Atmosphere of Venus." *Scientific American,* July 1981. This fine article compares the dynamics of Venus's atmosphere with Earth's.

THE MARTIAN INVASIONS

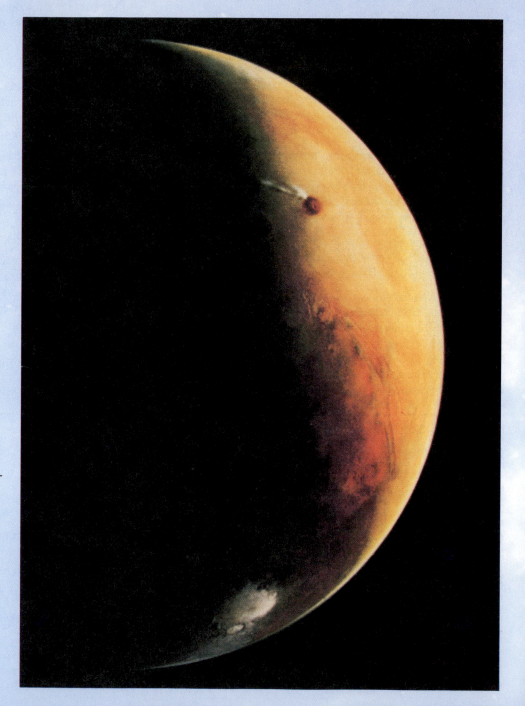

MARS Many of the major features of the Martian surface are seen in this view from *Viking 2* as the spacecraft approached the dawn side of the planet in August 1976. Clouds flank the western slopes of the huge volcano Olympus Mons near the top of the picture. In the middle is a vast canyon called Valles Marineris. This canyon stretches nearly 4000 km along the planet's equator. At the bottom of the photograph, carbon dioxide snow lines the floor of the Argyre Basin and surrounding craters. (NASA)

PEOPLE HAVE long speculated about the possibility of Martian life, because Mars has many Earthlike characteristics. Around 1900, some astronomers claimed to have seen networks of linear features on the Martian surface, perhaps "canals" built by an advanced civilization. Nearly a century later spacecraft reported many surprising facts about Mars—including the discovery of an enormous volcano and a huge canyon—but they found no canals, and no signs of life. Spacecraft that landed on the planet have made many discoveries about Mars, including the fact that water once flowed on this now-arid planet. Mars also has two small moons, possibly asteroids captured by Mars's gravity from the nearby asteroid belt. Many questions about Mars and its moons remain, questions that could be answered by future missions that place spacecraft in orbit about Mars and land on the planet's surface.

Mars is the only planet whose surface features can be seen through Earth-based telescopes. Early telescopic observers reported seeing seasonal variations that seemed to indicate vegetation on the Martian surface, and some reported geometric patterns that might represent a planetwide system of artificial canals. Scientists speculated about the possibility of life on Mars, and science fiction writers wove popular stories about invasions of Earth by hostile Martians. Ironically, the first actual invasion came in 1976, when automated spacecraft from the Earth landed on the Martian surface. These probes sent back pictures and data indicating that Mars is as sterile and lifeless as the Moon.

12-1 The best Earth-based observations of Mars are made during favorable oppositions

There are favorable and not-so-favorable times to observe Mars. Because Mars's orbit is noticeably elongated (Figure 12-1), the best views are obtained when Mars is simultaneously at opposition and near perihelion. This configuration is

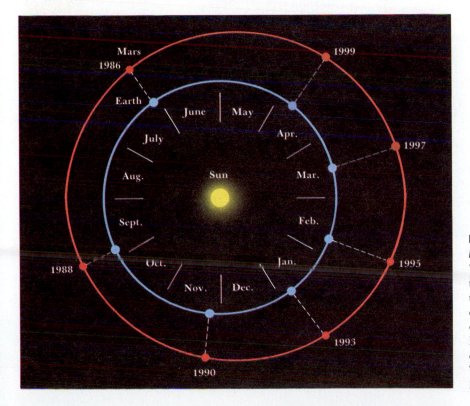

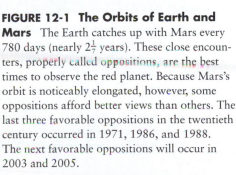

FIGURE 12-1 The Orbits of Earth and Mars The Earth catches up with Mars every 780 days (nearly $2\frac{1}{7}$ years). These close encounters, properly called oppositions, are the best times to observe the red planet. Because Mars's orbit is noticeably elongated, however, some oppositions afford better views than others. The last three favorable oppositions in the twentieth century occurred in 1971, 1986, and 1988. The next favorable oppositions will occur in 2003 and 2005.

BOX 12-1
Mars Data

Mean distance from Sun:	1.524 AU $= 2.279 \times 10^8$ km
Maximum distance from Sun:	1.666 AU $= 2.492 \times 10^8$ km
Minimum distance from Sun:	1.381 AU $= 2.067 \times 10^8$ km
Mean orbital velocity:	24.1 km/s
Sidereal period:	686.98 days $= 1.88$ years
Rotation period:	$24^h\ 37^m\ 22^s$
Inclination of equator to orbit:	25° 11′
Inclination of orbit to ecliptic:	1° 50′ 59″
Diameter (equatorial):	6786 km
Diameter (Earth = 1):	0.532
Mass:	6.42×10^{23} kg
Mass (Earth = 1):	0.107
Mean density:	3950 kg/m³
Surface gravity (Earth = 1):	0.380
Escape speed:	5.0 km/s
Surface temperatures:	Maximum: 20°C = 70°F = 293 K
	Minimum: −140°C = −220°F = 133 K

FIGURE 12-2 Mars Viewed from the Earth This high-quality Earth-based photograph of Mars was taken a few days after the opposition of 1971 when the Earth–Mars distance was only 56 million kilometers (35 million miles). At that time, Mars presented a disk nearly 26 arc sec in angular diameter. Circumstances as favorable as these will not be repeated for the rest of the century. This photograph portrays what you might typically see through a moderate-sized telescope under excellent observing conditions. Notice the prominent southern polar cap. (Courtesy of S. M. Larson)

called a **favorable opposition** because the Earth–Mars distance can be as small as 56 million kilometers (35 million miles). At such times, Mars appears brilliant and red in the sky, outshining all the stars and gleaming with a magnitude as bright as −2.8. Through a telescope, the ruddy planet presents a disk nearly 26 arc sec in diameter. That is the same angular diameter as a moderate-sized (50-km) lunar crater viewed from the Earth. The Earth-based photograph in Figure 12-2 was taken during the favorable opposition of 1971.

Four Martian oppositions will take place between 1993 and 1999 (Table 12-1). None will be particularly favorable. The previous favorable Martian opposition occurred in September 1988, when Mars and Earth were separated by only 58 million kilometers. The next favorable oppositions will occur in 2003 and 2005. A star chart showing the path of Mars during the 1995 opposition is given in Figure 4-1.

At unfavorable oppositions, such as those when Mars is near aphelion, the Earth–Mars distance can be as large as 101 million kilometers (63 million miles). And when Mars is not at opposition, the Earth–Mars distance can be much greater than 100 million kilometers. Mars then appears as a tiny reddish dot, virtually devoid of features even when viewed through a large telescope.

12-2 Earth-based observations originally suggested that Mars might have some form of extraterrestrial life

The first reliable record of surface features on Mars is found in observations recorded by the Dutch physicist Christian Huygens in November 1659. Using a refracting telescope of his own design, Huygens identified a prominent, dark, triangular feature that we now call Syrtis Major. After observing this feature for several weeks, Huygens concluded that the rotation period of Mars is approximately 24 hours. This observation was the first in a series that soon led to speculations about life on Mars, because the planet seemed similar to Earth. (Basic data about Mars are listed in Box 12-1.)

In 1666 the Italian astronomer Giovanni Domenico Cassini made the first accurate measurements of Mars's rotation period. He determined that a Martian day is nearly $37\frac{1}{2}$ minutes longer than an Earth solar day. Cassini was also the first to see the Martian polar caps, which bear a striking superficial resemblance to the Arctic and Antarctic polar caps on Earth (Figure 12-2). More than a century elapsed, however, before the famous German-born English astronomer William Herschel first suggested that the Martian polar caps might be made of ice or snow.

Herschel also determined the inclination of Mars's axis of rotation. Just as Earth's equatorial plane is tilted $23\frac{1}{2}°$ from the plane of its orbit, Mars's equator makes an angle of over $25°$ with its orbit. This striking coincidence means that Mars experiences Earthlike seasons. However, the Martian seasons last nearly twice as long as Earth's, because Mars takes nearly two years to orbit the Sun.

The Martian surface exhibits interesting seasonal variations. During spring and summer in a Martian hemisphere, the polar cap shrinks and the dark markings (which often look greenish) become very distinct. Half a Martian year later, with the approach of fall and winter, the dark markings fade and the polar cap grows. The implication that there is Martian vegetation was so strong that in 1802 the German mathematician Karl Friedrich Gauss proposed that we signal the Martian inhabitants by drawing huge geometric patterns in the Siberian snow. His plan was never carried out.

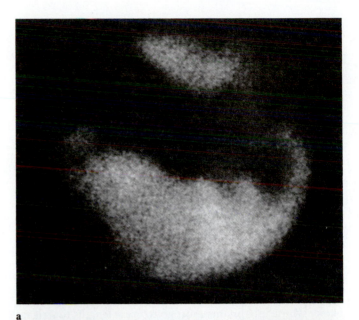
a

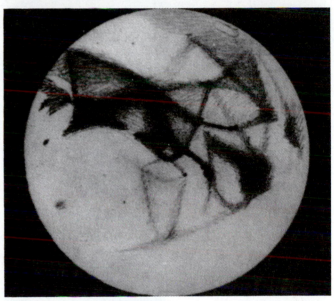
b

FIGURE 12-3 Martian Canals During moments of "good seeing," when the air is exceptionally calm and clear, some Earth-based observers reported seeing networks of lines crisscrossing the Martian surface. Some people suggested that these lines, called canals, are irrigation ditches constructed by intelligent creatures. This photograph (**a**) and the corresponding drawing (**b**) were made during the opposition of 1926. (Lick Observatory)

TABLE 12-1
Martian Oppositions, 1993–1999

Date of opposition		Earth–Mars distance (10^6 km)	Apparent diameter (arc sec)	Magnitude
1993	January 7	93	15	−1.2
1995	February 12	100	14	−1.0
1997	March 17	97	14	−1.1
1999	April 24	85	16	−1.5

Observations by the Italian astronomer Giovanni Virginio Schiaparelli during a favorable opposition in 1877 fueled further speculations about life on Mars. Schiaparelli reported seeing 40 straight-line features crisscrossing the Martian surface. He called these dark linear features *canali*, an Italian term meaning "natural water channels," which was soon mistranslated into English as *canals* (Figure 12-3). The alleged discovery of canals seemed to imply the existence on Mars of intelligent creatures capable of substantial engineering feats.

This speculation led Percival Lowell, who came from a wealthy Boston family, to finance a major new ob- servatory near Flagstaff, Arizona, primarily for additional studies of Mars. By the end of the nineteenth century, Lowell had reported observations of 160 Martian canals.

Not all astronomers saw the canals. In 1894, the American astronomer Edward Emerson Barnard, working at the Lick Observatory, complained that "to save my soul I can't believe in the canals as Schiaparelli draws them." Incidentally, Barnard was apparently the first to report seeing craters on Mars. However, he did not publish these observations for fear of ridicule.

The skepticism of cautious observers like Barnard was drowned out by the more fanciful pronouncements of Lowell and his colleagues. It soon became fashionable to speculate that the Martian canals formed an enormous planetwide irrigation network to transport water from melting polar caps to vegetation near the equator. In view of the planet's reddish, desertlike appearance, Mars was thought to be a dying planet whose inhabitants must go to great lengths to irrigate their farmlands. No doubt the Martians would readily aban-

don their arid ancestral homeland and invade the Earth for its abundant resources. Hundreds of science fiction stories and dozens of monster movies owe their existence to the canali of Schiaparelli.

12-3 Space probes to Mars found craters, volcanoes, and canyons—but no canals

Three American spacecraft journeyed past Mars in the 1960s and sent back pictures that clearly showed numerous flat-bottomed craters (Figure 12-4) but not one canal. Schiaparelli's canali had been an optical illusion.

From studies of moon rocks and the lunar surface, scientists knew that crater-making tapered off about 3.8 billion years ago, with mare-gouging impacts continuing to about 3 billion years ago. Like the Moon, the Martian surface must therefore be extremely ancient and have apparently experienced few changes over the past 3 billion years. The prospect of abundant Martian life suddenly seemed extremely remote.

The partially eroded, flat bottoms of the Martian craters probably result from dust storms that rage across the planet's

FIGURE 12-4 Martian Craters Numerous flat-bottomed craters are seen in this mosaic of about 100 images taken by a Viking orbiter in 1980. North is at the top. The large, heavily cratered circular area in the upper half of this view is known as Arabia. The dark area to the right of Arabia, called Syrtis Major, can be seen from Earth during favorable oppositions. Carbon dioxide snow covers the floors of craters near the Martian south pole. (NASA, USGS)

FIGURE 12-5 Olympus Mons This computer-enhanced view looks eastward across the largest volcano on Mars. Olympus Mons is nearly three times as tall as Mount Everest. Its cliff-ringed base measures 600 km (370 mi) in diameter. (NASA, USGS)

surface. Over the past two centuries, astronomers have seen faint surface markings disappear under a reddish-orange haze as thin Martian winds have stirred up finely powdered dust. Some storms actually obscure the entire planet. Over the ages, deposits of dust from these storms in the craters have filled in the crater bottoms.

However, these storms have not completely obliterated the craters because the Martian atmosphere is too thin to carry much dust. Measurements from spacecraft reveal an atmospheric pressure on the Martian surface of only 5 to 10 millibars, which is roughly the same as the air pressure at an altitude of 30 km (100,000 ft) above the Earth's surface. The material whipped up by the rarified Martian winds must be extremely fine-grained powder, because 3 billion years of sporadic dust storms have not wiped out the craters.

In 1971 the *Mariner 9* spacecraft went into orbit about Mars and began sending back detailed, close-up pictures that showed enormous volcanoes and vast canyons. The largest

FIGURE 12-6 The Olympus Caldera This view of the summit of Olympus Mons is based on a mosaic of six pictures taken by one of the Viking orbiters. The caldera consists of overlapping, volcanic craters and measures roughly 70 km across. The volcano is wreathed in mid-morning clouds formed from ice (H₂O) brought upslope by cool air currents. The cloudtops are about 8 km below the volcano's peak. (NASA)

Martian volcano, Olympus Mons, covers an area as big as the state of Missouri and rises 24 km (15 mi) above the surrounding plains—nearly three times the height of Mount Everest (Figure 12-5). The highest volcano on Earth, Mauna Loa in the Hawaiian Islands, has a summit only 8 km above the ocean floor. The Martian volcanoes and the Hawaiian Islands were both formed by hot-spot volcanism. Recall, however, that the process of plate tectonics also affects the Hawaiian Islands. The huge size of Olympus Mons strongly suggests an absence of plate tectonics on Mars. Instead, a hot spot in Mars's mantle probably kept pumping lava upward through the same vent for millions of years, producing one giant volcano rather than a long chain of smaller ones. The volcano's summit has collapsed to form a volcanic crater, called a **caldera,** large enough to contain the state of Rhode Island (Figure 12-6).

Olympus Mons is one of several large volcanoes clustered together on a huge bulge called the Tharsis Rise that covers an area 2500 kilometers in diameter centered just north of the Martian equator. Ground levels over this vast dome-shaped region are typically 5 to 6 kilometers higher than the average ground level for the rest of the planet. Another major though less dramatic grouping of volcanoes is located nearly on the opposite side of the planet. None of these volcanoes is active today.

For reasons we do not yet understand, most of the volcanoes on Mars are in the northern hemisphere, but most of the impact craters are in the southern hemisphere. Between the two hemispheres, *Mariner 9* discovered a vast canyon running roughly parallel to the Martian equator (Figure 12-7). In honor of the Mariner spacecraft that revealed so much of the Martian surface, this enormous chasm has been designated Valles Marineris.

Valles Marineris stretches over 3000 km, beginning with heavily fractured terrain in the west and ending with ancient cratered terrain in the east. If this canyon were located on Earth, it would stretch from New York to Los Angeles. Some geologists suspect Valles Marineris is a fracture in the Martian crust that suggests an attempt at plate tectonics.

Plate tectonics did not significantly affect the surface of Mars, possibly because the planet cooled rapidly when it was still young. In general, small bodies lose heat much more rapidly than large bodies. Mars's diameter is only half that of

FIGURE 12-7 A Segment of Valles Marineris This mosaic of Viking orbiter photographs shows a segment of Valles Marineris. The canyon is about 100 km wide in this region. The canyon floor has two major levels. The northern (upper) canyon floor is 8 km beneath the surrounding plateau, whereas the southern canyon floor is only 5 km below the plateau. (NASA, USGS)

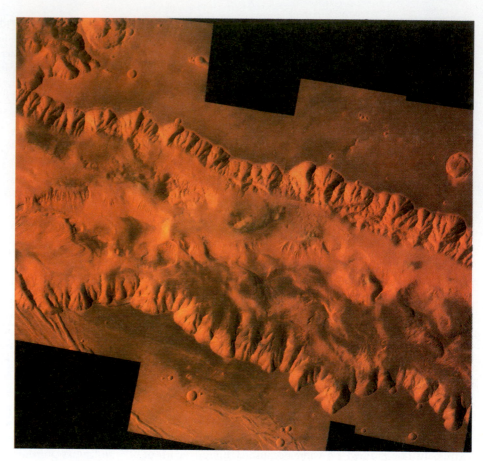

the Earth, and its mass is just one-tenth of the Earth's mass (only about twice the mass of Mercury). Plate tectonics on Mars bogged down early in the planet's history as the rapidly cooling lithosphere became thick and sluggish.

12-4 Surface features indicate that water once flowed on Mars

Valles Marineris was not formed by water erosion, but Mars-orbiting spacecraft did reveal many features that look like dried-up riverbeds (Figure 12-8). Intricate branched patterns and delicate channels meandering among flat-bottomed craters strongly suggest that water once flowed over the Martian surface. Because craters overlap many of the riverbeds, scientists estimate these erosional features to be older than the craters—perhaps about 4 billion years old.

In 1976 a pair of spacecraft called the Viking Orbiters sent back pictures that appear to show the results of flash floods, suggesting that water once raged across the Martian surface in great torrents. From the widths and depths of the Martian flash-flood channels, scientists estimate peak flood discharges must have been greater than anything known on Earth.

The Martian riverbeds were totally unexpected, because liquid water cannot exist today on Mars. Water is liquid over a limited range of temperatures and pressures: At low temperatures, water becomes ice; at high temperatures, it becomes steam. The atmospheric pressure above a body of

water also affects the state of the water. If the pressure is very low, molecules easily escape from the liquid's surface, causing the water to vaporize. Any liquid water on Mars today would furiously boil and rapidly evaporate into the thin Martian air.

Spacecraft data also indicate that the Martian atmosphere, which is 95% carbon dioxide, is very dry. It never rains on Mars. If all the water vapor could somehow be squeezed out of the Martian atmosphere, it would not fill one of the five Great Lakes in North America.

If water once flowed on Mars, where might that water be today? Scientists looked first to the Martian polar caps as a possible frozen reservoir. Because the winter temperature at the Martian poles is low enough ($-140°C = -220°F$) to freeze carbon dioxide as well as water, it was not clear how much of the polar caps consists of frozen carbon dioxide (dry ice) and how much is water ice.

In 1972 *Mariner 9* sent back photographs from orbit as summer came to the northern hemisphere of Mars. During the Martian spring, the polar cap receded rapidly, strongly suggesting that a thin layer of carbon dioxide frost was evaporating in the sunlight. However, with the arrival of summer, the rate of recession slowed abruptly, suggesting that a thicker layer of water ice had been exposed. Scientists concluded that the **residual polar caps** that survive the Martian summers contain a large quantity of frozen water (Figure 12-9). Calculating the volume of water ice in the residual caps is difficult, however, because we do not know

FIGURE 12-8 Ancient River Channels on Mars Many dried riverbeds cut across this cratered terrain, located only 5° south of the Martian equator. The existence of riverbeds on Mars suggests that the planet's atmosphere was thicker and its climate

more Earthlike long ago, when liquid water flowed across its surface. Any water exposed to the sparse Martian atmosphere today would rapidly boil away or freeze solid. (NASA, USGS)

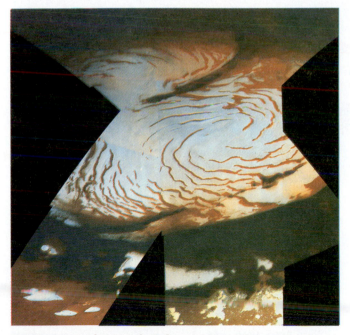

FIGURE 12-9 The North Polar Ice Cap This mosaic of Viking images shows the northern polar cap of Mars in mid-summer. Most of the carbon dioxide frost that covers northern latitudes during the winter has evaporated. The residual polar cap shown here is mostly water ice. The reddish streaks consist of dust, and dark bluish areas are sand dunes forming a huge "collar" that surrounds the polar cap. (NASA, USGS)

how thick the layer of ice is. Scientists are thus still debating the exact amount of frozen water stored in the Martian polar caps.

Close-up views show that flash-flood features usually emerge from craters or from collapsed, jumbled terrain. It therefore seems likely that frozen water also exists in a layer of **permafrost** under the Martian surface, similar to that beneath the tundra in the far northern regions on Earth. Apparently, heat from a meteoroid impact or from volcanic activity in the distant past occasionally melted this subsurface ice. The ground then collapsed, and millions of tons of rock pushed the water to the surface (Figure 12-10).

12-5 Earth and Mars began with similar atmospheres that evolved very differently

All the erosional features, from flash-flood channels to meandering stream beds, clearly indicate that the Martian climate must have been much warmer and more Earthlike 4 billion years ago. How could the Martian surface have been warm enough to permit substantial amounts of flowing water? The likeliest mechanism is the greenhouse effect (recall Figure 11-5).

Because the atmospheres of Earth and Mars are transparent to sunlight, they are not heated directly by the Sun. Instead, they are heated from below by energy reradiated from the ground at longer, infrared wavelengths. Water vapor and carbon dioxide absorb these wavelengths, trapping the heat

FIGURE 12-10 An Ancient Dried Lake? The layered terrain visible in this scene may be the residue of a huge, ancient lake. Note the erosional patterns that emanate from the cliffsides at the lower left and flow across the canyon floor. The area shown in this mosaic of Viking orbiter images is about the same size as the state of Ohio. (NASA, USGS)

and warming the atmosphere. In this way, the average worldwide temperature of the Earth's atmosphere is increased by about 33°C.

The greenhouse effect is not very efficient on Mars today because the sparse Martian atmosphere permits much of the infrared radiation to escape into space. If the greenhouse effect was once important on Mars, the carbon-dioxide content of the Martian atmosphere must have been much higher than it is today. Calculations indicate that 1000 millibars of carbon dioxide would have been needed to sustain a greenhouse effect capable of permitting free-flowing water. Thus, the primordial atmosphere of Mars would have been comparable in density and pressure to the Earth's atmosphere today.

Extinct Martian volcanoes suggest how such a primordial atmosphere could well have been created. The terrestrial planets obtained their atmospheres primarily through volcanic outgassing (see Figure 11-13). Water vapor, carbon dioxide, and nitrogen are among the most common gases in volcanic vapors. These three gases probably constituted the bulk of the primordial atmospheres of both Earth and Mars 4 billion years ago.

On Earth most of the water is in the oceans. Nitrogen, which is not very reactive, is still in the atmosphere. Carbon dioxide, however, dissolves in water; thus rain washed carbon dioxide out of the air. The dissolved carbon dioxide reacted with rocks in rivers and streams, and the residue was ultimately deposited on the ocean floors, where it turned into carbonate rocks such as limestone. Consequently, most of the Earth's carbon dioxide is tied up in the Earth's crust. Plate tectonics cycles carbonate rocks through volcanoes, which then continually reintroduce carbon dioxide back into the atmosphere. Without volcanoes, rain would wash all the carbon dioxide from the Earth's air in only a few thousand years. Of course, the oxygen in Earth's atmosphere is the result of photosynthesis by plant life (Figure 12-11a).

On Mars, rainfall 4 billion years ago washed much of the planet's carbon dioxide from its atmosphere, probably creating carbonate rocks in which the carbon dioxide is today chemically bound. Thus, on both Earth and Mars, carbonate rocks are significant reservoirs of the primordial carbon dioxide outgassed from ancient volcanoes. But plate tectonics did not occur on Mars, and its carbonate rocks were not recycled through volcanoes. The depletion of carbon dioxide from the Martian atmosphere was therefore permanent. According to calculations by James B. Pollack of NASA's Ames Research Center, a 1000-millibar carbon dioxide Martian atmosphere could have survived for only 10 to 100 million years. However, volcanic activity early in Mars's history may have recycled some carbon dioxide, thus prolonging the greenhouse effect for perhaps half a billion years. Ultimately, rainfall did succeed in destroying the greenhouse effect that had allowed water to exist in the first place. In effect, water on Mars led to its own disappearance.

As both the water vapor and the carbon dioxide became depleted, ultraviolet light from the Sun could penetrate the thinning Martian atmosphere to strip it of nitrogen. Nitrogen molecules (N_2) normally do not have enough thermal energy to escape from Mars, but they can acquire that energy from ultraviolet photons, which break the molecules in two.

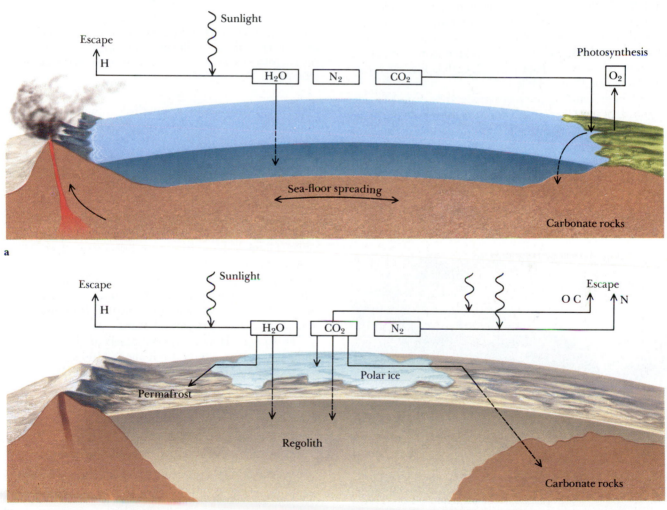

FIGURE 12-11 The Atmospheres of Earth and Mars The primitive atmospheres of Earth and Mars were probably similar, but they evolved differently. (a) On Earth, water condensed into oceans, and nitrogen remained in the air, while carbon dioxide was removed from the atmosphere by rainfall and then stored in carbonate rocks. (b) On Mars, rainfall destroyed the thick carbon dioxide atmosphere that kept the planet warm enough (through the greenhouse effect) for liquid water to exist. Today water ice is found in the polar caps and the Martian permafrost. Some water, along with carbon dioxide, may also be chemically bound in the fine-grained Martian regolith. Ultraviolet light from the Sun breaks apart both carbon dioxide and nitrogen molecules into their constituent atoms, which then escape into space. (Adapted from R. M. Haberle)

Ultraviolet photons can also split carbon dioxide and water molecules, giving their atoms enough energy to escape (Figure 12-11b). Perhaps in this way Mars came to have the sparse atmosphere we observe today.

12-6 The *Viking* landers sent back close-up views of a barren Martian surface

The discovery that water once existed on the surface of Mars rekindled speculation about Martian life. By the mid-1960s, although it was clear that Mars has neither civilizations nor fields of plants, some sort of microbial life forms still seemed possible. Searching for Martian microbes was one of the main objectives of the ambitious and highly successful Viking missions.

The two *Viking* spacecraft, which were launched from Earth during the summer of 1975, arrived at Mars almost a year later. Each spacecraft consisted of two modules: an orbiter and a lander. After entering orbit about Mars, each lander separated from its orbiter and descended to the Martian surface with the aid of a heat shield, retrorockets, and a parachute. As shown in Figure 12-12, each lander is roughly the size of a small automobile.

The *Viking 1* landing site was a moderately cratered rocky plain north of the Martian equator near the mouth of a large flash-flood channel. The upraised rim of a crater is visible on the horizon in Figure 12-13. Such craters are probably the source of the jagged rocks that litter the scene. Some light-colored bedrock is exposed near the horizon.

Some small sand drifts were seen at the *Viking 1* site. Careful scrutiny of pictures taken over several months

FIGURE 12-13 The *Viking 1* Site This view looks southward over the Chryse Planitia (the Golden Plains). The horizon is about 3 km (2 mi) from the spacecraft. Part of the upraised rim of a crater on the horizon is visible at the upper left. A few light-colored patches of exposed bedrock also appear in this picture. (NASA)

reveals that these drifts are pushed around slightly by the Martian winds. New rocks are occasionally exposed and others covered up, providing what may be the best explanation for the seasonal changes observed from Earth for the past two centuries. Seasonal winds transporting fine red dust alternately expose and cover dark rocks, mimicking color changes due to vegetation over the long Martian year.

What of the greenish hues reported for the dark surface markings by many Earth-based observers? Not life but a quirk of human color vision is probably responsible. When a neutral gray area is viewed next to a bright red-orange area, the eye perceives the gray to have a blue-green cast, the so-called complementary colors of red-orange. Perhaps fantasies of Martian vegetation also contributed to observations of greenish surface patterns on Mars.

The *Viking 2* lander set down nearly on the opposite side of the planet from *Viking 1* but about 1500 kilometers (930 miles) closer to Mars's north pole. As shown in Figure 12-14, the *Viking 2* site is remarkably crater-free, although many of the rocks in this view may have been ejected from a large crater about 200 kilometers east of the spacecraft.

12-7 Instruments on the Viking landers revealed a highly seasonal Martian climate

In all the pictures sent back from Mars, the sky is distinctly pinkish-orange. This remarkable color is probably caused by extremely fine-grained dust suspended in the thin Martian atmosphere.

Both Viking landers measured meteorological conditions on Mars. Their instruments promptly confirmed the thinness of the Martian atmosphere: Surface atmospheric pressure is in the range of 6 to 8 millibars, slightly less than 1% that of Earth's. Chemical analysis also confirmed a 95% abundance

FIGURE 12-12 The Viking Lander From the base of its footpads to the top of its antenna, each Viking lander is about 2 m (6 ft) tall and weighs 576 kg (1270 lb) without fuel. An extendable arm for retrieving soil samples is visible in the foreground. The two cylinders (one with an American flag) protruding from the top of the lander contain the two television cameras. The dish antenna was used to relay data to the Viking orbiter overhead, which in turn sent the information back to Earth. (NASA)

FIGURE 12-14 A Panorama of the *Viking 2* Site This mosaic of three summertime views shows about 180° of the Utopian plains. Northwest is at the left; southeast on the right. The flat, featureless horizon is approximately 3 km (2 mi) away from the spacecraft. (NASA)

of carbon dioxide; the remaining 5% is mostly nitrogen and argon.

After only a few weeks on the Martian surface, the Viking landers provided data clearly showing that the atmospheric pressure at both landing sites was dropping steadily. Mars seemed to be rapidly losing its atmosphere, leading some scientists to joke that soon all the air would be gone. A straightforward explanation was, however, readily available: Winter was coming to the southern hemisphere. At the Martian south pole, it became so cold that large amounts of carbon dioxide began condensing out of the atmosphere and covering the ground with flakes of dry ice (Figure 12-15). The formation of this "snow" removed gas from the Martian atmosphere, thereby lowering the atmospheric pressure.

Several months later, when spring came to the southern hemisphere, the dry-ice snow rapidly evaporated and the atmospheric pressure returned to its prewinter levels. When winter arrived in the northern hemisphere, another decrease in atmospheric pressure was observed there as dry-ice snow blanketed the northern latitudes (Figure 12-16). Therefore both temperature and atmospheric pressure change significantly with the Martian seasons.

FIGURE 12-15 The South Pole of Mars During the Martian winter, the temperature drops so low that carbon dioxide freezes out of the Martian atmosphere. A thin coating of carbon dioxide frost covers a heavily cratered region around Mars's south pole. Seasonal freezing and evaporation of carbon dioxide at both poles causes atmospheric pressure to vary planetwide. (NASA)

FIGURE 12-16 Winter in Mars's Northern Hemisphere This picture, taken during midwinter, shows a thin frost layer that lasted for about a hundred days at the *Viking 2* lander site. Freezing carbon dioxide adheres to water-ice crystals and dust grains in the atmosphere, causing them to fall to the ground. The sky is therefore not as pink as it is in the summertime view of Figure 12-14. (NASA)

FIGURE 12-17 Digging in the Martian Soil Viking's mechanical arm with its small scoop protrudes from the right side of this view of the *Viking 1* landing site. (Figure 12-12 shows the fully extended arm.) Several small trenches dug by the scoop in the Martian regolith appear near the left side of the picture. (NASA)

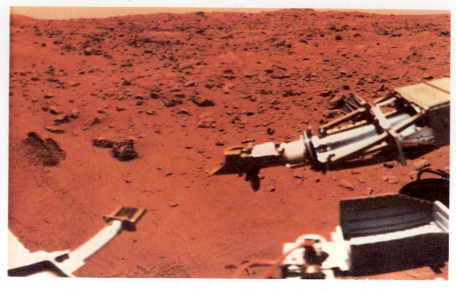

Summer comes to the southern hemisphere of Mars at the same time that the planet makes its closest approach to the Sun. Along the retreating edge of the polar cap, the sharp temperature contrast between the evaporating dry ice and the soil warmed by the sunshine gives rise to strong winds. These winds often produce swirling dust storms.

12-8 The Viking landers found abundant iron in the Martian regolith but failed to detect living organisms

Each Viking lander had a scoop at the end of a mechanical arm to dig in the Martian regolith and retrieve rock samples for analysis (Figure 12-17). Bits of the regolith were observed to cling to a magnet mounted on the scoop, indicating that the regolith contains iron. A device called an X-ray fluorescence spectrometer measured the chemical composition of the rocks at both lander sites and showed them to be rich in iron, silicon, and sulfur. The Martian regolith can best be described as an iron-rich clay. Indeed, the familiar reddish color of Mars may be caused by rust (iron oxides) in the regolith.

In spite of the high iron content of its crust, Mars has a lower average density (3950 kg/m^3) than the other terrestrial planets (more than 5000 kg/m^3 for Mercury, Venus, and Earth). Mars must therefore have a lower total iron content than these other planets. Furthermore, Mars does not have a measurable magnetic field. Perhaps Mars's iron is more uniformly distributed throughout the body of the planet rather than being concentrated in a dense core. Because a seismometer on one of the Viking landers failed, the structure of the Martian interior remains largely unknown.

In addition to instruments that analyzed rocks and the atmosphere, each Viking lander carried a compact biological laboratory designed to perform three different tests for microorganisms in the Martian soil. The three biological ex-

periments carried out by the Viking landers were based on the general proposition that living things must alter their environment. They eat, they breathe, and they give off waste products. In each of the three experiments, a sample of the Martian regolith was placed in a closed container, with or without a certain nutrient substance. The container was then examined for any changes in its contents.

The "gas-exchange experiment" was designed to detect any processes that might be broadly considered as respiration. A small sample of regolith was placed in a sealed container along with a controlled amount of gas and nutrients. The gases in the container were then monitored to see if their chemical composition changed.

The "labeled-release experiment" was designed to detect processes resembling metabolism. A small sample of regolith was moistened with nutrients containing radioactive carbon atoms. Scientists assumed that any organisms in the regolith would eat the food and then emit gases containing the telltale radioactive carbon.

The "pyrolytic-release experiment" was designed to detect photosynthesis, the biological process by which plants on Earth synthesize organic compounds from carbon dioxide, using sunlight as an energy source. In the Viking experiments, a regolith sample was placed in a container along with radioactive carbon dioxide and exposed to artificial sunlight. If plantlike photosynthesis were to occur, some of the radioactive carbon from the gas should become incorporated into the microorganisms in the regolith.

The first data returned by the Viking biological experiments caused great excitement: In almost every case, rapid and extensive changes were detected inside the sealed containers. However, further analysis of the data led to the conclusion that these changes were due solely to nonbiological chemical processes. It appears that the Martian regolith is rich in chemicals that effervesce (fizz) when moistened. A large amount of oxygen is apparently tied up in the regolith in the form of unstable chemicals called peroxides and superoxides, which break down in the presence of water to release oxygen gas.

The chemical reactivity of the Martian regolith probably comes from ultraviolet radiation that beats down on the planet's surface. Ultraviolet photons easily break apart molecules of carbon dioxide (CO_2) and water vapor (H_2O) by knocking off oxygen atoms, which then become loosely attached to chemicals in the regolith. Ultraviolet photons also produce ozone (O_3) and hydrogen peroxide (H_2O_2), which also become incorporated in the regolith. In all these cases, the loosely attached oxygen atom makes the regolith extremely reactive.

Here on Earth, hydrogen peroxide is commonly used as an antiseptic. When you pour this liquid on a wound, it fizzes and froths as the loosely attached oxygen atoms chemically combine with organic material, thereby destroying germs. The Viking landers may have failed to detect any organic compounds on Mars because the superoxides and peroxides in the regolith make it literally antiseptic.

12-9   The two Martian moons resemble asteroids

Two tiny moons move around Mars in orbits close to the planet's surface. These satellites are so small that they were not discovered until the favorable opposition of 1877. While Schiaparelli was seeing canals, the American astronomer Asaph Hall spotted the two moons, which orbit almost directly above its equator. He named them Phobos ("fear")

and Deimos ("panic"), after the mythical horses that drew the chariot of the Greek god of war.

Phobos is the inner and larger of the Martian moons. It circles Mars in only 7 hours 39 minutes, moving over the Martian surface at an average distance of only 6000 kilometers (3700 miles). (Recall that our Moon orbits at an average distance of 376,280 km above the Earth's surface.) This orbital period is much shorter than the Martian day; Phobos rises in the west and gallops across the sky in only $5\frac{1}{2}$ hours, as viewed by an observer near the Martian equator. During this time, Phobos appears several times brighter in the Martian sky than Venus does from Earth.

Deimos, which is farther from Mars and somewhat smaller than Phobos, appears about as bright from Mars as Venus does from Earth. Deimos's orbit, 20,000 kilometers (12,400 miles) above the Martian surface, is at almost the right distance for a synchronous orbit—an orbit in which a satellite seems to hover above a single location on the planet's equator. As seen from the Martian surface, Deimos rises in the east and takes about three full days to creep slowly from one horizon to the other.

The *Mariner 9* mission treated astronomers to their first close-up views of the Martian satellites. Numerous additional views provided by the Viking orbiters revealed Phobos and Deimos to be jagged, heavily cratered, football-shaped rocks (Figure 12-18). Phobos was found to be roughly 28 by 23 by 20 km, with Deimos being slightly smaller, at roughly 16 by 12 by 10 km. It was further revealed that both Phobos and Deimos rotate synchronously about Mars, always keeping their same sides facing the red planet.

a

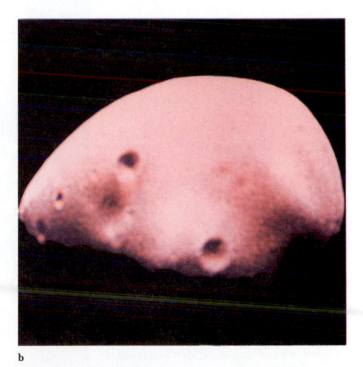

b

FIGURE 12-18 Phobos and Deimos Viking orbiters provided these images of Mars's tiny moons. (**a**) Phobos, the larger of Mars's two moons, is potato-shaped and measures roughly 28 by 23 by 20 km (about 17 by 14 by 12 mi). (**b**) Deimos seems to be less cratered than Phobos and measures roughly 16 by 12 by 10 km (about 10 by 7 by 6 mi). (NASA)

The origin of the Martian moons is unknown, but they may be captured asteroids. Mars is quite near the asteroid belt, where thousands of large rocks orbit the Sun. Like Phobos and Deimos, many asteroids have low albedos (typically less than 10%) because of a high carbon content. Perhaps two of these asteroids wandered close enough to Mars to become permanently trapped by the planet's gravitational field. Future measurements of the surface composition of the Martian moons may shed light on their origin.

12-10 Future missions to Mars will give us important new insights about the planet

In 1992 NASA launched the *Mars Observer* spacecraft on an eleven-month journey to the red planet (Figure 12-19). Just a few days before it was to go into orbit about Mars, all contact with the spacecraft was lost. The failure of this mission was a major disappointment to scientists around the world, some of whom had devoted their entire professional careers to the project. Instruments on board the spacecraft would have revealed the behavior of the Martian atmosphere over a full seasonal cycle. The *Mars Observer* would have also mapped the surface of Mars with unprecedented resolution. The Mariner and Viking spacecraft produced detailed maps of only about 15% of the Martian surface. The *Mars Observer* would have shown us much more of Mars, in some cases resolving features as small as 10 m (33 ft) across.

The next logical step in the exploration of Mars is to place a spacecraft in orbit about the planet and gather data over the course of a full Martian year. Science instruments, similar to those on Earth-orbiting satellites that study our planet, could provide a detailed portrait of Mars. A spectrometer could analyze the Martian surface by detecting gamma rays emitted by radioactive elements in the regolith. Cameras could take global pictures to show us daily weather changes. The same cameras could also take high-resolution images of the ice caps, volcanoes, and dried riverbeds. A laser altimeter could measure the heights of volcanoes, canyons, and other surface features with a precision of several meters.

Russian scientists plan to launch a spacecraft toward Mars in 1994 that will consist of an orbiter, two landers, and two "penetrators." The landers will carry instruments to measure surface properties and the weather. The spearlike penetrators will split in two when they hit the ground. The bottom half will bore several meters into the Martian regolith; the upper half, with its weather instruments and cameras, will remain above ground.

Mars beckons, not only as the most likely planet for human exploration, but also as an ideal laboratory—ancient and unchanged—for studying the formation and evolution of the inner solar system. Mars's hemispheres—one heavily cratered, the other smoothed by volcanic eruptions and flash floods—may soon tell us more about events that caused the planet to grow cold and nearly airless early in its history.

FIGURE 12-19 The *Mars Observer* All communication with this spacecraft was lost just before it was about to go into orbit about Mars in August 1993. Russian scientists will attempt to launch a spacecraft toward Mars in September 1994. (NASA)

KEY WORDS

caldera   favorable opposition   permafrost   residual polar cap

KEY IDEAS

• Earth-based observers found that the Martian solar day is nearly the same as that on the Earth, that Mars has polar caps that expand and shrink with the seasons, and that the Martian surface undergoes seasonal color changes.

The best Earth-based views of Mars are obtained at favorable oppositions, when Mars is simultaneously at opposition and near perihelion.

A few observers reported a network of linear features called canals; these observations led to many speculations about Martian life.

• Spacecraft have returned close-up views showing that the Martian surface has numerous flat-bottomed craters, several huge volcanoes, a vast canyon, and dried-up riverbeds—but no canals.

• The flash-flood features and dried riverbeds on the Martian surface indicate that water once flowed on Mars.

Liquid water would quickly boil away in Mars's thin atmosphere, but its polar caps contain frozen water, and a layer of permafrost may exist beneath the Martian regolith.

Mars's primordial atmosphere probably contained enough carbon dioxide to support a greenhouse effect that permitted liquid water to exist on the planet's surface.

The winter expansion of Mars's polar caps is caused by deposition of a thin layer of frozen carbon dioxide (dry ice).

• The Martian atmosphere is composed mostly of carbon dioxide; the atmospheric pressure on the surface there is about 1% that of Earth's and shows seasonal variations.

Great dust storms sometimes blanket Mars. Fine-grained dust in its atmosphere gives the Martian sky a pinkish-orange tint.

• Chemical reactions in the Martian regolith, together with ultraviolet radiation from the Sun, apparently act to sterilize the Martian surface.

• Mars has two small, football-shaped satellites that move in orbits close to the surface of the planet; they may be captured asteroids.

REVIEW QUESTIONS

1. Why is Mars red?

2. When is the best time to observe Mars from Earth?

3. Explain why Mars has the longest synodic period of all the planets, although its sidereal period is only 687 days.

4. Compare the cratered regions of Mercury, the Moon, and Mars. Assuming that the craters on all three worlds originally had equally sharp rims, what can you conclude about the environmental histories of these worlds?

5. Explain why the Earth's sky is blue, that of Venus is yellow-orange, and that of Mars is pink.

6. How would you tell which craters on Mars had been formed by meteoritic impacts and which by volcanic activity?

7. Compare the volcanoes of Venus, Earth, and Mars. Cite evidence that hot-spot volcanism is or was active on all three worlds.

8. What geologic features (or lack thereof) on Mars have convinced scientists that plate tectonics did not significantly shape the Martian surface?

9. What two types of Martian terrain suggest that Mars may have had a denser atmosphere in the past than it does today? What third type of terrain suggests that this episode of a denser atmosphere must have been comparatively short-lived?

10. Why is it reasonable to assume that the primordial atmospheres of Venus, Earth, and Mars were roughly the same?

11. Compare the evolution of the atmospheres of Venus, Earth, and Mars. Explain how the atmospheres of these three planets turned out so differently from each other, even though they all started with roughly the same chemical composition.

12. With carbon dioxide accounting for about 95% of the atmospheres of both Mars and Venus, why do you suppose there is no greenhouse effect on Mars today?

13. What is the current knowledge concerning life on Mars? Do you think Mars might be as barren and sterile as the Moon?

14. Briefly describe at least three types of remote-sensing observations that the *Mars Observer* is conducting.

ADVANCED QUESTIONS

Tips and tools . . .

You will need to remember the small angle formula, given in Box 1-1. Also, you may need to review the form of Kepler's third law, stated in Box 4-3, that explicitly includes mass.

15. Suppose you have a telescope with a resolving power of 1 arc sec. What is the smallest feature you could see on the Martian surface during the opposition of 1995? Suppose you had access to the Hubble Space Telescope, which has a resolving power of 0.1 arc sec. What is the smallest feature you could see on Mars with the HST during the 1995 opposition?

16. Imagine you are on Mars, looking back at the Earth when it is at greatest elongation from the Sun. Assuming the human eye has a resolving power of 50 arc sec, would you see the Earth and Moon as a "double star"?

17. Use the orbital data for Phobos to calculate the mass of Mars. How does your answer compare with the mass of Mars given in Box 12-1?

18. The synodic period of Mars is 780 days, but the dates given in Table 12-1 show that the intervals between successive oppositions vary. Explain this apparent contradiction.

19. Is it reasonable to suppose that the polar regions of Mars might harbor life forms, even though the Martian regolith is sterile at the Viking lander sites?

20. Why do you suppose that Phobos and Deimos are not round like our Moon?

21. Calculate the angular sizes of Phobos and Deimos as they pass overhead, seen by an observer standing on the Martian equator. How do these sizes compare with that of the Moon seen from the Earth's surface?

22. The *Mars Observer* arrived at Mars about the same time this book went to press. Consult such magazines as *Sky & Telescope* or *Science News* to learn about the current status of the mission. Is the *Mars Observer* working properly? What do the spacecraft's high-resolution images reveal that was not seen by *Viking* or *Mariner*? Have any surprises come from the mission?

DISCUSSION QUESTIONS

23. Suppose someone told you that the Viking mission failed to detect life on Mars simply because the tests were designed to detect terrestrial life forms, not Martian life forms. How would you respond?

24. Do you think sedimentary rocks might exist on Mars? Is it possible that sedimentary rocks might contain fossils of early Martian life forms? What sort of adaptations would these early life forms have had to make in order to survive to the present time?

25. Compare the scientific opportunities for long-term exploration offered by the Moon and Mars. What difficulties would be inherent in establishing a permanent base or colony on each of these two worlds?

26. Imagine you are an astronaut living at a base on Mars. Describe your day's activities, what you see, the weather, the spacesuit you are wearing, and so on.

OBSERVING PROJECT

27. Consult such magazines as *Sky & Telescope* or *Astronomy* to determine Mars's location among the constellations. If Mars is suitably placed for observation, arrange to view the planet through a telescope. Draw a picture of what you see. What magnifying power seems to give you the best image? Can you distinguish any surface features? Can you see a polar cap or dark markings? If not, can you offer an explanation for Mars's bland appearance?

FOR FURTHER READING

Ainsworth, D. "Mars Observer: Return to the Red Planet." *Astronomy,* September 1992. This comprehensive article describes the Mars Observer mission and what scientists hope to learn about the red planet in the next few years.

Batson, R., and others, eds. *The Atlas of Mars.* NASA SP-438, 1979. This NASA atlas contains photographs and maps of Mars produced by the U.S. Geological Survey.

Burgess, E. *To the Red Planet.* Columbia University Press, 1978. This book offers a fine summary of the Viking missions to Mars.

Carr, M. "The Surface of Mars: A Post-Viking View." *Mercury,* January/February 1983. This article by the leader of the Viking Orbiter Imaging Team gives an excellent summary of what we have learned about the Martian surface.

Chaikin, A. "Four Faces of Mars." *Sky & Telescope,* July 1992. This brief article includes three spectacular global views of Mars constructed from Viking images by a team of imaging wizards at the U.S. Geological Survey.

Cooper, H. *The Search for Life on Mars.* Holt, Rinehart & Winston, 1980. This book by a noted journalist profiles the Viking missions to Mars.

Gore, R. "Sifting for Life in the Sands of Mars." *National Geographic,* January 1977. This beautifully illustrated article summarizes the findings of the Viking missions.

Haberle, R. M. "The Climate of Mars." *Scientific American,* May 1986. This article argues that Mars's climate was once like Earth's early climate, but it evolved differently.

Hartmann, W. K. "What's New on Mars?" *Sky & Telescope,* May 1989. This fine article by a noted planetary scientist discusses the current state of our knowledge about Mars.

Horowitz, N. H. *To Utopia and Back*. W. H. Freeman and Company, 1986. This excellent book discusses the search for life on Mars.

Nichols, R. G. "Return to the Red Planet." *Sky & Telescope*, October 1992. This article by a former NASA contractor describes the Mars Observer program.

Parker, D., and others. "Mars' Grand Finale." *Sky & Telescope*, April 1989. This article by members of the Association of Lunar and Planetary Observers contains many fine photographs of Mars taken during the 1988 opposition, whose favorable circumstances will not be repeated until 2003.

Pollack, J. "Atmospheres of the Terrestrial Planets." In Beatty, J., and others, eds., *The New Solar System*, 3rd ed. Cambridge University Press, 1990. This chapter compares the chemical composition, circulation, and evolution of the atmospheres of Earth, Venus, and Mars.

Schiltz, P. "Polar Wandering on Mars." *Scientific American*, December 1985. The author of this intriguing article argues that Mars's entire lithosphere has shifted, moving polar regions toward the equator.

Spitzer, C., ed. *Viking Orbiter Views of Mars*. NASA SP-441. This magnificent album from NASA contains both color and black-and-white photographs of the Martian surface.

Toon, O. B. "How Climate Evolved on the Terrestrial Planets." *Scientific American*, February 1988. This superb article compares the evolution of the atmospheres of Venus, Earth, and Mars.

MARCIA NEUGEBAUER

Exploring the Planets

MARCIA NEUGEBAUER is a senior research scientist at the Jet Propulsion Laboratory of the California Institute of Technology, where her principal fields of interest are space plasma physics and comets. Early in her career she worked on an instrument that, while en route to Venus aboard the *Mariner 2* spacecraft, proved that the solar wind is a continuously blowing flow of plasma from the Sun. She also worked on an ion mass spectrometer that in 1986 flew on the European spacecraft through the coma of Halley's Comet.

Over the years Dr. Neugebauer has received substantial recognition for her work, including a NASA Medal for Exceptional Scientific Achievement, the Woman Scientist of the Year Medal from the California Museum of Science and Industry, and fellowship in the American Geophysical Union. She currently chairs the committee on solar and space physics of the National Research Council and is president-elect of the American Geophysical Union.

Humans are driven by the urge to explore. If no *terra incognita* remains on maps of the Earth, that can hardly be said of the rest of the solar system. No one has studied the polar regions of our own moon, where ice may yet lie at the bottom of deep valleys. Pluto and its satellite, Charon, are still known only as small, blurry disks seen through large telescopes. The surface (or oceans?) beneath the opaque atmosphere of Saturn's moon Titan, roughly half the surface of Mercury, as well as myriad asteroids and comets are all totally unknown.

Exploration is exhilarating, even when it must be done by remote control, using cameras as our eyes, spectrometers as noses, radio antennas as ears, thermometers instead of skin, and mechanical scoops in place of arms and hands. I have participated in space missions that flew past Venus in 1962, to the Moon in 1969,

through the coma, or gaseous head, of Comet Halley in 1986, and past Jupiter in 1992. Each time, other space scientists and I were at the terrestrial end of radio links to instruments that we had designed. In a sense we, too, were out there, exploring the solar system. Today one of our instruments is headed for the polar regions of the Sun.

Thus a major goal of the nation's planetary exploration program is to satisfy our curiosity. But the answers we seek to major scientific questions about the solar system require a much more sophisticated effort than just traveling to different bodies and taking their pictures.

One of those questions is the mystery of how the solar system itself originated. We believe it evolved from the collapse of a vast interstellar cloud of gas, ice, and grains of dust, which formed the solar nebula. To learn how the nebula evolved, we

look for clues in the present state of the solar system. And since the rocky inner planets have been pretty well baked out by the Sun's heat, those clues must be sought further from the Earth, beginning with Jupiter and Saturn.

The gravitational attraction of the two giant planets is so great that even the lightest, fastest-moving gas molecules cannot escape. Thus the present-day chemical compositions of Jupiter and Saturn show the ratios of different elements and isotopes in the parts of the solar nebula where those planets formed. Perhaps even more important, ratios of minerals found on planetary satellites, asteroids, and comets can provide us with detailed maps of the nebula—including the distribution of pressures and temperatures. Finally, from the molecules that now compose cometary ices, we can estimate what fraction of the atmospheres of the Earth, Venus, and Mars was delivered billions of years ago by a shower of comets, during the same period of intense bombardment that gave the Moon most of its craters.

We wish to understand not only how and from what the planets accreted, but also how life might have arisen on Earth. On this question, too, planetary exploration can help resolve a vigorous debate about the role that comets may have played. The young Earth's atmosphere might have been inhospitable to organic synthesis, the building up of the complex molecules on which life is based. Could organic synthesis have proceeded far enough in cometary ices to have supplied those compounds, and could such complex molecules have survived the impact of a comet with the Earth? Can the chemistry of Titan's present atmosphere tell us more about organic reactions on the early Earth? Might life have started on Mars as well, only to disappear with that planet's liquid water? Clearly biologists also will be exploring the solar system—alongside geophysicists.

Clues to the Earth's future may be found in the differing evolution of its sister planets—an aspect of space exploration called *comparative planetology*. We can learn the possible impact of human activity on Earth by studying Venus, the victim of a runaway greenhouse effect, and Mars, with its extreme desertification. Studies of Mars have already provided improved chemical data for computer models of Earth's atmosphere. If we are so foolish as to exhaust the Earth's critical resources, can we find relief by mining the Moon or some nearby asteroids or comets? I hope we never reach that point, but a survey of extraterrestrial resources can only help society make crucial decisions.

Still other questions require *ground-truth measurements*—basic data on astrophysical processes that cannot be observed in the laboratory.

Studies of the solar wind, for example, help in interpreting the flow of plasma from distant stars. Complex interactions of light, gases, plasma, and dust occurring throughout the universe can be studied directly in the tails of comets and in the magnetic field surrounding Saturn.

How does a program of interplanetary exploration go about solving mysteries like these? We begin with *flyby* missions, in which a spacecraft swings by a target on a trip somewhere else—or even nowhere else in particular. These missions are the easiest to launch and the least expensive to carry out. All early missions to Mars and Venus were flybys; and, in fact, flybys are all that we have had to date for five of the six other planets, asteroids, and comets.

Flyby missions, however, cannot offer a full range of studies of a planet and its surroundings. The next step is to *orbit* the target body. Orbiting spacecraft can observe an entire body from a wide selection of viewpoints, and Earth-based scientists can adjust the sequence of observations to take best advantage of data acquired on prior orbits. Spacecraft have now orbited the Moon, Mars, and Venus, where *Magellan* used radar to map the planet's surface through its clouds, which are opaque to visible light. The Galileo mission is headed for an orbit of Jupiter, and the Cassini mission is designed to orbit Saturn. *Cassini* will use Titan's gravity to adjust the spacecraft's trajectory. That "gravity assist" will give the orbiter close views of nearly all Saturn's moons and allow it to map Saturn's magnetosphere over a great range of longitude, latitude, and distance.

Next in scientific return are *probes*, *landers*, and *rovers*, which actually touch the planet or enter its atmosphere. *Viking* landed two stations on Mars, *Pioneer 12* sent probes deep into Venus's atmosphere, and instruments launched by

the former Soviet Union landed on Venus. Soon *Galileo* will deliver a probe into the atmosphere of Jupiter, and *Cassini* will launch the Huygens probe, built by the European Space Agency, into the atmosphere of Titan. These probes will finally determine the composition of the atmosphere and cloud layers, as well as measure atmospheric pressure, temperature, and wind speed. Plans are also being made for small rovers to explore the Martian surface.

Despite great strides in miniaturization, however, we cannot come close to flying the most sophisticated instruments on spacecraft. To analyze a planet's ratios of important elements and isotopes, for example, we must bring samples back to laboratories on Earth. This next step, *sample return*, may one day show us the specific asteroids from which many meteorites broke off. NASA's long-range plans include sample return from the Martian surface; American, European, and Japanese space agencies are studying how to collect samples of cometary dust and ice. Proving whether dust gathered in the Earth's stratosphere in fact matches material from comets would be a key to the origin of our atmosphere and perhaps of life itself.

But the ultimate scientific adventure is *human exploration*. No robot can yet begin to compete with a trained scientist's on-the-spot judgment, so essential in geophysical field work. The problem, of course, is cost: Humans in space require complex and expensive support.

I believe our curiosity will eventually lead people once again to the Moon—and beyond. In spite of the unfortunate failure of the *Mars Observer*, we should forge ahead with a full program of planetary exploration using flybys, orbiters, probes, rovers, landers, and sample return to address major scientific questions about our home planet, the solar system, and even faraway stars and galaxies.

JUPITER: LORD OF THE PLANETS

JUPITER WITH IO AND EUROPA   Jupiter is the largest and most massive planet in the solar system. Jupiter's diameter is $11\frac{1}{4}$ larger than Earth's, and its mass is 318 times greater. However, Jupiter is composed primarily of hydrogen and helium, in abundances similar to the Sun's chemical composition. This photograph from *Voyager 1* in 1979 shows Europa (right) and Io (in front of Jupiter). Both satellites are nearly the same size as our Moon. The Great Red Spot, a persistent storm in the Jovian atmosphere about twice the size of the Earth, is on the left. (NASA)

JUPITER IS an active, multicolored world, more massive than all the other planets combined. Jupiter's distinctive striping shows how its rapid rotation stretches its weather systems completely around the planet. Some features in the Jovian cloud cover are fleeting, whereas others, like the Great Red Spot, endure year after year. Four spacecraft have given us brief but revealing close-up views of Jupiter's colorful, turbulent clouds and probed its enormous magnetosphere. Jupiter's hydrogen-rich interior, so highly compressed by the planet's gravity that the gas has become a metal, is capable of conducting electricity and supporting the planet's vast magnetic field. The Galileo spacecraft, now on its way to Jupiter, will send a probe into the Jovian atmosphere to give us crucial information about the largest planet in the solar system.

More than any other single factor, it was the temperature distribution in the young solar nebula that dictated the natures of the planets that orbit the Sun. In the warm inner regions of this ancient nebula, the dust grains that survived the heat consisted primarily of rock-forming substances (metals, silicates, and oxides). As we saw in Chapter 7, the temperature was simply too high for volatile substances (water, methane, and ammonia) to condense significantly. The four planets that formed close to the Sun are therefore almost entirely made of rock. Their surface gravities were too low and their surface temperatures too high to retain any of the abundant but lightweight hydrogen and helium gases that made up most of the solar nebula.

The orbit of Jupiter, however, is about five times as large as the orbit of the Earth, and sunlight reaching Jupiter is only $\frac{1}{25}$ as bright as that reaching the Earth. In the past, as now, it was cold at this distance of $\frac{1}{2}$ billion miles from the Sun (recall Figure 7-17). In the young solar nebula, the dust grains at this distance from the protosun were covered with thick, frosty coatings of frozen water, methane, and ammonia. These volatile substances thus became important constituents of the planets that accreted in the outer reaches of our solar system.

As explained in Chapter 7, astronomers suspect that a planet like Jupiter formed in two stages. First, ice-coated dust grains accreted to make a protoplanet several times more massive than the Earth. The gravitational pull of this large protoplanet then attracted and retained substantial quantities of the hydrogen and helium existing in the cool outer reaches of the solar nebula. Supercomputer simulations show that these lightweight gases would have been gathered quite rapidly once the protoplanet had grown beyond a certain mass.

This process created the largest planet in our solar system—a planet whose bulk is approximately 82% hydrogen, 17% helium, and only 1% all other elements (by weight). This composition is probably close to that of the original solar nebula, and it is comparable to the Sun's atmosphere today (recall Table 7-4). Saturn—and to a lesser extent Uranus and Neptune—are thought to have formed in much the same way as Jupiter.

13-1 Huge, massive Jupiter is composed largely of lightweight gases

Jupiter is huge: Its diameter is $11\frac{1}{4}$ times bigger than Earth's. Jupiter is massive: Its mass is $2\frac{1}{2}$ times the combined mass of all the other planets, satellites, asteroids, meteoroids, and comets in the solar system. Basic data about Jupiter are listed in Box 13-1.

In addition to being the largest and most massive planet in our solar system, Jupiter also rotates the fastest. At its equator, Jupiter completes a full rotation in only 9 hours 50 minutes 28 seconds. However, Jupiter is not solid or rigid. By watching features in Jupiter's cloud cover, astronomers have discovered that the polar regions of the planet rotate a little more slowly than do the equatorial regions. Near the poles, the rotation period of Jupiter's atmosphere is about 9 hours 55 minutes 41 seconds. The first person to notice this **differential rotation** of Jupiter was the Italian astronomer Giovanni Domenico Cassini in 1690. (You may recall that Cassini is the same gifted observer who first determined Mars's rate of rotation.)

Through an Earth-based telescope, Jupiter appears as a colorful, intricately banded sphere, as shown in Figure 13-1.

BOX 13-1

Jupiter Data

Mean distance from Sun:	5.203 AU = 7.783 × 10⁸ km
Maximum distance from Sun:	5.454 AU = 8.159 × 10⁸ km
Minimum distance from Sun:	4.952 AU = 7.407 × 10⁸ km
Mean orbital velocity:	13.1 km/s
Sidereal period:	11.86 years
Rotation period:	Equatorial: $9^h 50^m 28^s$ Internal: $9^h 55^m 30^s$
Inclination of equator to orbit:	3° 07′
Inclination of orbit to ecliptic:	1° 18′
Diameter:	Equatorial: 142,980 km Polar: 133,700 km
Diameter (Earth = 1):	Equatorial: 11.21 Polar: 10.44
Mass:	1.90×10^{27} kg
Mass (Earth = 1):	317.9
Mean density:	1330 kg/m³
Surface gravity (Earth = 1):	2.34
Escape speed:	60 km/s
Mean surface temperature (at cloudtops):	−110°C = −166°F = 163 K

Figure 13-2, a close-up view from a spacecraft, shows the alternating dark and light bands parallel to Jupiter's equator in subtle tones of red, orange, brown, and yellow. The dark, reddish bands are called **belts**, the light-colored bands **zones**. In addition to these conspicuous stripes, a large, red-orange oval called the **Great Red Spot** is often visible in Jupiter's southern hemisphere. This remarkable feature, which has been observed since the mid-1600s, appears to be a long-lived storm in the planet's dynamic atmosphere. Many careful observers have also reported smaller spots and blemishes that last for only a few weeks or months in Jupiter's turbulent clouds.

The best time to observe Jupiter is, of course, when the planet is at opposition. At such times, Jupiter outshines all the stars in the sky (its brightest magnitude is −2.6). Through a telescope, Jupiter presents a disk nearly 50 arc sec in diameter—roughly twice the angular diameter of Mars under the most favorable conditions. Because Jupiter takes almost a dozen years to orbit the Sun, this planet appears to meander across the 12 constellations of the zodiac at the rate of approximately 1 per year. Successive oppositions occur at intervals of about 13 months (Table 13-1).

An important clue to Jupiter's composition came from its average density, which is only 1330 kg/m³. To explain this low average density, the Swiss-American astronomer Rupert

TABLE 13-1

Oppositions of Jupiter, 1993–1999

Date	Angular diameter (arc sec)	Magnitude
1993 March 30	44.2	−2.0
1994 April 30	44.5	−2.0
1995 June 1	45.6	−2.1
1996 July 4	47.0	−2.2
1997 August 9	48.6	−2.4
1998 September 16	49.7	−2.5
1999 October 23	49.8	−2.5

FIGURE 13-1 Jupiter from Earth Various belts and zones are easily identified in this Earth-based view of the largest planet in our solar system. The Great Red Spot was exceptionally prominent when this photograph was taken. (Courtesy of S. Larson)

FIGURE 13-2 Jupiter from a Spacecraft This view was sent back from *Voyager 1* in 1979, when the spacecraft was only 30 million kilometers from Jupiter. Features as small as 600 km across can be seen in the turbulent cloudtops. Complex cloud motions surround the Great Red Spot. (NASA)

Wildt suggested in the 1930s that Jupiter is composed mostly of hydrogen and helium compressed by its own gravity. Confirmation was slow in coming, however, because neither gas produces prominent spectral lines in the sunlight reflected from the planet.

Wildt did find prominent spectral lines of methane (CH_4) and ammonia (NH_3) in Jupiter's spectrum. Note that each of these molecules contains three or four hydrogen atoms—more evidence of hydrogen in Jupiter's atmosphere. At the low temperatures in Jupiter's upper atmosphere, hydrogen exists only as H_2 molecules, which produce very weak spectral lines that were first detected in 1960. Finally, infrared instruments on space probes passing near the planet in the 1970s detected collisions between hydrogen molecules and helium atoms in Jupiter's atmosphere. Today Jupiter's atmosphere is known to consist of approximately 86% hydrogen and 14% helium, with very small amounts of methane, ammonia, water vapor, and other gases. The planet as a whole has a slightly less hydrogen-rich composition: 82% hydrogen, 17% helium, and 1% for all heavier elements.

13-2 Pictures taken during flybys show many details in Jupiter's clouds

For many years astronomers realized that our understanding and appreciation of this extraordinary planet would be vastly improved if we could send spacecraft past Jupiter to examine its colorful atmosphere at close range. This program was carried out by four historic missions to Jupiter during the 1970s.

On the first Jupiter flyby in December 1973, *Pioneer 10* coasted over the planet's cloudtops (Figure 13-3). An identical spacecraft, *Pioneer 11*, followed in December 1974. In 1979, another pair of spacecraft, *Voyager 1* and *Voyager 2*, sailed past Jupiter. These spacecraft sent back spectacular close-up color pictures of Jupiter's dynamic atmosphere. The global views in Figure 13-4 show that Jupiter's atmosphere underwent some fundamental changes during the four years between the Pioneer and Voyager flybys, changes that are most apparent in the area surrounding the Great Red Spot. During the Pioneer flybys, the Great Red Spot was embedded in a broad white zone that dominated the planet's southern hemisphere. By the time of the Voyager missions, a dark belt had broadened and encroached on the Great Red Spot from the north, embroiling the entire region in greater turbulence.

Over the past three centuries, Earth-based observers have reported many long-term variations in the Great Red Spot's size and color. At its largest, it measured 40,000 by 14,000 km—so large that three Earths could fit side by side across it. At other times (as in 1976 and 1977), the spot almost faded from view. During the Voyager flybys of 1979, the Great Red Spot was only slightly larger than the Earth (Figure 13-5).

Careful examination of cloud motions in and around the Great Red Spot reveal that the spot rotates counterclockwise

FIGURE 13-3 The Pioneer Spacecraft This drawing shows one of four spacecraft that flew past Jupiter in the 1970s. Instead of using solar panels, these spacecraft drew power from electric generators containing radioactive isotopes. In addition to sending back pictures of Jupiter's clouds, instruments on the spacecraft analyzed Jupiter's magnetic environment. (NASA)

Pioneer 10, December 1973

Pioneer 11, December 1974

Voyager 1, March 1979

Voyager 2, July 1979

FIGURE 13-4 Four Views of Jupiter These four pictures of Jupiter span $5\frac{1}{2}$ years and show major changes in the planet's upper atmosphere. Noticeable changes are apparent even in the Voyager views, which were separated by only four months. All four pictures show the Great Red Spot. The black dot on the *Pioneer 10* view is the shadow of Jupiter's moon Io. (NASA)

FIGURE 13-5 The Great Red Spot Turbulence surrounding the Great Red Spot is clearly seen in this view from *Voyager 2*. When this picture was taken in 1979, the Great Red Spot measured about 20,000 km long and about 10,000 km wide. For comparison, the Earth's diameter is 12,756 km. The prominent white oval south of the Great Red Spot has been observed since 1938. (NASA)

with a period of about six days. Furthermore, winds to the north of the spot blow to the west, and winds south of the spot move toward the east. The circulation around the Great Red Spot is thus like a wheel spinning between two oppositely moving surfaces (Figure 13-6). This surprisingly stable wind pattern has survived for at least three centuries.

Voyager pictures also showed persistent features called **white ovals**, like the one seen near the Great Red Spot in Figure 13-5. Wind flow in these ovals is also counterclockwise. White ovals are apparently long-lived; Earth-based observers have reported seeing them in the same location since 1938.

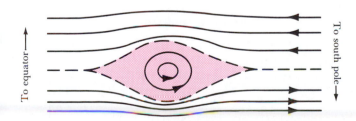

FIGURE 13-6 Circulation Around the Great Red Spot
The Great Red Spot spins counterclockwise, completing a full revolution in about six days. Meanwhile, the winds to the north and south of the spot blow in opposite directions. Consequently, circulation associated with the spot resembles a wheel spinning between two oppositely moving surfaces. (Adapted from A. P. Ingersoll)

Most of the white ovals are observed in Jupiter's southern hemisphere, whereas **brown ovals** are more common in Jupiter's northern hemisphere (Figure 13-7). Infrared observations indicate that brown ovals are holes in Jupiter's cloud cover that permit us to see down into warmer regions of the Jovian atmosphere.

The Voyager photographs might suggest that Jupiter's clouds are the result of incomprehensible turmoil. You might wonder if there is anything constant in the Jovian atmosphere. Surprisingly, there is. The computer-generated view in Figure 13-8 shows how Jupiter would look if you were located directly over either the planet's north pole or its south pole. Note the regular spacing of such cloud features as ripples, plumes, and light-colored wisps. The regular spacing of white ovals in the southern hemisphere is quite apparent. These regularities are probably the result of stable, large-scale weather patterns that encircle Jupiter.

13-3 Because of its rapid rotation, Jupiter's weather patterns are stretched around the planet

Infrared data from the Pioneer and Voyager spacecraft confirmed an extraordinary fact that had been deduced from previous Earth-based observations: Jupiter emits more energy than it receives from the Sun—in fact, nearly twice as much. Many scientists believe that this excess heat escaping from Jupiter is energy left over from the formation of the

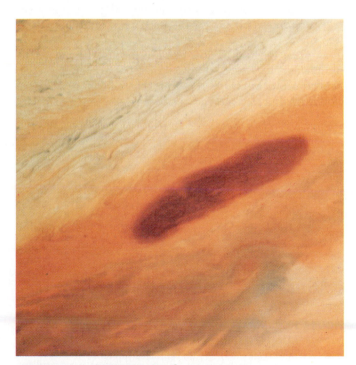

FIGURE 13-7 A Brown Oval Oval-shaped openings in Jupiter's main cloud layer reveal warm, brownish gases below. The length of this oval is roughly equal to the Earth's diameter. *Voyager 1* was 4 million kilometers from Jupiter when this picture was taken. (NASA)

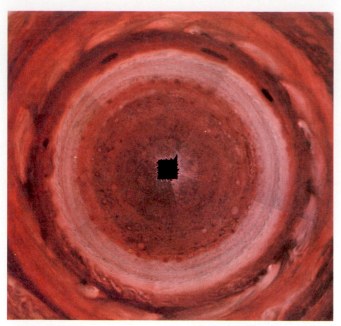

a North pole

b South pole

FIGURE 13-8 Jupiter's Northern and Southern Hemispheres   Computer processing was used to construct these views that look straight down onto Jupiter's north and south poles. (**a**) In this northern view, light-colored plumes are evenly spaced around the equatorial regions. Several brown ovals are visible. (**b**) In this southern view, the three biggest white ovals are separated by almost exactly 90° of longitude. In both views, the banded belt–zone structure is absent near the poles. The black spots are areas not photographed by the spacecraft. (NASA)

planet 4.5 billion years ago. As gases from the solar nebula fell into the protoplanet, vast amounts of gravitational energy were converted into the thermal energy that still radiates into space.

In order for heat to flow upward through the Jovian atmosphere and radiate out into space, the temperature must be warmer deep inside the atmosphere than it is at the cloudtops. Infrared measurements have confirmed that tem-perature rises with increasing depth in the Jovian cloud cover (Figure 13-9).

From spectroscopic observations and calculations of temperature and pressure in the Jovian clouds, scientists conclude that Jupiter has three main cloud layers of differing chemical composition. The uppermost cloud layer is com-posed of crystals of frozen ammonia. About 25 km deeper in the atmosphere, ammonia (NH_3) and hydrogen sulfide (H_2S)

FIGURE 13-9 The Vertical Structure of Jupiter's Upper Atmosphere   This graph dis-plays a temperature and pressure profile of Jupiter's upper atmosphere, as deduced from measurements at infrared and radio wavelengths. Three major cloud layers are shown, along with the colors that predominate at various depths.

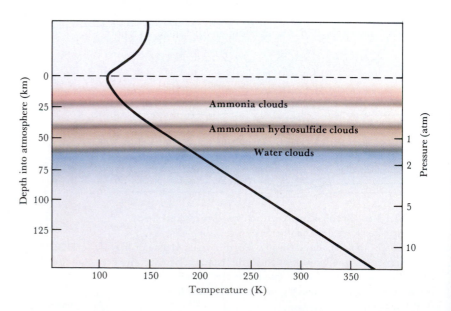

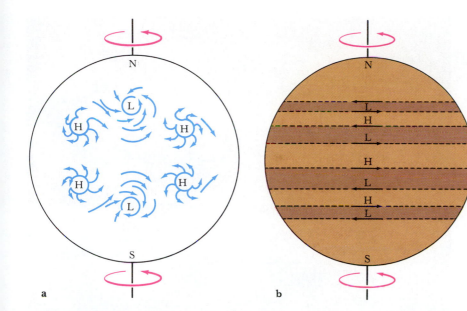

FIGURE 13-10 Wind Flow Patterns on Earth and Jupiter (a) Because of the Earth's rotation, cyclonic winds flowing into a low-pressure region or anticyclonic winds flowing out of a high-pressure region rotate either clockwise or counterclockwise, depending on the hemisphere in which the weather system is located. (b) Jupiter's rapid rotation stretches low- and high-pressure systems all the way around the planet. Light-colored zones are high-pressure systems, and dark-colored belts are low pressure.

combine to produce ammonium hydrosulfide (NH_4SH) crystals. Still farther down, the clouds are composed of water droplets and snowflakes of frozen water.

Colors in Jupiter's clouds correspond to differing temperatures and hence to different depths in the atmosphere. Brown clouds are the warmest and thus the deepest layers that we can see in the Jovian atmosphere. Whitish clouds form the next layer up, followed by red clouds in the highest layer. Jupiter's whitish zones are therefore somewhat higher than the brownish belts, while the Great Red Spot is a bulge in the Jovian cloud cover.

Anyone familiar with weather forecasting knows that the Earth's weather is dramatically affected by both high-pressure and low-pressure systems. A **high-pressure system** (commonly called a high) is simply a place where a more-than-usual amount of air happens to be located. This excess air weighs down on the Earth's surface to produce high barometric readings (recall Box 8-2). On the other hand, a **low-pressure system** (commonly called a low) is a place where there is a less-than-normal amount of air, thus producing low barometric readings. If we could see air, highs would look like bulges in the atmosphere (where extra amounts of air are piled up), and lows would look like depressions, or troughs.

Gravity strongly influences the basic dynamics of the Earth's weather. Gravity causes air to flow "downhill," from high-pressure bulges into low-pressure troughs. Because the Earth is rotating, however, the wind flowing from the highs toward the lows does not move along straight lines. The Earth's rotation deflects the winds, producing either a clockwise or a counterclockwise flow about the high- and low-pressure regions.

Figure 13-10a shows wind flow about highs and lows in the Earth's atmosphere. In the northern hemisphere, winds blowing toward a low-pressure region rotate counterclockwise about the low, forming a **cyclone**. Winds blowing away from a high-pressure region rotate clockwise about the high,

forming an **anticyclone**. In the southern hemisphere, the directions of rotation are reversed: Cyclonic winds about a low rotate clockwise, and anticyclonic winds about a high rotate counterclockwise.

Highs and lows also dominate the weather on Jupiter, but the planet's rapid rotation draws them into bands that completely encircle the globe, as shown in Figure 13-10b. The light-colored zones, where we see high-elevation clouds, are high-pressure systems stretched all the way around the planet. Similarly, dark-colored belts, where we see deeper into Jupiter's cloud cover, are low-pressure systems encircling the planet. Gases warmed by Jupiter's internal heat rise upward in the zones, while cool gases from the upper atmosphere descend in the belts (Figure 13-11). High-speed winds mark the boundaries between the cyclonic flow in the belts and the anticyclonic flow in the zones.

Superimposed on Jupiter's regular pattern of belts and zones are low- and high-pressure features that also exhibit

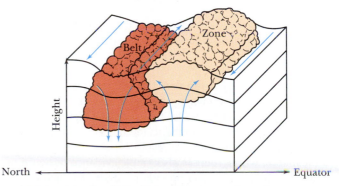

FIGURE 13-11 The Belt–Zone Structure of Jupiter's Clouds Gases warmed by heat from Jupiter's interior rise upward in the light-colored zones. Cool gases descend in the dark-colored belts. The horizontal surfaces indicate contours of constant pressure.

FIGURE 13-12 *Galileo's* **Atmospheric Probe** Just before arriving at Jupiter, *Galileo* will release an atmospheric probe that will make direct measurements and chemical analyses as it descends through the Jovian cloud cover. Scientists look forward to this historic event, which will occur in December 1995. (NASA)

cyclonic or anticyclonic flow. The six-day counterclockwise rotation of the Great Red Spot in Jupiter's southern hemisphere is consistent with infrared observations that the reddish cloudtops are cool and at a high elevation. The Great Red Spot is thus a long-lasting, high-pressure system in Jupiter's southern hemisphere.

In spite of all the data from the Pioneer and Voyager missions, we still do not know what gives Jupiter's clouds their colors. The crystals of ammonia, ammonium hydrosulfide, and frozen water in Jupiter's three main cloud layers are all white. Thus other chemicals must cause the browns, reds, and oranges. Some scientists believe that sulfur, which can assume many different colors depending on its temperature, plays an important role in Jupiter's clouds. Others think that phosphorus might be involved, especially in the Great Red Spot. The Sun's ultraviolet radiation, which can induce chemical reactions, may also play a role in coloring the Jovian clouds.

Perhaps the most intriguing proposal is that Jupiter's colors might be due to organic compounds, possibly resulting from biological activity. Many of Jupiter's colors are similar to the colors of organic material created in laboratory experiments that attempt to simulate the first steps in the creation of life. In these experiments, discussed in detail in the Afterword, electrical discharges (to simulate lightning) pass through a mixture of methane, ammonia, hydrogen, and water vapor. These gases are relatively abundant in Jupiter's atmosphere, where there is also ample evidence of lightning. Many of the organic compounds produced in such experiments may have been brewing for 4 billion years in Jupiter's clouds. Could those clouds harbor extraterrestrial life?

To explore these issues, *Galileo* will arrive at Jupiter in 1995. The spacecraft carries an atmospheric probe that will be released into Jupiter's atmosphere just before *Galileo* goes

into orbit around the planet. Using a heat shield and a parachute, the probe will descend through the Jovian clouds, making a wide range of measurements and chemical analyses (Figure 13-12). Just as the Voyager flybys revealed phenomena that stagger the imagination, the Galileo mission will undoubtedly provide a few answers and pose many new mysteries.

13-4 Jupiter's oblateness divulges the planet's rocky core

Jupiter's rapid rotation profoundly affects the overall shape of the planet. Even a casual glance through a small telescope shows that Jupiter is slightly flattened, or oblate. The diameter across Jupiter's equator (142,980 km) is 6.5% larger than its diameter from pole to pole (133,700 km). Thus Jupiter is said to have an **oblateness** of 6.5%, or 0.065.

If Jupiter did not rotate, it would be a perfect sphere. A massive, nonrotating object naturally settles into a spherical shape so that every atom on its surface experiences the same gravitational pull aimed directly at the object's center. However, because Jupiter is rotating, every part of the planet experiences an outward force, directed away from the axis of rotation. The size of this outward-directed **centrifugal force** is proportional to distance from the axis of rotation; the greater the distance, the greater the force. Jupiter's equatorial regions, which are farther from the planet's axis of rotation than the polar regions, are therefore subjected to a stronger centrifugal force than the polar regions. This centrifugal stretching of the equatorial regions gives Jupiter its characteristic oblate shape.

Jupiter's shape is an excellent indicator of its internal structure. Imagine two planets with the same mass, same average density, and same rotation. These planets will nevertheless have slightly different shapes if one has a compact core but the other does not. Because they have the same average density, one planet consists of less-dense material surrounding a compact core while the other is more homogeneous. This less-dense matter is more readily flung to a greater distance from the planet's axis of rotation than the material of a planet with a uniform distribution of matter. All other things being equal, therefore, a planet with a dense core will be more oblate than a planet without one.

The size of Jupiter's oblateness suggests that 4% of its mass is concentrated in a dense, rocky core. Jupiter's core is nearly 13 times more massive than the entire Earth. Much of this core was probably the original "seed" around which the proto-Jupiter accreted. However, additional amounts of rocky material were added later by meteoritic material falling onto the planet.

Although Jupiter's rocky core is 13 times more massive than the Earth, it is probably not much bigger than our planet. The tremendous crushing weight of the remaining bulk of Jupiter—equal to the mass of 305 Earths—compresses Jupiter's core down to a sphere 20,000 km in diameter (Earth's diameter is 12,800 km). The pressure at Jupiter's center is about 80 million atmospheres, meaning that the rocky material of Jupiter's core is squeezed to a density of about 20,000 kg/m³. The corresponding temperature at the planet's center is probably about 25,000 K. In contrast, the temperature at Jupiter's cloudtops is only 165 K.

13-5 A metallic hydrogen interior endows Jupiter with a powerful magnetic field

In the 1950s, astronomers began discovering radio emissions coming from Jupiter. Some of this radiation is **thermal**; that is, it comes from the planet's surface and is exactly what would be expected from a warm object. However, most of the radiation is **nonthermal**, and is found in two broad wavelength ranges. At wavelengths of a few meters are sporadic bursts of **decametric** ("ten-meter") **radiation**, probably caused by electrical discharges associated with powerful electric currents in Jupiter's ionosphere. As we shall see in Chapter 14, these discharges result from complex electromagnetic interactions between Jupiter and its large satellite, Io. At wavelengths of a few tenths of a meter (particularly at 3 to 75 cm) is **decimetric** ("tenth-meter") **radiation** that provided important clues about the hydrogen-rich bulk of Jupiter between its rocky core and its colorful cloudtops.

The first person to explain the source of decimetric radiation was the Soviet astrophysicist Iosif Shklovskii. In the 1950s, Shklovskii convincingly argued that this radio radiation is emitted by high-speed electrons moving through a magnetic field. Astronomers now realize that this process is very important in a wide range of phenomena throughout the universe, from pulsars and quasars to the radio emissions of entire galaxies. Whenever electrons traveling near the speed of light encounter a magnetic field, they spiral around that field and emit copious radio waves. Energy produced in this fashion is called **synchrotron radiation**.

If some of Jupiter's radiation is indeed synchrotron radiation, then Jupiter must have a magnetic field. From the intensity of this type of radiation, it was soon apparent that Jupiter's magnetic field must be extremely strong. At Jupiter's cloudtops, the Jovian magnetic field is roughly 19,000 times stronger than the magnetic field at the Earth's surface. A molten iron center inside Jupiter's rocky core could not have created such an enormous planetwide magnetic field. What, then, could be the source of Jupiter's magnetism?

The answer lies in the fact that Jupiter is composed largely of hydrogen. A hydrogen atom consists of a single proton orbited by a single electron. Deep inside Jupiter, pressures are so great that the electrons are stripped from their protons. The result is a mixture of protons and electrons. The electrons, no longer bound to their protons, are free to wander around on their own. Their motion, directed by Jupiter's rotation, creates an electric current, just as the ordered movement of electrons in a copper wire constitutes an electric current. In other words, the highly compressed hydrogen deep inside Jupiter behaves like a metal; thus it is called **liquid metallic hydrogen**.

Detailed calculations strongly suggest that hydrogen is transformed into a liquid metal when the pressure exceeds 3 million atmospheres. This transition occurs at a depth of about 20,000 km below Jupiter's cloudtops. Thus, the internal structure of Jupiter consists of three distinct regions (Figure 13-13): a rocky core, a 40,000-km-thick layer of liquid

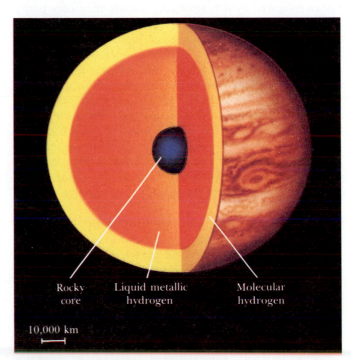

Rocky core   Liquid metallic hydrogen   Molecular hydrogen

10,000 km

FIGURE 13-13 The Structure of Jupiter Jupiter probably has a rocky core approximately 20,000 km in diameter. This core is surrounded by a 40,000-km-thick layer of liquid metallic hydrogen, which is responsible for Jupiter's powerful magnetic field. The outer, 20,000-km-thick layer is composed primarily of ordinary hydrogen.

FIGURE 13-14 Jupiter's Magnetosphere
Jupiter's magnetosphere is enveloped by a shock wave, where the supersonic solar wind is abruptly slowed to subsonic speeds. Most of the particles of the solar wind are deflected around Jupiter in a turbulent region colored purple in the scale drawing. Particles trapped inside Jupiter's magnetosphere are spewed out into a vast current sheet by the planet's rapid rotation. Jupiter's axis of rotation is inclined to its magnetic axis by about 11°.

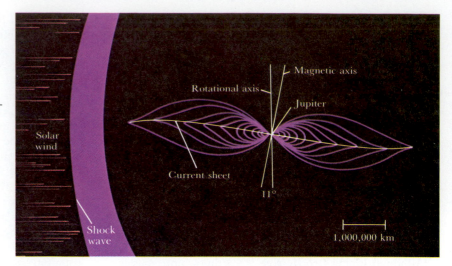

metallic hydrogen, and a 20,000-km-thick layer of ordinary molecular hydrogen. The colorful cloud patterns that we can see through telescopes are located in the outermost 100 km of the exterior layer of molecular hydrogen.

It is clear from Figure 13-13 that the preponderance of Jupiter's enormous bulk is electrically conductive liquid metal. Because of Jupiter's rapid rotation, electric currents in this thick layer of liquid metallic hydrogen generate a powerful magnetic field, in much the same way that liquid portions of the Earth's core produce the Earth's magnetic field. The Jovian magnetic field is so much stronger than ours partly because Jupiter's liquid metallic region is so much larger than the Earth's core and partly because Jupiter rotates so much faster than the Earth.

Jupiter's powerful magnetic field surrounds the planet with a magnetosphere so large that it envelops the orbits of many of its moons. Pioneer and Voyager spacecraft, which carried instruments to detect charged particles and magnetic fields, revealed the awesome dimensions of the Jovian magnetosphere. The shock wave that surrounds Jupiter's magnetosphere is nearly 30 million kilometers across. In other words, if you could see Jupiter's magnetosphere from Earth, it would cover an area in the sky 16 times larger than the full Moon.

Pioneer and Voyager spacecraft reported the outer boundary of Jupiter's magnetosphere at distances ranging from 3 to 7 million kilometers above the planet. Evidently Jupiter's magnetosphere is sensitive to fluctuations in the solar wind. It expands and contracts by a factor of 2, rapidly and frequently. In contrast, dramatic changes in the size of the Earth's magnetosphere are extremely rare in spite of occasional "gusts" in the solar wind.

The inner regions of our magnetosphere are dominated by two huge Van Allen belts (recall Figure 8-19) that are filled with charged particles. Jupiter also entraps charged particles in a similar manner. In addition, however, Jupiter's rapid rotation spews the particles out into a huge electrically charged **current sheet** (Figure 13-14). This current sheet lies in the plane of Jupiter's magnetic equator. Jupiter's magnetic axis is inclined 11° from the planet's axis of rotation, and the

orientation of Jupiter's magnetic field is the reverse of Earth's—a compass would point toward the south on Jupiter.

Radio radiation emitted by charged particles in the densest regions of Jupiter's magnetosphere is seen in Figure 13-15. This radio emission varies slightly with a period of 9 hours 55 minutes 30 seconds, as Jupiter's rotation wobbles the angle at which we view its magnetic field. Since the magnetic field is anchored deep within the planet, this fluctuation reveals Jupiter's **internal rotation period,** which is slightly slower than the atmospheric rotation that we see through a telescope.

The Voyager spacecraft discovered that the inner regions of Jupiter's magnetosphere contain a hot, gaslike mixture of charged particles called a **plasma.** A plasma is formed when a gas is heated to such extremely high temperatures that electrons are torn off the atoms of the gas. As a result, a plasma is an electrically neutral mixture of positively charged ions and negatively charged electrons. The plasma that envelops Jupiter consists primarily of electrons and protons, with some ions of helium, sulfur, and oxygen. These charged par-

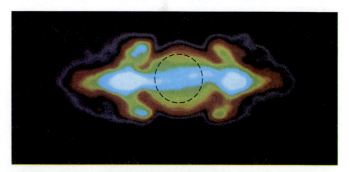

FIGURE 13-15 A Radio View of Jupiter The Very Large Array (VLA) was used to produce this map of synchrotron emission from Jupiter at a wavelength of 21 cm. The image is about five Jupiter diameters wide and is elongated parallel to the planet's magnetic equator. Jupiter is at the center of the image, indicated by the dashed circle. Strongest emission, shown in white, comes from electrons trapped in the densest regions of the current sheet that encircles Jupiter. (NRAO)

ticles are caught up in Jupiter's rapidly rotating magnetic field and accelerated to extremely high speeds.

The Voyager instruments relayed a startling discovery: The plasma surrounding Jupiter has an astoundingly high temperature of 300 million to 400 million kelvin. The Jovian plasma is the hottest place in the solar system! Even the center of our Sun is only about 20 million kelvin. However, the Jovian plasma is very thin: The average density in the hottest regions is about one charged particle per 100 cm³.

Because of its high temperature, Jupiter's huge plasma has a great deal of energy and exerts pressure that holds off the solar wind. The Voyager data suggest that the pressure balance between the solar wind and the hot plasma inside the Jovian magnetosphere is precarious. A gust in the solar wind can blow away some of the plasma, at which point the magnetosphere deflates rapidly to as little as one-half its original size. Charged-particle detectors carried by the Voyager spacecraft recorded several bursts of hot plasma that may have been associated with magnetospheric deflations. However, additional electrons and ions accelerated by Jupiter's rotating magnetic field soon replenish the plasma and the magnetosphere expands again.

KEY WORDS

anticyclone
belts
brown ovals
centrifugal force
current sheet
cyclone

decametric radiation
decimetric radiation
differential rotation
Great Red Spot
high-pressure system

internal rotation period
liquid metallic hydrogen
low-pressure system
nonthermal radiation
oblateness
plasma

synchrotron radiation
thermal radiation
white ovals
zones

KEY IDEAS

• Jupiter, whose mass is equivalent to the mass of 318 Earths, is composed of 82% hydrogen, 17% helium, and only 1% all other elements. This chemical composition is similar to that of the Sun.

Jupiter probably has a rocky core with a mass of about 13 Earth masses. This core is surrounded by a 40,000-km-thick layer of liquid metallic hydrogen and an outer layer of molecular hydrogen about 20,000 km thick.

The visible features of Jupiter—belts, zones, the Great Red Spot, ovals, and colored clouds—lie in the outermost 100 km of the molecular hydrogen layer.

The outer layers of the Jovian atmosphere show differential rotation, with the equatorial regions rotating slightly faster than the polar regions. The internal rotation rate, determined from the variations in radio emission, is nearly the same as the polar rotation rate.

Because of its rapid rotation, Jupiter is noticeably oblate.

• The colored ovals visible in the Jovian atmosphere represent gigantic storms; some such as the Great Red Spot, are quite stable and persist for many years.

• Jupiter emits more heat than it receives from the Sun; presumably the planet is still cooling.

There are three cloud layers in Jupiter's atmosphere. The reasons for the distinctive colors of these different layers are not yet known.

Some scientists speculate that life might be possible in the layers of the Jovian atmosphere where pressures and temperatures are not too different from those on Earth.

• Jupiter has a strong magnetic field created by currents in the metallic hydrogen layer. Its huge magnetosphere contains a vast current sheet of electrically charged particles.

Charged particles in the densest portions of Jupiter's magnetosphere emit synchrotron radiation at radio wavelengths.

The Jovian magnetosphere encloses a plasma of charged particles that is hotter than the center of the Sun but has very low density. The magnetosphere exists in a delicate balance between pressures from the plasma and from the solar wind, and its size fluctuates drastically.

REVIEW QUESTIONS

1. Which planet, Mars or Jupiter, is easier to observe with an Earth-based telescope? Explain your answer.

2. Compare Jupiter's chemical composition with that of the Sun. What does this tell us about Jupiter's formation?

3. If Jupiter does not have any observable solid surface and its atmosphere rotates differentially, how are astronomers able to determine the planet's internal rotation rate?

4. What is liquid metallic hydrogen?

5. Why do astronomers believe that Jupiter does not have a large iron-rich core, even though the planet possesses a strong magnetic field?

6. What is thought to be the source of Jupiter's excess internal heat?

7. What are the belts and zones in Jupiter's atmosphere? Is the Great Red Spot more like a belt or a zone?

8. What is the difference between a cyclone and an anticyclone?

9. Why is Jupiter oblate?

10. What data and techniques have been used to determine the internal structure of Jupiter?

11. What is a plasma?

12. Compare and contrast Jupiter's magnetosphere with the magnetosphere of a terrestrial planet like Earth. Why is the size of the Jovian magnetosphere variable, whereas the Earth's is not?

ADVANCED QUESTIONS

Tips and tools . . .

For a discussion of center of mass, see Box 4-3, which also contains a very useful form of Kepler's third law. Newton's universal law of gravitation is the basic equation from which a planet's surface gravity can be calculated.

13. Why do the magnitudes of Jupiter at opposition vary so little, compared to the oppositions of Mars (see Tables 12-1 and 13-1)?

14. Find the location of the center of mass of the Sun–Jupiter system. Is the center of mass inside or outside the Sun?

15. Using orbital data for a Jovian satellite of your choice (see Appendix 3), calculate the mass of Jupiter. How does your answer compare with the mass quoted in Box 13-1?

16. How much would a person who weighs 150 lb on Earth weigh on Jupiter's "surface"?

17. Estimate the wind velocities in the Great Red Spot, which rotates with a period of about six days.

18. What sort of experiment would you design in order to establish whether Jupiter has a rocky core?

DISCUSSION QUESTIONS

19. Describe some of the semipermanent features in Jupiter's atmosphere. Compare and contrast these long-lived features with some of the transient phenomena seen in Jupiter's clouds.

20. Suppose you were designing a mission to Jupiter involving an airplanelike vehicle that would spend many days (months?) flying through the Jovian clouds. What observations, measurements, and analyses should this aircraft be prepared to make? What dangers might the aircraft encounter, and what design problems would you have to overcome?

21. How is *Galileo* doing? Consult such magazines as *Sky & Telescope* and *Science News*. Where is the spacecraft right now? Are all its instruments functioning properly?

OBSERVING PROJECTS

22. Consult such magazines as *Sky & Telescope* and *Astronomy* to determine the visibility of Jupiter. If Jupiter is visible in the night sky, make arrangements to view the planet through a telescope. What magnifying power seems to give you the best view? Draw a picture of what you see. Can you see any belts and zones? How many? Can you see the Great Red Spot?

23. Make arrangements to view Jupiter's Great Red Spot through a telescope. The Calendar Notes section of *Sky & Telescope* lists the universal dates and times when the center of the Great Red Spot should pass across Jupiter's central meridian as seen from North America. The Great Red Spot is well placed for viewing for 50 minutes before and after meridian transit.

FOR FURTHER READING

Burgess, E. *By Jupiter.* Columbia University Press, 1982. This nontechnical book reviews our understanding of Jupiter in light of the Pioneer and Voyager flybys.

Carroll, M. "Project Galileo: The Phoenix Rises." *Sky & Telescope,* April 1987. This article describes the long-delayed Galileo mission to Jupiter.

Dyer, A. "*Ulysses* Meets a Giant." *Astronomy,* July 1992. This one-page article describes *Ulysses'* passage though the Jovian magnetosphere in February 1992, when Jupiter's gravity catapulted the tiny spacecraft into a polar orbit around the Sun.

Hunt, G., and Moore, P. *Jupiter.* Rand McNally, 1981. While occasionally technical, this book is a useful reference atlas for Jupiter and its satellites.

Ingersoll, A. "The Meteorology of Jupiter." *Scientific American,* March 1976. This article gives an exceptionally clear description of the dominant weather patterns on Jupiter.

Johnson, T., and Yeates, C. "Return to Jupiter: Project Galileo." *Sky & Telescope,* August 1983. This article by Jet Propulsion Laboratory scientists gives an excellent description of the Galileo spacecraft, but the dates quoted are wrong because the mission was delayed for many years.

Morrison, D., and Samz, J. *Voyage to Jupiter.* NASA SP-439, 1980. This superb book, which describes the Voyager missions to Jupiter, includes an excellent selection of color photographs.

Washburn, M. *Distant Encounters: The Exploration of Jupiter and Saturn.* Harcourt Brace Jovanovich, 1983. This entertaining book describes the Pioneer and Voyager missions to the outer solar system.

Wolfe, R. "Jupiter." *Scientific American,* 1975. Written shortly after the Pioneer flybys, this article describes Jupiter's interior with special emphasis on the role of liquid metallic hydrogen.

THE GALILEAN SATELLITES OF JUPITER

THE GALILEAN SATELLITES WITH OUR MOON AND MERCURY Jupiter's four large satellites are shown here along with our Moon and Mercury. All six worlds are reproduced to the same scale. Io has numerous active volcanoes and Europa has a smooth, icy surface; both are roughly the same size as our Moon. Ganymede and Callisto are each covered with a 1000-km-thick layer of ice; both are roughly the size of Mercury. Saturn's largest moon, Titan, is also about the size of Mercury. Neptune's largest moon, Triton, is slightly smaller than our Moon. (NASA)

IN 1610 GALILEO discovered four moons circling Jupiter, each about the same size as our Moon or Mercury. In 1979 two Voyager spacecraft flew by the Galilean satellites and found extraordinary panoramas. Io, the innermost of the four, is covered with colorful sulfur compounds deposited by active volcanoes. This geologically active satellite orbits deep within Jupiter's magnetosphere and affects radio emissions from the planet. The layer of ice that envelops Europa is crisscrossed with fine cracks. Ganymede and Callisto are encased in thick "oceans" of ice whose frozen surfaces bear the scars of meteoroid impacts. In 1995 the *Galileo* spacecraft will go into orbit about Jupiter and provide long-term, high-resolution observations of the four satellites. Our understanding of the physical processes that shape terrestrial worlds, including the Earth, is being advanced by the exploration of Jupiter's fascinating moons.

Galileo Galilei called them the "Medicean Stars" to attract the attention of a wealthy Florentine patron of the arts and sciences. Since their discovery in 1610, the four giant moons of Jupiter have played an important role in our understanding of the universe. To Galileo, they were observational evidence supporting the heretical Copernican cosmology. To the modern astronomer, the Voyager flybys of 1979 revealed four extraordinary terrestrial worlds, different from anything astronomers had ever seen or even imagined. We call them the **Galilean satellites**. They are named after the mythical lovers and companions of the Greek god Zeus: Io, Europa, Ganymede, and Callisto.

14-1 The Galilean satellites are easily seen with Earth-based telescopes

When viewed through an Earth-based telescope, the Galilean satellites look like pinpoints of light. With patience, you can follow these four worlds as they orbit Jupiter. Because their orbital periods are fairly short (ranging from 1.8 days for Io to 16.7 days for Callisto), major changes in the positions of the satellites are easily noticed from one night to the next. Their apparent magnitudes range from about 4.7 for Ganymede to nearly 6 for Callisto, so bright that you might expect all four satellites to be visible to the naked eye. Binoculars or a telescope are necessary, however, because of the overwhelming glare of Jupiter (Figure 14-1).

The brightness of each moon varies slightly as it revolves about Jupiter, since dark and light areas on the moons' surfaces are alternately exposed to and then hidden from our view. The apparent magnitude of each satellite varies with a period equal to the satellite's orbital period, which means that each Galilean satellite rotates exactly once on its axis during each trip around its orbit. Hence each Galilean satellite has synchronous rotation, caused by the tidal forces, that

keeps the same hemisphere perpetually facing Jupiter, just as our Moon keeps the same side facing the Earth.

By consulting a current issue of the *Astronomical Almanac* or a magazine like *Sky & Telescope*, you can schedule your observations to include transits, eclipses, and occultations. In a transit, one of the four moons passes between Jupiter and the Sun, and so the satellite's shadow is seen as a black dot against the planet's colorful cloudtops (examine Figure 13-4). In an eclipse, one of the moons moves behind

FIGURE 14-1 The Galilean Satellites This photograph, taken by an amateur astronomer with a small telescope, shows the four Galilean satellites alongside an overexposed image of Jupiter. Each Galilean satellite is bright enough to be seen with the unaided eye were it not overwhelmed by the glare of nearby Jupiter. (Courtesy of J. Jenkins)

Jupiter, suddenly seeming to disappear and then reappear, as it passes into and out of the planet's enormous shadow. In an **occultation,** a satellite passes in front of a star or another satellite, temporarily blocking its light from Earth-based viewers.

Observations of eclipses have been used to estimate the diameters of the Galilean satellites. When a satellite emerges from Jupiter's shadow, it does not blink on instantly. Instead, for a brief interval the satellite gets brighter and brighter as more of its surface becomes exposed to sunlight. The duration of this interval depends on the orbital speed of the satellite (which is known from Kepler's laws) and its diameter. Consequently, by accurately measuring the time it takes for a satellite to move into or out of Jupiter's umbra, we can calculate the satellite's diameter.

Occultations can also be used to determine the diameters of the four satellites. By measuring the time a satellite takes to pass in front of a background star we can calculate the satellite's diameter. In addition, the Galilean satellites occasionally block out each other, because all four orbit Jupiter in the plane of the planet's equator. Every six years, when the Earth passes through this equatorial plane, mutual occultations of the satellites occur for a few days. Once again, timing the occultations enables us to calculate the satellites' diameters.

14-2   Data from spacecraft yielded accurate sizes, masses, and densities of the Galilean satellites

More accurate data about the Galilean satellites came from photographs taken during the Voyager flybys (Figure 14-2). Direct measurements of the observed disks showed that the two inner Galilean satellites, Io and Europa, are approximately the same size as our Moon. The two outer satellites, Ganymede and Callisto, are comparable in size to Mercury.

The Voyager data (accurate to better than 10 km) are listed in Box 14-1.

As early as the 1920s, fairly accurate determinations of the masses of the Galilean satellites had been made from Earth. The satellites often pass near each other, their mutual gravitational attraction slightly shifting one another's orbit around Jupiter. From these tiny orbital perturbations and using Newtonian mechanics, astronomers were able to calculate the masses of the satellites. Europa was discovered to be the least massive (its mass is two-thirds that of our Moon). Ganymede is by far the most massive of the four, with at least twice the mass of our Moon. As you might expect, the Pioneer and Voyager flybys produced the best determinations, because the satellites' gravity also deflected spacecraft trajectories.

As soon as reliable mass and diameter measurements were available, it became apparent that the average densities of the four satellites are related to their distances from Jupiter. The innermost satellite, Io, has the highest average density (3570 kg/m^3), which is slightly greater than the density of our Moon (3340 kg/m^3). The next satellite out, Europa, has an average density of 2970 kg/m^3. Since typical rocks in the Earth's crust have densities around 3000 kg/m^3, it is reasonable to suppose that both Io and Europa are made primarily of rocky material. The two outer satellites are still less dense. Both Ganymede and Callisto have an average density of less than 2000 kg/m^3, suggesting that they are composed of rock and ice.

14-3   The formation of the Galilean satellites probably mimicked the formation of our solar system

The average densities of the Galilean satellites decline with distance from Jupiter in a way that mimics the solar system. Moving outward from the Sun, the planets' average density

FIGURE 14-2 Io and Europa The Galilean satellites look like pinpoints of light when viewed through an Earth-based telescope, but the Earth-orbiting Hubble Space Telescope produces views of Jupiter comparable to this photograph from *Voyager 1*. Surface features as small as 400 km (240 mi) across are visible. Io (left) and Europa (right) are each approximately the same size as our Moon. (NASA)

| | Io | Europa | Ganymede | Callisto |

FIGURE 14-3 The Galilean Satellites The four Galilean satellites are shown here to the same scale. Io and Europa have diameters and densities comparable to our Moon and are composed primarily of rocky material. Ganymede and Callisto are roughly as big as Mercury, but their low average densities indicate that each is covered with a thick layer of water and ice. (NASA)

steadily decreases, from more than 5000 kg/m^3 for Mercury to less than 1000 kg/m^3 for Saturn (recall Table 7-2). Scientists therefore suspect that the same processes that formed our solar system also shaped the Galilean satellites, though on a much smaller scale.

In the early 1970s, the American astronomer John Lewis pointed out that the low densities of Ganymede and Callisto are exactly what one would expect for objects formed by accretion of ice-covered dust grains. More recently, NASA scientists James Pollack and Fraser Fanale constructed theoretical models of the formation of the Galilean satellites. These models show that frozen water could be retained and incorporated into satellites at the distances of Ganymede and Callisto. Only rocky material would condense at the orbital distances of Io and Europa, because Jupiter emits twice as much energy as it receives from the Sun. Thus, Jupiter's gravity and warmth produced two distinct classes of Galilean satellites, just as warmth from the protosun caused a dichotomy between small, dense, rocky inner planets and huge, low-density, gaseous outer planets.

Spectroscopic observations confirm the presence of water ice on the Jovian satellites. Europa and Ganymede exhibit strong absorptions characteristic of water ice molecules. Since Europa is composed mostly of rock, its ice must be limited to a relatively thin surface layer. Callisto, the dimmest of the Galilean satellites, is covered with a dark, dusty coating that subdues the reflective properties of the underlying layer of ice. This dark coating can easily be seen in Figure 14-3, which shows all four Galilean satellites to the same scale.

BOX 14-1

The Galilean Satellites

Four of the sixteen known satellites that orbit Jupiter are comparable in size and mass to terrestrial planets. The table below lists basic data about these four worlds, with data about Mercury and our Moon for comparison.

	Mean distance from Jupiter (km)	Sidereal period (days)	Diameter (km)	Mass (kg)	Mass (Moon = 1)	Mean density (kg/m³)
Io	421,600	1.77	3630	8.94×10^{22}	1.22	3570
Europa	670,900	3.55	3138	4.80×10^{22}	0.65	2970
Ganymede	1,070,000	7.16	5262	1.48×10^{23}	2.01	1940
Callisto	1,883,000	16.69	4800	1.08×10^{23}	1.47	1860
Mercury	——	——	4878	3.30×10^{23}	4.49	5430
Moon	——	——	3476	7.35×10^{22}	1.00	3340

14-4 The Voyager spacecraft discovered several small moons and a ring around Jupiter

The first moon to be photographed at close range by *Voyager 1* was Amalthea, one of Jupiter's many small satellites (Box 14-2). Tiny, reddish Amalthea circles the planet in only 11.7 hours, along an orbit much closer to Jupiter than that of Io. As Figure 14-4 shows, Amalthea is irregularly shaped, much like a large asteroid. It measures only 270 km across— about ten times larger than the moons of Mars but ten times smaller than the smallest Galilean satellites. Like the Galilean satellites, Amalthea orbits Jupiter with synchronous rotation, keeping its longest axis pointed toward the planet.

Three of the 16 satellites now known to orbit Jupiter were discovered during the Voyager flybys. The three newly identified moons, along with Amalthea and the Galilean satellites, orbit close to Jupiter in the plane of the planet's equator. In contrast, the remaining eight moons all circle Jupiter along large orbits that are inclined at steep angles to the planet's equatorial plane. All are extremely tiny, with estimated diameters of less than 50 km. These eight outer moons may be wayward asteroids captured by Jupiter's powerful gravitational field. Indeed, the four outermost satellites all have retrograde orbits, and calculations show that it is easier for Jupiter to capture a satellite into a retrograde orbit than a direct one.

In addition to discovering several Jovian moons, the Voyager cameras revealed a faint ring around Jupiter (Figure 14-5). This ring, which is probably composed primarily of dust along with some small rock fragments, lies in Jupiter's equatorial plane, closer to the planet than even the orbit of Amalthea. The sharp outer boundary of the ring is only 1.81 Jupiter radii from the planet's center. As we shall see in the next two chapters, Saturn, Uranus, and Neptune also have

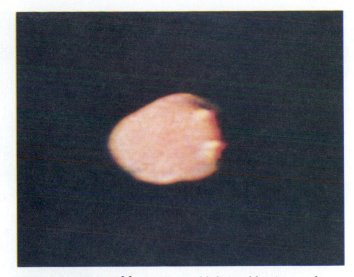

FIGURE 14-4 Amalthea Tiny, reddish Amalthea is one of Jupiter's four innermost satellites. It is an irregularly shaped, asteroidlike object measuring 270 km along its longest dimension. This photograph, taken by *Voyager 1*, has a resolution of 8 km. Amalthea was discovered in 1892 by the American astronomer E. E. Barnard at the Lick Observatory. (NASA)

rings and families of satellites whose orbits lie in their planets' equatorial planes.

14-5 Io is covered with colorful deposits of sulfur compounds ejected from active volcanoes

Only a few hours after *Voyager 1* passed Amalthea, Io loomed into view and the spacecraft began sending back a series of strange and unexpected pictures, like the one shown

FIGURE 14-5 Jupiter's Ring A portion of Jupiter's faint ring is seen in this photograph from *Voyager 2*. The ring, which is closer to Jupiter than any of the planet's satellites, is probably composed of tiny rock fragments. The brightest portion of the ring is about 6000 km in width. The outer edge of the ring is sharply defined, but the inner edge is somewhat fuzzy. A tenuous sheet of material extends from the ring's inner edge all the way down to the planet's cloudtops. (NASA)

FIGURE 14-6 Io This close-up view of Io was taken by *Voyager 1*. Notice the extraordinary range of colors, from white, yellow, and orange to black. Scientists believe that these brilliant colors are caused by surface deposits of sulfur ejected from Io's numerous volcanoes. (NASA)

BOX 14-2

Jupiter's Family of Moons

Sixteen confirmed satellites orbit Jupiter. Nearest Jupiter are four small moons. Next come the four giant Galilean satellites. The remaining eight moons, all very tiny, are located at great distances from Jupiter. The sixteen satellites are listed in the table with data about their orbits.

| | Average radius of orbit | | | |
	(km)	(Jupiter radii)	Orbital period (days)	Year of discovery
Metis	12,960	1.79	0.30	1979
Adrastea	128,980	1.81	0.30	1979
Amalthea	181,300	2.54	0.50	1892
Thebe	221,900	3.11	0.67	1979
Io	421,600	5.91	1.77	1610
Europa	670,900	9.40	3.55	1610
Ganymede	1,070,000	15.0	7.15	1610
Callisto	1,883,000	26.4	16.69	1610
Leda	11,094,000	155	238.7	1974
Himalia	11,480,000	161	250.6	1904
Lysithea	11,720,000	164	259.2	1938
Elara	11,737,000	164	259.6	1905
Ananke	21,200,000	297	631^R	1951
Carme	22,600,000	317	692^R	1938
Pasiphae	23,500,000	329	735^R	1908
Sinope	23,700,000	332	758^R	1914

Note: The outermost four satellites move in a retrograde direction about Jupiter, indicated by a superscript R with the orbital period.

in Figure 14-6. Baffled by what they were seeing, scientists at the Jet Propulsion Laboratory (JPL) in Pasadena (where the pictures were being received and enhanced by computer processing) jokingly compared Io to pizzas and rotten oranges. In fact, the scientists were seeing confirmation of a brilliant prediction made only a few days earlier.

Three days before *Voyager 1* flew past Io, Stanton Peale of the University of California at Santa Barbara and Patrick Cassen and Ray Reynolds of NASA's Ames Research Center reported their deduction that Io is acted upon by tremendous tidal forces. As it orbits Jupiter, Io repeatedly passes between Jupiter on one side and the remaining Galilean satellites on the other. The resulting gravitational tug-of-war distorts Io's orbit. As the distance between Io and Jupiter varies, tidal stresses on Io alternately squeeze and flex it. In turn, this constant tidal stressing causes frictional heating of Io's interior. Heat pumped into Io in this fashion could be as great as 10^{13} watts, which is equivalent to 2400 tons of TNT exploding every second. This energy must eventually make its way to the satellite's surface, so Peale, Cassen, and Reynolds predicted "widespread and recurrent volcanism" on Io.

No one expected to obtain photographs of erupting volcanoes on Io. After all, a spacecraft making a single trip past the Earth would be highly unlikely to catch a large volcano actually erupting. Yet that is exactly what happened at Io. Several days after the Jupiter flyby Linda Morabito, a navigation engineer at JPL, noticed a large umbrella-shaped cloud protruding from Io in one photograph. She had discovered an erupting volcano. Careful reexamination of the close-up photographs revealed eight giant eruptions. Thus was the brilliant Peale–Cassen–Reynolds prediction confirmed.

The volcanoes on Io are named after gods and goddesses traditionally associated with fire in Greek, Norse, Hawaiian, and other mythologies. For instance, Figure 14-7 shows two views of the symmetric plume of Prometheus.

The Voyager cameras also revealed numerous black dots on Io, which apparently are the volcanic vents through which the eruptions occur. These black spots, which are typically 10 to 50 km in diameter, cover 5% of Io's surface. Lava flows (Figure 14-8) radiate from many of these black dots, some of which are located at the origins of volcanic plumes. Some of the black spots have temperatures as high as $20°C$, in sharp contrast to the surface temperature surrounding them, which is only $-146°C$.

The plumes and fountains of material spewing from Io's volcanoes rise to astonishing heights of 70 to 280 km above the satellite's surface. To reach these altitudes, the material must emerge from the volcanoes' vents with speeds between 300 and 1000 m/s. Even the most violent terrestrial volcanoes, like Vesuvius, Krakatoa, and Mount St. Helens, have eruption velocities of only around 100 m/s. Scientists therefore began to suspect that Io's volcanoes must operate in a fundamentally different way from volcanoes here on Earth.

Infrared spectrometers on *Voyager 1* detected sulfur and sulfur dioxide in the material erupting from Io's volcanoes. Sulfur is normally bright yellow. If heated and suddenly cooled, however, it can assume a range of colors, from orange and red to black, which probably accounts for Io's

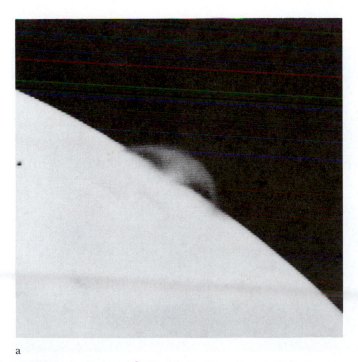

a

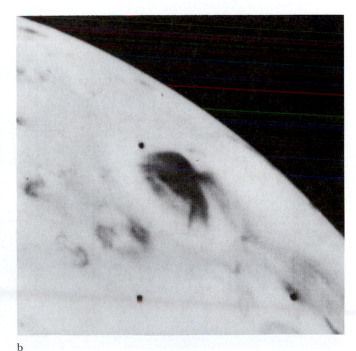

b

FIGURE 14-7 Prometheus on Io These two views from *Voyager 1*, taken two hours apart, show details of the plume of the volcano called Prometheus. (a) The plume's characteristic umbrella shape is silhouetted against the blackness of space.

(b) When viewed against the light background of Io's surface, jets of material give the plume a spiderlike appearance. The plume of Prometheus rises to an altitude of 100 km above Io's surface. (NASA)

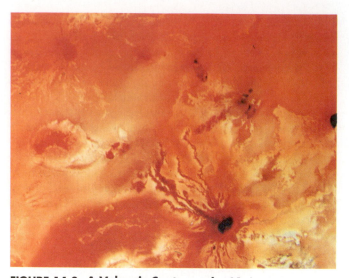

FIGURE 14-8 A Volcanic Center on Io No impact craters are seen in close-up pictures of Io such as this one taken by *Voyager 1*. Long, meandering lava flows radiate from many of the black dots that are the apparent sites of intense volcanic activity. This photograph covers an area measuring 1000 by 800 km, approximately twice the size of California. (NASA)

brilliant colors. Sulfur dioxide (SO_2) is an acrid gas commonly discharged from volcanic vents here on Earth. When this gas is released into the cold vacuum of space from eruptions on Io, it crystallizes into white snowflakes. It is likely that the whitish deposits on Io (examine Figure 14-6) are frozen sulfur dioxide (SO_2 frost or snow).

Abundant sulfur and sulfur dioxide on Io suggest an explanation for the mechanism of Io's volcanoes, which actually seem to be more like geysers here on Earth than volcanoes. In a geyser, water seeps down to volcanically heated rocks, where it suddenly changes to steam and erupts explosively through a vent. Geologist Susan Kieffer calculates that if geysers in Yellowstone Park in Wyoming were to erupt under the low gravity and vacuum that surround Io, Old Faithful would send a plume of water and ice to an altitude of 40 kilometers.

Planetary geologists Eugene Shoemaker and Bradford Smith have pointed out that sulfur dioxide could be the principal propulsive agent driving Io's eruptions. The compound is molten at depths of only a few kilometers below Io's surface, because of the heat generated by the tidal flexing of the satellite. Just as the explosive conversion of water into steam produces a geyser on Earth, a sudden conversion of liquid sulfur dioxide into a high-pressure gas could result in eruption velocities of up to 1000 m/s.

Each volcano on Io ejects an estimated 10,000 tons of material per second. This hot mixture of molten sulfur and sulfur dioxide gas is under high pressure as it first gushes from a volcanic vent, but it rapidly cools and solidifies in the cold, nearly perfect vacuum around the satellite. In about half an hour, most of the fine particles of sulfur dust and sulfur dioxide snow fall back down onto Io's surface.

Altogether, Io's volcanoes and vents eject roughly 1 trillion tons of matter each year. This amount is sufficient to cover the entire surface of Io to a depth of 10 meters in a thousand years, or to alter the color of an area of a thousand square kilometers in a few weeks. Because Io's surface is

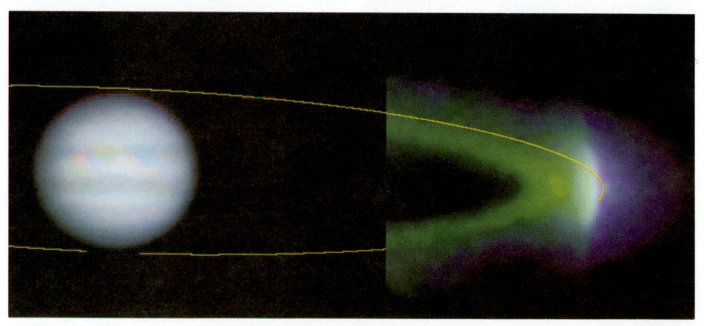

FIGURE 14-9 Io's Plasma Torus This view of Io's plasma torus was obtained with an Earth-based telescope. Sulfur ions that produce the purple and green hues are ejected from Io's volcanoes.

Because of the overwhelming glare of Jupiter, only the outer edge of the torus could be photographed. An artist added the drawing of Jupiter. (Courtesy of J. Trauger)

constantly changing, there are probably no long-lived features on this satellite. This rate of deposition of matter is easily sufficient to obliterate any impact craters rapidly.

Not all of the matter ejected by Io's volcanoes returns to the satellite's surface. Apparently a small fraction—perhaps 1 ton per second—manages to escape Io's gravity and become part of Jupiter's magnetosphere. High temperatures in the magnetosphere easily strip one or more electrons from each sulfur or oxygen atom. The result is a huge doughnut-shaped ring called the **Io torus**, circling Jupiter at the distance of Io's orbit, that contains a plasma composed primarily of electrons and ions of sulfur and oxygen. Although these ions are not plentiful, they produce radiation that is detected by Earth-based astronomers (Figure 14-9). The Io torus lies in the plane of Jupiter's magnetic equator, inclined by 11° to the plane of Io's orbit.

The Voyager flybys also illuminated Io's role in the bursts of the decametric radio radiation (radiation with wavelengths of tens of meters) that come from Jupiter. Jupiter rotates once every 10 hours, whereas Io takes 1.77 days to complete an orbit. As Jupiter's magnetic field sweeps past Io at high speed, Io develops a strong electric charge, which causes an electric current to flow between it and Jupiter. Although details are not yet fully understood, this electric current is believed to be responsible for the decametric radiation bursts.

14-6   Europa is covered with a smooth layer of ice that is crisscrossed with numerous cracks

Voyager 1 did not pass near Europa, but *Voyager 2* got close enough to capture the excellent view shown in Figure 14-10. Europa is a smooth-surfaced world with no mountains and very few craters, crisscrossed by spectacular streaks and cracks.

We have seen that Europa is only slightly less dense than our Moon. Spectroscopic observations from Earth indicate that Europa has frozen water on its surface. These two facts together suggest that Europa may be covered with ice. An ice layer would be consistent with the remarkable smoothness of Europa's surface, and such an "ocean" of ice 100 km deep would certainly hide mountain ranges and other topographic features. But what has caused the network of cracks? And why are impact craters so rare on Europa?

With Jupiter on one side and the two largest Galilean moons periodically passing on its opposite side, Europa is caught in a tidal tug-of-war similar to Io's. However, because it is farther from Jupiter, tidal effects on Europa are only about a quarter as strong as those on Io. NASA scientist Patrick Cassen has therefore suggested that tidal flexing has caused the network of cracks covering Europa's surface. Some of the darkest streaks do in fact follow paths along which the tidal stresses are strongest. Water gushed up and froze in these stress-induced cracks, leaving the streaks we

FIGURE 14-10 Europa Europa's icy surface is covered by numerous streaks and cracklike features that give this satellite a fractured appearance. The streaks are typically 20 to 40 km wide. Surface features as small as 5 km across can be seen in this picture taken by *Voyager 2*. (NASA)

see in the Voyager photographs. Tidal flexing might also supply enough energy to churn the satellite's icy coating. This movement would explain why only a few small impact craters have survived, though Europa's surface is much older than Io's.

There are various arguments against Cassen's theory. To create all the cracks—each one at least ten kilometers wide—Europa's surface area would have had to increase by 10% to 15%. Yet it seems unreasonable to suppose that Europa is expanding like an inflating balloon. Apparently, more is happening on Europa than meets the eye. Perhaps old surface material is somehow being pulled back down into the mushy layer below the ice coating and recycled, just as the Earth's crust is pulled back down into the Earth's mantle in subduction zones (recall Figure 8-14). Europa's surface may thus represent a water-and-ice version of plate tectonics.

14-7   Ganymede and Callisto have heavily cratered, icy surfaces

Only 100 km of ice and water on top of an otherwise rocky world suffices to explain Europa's density, but a much thicker water ice layer must surround a less dense moon. The average densities of Ganymede and Callisto suggest that their rocky cores are enveloped by mantles of water and ice nearly 1000 km thick.

The two outer Galilean satellites have ancient, cratered surfaces. As Figure 14-11 shows, Ganymede even looks somewhat like our Moon. Of course, the craters on both Ganymede and Callisto are made of ice rather than rock.

FIGURE 14-11 Ganymede This view from *Voyager 2*, taken at a distance of 1.2 million km, shows the hemisphere that always faces away from Jupiter. This hemisphere is dominated by a huge, dark circular area called Galileo Regio, which is the largest remnant of Ganymede's ancient crust. (NASA)

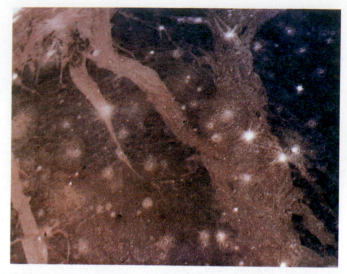

FIGURE 14-12 Young and Old Terrain on Ganymede This close-up view of Ganymede was taken by *Voyager 2* at a range of only 312,000 kilometers. Features as small as 5 km across can be seen. Dark, angular islands from Ganymede's ancient crust are separated by younger, light-colored, grooved terrain. The southwest edge of Galileo Regio appears at the right side of the picture. (NASA)

Ganymede is the largest satellite in our solar system. Its diameter is 5262 km, which is slightly greater than the diameter of Mercury. Like our Moon, Ganymede has two very different kinds of terrain (Figure 14-12). Dark, polygon-shaped regions are the oldest surface features on Ganymede, as judged by their high density of craters. Light-colored, heavily grooved terrain is found between the dark, angular islands. These lighter regions are much less cratered and therefore younger.

The largest single feature on Ganymede is the dark, circular island of ancient crust called Galileo Regio, seen in Figure 14-11. It measures 4000 kilometers in diameter and covers nearly one-third of the hemisphere of Ganymede that faces away from Jupiter. This dark region is the only surface feature on the Galilean satellites that can be detected with Earth-based telescopes.

Voyager photographs show noticeable differences between the young and the old craters on Ganymede. The youngest craters are surrounded by bright rays of freshly exposed ice (examine Figure 14-12). Older craters have been covered with deposits of dark meteoritic dust. The most ancient craters—sometimes called ghost craters or palimpsests, an archeological term for a parchment that was scraped clean and reused—are barely visible on Galileo Regio and other dark islands made of old crust. The near obliteration of the oldest craters was probably caused by a slow plastic flow of Ganymede's icy surface.

The dark, angular islands may be remnants of the ancient original crust. When Ganymede formed 4.5 billion years ago, it may have been completely covered with an ocean of water roughly 1000 km deep. During perhaps the next 200 million years, the water cooled and a coating of ice developed that is today about 100 km thick. Beneath it lies a 900-km-thick slushy mantle of water and ice.

Craters also tell us about the history of the younger, light-colored terrain, which is covered with numerous grooves. Cratering there varies in density, from about the same as on the ancient crust down to one-tenth that amount. We thus suspect that this grooved terrain formed over a long time. The process probably began quite early in Ganymede's history and continued through the period of intense meteoritic bombardment. The age of the grooved terrain thus ranges from about 4.5 to 3.5 billion years.

High-resolution photographs, such as Figure 14-13, show that this grooved terrain actually consists of parallel mountain ridges up to 1 km high, spaced 10 to 15 km apart. These features suggest that the process of plate tectonics may have dominated Ganymede's early history. Water seeping upward through cracks in the satellite's original crust froze and forced apart fragments of that crust, producing jagged, dark islands of old crust separated by bands of younger, light-colored, heavily grooved ice. The cracks may thus have played a role in Ganymedean plate tectonics, much like oceanic rifts on Earth. But unlike Europa, where tectonic activity may still occur today, tectonics on Ganymede bogged down 3 billion years ago, as the satellite's crust froze.

Callisto, Jupiter's outermost Galilean satellite, looks very much like Ganymede: Numerous impact craters are scattered over a dark, ancient, icy crust (Figure 14-14). There is one obvious difference, however: Callisto has no younger, grooved terrain. The absence of grooved terrain suggests that tectonic activity never began on Callisto. Perhaps

FIGURE 14-13 Grooved Terrain on Ganymede This picture, taken by *Voyager 1* at a range of 145,000 km, shows an area roughly as large as Pennsylvania. The smallest visible features are about 3 km across. The numerous parallel mountain ridges visible here are spaced 10 to 15 km apart and have heights up to 1000 m. (NASA)

FIGURE 14-14 Callisto Callisto, the outermost Galilean satellite, is almost exactly the same size as Mercury. Numerous craters pockmark Callisto's icy surface, as seen in this mosaic of views from *Voyager 1* taken at a distance of about 400,000 km. Note the series of faint, concentric rings that cover the left third of the image. These rings outline a huge impact basin called Valhalla that dominates the Jupiter-facing hemisphere of this frozen, geologically inactive world. (NASA)

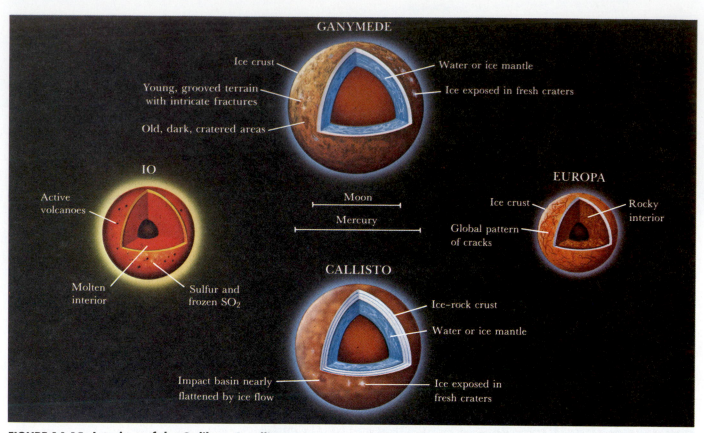

FIGURE 14-15 Interiors of the Galilean Satellites These cross-sectional diagrams show the probable internal structure of the four Galilean satellites, based on their average densities and on information from the Voyager flybys.

because of Callisto's greater distance from Jupiter, the ocean that enveloped the young satellite 4.5 billion years ago froze more rapidly and to a greater depth than on Ganymede, forever preventing tectonic processes. It is bitterly cold on Callisto: Voyager instruments measured a noontime temperature of -118°C (-180°F), and the nighttime temperature plunges to -193°C (-315°F).

Voyager 1 photographed a huge impact basin on Callisto, seen faintly in Figure 14-14. Located on Callisto's Jupiter-facing hemisphere, this Valhalla Basin was produced by the impact of an asteroid-sized object early in the satellite's history. Valhalla consists of a large number of concentric rings having diameters up to 3000 km.

The great age of Valhalla can be inferred both from the presence of impact craters there and from the absence of substantial vertical relief in the concentric rings. The Valhalla impact occurred probably around 4 billion years ago, when the satellite's young, relatively thin crust still let plastic flow reduce the height of the upraised rings in the ice.

In looking at the Galilean satellites, we see a neat, orderly progression from a high-density, geologically active world (Io) nearest Jupiter to a low-density, geologically dead world (Callisto) farthest from Jupiter. The diagrams in Figure 14-15 show their probable interior structures. These terrestrial worlds demonstrate the remarkable variety of which nature is capable.

In December 1995, *Galileo* will arrive at Jupiter. After releasing a probe into Jupiter's atmosphere, the spacecraft will make a single, inbound pass by Io, which lies deep in Jupiter's magnetosphere. Scientists fear that the spacecraft's electronics might suffer significant radiation damage if more than one flyby of Io is attempted. *Galileo* will then begin long-term surveillance of the remaining Galilean satellites, using their gravity to redirect the spacecraft's orbit. Such maneuvers will permit the spacecraft to go from one Galilean satellite to another. The resulting flybys will give scientists the opportunity to study three icy worlds that were glimpsed only briefly by the Voyagers.

KEY WORDS

Galilean satellites   Io torus   occultation

KEY IDEAS

• The four Galilean satellites of Jupiter orbit the planet in the plane of its equator, with synchronous rotation. Their periods of rotation and revolution about the planet are relatively short (from about 2 to 17 days).

The two inner Galilean moons, Io and Europa, have roughly the same size and density as our Moon. The two outer Galilean moons, Ganymede and Callisto, are roughly the size of Mercury and are lower in density than either the Moon or Mercury.

Several small moons and a faint ring of mostly dust also orbit in the plane of Jupiter's equator.

Eight small moons move in much larger orbits that are noticeably inclined to the plane of Jupiter's equator; some have retrograde orbits.

• The inner moons, Galilean satellites, and ring of Jupiter probably all formed through a process of accretion, similar to that which formed our solar system but on a smaller scale; the outer moons were possibly asteroids later captured by Jupiter's gravity.

• The Galilean satellites help us understand physical processes on the terrestrial planets, because they provide examples of how these processes operate under conditions quite different from those prevailing on the inner planets of our solar system.

• Io is covered with a colorful layer of sulfur compounds deposited by frequent explosive eruptions from volcanic vents.

Io's volcanic eruptions resemble terrestrial geysers. The energy heating Io's interior comes from tidal forces that flex the moon.

The Io torus is a ring of electrically charged particles circling Jupiter at the distance of Io's orbit. Interactions between this ring and Jupiter's magnetic field produce the strong decametric radio emissions associated with Jupiter.

• Europa is covered with a smooth layer of frozen water crisscrossed by an intricate pattern of long cracks that are probably produced by tidal flexing of the moon.

• The heavily cratered surface of Ganymede is composed of frozen water with large polygons of dark, ancient surface separated by regions of heavily grooved, lighter-colored, younger terrain.

Plate tectonics apparently operated during the early history of Ganymede.

• Callisto has a heavily cratered crust of frozen water, but plate tectonics apparently never operated on this moon, presumably because it quickly developed a thick, solid crust.

The impact basin called Valhalla on Callisto provides evidence for plastic flow in the icy crust of this moon.

REVIEW QUESTIONS

1. How does the Galilean satellite system resemble our solar system? How is it different?

2. Compare and contrast the surface features of the four Galilean satellites, discussing their relative geological activity and the evolution of these four satellites.

3. How would you account for the existence of the non-Galilean satellites of Jupiter as well as for Jupiter's ring?

4. What is the source of energy that powers Io's volcanoes?

5. With all its volcanic activity, why doesn't Io possess a thick atmosphere?

6. Long before the Voyager flybys, Earth-based astronomers reported that Io appeared brighter than usual for the few hours after it emerged from Jupiter's shadow. From what we know about the material ejected from Io's volcanoes, explain this brief brightening of Io.

7. Why are numerous impact craters found on Ganymede and Callisto but not on Io or Europa?

8. What is the Io torus and what is its source?

9. Compare and contrast Valhalla (see Figure 14-14) with Mare Orientale on our Moon and the Caloris Basin on Mercury.

10. Describe the plans for the *Galileo* orbiter to conduct long-term surveillance of the Galilean satellites. Why will this spacecraft make many close approaches to Europa, Ganymede, and Callisto but not to Io?

ADVANCED QUESTIONS

Tips and tools . . .

Kepler's third law tells us that P^2 divided by a^3 is a constant. The small-angle formula is discussed in Box 1-1. The best seeing conditions on Earth give a seeing disk $\frac{1}{4}$ arc second in diameter. Because the orbits of the Galilean satellites are almost perfect circles, the orbital speeds of these satellites can easily be calculated from the data listed in Box 14-1.

11. Using orbital data given in Box 14-1, demonstrate that the Galilean satellites obey Kepler's third law.

12. Invent a system of units in which $P^2 = a^3$ for Jupiter's satellites.

13. What is the size of the smallest feature you should be able to see on a Galilean satellite through a large telescope under conditions of excellent seeing when Jupiter is near opposition?

14. Astronomers who took the accompanying photograph of Io with the Hubble Space Telescope in 1992 claim that they can see features as small as 150 km across. When the picture was taken, Io was 4.45 AU from Earth. What is the angular resolution of the Hubble Space Telescope? (For

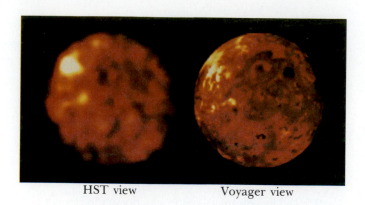

HST view   Voyager view

comparison, a Voyager image taken from a distance of about 400,000 km is also shown.)

15. Using the diameter of Io (3630 km) as a scale, estimate the height to which the plume of Prometheus rises above the surface of Io in Figure 14-7.

16. If material is ejected into space from Io's volcanoes at the rate of 1 ton/s, how long will it be before Io loses 10% of its mass? How does your answer compare with the age of the solar system?

17. How long does it take for Ganymede to enter or leave Jupiter's shadow?

18. Consult magazines like *Science News* or *Sky & Telescope* to learn about the current status of the Galileo mission. Is all well with the spacecraft? When is it scheduled to arrive at Jupiter?

DISCUSSION QUESTIONS

19. Speculate on the possibility that Europa, Ganymede, or Callisto might harbor some sort of marine life.

20. Suppose you were planning four missions to the moons of Jupiter that would land a spacecraft on each of the Galilean satellites. What kinds of questions would you want these missions to answer, and what kinds of data would you want your spacecraft to send back? In view of the different environments on the four satellites, how would the designs of the four spacecraft differ? Be specific about the possible hazards and problems each spacecraft might encounter in landing on the four satellites.

OBSERVING PROJECTS

21. Observe Jupiter through a pair of binoculars. Can you see all four Galilean satellites? Make a drawing of what you observe.

22. Observe Jupiter through a small telescope on three or four consecutive nights. Make a drawing each night showing the positions of the satellites relative to Jupiter. Record the time and date of each observation. Consult the section called "Satellites of Jupiter" in the *Astronomical Almanac* for the current year to see if you can identify the satellites by name. This section of the *Astronomical*

Almanac shows the apparent paths of the Galilean satellites as they move back and forth, from one side of Jupiter to the other. Note that universal time (UT) is used in the *Astronomical Almanac,* and so you must convert your local time to UT. Universal time, which is the same as Greenwich mean time, is counted from 0 hours beginning at midnight in Greenwich, England. Thus, for instance, to convert eastern standard time (EST) to universal time, you must add five hours. For example, 10:30 PM EST on January 3 is the same as 3:30 UT on January 4.

23. Make arrangements to observe an eclipse, a transit, or an occultation of one of the Galilean satellites. Consult a listing of such phenomena in the section called "Satellites of Jupiter" in the *Astronomical Almanac* for the current year. Choose the phenomenon you would like to see and calculate its scheduled time by converting the universal time given in the *Astronomical Almanac* to your local time.

24. If you are fortunate enough to have access to a large telescope with a primary mirror 1 meter or more in diameter, make arrangements to view Jupiter through that telescope. Describe what you see. Under conditions of excellent seeing when Jupiter is near opposition, the Galilean satellites should look like tiny discs rather than starlike pinpoints of light.

FOR FURTHER READING

Carroll, M. "Project Galileo: The Phoenix Rises." *Sky & Telescope,* April 1987. This brief article discusses the Galileo mission.

Croswell, K. "Io: Jupiter's Fiery Satellite." *Space World,* July 1988. This well-written article surveys current understanding of Io.

Griffin, R. "Barnard and His Observations of Io." *Sky & Telescope,* November 1982. This brief article discusses pre-Voyager visual observation of Io, including some attempts to map its surface.

Johnson, T. "The Galilean Satellites." In Beatty, J., and others, eds., *The New Solar System.* 3rd ed. Sky Publishing and Cambridge University Press, 1990. A well-written, up-to-date summary of our current understanding of the Galilean satellites.

Johnson, T., and Soderblom, L. "Io." *Scientific American,* December 1983. This article, which is devoted to volcanism on Io, includes an enlightening discussion of the properties of sulfur and sulfur dioxide.

Morrison, D. "Four New Worlds: The Voyager Exploration of Jupiter's Satellites." *Mercury,* May/June 1980. This article by a noted planetary scientist discusses what we have learned from the Voyager flybys.

Soderblom, L. "The Galilean Moons of Jupiter." *Scientific American,* January 1980. An excellent selection of photographs accompany this fine article, which summarizes the discoveries made during the Voyager flybys.

THE SPECTACULAR SATURNIAN SYSTEM

SATURN The second-largest planet in the solar system is surrounded by broad, bright rings. Note that the cloud features on Saturn are much less distinct than those on Jupiter. This photograph, taken by *Voyager 1* in 1980, also shows two of Saturn's medium-sized satellites: Enceladus (off the left edge of rings) and Dione (just below Saturn). The rings and most of Saturn's 18 known moons orbit the planet in the plane of Saturn's equator. (NASA)

SATURN IS a fascinating and beautiful world primarily because of the system of thin, flat rings that encircles it. In the early 1980s, two Voyager flybys led to many important discoveries about Saturn, its rings, and its many satellites. These explorations of Saturn's hydrogen-rich atmosphere and cloud patterns suggest differences in how Saturn and neighboring Jupiter evolved. The Voyager spacecraft glimpsed surface features on many of Saturn's satellites and showed that the gravity of some of these satellites profoundly affects the structure and appearance of the rings. Most interesting is Titan, Saturn's largest moon, which is enveloped by a thick nitrogen-rich atmosphere where methane snow and rain may fall beneath a smog of carbon–hydrogen compounds. These worlds pose many new puzzles for planetary scientists that could be answered by the Cassini spacecraft, tentatively scheduled to orbit the ringed planet early in the twenty-first century.

The magnificent rings of Saturn make this planet one of the most spectacular objects you can see in the nighttime sky with a small telescope. Saturn is so far away, however, that Earth-based telescopes reveal only coarse, large-scale features of its rings and cloudtops (Figure 15-1). Most of our detailed knowledge of this planet came from the Voyager spacecraft that sped past Saturn in the early 1980s. (Box 15-1 lists basic data about Saturn.)

15-1 Earth-based observations reveal three broad rings encircling Saturn

When Galileo focused his telescope on Saturn in the early 1600s, he saw few details, but he did notice two puzzling lumps protruding from opposite edges of the planet's disk. In 1655 the Dutch astronomer Christian Huygens observed Saturn with a better telescope and noted no protrusions. Having faith in Galileo's observations, Huygens suggested that Saturn might be surrounded by a thin, flattened ring that just happened to be edge-on as viewed from Earth in 1655, making it temporarily almost impossible to see. This brilliant deduction was confirmed ten years later, by which time Saturn had moved far enough along its orbit for the ring to be seen again.

As the quality of telescopes improved, astronomers realized that Saturn's "ring" is actually a *system* of rings. In 1675, G. D. Cassini discovered a dark division in the ring, an apparent gap about 5000 kilometers wide. This **Cassini division** separates the outer **A ring** from the brighter **B ring** closer to the planet. By the mid-1800s, astronomers managed to identify the faint **C ring**, or crepe ring, that lies just inside the B ring. The diameters of these rings are superimposed on a *Voyager* photograph in Figure 15-2.

FIGURE 15-1 Saturn from the Earth This view of Saturn is one of the best ever produced by an Earth-based observatory. Sixteen original color images taken on the same night with the 1.5-m telescope at the Catalina Observatory were combined to make this photograph. Note the faint stripes in Saturn's cloudtops as well as a prominent gap in the rings. (NASA)

Saturn's rings are best seen when the planet is at or near opposition (Table 15-1). Whereas a modest telescope gives a good view of the A and B rings, a large telescope and excellent observing conditions are needed to see the C ring.

Earth-based views of the Saturnian ring system change as Saturn slowly orbits the Sun. This change is observed because the rings, which lie in the plane of Saturn's equator, are tilted 27° from the plane of Saturn's orbit. Thus, over the course of a Saturnian year, the rings are viewed from various angles by an Earth-based observer (Figure 15-3). At one time, the observer looks "down" on the rings, but half a Saturnian year later the "underside" of the rings is exposed to our Earth-

BOX 15-1
Saturn Data

Mean distance from Sun:	9.529 AU = 1.427×10^9 km
Maximum distance from Sun:	10.073 AU = 1.507×10^9 km
Minimum distance from Sun:	9.005 AU = 1.347×10^9 km
Mean orbital velocity:	9.64 km/s
Sidereal period:	29.458 years
Rotation period:	Equatorial: $10^h\,13^m\,59^s$ Internal: $10^h\,39^m\,25^s$
Inclination of equator to orbit:	26.7°
Diameter (equatorial):	120,540 km
Diameter (Earth = 1):	9.45
Mass:	5.69×10^{26} kg
Mass (Earth = 1):	95.2
Mean density:	690 kg/m³
Surface gravity (Earth = 1):	0.93
Escape speed:	35.5 km/s
Mean surface temperature (at cloudtops):	−180°C = −292°F = 93 K

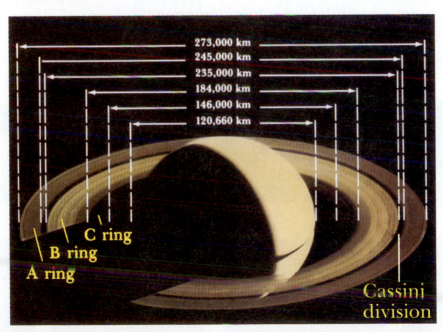

FIGURE 15-2 Saturn's Classic Rings Details of Saturn's rings are visible in this photograph sent back by *Voyager 1*. The equatorial diameter of Saturn as well as the diameters of the inner and outer edges of the rings are given in the overlay. Closest to the planet is the 19,000-km-wide C ring, so faint that it is almost invisible in this view. Next outward from Saturn is the broad, bright B ring, whose width is 25,500 km. The outermost ring that can be seen from Earth is the A ring, which is 14,000 km wide. The 5000-km-wide Cassini division lies between the B ring and the A ring. (NASA)

TABLE 15-1

Oppositions of Saturn, 1993–1999

Date of opposition	Magnitude
1993 August 19	+0.5
1994 September 1	+0.7
1995 September 14	+0.8
1996 September 26	+0.7
1997 October 10	+0.4
1998 October 23	+0.2
1999 November 6	0.0

based view. In between, the rings are viewed edge-on, when they seem to disappear entirely. This disappearance indicates that the rings are very thin—less than 2 km thick, according to recent estimates. The last edge-on presentation of Saturn's rings was in 1980, and the next will occur in 1995.

15-2 Saturn's rings are composed of numerous fragments of ice and ice-coated rock

Astronomers have long known that Saturn's rings could not possibly be solid, rigid, thin sheets of matter. In 1857 the Scottish physicist James Clerk Maxwell mathematically proved that such a broad, thin sheet would break apart. He concluded that Saturn's rings are composed of "an indefinite number of unconnected particles."

Proof that rings are not rigid came in 1895 when James Keeler at the Allegheny Observatory in Pittsburgh photographed the spectrum of sunlight reflected from Saturn's rings. Because the rings orbit Saturn, spectral lines on the side

approaching us are blueshifted, while spectral lines on the receding side are redshifted, as demanded by the Doppler effect (recall Section 5-8 and Figure 5-19). More importantly, Keeler noted that the size of the wavelength shift increased inward across the rings; the closer to the planet, the greater the shift. This variation in Doppler shift proves that the inner portions of Saturn's rings are moving about the planet more rapidly than the outer portions. Indeed, the orbital speeds across the rings are in complete agreement with Kepler's third law: The square of the orbit period about Saturn at any place in the rings is proportional to the cube of the distance from Saturn's center (recall Section 4-7 and Box 4-3 for a discussion of Kepler's laws). This is exactly what would be expected if the rings consisted of numerous tiny moonlets, each individually circling Saturn.

Saturn's rings are quite bright; they reflect 80% of the sunlight that falls on them. Astronomers therefore long suspected that the rings are made of ice and ice-coated rocks. This hunch was confirmed in the 1970s, when American astronomers Gerard Kuiper and Carl Pilcher identified absorption features of frozen water in the ring's near-infrared spectrum. Infrared measurements by Voyager spacecraft tell us that the temperature of the rings ranges from −180°C (−290°F) in the sunshine to less than −200°C (−300°F) in Saturn's shadow. Water ice is in no danger of melting or evaporating at these temperatures.

Changes in the radio signals received as the Voyager spacecraft passed behind the rings enabled astronomers to estimate the sizes of the particles. They range from snowflakes less than 1 mm in diameter to icy boulders tens of meters across. Most abundant are snowball-sized particles about 10 cm in diameter. It seems reasonable to suppose that all this material is ancient debris that failed to accrete (fall together) into satellites. As a matter of fact, the ring particles are so close to Saturn that they will never be able to form moons.

FIGURE 15-3 The Changing Appearance of Saturn's Rings Saturn's rings are tilted 27° from the plane of Saturn's orbit. Earth-based observers therefore see the rings at various angles as Saturn moves around its orbit. Note that the rings seem to disappear entirely when viewed edge-on. (Lowell Observatory)

To see why, imagine a collection of small fragments of rock orbiting a planet. Gravitational attraction between neighboring rocks tends to pull the rock fragments together. However, because the various rock fragments are at differing distances from the parent planet, they also experience different amounts of gravitational pull from the planet. This difference in gravitational pull is a **tidal force** that tends to keep the fragments separated. At a certain distance from the planet's center called the **Roche limit**, the attractive gravitational pull between neighboring rocks is exactly equal to the disruptive tidal force of the planet. Inside the Roche limit, fragments will not accrete to form a larger body; instead, they tend to spread out into a ring around the planet. All large satellites are therefore found only outside a planet's Roche limit. The Roche limit for Saturn lies just outside the outer edge of the A ring. Moreover, if any large moon were to come inside a planet's Roche limit, the planet's tidal forces would cause the satellite to break up into fragments. As we shall see in the next chapter, such a tidal disruption may be the catastrophic fate of Neptune's large satellite, Triton, whose orbit is gradually decaying.

Rock fragments survive in Saturn's rings because the Roche limit applies only to objects held together by gravity. Chemical bonds between atoms and molecules hold a rock together. Because these chemical forces are much stronger than the disruptive tidal force of a nearby planet, the rock does not break apart. In the same way, people walking around on the Earth's surface (which is inside the Earth's Roche limit) are in no danger of coming apart, because we are held together by comparatively strong intermolecular forces rather than gravity.

15-3 Saturn's rings consist of thousands of narrow, closely spaced ringlets

During the Voyager flybys of 1980 and 1981, spacecraft cameras sent back pictures showing the detailed structure of Saturn's rings. The broad rings were seen to consist of hundreds upon hundreds of closely spaced bands, or **ringlets,** of particles (Figure 15-4). These ringlets are orderly orbits of rock and ice fragments arranged by the continual gravitational pull of neighboring fragments, Saturn's moons, and the planet itself over millions of years.

The Voyager pictures revealed many additional details of Saturn's rings. For instance, they showed another division in the outer half of the A ring (examine Figure 15-4). This 270-km-wide gap is named the **Encke division,** after the German astronomer Johann Franz Encke, who reported seeing it in 1838. Many astronomers doubted Encke's report, however, because of the poor resolving power of his telescope. The first undisputed observation of the Encke division was made in the late 1880s by the American astronomer James Keeler, using the newly constructed 36-inch refractor at the Lick Observatory. For this reason, the Encke division is called the Keeler gap by some astronomers.

The Voyager cameras also sent back high-quality pictures of the **F ring**, which was first detected during the 1979 flyby of *Pioneer 11*. The F ring lies just beyond the outer edge of the A ring (examine Figure 15-4). Close-up views revealed a startling and mysterious fact: The F ring is kinky and braided. It actually consists of several intertwined strands, as shown in Figure 15-5. One *Voyager 1* image displayed a total of five strands, each about 10 km across. Astronomers are at

Cassini division
Encke division   F ring

FIGURE 15-4 Details of Saturn's Rings This view from *Voyager 1* shows that Saturn's rings are actually composed of thousands of closely spaced ringlets. The broad Cassini division is clearly visible, as is the narrow Encke division in the outer A ring. The very thin F ring is visible just beyond the outer edge of the A ring. (NASA)

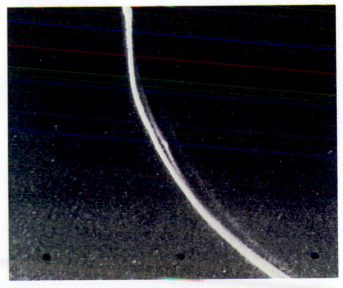

FIGURE 15-5 Details of the F Ring This photograph from *Voyager 1* shows several strands, each measuring roughly 10 km across, that together comprise the F ring. The total width of the F ring is about 100 km. (NASA)

a loss to explain this complex structure. The braids, kinks, knots, and twists in the F ring pose a challenging puzzle.

Through Earth-based telescopes, we can see only the sunlit side of Saturn's rings. From this perspective, the B ring appears very bright, the A ring moderately bright, the C ring dim, and the Cassini division dark. The proportion of sunlight reflected back toward the Sun—the albedo—is directly related to the concentration of the fragments or particles in the ring. The B ring is bright because it has a high concentration of ice and rock fragments, whereas the darker Cassini division has a lower concentration of such fragments.

Eight hours after it crossed from the sunlit side to the shaded side of the rings, *Voyager 1* took the photograph that appears in Figure 15-6. Since the spacecraft was looking back toward the Sun, this picture shows sunlight that has passed through the rings. The B ring looks darkest because little sunlight gets through its dense collection of fragments, but the Cassini division looks bright because sunlight passes relatively freely through its more widely spaced fragments. Nevertheless, the fact that the Cassini division does appear bright is clear evidence that it contains some fragments. If it contained no fragments at all to scatter the sunlight, we would see only the blackness of space through it.

The process by which light bounces off particles or fragments in its path is called **scattering**. The proportion of light scattered in various directions depends upon both the size of the particles and the wavelength of the light. By measuring the albedo of the rings at various wavelengths as the Voyager spacecraft sped past Saturn, scientists determined that the F ring consists primarily of tiny, micrometer-sized particles. The A and B rings have a mixture of particles of various sizes,

ranging from snowflakes to boulders. In contrast, the Cassini division and the C ring each contain relatively few tiny particles.

Subtle differences in color from one ring to the next also give important clues about the composition of the particles in the different rings. These variations are clearly visible in Figure 15-7, in which the colors have been exaggerated by computer processing. Although the main chemical constituent is frozen water, trace amounts of other chemicals—perhaps coating the surfaces of the ice particles—are probably responsible for the colors seen in the computer-enhanced view. These trace chemicals have not yet been identified. Nevertheless, the existence of color variations and their probable persistence over millions of years suggest that the icy particles do not migrate substantially from one ringlet to another.

In addition to revealing new details about the A, B, C, and F rings, the Voyager cameras also discovered three new ring systems: the D, E, and G rings. The **D ring** is Saturn's innermost ring system. It consists of a series of extremely faint ringlets located between the inner edge of the C ring and the Saturnian cloudtops. The **E ring** and the **G ring** both lie far from the planet, well beyond the outer edge of the A ring. Both of these outer ring systems are extremely faint, fuzzy, and tenuous. Each lacks the ringlet structure so prominent in the main ring systems. The E ring lies along the orbit of Enceladus, one of Saturn's icy satellites. Some scientists suspect that water geysers on Enceladus are the source of ice particles in the E ring, much as Io's volcanoes produce a torus of material along its orbit around Jupiter. The drawing in Figure 15-8 shows the layout of Saturn's rings along with the orbits of some of Saturn's 18 known moons.

FIGURE 15-6 Details of Saturn's A Ring This view of the underside of Saturn's rings was taken by *Voyager 1*. Both the Encke division and the thin F ring are clearly visible toward the right side of the picture. The Cassini division and the outer edge of the B ring are on the left side of the picture. The Cassini division appears bright in this view of the shaded side of the rings. (NASA)

FIGURE 15-7 An Enhanced-Color View of Ring Details Computer processing has severely exaggerated the subtle color variations in this view of the sunlit side of the rings from *Voyager 2*. Note that the C ring and Cassini's division appear bluish. Also note the distinct color variations across both the A and the B rings. (NASA)

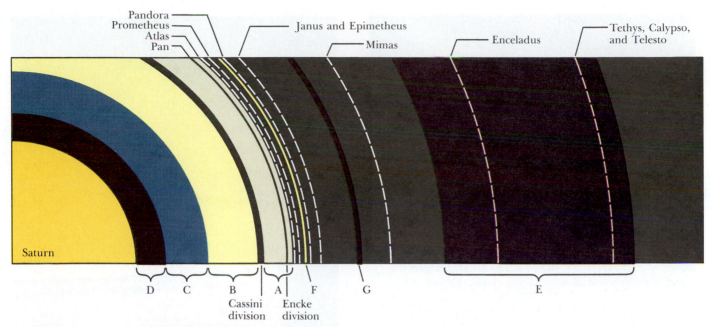

Pandora
Prometheus
Atlas
Pan
Janus and Epimetheus
Mimas
Enceladus
Tethys, Calypso,
and Telesto
Saturn

D   C   B   A   F   G   E

Cassini
division   Encke
division

FIGURE 15-8 The Arrangement of Saturn's Rings This scale drawing shows the location of Saturn's rings along with the orbits of the inner satellites. Only the A, B, and C rings can be seen from Earth. The remaining rings, which are very faint, were discovered during spacecraft flybys.

15-4 Saturn's innermost satellites affect the appearance and structure of its rings

Astronomers have long known that one of Saturn's moons, Mimas, has a gravitational effect on its ring system. Mimas is a moderate-sized satellite that orbits Saturn every 22.6 hours. According to Kepler's third law, particles in Cassini's division should orbit Saturn approximately every 11.3 hours. Consequently, on every second orbit, the particles in Cassini's division line up between Saturn and Mimas. Because of these repeated alignments, the combined gravitational forces of Saturn and Mimas cause small fragments to deviate from their original orbits. This "2-to-1 resonance" with Mimas depletes Cassini's division of the particles and fragments that would otherwise scatter sunlight back toward Earth. Earth-based astronomers therefore see the Cassini division as a relatively dark band between neighboring rings.

Similar effects are produced by two tiny satellites that follow orbits on either side of the F ring (Figure 15-9). Peter Goldreich of the California Institute of Technology (Caltech) and Scott Tremaine at Princeton pointed out that the gravitational forces of these two satellites keep the F ring particles in place. The outer of the two satellites moves around Saturn at a slightly slower speed than do the ice particles in the ring. As the ring particles pass the outer satellite, they experience a tiny gravitational tug, which tends to slow them down. As a result, these particles lose a little energy, causing them to fall into orbits a bit closer to Saturn. Meanwhile, the inner satellite is orbiting the planet somewhat faster than are the F ring particles. Its gravitational pull tends to nudge the particles into a slightly higher orbit. The combined effect of these two satellites therefore focuses the icy particles into a well-defined, narrow band about 100 km wide. Because of their confining influence, these two moons, Prometheus and Pandora, are called **shepherd satellites.**

FIGURE 15-9 The F Ring and Its Two Shepherds Two tiny satellites (Prometheus and Pandora), each measuring only about 50 km across, orbit Saturn on either side of its F ring. The gravitational effects of these two shepherd satellites focus and confine the particles in the F ring to a band about 100 km wide. (NASA)

FIGURE 15-10 Saturn from *Voyager 2* *Voyager 2* took this picture when the spacecraft was 34 million kilometers from the planet. Faint belts and zones are clearly visible. (NASA)

15-5 Saturn's atmosphere extends to a greater depth and has higher wind speeds than Jupiter's atmosphere

Earth-based spectroscopic observations along with data from spacecraft confirm that Saturn has a hydrogen-rich atmosphere much like Jupiter's, with trace amounts of methane (CH_4), ammonia (NH_3), and water vapor (H_2O). These compounds are the simplest combinations of hydrogen with carbon, nitrogen, and oxygen. Like Jupiter, Saturn's atmosphere also has three distinct cloud layers: an upper layer of frozen ammonia crystals, a middle layer of crystals of ammonium hydrosulfide (NH_4SH), and a lower layer of frozen water crystals.

Although their atmospheres have similar structures and compositions, Saturn and Jupiter are by no means identical in appearance. Saturn's clouds lack the colorful contrast of Jupiter's (Figure 15-10). Nevertheless, many Earth-based photographs and most spacecraft views do show faint belts and zones (examine Figure 15-11).

On rare occasions, Earth-based observers have reported seeing markings in Saturn's clouds that last for several days. One such "white spot" was discovered near Saturn's equator in 1990. The Hubble Space Telescope photographed the spot as it grew and spread along the equator (Figure 15-12). About 20 temporary white spots have been seen on Saturn over the past two centuries.

The different appearances of Saturn and Jupiter are related to the different masses of the two planets. Jupiter's strong surface gravity compresses its three cloud layers into a range of 75 km in the planet's upper atmosphere. Because Saturn's weaker surface gravity subjects its atmosphere to less compression, the same three cloud layers are spread out over a range of nearly 300 km (Figure 15-13). The colors of Saturn's clouds are less dramatic than Jupiter's because

FIGURE 15-11 Saturn's Cloudtops from *Voyager 1*
This view of Saturn's cloudtops was taken by *Voyager 1* at a range of $1\frac{3}{4}$ million kilometers. Note that there is much less contrast between the belts and zones on Saturn than there is on Jupiter. (NASA)

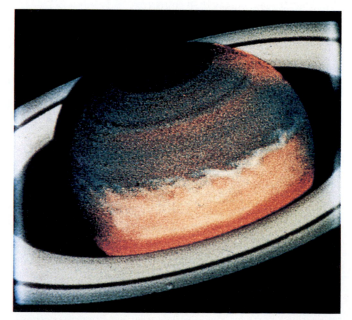

FIGURE 15-12 Saturn from HST This picture, a combination of blue and infrared views, shows Saturn's temporary "Great White Spot" seven weeks after it was discovered in September 1990. By that time, the spot had spread out along Saturn's equator. Note the prominent waves along the northern edge of the spot. (NASA; ESA)

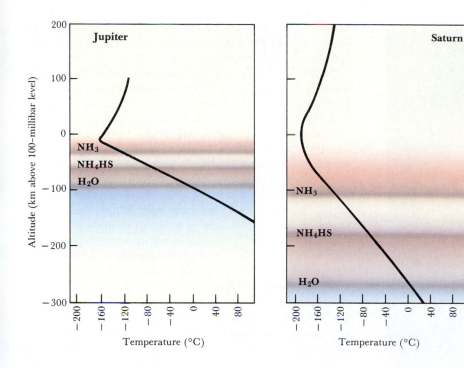

FIGURE 15-13 **The Temperature Profiles of Jupiter and Saturn** The structures of the upper atmospheres of Jupiter and Saturn are displayed on these graphs of temperature versus depth. Note that Saturn's atmosphere is more spread out than Jupiter's, as a direct result of Saturn's weaker surface gravity. (Adapted from A. P. Ingersoll)

deeper layers are partly obscured by the hazy atmosphere above them.

By following features in the Saturnian clouds, scientists have determined the wind speeds in the planet's upper atmosphere. As in Jupiter's atmosphere, there are counterflowing easterly and westerly currents in Saturn's upper atmosphere. However, Saturn's equatorial jet is much broader—and much faster—than Jupiter's. In fact, wind speeds near Saturn's equator approach 500 m/s (1100 mph), which is approximately two-thirds the speed of sound in Saturn's atmosphere.

15-6 Saturn's internal structure is quite similar to Jupiter's and includes a layer of liquid metallic hydrogen

Although Saturn's atmosphere has a composition and structure similar to Jupiter's, there are significant differences between the two planets. The average density of Saturn—690 kg/m³— is only about half that of Jupiter. Such a low density means that Saturn is composed mostly of hydrogen and helium.

Saturn is even more oblate than Jupiter (examine Figure 15-10). Saturn's equatorial diameter is about 10% larger than its polar diameter—that is, its oblateness is about 0.11. As explained in Chapter 13, the oblateness of a planet is directly related to its speed of rotation and the degree to which its mass is concentrated near its center. Because Saturn rotates a little more slowly than Jupiter does (a solar day near Saturn's equator is about 24 minutes longer than a solar day near Jupiter's equator), the greater oblateness of Saturn can-

not be caused by faster rotation. Consequently, Saturn's oblateness must result from a greater concentration of mass at its center. This concentration cannot be caused by greater compression, because Saturn is smaller and less dense than Jupiter. We may conclude, therefore, that Saturn's rocky core must be larger and more massive than Jupiter's. Detailed calculations suggest that about 26% of Saturn's mass is contained in its rocky core, whereas Jupiter's core contains only about 4% of its entire mass.

Saturn's massive core and low average density means that its density must decrease rapidly outward from the core toward the surface. This suggests that, in a general way, Saturn resembles Jupiter in structure: a solid, rocky core surrounded by a mantle of liquid metallic hydrogen, which in turn is surrounded by a layer of liquid and gaseous molecular hydrogen (Figure 15-14).

Saturn's mantle of liquid metallic hydrogen, like Jupiter's, is thought to be the source of the planet's magnetic field. Saturn's magnetic field turns out to be somewhat weaker than Jupiter's, probably because of Saturn's slightly slower rotation and much smaller volume of liquid metallic hydrogen.

Saturn's magnetosphere contains far fewer charged particles than Jupiter's. There are two reasons for this deficiency. First, Saturn lacks a continuous source of particles, like the volcanoes of Io that dump one ton of material into Jupiter's magnetosphere every second. Second, many charged particles are absorbed by the rock and ice fragments in Saturn's rings, thus depleting the concentration of particles in the inner magnetosphere. The charged particles that do exist in Saturn's magnetosphere are concentrated in radiation belts similar to the Van Allen belts in the Earth's magnetosphere.

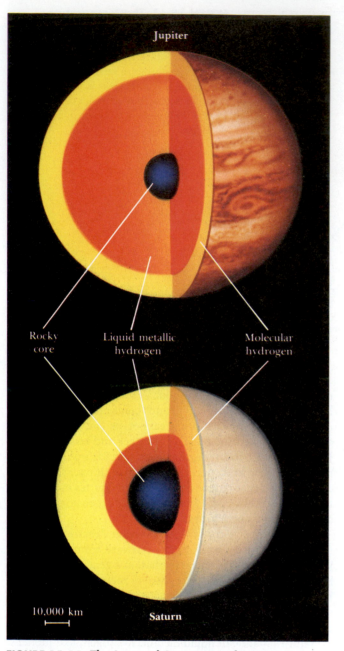

Jupiter

Rocky core   Liquid metallic hydrogen   Molecular hydrogen

10,000 km

Saturn

FIGURE 15-14 The Internal Structures of Saturn and Jupiter There are three distinct layers in Saturn's interior, as in Jupiter's. Each planet's rocky core is surrounded by a layer of liquid metallic hydrogen, which in turn is enveloped in a thick layer of liquid and gaseous molecular hydrogen. Both diagrams are drawn to the same scale.

15-7   Like Jupiter, Saturn emits more radiation than it receives from the Sun

Both Jupiter and Saturn have internal sources of energy. Each planet radiates more energy than it receives as sunlight. As we saw in Chapter 13, Jupiter's internal heat is thought to be the residual energy trapped as the planet accreted from the

original solar nebula 4.5 billion years ago. Jupiter has been slowly cooling off ever since, as this energy escapes in the form of infrared radiation.

Saturn is both smaller and less massive than Jupiter. One would thus expect Saturn to have cooled more rapidly than Jupiter and hence to emit less energy today. But in fact Saturn radiates about $2\frac{1}{2}$ times as much energy as it receives from the Sun, whereas Jupiter emits only about $1\frac{1}{2}$ times the heat it absorbs from sunlight. Why does Saturn emit so much more heat than does Jupiter?

A second puzzle is Saturn's apparent deficiency of helium. Before the Voyager flybys in the 1980s, astronomers suspected both Jupiter and Saturn to have compositions similar to that of the original solar nebula, which is assumed to be the same as the Sun's composition. Each of these giant planets is both massive enough and cool enough to have retained all the gases that originally accreted from the solar nebula. The Voyager flybys did confirm that Jupiter has an abundance of elements (82% hydrogen, 17% helium, and 1% all other elements, by weight) that is similar to the solar abundance. Surprisingly, however, the Voyager spacecraft reported that Saturn's atmosphere has less helium than expected. The chemical composition of Saturn's atmosphere is, by weight, 88% hydrogen, 11% helium, and 1% all other elements.

Edwin E. Salpeter of Cornell University and David J. Stevenson of Caltech offered a brilliant hypothesis that explains both Saturn's excess heat output and its helium deficiency. According to this theory, Saturn did indeed cool more rapidly than Jupiter, triggering a process analogous to the development of a rainstorm here on Earth. When the air is cool enough, humidity in the Earth's atmosphere condenses into raindrops, which fall to the ground. On Saturn, though, it is helium droplets that rain downward through the planet's atmosphere toward its core. As the helium droplets descend, their gravitational energy is converted into thermal energy (heat) that eventually escapes from Saturn's surface.

Helium is deficient in Saturn's upper atmosphere simply because it has fallen farther down into the planet. The precipitation of helium from Saturn's clouds is calculated to have begun 2 billion years ago. The energy released adequately accounts for the extra heat radiated by Saturn since that time. Similar calculations for Jupiter indicate that it is only now reaching the stage where a significant amount of helium precipitation can begin in its outer layers. Saturn has therefore given us important clues about the probable course of Jupiter's future evolution.

15-8   Titan has a thick, opaque atmosphere rich in methane, nitrogen, and hydrocarbons

Saturn's largest moon, Titan, was discovered by Christian Huygens in 1655, the same year that he proposed that Saturn has rings. By the early 1900s, several scientists had begun to suspect that Titan might have an atmosphere, because it is cool enough and massive enough to retain heavy gases. These

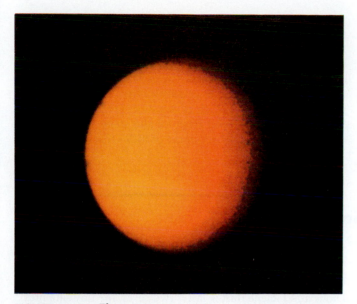

FIGURE 15-15 Titan This view of Titan was taken by *Voyager* 2 from a distance of $4\frac{1}{2}$ million kilometers. Hardly any features are visible in the thick, unbroken haze that surrounds this large satellite. The main haze layer is located nearly 300 kilometers above Titan's surface. (NASA)

suspicions were confirmed in 1944, when the American astronomer Gerard P. Kuiper discovered the spectral lines of methane in sunlight reflected from Titan. Titan is now known to be the only satellite in the solar system to have an appreciable atmosphere.

Because of its atmosphere, Titan was a primary target for the Voyager missions. To everyone's chagrin, however, the Voyagers spent hour after precious hour sending back featureless images (Figure 15-15). Titan's cloud cover is so thick that it blocks any view of the surface. The haze surrounding Titan is so dense that little sunlight penetrates to the ground; its surface must be a dark, gloomy place.

Titan's thick atmosphere distinguishes it from all other satellites in the solar system. The atmospheric pressure at Titan's surface is 1.6 bars, which is 60% greater than the atmospheric pressure at sea level on Earth, even though Titan's surface gravity is weaker than the Earth's. Thus, Titan must have considerably more gas in its atmosphere than does Earth. Indeed, calculations show that about ten times more gas lies above each square meter of Titan's surface than lies above each square meter of Earth's.

Voyager data suggest that roughly 90% of Titan's atmosphere is nitrogen. Most of this nitrogen probably came from ammonia (NH_3), which is easily broken into hydrogen and nitrogen atoms by the Sun's ultraviolet radiation. Because Titan's gravity is too weak to retain hydrogen, it escapes into space, leaving behind ample nitrogen.

The second-most-abundant gas on Titan is methane, a major component of "natural gas," which is used as a fuel on Earth. The interaction of sunlight with methane induces chemical reactions that produce a variety of other carbon–hydrogen compounds, or **hydrocarbons.** For example, the Voyager infrared spectrometers detected small amounts of ethane (C_2H_6), acetylene (C_2H_2), ethylene (C_2H_4), and propane (C_3H_8) in Titan's atmosphere. Ethane is the most abundant of these by-products. It condenses into droplets as it is produced and falls to Titan's surface to form a liquid. Over Titan's entire history, enough ethane may have been produced to create rivers and lakes.

Furthermore, nitrogen combines with these hydrocarbons to produce other compounds, such as hydrogen cyanide (HCN). Although hydrogen cyanide is a poison, some of the other nitrogen compounds formed are the building blocks of the organic molecules on which life is based. There is little reason to suspect that life exists on Titan, however; its surface temperature of 95 K ($-178°C = -288°F$) is prohibitively cold. Nevertheless, a more detailed study of the chemistry of Titan may shed light on the origins of life on Earth.

Some molecules can join together in long, repeating molecular chains to form substances called **polymers.** Many of the hydrocarbons and carbon–nitrogen compounds in Titan's atmosphere can form such polymers. Droplets of some polymers remain suspended in the atmosphere to form the kind of mixture called an **aerosol,** but the heavier polymer particles settle down onto Titan's surface. As a result, any land that protrudes above Titan's ethane oceans is probably covered with a thick layer of sticky, tarlike goo.

15-9 The features of the icy surfaces of Saturn's six moderate-sized moons provide clues to their histories

Eighteen moons are known to orbit Saturn (Box 15-2). They can be placed in three groups according to size. First, planet-sized Titan is one of the terrestrial worlds of our solar system. With its diameter of 5150 km, Titan is intermediate in size between Mercury and Mars. Second, six moderate-sized satellites were all discovered before 1800. These six satellites have properties and diameters that lead us to group them in pairs: Mimas and Enceladus (400 to 500 km in diameter), Tethys and Dione (roughly 1000 km in diameter), and Rhea and Iapetus (about 1500 km in diameter). Third, several tiny moons may be either captured asteroids (as in the case of Phoebe, Saturn's outermost satellite) or jagged fragments of ice and rock produced by impacts and collisions. The shepherd satellites, for instance, may be such fragments.

Saturn's six moderate-sized satellites share several characteristics. All six circle the planet along **regular orbits,** which is to say, in the plane of Saturn's equator and in the direction of the planet's rotation. These six satellites also exhibit synchronous rotation. Finally, all six have average densities between 1200 and 1400 kg/m^3, a value consistent with a composition largely of ice.

Mimas, the innermost of the six middle-sized moons, is heavily cratered. In fact, one of the craters (Figure 15-16) is so large that the impact forming it must have come close to shattering Mimas into fragments.

In contrast to Mimas's heavily cratered surface, its neighbor and near twin, Enceladus, has extensive crater-free regions (Figure 15-17). These areas apparently have been resurfaced within the past 100 million years by some form of geological activity. Furthermore, Enceladus is the most highly reflective large object in our solar system, even more reflective than newly fallen snow. Its cover of extremely pure ice is free of the rock and dust thought to explain the lower albedos of the other icy satellites.

Some astronomers theorize that Enceladus, like Jupiter's moon Io, is heated by tidal forces. The satellite Dione orbits Saturn in 65.7 hours, almost exactly twice Enceladus's orbital period of 32.9 hours. Enceladus, therefore, is repeatedly caught in a tidal tug-of-war between Saturn and Dione. This tidal flexing may melt Enceladus's interior, forming water geysers on its surface, much as volcanoes are produced on Io. Such geysers may deposit fresh ice and snow on the surface and may also be the source of the particles that form the thin

---

BOX 15-2

Saturn's Satellites

The table below lists some facts about Saturn's 18 known satellites. All the larger satellites are spherical, and their diameters are listed in the "Size" column. Because the smaller satellites are not spheres, three dimensions—width, length, and height—are given for each of these satellites. The masses, and hence the densities, of the smaller satellites are not yet known.

The F ring shepherds are Prometheus and Pandora.

Janus and Epimetheus are called **coorbital satellites,** because they move in almost the same orbit. Tethys, Calypso, and Telesto are also coorbital satellites, as are Dione and Helene. In the latter two cases, the tiny satellites occupy specific locations along the orbits of the larger moons, where a balance exists between the gravitational pulls of Saturn and the larger moon. These locations, called the Lagrangian points, are discussed in Box 17-1.

	Distance from center of Saturn		Orbital period	Size	Density
	(km)	(Saturn radii)	(days)	(km)	(kg/m³)
Pan	133,570	2.22	0.573	20	——
Atlas	137,640	2.28	0.602	20 × 20 × 40	——
Prometheus	139,350	2.31	0.613	140 × 100 × 80	——
Pandora	141,700	2.35	0.629	110 × 90 × 70	——
Epimetheus	151,422	2.51	0.694	140 × 120 × 100	——
Janus	151,472	2.51	0.695	220 × 200 × 160	——
Mimas	185,520	3.08	0.942	392	1400
Enceladus	238,020	3.95	1.370	500	1200
Tethys	294,660	4.89	1.888	1060	1200
Calypso	294,660	4.89	1.888	34 × 28 × 26	——
Telesto	294,660	4.89	1.888	24 × 22 × 22	——
Dione	377,400	6.26	2.737	1120	1400
Helene	377,400	6.26	2.737	36 × 32 × 30	——
Rhea	527,040	8.74	4.518	1530	1300
Titan	1,221,850	20.25	15.945	5150	1880
Hyperion	1,481,000	24.55	21.277	410 × 260 × 220	——
Iapetus	3,561,300	59.02	79.331	1460	1200
Phoebe	12,952,000	214.7	550.48^R	220	——

Note: Saturn's outermost satellite, Phoebe, moves in a retrograde direction about the planet, indicated by a superscript R with the orbital period.

FIGURE 15-16 Mimas Mimas is the smallest and innermost of Saturn's six moderate-sized satellites. This view was taken by *Voyager 1* at a range of nearly 500,000 km. The huge impact crater, named Herschel after a famous astronomer, is 130 km in diameter; Mimas itself is only 400 km in diameter. (NASA)

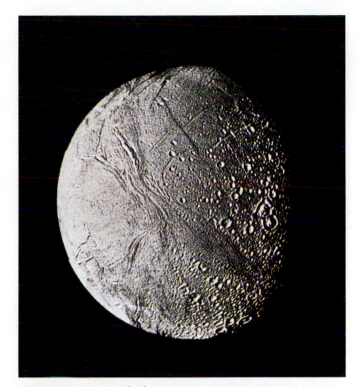

FIGURE 15-17 Enceladus This high-resolution image of Enceladus was obtained by *Voyager 2* from a distance of 191,000 km. Ice flows and cracks strongly suggest that Enceladus's surface has been subjected to recent geological activity. The youngest, crater-free ice flows are estimated to be less than 100,000 years old. (NASA)

E ring. No direct evidence for such geysers or of a molten interior was obtained by the Voyager spacecraft, however.

Tethys and Dione are the next-largest pair of Saturn's six moderate-sized satellites. Tethys, like Mimas, is heavily cratered and has one exceptionally huge impact crater. Dione is slightly larger than Tethys. The surfaces of its leading hemisphere (that is, the hemisphere pointing toward its direction of orbital motion) and trailing hemisphere are quite different. Dione's leading hemisphere is heavily cratered, but its trailing hemisphere is covered by a network of strange, wispy markings (Figure 15-18). These wisps may be troughs and valleys in the icy surface, or they may be fresh deposits of frozen water formed by outgassings from Dione's interior.

Rhea and Iapetus are the largest and most distant of Saturn's moderate-sized moons. Rhea resembles Dione in that its trailing hemisphere is covered by a network of wispy markings and its leading hemisphere is heavily cratered (Figure 15-19). Whatever process produced the wisps on the one side of Dione has apparently also been active on the corresponding side of Rhea.

Iapetus is the most distant of Saturn's moderate-sized satellites, orbiting $3\frac{1}{2}$ million kilometers from the planet. Iapetus was known to be unusual ever since its discovery, because its brightness varies greatly as it orbits Saturn. Pictures from the Voyager spacecraft showed extreme differences between the leading and trailing hemispheres of Iapetus (Figure 15-20). The leading hemisphere is as black as asphalt, but the trailing hemisphere is highly reflective, like the surfaces of the other moderate-sized moons.

FIGURE 15-18 Dione Bright, wispy streaks crisscross Dione's trailing hemisphere. (A similar network of light-colored wisps covers the trailing hemisphere of Rhea.) The nature and cause of these streaks are not known. This view of Dione was obtained by *Voyager 1* from a distance of 695,000 km. (NASA)

FIGURE 15-19 Rhea The leading hemisphere of Rhea is heavily cratered, like the leading hemisphere of Dione. This view was taken by *Voyager 1* at a distance of 128,000 km. Surface features as small as 3 km across are visible. (NASA)

The dark material covering Iapetus's leading hemisphere may have come from Phoebe, the most distant of Saturn's known satellites. Phoebe moves in a retrograde direction along an orbit tilted well away from the plane of Saturn's equator. This motion would lead us to suspect that Phoebe is a captured asteroid. Its dark surface does indeed resemble the surfaces of a particular class of carbon-rich asteroids. Bits

FIGURE 15-20 Iapetus *Voyager 2* took this photograph of Iapetus from a distance of 1.1 million kilometers. The leading hemisphere of Iapetus is extremely dark. One theory suggests that this hemisphere may have been coated with material drifting inward from Saturn's outermost satellite, Phoebe. (NASA)

and pieces of this dark charcoal-like material drifting toward Saturn may have been swept up onto the leading hemisphere of Iapetus.

The Voyager spacecraft revealed many new and unsuspected facts about Saturn's complex system of rings and satellites. There is much yet to be learned about the Saturnian system, information that should increase our understanding of the creation and evolution of other planets, including the Earth. The giant moon Titan may even harbor important clues about the origin of life on Earth. Scientists are designing a new spacecraft, called *Cassini,* for a return mission to Saturn and Titan (Figure 15-21). *Cassini* will carry a cone-shaped probe that will descend through Titan's dense atmosphere and plop onto its surface—be it solid, goo, or liquid. After dispatching the probe, *Cassini* will go into orbit about Saturn to conduct long-term observations. The spacecraft could be launched toward Saturn as early as 1996 and arrive at the planet $6\frac{1}{2}$ years later, but delays in funding may postpone the mission.

FIGURE 15-21 Cassini If funding is approved, this spacecraft will explore Saturn and its largest moon, Titan, early next century. After dropping a probe into Titan's thick atmosphere, *Cassini* will go into orbit about Saturn to perform long-term observations of the planet. (NASA)

KEY WORDS

Terms preceded by an asterisk are discussed in the boxes.

A ring	B ring	Cassini division	D ring
aerosol	C ring	*coorbital satellites	E ring

Encke division   hydrocarbon   ringlets   shepherd satellite
F ring   polymer   Roche limit   tidal force
G ring   regular orbit   scattering (of light)

KEY IDEAS

- Saturn is circled by a system of thin, broad rings lying in the plane of the planet's equator. This system is tilted away from the plane of Saturn's orbit, which causes the rings to be seen at various angles by an Earth-based observer over the course of a Saturnian year.

  Three major, broad rings can be seen from Earth. The faint C ring lies nearest Saturn. Just outside it is the much brighter B ring, then a dark gap called the Cassini division, and then the moderately bright A ring. Other, fainter rings were observed by the Voyager spacecraft.

  The principal rings of Saturn are composed of numerous fragments of ice and ice-coated rock ranging in size from a few micrometers to around 10 m.

  The rings exist inside the Roche limit of Saturn, where disruptive tidal forces are stronger than the gravitational forces attracting the fragments to each other.

  Each of the major rings is composed of a great many narrow ringlets.

  The faint, narrow F ring, which is just outside the A ring, has an intricately braided structure of unknown origin.

  Some of the ring boundaries are produced by so-called shepherd satellites, whose gravitational pull restricts the orbits of the ring fragments.

- Saturn's internal structure and atmosphere are similar to those of Jupiter, but Saturn has a larger rocky core and a thinner mantle of liquid metallic hydrogen; therefore Saturn has a more oblate shape and a weaker magnetic field.

  Saturn's atmosphere contains less helium than does Jupiter's. This lower abundance may be the result of precipitation of helium downward into the planet.

  If helium rain does fall, the resulting conversion of gravitational into thermal energy would account for Saturn's surprisingly strong heat output, which is $2\frac{1}{2}$ times as great as the incoming solar energy.

  Saturn has belts, zones, and ovals like those of Jupiter.

  The cloud layers in Saturn's atmosphere are spread out over a greater range of altitude than those of Jupiter.

- The largest Saturnian satellite, Titan, is a terrestrial world with a dense nitrogen atmosphere. A variety of hydrocarbons are produced there by the interaction of sunlight with methane. These compounds form an aerosol layer in Titan's atmosphere and possibly cover some of its surface with lakes of ethane.

- Six moderate-sized moons circle Saturn in regular orbits:

Mimas, Enceladus, Tethys, Dione, Rhea, and Iapetus. They are probably composed largely of ice, but their surface features and histories vary significantly.

- Saturn has more than a dozen much smaller satellites that may be captured asteroids.

REVIEW QUESTIONS

1. Is there any way we can infer from naked-eye observations that Saturn is the most distant of the planets visible without a telescope? Explain.

2. Why do you suppose there are fewer transits, eclipses, and occultations of the Saturnian satellites than of the Galilean satellites?

3. Describe the structure of Saturn's rings. What evidence is there that ring particles do not migrate significantly between ringlets?

4. It has been claimed that Saturn would float if one had a large enough bathtub. Using the mass and size of Saturn given in Box 15-1, confirm that the planet's average density is 690 kg/m^3, and comment on this claim.

5. Compare the atmospheres of Jupiter and Saturn. Why does Saturn's atmosphere look "washed out" compared to Jupiter's?

6. Compare the interiors of Jupiter and Saturn. Why does Saturn have a larger core than Jupiter, even though Saturn is less massive?

7. Explain why Saturn is more oblate than Jupiter, even though Saturn rotates more slowly.

8. Compare the internal sources of energy in Jupiter and Saturn that cause both planets to emit more energy than they receive from the Sun in the form of sunlight.

9. Why is the term "shepherd satellite" appropriate for the objects so named? Explain how a shepherd satellite operates.

10. Describe Titan's atmosphere. What effect has the Sun's ultraviolet radiation had on Titan's atmosphere?

11. Why do astronomers suspect that there may be liquid ethane on Titan?

ADVANCED QUESTIONS

Tips and tools . . .

The small-angle formula is given in Box 1-1. Saturn's satellites, whose orbital parameters are given in Box 15-2, obey Kepler's third law. A discussion of the retention of an atmosphere is found in Box 7-2.

12. It is well known that Cassini's division involves a 2-to-1 resonance with Mimas. Does the location of the Encke division—133,500 km from Saturn's center—correspond to a resonance with one of the other satellites? If so, which one?

13. Using the answer to Question 14 in Chapter 14, calculate the size of the smallest feature that can be seen in HST images of Saturn like Figure 15-12.

14. Computer processing of images from the Hubble Space Telescope can yield a resolving power of 0.05 arc sec. Is this sufficient to produce resolved images of any of Saturn's satellites? If so, which ones?

15. Find the escape velocity on Titan. Assuming an average atmospheric temperature comparable to that of Saturn, what is the limiting molecular weight of gases that could be retained by Titan's gravity?

16. Consult such magazines as *Sky & Telescope* or *Science News* to learn the current status of the Cassini mission. Has the U.S. Congress approved funds for the mission? What is the estimated launch date? When will *Cassini* arrive at Saturn?

17. Use Kepler's third law to calculate the orbital speeds of particles at the outer edge of the A ring and at the inner edge of the B ring. Use your answers to calculate the size of the Doppler shift of sunlight (wavelength $\cong$ 500 nm) reflected from Saturn's brightest rings.

18. As seen by Earth-based observers, the intervals between successive edge-on presentations of Saturn's rings alternate between 13 years 9 months and 15 years 9 months. Why do you think these two intervals are not equal?

DISCUSSION QUESTIONS

19. Compare and contrast the internal structures and atmospheres of Jupiter and Saturn. Wherever possible, describe the reasons for these differences or similarities in terms of such physical parameters as their mass, chemical composition, and surface gravity.

20. Comment on the suggestion that Titan may harbor life forms.

21. NASA, the Jet Propulsion Laboratory, and the European Space Agency are currently planning a mission to Saturn that will place the Cassini spacecraft in orbit about the planet. In your opinion, what issues should be examined, what data should be collected, and what kinds of questions might such a mission answer?

OBSERVING PROJECTS

22. View Saturn through a small telescope. Make a sketch of what you see. Estimate the angle at which the rings are tilted to your line of sight. Can you see the Cassini division? Can you see any belts or zones in Saturn's clouds? Is there a faint, starlike object near Saturn that might be Titan? What observations could you perform to test whether the starlike object is a Saturnian satellite?

23. If you have access to a moderately large telescope, make arrangements to observe several of Saturn's satellites. Consult the section entitled "Satellites of Saturn" in the current *Astronomical Almanac,* which includes a diagram showing the orbits of Mimas, Enceladus, Tethys, Dione, Rhea, Titan, and Hyperion. Plan your observing session by looking up the dates and times of the most recent greatest eastern elongations of the various satellites. You will have to convert from universal time (UT), which is the same as Greenwich mean time, to your local time zone. Then, using the tick marks along the orbits in the diagram, estimate the positions of the satellites relative to Saturn at the time you will be at the telescope. At the telescope, you should have no trouble identifying Titan, which reaches a magnitude of +8 at opposition. Tethys, Dione, and Rhea should be the next easiest satellites to find because they each have a magnitude of +10 at opposition. Can you confidently identify any of the other satellites?

FOR FURTHER READING

Beatty, J. K. "Report on the Voyager Encounters with Saturn." *Sky & Telescope,* January, October, and November 1981. This series of articles boasts a magnificent collection of the best images from the Voyager flybys of Saturn.

Gore, R. "Saturn: Riddle of the Rings." *National Geographic,* July 1981. A discussion of some of the puzzling aspects of Saturn's rings is lavishly illustrated in the tradition of this famous periodical.

Ingersoll, A. "Jupiter and Saturn." *Scientific American,* December 1981. Written shortly after the Voyager flybys, this enlightening article compares and contrasts the two largest planets.

Owen, T. "Titan." In Beatty, J. K., and others, eds., *The New Solar System.* 3rd ed. Sky Publishing and Cambridge University Press, 1990. This brief, six-page chapter presents the latest ideas about Titan's atmosphere and surface.

Pollack, J., and Cuzzi, J. "Rings in the Solar System." *Scientific American,* November 1981. This excellent article on the structure and evolution of planetary rings focuses primarily on Saturn, but also discusses the rings around Jupiter and Uranus.

Soderblom, L., and Johnson, T. "The Moons of Saturn." *Scientific American,* January 1982. This article presents a coherent, comprehensive survey of Saturn's moons as revealed by the Voyager spacecraft.

Washburn, M. *Distant Encounters:* Voyagers 1 *and* 2 *at Jupiter and Saturn.* Harcourt, Brace, Jovanovich, 1983. This eloquent summary of the Voyager missions, written by a journalist, is filled with awe and enthusiasm.

THE OUTER WORLDS

URANUS, EARTH, AND NEPTUNE These images of Uranus, the Earth, and Neptune are to the same scale. Uranus and Neptune are quite similar in mass, size, and chemical composition. Both planets are surrounded by thin, dark rings, quite unlike Saturn's, which are broad and bright. Almost everything we know about the outer two Jovian planets was gleaned from *Voyager 2*, which flew past Uranus in 1986 and Neptune in 1989. (NASA)

SHROUDED IN MYSTERY because of their remoteness, Uranus and Neptune were unveiled by *Voyager 2*. Surrounded by a system of small moons and thin, dark rings, Uranus is tipped on its side so that its axis of rotation lies nearly in its orbital plane. This remarkable orientation suggests that Uranus may have been the victim of a staggering impact by a massive planetesimal. Neptune is a bluish world with markings that give it a Jupiterlike appearance. Neptune even has a huge storm reminiscent of the Great Red Spot. Most intriguing of all is Neptune's giant moon, Triton. Nearly devoid of impact craters, this geologically active world is covered with puzzling features, including frozen lakes and geysers that squirt nitrogen-rich vapors. Triton's retrograde orbit suggests that this strange world may have been gravitationally captured by Neptune. Pluto, which shares many of Triton's general characteristics, may be just one of a thousand small, icy worlds that orbit the Sun at the outskirts of the solar system.

At first glance, Uranus and Neptune appear to be twins. They have nearly the same size, the same mass, the same density, and the same chemical composition. When *Voyager 2* flew past both planets in the late 1980s, however, astronomers discovered that Uranus and Neptune are really quite different, despite their superficial similarities. The brief glimpses of Uranus and Neptune provided by *Voyager 2* gave us profound insights into these dim and distant worlds.

16-1 Uranus was discovered by chance, but Neptune's existence was predicted by applying Newtonian mechanics

Uranus was discovered by chance on March 13, 1781, by the then-little-known astronomer William Herschel. This German-born musician emigrated to England and became fascinated by astronomy. Using a telescope that he built himself, Herschel was systematically surveying the sky when he noticed a faint, fuzzy object that he first thought to be a distant comet. By the end of 1781, however, he knew that the object had a planetlike orbit far out beyond Saturn, beyond where comets can normally be seen. Herschel had not only discovered the seventh planet from the Sun but in doing so had doubled the diameter of the known solar system.

Herschel acquired instant fame. He was appointed court astronomer to King George III and given the financial resources to pursue his scientific interests. During the next few decades, Herschel made discoveries that established him as one of the great astronomers of all time.

Although Herschel received the credit for discovering Uranus, many other astronomers had sighted this planet in earlier times but had mistaken it for a dim star. It is plotted on at least 20 star charts drawn between 1690 and 1781. At opposition, Uranus reaches a magnitude of +5.6, which means that it can just be seen with the naked eye under good observing conditions.

By the beginning of the nineteenth century, it had become painfully clear to astronomers that they could not accurately predict the orbit of Uranus using Newtonian mechanics. By 1830 the discrepancy between the planet's predicted and observed positions had become so great (2 arc min) that some scientists had begun to suspect a flaw in Newton's laws. The Astronomer Royal of Great Britain, George Airy, was one of those who thought that Newton's law of gravitation might not be accurate at great distances from the Sun.

In 1843, a 24-year-old student at Cambridge University in England began to explore an earlier suggestion that the gravitational pull of some more-distant planet was causing Uranus to deviate slightly from its predicted orbit. John Couch Adams began to calculate the orbit of this unknown object. After two years of hard work, Adams reached the conclusion that Uranus had caught up with and had passed a more distant planet. Uranus had accelerated slightly as it approached the unknown planet and then decelerated slightly as it receded from the planet. Adams predicted that the unknown planet would be found at a certain position in the constellation of Aquarius.

In October 1845, Adams submitted his calculations to Airy. The Astronomer Royal was unconvinced, and the matter was dropped. A few months later, however, the French

BOX 16-1

Uranus Data

Mean distance from Sun:	19.19 AU = 2.871×10^9 km
Maximum distance from Sun:	20.08 AU = 3.003×10^9 km
Minimum distance from Sun:	18.31 AU = 2.739×10^9 km
Mean orbital velocity:	6.8 km/s
Sidereal period:	84.01 years
Rotation period (internal):	17.24 hours
Inclination of equator to orbit:	97.9°
Inclination of orbit to ecliptic:	0.77°
Orbital eccentricity:	0.046
Diameter:	51,118 km
Diameter (Earth = 1):	4.01
Mass:	8.68×10^{25} kg
Mass (Earth = 1):	14.6
Mean density:	1290 kg/m³
Surface gravity (Earth = 1):	0.79
Escape speed:	21.3 km/s
Mean surface temperature (at cloudtops):	−216°C = −357°F = 57 K

astronomer Urbain Leverrier independently made the same calculations and came to the same conclusion as Adams. Their predicted positions for the unknown planet differed by less than 1°.

Shocked into action, Airy quickly instructed James Challis, the director of the Cambridge Observatory, to begin a thorough search. The search was hampered by the lack of an up-to-date star map at Cambridge Observatory for the constellation of Aquarius. Many uncharted stars had to be observed, then followed over many nights to see whether any of them showed a planetlike motion with respect to other stars.

Meanwhile, Leverrier wrote to Johann Gottfried Galle at the Berlin Observatory. The letter was received on September 23, 1846, and the planet was sighted that very night. Because he had excellent star charts, Galle easily located an uncharted star with the expected brightness in the predicted location.

Leverrier proposed that the planet be called Neptune. After years of debate between English and French astronomers, the credit for its discovery came to be divided equally between Adams and Leverrier.

An interesting postscript to this story is that Galileo may have in fact sighted Neptune over two hundred years before anyone else, in January 1613. Galileo had been regularly observing the motions of the four large satellites of Jupiter. Galileo's drawings of Jupiter and its satellites show a "star" less than 1 arc min from Neptune's location during those winter nights. Neptune can be as bright as magnitude +7.7, making it visible through small telescopes. Galileo even noted in his observation log that on one night this star seemed to have moved in relation to the other stars.

Even today through a large telescope, both Uranus and Neptune are dim, uninspiring sights. Each planet appears as a small, hazy, featureless disk with a faint greenish-blue tinge (Figures 16-1 and 16-2). Throughout the 1990s, Uranus and Neptune are together in the same part of the sky, moving slowly from Sagittarius into Capricornus. Both planets are at opposition in either July or August, which are therefore the best months to view these worlds with a telescope for the rest of the twentieth century. Basic data about Uranus and Neptune are found in Boxes 16-1 and 16-2.

BOX 16-2
Neptune Data

Mean distance from Sun:	30.06 AU = 4.497×10^9 km
Maximum distance from Sun:	30.35 AU = 4.541×10^9 km
Minimum distance from Sun:	29.77 AU = 4.453×10^9 km
Mean orbital velocity:	5.4 km/s
Sidereal period:	164.8 years
Rotation period (internal):	16.11 hours
Inclination of equator to orbit:	29.6°
Inclination of orbit to ecliptic:	1.77°
Diameter:	49,528 km
Diameter (Earth = 1):	3.88
Mass:	1.02×10^{26} kg
Mass (Earth = 1):	17.23
Mean density:	1640 kg/m³
Surface gravity (Earth = 1):	1.12
Escape speed:	23.3 km/s
Mean surface temperature (at cloudtops):	–216°C = –357°F = 57 K

FIGURE 16-1 Uranus Nearly 3 billion kilometers from the Sun, Uranus receives only $\frac{1}{400}$th the intensity of sunlight that we experience here on Earth. Uranus is therefore a dim, frigid world that never shines brighter than magnitude +5.6 in our skies. Even though its diameter is about four times that of Earth, Uranus always shows us a disk less than 4 arc sec across, which is roughly the size of a golf ball seen from a distance of 1 kilometer. (New Mexico State University Observatory)

FIGURE 16-2 Neptune At $4\frac{1}{2}$ billion kilometers from the Sun, Neptune's surface receives only $\frac{1}{900}$th the intensity of sunlight we receive on Earth. Neptune is therefore dimmer than Uranus; it never shines brighter than magnitude +7.7 in our skies. Although it is about the same size as Uranus, Neptune looks smaller through Earth-based telescopes simply because it is farther away. The maximum possible angular diameter of Neptune's disk is 2.2 arc sec, which is the same size as a dime seen from a distance of 1 km. (New Mexico State University Observatory)

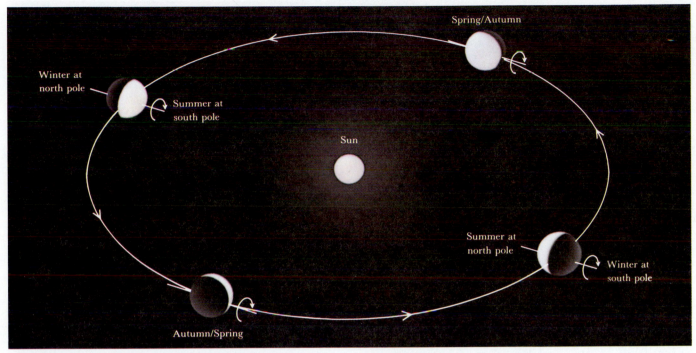

Winter at
north pole

Summer at
south pole

Spring/Autumn

Sun

Summer at
north pole

Winter at
south pole

Autumn/Spring

FIGURE 16-3 Exaggerated Seasons on Uranus Uranus's axis of rotation is tilted so steeply that it lies nearly in the plane of the planet's orbit. The seasonal changes on Uranus are thus severely exaggerated. For instance, during midsummer at Uranus's south pole, the Sun appears nearly overhead for many Earth years, while the planet's northern regions are subjected to a long, continuous winter night.

16-2 The large tilt of Uranus's axis of rotation produces extreme seasonal changes on this nearly featureless planet

Uranus is unique among the planets because its axis of rotation lies very nearly in the plane of its orbit (Figure 16-3). As a result, when Uranus moves along its 84-year orbit, its north and south poles alternately point toward or away from the Sun, producing highly exaggerated seasonal changes on the planet. For instance, near Uranus's south pole during the summer, the Sun remains above the horizon while northern latitudes are subjected to a continuous, frigid winter night. Half a Uranian year later (42 of our years), the situation is reversed.

When *Voyager 2* flew by in 1986, Uranus's south pole was aimed almost directly at the Sun. Thus the face-on, fully illuminated view of Uranus's daytime hemisphere seen in Figure 16-4 looks nearly straight down onto Uranus's south pole. The Voyager pictures showed that Uranus is remarkably featureless. With severe computer enhancement, however, some faint cloud markings do become visible. By following these markings, scientists determined that the winds on Uranus blow in the same direction in which the planet rotates, just as they do on Venus, Earth, Jupiter, and Saturn. The Uranian winds move at speeds of 40 to 160 meters per second (90 to 360 miles per hour), whereas on Earth the jet streams blow at about 50 meters per second (110 miles per hour).

FIGURE 16-4 Uranus from Voyager 2 No distinctive cloud patterns are to be seen in any of the Voyager views of Uranus. The blue-green appearance of Uranus comes from methane in the planet's atmosphere, which absorbs red wavelengths from the sunlight. Severe computer enhancement of the Voyager pictures does reveal faint cloud features, from which the rotation period of Uranus's atmosphere was determined to be about $16\frac{1}{2}$ hours. (NASA)

Miranda   Ariel   Umbriel   Titania   Oberon

FIGURE 16-5 The Principal Uranian Satellites This "family portrait" of Uranus's largest moons shows the five satellites to the same scale and correctly displays their relative reflectivities. The surfaces of all five satellites vary only slightly in color, with the average color being nearly gray. Note how dark Umbriel is. (NASA)

The Uranian winds arrange the planet's cloud features into east–west bands similar to those on Jupiter and Saturn. The equatorial region receives little sunlight, but its temperature (about 56 K = −217°C = −359°F) is about the same as that at the sunlit pole. Heat must therefore be efficiently transported from the poles to the equator. Perhaps this north–south heat transport, which is perpendicular to the wind flow, mixes and homogenizes the atmosphere to make Uranus nearly featureless.

Uranus rotates differentially, as do Jupiter and Saturn. That is, a day on Uranus is between 14.2 and 16.5 hours long, depending on the latitude. This rotation period is only for the cloudtops, however. To determine the rotation period for the underlying body of the planet, scientists looked to Uranus's magnetic field, which is presumably anchored in the planet's core, or at least in the deeper and denser layers of its atmosphere. Data from *Voyager 2* indicate that Uranus's internal period of rotation is 17.24 hours.

16-3   Uranus is orbited by satellites that bear the scars of many shattering collisions

Prior to the *Voyager 2* flyby, five moderate-sized satellites were known to orbit Uranus. All five are named after sprites and spirits in Shakespearian literature. The two largest moons, Titania and Oberon, were found by William Herschel in 1789. Another English astronomer, William Lassell, discovered Ariel and Umbriel in 1851. The smallest of the five, Miranda, was first detected by Gerard P. Kuiper in 1948 at the McDonald Observatory in Texas.

Photographs from *Voyager 2* (Figure 16-5) gave accurate information about the sizes of the five principal Uranian moons. They range in diameter from about 1600 km (1000 mi) for Titania and Oberon to only about 500 km (300 mi) for Miranda. Box 16-3 lists 10 very small Uranian satellites discovered by *Voyager 2*, most of which have diameters of less than 100 km (60 mi).

The average density of the larger Uranian moons is about 1500 kg/m^3, which is consistent with a mixture of ice and rock. Specifically, Voyager data suggest a composition of 50% water, 20% carbon- and nitrogen-based materials, and 30% rock.

All of the Uranian moons are quite dark. Umbriel is one of the darkest moons in our solar system, but craters on Oberon and Titania appear to have exposed fresh, undarkened ice there. Thus, the low albedo of the Uranian moons is probably caused by a coating of dark, sootlike material. Perhaps Umbriel simply had the luck to escape recent impacts.

Ariel exhibits mysterious volcanic features shown in Figure 16-6. Apparently some sort of thick, viscous fluid flooded a low-lying area. Is this evidence for a warmer climate there in the past? No one knows.

Miranda has a landscape unlike that of any other world. From afar, Miranda appears to consist of several large, angular, blocklike segments (Figure 16-7). From close range, Voyager's cameras revealed dramatic topography, including a cliff twice as high as Mount Everest (Figure 16-8).

FIGURE 16-6 Volcanism on Ariel? This close-up view of Ariel shows a channel that once carried a viscous fluid. As the material flowed down the channel (from left to right in this view), it partly buried a crater when it inundated a low-lying area at the right of the picture. Note that the large impact crater near the upper left corner has uncovered fresh, undarkened ice. (NASA)

FIGURE 16-7 Miranda The patchwork appearance of Miranda suggests that this satellite consists of huge chunks of rock and ice that came back together after an ancient, shattering impact by a large asteroid. The curious banded features that cover much of Miranda are paralleled by numerous closely spaced valleys and ridges, which may have formed as dense, rocky material sank toward the satellite's core. (NASA)

FIGURE 16-8 Cliffs on Miranda This close-up view of Miranda shows impact craters and parallel networks of valleys and ridges. In the upper right corner are enormous cliffs that jut upward 20 km to an elevation twice as high as Mount Everest. These cliffs may consist of low-density ice that was thrust upward as higher-density rock sank toward Miranda's core. (NASA)

Some planetary scientists theorize that Miranda's extraordinary surface may be the result of an impact that shattered the satellite, leaving a chaotic mix of rock and ice. The landscape we see today on Miranda is the result of huge, dense rocks trying to settle toward the satellite's center as blocks of less dense ice are forced upward toward the surface.

Some scientists speculate that a catastrophic collision that knocked Uranus on its side was responsible for the unique features of the planet and its satellites. Prior to the impact of an Earth-sized object, Uranus's satellites presumably orbited the upright planet in the plane of its equator, just like the regular satellites of the other Jovian planets. After the impact, tidal forces exerted by its equatorial bulge disrupted the orbits of the satellites, causing them to collide with one another. Only after many years did the remaining chunks of rock and ice settle into the highly tilted plane of Uranus's equator, where we find them today.

16-4 Uranus is circled by a system of thin, dark rings

Like the planet itself, Uranus's rings were discovered by accident. On March 10, 1977, Uranus was scheduled to occult (move in front of) a faint star, as seen from the Indian Ocean. This event afforded an excellent opportunity to measure Uranus's diameter and study its upper atmosphere. Because Uranus's position and speed along its orbit were already known, astronomers needed only to measure how long the star was hidden to deduce Uranus's size. In addition, by carefully measuring how the starlight faded when Uranus passed

in front of the star, they planned to deduce important properties of Uranus's upper atmosphere.

With these goals in mind, a team of astronomers headed by James L. Elliot of Cornell University observed the occultation from a NASA airplane equipped with a telescope. To everyone's surprise, the background star briefly blinked on and off several times just before the occultation and again immediately after it. The astronomers concluded that Uranus must be surrounded by a series of narrow rings—nine in all. During the Voyager flyby, two more rings were discovered.

Unlike Saturn's rings, Uranus's are dark and narrow, most less than 10 km wide. Typical particles in Saturn's rings are chunks of ice with the reflectivity and dimensions of snowballs, but typical particles in Uranus's rings could be compared with large lumps of coal roughly 1 m in size. The Uranian rings reflect only about 1% of the sunlight that falls on them. Two contrasting views of the Uranian rings are shown in Figures 16-9 and 16-10.

All of Uranus's rings are located less than 2 Uranian radii from the planet's center, well within the planet's Roche limit. Some sort of mechanism, possibly one involving shepherd satellites, efficiently confines particles to their narrow orbits. Voyager 2 searched for shepherd satellites but found only two. The others may be so small and black that they have simply escaped detection. Or perhaps they aren't there, and our ideas need revision.

Figure 16-10 was obtained when Voyager 2 passed into Uranus's shadow and looked back toward the Sun. This picture demonstrates that the gaps between the rings are not empty. Broad bands of tiny dust grains gleam in the sunlight.

FIGURE 16-9 The Sunlit Side of Uranus's Rings This view looks down onto the sunlit side of Uranus's rings. The nine principal rings that were discovered by Earth-based observers can be easily seen. Prominent in the upper-right corner is the ε ring, which is about 100 km wide at this location. The ring particles are so dark that they reflect only 1% of the light that falls on them. To make the rings visible, the photograph's contrast had to be enhanced so much that the dark sky appears mottled. (NASA)

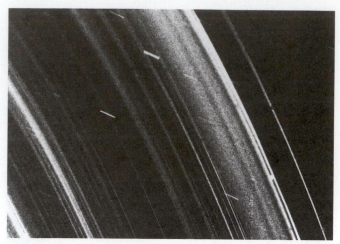

FIGURE 16-10 The Shaded Side of Uranus's Rings This view, taken when *Voyager 2* was in Uranus's shadow, looks back toward the Sun. Numerous fine dust particles between the main rings gleam in the sunlight. This dust is probably fine-grained debris from collisions between larger particles in the main rings. The short streaks are star images, blurred because of the spacecraft's motion during the exposure. (NASA)

BOX 16-3

Uranus's Satellites

Ten of Uranus's 15 known satellites were discovered during the 1986 *Voyager 2* flyby. These ten tiny moons (listed first in the following table) orbit closer to the planet than the five moderate-sized satellites visible to Earth-based observers.

	Distance from center of Uranus (km)	Orbital period (d)	Diameter or size (km)	Density (kg/m³)
Cordelia	49,750	0.335	30	——
Ophelia	53,760	0.376	30	——
Bianca	59,160	0.435	40	——
Cressida	61,770	0.464	70	——
Desdemona	62,660	0.474	60	——
Juliet	64,360	0.493	80	——
Portia	66,100	0.513	110	——
Rosalind	69,930	0.558	60	——
Belinda	75,260	0.624	70	——
Puck	86,010	0.762	150	——
Miranda	129,780	1.414	470	1350
Ariel	191,240	2.520	1160	1660
Umbriel	265,970	4.144	1170	1510
Titania	435,840	8.706	1610	1580
Oberon	582,600	13.463	1550	1520

Note: The sizes given for the 10 smallest satellites are estimates.

Collisions among ring particles are probably a continuous source of this dust. Some planetary scientists suggest that the Uranian rings are eroding before our eyes and will probably be gone in only a few hundred million years.

16-5 Neptune is a cold, bluish world with Jupiterlike atmospheric features and a Uranuslike interior

The arrival of *Voyager 2* at Neptune in August 1989 capped one of NASA's most ambitious and successful missions. Scientists were overjoyed at the detailed, close-up pictures and wealth of data sent back by the spacecraft.

At first glance, Neptune looks like a bluish Jupiter (Figure 16-11). Because methane in Neptune's atmosphere absorbs longer visible wavelengths, sunlight reflected from Neptune's clouds is depleted of reds and yellows, making the planet appear quite blue.

The most prominent feature in Neptune's atmosphere, called the **Great Dark Spot**, gives the planet its distinctly Jupiterlike appearance. It falls at about the same southern latitude on Neptune as the Great Red Spot does on Jupiter. The Great Dark Spot is roughly the same size as the Earth, measuring about 12,000 by 8000 km.

Like Jupiter's Great Red Spot, the Great Dark Spot is a high-pressure system in which the winds flow counterclock-wise. This **anticyclonic flow** on Neptune is difficult to observe from the few conspicuous whitish clouds that hover over the Great Dark Spot. These clouds are produced when winds carry methane gas into the cool, upper atmosphere, where it condenses into visible methane-ice crystals. The high elevation of Neptune's whitish clouds was confirmed in photographs, which show these clouds casting shadows on the main cloud deck (Figure 16-12).

The Jupiterlike appearance of Neptune is emphasized by faint belts and zones parallel to the planet's equator. Most prominent is a broad, darkish band at high southern latitudes. Embedded in this band is a smaller dark spot, about a third the size of the Great Dark Spot. White, wispy clouds seem to hover over the smaller dark spot just as they do over the Great Dark Spot. Temperatures in the Neptunian atmosphere range from about 60 to 50 K.

The dominant wind flow in Neptune's clouds is quite unlike that on any other Jovian planet. On Jupiter, Saturn, and Uranus, the clouds revolve around the planets more rapidly than the planets themselves rotate. In contrast, most of Neptune's cloud cover revolves about the planet more slowly than the planet rotates. This was deduced from *Voyager 2*'s discovery that Neptune's radio emission varies with a period of 16 hours 7 minutes. It is reasonable to suppose that these emissions reveal the underlying rotation of Neptune's bulk, because they are driven by the planet's magnetic field which is embedded in its interior. Because some cloud features take

FIGURE 16-11 Neptune This view from *Voyager 2* looks down on the southern hemisphere of Neptune. The Great Dark Spot, which is about the same size as the Earth, is near the center of this picture. Note the white, wispy methane clouds. Toward the lower left is a smaller, dark spot, located 54° south of Neptune's equator. (NASA)

FIGURE 16-12 Cirrus Clouds over Neptune This picture shows vertical relief in Neptune's bright, methane cloud streaks. *Voyager 2* photographed these clouds north of Neptune's equator near the terminator (the border between day and night on the planet). Note the shadows cast by the clouds onto the main cloud deck. (NASA)

FIGURE 16-13 Neptune's Rings Two main rings are easily seen in this view, which blocks out an overexposed image of Neptune. This picture also reveals a faint, inner ring. A sheet of particles, whose outer edge is located between the two main rings, extends inward toward the planet. (NASA)

as long as 18 hours to go once around the planet, Neptune's clouds fail to keep up with the underlying body of the planet.

Like Uranus, Neptune is surrounded by a system of thin, dark rings (Figure 16-13). It is so cold at Uranus and Neptune that the ring particles can retain methane ice. Scientists speculate that eons of radiation damage have converted this methane ice into darkish carbon compounds, thus accounting for the low reflectivity of the rings.

Neptune radiates more energy than it receives from the Sun, which means that the planet has an internal heat source—yet another resemblance to Jupiter. Computer models suggest that Neptune could still be radiating primordial heat, just as Jupiter does. Uranus may have a similar heat source that is masked by the greater amount of energy that Uranus absorbs from sunlight.

As we saw in previous chapters, Jupiter and Saturn are composed primarily of hydrogen and helium, both in nearly the same abundance as for the Sun. That composition, however, is incompatible with the high average densities of Uranus and Neptune. Both of these outer planets are distinctly smaller and less massive than either Jupiter or Saturn. If Uranus and Neptune also had solar abundances of the elements, the planets' smaller masses would produce less compression, and thus lower average densities than those of Jupiter and Saturn. In fact, though, Uranus and Neptune have average densities comparable to or greater than those of Jupiter or Saturn. We may therefore conclude that Uranus and Neptune contain greater proportions of the heavier elements. Each planet probably has a rocky core, composed primarily of iron and silicon, that is roughly the size of the Earth.

This picture is not what we might expect. According to the simplest model of how the solar system formed, hydrogen and helium should be more abundant as distance from the vaporizing heat of the Sun increases. Yet Uranus and Neptune contain a greater percentage of heavy elements than do Jupiter and Saturn. If we knew the origin of the heavy material in Uranus and Neptune, we would have a better understanding of the prevailing conditions far from the Sun as our solar system was being formed.

16-6 The magnetic fields of both Uranus and Neptune are oriented at unusual angles

Astronomers were quite surprised by data from *Voyager 2*'s magnetometer, sent back as the spacecraft sped past Uranus and Neptune. These data, as well as radio emissions from charged particles in their magnetospheres, showed that the magnetic fields of both Uranus and Neptune are tilted at steep angles to their axes of rotation. Uranus's **magnetic axis,** the line connecting its north and south magnetic poles, is inclined by 59° from its axis of rotation; Neptune's is tilted by 47°.

The magnetic axes of Earth, Jupiter, and Saturn are all nearly aligned with their rotation axes; the angle between their magnetic and rotation axes is 12° or less (Figure 16-14). Because of this small angle, the magnetospheres of Earth, Jupiter, and Saturn do not wobble very much as the planets rotate. As a result, these magnetospheres maintain their orientations in the solar wind. In contrast, the magnetospheres of Uranus and Neptune wobble considerably. Because of the large angle between their magnetic and rotation axes, the magnetospheres of Uranus and Neptune can change dramatically over a single rotation. For example, during half a day on Neptune, its magnetic field goes from being nearly per-

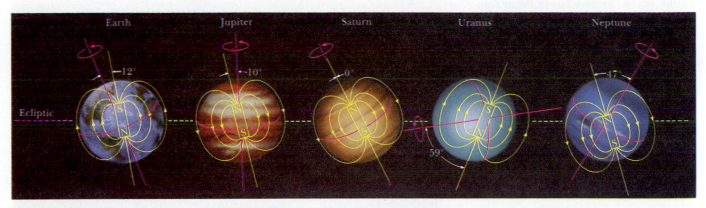

FIGURE 16-14 The Magnetic Fields of Five Planets This drawing shows how the magnetic fields of Earth, Jupiter, Saturn, Uranus, and Neptune are tilted relative to their rotation axes. Note that the magnetic fields of Uranus and Neptune are offset from the center of the planets and steeply inclined to their rotation axes. Also note that the magnetic fields of all four Jovian planets are oriented opposite to that of Earth. A compass on a Jovian planet would point southward.

pendicular to the solar wind to pointing its south magnetic pole almost directly toward the Sun. *Voyager 2* did not remain within Neptune's magnetosphere long enough to examine the effects of such a dramatic daily change.

What could have caused the misalignment on Uranus and Neptune? One possibility is that their magnetic fields might be undergoing a reversal; geological data show that the Earth's magnetic field has switched north to south and back again many times in the past. Another possibility is that the misalignments resulted from catastrophic collisions with planet-sized bodies. As we noted, the tilt of Uranus's rotation axis and its system of moons suggest that just such collisions occurred long ago. As we shall see shortly, Neptune may have gravitationally captured its largest moon, Triton. Such a near-collision would have significantly perturbed Neptune, but no one knows if that incident was responsible for offsetting Neptune's magnetic axis.

The source of the magnetic fields of Uranus and Neptune is not known. Because neither planet is massive enough to compress hydrogen to a metallic state, their magnetic fields are not generated in the same way as those of Jupiter and Saturn. Some scientists speculate that the thick atmospheres of Uranus and Neptune, although primarily composed of hydrogen and helium, also contain a slushy mixture of methane, water, and ammonia that extends all the way down to the planets' rocky cores. Perhaps electric currents in this fluid give rise to the planets' magnetic fields.

16-7 Triton is a frigid, icy world with a young surface and a tenuous atmosphere

Triton was discovered in 1846 by the English astronomer William Lassell. Even through a large telescope, this outermost giant satellite of the solar system is a faint, starlike pinpoint of light. In 1989, *Voyager 2* revealed Triton to be an ice-bound world whose puzzling surface features present a challenge to planetary scientists (Figure 16-15).

An important property of Triton, known since its discovery, is its retrograde orbit. Triton goes around Neptune opposite to the direction in which the planet rotates. No other giant satellite has such an orbit, and it is difficult to imagine how any satellite might form near a planet in an orbit opposing the direction of the planet's rotation. Only a few of the outer satellites of Jupiter and Saturn have retrograde orbits, and these bodies are probably captured asteroids. Some scientists have therefore hypothesized that Triton may have been captured long ago by Neptune's gravity.

FIGURE 16-15 Triton Features as small as 100 km across can be seen in this image of Triton, photographed by *Voyager 2* from a distance of about 5 million kilometers. The dark areas near the top of this image are part of a belt of dark markings near Triton's equator. (NASA)

FIGURE 16-16 Triton's South Polar Cap
Approximately a dozen high-resolution images were combined to produce this view of Triton's southern hemisphere. The pinkish polar cap is probably made of nitrogen frost. A notable scarcity of craters suggests that Triton's surface was either melted or flooded by icy lava after the era of bombardment that characterized the early history of the solar system. (NASA)

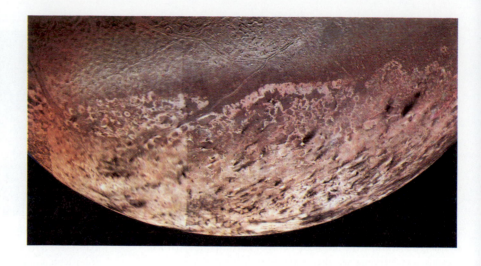

Triton is gradually spiraling in toward Neptune. Triton's orbit is decaying because of tidal interaction between the satellite and the planet. Triton raises a tidal bulge on Neptune just as our Moon distorts the Earth. In the case of the Earth–Moon system, the gravitational pull of the Earth's tidal bulge causes the Moon to spiral away from the Earth (recall Figure 9-14). However, Triton's orbit is retrograde, which makes the tidal bulge on Neptune lag behind the satellite's position in the sky. This is causing Triton to spiral gradually in toward Neptune. In roughly 100 million years, Triton will be inside Neptune's Roche limit. The giant satellite will eventually be torn to pieces by Neptune's tidal forces, and the planet will develop a spectacular ring system as rock fragments gradually spread out along Triton's former orbit.

A high-resolution image of Triton's south polar region shows very few craters (Figure 16-16). Evidently, some process obliterated the ancient scars of numerous impacts from the first billion years of the solar system. One plausible explanation is that Triton once orbited the Sun on its own but was captured by Neptune three or four billion years ago. After its capture, Triton most likely started off in a highly elliptical orbit, but today the satellite's orbit is quite circular. Triton's original elliptical orbit would have been made circular by tidal forces exerted on the satellite by Neptune's gravity. These tidal forces, which stretched and flexed Triton, could also have supplied enough energy to melt much of its interior and produce volcanism that obliterated Triton's original surface features, including craters.

The pinkish ice that constitutes Triton's south polar cap is probably a layer of nitrogen frost. When summer comes to Triton's south pole, this frost layer slowly evaporates, and many of the surface markings seen in Figure 16-16 may be the result of the northward flow of the resulting nitrogen gas.

Some of Triton's surface features resemble the long cracks seen on Europa and Ganymede. Other features are unique to Triton and are quite puzzling. For example, near the top of Figure 16-16 you can see a wrinkled terrain that resembles the skin of a cantaloupe. Triton also has a few frozen lakes like the one shown in Figure 16-17. Some scientists speculate that these lakelike features are the calderas of extinct ice

volcanoes. It is unlikely that any lava is flowing on Triton today, however, because the satellite is so very cold. Voyager instruments measured a surface temperature of 37 K (−236°C = −395°F), making Triton the coldest world we have ever visited. Nevertheless, Voyager cameras did glimpse two towering plumes of gas extending up to 8 km above the satellite's surface. Perhaps these plumes are nitrogen gas escaping through vents or fissures warmed by the feeble summer Sun.

FIGURE 16-17 A Frozen Lake on Triton? Some scientists believe that this lakelike feature is the caldera of an ice volcano. The flooded basin, about 200 km wide and 400 km long, covers an area about the size of the state of West Virginia. (NASA)

─────────────── **BOX 16-4** ───────────────

Neptune's Satellites

Eight satellites are known to orbit Neptune. Triton and Nereid were discovered from Earth; the rest were discovered during the *Voyager 2* flyby in August 1989. Three of these tiny moons are shown alongside an overexposed

Neptune in the accompanying photograph. Unlike Triton and Nereid, the orbits of all of the newly discovered moons lie nearly in the plane of Neptune's equator.

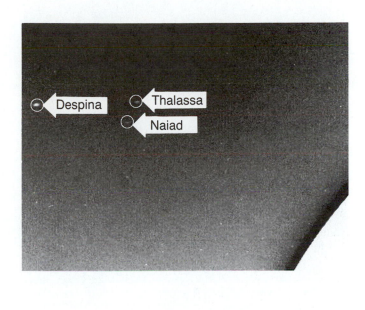

	Distance from center of Neptune (km)	Orbital period (d)	Diameter (km)
Naiad	48,230	0.296	60
Thalassa	50,070	0.312	80
Despina	52,530	0.333	150
Galatea	61,950	0.429	180
Larissa	73,550	0.554	190
Proteus	117,640	1.121	415
Triton	354,800	5.877^R	2700
Nereid	5,513,400	359.2	340

Note: Triton moves in a retrograde direction about Neptune, indicated by a superscript R with the orbital period. The semimajor axis of Nereid's highly elliptical orbit is given as the distance from Neptune's center.

───────────────────────────────

Prior to *Voyager 2,* only one other satellite was known to orbit Neptune. Nereid, which was first sighted in 1949, has the most eccentric orbit of any satellite in the solar system; its distance from Neptune varies from 1.4 million to 9.7 million kilometers. During its brief flyby, *Voyager 2* discovered six more moons (see Box 16-4), all of which escaped detection from Earth because they are small and orbit quite close to Neptune.

16-8   Pluto was discovered after a laborious search of the heavens

Speculations about a ninth planet date back to the late 1800s, when a few astronomers suggested that Neptune's orbit was being perturbed by an unknown object. Encouraged by the fame of Adams and Leverrier, several people set out to become the discoverer of "Planet X." Two Boston gentlemen, William Pickering and Percival Lowell, were prominent in this effort. Modern calculations show that there are, in fact, no significant perturbations of Neptune's orbit. It is thus not surprising that no planet was found at the positions predicted by Pickering, Lowell, and others, but the search continued.

Before he died in 1916, Percival Lowell urged that a special wide-field camera be constructed to help search for

Planet X. After many delays, the camera was finished in 1929 and installed at the Lowell Observatory in Arizona, where a young astronomer, Clyde W. Tombaugh, had joined the staff to carry on the project. On February 18, 1930, Tombaugh finally discovered the long-sought planet. It was disappointingly faint—fifteenth magnitude—and presented no discernible disk. The planet was named for Pluto, the mythological god of the underworld, whose name has Percival Lowell's initials as its first two letters. The discovery was publicly announced on March 13, 1930, the 149th anniversary of the discovery of Uranus. Two photographs showing one day's motion of Pluto appear in Figure 16-18.

Pluto's orbit about the Sun is more elliptical and more steeply inclined to the plane of the ecliptic than is the orbit of any other planet. In fact, Pluto's orbit is so eccentric that this planet is sometimes closer to the Sun than Neptune is. For example, when Pluto was at perihelion in 1989, it was more than 100 million kilometers closer to the Sun than was Neptune. (Additional data about Pluto are listed in Box 16-5.)

A solar day on Pluto lasts 6 days 9 hours 18 minutes. Although we cannot see any surface details on the planet, we can observe Pluto's magnitude varying with this period. Features on Pluto—perhaps a large frozen lake—are periodically exposed to our Earth-based view, causing these variations in brightness.

BOX 16-5

Pluto Data

Mean distance from Sun:	39.53 AU = 5.914×10^9 km
Maximum distance from Sun:	49.34 AU = 7.381×10^9 km
Minimum distance from Sun:	29.72 AU = 4.446×10^9 km
Mean orbital velocity:	4.7 km/s
Sidereal period:	248.5 years
Rotation period:	6.387 days
Inclination of equator to orbit:	122°
Inclination of orbit to ecliptic:	17.1°
Diameter:	2300 km
Diameter (Earth = 1):	0.18
Mass:	1.29×10^{22} kg
Mass (Earth = 1):	0.002
Mean density:	2030 kg/m³
Surface gravity (Earth = 1):	0.04
Escape speed:	1.1 km/s
Mean surface temperature:	−223°C = −369°F = 50 K

16-9   Pluto and its moon Charon may be typical of a thousand "ice dwarfs" that orbit the Sun at the outskirts of the solar system

In 1978, while examining some photographs of Pluto, James W. Christy of the U.S. Naval Observatory noticed that the image of the planet on a photographic plate was slightly elongated. Pluto seemed to have a lump on one side. Numer-ous other photographs of Pluto were promptly examined. Some showed this lump, but others did not. It soon became evident that the lump was actually a satellite that is occasion-ally far enough away from Pluto to show up as a slight elongation in the planet's image. Christy proposed that the newly discovered moon be christened Charon (pronounced KAR-en), after the mythical boatman who ferried souls across the River Styx to Hades, the domain ruled by Pluto. The best

FIGURE 16-18 Pluto Pluto was discovered in 1930 by searching for a dim, starlike object that moves slowly in relation to the background stars. These two photographs were taken one day apart. Pluto's moon was discovered in 1978 when an astronomer noted a "lump" protruding from one side of a greatly magnified image of the planet. (Lick Observatory)

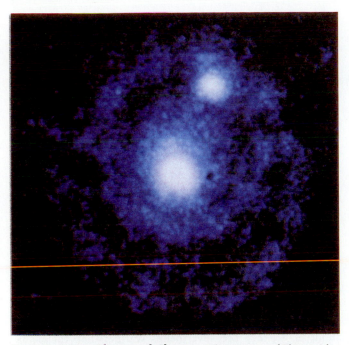

FIGURE 16-19 Pluto and Charon This picture of Pluto with its moon, Charon, was taken by the Hubble Space Telescope. Pluto and Charon are separated by only 19,700 km. The Pluto–Charon system could be called a "double planet," because these two objects resemble each other in mass and size more closely than do any other planet–satellite pair in the solar system. (NASA; ESA)

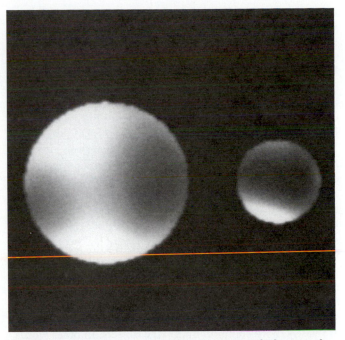

FIGURE 16-20 Computer-Modeled Image of Pluto and Charon A supercomputer was used to deduce variations on the surfaces of Pluto and Charon that could account for the variations in brightness observed as the two bodies transited and eclipsed each other. Pluto apparently has bright polar caps and a large, dark equatorial spot, turned toward Charon in this view. (Courtesy of M. Buie, K. Horne, and D. Tholen)

pictures of Pluto and Charon have recently come from the Hubble Space Telescope (Figure 16-19).

The average distance between Charon and Pluto is a scant 19,700 km—less than $\frac{1}{20}$th the distance between the Earth and our Moon. Furthermore, Charon's orbital period of 6.3874 days is the same as the rotational period of Pluto. In other words, Pluto rotates synchronously with the revolution of its satellite. As seen from the satellite-facing side of Pluto, Charon neither rises nor sets but instead seems to hover in the sky, as if perpetually suspended above the horizon.

Soon after the discovery of Charon, astronomers realized that they were about to witness an alignment that occurs only once every 124 years. From 1985 through 1990, Charon's orbital plane swept across the inner solar system, allowing astronomers to view mutual eclipses of Pluto and its moon. In this **eclipse season,** as the bodies passed in front of each other, their combined brightness diminished in ways that revealed their sizes and surface characteristics. Data from the eclipses gave Pluto's diameter as 2300 km and Charon's as 1190 km. For comparison, our Moon's diameter (3476 km) nearly equals the diameters of Pluto and Charon added together.

During this last eclipse season, astronomers carefully measured changes in brightness as Pluto and Charon passed in front of each other. Marc Buie at the Space Telescope Science Institute and his colleagues then used a supercomputer to calculate what sort of surface features would produce the observed variations. The simulations suggest that

Pluto has bright polar caps and a large, dark equatorial spot (Figure 16-20). Charon, like Triton, apparently has a bright southern polar cap.

The average densities of Pluto and Charon, at about 2000 kg/m^3, are essentially the same as that of Triton (2070 kg/m^3). All three worlds are therefore probably composed of a mixture of rock and ice. These average densities, notably higher than those of either Jupiter or Saturn, support the idea that some process favored the formation of rocklike compounds in the outer solar system.

Pluto's spectrum shows absorption lines of methane, nitrogen, and carbon monoxide, which are ices that cover the planet's surface. Exposed to a daytime temperature of 58 K, nitrogen and carbon monoxide ices turn to gas more easily than frozen methane does. For this reason, most of Pluto's tenuous atmosphere probably consists of these two gases. Charon's weaker gravity has allowed these three substances to escape, exposing a layer of water ice that covers its surface.

Pluto and Charon are remarkably like each other in mass, size, density, and quite possibly chemical composition. Throughout the rest of the solar system, planets are always many times larger and more massive than any of their satellites. The exceptional similarities between Pluto and Charon suggest to some astronomers that this binary system may have formed when Pluto collided with a similar body. Perhaps chunks of matter were stripped from the second body, leaving behind a mass, now called Charon, that was captured into orbit by Pluto's gravity. Alternatively, perhaps Charon was simply captured by Pluto during a close encounter between the two worlds.

For either of these scenarios to be feasible, there must have been many Plutolike objects in the outer regions of the solar system. Some astronomers estimate that there must have been at least a thousand Plutos in order for a collision or close encounter between two of them to have occurred at least once since the solar system formed 4.5 billion years ago.

Triton seems to offer evidence supporting the idea that many Plutos existed at the outskirts of the solar system. Recall that, in addition to being quite like Pluto and Charon, Triton has a retrograde orbit, suggesting that it was gravitationally captured by Neptune. A gravitational capture of one body by another requires a very special set of orbital circumstances; the vast majority of close encounters between passing bodies serves only to deflect them from their original paths. For one successful capture of Triton by Neptune, there must have been many near misses, which implies the existence of many Plutolike objects.

An overview of the solar system reveals that small objects are vastly more common than large objects. In other words, the number of objects in the solar system increases dramatically with decreasing mass. This trend suggests that for a thousand Plutolike worlds, there also must have been 10 to 50 icy, Earth-sized worlds in the vicinity of Uranus and Neptune. A direct hit by an object with roughly the mass of the Earth could have knocked Uranus on its side. None of these objects remain near the orbits of Neptune or Uranus today, because gravitational deflections by these two giant planets long ago catapulted all the remaining Plutolike worlds far from the Sun.

In 1992 astronomers David Jewitt and Jane Luu discovered a small object 42 AU from the Sun. Named 1992 QB1, its diameter is estimated to be only 240 km. Like Pluto, 1992 QB1 has a reddish color, possibly because frozen methane has been degraded by eons of radiation exposure. Some astronomers suspect that this object may be one of thousands of tiny worlds yet to be discovered in the outer reaches of the solar system.

Other astronomers take issue with the idea of many Plutos. There really is no complete theory for the formation of Uranus and Neptune, they argue, and so speculation about Pluto and Charon seems pointless. Observers, however, may soon have the chance to test the theory by looking for these Plutolike worlds. Attempts to find distant Plutos with ordinary telescopes is frustrating because these tiny, remote bodies do not reflect enough sunlight to be seen. They would, however, be weak sources of infrared radiation. Even though these icy objects probably have surface temperatures of only about 60 K, they are nevertheless warmer than the blank background sky. Highly sensitive, wide-angle infrared telescopes may therefore be able to detect a population of Plutos at distances of about 100 AU from the Sun. An 8-m infrared telescope currently planned for the summit of Mauna Kea in Hawaii will be an excellent tool for the search. If successful, astronomers will have to account for a new class of objects in their theories about the formation of the solar system.

KEY WORDS

anticyclonic flow	eclipse season	Great Dark Spot	magnetic axis

KEY IDEAS

• Uranus was discovered by chance, Neptune at a location predicted by applying Newtonian mechanics, and Pluto after a long search. If another planet exists beyond Pluto, it is either very small or very far away.

• Both Uranus and Neptune have a rocky core surrounded by a liquid mantle of water, ammonia, and methane, with an outer gaseous envelope composed predominantly of hydrogen and helium.

• Uranus is unique among the planets and satellites of our solar system in that its axis of rotation lies nearly in the plane of its orbit, producing greatly exaggerated seasonal changes on the planet.

Uranus's magnetic axis is inclined by 59° from its axis of rotation, while Neptune's is inclined by 47°. The magnetic and rotational axes of all the other planets are roughly parallel.

Uranus has a system of thin, dark rings and five satellites similar to the moderate-sized moons of Saturn; 10 more small Uranian satellites were discovered during the *Voyager 2* flyby.

The unusual orientation of Uranus's rotational and magnetic axes, as well as the surface features of some of its satellites (especially Miranda) suggest that Uranus may have been knocked on its side by a collision with a planetlike object early in the history of our solar system.

• Neptune's surface features, which include the Great Dark Spot, resemble those of a bluish Jupiter.

Like Uranus, Neptune is surrounded by a system of thin, dark rings. The low reflectivity of the ring particles may be due to radiation-damaged methane ice.

• Neptune has an icy, terrestrial-sized satellite, Triton, with a tenuous nitrogen atmosphere.

The scarcity of craters on Triton suggests that its surface was either molten or flooded with icy lava sometime after the era of bombardment, which left numerous craters on such worlds as Mercury and our Moon.

- Pluto and its moon Charon have roughly comparable masses and move in a highly elliptical orbit steeply inclined to the plane of the ecliptic.

Pluto, Charon, and Triton may be typical of a thousand small, icy worlds that orbited far from the Sun shortly after the solar system was formed.

REVIEW QUESTIONS

1. Could astronomers in antiquity see Uranus? If so, why was it not recognized as a planet?

2. Why do you suppose that the discovery of Neptune is rated as one of the great triumphs of science, whereas the discoveries of Uranus and Pluto are not?

3. Describe the seasons on Uranus. Why are the Uranian seasons different from those on any other planet?

4. Why is Neptune distinctly bluer than Jupiter or Saturn?

5. How does the orientation of Uranus's and Neptune's magnetic axes differ from those of other planets?

6. Briefly describe the evidence supporting the idea that Uranus was struck by a large planetlike object several billion years ago.

7. Briefly describe the evidence supporting the idea that Triton was captured by Neptune.

8. Compare the rings that surround Jupiter, Saturn, Uranus, and Neptune. Briefly discuss their similarities and differences.

9. Why is it reasonable to suppose that Neptune will someday be surrounded by a broad system of rings, perhaps similar to those that surround Saturn?

10. How can astronomers distinguish a faint solar system object like Pluto from background stars within the same field of view?

11. Describe the circumstantial evidence supporting the idea that Pluto is one of a thousand similar icy worlds that once occupied the outer regions of the solar system.

ADVANCED QUESTIONS

Tips and tools . . .

The small-angle formula is given in Box 1-1. After its optical problems are fixed, the Hubble Space Telescope will have a resolving power of about 0.1 arc sec. Problem 14 involves some geometry, and you need to remember that the circumference of a circle is π times its diameter. Newton's formula for the gravitational force between two objects was given in Chapter 4.

12. If Earth-based telescopes can resolve angles down to 0.25 arc sec, how large could an object be at Pluto's average distance from the Sun and still not present a resolvable disk?

13. Can the Hubble Space Telescope resolve the Great Dark Spot? Is it capable of seeing any features on Pluto?

14. Using the orbital data given in Box 16-1, estimate the maximum duration of a stellar occultation by Uranus.

15. At what planetary configuration is the gravitational force of Neptune on Uranus at a maximum? Calculate the fraction by which the sunward gravitational pull on Uranus is reduced by Neptune at that configuration.

16. Suppose you were standing on Pluto. Describe the motions of Charon relative to the Sun, the stars, and your own horizon. Would you ever be able to see a total eclipse of the Sun? Under what circumstances would you never see Charon?

17. Suppose you wanted to search for planets beyond Pluto. Why might it be advantageous to do your observations at infrared rather than visible wavelengths? (Use Wien's law to calculate the wavelength range best suited for your search.) Could such observations be done at an Earth-based observatory?

18. Calculate the ratio of the diameters of the Great Red Spot and the Great Dark Spot and compare it to the ratio of the diameters of Jupiter and Neptune. Does this comparison strengthen or weaken the similarities between the Great Red Spot and the Great Dark Spot?

19. Consult recent issues of *Sky & Telescope* and *Science News* to learn about the current status of 1992 QB1. Has its orbit been determined? If so, how does that orbit compare with Pluto's? Have any more objects like 1992 QB1 been found?

DISCUSSION QUESTIONS

20. Discuss the evidence presented by the outer planets that catastrophic impacts of planetlike objects occurred during the early history of our solar system.

21. Would you expect the surfaces of Pluto and Charon to be heavily cratered? Explain.

22. At NASA and the Jet Propulsion Laboratory, some scientists are discussing the possibility of placing spacecraft in orbit about Uranus and Neptune early in the twenty-first century. What kinds of data should be collected, and what questions would you like to see answered by these missions?

OBSERVING PROJECTS

23. Make arrangements to view Uranus through a telescope. Throughout the 1990s, Uranus is best seen in July and August. To help you find the planet, a star chart showing the path of Uranus against the background stars is

published each January in *Sky & Telescope.* Use that star chart at the telescope to find the planet. Are you certain that you have found Uranus? Can you see a disk? What is its color?

24. If you have access to a large telescope, make arrangements to view Neptune. Throughout the 1990s, Neptune is in Sagittarius, so the planet is best seen in July and August. Neptune is most easily found if you know its right ascension and declination, which are given in the *Astronomical Almanac.* Alternatively, you could consult the star chart published in the January issue of *Sky & Telescope.* Can you see a disk? What is its color?

FOR FURTHER READING

Bennett, J. "The Discovery of Uranus." *Sky & Telescope,* March 1981. This fascinating historical article describes Herschel's discovery of Uranus.

Berry, R. "Neptune Revealed." *Astronomy,* December 1989. This summary of Voyager results at Neptune includes an outstanding collection of photographs.

Binzel, R. "Pluto." *Scientific American,* June 1990. This superb article summarizes our current understanding of Pluto and Charon.

Brown, R., and Cruikshank, D. "The Moons of Uranus, Neptune, and Pluto." *Scientific American,* July 1985. This article, written before the Voyager flybys, contains many insights gleaned from Earth-based observations.

Cuzzi, J., and Esposito, L. "The Rings of Uranus." *Scientific American,* July 1987. The authors of this article argue that Uranus's rings are actually a fleeting phenomenon.

Elliot, J., and others. "Discovering the Rings of Uranus." *Sky & Telescope,* June 1977. The team of astronomers who discovered Uranus's rings describe their airborne observations.

Harrington, R., and Harrington, B. "The Discovery of Pluto's Moon." *Mercury,* January/February 1979. This article gives authoritative insights into the discovery of Pluto's moon.

Ingersoll, A. P. "Uranus." *Scientific American,* January 1987. This superb article summarizes the results of the *Voyager 2* flyby.

Johnson, T., and others. "The Moons of Uranus." *Scientific American,* April 1987. This beautifully illustrated article asserts that Uranus's moons had a geologically vigorous history.

Kinoshita, J. "Neptune." *Scientific American,* November 1989. Pictures of Neptune and Triton grace this article about the Voyager flyby.

Moore, P. "The Discovery of Neptune." *Mercury,* July/August, 1989. A prolific popularizer of astronomy gives a detailed history of how Neptune was found.

Tombaugh, C. "The Search for the Ninth Planet: Pluto." *Mercury,* January/February 1979. The discoverer of Pluto describes his historic observations.

INTERPLANETARY VAGABONDS

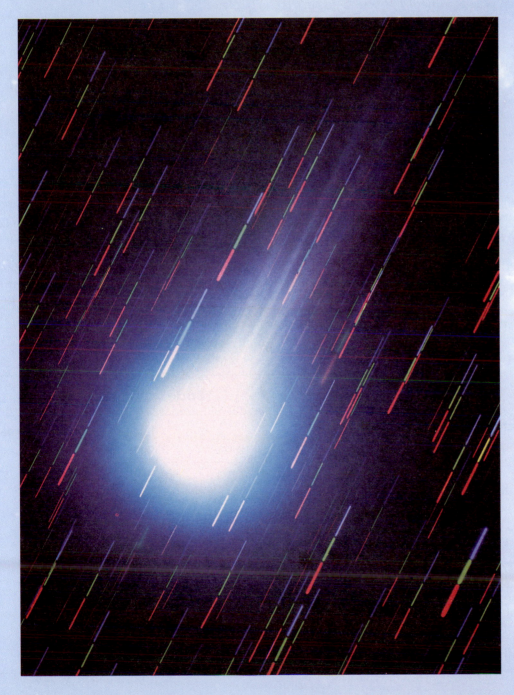

THE HEAD OF COMET HALLEY This photograph shows the bluish head of Halley's Comet as it approached the Sun in December 1985. Comet Halley orbits the Sun with an average period of 76 years along a highly elliptical path that stretches from just inside the Earth's orbit to slightly beyond the orbit of Neptune. This color photograph was constructed from three black-and-white photographs taken in rapid succession with red, blue, and green filters on the same telescope. (Anglo-Australian Observatory)

THE PLANETS are not the only objects that orbit the Sun. Asteroids, meteoroids, and comets are small but significant members of our solar system, providing valuable information about the early history of the planets. Some meteoroids are fragments of asteroids that have undergone significant melting and chemical differentiation. Others seem to have escaped such processing and thus contain important chemical clues about conditions in the solar nebula 4.5 billion years ago. Likewise, comets are dusty chunks of ice that contain frozen samples of the solar nebula. When one of these "dirty snowballs" passes near the Sun, solar radiation vaporizes some of its ice, producing a long, flowing tail. After many such passages near the Sun, a comet's ice is depleted, leaving only a swarm of dust particles that rain down on Earth in a meteor shower. Meteoroids and asteroids occasionally strike the Earth. Indeed, there is substantial evidence that an asteroid was responsible for the extinction of the dinosaurs some 65 million years ago. Even today, life on Earth is threatened by the devastation that could be wrought by a wayward asteroid.

When William Herschel discovered a new planet beyond Saturn in 1781, he proposed that it be named Georgium Sidus, in honor of King George III of Great Britain. Astronomers in other countries rejected this suggestion and adopted the name Uranus, which had been proposed by Johann Elert Bode. This young German astronomer was already far more famous, however, for popularizing a simple rule that describes the average distances of the planets from the Sun. Although Bode did not discover this rule, it is generally known today as Bode's law. Most astronomers now regard this "law" as merely a coincidence, but it did lead directly to the discovery of a large number of previously unknown objects that orbit the Sun.

17-1 Bode's law led to the discovery of asteroids between the orbits of Mars and Jupiter

Bode's rule for remembering the distances of planets from the Sun goes like this:

1. Write down the sequence of numbers 0, 3, 6, 12, 24, 48, 96, ... (Note that each number after the 3 is simply twice the preceding number.)

2. Add 4 to each number in the sequence.

3. Divide each of the resulting numbers by 10.

As shown in Table 17-1, the result is a series of numbers that corresponds remarkably well with the distances of most of the planets from the Sun measured in astronomical units.

Astronomers regarded Bode's rule as merely a useful trick for remembering the planetary distances—until Herschel's unexpected discovery of Uranus, very near the orbit for a planet beyond Saturn predicted by Bode's scheme. For a time, it seemed likely that Bode's rule might actually represent some physical property of the solar system. The numerical sequence soon became known as **Bode's law**, an unfortunate name because it is not a physical law. In fact, there is no

---

TABLE 17-1

Bode's Law

Bode's progression	Planet	Actual distance (AU)
(0 + 4)/10 = 0.4	Mercury	0.39
(3 + 4)/10 = 0.7	Venus	0.72
(6 + 4)/10 = 1.0	Earth	1.00
(12 + 4)/10 = 1.6	Mars	1.52
(24 + 4)/10 = 2.8	?	——
(48 + 4)/10 = 5.2	Jupiter	5.20
(96 + 4)/10 = 10.0	Saturn	9.54
(192 + 4)/10 = 19.6	Uranus	19.18
(384 + 4)/10 = 38.8	Neptune	30.06
(768 + 4)/10 = 77.2	Pluto	39.44

reason why this particular numerical sequence should have any physical significance. Nor was it invented by Bode. Johann Titius, a German physicist and mathematician, first published it in 1766. A better name for this numerical sequence might therefore be the "Titius-Bode rule."

Astronomers looked with new interest at the location of the "missing planet" in the sequence of Bode's law—the gap between the orbits of Mars and Jupiter (see Table 17-1). Six German astronomers who jokingly called themselves the "Celestial Police" organized an international group to begin a careful search for the missing planet. Before their search got under way, however, the anticipated announcement came from Sicily.

The Sicilian astronomer Giuseppe Piazzi had been carefully preparing a map of faint stars in the constellation of Taurus. On January 1, 1801, he noticed a dim, previously uncharted star whose position shifted slightly over the next several nights. Suspecting that he might have found the missing planet suggested by Bode's law, Piazzi excitedly wrote to Bode in Berlin.

Unfortunately, Piazzi's letter did not reach Bode until late March. By that time, Piazzi's object was too near the Sun to be seen. To make matters worse, Piazzi had made only a limited number of observations, and the best mathematical techniques of the day could not predict the object's position from so few data. Astronomers feared that Piazzi's object might have been lost.

Upon hearing of the plight of these astronomers, the brilliant young mathematician Karl Friedrich Gauss took up the challenge and developed a general method of computing orbits from only three separate measurements of right ascension and declination. In November 1801, Gauss used this method to predict that Piazzi's object would be found in the constellation of Virgo. One of the self-styled "Celestial Police," Baron Franz Xavier von Zach, sighted Piazzi's object on December 31, 1801, only a short distance from the position calculated by Gauss. At Piazzi's request, the object was named Ceres (pronounced SEE-reez), after the Roman goddess of agriculture, popular in Sicily.

Ceres orbits the Sun once every 4.6 years at an average distance of 2.77 AU, which is in remarkable agreement with the distance given by Bode's law for the missing planet. Ceres is very small, however. At opposition, its magnitude is only +7.4, and its diameter is estimated to be a scant 900 km. Ceres thus did not qualify as a full-fledged planet, and astronomers continued the search. On March 28, 1802, Heinrich Olbers discovered another faint, starlike object that moved against the background stars. He called it Pallas, after the Greek goddess of wisdom. Like Ceres, Pallas orbits the Sun every 4.6 years at an average distance of 2.77 AU. Pallas is even dimmer and smaller than Ceres, reaching a magnitude of only +8.0 at opposition and having an estimated diameter of only 600 km. Obviously, Pallas was not the missing planet either.

The discovery of two small objects with similar orbits at the distance predicted for Bode's missing planet led astronomers to suspect that the planet might have somehow broken apart or exploded. The search for other small objects went on. Only two more were found, Juno and Vesta, until the mid-1800s, by which time telescopic equipment and techniques had improved. Astronomers then began to stumble across many more such objects orbiting the Sun between the orbits of Mars and Jupiter. These objects are known today as **asteroids** or **minor planets**.

17-2   Many asteroids orbit the Sun between the orbits of Mars and Jupiter

The next major breakthrough came in 1891 when the German astronomer Max Wolf began using photographic techniques to search for asteroids. Some three hundred asteroids had been found up to that time, each painstakingly discovered by scrutinizing the skies for faint, uncharted, starlike objects whose positions move slightly from one night to the next. With photography, however, the floodgates were opened. An astronomer simply aims a camera-equipped telescope at the stars and takes a long exposure. If an asteroid happens to be in the field of view, it leaves a distinctive, blurred trail on the photographic plate, because of its movement along its orbit during the long exposure (Figure 17-1). Using this technique, Wolf alone discovered 228 asteroids.

FIGURE 17-1 Two Asteroids Asteroids can be detected by their blurred trails on time exposure photographs of the stars. The images of two asteroids are seen in this picture. Astronomers sometimes find asteroids accidentally while photographing various portions of the sky for other purposes. (Yerkes Observatory)

FIGURE 17-2 The Asteroid Belt Most asteroids orbit the Sun in a belt $1\frac{1}{2}$ AU wide between the orbits of Mars and Jupiter. The orbits of Ceres, Pallas, and Juno are indicated. The orbits of the asteroids Apollo and Icarus are also shown.

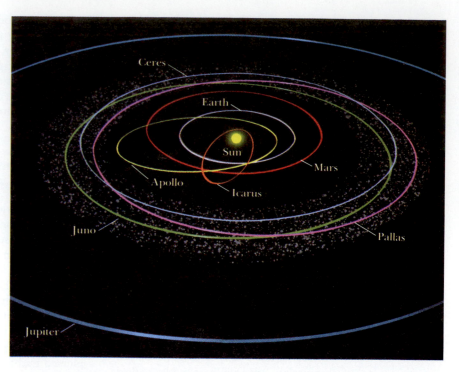

To become the official discoverer of an asteroid, you must do a lot more than merely photograph a blurred trail that matches no known orbit. You must also track the asteroid long enough to compute an accurate, reliable orbit. (Important data may also be available from colleagues, who may have inadvertently photographed your asteroid.) Then you must prove the accuracy of your orbit by locating the asteroid again on at least one succeeding opposition. At that time, an official number is assigned to your asteroid (Ceres is 1, Pallas is 2, and so forth), and you are given the privilege of naming your asteroid. The orbits of all officially discovered asteroids are published annually in the catalogue *Ephemerides of Minor Planets*.

Astronomers estimate that roughly 100,000 asteroids are bright enough to appear on Earth-based photographs. The vast majority are less than 1 km across. Like Ceres, Pallas, Vesta, and Juno, most asteroids orbit the Sun at distances between 2 and $3\frac{1}{2}$ AU. This region of our solar system between the orbits of Mars and Jupiter is called the **asteroid belt** (Figure 17-2). The asteroids whose orbits lie entirely within this region are called **belt asteroids.**

Ceres is unquestionably the largest asteroid. With a diameter of about 900 km, Ceres accounts for about 30% of the mass of all the asteroids combined. Only three asteroids —Ceres, Pallas, and Vesta—have diameters greater than 300 km. Thirty other asteroids have diameters between 200 and 300 km, and 200 more are bigger than 100 km across.

The combined matter of all the asteroids (including an estimate for those not yet officially known) would produce an object barely 1500 km in diameter, considerably smaller than anything that could be considered a missing planet. It seems more reasonable to believe that asteroids are debris left over from the formation of the solar system.

17-3 Jupiter's gravity affects the structure of the asteroid belt and captures asteroids along its orbit

Supercomputer simulations of the formation of the terrestrial planets, like the one we examined in Figure 7-18, give important insights into why asteroids rather than a planet formed between Mars and Jupiter. A typical simulation starts off with a billion planetesimals, each with a mass of 10^{15} kg, so that their combined mass equals that of the terrestrial planets. The computer simply follows these planetesimals as they collide and coalesce to form the planets. By selecting different speeds and positions for the planetesimals at the start of the simulation, a scientist can study various ways in which the inner solar system might have turned out.

If the effects of Jupiter's gravity are not included in a simulation, an Earth-sized planet usually forms in the asteroid belt, giving us five terrestrial planets instead of four. When Jupiter's gravity is added, however, no fifth terrestrial planet is created. Jupiter's strong pull disrupts the orbits of planetesimals in the asteroid belt, ejecting most of them from the solar system altogether. As a result, the asteroid belt becomes depleted before a planet has a chance to form. Because the surviving asteroids are but a fraction of those that originally orbited the Sun, astronomers say that Jupiter "cleared" the asteroid belt as the solar system formed.

In addition to clearing the asteroid belt, Jupiter affects the distribution of the remaining asteroids. In 1867 the American astronomer Daniel Kirkwood called attention to gaps in the asteroid belt. These features, today called **Kirkwood gaps,** are best seen in a histogram of asteroid orbital periods, like the one in Figure 17-3. Note the gaps at simple fractions $(\frac{1}{3}, \frac{2}{5}, \frac{3}{7},$ and $\frac{1}{2})$ of Jupiter's orbital period.

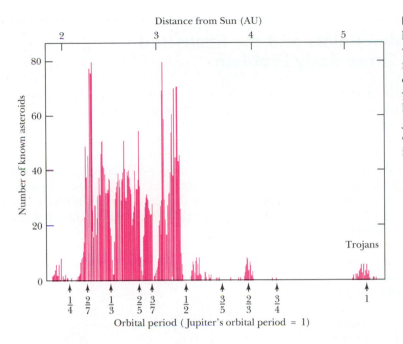

FIGURE 17-3 The Kirkwood Gaps This histogram displays the numbers of asteroids at various distances from the Sun. Notice that very few asteroids have orbits whose orbital periods correspond to such simple fractions as $\frac{1}{3}$, $\frac{2}{5}$, $\frac{3}{7}$, and $\frac{1}{2}$ of Jupiter's orbital period. Gravitational perturbations resulting from repeated alignments with Jupiter have deflected asteroids away from these orbits. The Trojan asteroids accompany Jupiter as it orbits the Sun.

To understand why the Kirkwood gaps exist, imagine a belt asteroid circling the Sun once every 5.93 years, exactly half of Jupiter's orbital period. On every second trip around the Sun, the asteroid finds itself lined up again and again between Jupiter and the Sun, always at the same location and with the same orientation. Because of these repeated alignments, Jupiter's gravity deflects the asteroid from its original 5.93-year orbit, ultimately ejecting it from the asteroid belt.

According to Kepler's third law, a period of 5.93 years corresponds to a semimajor axis of 3.28 AU. Because of Jupiter, there are no asteroids that orbit the Sun at this average distance. Similarly, there is a gap corresponding to an orbital period of one-third Jupiter's period, or 3.95 years. Additional gaps exist for other simple ratios between the periods of asteroids and Jupiter. Major divisions in Saturn's rings were created by similar gravitational effects of Saturnian moons on the icy fragments in the rings.

While Jupiter's gravitational pull depletes certain orbits in the asteroid belt, it also captures asteroids at two locations much farther from the Sun. As explained in Box 17-1, the gravitational forces of the Sun and Jupiter work together to hold asteroids at locations called the **Lagrangian points** L_4 and L_5. Point L_4 is located one-sixth of the way around Jupiter's orbit ahead of the planet, and point L_5 occupies a similar position behind the planet (Figure 17-4).

The asteroids trapped at Jupiter's Lagrangian points are called **Trojan asteroids,** named individually after heroes of the Trojan War. Approximately four dozen Trojan asteroids have been catalogued so far, and some astronomers suspect that several hundred rock fragments orbit near each Lagrangian point.

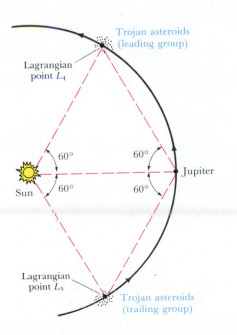

◀ **FIGURE 17-4 The Trojan Asteroids** The combined gravitational forces of Jupiter and the Sun trap asteroids at the two Lagrangian points along Jupiter's orbit. The asteroids at these locations are named after Homeric heroes of the Trojan War.

BOX 17-1

Lagrangian Points and the Restricted Three-Body Problem

In the late 1700s the great French mathematician Joseph Louis Lagrange (pronounced la-GRAHNZH) tackled the famous three-body problem: the task of predicting the motions of three objects moving freely in space under the mutual influences of their gravitational forces. Lagrange succeeded in solving a restricted form of the problem, in which one of the objects is assumed to be small enough that its gravity has no effect on the other two, more massive, objects. (Mathematicians later proved that the general three-body problem cannot be solved precisely.) Among the practical applications of a solution to the restricted three-body problem are plotting a course for a spaceship from the Earth to the Moon and calculating the path of an asteroid affected chiefly by the gravitational pulls of Jupiter and the Sun.

Lagrange discovered some interesting facts. The easiest way to display the combined gravitational fields of the two massive objects (we shall call their masses M_1 and M_2, where $M_1 > M_2$) is to draw so-called **equipotential contours**. We can think of an equipotential contour map as a sort of topographic map showing "hills" and "valleys" in the gravitational field, as shown in the accompanying diagram. Any small object in this field will feel a force pulling it in the "downhill" direction. The massive objects M_1 and M_2

(along with the entire gravitational field) revolve about the center of mass of the system, which is nearer M_1 than M_2 because $M_1 > M_2$.

Note that there are three locations along a line through M_1 and M_2 where contour lines cross. A tiny object placed at one of these locations would be balanced unsteadily at a local "flat spot" in the gravitational field. These three points are the unstable Lagrangian points L_1, L_2, and L_3 (see the diagram below on the right). These points are said

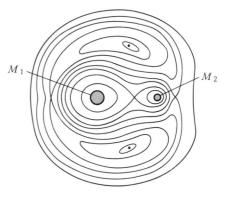

The equipotential contours

17-4   Asteroids occasionally collide with each other and with the inner planets

Some asteroids have highly elliptical orbits that bring them into the inner regions of the solar system. Asteroids that cross Mars's orbit are called **Amor asteroids** (after their prototype, Amor), and asteroids that cross Earth's orbit are called **Apollo asteroids** (after their prototype, Apollo).

It is probable that the Amor and Apollo asteroids are somehow related. Edward Anders of the University of Chicago argues that gravitational perturbations by Mars deflect Amor asteroids into Earth-crossing orbits, causing Amor asteroids to become Apollo asteroids.

Occasionally an Apollo asteroid passes quite close to Earth. Figure 17-5 shows Eros as it passed within 23 million kilometers of our planet in 1931. On June 14, 1968, Icarus passed Earth at a distance of only 6 million kilometers. One of the closest near-misses in recent history occurred on October 30, 1937, when Hermes passed Earth at a distance of 900,000 km—only a little more than twice the distance to the Moon. A similar close call occurred on March 23, 1989, when an asteroid called 1989FC passed within 800,000 km of the Earth. If this asteroid had struck the Earth, the impact

FIGURE 17-5   Eros   Like all known Apollo asteroids, Eros occasionally passes near the Earth. This photograph was taken in February 1931, when the distance between Eros and the Earth was only 23 million kilometers. The dimensions of Eros are roughly $10 \times 20 \times 30$ km. This asteroid rotates with a period of 5.27 hours. (Yerkes Observatory)

to be unstable because if the small object moves ever so slightly away from the exact Lagrangian point, it will experience a gravitational pull in the "downhill" direction, away from the Lagrangian point. An object such as an asteroid cannot therefore be permanently trapped at one of the unstable Lagrangian points.

Notice that there are two points at the bottoms of the teardrop-shaped "valleys" on either side of the line through M_1 and M_2. These are the stable Lagrangian points L_4 and

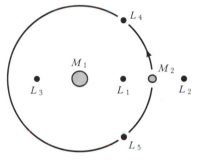

The five Lagrange points

L_5. If a small object is moved slightly away from one of these points, it will experience a gravitational pull in the "downhill" direction, back toward the Lagrangian point. A small object like an asteroid can thus be permanently trapped at one of these stable Lagrangian points. In fact, the Trojan asteroids do move in small orbits around the two stable Lagrangian points, each of which is located at a corner of an equilateral triangle, with Jupiter and the Sun at the other corners (see Figure 17-4).

The Trojan asteroids at L_4 and L_5 along Jupiter's orbit have been known for many years. (The discovery of the first one, in 1906, provided the first practical proof of Lagrange's theoretical ideas about these points.) Of course, stable Lagrangian points exist at many places in the solar system. While passing Saturn, the Voyager spacecraft discovered tiny satellites at the L_4 and L_5 points of the Saturn–Tethys and Saturn–Dione systems. Small clouds of dust-grain–sized particles have been observed at the L_4 and L_5 points of the Earth–Moon system. A group called the L_5 Society argues that the L_5 point of the Earth–Moon system would be the ideal location for a huge space station with a permanent human population. Despite careful searches, no asteroids have been found at the stable Lagrangian points of the Earth–Sun and the Saturn–Sun systems.

would have been equivalent to the explosion of a thousand 20-megaton hydrogen bombs.

During these close encounters, astronomers can examine details about the asteroids. For example, an asteroid's magnitude often varies periodically, presumably because different surfaces turn toward us as the asteroid rotates. These periodic variations in brightness therefore reveal an asteroid's rate of rotation. Typical rotation periods for asteroids are in the range of 5 to 20 hours.

These variations in brightness can also reveal an asteroid's shape. For instance, a small asteroid looks dim when seen end-on but appears brighter when seen broadside. Only the largest asteroids, like Ceres, Pallas, and Vesta, have enough gravity to pull themselves into a spherical shape. Smaller asteroids retain the odd shapes produced by previous collisions with other asteroids.

In October 1991 the Jupiter-bound *Galileo* spacecraft passed near the asteroid Gaspra and sent back close-up views (Figure 17-6). Gaspra appears to be a survivor of numerous catastrophic collisions in which a parent body was broken down into smaller and smaller pieces. Indeed, a series of grooves on Gaspra's surface suggest that the asteroid was hit with nearly enough force to smash it to bits.

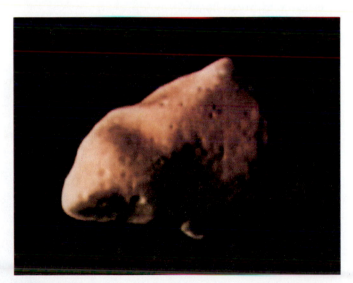

FIGURE 17-6 The Asteroid Gaspra This photograph of Gaspra was taken in October 1991, when *Galileo* passed within 1600 km of the asteroid. Gaspra's cratered, irregular shape measures 16×12 km in this view. However, its overall dimensions are an estimated $12 \times 20 \times 11$ km. Craters as small as 160 m across are visible on Gaspra's surface. (NASA)

There is ample evidence of interasteroid collisions. In 1918 the Japanese astronomer Kiyotsugu Hirayama drew attention to groups of asteroids that share nearly identical orbits. These groupings, now called **Hirayama families,** presumably resulted from fragmentation of parent asteroids. Only a high-velocity collision has enough energy to shatter an asteroid permanently and create a Hirayama family. Of the three dozen known Amor asteroids, 20 can be grouped into four Hirayama families.

A collision between kilometer-sized asteroids must be an awesome event. Typical collision velocities are estimated to be 1 to 5 km/s (2000 to 11,000 mi/h), which is more than sufficient to shatter rock. In some collisions, the resulting fragments may not have enough speed to escape from each other and thus reassemble because of their mutual gravitational attraction. Alternatively, several large fragments may end up orbiting each other. This is probably what happened to Pallas, which actually consists of a main asteroid and a large satellite.

Interasteroid collisions produce numerous chunks of rock, many of which eventually rain down on Venus, Earth, and Mars. Fortunately for us, the vast majority of these asteroid fragments, usually called **meteoroids,** are quite small. On rare occasions, however, a large fragment does collide with our planet. The result is an **impact crater,** whose diameter depends on the mass and the speed of the impinging object.

One of the most impressive, best-preserved terrestrial impact craters is the famous Barringer Crater near Winslow, Arizona (Figure 17-7). The crater measures 1.2 km across and is 200 m deep. The crater was formed 50,000 years ago when an iron-rich meteoroid measuring roughly 50 m across struck the ground with a speed estimated at 11 km/s (25,000 mi/h). The resulting blast was equal to the detonation of a 20-megaton hydrogen bomb.

Iron is one of the more abundant elements in the universe (recall Table 7-4), as well as being one of the most common

FIGURE 17-7 The Barringer Crater   An iron meteoroid measuring about 50 m across struck the ground in Arizona 50,000 years ago. The result was this impact crater, 1.2 km in diameter and 200 m deep at its center. (Meteor Crater Enterprises)

rock-forming elements, so it is not surprising that iron is an important constituent of asteroids and meteoroids. Another element, iridium, is common in iron-rich minerals but rare in ordinary rocks. Measurements of iridium in the Earth's crust can thus tell us the rate at which meteoritic material has been deposited on the Earth over the ages. Geologist Walter Alvarez and his physicist father, Luis Alvarez, from the University of California at Berkeley, made such measurements in the late 1970s.

Working at a site of exposed marine limestone in the Apennine Mountains in Italy, the Alvarez team discovered an exceptionally high abundance of iridium in a dark-colored layer of clay between limestone strata (Figure 17-8). Since

FIGURE 17-8 Iridium-Rich Clay This photograph of strata in the Apennine Mountains of Italy shows a dark-colored layer of iridium-rich clay sandwiched between white limestone (below) from the late Mesozoic era and grayish limestone (above) from the early Cenozoic era. This iridium-rich layer may be the result of an asteroid whose impact caused the extinction of the dinosaurs. The coin is the size of a U.S. quarter. (Courtesy of W. Alvarez)

this discovery in 1979, a comparable layer of iridium-rich material has been uncovered at various sites around the world. In every case, geological dating reveals that this apparently worldwide layer of iridium-rich clay was deposited about 65 million years ago.

Paleontologists were quick to realize the significance of this date, for it was 65 million years ago that all the dinosaurs rather suddenly became extinct. In fact, two-thirds of all the species on Earth disappeared within a relatively brief span of time.

The Alvarez discovery suggests a startling explanation for the dramatic extinction of more than half the life forms that inhabited our planet at the end of the Mesozoic era. Perhaps an asteroid hit the Earth at that time. An asteroid 10 km in diameter slamming into the Earth could have thrown enough dust into the atmosphere to block out sunlight for several years. As the temperature dropped drastically and plants died for lack of sunshine, the dinosaurs would have perished, along with many other creatures in the food chain that were highly dependent on vegetation. The dust eventually settled, depositing an iridium-rich layer around the world. Tiny, rodentlike creatures capable of ferreting out seeds and nuts were among the animals that managed to survive this holocaust, setting the stage for the rise of mammals in the Cenozoic era.

In 1992 a team of geologists suggested that this asteroid crashed into a site in Mexico. They based this conclusion on glassy debris and violently shocked grains of rock ejected from the 180-km-diameter Chicxulub Crater on the Yucatán Peninsula. From the known rate at which radioactive potassium decays, the scientists claim to have pinpointed the date when the asteroid struck: 64.98 million years ago.

Some geologists and paleontologists are not yet convinced that an asteroid impact led to the extinction of the dinosaurs at the end of the Mesozoic era, but many scientists agree that this hypothesis fits the available evidence better than any other explanation that has been offered so far.

17-5 Meteorites are classified as stones, stony irons, or irons, depending on their composition

A meteoroid, like an asteroid, is a chunk of rock in space. There is no official dividing line between meteoroids and asteroids, but the term *asteroid* is generally applied only to objects larger than a few hundred meters across.

A **meteor** is the brief flash of light (sometimes called a *shooting star*) that is visible at night when a meteoroid enters the Earth's atmosphere (Figure 17-9). Frictional heat is generated as the meteoroid plunges through the atmosphere, the result being a fiery trail across the night sky.

If a piece of rock survives its fiery descent through the atmosphere and reaches the ground, it is called a **meteorite**. People have been finding specimens of meteorites for thousands of years, and descriptions of them appear in ancient

FIGURE 17-9 A Meteor A meteor is produced when a piece of interplanetary rock or dust strikes the Earth's atmosphere at high speed. Exceptionally bright meteors, such as the one shown in this long exposure (notice the star trails), are usually called fireballs. (Courtesy of R. A. Oriti)

Chinese, Greek, and Roman literature. Our ancestors placed special significance on these "rocks from heaven." There have been numerous examples of the veneration of meteorites, such as the sacred black stone of Kaaba enshrined at Mecca.

The extraterrestrial origin of meteorites was hotly debated until as late as the eighteenth century. Upon hearing a lecture by two Yale professors, President Thomas Jefferson is said to have remarked, "I could more easily believe that two Yankee professors could lie than that stones could fall from Heaven." Although several falling meteorites had been widely witnessed and specimens from them had been collected (as, for example, in 1751 near Zagreb, Yugoslavia), many scientists were reluctant to accept the fact that rocks could fall to Earth from outer space.

Conclusive evidence for the extraterrestrial origin of meteorites came on April 26, 1803, when fragments pelted the French town of L'Aigle. The austere French Academy, whose members were among the last holdouts, sent the noted physicist J. B. Biot to investigate. His exhaustive report finally convinced the scientific community that meteorites were in fact extraterrestrial.

Meteorites are classified into three broad categories: stones, stony irons, and irons. As their name suggests, **stony meteorites,** or **stones,** look like ordinary rocks at first glance, but they are sometimes covered with a **fusion crust** (Figure 17-10). This crust is produced by a momentary melting of the meteorite's outer layers during its fiery descent through the atmosphere. When a stony meteorite is cut in two and polished, tiny flecks of iron can sometimes be found in the rock (Figure 17-11).

Although stony meteorites account for about 95% of all the meteoritic material that falls to Earth, stones are the most

FIGURE 17-10 A Stony Meteorite Of all meteorites that fall on the Earth, 95% are stones. Many freshly discovered specimens, like the one shown here, are coated with dark fusion crusts. This particular stone fell near Plainview, Texas. (From the collection of R. A. Oriti)

FIGURE 17-11 A Cut and Polished Stone When stony meteorites are cut and polished, some are found to contain tiny specks of iron mixed in the rock. This specimen was discovered near Neenach, California. (From the collection of R. A. Oriti)

difficult specimens to find. If they lie undiscovered and become exposed to the weather for a few years, they look almost indistinguishable from common terrestrial rocks. Meteorites with a high iron content are much easier to find, because they can be located with a metal detector. Consequently, iron and stony iron meteorites dominate most museum collections.

Stony iron meteorites consist of roughly equal amounts of rock and iron. Figure 17-12, for example, shows the mineral olivine suspended in a matrix of iron. Only about 1% of the meteorites that fall to Earth are stony irons.

Iron meteorites (Figure 17-13), or **irons,** account for about 4% of the material that falls on the Earth. Iron meteorites usually contain no stone, but many contain from 10% to 20% nickel.

In 1808 Count Alois von Widmanstätten, director of the Imperial Porcelain works in Vienna, discovered a conclusive test for the most common type of iron meteorite. About 75% of all iron meteorites have a unique crystalline structure. These **Widmanstätten patterns** become visible when the meteorite is cut, polished, and briefly dipped into a dilute solution of acid (Figure 17-14). Nickel–iron crystals can grow to lengths of several centimeters only if the molten metal cools slowly over many millions of years. So Widmanstätten patterns are never found in counterfeit meteorites, or "meteor-wrongs," as they are humorously called.

The Widmanstätten patterns in meteorites suggest that some asteroids remained partly molten long after they were formed. A parent asteroid must have been large enough to insulate its molten nickel–iron interior and slow its cooling. Typical meteorites are probably fragments of parent asteroids that were 200 to 400 km in diameter.

As soon as an asteroid accreted from planetesimals 4.5 billion years ago, rapid decay of short-lived radioactive iso-

topes heated the asteroid's interior to temperatures above the melting point of rock. Over the next few million years, a process called *chemical differentiation* occurred: Iron sank toward the asteroid's center, which forced the lighter rock upward toward the asteroid's surface. After the asteroid cooled and its core solidified, interasteroid collisions fragmented the parent body into meteoroids. Iron meteorites are

FIGURE 17-12 A Stony Iron Meteorite Stony irons account for about 1% of all the meteorites that fall to Earth. This particular specimen is a variety of stony iron called a pallasite. It fell near Antofagasta, Chile. (From the collection of R. A. Oriti)

FIGURE 17-13 An Iron Meteorite Irons are composed almost entirely of nickel–iron minerals. The surface of a typical iron is covered with thumbprintlike depressions caused by ablation (removal by melting) during its high-speed descent through the atmosphere. This specimen was found near Henbury, Australia. (From the collection of R. A. Oriti)

FIGURE 17-14 Widmanstätten Patterns When cut, polished, and etched with a weak acid solution, most iron meteorites exhibit interlocking crystals in designs called Widmanstätten patterns. These patterns appear only in the type of iron meteorite called octahedrites. This octahedrite was found near Henbury, Australia. (From the collection of R. A. Oriti)

therefore specimens from an asteroid's core, and stones are samples of its crust. Stony irons presumably come from regions between an asteroid's core and its crust.

17-6 Some meteorites retain traces of the early solar system

A class of rare meteorites, called **carbonaceous chondrites**, show no evidence of ever having been subjected to heating and melting, which affect the structure and composition of most other meteorites. Carbonaceous chondrites may therefore be samples of the primordial material from which our solar system was created. They contain complex organic compounds and as much as 20% water. These compounds would have been broken down and the water driven out if these meteorites had been significantly heated.

Amino acids, the building blocks of proteins upon which terrestrial life is based, are among the organic compounds occasionally found inside carbonaceous chondrites. Interstellar organic material has therefore probably been falling on our planet since it formed 4.5 billion years ago. Some scientists suspect that carbonaceous chondrites may have played a role in the origin of life on Earth.

Shortly after midnight on February 8, 1969, the night sky around Chihuahua, Mexico, was illuminated by a brilliant blue-white light moving across the heavens. The dazzling display was witnessed by hundreds of people, many of whom thought the world was coming to an end. As the light crossed the sky, it disintegrated in a spectacular, noisy explosion that dropped thousands of rocks and pebbles over the terrified onlookers. Within hours, teams of scientists were on their way to collect specimens of a carbonaceous chondrite, collectively named the Allende meteorite, after the locality (Figure 17-15).

Specimens were scattered in an elongated ellipse approximately 50 km long by 10 km wide. Larger specimens, which the atmosphere had slowed down the least, were found at one end of the ellipse, with the largest piece (110 kg) located at the very tip. Most fragments were coated with a fusion crust, but minerals immediately beneath the crust showed no signs of damage. Surface material is peeled away as it becomes heated during flight through the atmosphere; this

FIGURE 17-15 A Piece of the Allende Meteorite This carbonaceous chondrite fell near Chihuahua, Mexico, in February 1969. Note the meteorite's dark color, caused by a high abundance of carbon. The amount of argon-40 found in the meteorite corresponds to that which would have been generated by the decay of radioactive potassium-40 in about 4.6 billion years. Geologists therefore believe that this meteorite is a specimen of primitive planetary material. The ruler is 6 inches long. (Courtesy of J. A. Wood)

process forms the fusion crust. Because the heat has little chance to penetrate the meteorite's interior, compounds there are left intact.

One of the most significant discoveries to come from the Allende meteorite was evidence of the detonation of a nearby supernova 4.5 billion years ago. One of nature's most violent and spectacular phenomena, a **supernova explosion**, occurs when a massive star dies. During a supernova explosion, the doomed star blows itself apart in a cataclysm that hurls matter outward at tremendous speeds, as we shall see in Chapter 22. During this detonation, violent collisions between nuclei produce a host of radioactive elements, including a short-lived radioactive isotope of aluminum, ^{26}Al.

Gerald J. Wasserburg and his colleagues at the California Institute of Technology found clear evidence for the former presence of ^{26}Al in the Allende meteorite. Chemical analyses revealed a high abundance of a stable isotope of magnesium (^{26}Mg), which is produced by the radioactive decay of ^{26}Al. Some astronomers interpret this as evidence for a supernova in our vicinity at about the time the Sun was born. Indeed, by compressing interstellar gas and dust, the supernova's shock wave may have triggered the birth of our solar system.

17-7   A comet is a dusty chunk of ice that becomes partly vaporized as it passes near the Sun

Many rocks and chunks of ice that originally condensed out of the primordial solar nebula still orbit the Sun. Just as heat from the protosun produced two classes of planets, the terrestrial and the Jovian, two main types of interplanetary material were created. Near the Sun, interplanetary debris consists of the rocks called asteroids or meteoroids. Far from the Sun are the loose collections of ice and dust called **comets.**

Asteroids travel around the Sun along roughly circular orbits that are largely confined to the asteroid belt and to the plane of the ecliptic. In sharp contrast, comets travel around the Sun along highly elliptical orbits inclined at random angles to the ecliptic. As a comet approaches the Sun, solar heat begins to vaporize the ices. The liberated gases begin to glow, producing a fuzzy, luminous ball called a **coma.** The solar wind and sunlight blow these luminous gases outward into a long, flowing **tail,** one of the most awesome sights that can be seen in the night sky (Figure 17-16).

The solid part of a comet, called the **nucleus,** is a roughly half-and-half mixture of ice and dust, typically measuring a few kilometers across. To emphasize this composition, Harvard astronomer Fred L. Whipple, a pioneer in comet research, describes comets as "dirty snowballs."

The first pictures of a comet's nucleus were obtained by several spacecraft that flew past Comet Halley in 1986 (Figure 17-17). Halley's potato-shaped nucleus is darker than coal, reflecting only about 4% of the light that strikes it. This

FIGURE 17-16 Comet West   A comet is always named after the person who first sees it. Astronomer Richard M. West first noticed this comet in mid-1975 on a photographic plate taken with a telescope. After passing near the Sun, Comet West became one of the brightest comets in recent years. This photograph shows the comet gracing the predawn sky in the spring of 1976. (Courtesy of H. Vehrenberg)

dark color is probably caused by a layer of carbon-rich compounds and dust left behind as the comet's ice evaporated.

Several jets about 15 km long protrude from Halley's nucleus where "vents" squirt fountains of dust toward the Sun. Vents, which probably cover about 10% of the surface of the nucleus, seem to be active only when exposed to the Sun. When the nucleus's rotation brings them into darkness, away from the Sun, the vents shut off.

The structure of a comet is outlined in Figure 17-18. The nucleus of a comet is never visible to Earth-based astronomers, because it is small, dim, and buried in the glare of the coma. A coma is typically 1 million km in diameter, and a comet's tail can stretch to over 100 million km in length. Also not visible to the human eye is the **hydrogen envelope,** a huge sphere of sparse gas surrounding the comet's nucleus. This

FIGURE 17-17 The Nucleus of Comet Halley In March 1986 five spacecraft passed near Comet Halley. This close-up picture, taken by a camera on board *Giotto*, shows the potato-shaped nucleus of the comet. The nucleus is darker than coal and measures 15 km in the longest dimension and about 8 km in the shortest. The Sun illuminates the comet from the left. Two bright jets of dust extend 15 km from the nucleus toward the Sun, suggesting that most of the vaporization takes place on the sunlit side. (Max Planck Institut für Aeronomie)

hydrogen comes from water molecules (H_2O) escaping from the comet's evaporating ice. Figure 17-19 shows two views of Comet Kohoutek: as it appeared to Earth-based observers and as photographed by an ultraviolet camera from a rocket. This ultraviolet photograph showed astronomers the enormous extent of the hydrogen envelope, which can span 10 million kilometers.

Comets come in a wide range of shapes and sizes. For example, the comet shown in Figure 17-20 had a large, bright coma but a short, stubby tail. In contrast, the comet seen in Figure 17-21 had an inconspicuous coma, but its tail had an astonishing length of 1 AU, long enough to stretch all the way from the Earth to the Sun.

It has long been known that comets' tails always point away from the Sun (Figure 17-22), regardless of the direction of the comet's motion. The implication that something from the Sun was "blowing" the comet's gases radially outward led Ludwig Biermann to predict the existence of the solar wind a full decade before it was actually discovered in 1962 by instruments on a spacecraft.

In fact, the Sun usually produces two comet tails: an **ion tail** and a **dust tail.** Ionized atoms (that is, atoms missing one or more electrons) are swept directly away from the Sun by the solar wind to form the ion tail. The dust tail is formed when photons of light strike dust particles freed from the evaporating nucleus. Light exerts a pressure on any object that absorbs or reflects it. This pressure, called **radiation pressure,** is quite weak, but fine-grained dust particles in a

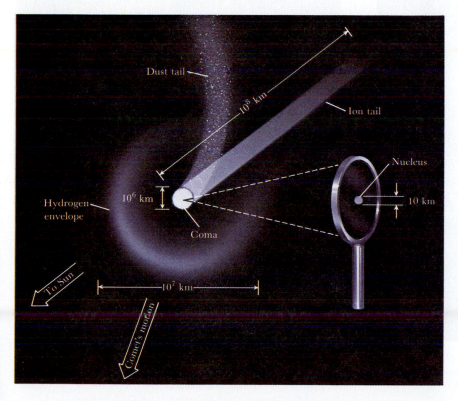

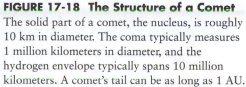

FIGURE 17-18 The Structure of a Comet The solid part of a comet, the nucleus, is roughly 10 km in diameter. The coma typically measures 1 million kilometers in diameter, and the hydrogen envelope typically spans 10 million kilometers. A comet's tail can be as long as 1 AU.

**FIGURE 17-19 Comet Kohoutek and Its
Hydrogen Envelope** These two photographs of Comet Kohoutek are reproduced to
the same scale. The picture on the left shows the
comet in visible light. The ultraviolet view on
the right reveals a huge hydrogen cloud surrounding the comet's head. (Johns Hopkins
University; Naval Research Laboratory)

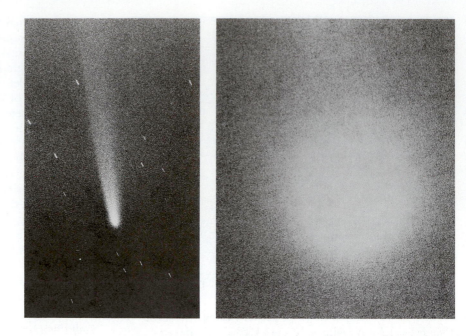

FIGURE 17-20 The Head of Comet Brooks With an
exceptionally large, bright coma, this comet dominated the night
skies during October 1911. It was named Comet Brooks after its
discoverer. (Lick Observatory)

FIGURE 17-21 Comet Ikeya-Seki Although the head of
Comet Ikeya-Seki was tiny, its tail was 1 AU long and dominated
the predawn sky during late October 1965. This comet was named
after its two Japanese codiscoverers. (Lick Observatory)

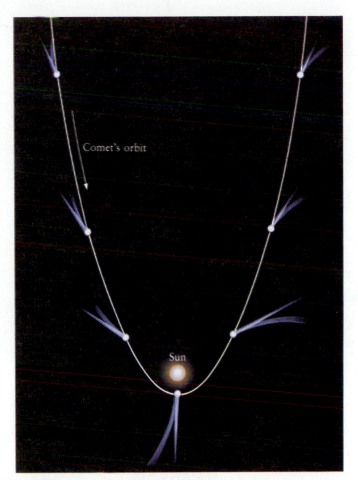

FIGURE 17-22 The Orbit and Tail of a Comet The solar wind and radiation pressure from sunlight blow a comet's dust particles and ionized atoms away from the Sun. Consequently, a comet's tail always points away from the Sun.

comet's coma offer little resistance and are blown away from the comet, thus producing a dust tail. The relatively straight ion tail can exhibit a dramatic structure that changes from night to night (Figure 17-23). The more amorphous dust tail is usually arched. On rare occasions the geometry of the Earth, the comet, and its arched dust tail is such that the dust tail appears to be sticking out the front of the comet (Figure 17-24).

Astronomers discover at least a dozen new comets in a typical year. Some are **short-period comets**, which circle the Sun in less than 200 years. Like Halley's Comet, they appear again and again at predictable intervals. However, the majority of comets discovered each year are long-period comets, which take roughly 1 to 30 million years to complete one orbit of the Sun. These comets travel along extremely elongated orbits and consequently spend most of their time at distances of roughly 10^4 to 10^5 AU from the Sun, or about one-fifth of the way to the nearest star.

FIGURE 17-24 The Antitail of Comet Arend-Roland Comet Arend-Roland, seen in April 1957, exhibited an "antitail." Actually, this antitail was merely the end of the dust tail. The Earth and the comet were oriented in such a way that the end of the arched dust tail, as seen past the comet's head, looked like a spike sticking out of the comet's head. (Lick Observatory)

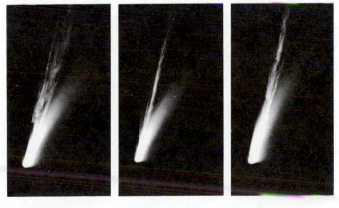

August 22   August 24   August 26

FIGURE 17-23 The Two Tails of Comet Mrkos Comet Mrkos dominated the evening sky during August 1957. These three views, taken at two-day intervals, show dramatic changes in the comet's ion (Type I) tail. In contrast, the slightly curved dust (Type II) tail remained fuzzy and featureless. (Palomar Observatory)

Because astronomers discover long-period comets at the rate of roughly one per month, it is reasonable to suppose that there is an enormous population of comets out there around 50,000 AU from the Sun. This reservoir of cometary nuclei surrounding the Sun is called the **Oort cloud,** after the Dutch astronomer Jan Oort, who first proposed its existence in the 1950s. Estimates of the number of "dirty snowballs" in the Oort cloud range as high as 12 billion. Only such a large reservoir of cometary nuclei would explain why we see so many long-period comets, even though each one takes several million years to travel once around its orbit.

The Oort cloud was probably created 4.5 billion years ago from numerous icy planetesimals that orbited the Sun in the vicinity of the newly formed Jovian planets. During near-collisions with the giant planets, many of these chunks of ice

and dust were catapulted by gravity into the highly elliptical orbits that they now occupy, much in the same way that Pioneer and Voyager spacecraft were flung far from the Sun during their flybys of the Jovian planets. During a return trip toward the Sun, an encounter with a Jovian planet may force a long-period comet into a short-period orbit. There the comet will eventually be destroyed by its frequent passages near the Sun.

17-8 Cometary dust and debris rain down on the Earth during meteor showers

A typical comet loses about 0.1% of its ice each time it passes near the Sun. Its ice thus completely vaporizes after about 1000 perihelion passages, leaving only a swarm of meteoritic

---

BOX 17-2

Meteor Showers

At least ten notable meteor showers occur each year. They are listed with relevant information in the table below. The date of maximum is the best time to observe a particular shower, although good displays can often be seen a day or two before or after the maximum. The radiant is the location among the stars from which meteors seem to be coming. The hourly rate is given for a single observer under excellent conditions. The velocity is the average speed of the meteoritic material as it strikes the atmosphere.

In order to see a fine meteor display, you need a clear, moonless sky. The Moon's presence above the horizon can significantly detract from the number of faint meteors you will be able to see. In addition, the early morning hours (between roughly 2 AM and dawn) are the best times to make your observations. As sketched in the diagram, you are on the leading side of the Earth during the early morning hours, and all meteor-producing particles in the

Earth's path are swept into the atmosphere above you. On the other hand, you are on the trailing side of the Earth during the evening hours, where only high-speed particles manage to catch up with our planet to produce meteors.

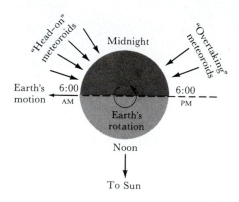

| Shower | Date of maximum | Radiant | | Hourly rate | Velocity (km/s) |
		Right ascension	Declination		
Quadrantids	January 3	15^h 28^m	+50°	40	40
Lyrids	April 22	18 16	+34	15	50
Aquarids	May 4	22 24	0	20	64
Aquarids	July 30	22 36	−17	20	40
Perseids	August 12	3 4	+58	50	60
Orionids	October 21	6 20	+15	20	66
Taurids	November 4	3 32	+14	15	30
Leonids	November 16	10 8	+22	15	70
Geminids	December 13	7 32	+32	50	35
Ursids	December 22	14 28	+76	15	35

| March 8 | March 12 | March 14 | March 18 | March 24 |

FIGURE 17-25 The Fragmentation of Comet West
Shortly after passing near the Sun in 1976, the nucleus of Comet West broke into four pieces. This series of five photographs clearly shows that disintegration. (New Mexico State University Observatory)

dust and pebbles. A comet's nucleus will disintegrate more quickly if it happens to pass very near the Sun as what is called a **Sun-grazing comet.** On rare occasions, a comet's nucleus has been observed to fragment (Figure 17-25).

As a comet's nucleus evaporates, residual dust and rock fragments spread out to form a **meteoritic swarm,** a loose collection of debris that continues to circle the Sun along the comet's orbit (Figure 17-26). If the Earth's orbit happens to pass through this swarm, a **meteor shower** is seen as the dust particles strike Earth's upper atmosphere.

Nearly a dozen meteor showers can be seen each year (see Box 17-2). Note that the **radiants** for these showers (that is, the places among the stars from which the meteors appear to come) are not confined to the constellations of the zodiac. Meteor showers can come from various parts of the sky, because the orbits of their parent comets are inclined at random angles to the plane of the ecliptic.

As incredible as it may seem, an estimated total of 300 tons of extraterrestrial rock and dust fall on the Earth each day. The low-density material from comets and stony meteoroids burns up in the atmosphere, so only the denser matter originating from iron-rich asteroids typically reaches the ground.

On June 30, 1908, a spectacular explosion occurred over the Tunguska region of Siberia that released about 10^{15} joules of energy, equivalent to a nuclear detonation of several hundred kilotons. The blast knocked a man off his porch some 60 km away and was audible more than 1000 km away. Millions of tons of dust were injected into the atmosphere, causing a decrease in air transparency detected as far away as California.

Preoccupied with political upheaval and World War I, Russia did not send a scientific expedition to the site until 1927. At that time, Soviet researchers found that trees had

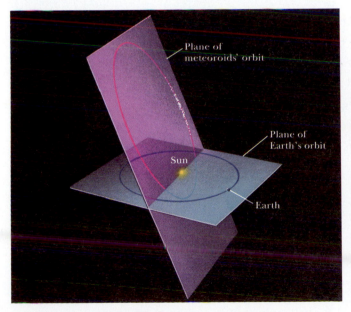

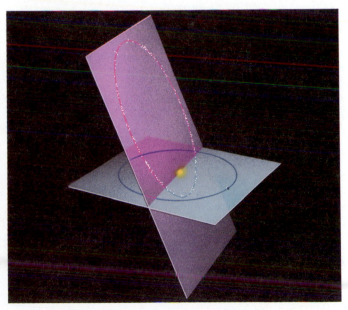

FIGURE 17-26 Meteoritic Swarms Rock fragments and dust from "burned out" comets continue to circle the Sun. (a) If the comet is only recently extinct, the particles will still be tightly concentrated in a compact swarm. The most spectacular meteor showers occur when the Earth happens to pass through such a swarm. (b) Over the ages, the particles gradually spread out along the old comet's elliptical orbit. This configuration produces the most predictable meteor showers, because the Earth must pass through the evenly distributed swarm on each trip around the Sun.

FIGURE 17-27 Aftermath of the Tunguska Event In 1908 a small stony asteroid struck the Earth's atmosphere over the Tunguska region of Siberia. Trees were blown down for many kilometers in all directions around the impact site. (Courtesy of Sovfoto)

been seared and felled radially outward in an area about 30 km in diameter (Figure 17-27). There was no clear evidence of a crater. In fact, the trees at "ground zero" were left standing upright, although they were completely stripped of branches and leaves. Because no significant meteorite samples were found, for many years it was assumed that a small comet had struck the Earth.

Recently, however, several teams of astronomers have argued that the Tunguska explosion was actually caused by a stony asteroid traveling at supersonic speed. They arrived at this conclusion after assessing the effects of various impactor sizes, speeds, and compositions. Even data from aboveground nuclear detonations of the 1940s and 1950s were worked into the calculations. The Tunguska event is well-matched by a stony asteroid about 80 m (260 ft) in diameter entering the Earth's atmosphere at 22 km/s (50,000 mph). A small comet breaks up too high in the atmosphere to cause significant damage on the ground.

Orbital calculations suggest a small chance that a piece of Comet Swift-Tuttle might strike the Earth in 2126. This comet, discovered in 1862, is the parent of the Perseid meteor shower. If gravity alone determines the comet's orbit, it will return to pass nearest the Sun on July 11, 2126, and then miss the Earth by a wide margin in August of that year. However, jets were observed erupting from the comet's nucleus during its 1992 perihelion passage (Figure 17-28). Much like rocket thrusters on a spacecraft, they can change the comet's orbit slightly. Furthermore, past behavior of Swift-Tuttle suggests that it sheds large chunks of material. An unlucky combination of orbit deflection and comet fragmentation could send a piece of Swift-Tuttle slamming into the Earth, possibly on August 14, 2126.

Large objects occasionally strike the Earth with a destructive force of a nuclear weapon. A comet or small asteroid approaching our planet in the daytime sky from the general direction of the Sun would probably not be discovered by astronomers, so its impact would occur without warning. In spite of safeguards and diplomacy, Armageddon could be triggered by an entirely natural phenomenon.

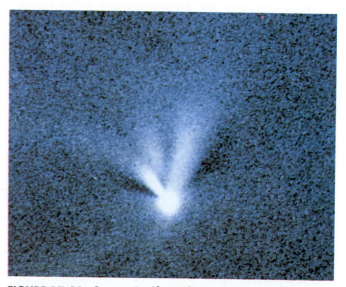

FIGURE 17-28 Comet Swift-Tuttle This CCD image of Comet Swift-Tuttle shows several jets erupting from its nucleus six weeks before perihelion passage in December 1992. These jets alter the comet's path, making it difficult to predict the comet's orbit precisely. (Courtesy of J. DeYoung and R. Schmidt: USNO)

KEY WORDS

Terms preceded by an asterisk are discussed in the boxes.

amino acids	asteroid	Bode's law	comet
Amor asteroid	asteroid belt	carbonaceous chondrite	dust tail
Apollo asteroid	belt asteroid	coma (of a comet)	*equipotential contour

fusion crust

Hirayama families

hydrogen envelope

impact crater

ion tail

iron meteorite

Kirkwood gaps

Lagrangian points

meteor

meteorite

meteoritic swarm

meteoroid

meteor shower

minor planet

nucleus (of a comet)

Oort cloud

radiant (of a meteor shower)

radiation pressure

short-period comet

stony iron meteorite

stony meteorite (stone)

Sun-grazing comet

supernova explosion

tail (of a comet)

Trojan asteroid

Widmanstätten patterns

KEY IDEAS

- Bode's law is a numerical sequence that gives the distances of most of the planets (Mercury through Uranus) from the Sun in astronomical units. This "law" inspired nineteenth-century astronomers to search for a planet in a gap between the orbits of Mars and Jupiter.

- Thousands of belt asteroids with diameters ranging from a few kilometers up to 1000 kilometers circle the Sun between the orbits of Mars and Jupiter.

    Gravitational perturbations by Jupiter deplete certain orbits within the asteroid belt. The resulting gaps, called Kirkwood gaps, occur at simple fractions of Jupiter's orbital period.

    Jupiter's gravity also captures asteroids in two locations, called Lagrangian points, along Jupiter's orbit.

    Some asteroids, called Amor and Apollo asteroids, move in elliptical orbits that cross the orbits of Mars and Earth, respectively.

- Small rocks in space are called meteoroids. If a meteoroid enters the Earth's atmosphere, it produces a fiery trail called a meteor. If part of the object survives the fall, the fragment that reaches the Earth's surface is called a meteorite.

    Meteorites are grouped into three major classes, according to composition: iron, stony iron, and stony meteorites.

    When a large meteoroid strikes the Earth, it forms an impact crater whose diameter depends on both the mass and the speed of the impinging object.

    An asteroid may have struck the Earth 65 million years ago, probably causing the extinction of the dinosaurs and many other species.

- Rare stony meteorites called carbonaceous chondrites may be relatively unmodified material from the solar nebula; these meteorites often contain organic material and may have played a role in the origin of life on Earth.

- Analysis of isotopes in certain meteorites suggests that a nearby supernova explosion may have triggered the formation of the solar system 4.5 billion years ago.

- A comet is a chunk of ice and rock fragments that generally moves in a highly elliptical orbit about the Sun.

As a comet approaches the Sun, its icy nucleus develops a luminous coma, surrounded by a vast hydrogen envelope; an ion tail and a dust tail extend from the comet, pushed away from the Sun by the solar wind and radiation pressure, respectively.

Fragments of "burned out" comets produce meteoritic swarms; a meteor shower is seen when the Earth passes through a meteoritic swarm.

Millions, perhaps billions, of cometary nuclei probably exist in the Oort cloud located between 10,000 and 100,000 AU from the Sun.

REVIEW QUESTIONS

1. What is Bode's law, and why is it not really a "law"?

2. Why are asteroids, meteorites, and comets of special interest to astronomers who want to understand the early history of the solar system?

3. Describe the asteroid belt.

4. What are Kirkwood gaps and what causes them?

5. Can you think of another place in our solar system where a phenomenon similar to Kirkwood gaps occurs? Explain.

6. What are the Trojan asteroids, and where are they located?

7. Where on Earth might you find large numbers of stony meteorites that are not significantly weathered?

8. Suppose you found a rock you suspect might be a meteorite. Describe some of the things you could do to determine whether it was a meteorite or a "meteorwrong."

9. Why do astronomers believe that meteorites come from asteroids, whereas meteor showers are related to comets?

10. With the aid of a drawing, describe the structure of a comet.

11. Why is the phrase "dirty snowball" an appropriate characterization of a comet's nucleus?

12. Explain why comets are generally brighter after passing perihelion.

13. What is the Oort cloud, and how might it be related to planetesimals left over from the formation of the solar system?

ADVANCED QUESTIONS

Tips and tools . . .

You will need to use Kepler's third law in two of the problems below. A spherical object intercepts an amount of sunlight proportional to the square of its radius. The volume of a sphere of radius r is $\frac{4}{3}\pi r^3$.

14. Suppose that a planet were indeed 77.2 AU from the Sun, as predicted by Bode's law. How long would it take such an object to orbit the Sun? At what rate (in degrees per year) would this object appear to move in relation to the background stars? Would this motion be detectable with Earth-based telescopes?

15. Suppose that a double asteroid is observed in which one member is 16 times brighter than the other. Suppose that both members have the same albedo and that the larger of the two is 200 km in diameter. What is the diameter of the other member?

16. Why do you suppose there aren't any asteroids at the L_1, L_2, or L_3 Lagrangian points in the Sun–Jupiter system?

17. Are there any examples in the solar system of objects being trapped at the L_4 and L_5 Lagrangian points other than in the Sun–Jupiter system?

18. Find the orbital periods of Sun-grazing comets whose aphelion distances are (**a**) 100 AU, (**b**) 1000 AU, (**c**) 10,000 AU, and (**d**) 100,000 AU. Assuming that these comets can survive only a hundred perihelion passages, calculate their lifetimes.

19. Use the percentages of stones, irons, and stony iron meteorites that fall to Earth to estimate what fraction of their parent asteroids originally consisted of an iron core and a stony mantle. How do these percentages compare with that for the Earth? (*Note:* Assume that the percentages of stones and irons that fall to Earth indicate the fractions of parent asteroids occupied by rock and iron, respectively.)

20. In 1992 Comet Shoemaker-Levy passed within 10 million kilometers of Jupiter and broke into at least a dozen pieces that were captured into orbit about the planet. As this textbook went to press, astronomers calculated that the fragmented comet should plunge into Jupiter in August 1994. Consult such magazines as *Sky & Telescope, Astronomy,* and *Science News* to learn the latest word on this extraordinary event.

DISCUSSION QUESTIONS

21. Suppose it were discovered that the asteroid Hermes had been disturbed in such a way as to put it on a collision course with Earth. Describe what humanity could do within the framework of present technology to counter such a catastrophe.

22. From the abundance of craters on the Moon and Mercury, we know that numerous asteroids and meteoroids struck the inner planets early in the history of our solar system. Is it reasonable to suppose that numerous comets also pelted the planets 3.5 to 4.5 billion years ago? Speculate about the effects of such a cometary bombardment, especially with regard to the evolution of the primordial atmospheres on the terrestrial planets.

23. Compare the consequences of a global thermonuclear war with that of an asteroid hitting the Earth.

OBSERVING PROJECTS

24. Make arrangements to view a meteor shower. Dates of major meteor showers are given in Box 17-2. Choose a shower that will occur near the time of a new moon. Informative details concerning upcoming meteor showers are often published a month ahead of time in such magazines as *Sky & Telescope* and *Astronomy.* Set your alarm clock for the early morning hours (1 to 3 AM). Get comfortable in a reclining chair or lie on your back so that you can view a large portion of the sky. Make note of how long you observe, how many meteors you see, and what location in the sky they seem to come from. How well does your observed hourly rate agree with published estimates such as those in Box 17-2? Is the radiant of the meteor shower apparent from your observations?

25. Make arrangements to view a comet through a telescope. Since astronomers discover roughly a dozen comets each year, there is usually a comet visible somewhere in the sky. Unfortunately, they are often quite dim, and so you will need to have access to a moderately large telescope. Consult recent issues of the IAU *Circular,* published by the International Astronomical Union's Central Bureau for Astronomical Telegrams, which contains the latest word on predicted positions and anticipated brightness of comets in the sky. Also, if there is an especially bright comet in the sky, useful information about it might be found in the latest issue of *Sky & Telescope.* Is a comet visible with a telescope at your disposal? If so, can you distinguish the comet from background stars? Can you see its coma? Can you see a tail?

26. Make arrangements to view an asteroid. At opposition, some of the largest asteroids are bright enough to be seen through a modest telescope. Check the "Minor Planets" section of the current issue of the *Astronomical Almanac* to see if any bright asteroids are near opposition. If so, check the current issue as well as the most recent January issue of *Sky & Telescope* for a star chart showing the asteroid's path among the constellations. You will need such a chart to distinguish the asteroid from background stars. Also, the right ascensions and declinations of Ceres, Pallas, Juno, and Vesta are listed in the *Astronomical Almanac.* Observe the asteroid on at least two occasions separated by a few days. On each night, draw a star chart of the objects in your telescope's field of view. Has the position of one starlike object shifted between observing sessions? Does the position of the moving object agree with the path plotted on published star charts? Do you feel confident that you have in fact seen the asteroid?

FOR FURTHER READING

Beatty, J. K. "An Inside Look at Halley's Comet." *Sky & Telescope*, May 1986. Published only a few months after the Comet Halley flybys, this article contains some fine photographs of the comet's nucleus as well as a "first look" at the data sent back by the spacecraft.

———. "Killer Crater in the Yucatán?" *Sky & Telescope*, July 1991. This well-written article summarizes evidence that the Chicxulub Crater is the impact site of the asteroid that killed off the dinosaurs 65 million years ago.

Binzel, R. P., Barucci, M. A., and Fulchignoni, M. "The Origins of the Asteroids." *Scientific American*, October 1991. This article discusses the important clues about the birth of the solar system that asteroids can give us.

Cunningham, C. J. "Giuseppe Piazzi and the 'Missing Planet.'" *Sky & Telescope*, September 1992. This fascinating two-page article tells the story of Piazzi's discovery of Ceres.

Dodd, R. *Thunderstones and Shooting Stars: The Meaning of Meteorites*. Harvard University Press, 1986. This clear introduction to the recovery, classification, and study of meteorites was written by a noted geologist.

Durda, D. "All in the Family." *Astronomy*, February 1993. This well-illustrated article shows how families of asteroids arise from interasteroid collisions.

Gingerich, O. "Newton, Halley, and the Comet." *Sky & Telescope*, March 1986. This fascinating article by a noted science historian describes the relationship between Newton and Halley.

Hartmann, W. "The Smaller Bodies of the Solar System." *Scientific American*, September 1975. This article by a noted planetary scientist discusses asteroids and meteorites.

Knacke, R. "Sampling the Stuff of a Comet." *Sky & Telescope*, March 1987. This article, last of three in a special issue of *Sky & Telescope* describing results from the Comet Halley flybys, discusses the chemical composition of comets.

Kowal, C. *Asteroids*. Ellis Horwood/John Wiley, 1988. This fascinating book covers the history and science of aster-oids and speculates about future space missions to take advantage of asteroid resources.

Marsden, B. G., "Comet Swift-Tuttle: Does it Threaten Earth?" *Sky & Telescope*, January 1993. In this article, a renowned comet and asteroid pundit discusses his predictions for the possibility of a catastrophic return of Comet Swift-Tuttle in 2126.

Olson, R. "Giotto's Portrait of Halley's Comet." *Scientific American*, July 1979. This article describes historical sightings of Comet Halley, including that by Giotto di Bondone, the Renaissance painter after whom a spacecraft was named.

Russell, D. "The Mass Extinctions of the Late Mesozoic." *Scientific American*, January 1982. This article discusses evidence that an asteroid hit the Earth 65 million years ago, causing the extinction of the dinosaurs and many other species.

Sagan, C., and Druyan, A. *Comet*. Random House, 1985. This beautiful volume on comet science and lore is stylishly written, well organized, and contains excellent illustrations.

Weissman, P. "Comets at the Solar System's Edge." *Sky & Telescope*, January 1993. This article discusses the Oort cloud and a second, less numerous band of comets closer to the Sun called the Kuiper Belt.

———. "Realm of the Comet." *Sky & Telescope*, March 1987. This article, the first of three in a special issue of *Sky & Telescope* describing results from the Comet Halley flybys, discusses the Oort cloud.

Whipple, F. "The Black Heart of Comet Halley." *Sky & Telescope*, March 1987. This article, the second of three in a special issue of *Sky & Telescope* describing results from the Comet Halley flybys, discusses the structure and properties of the comet's nucleus.

———. *The Mystery of Comets*. Smithsonian Institution Press, 1985. This appealing book for the layperson includes personal reminiscences by a distinguished astronomer who has spent most of his life studying comets.

———. "The Nature of Comets." *Scientific American*, February 1974. The scientist who invented the "dirty snowball" model discusses the structure and behavior of comets.

———. "The Spin of Comets." *Scientific American*, March 1980. This article, which focuses on Comet Encke, describes the rotation of a cometary nucleus and the thrust of gases evaporating from it.

OUR STAR

THE X-RAY SUN AND THE SOLAR CORONA This composite view of the Sun's outer atmosphere combines a white-light photograph taken during a solar eclipse visible in Hawaii with an X-ray image taken concurrently by a camera on board a rocket launched from New Mexico. The white-light photograph shows streamers extending far above the solar surface. The X-ray view shows hot gases near the solar surface in shades of yellow, orange, and red. By matching the two views, scientists can study the Sun's outer atmosphere over a range of wavelengths and construct a three-dimensional picture of solar activity. (Courtesy of L. Golub and S. Koutchmy)

THE SUN is a typical star. Because we can view it from close range, studying the Sun gives us important insights into the nature of stars in general. The solar atmosphere exhibits a host of bewildering phenomena, as ionized gases, photons, and magnetic fields interact on a colossal scale. Most conspicuous are sunspots, which are sites of concentrated magnetic field. The number of sunspots varies with an 11-year period, part of a more general 22-year solar cycle that affects the entire solar atmosphere. The energy source that powers all these phenomena—and causes the Sun to shine—lies buried at the Sun's center, where thermonuclear reactions convert hydrogen into helium. Astrophysicists use the laws of physics to construct a model of the Sun, which fully describes the Sun's structure. Nevertheless, astronomers are still puzzled by many basic questions. For instance, why do sunspots exist in the first place? There is also much contention over solar neutrinos, particles emanating from the nuclear reactions in Sun's core. For the clues that it can give us about the universe as well as for its importance to life on Earth, solar astronomy is an exciting and rewarding field of research.

The Sun is the source of our heat and light. Life would not be possible on Earth without the energy provided by the Sun. Even a small change in its size or surface temperature could dramatically alter conditions on the Earth, either melting the polar caps or producing another ice age.

The Sun is also a dramatic arena where we can observe the beautiful and bewildering interaction of matter and energy on an immense scale. Columns of hot gases gush up to the solar surface, where they interact with the Sun's magnetic field and dissipate energy into its outer atmosphere.

Despite its remarkable activity, the Sun is an average star. Its mass, size, surface temperature, and chemical composition lie roughly midway between the extremes exhibited by other stars. (Data about the Sun are listed in Box 18-1.) Studying the Sun therefore gives important insights into the nature of stars in general, for unlike other stars, our Sun is available for detailed, close-up examination.

18-1 The photosphere is the lowest of three main layers in the Sun's atmosphere

Although astronomers often speak of the solar surface, the Sun really has no surface at all. As you move in toward the Sun, you encounter ever denser gases, but there is no sharp boundary comparable to the surfaces of the Earth or the Moon. The Sun appears to have a kind of surface (Figure 18-1), however, because there is a specific layer in the Sun's atmosphere from which most of the visible light comes. It is an *optical*, rather than a physical surface. This layer, which is about 300 to 400 km thick, is appropriately called the Sun's **photosphere** ("sphere of light").

The photosphere is the lowest of three layers that together constitute the **solar atmosphere.** Above the photosphere are two additional layers, the chromosphere and the corona, which are discussed later in this chapter. These upper layers are transparent to visible light; we can see through them

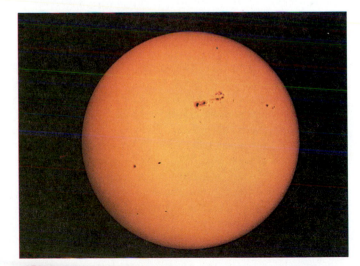

FIGURE 18-1 The Photosphere The Sun is the only star whose surface details can be examined through Earth-based telescopes. When viewing the Sun, astronomers always take great care to avoid severe damage to their eyes by using extremely dark filters or by projecting the Sun's image onto a screen. The Sun is so bright that its rays, focused by the lens of an unprotected eyeball, can destroy the retina. Therefore, **never look directly at the Sun.** (Celestron International)

BOX 18-1

Sun Data

Distance from Earth:	Mean: 1 AU = 149,598,000 km Maximum: 152,000,000 km Minimum: 147,000,000 km
Light travel time to Earth:	8.3 min
Mean angular diameter:	32 arc min
Radius:	696,000 km = 109 Earth radii
Mass:	1.99×10^{30} kg = 3.33×10^5 Earth masses
Composition (by mass):	74% hydrogen 25% helium 1% other elements
Mean density:	1410 kg/m^3
Mean temperatures:	Surface: 5800 K Center: 15.5×10^6 K
Spectral type:	G2
Luminosity:	3.90×10^{26} W
Apparent magnitude:	−26.8
Absolute magnitude:	+4.8
Distance from center of Galaxy:	7.7 kpc = 25,000 ly
Orbital period around center of Galaxy:	200,000,000 years
Orbital velocity around center of Galaxy:	230 km/s

down to the photosphere. We cannot, however, see through the shimmering gases of the photosphere, and so everything below the photosphere is called the **solar interior.**

Under good observing conditions with a telescope—and special dark filters to protect their eyes from damage—astronomers can often see a blotchy pattern in the photosphere called **granulation** (Figure 18-2). Each light-colored **granule** measures about 1000 km across and is surrounded by a darkish boundary. The difference in brightness between the center and the edge of a granule corresponds to a temperature drop of about 300 K.

Time lapse photography shows that granules form, disappear, and re-form in cycles lasting only a few minutes. At any one time, roughly four million granules cover the solar surface. Each granule occupies an area equal to Texas and Oklahoma combined (about a million square kilometers).

High-resolution spectroscopy reveals a blueshifting of spectral lines in the central, bright regions of granules and a redshifting along the dark boundaries between them. These Doppler shifts show that gas is rising upward in the centers of granules and descending downward along "intergranule boundaries."

Granulation is caused by **convection,** the transport of heat by the movement of a gas or fluid. Convection is easily observed in a simmering pot of soup on a warm stove. In the solar photosphere, convection causes hot gases to rise upward in granules, cool off, spill over the edges of the granules, and then plunge back down into the Sun (Figure 18-3).

The photosphere appears darker around the edge, or **limb**, of the Sun than it does toward the center of the solar disk (examine Figure 18-1). This **limb darkening** arises because we are looking obliquely at the photosphere near the edge of the Sun. We do not see as deeply into the Sun there as we do when we look near the center of the disk. Near the limb, the gas we observe is cooler and thus appears dimmer than the deeper, warmer gas seen near the disk center.

The Sun's photosphere shines with a continuous, nearly perfect blackbody spectrum that corresponds to an average

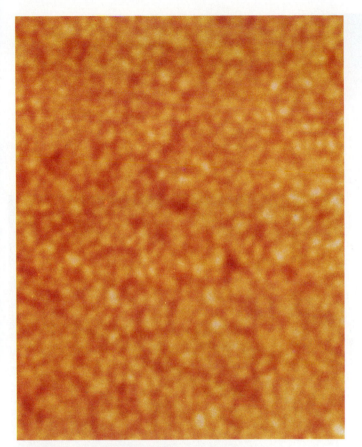

FIGURE 18-2 Solar Granulation High-resolution photographs of the Sun's surface reveal a blotchy pattern called granulation. The granules, each measuring about 1000 km across, are convection cells in the Sun's outer layers. (NOAO)

temperature of about 5800 K (recall Figure 5-9). The temperature declines to about 4000 K in the uppermost layers of the photosphere. All the absorption lines in the Sun's spectrum are produced in this relatively cool layer by atoms extracting energy from the sunlight at various wavelengths.

18-2 The chromosphere is located between the photosphere and the Sun's outermost atmosphere

Immediately above the photosphere is a relatively cool, dim layer called the **chromosphere** ("sphere of color"), which is the second of the three major levels in the Sun's atmosphere. When the Moon blocks out the photosphere during a total solar eclipse, the chromosphere is visible as a pinkish strip around the edge of the dark Moon.

High-resolution images of the Sun's limb reveal that the chromosphere is composed of numerous spikes called **spicules** (Figure 18-4). Spicules are jets of gas that surge upward from the Sun. A typical spicule rises at the rate of 20 km/s, reaches a height of near 10,000 km, then collapses and fades away after a few minutes. A schematic diagram of the solar atmosphere with a spicule is shown in Figure 18-5. Roughly 300,000 spicules exist at any one time, covering a few percent of the Sun's surface.

The spectrum of the chromosphere is dominated by emission lines. As we saw in Chapter 5, emission lines are produced by atoms when their electrons fall from higher to lower energy levels. The characteristic reddish-pink color of the chromosphere comes from the Balmer line H_α, one of the brightest emission lines in the chromosphere's spectrum. This spectral line is produced as electrons fall from the $n = 3$ to the $n = 2$ levels in the hydrogen atom (recall Figure 5-18b).

The blue side of the chromosphere's spectrum is dominated by two bright emission lines, the H and K lines of singly ionized calcium. The spectrum of the chromosphere also exhibits several emission lines not seen in the photospheric spectrum, including lines due to ionized helium and ionized metals. These lines must therefore originate in areas of the chromosphere that are hotter than the photosphere.

The spectrum of the photosphere is dominated by numerous absorption lines (recall Figure 5-11), including broad, dark absorption lines of hydrogen and ionized calcium. These lines are produced by hydrogen atoms and calcium ions in the upper photosphere that efficiently absorb photons

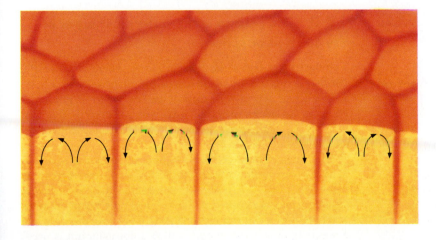

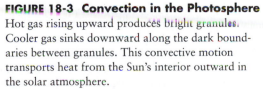

FIGURE 18-3 Convection in the Photosphere Hot gas rising upward produces bright granules. Cooler gas sinks downward along the dark boundaries between granules. This convective motion transports heat from the Sun's interior outward in the solar atmosphere.

FIGURE 18-4 Spicules and the Chromosphere
Numerous spicules are seen in this photograph of the Sun's limb. The chromosphere consists of spicules, which are jets of cool gas that surge upward into the warmer regions of the Sun's outer atmosphere. This photograph was taken through a filter that is transparent only to the wavelength of the Balmer line H_α at 656.3 nm. (NOAO)

tons coming from deeper layers within the Sun. As a result, the photosphere is rather dim at the wavelengths of H_α and calcium's H and K lines. But the chromosphere is bright at these wavelengths. Astronomers can thus study details of the chromosphere by viewing the Sun through special filters transparent to light only at the wavelengths of H_α or the calcium lines.

Figure 18-6 is a photograph of the solar surface taken through an H_α filter, which shows what is going on in the chromosphere. In this picture, spicules appear as numerous dark, brushlike spikes protruding upward.

Spicules are generally located on boundaries between large, organized regions called **supergranules**, which are enormous convective cells in the photosphere. A typical supergranule is about 30,000 km in diameter and contains many hundreds of ordinary granules. Gases rise upward in the middle of a supergranule, move horizontally outward toward its edge, and descend back into the Sun. This large-

FIGURE 18-6 Spicules and Supergranulation Numerous spicules are visible in this H_α image, which also shows many details about the chromosphere. Spicules are located along the irregularly shaped boundaries between large organized cells called supergranules. (NOAO)

scale convection moves with the speed of roughly 0.5 km/s, which is about a tenth of the speed of gases churning in a granule.

18-3 The corona is the outermost layer of the Sun's atmosphere

The outermost region of the Sun's atmosphere is called the **corona**. It extends from the top of the chromosphere out to a distance of several million kilometers, where it gradually becomes the solar wind. (The solar wind, as described earlier, consists of high-speed protons and electrons constantly escaping from the Sun.)

The total amount of visible light emitted by the solar corona is comparable to the brightness of the Moon when it is full—only about one-millionth as bright as the photosphere. Consequently, the corona can be viewed only when the photosphere is blocked out either during a total eclipse or when a specially designed telescope called a **coronagraph** is used.

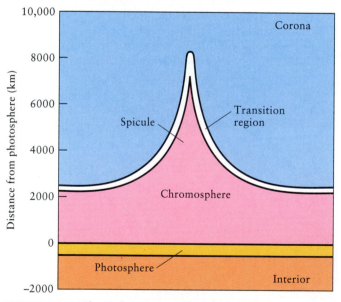
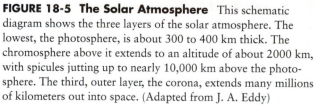

FIGURE 18-5 The Solar Atmosphere This schematic diagram shows the three layers of the solar atmosphere. The lowest, the photosphere, is about 300 to 400 km thick. The chromosphere above it extends to an altitude of about 2000 km, with spicules jutting up to nearly 10,000 km above the photosphere. The third, outer layer, the corona, extends many millions of kilometers out into space. (Adapted from J. A. Eddy)

FIGURE 18-7 The Solar Corona This extraordinary photograph was taken during the total solar eclipse of July 11, 1991. Numerous streamers are visible, extending millions of kilometers above the solar surface. (Courtesy of R. Christen and M. Christen: Astro-Physics Inc.)

Figure 18-7 is an exceptionally detailed photograph of the Sun's corona taken during a solar eclipse. Numerous streamers extend far above the solar surface.

Around 1940 astronomers realized that the corona is extraordinarily hot. Its spectrum contains the emission lines of certain highly ionized elements. For example, there is a prominent green line at 530.3 nm that is caused by iron atoms, each stripped of 13 electrons. Temperatures in the range of 1 to 2 million K are required to strip that many electrons from atoms. Figure 18-8 shows how temperature varies with distance above the photosphere.

Despite these high temperatures, the corona is nearly a vacuum. The typical density of coronal gases is only about 10^{11} particles per cubic meter. In comparison, the density of the photosphere is about 10^{23} particles per cubic meter, and the density of the air we breathe is about 10^{25} particles per cubic meter. Consequently, the corona emits relatively little light because of its low density.

Astronomers do not fully understand why the corona is so hot. The corona may be heated by the churning motions of gases in the photosphere, but no one has been able to explain how that energy is transported to the corona. Some researchers suspect that sound waves or waves in the Sun's magnetic field carry that energy from the photosphere to the

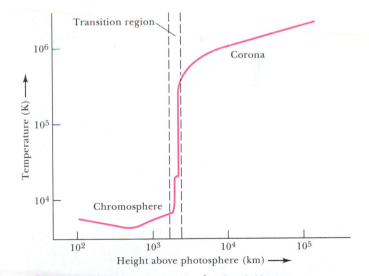

FIGURE 18-8 Temperatures in the Sun's Upper Atmosphere In a narrow region about 11,000 km above the Sun's photosphere, the temperature rises abruptly to about 1 million K. Sound or magnetic waves propagating outward from the Sun's convective zone may provide the energy that heats the Sun's tenuous outer layers to such extraordinarily high temperatures. (Adapted from A. Gabriel)

corona. The essay that follows this chapter discusses recent advances in our understanding of coronal heating.

Observations from space show that the corona has a complicated structure that is the site of considerable activity. For example, Figure 18-9 shows a huge bubble called a **solar transient**. These short-lived disturbances erupt suddenly and expand rapidly outward through the corona. Solar transients probably occur as often as once a day, but they were unknown until the Sun could be examined with a coronagraph carried aloft on board Skylab in the 1970s. Solar transients were later identified as the source of bursts of radio noise that are easily detected on Earth.

Because of the corona's high temperature, its ionized atoms are traveling very rapidly. High speed collisions between these ions boost their electrons to upper energy levels. As the electrons fall back to lower levels, they emit radiation, primarily at X-ray wavelengths. The Skylab astronauts as well as brief rocket flights and Earth-orbiting satellites have carried cameras above the Earth's atmosphere to obtain X-ray photographs of the corona. These pictures reveal a blotchy, irregular inner corona (Figure 18-10). Note the large dark area, called a **coronal hole** because it is devoid of the usual hot, glowing coronal gases. Many astronomers suspect that coronal holes are the main corridors through which particles of the solar wind escape from the Sun.

In addition to dark coronal holes, X-ray photographs also reveal numerous bright spots that are hotter than the surrounding corona. The temperatures in these bright points occasionally reach 4 million K. Many of the hot spots seen in X-ray pictures are located over one of the Sun's most familiar features—sunspots.

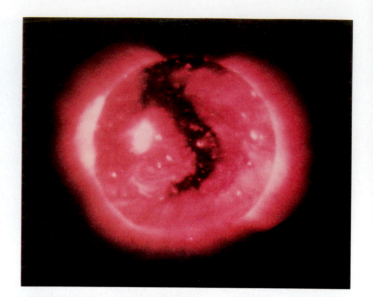

FIGURE 18-10 A Coronal Hole in an X-Ray Photograph This X-ray picture of the Sun was taken by Skylab astronauts on August 21, 1973. A huge, dark, boot-shaped coronal hole dominates this view of the inner corona. Numerous bright points are also visible. (NASA; Harvard College Observatory)

18-4 The number of sunspots varies with an 11-year period

Superimposed on the basic structure of the solar atmosphere are a host of phenomena that recur with an 11-year period and that provide clues to a more fundamental 22-year solar cycle. Irregularly shaped dark regions called **sunspots** are the most easily recognized of these phenomena, because they occur in the photosphere (Figure 18-11). The dark central core of a sunspot, called the **umbra**, is usually surrounded by a lighter border called the **penumbra**. Although sunspots vary greatly in size, typical ones measure a few tens of thousands of kilometers across.

The darkened appearance of sunspots means they are cooler than the surrounding photosphere. As we saw in Chapter 5, the Stefan–Boltzmann law relates the energy output to temperature: The energy flux from a blackbody is proportional to the fourth power of its temperature. Using this relationship, astronomers find that the reduced brightness of sunspots corresponds to temperatures of 4000 to 4500 K, or more than 1000 K cooler than the surrounding, undisturbed photosphere.

On rare occasions, a sunspot group is large enough to be seen with the naked eye. Chinese astronomers recorded such sightings 2000 years ago, and a huge sunspot group visible to the naked eye was seen in 1989. **ALWAYS USE SPECIAL DARK FILTERS OR OTHER MEANS TO PROTECT YOUR EYES WHEN OBSERVING THE SUN! Looking directly at the Sun can cause blindness.** Of course, a telescope gives a much better view, and so it was not until Galileo that anyone was able to examine sunspots in detail. In fact,

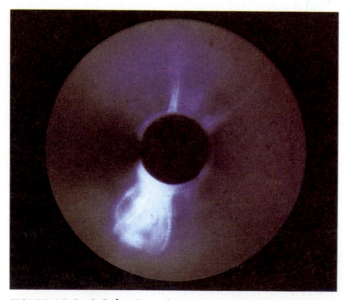

FIGURE 18-9 A Solar Transient During the early 1970s, astronauts on Skylab discovered huge bubbles erupting outward through the Sun's corona. The leading edge of this bubble moved outward from the Sun at a speed of 500 km/s. This solar transient was photographed on June 10, 1973. (NASA; High Altitude Observatory)

Galileo discovered that he could determine the Sun's rotation rate by tracking sunspots as they moved across the solar disk (Figure 18-12). He found that the Sun rotates once in about four weeks. A typical sunspot group lasts about two months, so a specific one can be followed for two solar rotations.

More careful observations by the British astronomer Richard Carrington in 1859 demonstrated that the Sun does not rotate as a rigid body. Carrington found that the equatorial regions rotate more rapidly than the polar regions. A sunspot near the solar equator takes only 25 days to go once around the Sun, while at 30° north or south of the equator a sunspot takes $27\frac{1}{2}$ days to complete a rotation. The rotation period at 75° north or south of the solar equator is about 33 days, and at the poles it may be as long as 35 days. This phenomenon is known as **differential rotation.**

Careful observations over many years revealed that the numbers of sunspots change periodically. In some years there are many sunspots, in other years almost none. This phenomenon, first reported by the German astronomer Heinrich Schwabe in 1843 after many years of observing, is called the **sunspot cycle.** As shown in Figure 18-13, the average number of sunspots varies with a period of about 11 years. A time of exceptionally many sunspots is a **sunspot maximum,** as

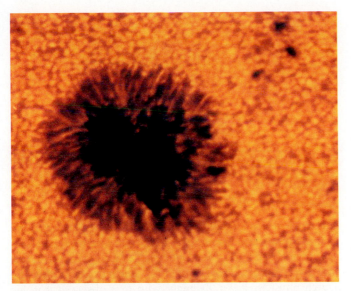

FIGURE 18-11 A Sunspot Group This high-resolution photograph shows a mature sunspot. The dark center of the spot is the umbra; the penumbra is less dark and has a featherlike appearance. Granulation is visible in the surrounding, undisturbed photosphere. (NOAO)

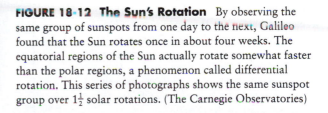

FIGURE 18-12 The Sun's Rotation By observing the same group of sunspots from one day to the next, Galileo found that the Sun rotates once in about four weeks. The equatorial regions of the Sun actually rotate somewhat faster than the polar regions, a phenomenon called differential rotation. This series of photographs shows the same sunspot group over $1\frac{1}{2}$ solar rotations. (The Carnegie Observatories)

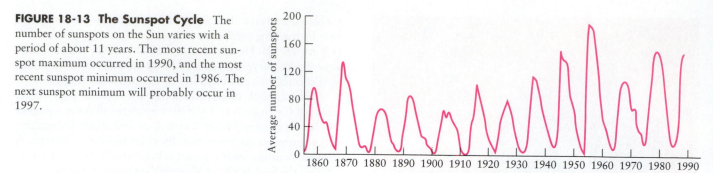

FIGURE 18-13 The Sunspot Cycle The number of sunspots on the Sun varies with a period of about 11 years. The most recent sunspot maximum occurred in 1990, and the most recent sunspot minimum occurred in 1986. The next sunspot minimum will probably occur in 1997.

occurred in 1968, 1979, and 1990. Conversely, the Sun was almost devoid of sunspots at times of **sunspot minimum** in 1964, 1975, and 1986.

The locations of the sunspots also vary over a sunspot cycle. At the beginning of a cycle, just after sunspot minimum, sunspots first start to appear at latitudes 30° north and south of the solar equator. Over the succeeding years, the sunspots occur closer and closer to the equator, until at the end of the cycle they are virtually all on the solar equator. Observations showing this "equatorial migration" of sunspots are displayed in Figure 18-14. Because the data on this graph cover areas shaped somewhat like butterflies, it is called a **Maunder butterfly diagram,** after E. Walter Maunder, who discovered this phenomenon of sunspot migration in 1904.

18-5 Sunspots are one of many phenomena produced by the Sun's magnetic field and a 22-year solar cycle

In 1908 the American astronomer George Ellery Hale made the important discovery that sunspots are associated with intense magnetic fields on the Sun. When Hale focused a spectroscope on sunlight coming from a sunspot, he found that many spectral lines appear to be split into several closely spaced spectral lines. This "splitting" of a single spectral line into two or more lines is called the **Zeeman effect** after the Dutch physicist Pieter Zeeman, who first observed it in his laboratory in 1896. Zeeman showed that the spectral line is split when the atoms are subjected to an intense magnetic field. The more intense the magnetic field, the wider the separation of the split lines in the Zeeman effect (see Box 18-2).

Hale's discovery proved that sunspots are places where a powerful, concentrated magnetic field protrudes through the hot gases of the photosphere. Because of the temperature in the photosphere, many of the atoms there are ionized. The photosphere is thus a hot mixture of electrically charged ions and electrons, technically called a **plasma.** Because it is an extremely good conductor of electricity, a plasma interacts vigorously with magnetic fields, which are able to constrain the plasma's motions.

A dramatic example of how magnetic fields can affect a plasma are huge arching columns of gas called **prominences,** which often appear above sunspots (Figure 18-15). Some of these formations, called *quiescent prominences,* may hang suspended for days above the solar surface. In contrast, *erup-*

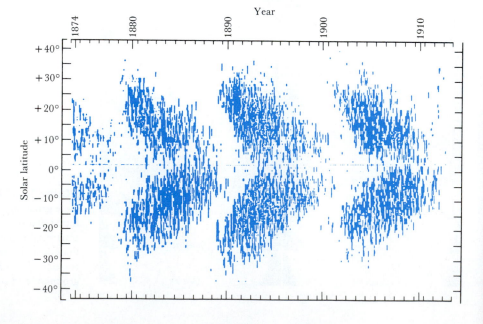

FIGURE 18-14 The Maunder Butterfly Diagram The location of sunspots varies in a regular fashion over the sunspot cycle. The first sunspots in a cycle appear at great distances from the solar equator, but the last spots are formed very near the equator. At sunspot maximum, in the middle of the cycle, most sunspots occur at latitudes of 10° to 15° north and south of the equator.

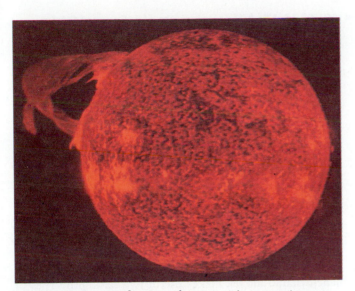

FIGURE 18-15 A Solar Prominence A huge prominence arches above the solar surface in this Skylab photograph taken December 19, 1973. The radiation that exposed this picture is from singly ionized helium (He II) at a wavelength of 30.4 nm, which corresponds to a temperature of about 50,000 K. (Naval Research Laboratory)

tive prominences blast material outward from the Sun at speeds of roughly 1000 km/s.

The most violent, eruptive events on the Sun, called **solar flares,** occur in complex sunspot groups. Within only a few minutes, temperatures in a compact region may soar to 5 million K. Vast quantities of particles and radiation are blasted out into space. Most astronomers believe that flares involve sudden changes where the Sun's magnetic field is strongly concentrated.

The magnetic effects of a sunspot also reach into the corona, where they cause distinctive emission lines in the spectra of highly ionized atoms. This high-altitude activity allows astronomers to study the solar cycle toward the polar latitudes, where the angle of Earth-based observation prohibits

a clear view of the solar surface. In 1987, for instance, Richard C. Altrock reported that his long-term observations of a green emission line of Fe XIV showed that each sunspot cycle begins much earlier than had previously been believed. The magnetic effects of a new cycle actually start at latitudes of 75° north and south of the solar equator several years before the sunspots of the previous cycle have faded away. The locations of these magnetic disturbances gradually migrate toward the equator. Their paths merge with the Maunder butterfly diagram (see Figure 18-14) at about 30° from the equator when the first sunspots of a traditional cycle appear at that latitude. Thus the sunspot cycle is actually a series of overlapping cycles.

Astronomers construct artificial pictures known as **magnetograms,** which display the magnetic fields in the solar atmosphere, by combining two photographs taken at wavelengths on either side of a magnetically split spectral line. The magnetogram in Figure 18-16 shows the entire solar surface, along with an ordinary photograph of the Sun taken at the same time. Dark blue indicates the areas of the photosphere covered by one magnetic polarity (north), and yellow indicates the area covered by the opposite (south) magnetic polarity.

Many sunspot groups are said to be **bipolar,** meaning that they have roughly comparable areas covered by north and south magnetic polarities (Figure 18-17). The sunspots on the side of the group toward which the Sun is rotating are called the "preceding members" of the group. The spots that follow behind are referred to as the "following members."

After years of studying the solar magnetic field, George Ellery Hale was able to piece together a remarkable magnetic description of what we now call the solar cycle. First of all, Hale discovered that the preceding members of all sunspot groups in one solar hemisphere have the same magnetic polarity. In fact, this polarity is the same as that of the hemisphere in which the group is located. In other words, in the hemisphere of the Sun's north magnetic pole, the preceding members of all sunspot groups have north magnetic polarity. Likewise, in the south magnetic hemisphere all the preceding members have south magnetic polarity.

FIGURE 18-16 Two Views of the Sun The optical view of the Sun (right) was taken at the same time as the magnetogram (left). Notice how regions with strong magnetic fields are related to the locations of sunspots. Careful examination of the magnetogram also reveals that the magnetic polarity of the sunspots in the northern hemisphere is opposite that of those in the southern hemisphere. (NOAO)

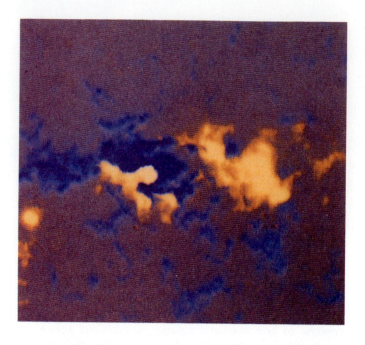

◄ FIGURE 18-17 A Magnetogram of a Sunspot Group
This artificially colored picture displays the intensity and polarity of the magnetic field associated with a large sunspot group. In a typical sunspot group, one side has one magnetic polarity, and the other side has the opposite polarity. (NOAO)

In addition, Hale found that the polarity pattern completely reverses itself every 11 years. In other words, the hemisphere with a north magnetic polarity at one solar maximum has a south magnetic polarity at the next solar maximum. For this reason astronomers prefer to speak of the entire **solar cycle** as having a period of 22 years, instead of the sunspot cycle of 11 years.

In 1960 the American astronomer Horace Babcock proposed a description that seems to account for many features of the 22-year solar cycle. Babcock's scenario, called a **magnetic-dynamo model**, makes use of two basic properties of the Sun: its differential rotation and its convective envelope. Differential rotation causes the magnetic field in the photo-

---

BOX 18-2

The Zeeman Effect

In the 1890s the Dutch physicist Pieter Zeeman placed some hot gases between the poles of a powerful magnet and examined the spectra of the glowing gases through a spectroscope. He found that each spectral line was split into two, three, or more closely spaced lines, depending on the particular gases used. Electrons circle the nuclei of atoms at specific energy levels that are dictated by the laws of quantum mechanics. As we saw in Chapter 5, only certain energy levels are allowed. Spectral lines are produced when electrons jump from one energy level to another.

An electron moving around a nucleus is actually a miniature electric current, which produces a tiny magnetic field. When an atom is placed in a magnetic field, the tiny magnetic fields caused by the atom's moving electrons try to line up with the overall field. According to the principles of quantum mechanics, this alignment occurs only at specific angles. Each one of these angles corresponds to a slightly different energy level for the electron. The number of allowed inclinations increases for higher energy levels.

The diagram shows an example of how the Zeeman effect is produced. When the magnetic field is not present, electrons jumping from a lower energy level to a higher level absorb photons of a specific wavelength, producing a single absorption line in the spectrum. However, when the magnetic field is turned on around the light source, the lower energy level is replaced by two slightly different energy levels, corresponding to two different alignments of a particular electron orbit in the outside magnetic field. Similarly, the higher energy level is replaced by four slightly different energy levels. Eight different electron jumps are now possible, corresponding to the absorption of photons

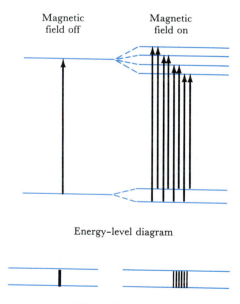

Magnetic field off   Magnetic field on

Energy-level diagram

Observed spectrum

having eight slightly different wavelengths. In this case, some of the jumps involve nearly the same energy, so only six distinct spectral lines can be resolved in the spectrum with the magnetic field turned on.

The stronger the magnetic field, the greater the difference in the energy levels. Thus, astronomers can deduce the strength of the magnetic field surrounding the atoms or ions that emit the light by measuring the separations of the lines. The strength of a magnetic field is usually expressed

sphere to become wrapped around the Sun (Figure 18-18). As a result, the magnetic field then becomes concentrated at latitudes on either side of the solar equator. Convection in the photosphere causes the concentrated magnetic field to become tangled, and "kinks" erupt through the solar surface. Sunspots appear where the magnetic field protrudes through the photosphere. Careful examination of the orientation of the magnetic field in Figure 18-18 reveals that the preceding members of a sunspot group should indeed have the same magnetic polarity as the hemisphere in which they are located.

There are many things that the magnetic-dynamo model fails to explain, however. Most embarrassing is that it does not tell us what holds a sunspot together week after week. Our best calculations predict that a sunspot should break up and disperse almost as soon as it forms.

Our understanding of sunspots is further confounded by irregularities in the solar cycle. For example, the overall reversal of the Sun's magnetic field is often piecemeal and haphazard. One pole may reverse polarity long before the other.

For several weeks the Sun's surface may show two north poles and no south pole at all. To make matters worse, there seem to be times when all traces of sunspots and the sunspot cycle vanish for many years. For example, virtually no sunspots were seen from 1645 through 1715, a period called the **Maunder minimum.** Similar sunspot-free periods apparently occurred at irregular intervals in earlier times as well.

The Maunder minimum suggests a connection between the Sun and the Earth's weather. During this particular sunspot-free period, Europe experienced years of record low temperatures often referred to as the "Little Ice Age," whereas the western United States was subjected to severe drought. Some scientists speculate that solar activity may have an ongoing, though subtle, effect on terrestrial climates. Recent research demonstrates that the Sun is actually fainter during sunspot minimum than it is at sunspot maximum. During sunspot maximum, regions of bright emission in the chromosphere, called **plages**, seem to more than compensate for the energy output lost because of sunspots in the photosphere.

---

in terms of a unit called the **gauss** (G), named after Karl Friedrich Gauss, a German mathematician who experimented with magnets in the early 1800s. For example, the strength of the Earth's magnetic field at its north and south magnetic poles is about 0.7 G.

The two pictures show the Zeeman splitting of a spectral line of a sunspot's magnetic field. On the left, a

black line drawn across the sunspot indicates the location toward which the slit of the spectrograph was aimed. On the right is the resulting spectrogram, which shows an iron line split into three components. The separation between the three lines corresponds to a magnetic field strength in excess of 4130 G.

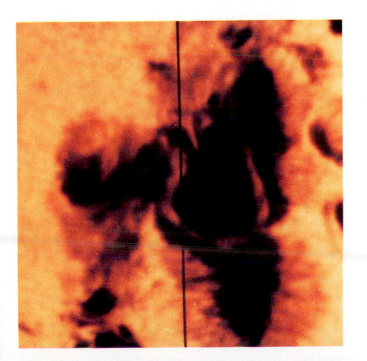

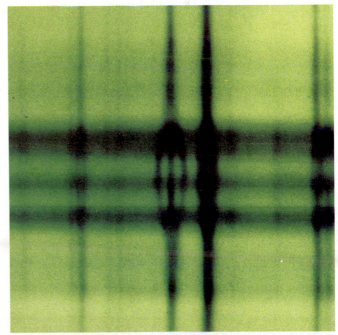

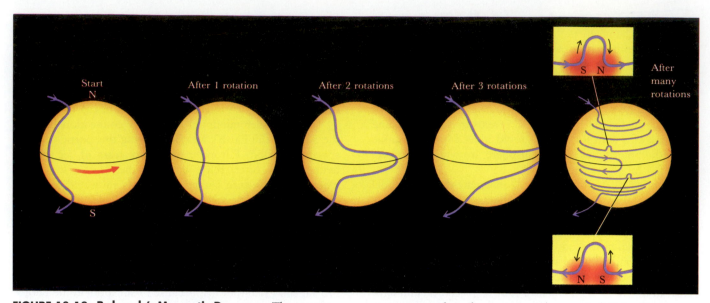

FIGURE 18-18 Babcock's Magnetic Dynamo The sunspot cycle may in part result from the way the Sun's magnetic field becomes wrapped around the Sun by differential rotation. Sun- spots appear where the concentrated magnetic field breaks through the solar surface.

Not so subtle would be the effects of a varying solar diameter. Jack Eddy of the High Altitude Observatory in Colorado and other astronomers have recently reported controversial observations suggesting that the Sun may be shrinking slightly. This topic is currently a matter for heated debate, because astronomers are at a loss to explain these reported changes in the Sun's size.

18-6   The Sun's energy is produced by thermonuclear reactions in the core of the Sun

During the nineteenth century, geologists and biologists found convincing evidence that the Earth had existed for hundreds of millions of years. This posed a severe problem for physicists: How could the Sun have been shining for so long, radiating immense amounts of energy into space? If the Sun were made of coal, for example, which produces its heat and light by chemical burning, it could last for only 5000 years.

One attempt to explain solar energy was made in the mid-1800s by Lord Kelvin, after whom the temperature scale is named, and Hermann von Helmholtz. They argued that the tremendous weight of the Sun's outer layers should cause the Sun to contract gradually, compressing its interior gases. Whenever a gas is compressed, its temperature rises. (You can demonstrate this effect yourself with a bicycle pump; air becomes warm as you compress it and push it into a tire.) Kelvin and Helmholtz thus suggested that gravitational con- traction could cause the Sun's gases to become hot enough to radiate energy out into space.

This process, called **Kelvin–Helmholtz contraction,** actu- ally does occur during the earliest stages of the birth of a star. As a large ball of interstellar gas shrinks in response to the inward pull of its own gravity, gravitational energy (the en- ergy associated with gravitational forces) is converted into thermal energy (the energy associated with heat). The gases then glow with a spectrum much like that of a blackbody.

Kelvin–Helmholtz contraction cannot, however, be the major source of the Sun's energy. Helmholtz's own calcula- tions showed that in order to be as bright as it is today, the Sun would have extended beyond the Earth's orbit only 25 million years ago! Yet the Earth has taken hundreds of mil- lions of years to produce its present surface and life forms.

The key to the source of the Sun's energy was discovered in 1905 by Albert Einstein. According to his special theory of relativity, matter can be converted into energy according to the simple equation

$$E = mc^2$$

In words, a mass *(m)* can be converted into an amount of energy *(E)* equivalent to mc^2, where *c* is the speed of light. Because *c* is a large number, and thus c^2 is huge, a small amount of matter can release an awesome amount of energy.

Astronomers then began to wonder if the Sun's energy output might come from the conversion of matter into en- ergy. In the 1920s the British astronomer Arthur Eddington showed that temperatures near the center of the Sun must be much greater than had previously been thought. Another British astronomer, Robert Atkinson, suggested hydrogen nuclei near the Sun's center might fuse together to produce helium nuclei in a reaction that transforms a tiny amount of mass into a large amount of energy. This process of **nuclear fusion,** by which hydrogen is converted into helium at the Sun's center, is called **hydrogen burning,** even though nothing is actually burned in the conventional sense. (Ordinary burn- ing, like a log in a fireplace, involves chemical reactions that rearrange the outer electrons of atoms but have no effect on their nuclei.)

Recall that the nucleus of a hydrogen atom (H) consists of a single proton. The nucleus of a helium atom (He) consists of two protons and two neutrons. In the nuclear process described by Atkinson, four hydrogen nuclei would combine to form one helium nucleus, with a concurrent release of energy:

$$4 \, H \rightarrow He + energy$$

In several separate reactions, two of the four protons from hydrogen are changed into neutrons and eventually combine to produce a single helium nucleus. These reactions also release two positively charged electrons, called **positrons,** that carry off the electric charges relinquished by the protons that changed into neutrons. In addition, two seemingly massless particles, called **neutrinos,** carry off some energy and momentum. All together, the release of these four supplementary particles ensures that the reactions conserve electric charge, energy, and momentum.

When hydrogen is converted into helium, matter is lost. The ingredients (four hydrogen nuclei) have a combined mass just slightly more than the final product (one helium nucleus):

4 hydrogen atoms =	6.693×10^{-27} kg
−1 helium atom =	-6.645×10^{-27} kg
Mass lost =	0.048×10^{-27} kg

Thus, a small fraction (0.7%) of the mass of the hydrogen going into the nuclear reaction does not show up in the mass of the helium. This lost mass is converted into energy predicted by the equation $E = mc^2$:

$$E = mc^2 = (0.048 \times 10^{-27} \text{ kg})(3 \times 10^8 \text{ m/s})^2$$

$$= 4.3 \times 10^{-12} \text{ joule}$$

This is only a tiny amount of energy, because it results from the creation of just a single helium atom. However, consider the conversion of 1 kg of hydrogen into helium. Although a kilogram of hydrogen goes into this reaction, only 0.993 kg of helium comes out. Using Einstein's equation, we find that the missing 0.007 kg of matter has been transformed into 6.3×10^{14} joules of energy. For comparison, this equals the energy released by burning 20,000 metric tons (2×10^7 kg) of coal.

The Sun's total power output, called its **luminosity** ($L_\odot$), is 3.9×10^{26} watts. To produce this luminosity, 6×10^{11} kg of hydrogen must be converted into helium within the Sun each second. This rate is prodigious, but the Sun contains a vast amount of hydrogen. The Sun's total mass, usually designated $M_\odot$, is 2×10^{30} kg (333,000 Earth masses). The Sun's core contains enough hydrogen to continue to give off energy at the present rate for another five billion years.

The Sun's energy actually comes from a *series* of nuclear reactions, called the **proton–proton chain** in which hydro-gen nuclei combine to form helium. In stars whose central temperatures are hotter than the Sun's, however, hydrogen burning also proceeds according to a different set of nuclear reactions, called the **CNO cycle,** in which carbon, nitrogen, and oxygen nuclei absorb protons to produce helium nuclei. Details of the proton–proton cycle and the CNO cycle are discussed in Box 18-3.

All of these nuclear reactions require the extremely high temperatures of a star's center. Normally at lower temperatures, the positive electric charges on protons and nuclei keep these particles separated, because like charges repel each other. But in the extreme heat of a star's center, the particles move so fast that they can penetrate each other's electric repulsion so deeply that a powerful short-range nuclear force takes over and causes the particles to stick together. Hydrogen burning is called a **thermonuclear reaction,** or **thermonuclear fusion,** because it can occur only at high temperatures. In later chapters, we will see that other thermonuclear reactions, such as helium burning, carbon burning, and oxygen burning, occur late in the lives of many stars.

18-7 A theoretical model of the Sun shows how energy gets from the Sun's center to its surface

Although the Sun's interior is hidden from our view, we can use the laws of physics to calculate what is going on below the solar surface. The results of such computations constitute a **model** of the Sun, which tells us physical characteristics at various depths inside the star. A model of the Sun can be represented as a table or a graph, to show how such quantities as pressure, temperature, and density change with depths beneath the solar surface.

To develop a model of the Sun or any other stable star, we first note that the Sun is not undergoing any dramatic changes. The Sun is not exploding or collapsing; nor is it significantly heating or cooling. The Sun is thus in balance both mechanically and thermally.

Mechanical balance, often called **hydrostatic equilibrium,** simply means that each layer of a star can support the weight of the layers above it. Because of gravity, the tremendous weight of the Sun's outer layers pressing inward from all sides tries to make the gases contract. However, as gravity compresses the Sun, gas pressure and nuclear fusion both increase inside the star at just the right rate to support the weight of the overlying layers. This condition governs the rate at which pressure increases toward deeper layers.

Thermal balance, often called **thermal equilibrium,** simply means that a star keeps on shining. Vast amounts of energy escape from the Sun's surface each second. In a state of thermal equilibrium, this energy is constantly resupplied from the Sun's interior so that temperatures throughout the star do not fluctuate.

But exactly how is energy transported from the Sun's center to its surface? There are three methods of energy transport: conduction, convection, and radiative diffusion. Only the last two are operating inside the Sun.

BOX 18-3

The Proton–Proton Chain and the CNO Cycle

There are two avenues by which hydrogen burning proceeds inside a star. Both yield the same result: Four hydrogen protons combine to form one helium nucleus, and a slight amount of mass is converted into energy. The temperature at the star's core determines which avenue is taken.

For stars with masses not greater than the Sun's mass, the central temperature does not exceed 16 million K, and hydrogen burning proceeds predominantly via the proton–proton chain. For stars more massive than the Sun, the central temperature is above 16 million K, and hydrogen burning occurs primarily through a series of reactions called the CNO cycle.

The proton–proton chain has four branches. The primary branch, which produces 85% of the Sun's energy, occurs in three steps (see accompanying diagram). First, two protons (each denoted by ^{1}H) combine to form an isotope of hydrogen called **deuterium** (^{2}H), which consists of one proton and one neutron bound together. In this step, one of the two protons turns into a neutron, releasing a positron (e^+) and a neutrino (ν). Thus, the first step is

$$^1\text{H} + {}^1\text{H} \rightarrow {}^2\text{H} + e^+ + \nu$$

A positron (e^+) is just like an ordinary electron (e^-) except that it has a positive rather than a negative electric charge.

Most matter is virtually transparent to a neutrino. Consequently, neutrinos are very difficult to detect, and their properties are currently a topic of research and debate. Neutrinos have no charge, but they may have a small mass. A neutrino is produced whenever a proton is converted into a neutron. Conversely, an antineutrino ($\bar{\nu}$) is liberated when a neutron turns into a proton.

In the second step of the proton–proton chain, a third proton combines with the deuterium nucleus to produce a low-mass isotope of helium (^{3}He), whose nucleus contains two protons and one neutron. This reaction releases energy (γ stands for a gamma-ray photon):

$$^1\text{H} + {}^2\text{H} \rightarrow {}^3\text{He} + \gamma$$

In the final step, two ^{3}He nuclei combine to produce an ordinary nucleus of helium (^{4}He = 2 protons + 2 neutrons), with the release of two protons:

$$^3\text{He} + {}^3\text{He} \rightarrow {}^4\text{He} + {}^1\text{H} + {}^1\text{H}$$

In the other branches of the proton–proton chain, the ^{3}He nucleus follows different fates. About 15% of the Sun's energy comes from a branch of the proton–proton chain that temporarily creates beryllium (Be) and lithium (Li). An ^{3}He combines with an ^{4}He nucleus to produce ^{7}Be:

$$^3\text{He} + {}^4\text{He} \rightarrow {}^7\text{Be} + \gamma$$

The beryllium nucleus then captures an electron and becomes ^{7}Li:

$$^7\text{Be} + e^- \rightarrow {}^7\text{Li} + \nu$$

Finally, the lithium nucleus captures a proton and breaks apart into two helium nuclei:

$$^7\text{Li} + {}^1\text{H} \rightarrow 2\ {}^4\text{He}$$

The remaining branch of the proton–proton chain, which together produces only 0.02% of the Sun's energy, involves similar reactions. Although they account for only a tiny fraction of the Sun's luminosity, these branches are important because they also produce neutrinos. Teams of physicists around the world are currently conducting experiments to detect neutrinos from the various branches of the proton–proton chain.

In massive stars, where the central temperatures exceed 16 million K, hydrogen burning proceeds with carbon as a catalyst. Cornell University physicist Hans Bethe was instrumental in discovering the steps by which a carbon nucleus absorbs protons and finally emits a helium nucleus. Along the way, the carbon is transformed into nitrogen and oxygen; hence this process is called the CNO cycle. There are six steps in the CNO cycle:

$$^{12}\text{C} + {}^1\text{H} \rightarrow {}^{13}\text{N} + \gamma$$

$$^{13}\text{N} \rightarrow {}^{13}\text{C} + e^+ + \nu$$

$$^{13}\text{C} + {}^1\text{H} \rightarrow {}^{14}\text{N} + \gamma$$

$$^{14}\text{N} + {}^1\text{H} \rightarrow {}^{15}\text{O} + \gamma$$

$$^{15}\text{O} \rightarrow {}^{15}\text{N} + e^+ + \nu$$

$$^{15}\text{N} + {}^1\text{H} \rightarrow {}^{12}\text{C} + {}^4\text{He}$$

For the CNO cycle to proceed, the ^{12}C nucleus, which acts as a nuclear catalyst, must be present. Because it is restored at the end of the cycle, however, the process does not use up any carbon.

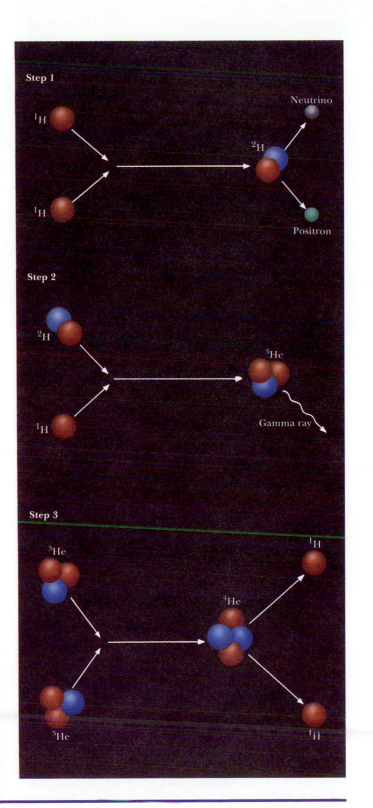

Step 1

^{1}H
^{1}H
^{2}H
Neutrino
Positron

Step 2

^{2}H
^{1}H
^{3}He
Gamma ray

Step 3

^{3}He
^{3}He
^{4}He
^{1}H
^{1}H

Experience teaches us that energy always flows from hot regions to cooler ones. For example, if you heat one end of a metal bar with a blowtorch, the other end of the bar eventually becomes warm. The efficiency of this method of energy transport, called **conduction**, varies significantly from one substance to another. For example, copper is a good heat conductor, but wood is not. Conduction is not an efficient means of energy transport inside stars like the Sun. But as we shall see in a later chapter, conduction is important in compact stars called white dwarfs.

Inside stars like our Sun, energy moves from center to surface by two other means: convection and radiative diffusion. As we saw earlier in this chapter, convection is the circulation of gases between hot and cool regions. Hot gases rise toward a star's surface, while cool gases sink back down toward the star's center. This physical movement of gases transports heat energy outward in a star.

In **radiative diffusion**, photons created in the thermonuclear inferno at a star's center diffuse outward toward the star's surface. Individual photons are absorbed and re-emitted by atoms and electrons inside the star. The overall result is an outward migration from the hot core, where photons are constantly created, toward the cooler surface, where they escape into space. In all, it takes roughly a million years for energy created at the Sun's center to reach the solar surface and finally escape as sunlight.

To construct a model of the Sun, astrophysicists bring together the concepts of hydrostatic equilibrium, thermal equilibrium, and energy transport. These concepts can be expressed as a set of equations, collectively called the **equations of stellar structure**. The solutions to these equations tell us about the conditions that must exist inside the star.

Astrophysicists use high-speed computers to solve the equations of stellar structure and develop a detailed theoretical model for the structure of a given star. The astrophysicist begins with astronomical data about the star's surface. (For example, the Sun's surface temperature is 5800 K, its luminosity is 3.9×10^{26} W, and the gas pressure and density at the surface are almost zero.) Using the equations of stellar structure, the astrophysicist then calculates conditions layer by layer in toward the star's center. The results tell us how temperature, pressure, and density increase with increasing depth below a star's surface. In this way, we have learned that the temperature at the Sun's center is 15.5 million K, the pressure is 3.4×10^{11} atm, and the density is 150,000 kg/m^3.

As a result of solving the equations of stellar structure, we know various physical quantities throughout a star, such as pressure, density, and temperature. This description constitutes the **stellar model**, which can be presented in tables or graphs. Table 18-1 and Figure 18-19 present a theoretical model of the Sun.

Figure 18-19 shows that the luminosity rises to 100% at about one-quarter of the way from the Sun's center to its surface. In other words, the Sun's energy production occurs within a volume extending out to $\frac{1}{4}$ solar radius (recall that

---

TABLE 18-1

A Theoretical Model of the Sun

Fraction of radius	Fraction of luminosity	Fraction of mass	Temperature ($\times 10^6$ K)	Density (kg/m³)	Fraction of central pressure
Center	0.00	0.00	15.5	160,000	1.00
0.1	0.42	0.07	13.0	90,000	0.46
0.2	0.94	0.35	9.5	40,000	0.15
0.3	1.00	0.64	6.7	13,000	0.04
0.4	1.00	0.85	4.8	4,000	0.007
0.5	1.00	0.94	3.4	1,000	0.001
0.6	1.00	0.98	2.2	400	0.003
0.7	1.00	0.99	1.2	80	4×10^{-5}
0.8	1.00	1.00	0.7	20	5×10^{-6}
0.9	1.00	1.00	0.3	2	3×10^{-7}
Surface	1.00	1.00	0.006	0.00030	4×10^{-13}

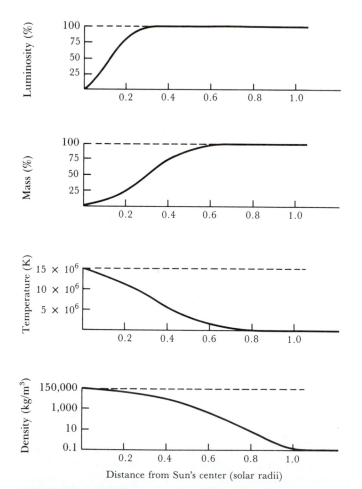

FIGURE 18-19 A Theoretical Model of the Sun The Sun's internal structure is displayed here with graphs that show how the density, temperature, mass, and luminosity vary with the distance from the Sun's center.

1 solar radius is nearly 700,000 km). Also note that the mass rises to nearly 100% at 0.6 solar radius. Consequently, the outer 0.4 solar radius contains very little matter.

The energy flow from the Sun's center toward its surface depends on how easily photons move through the gas. If the solar gases are comparatively transparent, photons can travel moderate distances before being scattered or absorbed, and so energy is transported by radiative diffusion. If the gases are rather opaque, photons cannot get through the gas easily and heat builds up. The gases start to churn as convection carries hot regions upward while cooler regions sink downward. From the center of the Sun out to about three-quarter solar radius, energy is transported by radiative diffusion, and so this region is called the **radiative zone.** Convection dominates the energy flow in the Sun's tenuous outer layer. We thus say that the Sun has a **convective zone,** or **convective envelope.** These aspects of the Sun's internal structure are sketched in Figure 18-20. In all, it takes about a million years for a photon to make its way from the Sun's center to its surface.

18-8 The mystery of the missing neutrinos inspires speculation about the Sun's interior

During hydrogen burning, protons change into neutrons and release a neutrino in the process. We should therefore expect a great outflow of neutrinos from the Sun's center.

Neutrinos have no electric charge and are believed to have little or no mass. If they are massless, they would travel through space at the speed of light, just as photons do. Some physicists theorize that neutrinos might have a tiny mass, probably less than one ten-thousandth the mass of an electron. If neutrinos do have mass, they would travel at speeds slightly less than that of light.

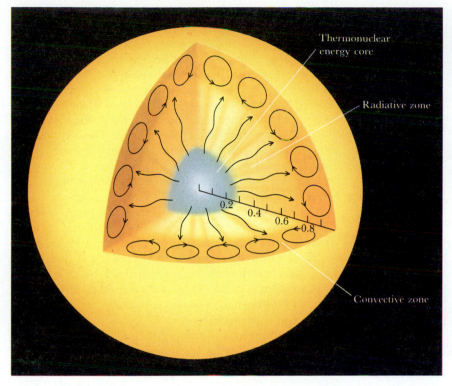

FIGURE 18-20 The Sun's Internal Structure Thermonuclear reactions occur in the Sun's core, which extends out to a distance of 0.25 solar radius from the center. Energy is transported outward, via radiative diffusion, to a distance of about three-quarter solar radius. Convection is responsible for energy transport in the Sun's outer layers.

Neutrinos are difficult to detect because they do not interact much with ordinary matter. In fact, neutrinos interact so weakly with ordinary matter that they pass through the Earth as if it were not there. On rare occasions, a neutrino might strike a neutron and convert it into a proton. Such collisions are highly improbable, however, so even objects as big and massive as the Earth are virtually transparent to neutrinos.

The Sun is also largely transparent to neutrinos, allowing these particles to stream outward, unimpeded, from its core. If astronomers could detect these particles, they would have an excellent means of probing the Sun's center. The conversion of hydrogen into helium at the Sun's center produces 10^{38} neutrinos each second. This output is so enormous that the number of neutrinos passing through each square meter of the Earth is quite large: about 10^{14} neutrinos per square meter every second. In other words, roughly 100 billion neutrinos pass through every square inch of your body every second! If astronomers could detect these particles, they might be able build "neutrino telescopes" that could "see" the thermonuclear inferno at the Sun's core.

Inspired by such possibilities, Raymond Davis of the Brookhaven National Laboratory designed and built a large neutrino detector. This device uses 100,000 gallons of perchloroethylene cleaning fluid (C_2Cl_4) in a huge tank buried deep in a mine in South Dakota (Figure 18-21). Because matter is virtually transparent to neutrinos, most of the neutrinos from the Sun pass right through Davis's tank, with no effect whatsoever. On rare occasions, though, a neutrino strikes the nucleus of one of the chlorine atoms (^{37}Cl) in the cleaning fluid and converts one of its neutrons into a proton, creating a radioactive atom of argon (^{37}A). The rate at which this argon is produced is related to the **flux** of solar neutrinos (i.e., the number of neutrinos from the Sun arriving at the Earth per square meter per second). By regularly flushing the entire 100,000 gallons of cleaning fluid though detectors that count the number of newly created argon atoms, Davis was able to measure the neutrino flux from the Sun.

This experiment, which began in the mid-1960s, has been repeated with extreme care for almost 30 years. On the average, solar neutrinos create one radioactive argon atom every three days in Davis's tank. To the continuing consternation of astronomers, this rate corresponds to only one-third of the neutrino flux predicted from standard models of the Sun. This troubling discrepancy between theory and observation has motivated teams of physicists around the world to conduct new experiments to measure the flux of solar neutrinos.

The various nuclear reactions in the proton–proton chain produce neutrinos of differing energies. The vast majority of the neutrinos from the Sun are created during the first reaction, in which two protons combine to form a heavy isotope of hydrogen (see Box 18-3 for details). Because these neutrinos have too little energy to convert chlorine into argon, Davis's experiment does not detect them. Davis's equipment instead responds only to high-energy neutrinos, produced by reactions that occur only part of the time near the end of the proton–proton chain.

To measure the flux of the low-energy neutrinos produced by the first reaction of the proton–proton chain, teams of physicists have recently constructed neutrino detectors in

FIGURE 18-21 The Solar Neutrino Experiment This tank, buried $1\frac{1}{2}$ km below ground at the Homestake Gold Mine in South Dakota, contains 100,000 gallons of perchloroethylene (C_2Cl_4). The Earth shields the tank from stray particles so that only solar neutrinos can convert chlorine atoms in the fluid into radioactive argon atoms. The number of argon atoms created in the tank is a direct measure of the flux of neutrinos from the Sun. (Courtesy of R. Davis, Brookhaven National Laboratory)

Italy and Russia. In both experiments, low-energy neutrinos convert gallium (^{71}Ga) into a radioactive isotope of germanium (^{71}Ge). By chemically separating the germanium from the gallium and counting the radioactive atoms, scientists hope to deduce the low-energy neutrino flux from the Sun.

In the Italian experiment, called GALLEX, 30 tons of gallium chloride are buried in a tunnel under a mountain in central Italy. In 1992, after a year of taking data, the GALLEX team announced they had detected only 60% of the expected neutrino flux from the Sun. The Russian experiment, anachronistically called SAGE (for Soviet-American Gallium Experiment), uses 60 tons of gallium metal. Data from the first year of operation indicated about 50% of the expected flux. Both experiments therefore seem to confirm Davis's results that the Sun is emitting fewer neutrinos than standard solar models predict.

This continuing discrepancy between theory and observation is so troubling that many scientists suspect that either we do not fully understand neutrinos or there is something wrong with standard solar models. To solve the "mystery of the missing solar neutrinos," some scientists have theorized that neutrinos change as they travel through space; other scientists have proposed nonstandard solar models, and still others have suggested that the whole problem would vanish if certain kinds of hypothetical particles exist.

During the late 1980s, experiments in high-energy physics revealed that there are three different kinds of neutrinos. Only one of these, called the electron-type neutrino, is produced in the Sun; the other two types are created during experiments with high-energy particle accelerators. Some theorists argue that if neutrinos have mass, they may continually change from one type to another. This hypothetical transformation, called **neutrino oscillation,** could explain the low measurements of neutrino flux: by the time solar neutrinos reach the Earth, many of them have changed into the two exotic types, which the experiments cannot detect.

Another possible solution to the neutrino problem is that the Sun's core is cooler than solar models predict. If the Sun's center is only 10% cooler than the current estimate, fewer neutrinos would be produced and the neutrino flux would agree with experiments. However, if the Sun's central temperature were a million degrees cooler, other obvious features, such as the Sun's size and surface temperature, would differ from what we observe.

In 1985 American astrophysicists John Faulkner and Ron Gilliland proposed a third, controversial explanation for the missing solar neutrinos. They suggested that the universe may contain numerous "weakly interacting massive particles," or **WIMPS.** Like neutrinos, WIMPS can easily pass through vast amounts of matter. Unlike neutrinos, however, WIMPS are quite massive, perhaps several times more so than a proton or neutron. Over the ages, many billions of these hypothetical WIMPS could have been gravitationally captured by the Sun, so that today they are deep inside the Sun's thermonuclear core, freely orbiting its center. As they move along their orbits, these particles could transport energy from the Sun's center to the rest of the hydrogen-burning core. This simple movement of energy could lower the Sun's central temperature just enough to bring the neutrino flux into line with experimental results without significantly altering any other observable aspects of the Sun.

Although most astronomers doubt the existence of WIMPS, the Faulkner–Gilliland proposal is perhaps not as outlandish as it might at first seem. As we shall see when we study galaxies, most of the matter in the universe seems to be dark, cold, and quite invisible. Some studies suggest that ordinary matter constitutes only 10% of the mass of all the matter in the universe. The remaining 90%, which seems to be made of something quite different from ordinary atoms and elements, is very hard to detect. Perhaps solving the mystery of the Sun's center will be our first clue toward discovering what the universe is really made of.

18-9 Solar seismology and new satellites are among the latest tools for solar research

Many clues about the Sun's behavior lies buried deep within the Sun's interior. For instance, the Sun's magnetic field might have started as a "primordial field," which was compressed and trapped deep within the Sun as the solar nebula con-

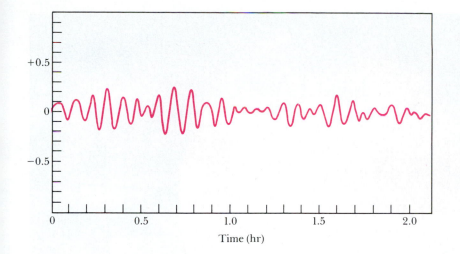

FIGURE 18-22 Vibrations of the Sun's Surface Doppler shift measurements give the speed at which the Sun's photosphere bobs up and down. These data extend over 2 hours and show numerous oscillations with a period of about 5 minutes. The gradual variation in the sizes of the oscillations results from interference between many vibrations and wavelengths. (Adapted from O. R. White)

tracted 4.5 billion years ago. To determine the roots of the Sun's magnetic field and to study other deeply buried phenomena, astronomers rely on vibrations within the Sun to probe the solar interior.

Geologists can determine the structure of the Earth's interior by using seismographs to record vibrations during earthquakes. Although there are no true sunquakes, the Sun does vibrate at a variety of frequencies, somewhat like a ringing bell. These vibrations were first noticed in 1960 by Robert Leighton at Caltech, who made high-precision Doppler shift observations of the photosphere. These measurements revealed that the Sun's surface moves up and down by about 10 km every five minutes (Figure 18-22). Since the mid-1970s, several astronomers have reported slower vibrations, having periods ranging from 20 to 160 minutes. The detection of extremely slow vibrations has inspired astronomers calling themselves "helioseismologists" to set up telescopes at the South Pole, where the Sun can be observed continuously for many days. This new field of solar research is called **solar seismology.**

The vibrations of the Sun's surface may be compared with sound waves. If you were within the Sun's atmosphere, you would first notice a deafening roar, somewhat like a jet engine, produced by the turbulence associated with granulation. Superimposed on this noise would be a variety of nearly pure tones. You would be unable to hear these tones, however, for they would be 16 octaves below the lowest note on the piano keyboard.

In 1970 Roger Ulrich at UCLA pointed out that sound waves moving upward from the solar interior would be reflected back in after reaching the solar surface. However, as a reflected sound wave descended back into the Sun, the increasing density and pressure would bend the wave so severely that it would be turned around and aimed back out again toward the solar surface. In other words, sound waves bounce back and forth between the solar surface and layers deep within the Sun. These sound waves can reinforce each other if their wavelength is the right size, just as sound waves of a particular wavelength resonate inside an organ pipe.

The Sun oscillates in millions of ways as a result of waves resonating in its interior. Figure 18-23 is a computer-generated illustration of one such vibrational mode. Helioseismologists can deduce information about the solar interior from measurements of these oscillations. For instance, it is possible to infer the rotation rate of the Sun's interior. By comparing the speeds of sound waves that travel east to west with those that travel west to east, helioseismologists have

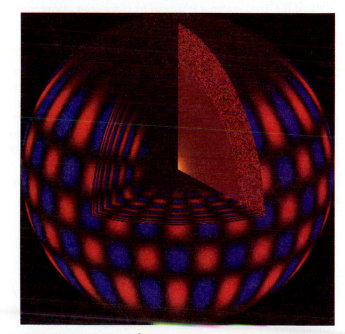

FIGURE 18-23 A Sound Wave Resonating in the Sun This computer-generated image shows one of the millions of ways in which the Sun vibrates because of sound waves resonating in its interior. The regions that are moving outward are colored blue, those moving inward, red. The cutaway shows how deep these oscillations are believed to extend. (National Solar Observatory)

FIGURE 18-24 Rotation Rates in the Solar Interior This cutaway picture of the Sun shows how the solar rotation rate varies with depth and latitude. Colors represent rotation periods according to the scale given at the right. Note that the surface rotation pattern, which varies from 25 days at the equator to 35 days near the poles, persists throughout the Sun's convective envelope. The Sun's radiative core seems to rotate like a rigid body. (Courtesy of K. Libbrecht; Big Bear Solar Observatory)

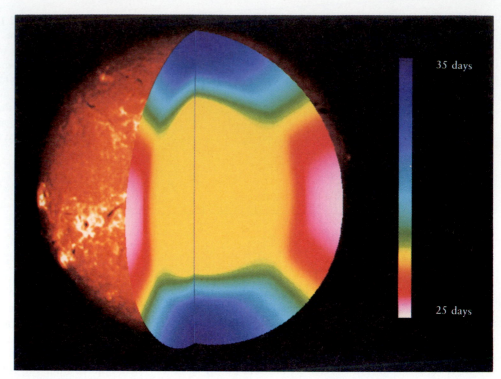

attempted to figure out the Sun's rotation rate at different depths and latitudes. As shown in Figure 18-24, the Sun's surface rotation pattern may persist throughout the outer 20% of the Sun, where it is driven by large-scale convection. Further in, convection ceases and the Sun seems to rotate like a rigid object with a period of about 27 days.

While some solar astronomers have been probing the Sun's core, others have observed the solar atmosphere from Skylab and with Earth-orbiting satellites. During the 1970s, a series of satellites called the Orbiting Solar Observatories provided a wealth of data about the Sun at ultraviolet, X-ray, and gamma-ray wavelengths. In 1980 the Solar Maximum Mission (SMM) spacecraft was launched to study solar flares. After being repaired by Space Shuttle astronauts in 1984, the satellite continued to send back spectra and pictures of violent solar phenomena until it fell to Earth in 1989. In late 1991, Japanese scientists launched *Yohkoh,* an Earth-orbiting satellite that is conducting a three-year study of solar flares in visible light and X rays. Meanwhile, the spacecraft *Ulysses,* a joint venture between NASA and the European Space Agency (ESA), is on its way to a polar orbit about the Sun. In 1994 and 1995, this spacecraft will give astronomers their first look at the Sun's polar regions. These missions will help unravel some of the mysteries that still surround the Sun.

KEY WORDS

Terms preceded by an asterisk are discussed in the boxes.

bipolar sunspot group	coronal hole	granule	magnetogram
chromosphere	*deuterium	hydrogen burning	Maunder butterfly diagram
CNO cycle	differential rotation (of the Sun)	hydrostatic equilibrium	
conduction		Kelvin–Helmholtz contraction	Maunder minimum
convection	equations of stellar structure		model (of the Sun)
convective envelope	flux	limb	neutrino
convective zone	*gauss	limb darkening	neutrino oscillation
corona	granulation	luminosity (of the Sun)	nuclear fusion
coronagraph		magnetic-dynamo model	penumbra (of a sunspot)

photosphere	radiative zone	spicule	thermal equilibrium
plage	solar atmosphere	stellar model	thermonuclear fusion
plasma	solar cycle	sunspot	thermonuclear reaction
positron	solar flare	sunspot cycle	umbra (of a sunspot)
prominence	solar interior	sunspot maximum	WIMPS
proton–proton chain	solar seismology	sunspot minimum	Zeeman effect
radiative diffusion	solar transient	supergranule	

KEY IDEAS

• The Sun's atmosphere has three main layers: the photosphere, the chromosphere, and the corona. Everything below the solar atmosphere is called the solar interior.

The visible surface of the Sun, the photosphere, is the lowest layer in the solar atmosphere. The gases in the photosphere shine with almost perfect blackbody radiation corresponding to a temperature of 5800 K; convection in the photosphere produces granules.

Above the photosphere is a layer of cooler, less dense gases called the chromosphere; spicules extend upward into the chromosphere from the photosphere along the boundaries of supergranules.

The outermost layer of the solar atmosphere, the corona, is made of very hot gases at a very low density. Activity in the corona includes solar transients, coronal streamers, and coronal holes. The solar corona blends into the solar wind at great distances from the Sun.

• The solar surface's features vary periodically in a 22-year solar cycle.

Sunspots are relatively cool regions produced by local concentrations of the Sun's magnetic field; the average number of sunspots increases and decreases in a regular cycle of approximately 11 years.

A solar flare is a brief eruption of hot, ionized gases from a sunspot group.

The magnetic-dynamo model suggests that many features of the solar cycle are caused by the effects of the Sun's differential rotation and convection on the Sun's magnetic field.

• The Sun's energy is produced by a thermonuclear process called hydrogen burning, in which four hydrogen nuclei combine to produce a single helium nucleus, releasing energy in the process.

The energy released in a nuclear reaction corresponds to a slight reduction of mass according to Einstein's equation $E = mc^2$.

A thermonuclear reaction is a nuclear reaction that occurs only at very high temperatures; for instance, hydrogen-burning reactions occur only at temperatures of more than about 8 million K.

• A stellar model is a theoretical description of a star's interior derived from calculations based upon the laws of physics.

The solar model suggests that hydrogen burning takes place in a core extending from the Sun's center to about 0.25 solar radius.

The solar core is surrounded by a radiative zone extending to about 0.8 solar radius, in which energy travels outward through radiative diffusion.

The radiative zone is surrounded by a convective zone of gases at relatively low temperatures and pressures in which energy travels outward primarily through convection.

• Space missions to observe the Sun from outside the Earth's atmosphere, along with studies of solar seismology, are expected to yield further understanding of the Sun—and hence of stellar processes in general—in coming years.

REVIEW QUESTIONS

1. Describe the dangers in attempting to observe the Sun. How have astronomers learned to circumvent these observational problems?

2. Briefly describe the three layers that make up the Sun's atmosphere. In what ways do they differ from each other?

3. What is solar granulation? Describe how convection gives rise to granules.

4. What is a spicule? How can you observe spicules?

5. What is the difference between granules and supergranules?

6. How do astronomers know that the corona is hot?

7. When do you think the next sunspot maximum and minimum will occur? Explain.

8. Why do astronomers say that the solar cycle is really 22 years long, even though the number of sunspots varies over an 11-year period?

9. Explain how the magnetic-dynamo model accounts for the solar cycle.

10. Give some everyday examples of conduction, convection, and radiative diffusion.

11. Give an everyday example of hydrostatic equilibrium. Give an example of thermal equilibrium.

12. Why do thermonuclear reactions occur only in the Sun's core?

13. Describe the Sun's interior. Include references to the main physical processes that occur at various depths within the Sun.

14. What is a neutrino, and why are astronomers so interested in detecting neutrinos emanating from the Sun?

15. What is hydrogen burning? Why is hydrogen burning fundamentally unlike the burning of a log in a fireplace?

16. Why do you suppose the Sun generates energy only around its center?

17. Explain why studying the oscillations of the Sun's surface can give important, detailed information about physical conditions deep within the Sun.

ADVANCED QUESTIONS

Tips and tools . . .

You may have to review Wien's law and the Stefan–Boltzmann law, which were discussed in Chapter 5. By taking ratios, such as the ratio of the flux from a sunspot to that from the undisturbed photosphere, all the cumbersome constants cancel out, thereby simplifying your calculation.

18. Using the mass and size of the Sun given in Box 18-1, verify that the average density of the Sun is 1410 kg/m^3. Compare your answer with the average densities of the Jovian planets.

19. Suppose that you want to determine the Sun's rotation rate by observing its sunspots. Is it necessary to take the Earth's orbital motion into account? Why or why not?

20. What would happen if the Sun were not in a state of both hydrostatic and thermal equilibrium?

21. How much energy would be released if each of the following masses were converted *entirely* into its equivalent energy: (**a**) a carbon atom with a mass of 2×10^{-26} kg (**b**) one kilogram, and (**c**) a planet as massive as the Earth (6×10^{24} kg)?

22. Use the answers to the previous question to calculate how long the Sun must shine in order to release an amount of energy equal to that produced by the complete mass-to-energy conversion of (**a**) a carbon atom, (**b**) one kilogram, and (**c**) the Earth.

23. The brightest-appearing star in the sky, Sirius, has a luminosity of 40 L$_\odot$, which means that it is 40 times as bright as the Sun and burns hydrogen at a rate 40 times greater than the Sun does. How much hydrogen does Sirius burn each second?

24. Assuming that the current rate of hydrogen burning in the Sun remains constant, what fraction of the Sun's mass will be converted into helium over the next five billion years? How will this affect the chemical composition of the Sun?

25. Calculate the wavelengths at which the photosphere, chromosphere, and corona emit the most radiation. Explain how the results of your calculations suggest the best way to observe these regions of the solar atmosphere. (*Hint:* Assume average temperatures of 50,000 K and 1.5×10^6 K for the chromosphere and corona, respectively.)

26. Assuming that a sunspot is 1500 K cooler than the undisturbed photosphere, calculate how much less intense radiation from the sunspot is compared to the surrounding photosphere.

DISCUSSION QUESTIONS

27. Discuss the extent to which cultures around the world have worshiped the Sun as a deity throughout history. Why do you suppose there has been such widespread veneration?

28. Discuss some of the difficulties in correlating solar activity with changes in the terrestrial climate.

29. Describe some of the advantages and disadvantages of observing the Sun (**a**) from space and (**b**) from the South Pole. What kinds of phenomena and issues do solar astronomers want to explore from both the Earth-orbiting and the Antarctic observatories?

OBSERVING PROJECTS

30. Use a telescope to view the Sun by projecting the Sun's image onto a screen or sheet of white paper. **DO NOT LOOK AT THE SUN! Looking directly at the Sun can cause blindness.** Do you see any sunspots? Sketch their appearance. Can you distinguish between the umbrae and penumbrae of the sunspots? Can you see limb darkening? Can you see any granulation?

31. If you have access to an H$_\alpha$ filter attached to a telescope especially designed for viewing the Sun safely, use this instrument to examine the solar surface. How does the appearance of the Sun differ from that in white light? What do sunspots look like in H$_\alpha$? Can you see any prominences? Can you see any filaments? Are the filaments in the H$_\alpha$ image near any sunspots seen in white light?

FOR FURTHER READING

Bahcall, J. N. "The Solar-Neutrino Problem." *Scientific American,* May 1990. This article describes a variety of experiments designed to measure the flux of neutrinos from the Sun.

Eddy, J. "The Case of the Missing Sunspots." *Scientific American,* May 1977. This article by a noted solar astronomer discusses evidence that there have been long periods when no sunspots were visible.

———. *A New Sun.* NASA SP-402, 1979. This lavishly illustrated book gives a superb summary of our understanding of the Sun.

Foukal, P. "The Variable Sun." *Scientific American,* February 1990. This up-to-date article discusses recent observations of the Sun's variability and its effects on Earth.

Frazier, K. *Our Turbulent Sun.* Prentice-Hall, 1983. This well-researched report by a noted science writer describes the irregularity and variability of the Sun's energy output.

Friedman, H. *Sun and Earth.* Scientific American Library, 1986. This excellent, beautifully illustrated book conveys the excitement and adventure of solar research.

Harvey, J., and others. "GONG: To See inside Our Sun." *Sky & Telescope,* November 1987. This article describes the Global Oscillation Network Group (GONG), a worldwide program of scientists exploring the solar interior by means of the Sun's naturally occurring oscillations.

Leibacher, J. W., and others. "Helioseismology." *Scientific American,* September 1985. This lucid article explains how astronomers can learn about the structure, composition, and dynamics of the Sun's interior from oscillations visible on its surface.

Levine, R. "The New Sun." In J. Cornell and P. Gorenstein, eds. *Astronomy from Space: Sputnik to Space Telescope.* MIT Press, 1983. This article surveys the advantages of solar astronomy from space.

Mitton, S. *Daytime Star.* Scribner's, 1981. This easily readable book presents a completely nontechnical introduction to the Sun.

Nichols, R. "Solar Max: 1980–89." *Sky & Telescope,* December 1989. This brief article describes the Solar Maximum Mission (SMM) satellite, which made many important contributions to our understanding of the Sun during the 1980s.

Noyes, R. *The Sun, Our Star.* Harvard University Press, 1982. This clear, well-presented introduction to the Sun includes interesting chapters on the climate and solar energy.

Robinson, L. "The Disquieting Sun: How Big, How Steady?" *Sky & Telescope,* April 1982. This article describes controversial observations that the Sun is shrinking.

Wallenhorst, S. "Sunspot Numbers and Solar Cycles." *Sky & Telescope,* September 1982. This article discusses the counting of sunspots and the plotting of the solar cycle.

Wentzel, D. *The Restless Sun.* Smithsonian Institution Press, 1989. This lucid, up-to-date, stimulating account of the Sun emphasizes the wealth of activity—from sunspots to flares—that our star so vividly displays.

Wolfson, R. "The Active Solar Corona." *Scientific American,* February 1983. This article describes the dynamic processes that occur in the solar corona as rarified gases and the magnetic field of the Sun interact on a grand scale.

ARTHUR B. C. WALKER, JR.

The Solar Corona

ARTHUR WALKER, after learning at age 10 that it was extremely unlikely that he could follow in the footsteps of New York Yankee center fielder Joe DiMaggio, decided instead to emulate his number two idol, Albert Einstein. Today Dr. Walker is professor of applied physics at Stanford University, where he has conducted pioneering studies of the solar corona by applying X-ray optics to the corona's fine-scale structure.

A graduate of Case Institute of Technology, Dr. Walker received his Ph.D. in physics from the University of Michigan in 1962. His first experience with space observations came while serving at the Air Force Weapons Laboratory in Albuquerque, New Mexico. He began probing the X-ray spectrum of the solar corona while at the Aerospace Corporation, before moving to Stanford's Center for Space Science and Astrophysics in 1974. Among his many research assignments for NASA and other federal agencies, the most demanding was his service on the commission appointed to investigate the accident of the Space Shuttle *Challenger*.

The solar corona is the tenuous, hot outer atmosphere of the Sun where temperatures range from 1 to 10 million kelvin. Because it is so hot, the corona is best observed at X-ray wavelengths. The photograph on the next page, which was taken with an X-ray telescope during a NASA rocket flight in 1991, shows two views of the solar corona in the light of 11-times ionized iron (Fe XII) at 19.2 µm. Much of the coronal gas is confined by magnetic fields to thin, threadlike loops. Occasionally, tangled magnetic fields can break and reconnect, impulsive releasing magnetically stored energy in a solar flare that drives temperatures up to 100 million kelvin.

The existence of the corona seems paradoxical. As you saw in Chapter 18, temperatures inside the Sun decline from 15 million kelvin at its center to 5800 K at the photosphere. How can the corona, in contact only with the 5800-K photosphere, be heated to million-degree temperatures, and how can it maintain those temperatures indefinitely?

Until 50 years ago, astronomers erroneously believed that the temperature of the corona was 5800 K, primarily because they did not understand what the corona's spectrum was telling them. To account for spectral lines they could not identify, some astronomers proposed that the solar corona is composed of an undiscovered element called "coronium," which is even lighter than hydrogen. To explain why coronium is not found on Earth, they argued that our planet's gravitational field is simply not strong enough to retain any, so all our coronium was lost soon after the Earth was formed.

The corona's spectrum was correctly interpreted in the 1940s by Walter Grotrian of the Potsdam Observatory. Using measurements made by the Swedish spectroscopist Edlén, Grotrian proved that the visible coronal lines are in fact caused by highly ionized ions of iron, with from 9 electrons (Fe X) to 13 electrons (Fe XIV) removed. These ions can be present only if temperatures exceed one million kelvin. This discovery challenged astronomers to explain why the corona is so hot.

The unusual conditions that prevail in the corona give important clues to the mystery of its high temperature. First, magnetic fields in the corona are so strong that they completely dominate the behavior of the thin coronal gas. Astronomers have recently proposed several mechanisms by which energy stored in these magnetic fields can be transferred directly into the coronal gas. Second, the density of the corona is so low that it does not lose energy easily; energy loss by radiation and conduction are strongly inhibited. This combination of efficient heating and ineffective cooling is the main reason why the corona stays so hot.

The best-documented heating process is magnetic reconnection, which is responsible for solar flares. When tangled magnetic fields break and reconnect, they generate powerful electric fields that accelerate electrons and protons to very high speeds. This transient population of energetic particles, which can produce gamma rays and radio bursts, heats the coronal gases to 100 million kelvin. Eugene Parker at the University of Chicago has recently proposed that many small flares, called "nanoflares," occur many times per second all over the solar surface. This continual small-scale release of energy provides a steady input of heat to the corona.

The corona can be heated in at least two other ways. One involves waves, called *magnetoacoustic waves*, that propagate in ionized gases and magnetic fields. These waves, which consist of both pressure and electromagnetic disturbances, are generated in the photosphere and the chromosphere. As they travel upward into the corona,

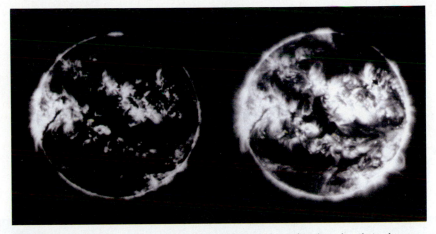

These two views of the solar corona, taken in 1991, show fine-line details in the core (left) and faint X-ray bright points, coronal loops, and loop arcades (right). (Courtesy of A. B. C. Walker, Jr., Stanford University)

they become shock waves that heat the coronal gases. A second mechanism involves electric currents flowing in the corona. These powerful currents, which result when magnetic fields and the ionized gas interact, can also heat the corona as they dissipate energy.

The fundamental mystery of the corona has been solved: Nonthermal magnetic processes heat the low-density corona. This conclusion is reinforced by the discovery that many stars have coronas that emit X rays. Ultimately, the source of a star's magnetic energy is its rotation. Observations by the Einstein Observatory and other spacecraft have shown a clear tendency: The faster a star rotates, the stronger the X-ray emission from its corona.

It is not known which combination of the three mechanisms—nanoflares, magnetoacoustic waves, or coronal currents—is responsible for steady heating of the corona. To discover the relative importance of these processes, we shall continue to

build and launch new generations of X-ray telescopes.

Focusing X rays is difficult because these high-energy photons are not easily reflected. The first X-ray astronomical pictures were obtained in the 1960s with cumbersome and expensive "grazing-incidence" mirrors that reflected X rays at very shallow angles, much like skipping a stone on the surface of a lake. Recently, however, a technology has been developed that uses ultra-thin films arranged in layers of near-atomic dimensions to reflect X rays over a narrow range of wavelengths. My colleagues and I used this technology in the 1991 rocket flight that took the picture of the corona shown above. In addition to providing the superb X-ray images of the Sun, this technology may be very useful in fabricating microelectronics and imaging very small structures such as biological cells. Our quest to understand the tenuous outer layers of the Sun may therefore yield some important practical applications.

THE NATURE OF STARS

A CLUSTER OF STARS By analyzing starlight, we can determine a star's surface temperature, chemical composition, and luminosity. This photograph shows color differences in the star cluster NGC 3293. Reddish stars are comparatively cool, with surface temperatures around 3000 K. They are also quite luminous and have large diameters, typically 100 times larger than our Sun. Bluish and blue-white stars have much higher surface temperatures (15,000 to 30,000 K) but are roughly the same size as the Sun. (Anglo-Australian Observatory)

ALTHOUGH THEY appear only as brilliant pinpoints of light, stars are massive spheres of glowing gas, much like our Sun. Using straightforward techniques, astronomers have measured the distances to many of the nearer stars. With such data, we learn how to calculate the energy output of the stars. We also see that the colors and spectra of stars indicate a wide range of stellar properties such as surface temperature, chemical composition, and luminosity. Stellar luminosities and surface temperatures come together in the all-important Hertzsprung–Russell diagram, which reveals the fundamental types of stars. Most common are stars like the Sun, called main sequence stars. There are also huge, bright, cool stars called red giants and small, dim, hot stars called white dwarfs. Binary stars, which consist of two stars orbiting each other, are also quite common. By observing the motions of the stars in nearby binary systems, astronomers can determine stellar masses. All this knowledge about the nature of stars is essential to understanding stellar evolution and serves as an indispensable tool in the astronomer's quest to probe the distant reaches of the universe.

To the unaided eye, the night sky is spangled with thousands of stars, each appearing as a bright pinpoint of light. A telescope reveals many more thousands of stars too faint to be seen with the naked eye. Each star is a huge, massive ball of hot gas much like our Sun, held together by its own gravity. Astronomers have discovered that some stars are larger than our Sun and some smaller; some stars are brighter than the Sun and some dimmer. Some stars are hotter than the Sun, but others are cooler.

Over the past century, astronomers have gathered basic information about the masses, luminosities, temperatures, and chemical compositions of the stars. In recent years, a remarkably complete picture has emerged, offering insight into our relationship to the universe as a whole and our place in the cosmic scope of space and time.

19-1 Careful measurements of the positions and motions of stars reveal stellar distances and space velocities

Looking up at the nighttime sky, you can see thousand of stars. Some appear quite bright, but most are rather dim. As you gaze up at this starry panorama, one of the questions you might ask is: How far away are the stars? Are they relatively nearby or inconceivably remote? The apparent brightness of the stars do not indicate their distances. A star that looks dim might actually be a brilliant star that happens to be extremely far away. To determine the distances to the stars, astronomers use geometric techniques that involve painstaking measurements of stellar positions and motions.

The most straightforward way of measuring stellar distances is based on **parallax**, which is the apparent displacement of an observed object because of a change in the observer's point of view. You experience parallax when nearby objects appear to shift their positions against a distant background as you move from one place to another (Figure 19-1). Stars exhibit the same phenomenon. As Earth

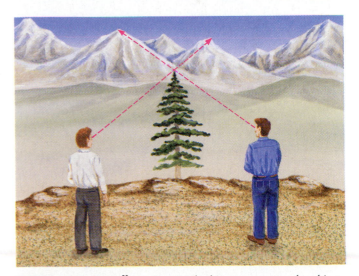

FIGURE 19-1 Parallax Imagine looking at some nearby object (a tree) as seen against a distant background (mountains). When you move from one location to another, the nearby object appears to shift with respect to the distant background scenery. This familiar phenomenon is called parallax.

orbits the Sun, nearby stars appear to move back and forth against the background of more distant stars.

The distance to a star can be determined by measuring the star's parallax. Mathematically, the parallax (p) of a star is half the angle through which the star's apparent position shifts as the Earth moves from one side of its orbit to the other (Figure 19-2). If the angle p is measured in seconds of arc, then the distance d to the star in parsecs is given by

$$d = \frac{1}{p}$$

For example, a star whose parallax is $\frac{1}{2}$ arc sec is 2 parsecs from Earth. This simple relationship between parallax and distance in parsecs is one of the main reasons that astronomers usually measure cosmic distances in parsecs rather than light years. (Recall that 1 parsec equals 3.26 light-years, and 1 light-year is nearly 10^{13} kilometers. See Figure 1-14.)

The German astronomer-mathematician Friedrich Wilhelm Bessel made the first parallax measurement. He found the parallax of 61 Cygni to be $\frac{1}{3}$ arc sec, and so he determined

that its distance is about 3 pc from Earth. The nearest star, Proxima Centauri, has a parallax of 0.772 arc sec, and thus its distance is 1.30 pc. In the parallax method, extremely tiny angles must be measured. For instance, the parallax of Proxima Centauri is comparable to the angular diameter of a dime seen from a distance of two miles. Because of such tiny angles, measuring stellar distances is one of the most difficult and challenging tasks that astronomers tackle.

Because parallaxes smaller than about $\frac{1}{50}$ arc sec are extremely difficult to measure from Earth, the parallax method gives reliable distances only for those stars nearer than about 50 pc. Most of these nearby stars are far too dim to be seen with the naked eye. The majority of the familiar, bright stars in the nighttime sky are so far away that their parallaxes cannot be measured from the Earth's surface.

Parallax measurements made by an Earth-orbiting satellite would be unhampered by our atmosphere, permitting astronomers to determine the distances to stars well beyond the reach of ground-based observations. In 1989 the European Space Agency (ESA) launched a satellite called *Hipparcos* (an acronym for *High Precision Parallax Collecting Satellite*), specifically designed to measure parallaxes as small as 0.002 arc second. Although the satellite failed to achieve its proper orbit, astronomers are still hopeful that *Hipparcos* will be able to complete much of its mission of measuring the distances to stars as far away as 500 pc. During the next century, astronomers will increasingly turn to space-based observations to determine stellar distances.

In addition to annual cyclic motion due to parallax, many stars change their apparent positions very slowly as they move through space. The component of their true motion perpendicular to our line of sight is called **proper motion**. Because proper motion does not repeat itself yearly, it can be distinguished from the apparent back-and-forth motion due to parallax. The proper motion of a star, usually designated by the Greek letter μ (mu), is measured in arc sec per year.

The star with the largest proper motion, called Barnard's star after the American astronomer Edward E. Barnard, exhibits a proper motion of 10.3 seconds of arc per year (Figure 19-3). It takes Barnard's star only 200 years to travel the apparent angular diameter of the full moon. This huge proper motion is far greater than that of any other star. In fact, the relative positions of stars are changing so slowly that the constellations look virtually the same today as they did thousands of years ago. Over a few million years, however, star positions shift by several degrees, noticeably altering constellations.

In order to determine the proper motion of a star, an astronomer must carefully measure the star's position for many years. During a single night at a telescope, however, an astronomer can use the Doppler shift to determine its motion along our line of sight. As discussed in Chapter 5 (recall Figure 5-20), the Doppler shift is a change in wavelength caused by the relative motion between a light source and an observer. If a star is approaching you, the wavelengths of all of its spectral lines are decreased (blueshifted); if the star is receding, the wavelengths are increased (redshifted). The size

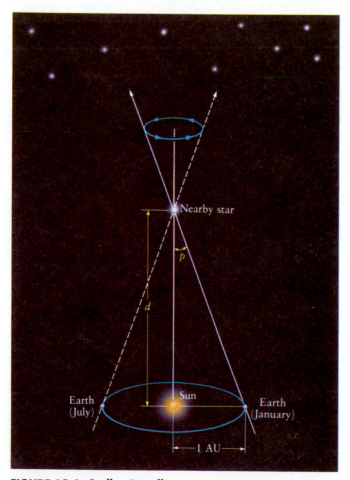

FIGURE 19-2 Stellar Parallax As the Earth orbits the Sun, a nearby star appears to shift its position against the background of distant stars. The parallax (p) of the star is equal to the angular radius of the Earth's orbit as seen from the star.

August 24, 1894

May 30, 1916

FIGURE 19-3 Barnard's Star These two photographs, taken 22 years apart, show the proper motion of Barnard's star in the constellation of Ophiuchus. In addition to having the largest

known proper motion (10.3″ per year), Barnard's star is one of the nearest stars, only 5.9 light-years from Earth. (Yerkes Observatory)

of the wavelength shift ($\Delta\lambda$) of a specific spectral line whose rest wavelength is λ_0 is given by

$$\frac{\Delta\lambda}{\lambda_0} = \frac{v_r}{c}$$

where v_r, called the star's **radial velocity,** is the component of the star's velocity parallel to our line of sight and c is the speed of light. A specific example showing how this equation is used to determine the radial velocity of the star Vega was given in Chapter 5 (recall Figure 5-20).

Together, the proper motion and radial velocity of a star give the size and direction of the total velocity of the star in space. As shown in Figure 19-4, a star's total velocity v consists of components parallel and perpendicular to our line of sight. The component parallel to our line of sight is just the radial velocity (v_r), determined from measurements of the Doppler shifts of the star's spectral lines. The component perpendicular to our line of sight, usually called the star's **tangential velocity** (v_t), can be directly related to the star's proper motion (μ) if the distance to the star is known. If the star's distance d is measured in parsecs and its proper motion is measured in arc seconds per year, then the tangential velocity in kilometers per second is given by

$$v_t = 4.74\mu d$$

Finally, we note that the tangential and radial components of a star's velocity form a right triangle. We can therefore use the Pythagorean theorem to write a concise formula for the star's total velocity in terms of its radial velocity and proper motion as follows:

$$v = \sqrt{v_r^2 + v_t^2} = \sqrt{v_r^2 + 22.5\mu^2 d^2}$$

As an example, consider Barnard's star: From Earth-based observation, its parallax (p) is 0.545 arc sec, its proper motion (μ) is 10.3 arc sec per year, and its radial velocity (v_r) is −108 km/s. (Recall that a minus sign means that the star

is approaching us.) We first calculate its distance from its parallax:

$$d = \frac{1}{p} = \frac{1}{0.545} = 1.83 \text{ pc}$$

We then use this result to calculate the star's tangential velocity:

$$v_t = 4.74d = 4.74 \times 10.3 \times 1.83 = 89.3 \text{ km/s}$$

Finally, we combine the observed radial velocity with our calculated tangential velocity to obtain the star's space velocity:

$$v = \sqrt{v_r^2 + v_t^2} = \sqrt{108^2 + 89.3^2} = 140 \text{ km/s}$$

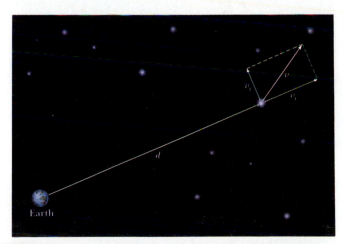

FIGURE 19-4 Components of a Star's Space Velocity It is useful to separate a star's velocity in space (v) into a component parallel to our line of sight plus a component perpendicular to our line of sight. The radial velocity (v_r) is determined from the Doppler shift of the star's spectral lines. The tangential velocity (v_t) can be determined from the star's proper motion if its distance from Earth is known. By combining these two components, the star's speed and direction of motion can be deduced.

Barnard's star is therefore moving though space at a speed of 140 km/s.

Parallax is a straightforward method that involves simple geometry and the idea that the Earth goes around the Sun. A second technique of determining distance, called the **moving cluster method,** applies to a cluster of stars. Because all the stars in a cluster move together as a group, they seem to be headed toward the same point on the celestial sphere. By knowing this point and by measuring the proper motions and radial velocities of stars in a cluster, the cluster's distance from Earth can be determined as described in Box 19-1. The moving cluster method is an important technique because it gives reliable stellar distances out to about 100 pc. Unfortunately, there is only one cluster (the Hyades) near enough so that the moving cluster method gives an accurate distance. Attempts to use this method with more distant clusters has met with mixed results.

The measurement of the distances and motions of stars is a cornerstone of modern astronomy. With such measurements, astronomers can go on to chart the size, shape, and behavior of the entire Galaxy. Even our knowledge of the overall size, age, and structure of the entire universe is based on measurements of stellar distances. Small inaccuracies in the measurements of distances to nearby stars can translate into huge errors in measurement for the whole universe. For this reason, astronomers are continually trying to perfect their distance-measuring techniques.

19-2   A star's luminosity can be determined from its apparent magnitude and distance

The system of magnitudes that astronomers use to denote the brightness of stars was invented in ancient Greece by the astronomer Hipparchus. Hipparchus called the brightest stars first-magnitude stars, those about one-half as bright second-magnitude stars, and so forth, down to sixth-magnitude stars, the dimmest ones he could see. After telescopes came into use, astronomers extended Hipparchus's magni-

---

BOX 19-1

The Moving Cluster Method

An **open cluster** is a loose collection of a few dozen to a few hundred stars. The Pleiades in the constellation of Taurus (the Bull) is a fine example of an open cluster easily seen with the naked eye. The nearest open cluster is the Hyades, whose stars comprise the V-shaped outline of the head of Taurus. The accompanying photograph shows the Pleiades and the Hyades. You should note that the brightest star in the V, Aldebaran, is not actually a member of the cluster.

The distances to the nearest open clusters can be determined by the moving cluster method, which makes use of the fact that the stars in a cluster all move together in nearly the same direction in space. For instance, the star

chart below shows the proper motions of the stars in the Hyades. The lengths of the arrows indicate how far each star will move over the next 10,000 years. Note that all the stars seem to be headed toward the same point. This

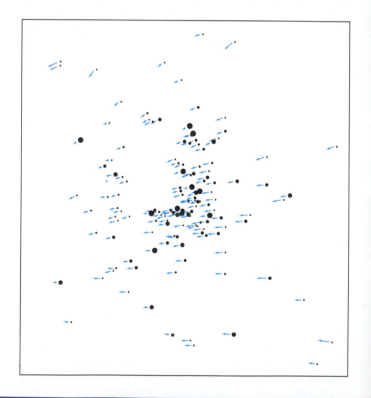

tude scale to include the dimmer stars now visible through their instruments.

These magnitudes are properly called **apparent magnitudes**, because they describe how bright an object *appears* to an Earth-based observer. Apparent magnitude is a measure of the light energy arriving at the Earth.

In the nineteenth century, better techniques were developed for measuring the light energy arriving from a star. Astronomers then set out to define the magnitude scale more precisely. Their measurements showed that a first-magnitude star is about 100 times brighter than a sixth-magnitude star. In other words, it would take 100 stars of magnitude +6 to provide as much light energy as we receive from a single star of magnitude +1. As a result, the magnitude scale was redefined so that a magnitude difference of 5 corresponds exactly to a factor of 100 in the amount of light energy received. A magnitude difference of 1 corresponds to a factor of 2.512 in light energy, because

$$2.512 \times 2.512 \times 2.512 \times 2.512 \times 2.512 = (2.512)^5 = 100$$

Thus, it takes about $2\frac{1}{2}$ third-magnitude stars to provide as much light as we receive from a single second-magnitude star.

Figure 19-5 illustrates the modern apparent magnitude scale. Note, for example, that the dimmest stars visible through a pair of binoculars have a magnitude of +10 and the dimmest stars that can be photographed in a one-hour exposure with a large telescope are magnitude +25.

Apparent magnitude tells us how bright a star appears to our eyes. To better understand the stars, we need to know how bright they really are. For this purpose, astronomers have invented a true measure of a star's energy output called absolute magnitude.

The **absolute magnitude** of a star is defined as the apparent magnitude it would have if it were located at a distance of exactly 10 parsecs from the Earth. For example, if the Sun were moved to a distance of 10 parsecs from the Earth, it would have an apparent magnitude of +4.8. The absolute magnitude of the Sun is thus +4.8. The absolute magnitudes of other stars range from roughly −10 for the brightest to +15 for the dimmest. The Sun's absolute magnitude is about

---

convergence is the result of perspective, just as parallel train tracks stretching across a prairie seem to meet at the horizon.

Let θ be the angular distance between a particular star and the cluster's convergent point K, as shown in the diagram on the right. Also shown are the star's space velocity v, radial velocity v_r, and tangential velocity v_t. The star's space velocity is parallel to our line of sight toward point K, and thus the angle between v and v_r is also θ. We can therefore write

$$v_r = v \cos \theta$$

$$v_t = v \sin \theta$$

As noted in the text, the radial velocity v_r can be measured from the Doppler shift of spectral lines in the star's spectrum. The tangential velocity v_t is related to the proper motion μ and the star's distance d from Earth by the equation

$$v_t = 4.74 \mu d$$

where μ is measured in arc sec per year and d is measured in parsecs. A formula for the distance to the star in terms of the observable quantities v_r, μ, and θ can be derived as follows:

$$d = \frac{v_t}{4.74\mu} = \frac{v \sin \theta}{4.74\mu} = \frac{0.211 v_r \tan \theta}{\mu}$$

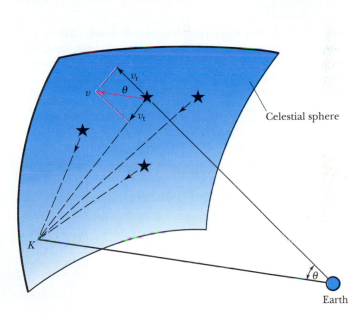

By this method, the distances of individual stars can be determined from the motion of the cluster as a whole. The distance to the Hyades obtained in this way is about 40 pc. Because the distance to the Hyades cluster is the most accurately determined of all stellar distances, it provides the basis upon which all other astronomical distances are determined.

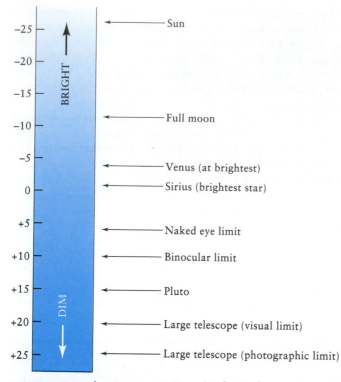

FIGURE 19-5 The Apparent Magnitude Scale Astronomers denote the brightness of objects in the sky by their apparent magnitudes. Most stars visible to the naked eye have magnitudes in the range +1 to +6. Photography through large telescopes can reveal stars as faint as magnitude +24.

in the middle of this range, providing us with our first hint that the Sun is an average star.

To appreciate how astronomers determine the absolute magnitude of a star, you must first realize that the farther away a source of light is, the dimmer it appears. Imagine a source of light, such as a light bulb or a star. Suppose that L is the amount of energy emitted by the light source each second, a quantity called the **luminosity**. The value of L is usually measured in joules per second or watts. For example, the Sun's luminosity is 3.90×10^{26} W.

As the light energy moves away from its source, it spreads out over increasingly larger regions of space (Figure 19-6). Imagine a sphere of radius d centered on the light source. The amount of energy passing through each square meter of the sphere each second is simply the total luminosity of the source (L) divided by the sphere's total surface area ($4\pi d^2$). The resulting quantity, called the **apparent brightness** of the light (b), is measured in joules per square meter per second:

$$b = \frac{L}{4\pi d^2}$$

This relationship is called the **inverse-square law**, because the apparent brightness of light that an observer can see or measure is inversely proportional to the square of the observer's distance (d) from the source. If you double your distance from a light source, its radiation is spread out over an area four times larger, so the apparent brightness you see is decreased by a factor of 4. Similarly, at triple the distance

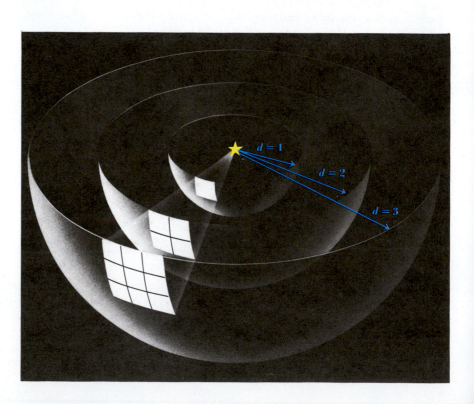

FIGURE 19-6 The Inverse-Square Law
This drawing shows how the same amount of radiation from a light source must illuminate an ever-increasing area as distance from the light source increases. Because this area increases as the square of the distance from the source, the apparent brightness decreases as the square of the distance.

TABLE 19-1
Distance Moduli

m – M	Distance (pc)
0	10
1	16
2	25
3	40
4	63
5	100
10	10^3
15	10^4
20	10^5

the apparent brightness decreases by a factor of 9, as Figure 19-6 demonstrates.

Using the inverse-square law, astronomers have derived an equation that reflects the special character of the magnitude scale. It relates a star's apparent magnitude (m), its absolute magnitude (M), and its distance (d, measured in parsecs) from the Earth:

$$m - M = 5 \log d - 5$$

It is important to realize that if you know any two of these quantities, such as apparent magnitude and distance, you can calculate the third one (in this case, absolute magnitude). For instance, astronomers might measure the apparent magnitude of a nearby star, find its distance by measuring its parallax, and then calculate its absolute magnitude. (For the reader familiar with logarithms, Box 19-2 shows how we obtain the equation relating m, M, and d, with several examples of how it is used.)

It is useful to write the previous equation in the following equivalent form:

$$d = 10^{(m - M + 5)/5}$$

The quantity $m - M$, called the **distance modulus**, is a measure of a star's distance from Earth. The larger the distance modulus, the greater the distance. We can use the above equation to construct Table 19-1. If you know apparent and absolute magnitudes of a star, you can use this table to estimate its distance.

Many scientists prefer to speak of a star's luminosity rather than its absolute magnitude, because luminosity is a direct measure of the star's energy output. There is a simple relationship between absolute magnitude and luminosity (Box 19-3), which astronomers use to convert one to the other as they see fit. For convenience, stellar luminosities are

expressed as multiples of the Sun's luminosity ($L_\odot$), which equals 3.90×10^{26} watts. The most luminous stars (those with an absolute magnitude of –10) have luminosities of 10^6 $L_\odot$, which means that each of these stars has the energy output of a million Suns. The least luminous stars (absolute magnitude = +15) have luminosities of 10^{-4} $L_\odot$.

19-3   Dim stars are more common than bright ones

Bright stars may attract your attention as you gaze up at the nighttime sky, but they are not typical of stars near our Sun. Appendix 4 at the end of this book lists all the known stars within about 13 light-years of the Earth. Of the 43 stars on this list, only three (α Cenaturi, Sirius, and Procyon) are brighter than the Sun. This is our first clue that the majority of stars are dimmer than the Sun.

To better characterize a typical population of stars, astronomers count the stars out to a certain distance from the Sun and plot the number at different brightnesses. The resulting graph, called the **luminosity function,** shows how the number of stars depends on their luminosity.

Figure 19-7 shows the luminosity function for stars in a 1000-pc^3 volume centered on the Sun. Note how steeply the curve declines for the brightest stars toward the left side of the graph, indicating that they are much rarer than dimmer ones. For instance, this graph shows that stars like the Sun ($M = +4.8$) are about ten thousand times more common than star's like Spica ($M = -3.5$).

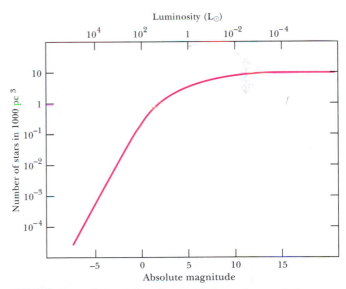

FIGURE 19-7 The Luminosity Function This graph shows how the number of stars in the solar neighborhood depends on brightness. Note that dim stars are far more numerous than bright stars. The steep decline of the curve toward the left side of the graph demonstrates that the most luminous stars are indeed quite rare. (Adapted from J. Bachall and R. Soneira)

BOX 19-2

Magnitude and Brightness

The concept of a star's magnitude dates back to the days of ancient Greek astronomy. Modern astronomers often prefer to talk instead about a star's apparent brightness. The apparent brightness of a star is the number of joules of starlight energy arriving at each square meter on Earth each second; it is usually expressed in joules per square meter per second ($J\ m^{-2}\ s^{-1}$). As explained in this chapter, each step in magnitude corresponds to a factor of 2.512 in brightness. In other words, your eyes receive 2.512 times more energy per square meter per second from a third-magnitude star than they do from a fourth-magnitude star.

Consider two stars with magnitudes m_1 and m_2 and brightnesses b_1 and b_2, respectively. The ratio of their apparent brightnesses (b_1/b_2) corresponds to a difference in their magnitudes ($m_2 - m_1$). Because each step in magnitude corresponds to a factor of 2.512 in brightness, we can construct the table shown below.

In general, the relationship between ($m_2 - m_1$) and (b_1/b_2) can be written as

$$\frac{b_1}{b_2} = 100^{\,(m_2 - m_1)/5}$$

Taking the logarithm of both sides of this equation and rearranging terms, we obtain

$$m_2 - m_1 = 2.5 \log\left(\frac{b_1}{b_2}\right)$$

The usefulness of these relationships and equations is best illustrated by examples.

EXAMPLE: At greatest brilliance, Venus has a magnitude of −4. Compare the apparent brightness of Venus with that of the dimmest stars visible to the naked eye, which have a magnitude of +6. The magnitude difference is therefore +6 − (−4) = 10. This difference corresponds to a brightness ratio of $(2.512)^{10} = 10^4$. It would thus take 10,000 sixth-magnitude stars to shine as brilliantly as Venus.

Magnitude difference $(m_2 - m_1)$	Ratio of apparent brightness (b_1/b_2)
1	2.512
2	$(2.512)^2 = 6.31$
3	$(2.512)^3 = 15.85$
4	$(2.512)^4 = 39.82$
5	$(2.512)^5 = 100$
10	$(2.512)^{10} = 10^4$
15	$(2.512)^{15} = 10^6$
20	$(2.512)^{20} = 10^8$

EXAMPLE: In August 1975 a nova appeared in the constellation of Cygnus (the Swan). A nova is a violent outburst of a certain kind of star. The magnitude of this nova changed in just two days from +15 to +2. By what factor did its brightness increase?

The change in magnitude ($m_2 - m_1$) is 13. Thus the ratio of brightness is $b_1/b_2 = 100^{13/5} = 158,500$. In two days, the nova's brightness therefore increased by a factor of nearly 160,000.

EXAMPLE: The variable star RR Lyrae periodically doubles its light output. How much does its magnitude change?

The brightness of the star varies by a factor of 2, so $b_1/b_2 = 2$. Thus $m_2 - m_1 = 2.5 \log(2) = 0.7$. RR Lyrae therefore varies periodically by seven-tenths of a magnitude.

With these relationships, we can derive a useful equation between a star's apparent magnitude, its absolute magnitude, and its distance from Earth. Let m and b be the apparent magnitude and the apparent brightness, respectively, of a star at a distance d from Earth. Let M and B be the apparent magnitude and the apparent brightness, respectively, of the star if it were 10 parsecs from Earth. By definition, M is the star's absolute magnitude. From the general relationship we have established between apparent brightness and apparent magnitude,

$$m - M = 2.5 \log\left(\frac{B}{b}\right)$$

According to the inverse-square law, the apparent brightness of a light is inversely proportional to the square of the distance between the light and the observer. Thus,

$$\frac{B}{b} = \left(\frac{d}{10}\right)^2$$

Combining these two equations and rearranging terms, we obtain

$$M = m - 5 \log\left(\frac{d}{10}\right)$$

where d is measured in parsecs. This is often written as

$$m - M = 5 \log d - 5$$

where ($m - M$) is called the distance modulus.

EXAMPLE: Consider Capella, a bright, nearby star. Its apparent magnitude is +0.05, and its distance is 14 parsecs. Thus, its absolute magnitude is $M = 0.05 - 5 \log(1.4) = -0.7$. Comparing this value to the Sun's absolute magnitude (+4.8), we see that Capella is actually 5.5 magnitudes brighter than the Sun. From $m_2 - m_1 = 5.5 = 2.5 \log(b_1/b_2)$, we find $b_1/b_2 = 158$. Thus, Capella emits about 160 times the light energy of our Sun.

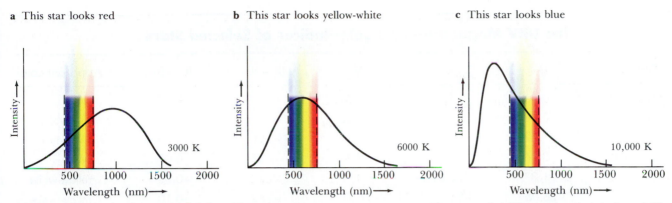

a This star looks red

b This star looks yellow-white

c This star looks blue

3000 K

6000 K

10,000 K

FIGURE 19-8 Temperature and Color This schematic diagram shows the relationship between the color of a star and its surface temperature. The intensity of light emitted by three hypothetical stars is plotted here against wavelengths (compare Figure 5-8). The range of visible wavelengths is indicated. The slope of a star's intensity curve across the visible wavelengths determines the star's apparent color.

The exact shape of the curve in Figure 19-7 applies only to the vicinity of the Sun and similar regions in our Milky Way Galaxy. Other locations have somewhat different luminosity functions, but the overall tendency for faint stars to be much more common than bright ones generally prevails in all stellar populations.

19-4 A star's color reveals its surface temperature

One of the first things you notice when comparing stars in the nighttime sky is their differences in apparent magnitude. More careful examination, even with the naked eye, reveals that the stars also have different colors. For example, in the constellation of Orion you can easily note the difference between reddish Betelgeuse and bluish Rigel (examine Figure 2-1).

As we saw in Chapter 5, a star's color is directly related to its surface temperature by Wien's law. The intensity of light from a cool star peaks at long wavelengths, making the star look red (Figure 19-8*a*). A hot star's intensity curve peaks instead at shorter wavelengths, so the star looks blue (Figure 19-8*c*). The maximum intensity of a star with an intermediate temperature, such as the Sun, occurs near the middle of the visible spectrum, giving the star a yellowish color (Figure 19-8*b*).

To measure accurately the colors of the stars, astronomers invented a technique called **photometry**, which uses a light-sensitive device (such as a CCD) at the focus of a telescope behind a standardized set of colored filters. The most commonly used filters are the **UBV filters**. Each of the three UBV filters is transparent in one of three broad wavelength bands: the ultraviolet (U), the blue (B), and the central yellow (V, for *visual*) region of the visible spectrum (Figure 19-9). (The transparency of the V filter mimics the sensitivity of the human eye.)

To do photometry, the astronomer aims a telescope at a star and measures the intensity of the starlight that passes through each of the filters. This procedure gives three apparent magnitudes for the star, usually designated by the capital letters U, B, and V. (As with all magnitudes, the lower the number, the brighter the star.) The astronomer then compares the intensity of starlight in neighboring wavelength bands by subtracting one magnitude from another to form the combinations (B – V) and (U – B), which are called the star's **color indices**.

A color index tells you how much brighter or dimmer a star is in one wavelength band than in another. For example, the (B – V) color index tells you how much brighter or dimmer a star appears through the B filter than through the V filter. The UBV magnitudes and color indices for several representative stars are given in Table 19-2.

The color index of a star is directly related to the star's surface temperature. If a star is very hot, its radiation is skewed toward the short-wavelength ultraviolet, which makes the star bright through the U filter, dimmer through

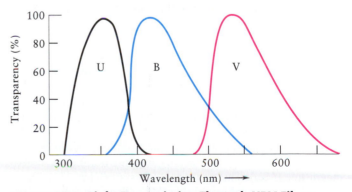

FIGURE 19-9 Light Transmission Through UBV Filters This graph shows the wavelength ranges over which the standardized U, B, and V filters are transparent to light. The U filter is transparent to the near-ultraviolet. The B filter is transparent from about 380 to 550 nm, and the V filter is transparent from about 500 to 650 nm.

TABLE 19-2
The UBV Magnitudes and Color Indices of Selected Stars

Star	U	B	V	(B – V)	(U – B)	Apparent color
Bellatrix (γ Ori)	0.55	1.42	1.64	−0.22	−0.87	Blue
Regulus (α Leo)	0.88	1.24	1.35	−0.11	−0.36	Blue-white
Sirius (α CMa)	−1.52	−1.46	−1.46	0.00	−0.06	Blue-white
Megrez (δ UMa)	3.46	3.39	3.31	+0.08	+0.07	White
Altair (α Aql)	1.07	0.99	0.77	+0.22	+0.08	Yellow-white
Sun	−26.06	−26.16	−26.78	+0.62	+0.10	Yellow-white
Aldebaran (α Tau)	4.29	2.39	0.85	+1.54	+1.90	Orange
Betelgeuse (α Ori)	4.41	2.35	0.50	+1.85	+2.06	Red

the B filter, and dimmest through the V filter. Regulus (see Table 19-2) is such a star. In contrast, if the star is cool, its radiation peaks at long wavelengths, making the star brightest through the V filter, dimmer through the B filter, and dimmest through the U filter. The stars Aldebaran and Betelgeuse are examples.

The graph in Figure 19-10 gives the relationship between a star's (B – V) color index and its temperature. If you know a star's (B – V) color index, you can use this graph to find the star's surface temperature. For example, the Sun's (B – V) index is +0.62, which corresponds to a surface temperature of 5800 K.

BOX 19-3

Bolometric Magnitude and Luminosity

In determining a star's absolute magnitude, astronomers must make allowances for nonvisible light and for the Earth's atmosphere. The apparent magnitude of a star could be misleading if the star happens to emit a significant fraction of its radiation at nonvisible wavelengths. For example, a luminous hot star with a surface temperature of 35,000 K appears deceptively dim to our eyes simply because most of the star's light is emitted at ultraviolet wavelengths. Furthermore, because the Earth's atmosphere is opaque to many nonvisible wavelengths, a sizable fraction of the light from both the hottest and the coolest stars does not penetrate the air to reach our eyes or telescopes.

To cope with this difficulty, astronomers have defined the **bolometric magnitude** of a star, the star's apparent magnitude as measured above the Earth's atmosphere and over all wavelengths. In recent years astronomical satellites have allowed us to determine the bolometric magnitudes of many stars.

The absolute magnitude of a star as deduced from its bolometric magnitude is called its **absolute bolometric magnitude** (M_{bol}). This quantity is always brighter than the star's absolute visual magnitude (M) as deduced from ground-based observations at visible wavelengths alone. By comparing satellite and ground-based data, astronomers have figured out how much they must add to a star's absolute visual magnitude to get its absolute bolometric

magnitude. This factor, called the **bolometric correction** (BC), is given in the simple equation

$$M_{bol} = M + BC$$

The bolometric correction is quite large for both the hottest stars and the coolest stars. For example, for stars hotter than 20,000 K or cooler than 3000 K, the bolometric correction is 3 magnitudes or greater. For stars such as the Sun, which emit an overwhelming percentage of their radiation at visible wavelengths, the bolometric correction is almost zero. The Sun's absolute bolometric magnitude is +4.72, whereas its absolute visual magnitude is +4.83.

A star's absolute bolometric magnitude is directly related to the star's luminosity (L). From the equations in Box 19-2, it is possible to show that

$$M_{bol} = 4.72 - 2.5 \log(L/L_\odot)$$

By knowing a star's M_{bol} and using this equation, astronomers can calculate exactly how much energy is being released from the star's surface each second.

EXAMPLE: Consider Sirius, which has an apparent visual magnitude of −1.46, making it the brightest star in the night sky. The distance to Sirius is 2.7 parsecs, and the surface temperature of Sirius is about 10,000 K, corre-

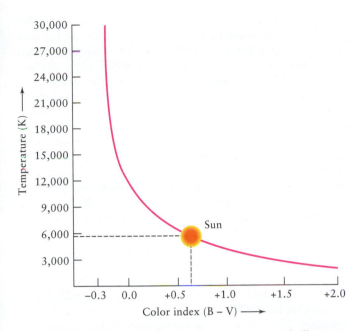

◄ **FIGURE 19-10 Blackbody Temperature Versus Color Index** The (B − V) color index is the difference between the B and V magnitudes of a star. If the star is hotter than about 10,000 K, it is a very bluish star with a (B − V) index of less than zero. If a star is cooler than about 10,000 K, its (B - V) index is greater than zero. The Sun's (B − V) index is about 0.62, which corresponds to a temperature of 5800 K. After measuring a star's B and V magnitudes, an astronomer can estimate the star's surface temperature from a graph like this one.

A word of caution here is in order. As we shall see in Chapter 20, dust and gas that pervade interstellar space cause distant stars to appear redder than they really are. Astronomers must therefore take this reddening into account whenever they attempt to determine a star's surface temperature from its color indices. A star's spectrum provides a more direct measure of a star's surface temperature, as we shall see next. Nevertheless, it is quicker and easier to observe a star's colors with a set of UBV filters than it is to take the star's spectrum with a spectrograph.

sponding to a bolometric correction of −0.6 magnitude. Suppose that you want to know the luminosity of Sirius.

First, calculate the absolute visual magnitude of Sirius:

$$M = m - 5 \log \left(\frac{d}{10} \right)$$

$$= -1.46 - 5 \log 0.27$$

$$= -1.46 - 5(9.4314 - 10)$$

$$= 1.4$$

The absolute magnitude of Sirius is 1.4 and the bolometric correction is −0.6, so we find that the absolute bolometric magnitude of Sirius is $M_{bol} = M + BC = 1.4 - 0.6 = 0.8$. We know that $M_{bol} = 4.72 - 2.5 \log (L/L_\odot)$. This expression is the same as

$$\log(L/L_\odot) = \frac{4.72 - M_{bol}}{2.5}$$

$$= \frac{4.72 - 0.8}{2.5}$$

$$= 1.6$$

Finally, we obtain $L/L_\odot = 10^{1.6} = 40$. Thus, we have determined that Sirius is about 40 times more luminous than the Sun.

19-5 Stars are classified according to the appearance of their spectra in a way that reveals their surface temperatures

The field of stellar spectroscopy was born in the 1860s, when the Italian Jesuit astronomer Angelo Secchi attached a spectroscope to his telescope and pointed it toward the stars. Secchi promptly discovered that many stars have absorption lines in their spectra. He was also the first astronomer to classify stellar spectra into **spectral types**, according to their appearance.

Astronomers who pioneered stellar spectroscopy were aware that stellar spectra must contain important information about stars, because chemists Robert Bunsen and Gustav Kirchhoff had demonstrated that spectral lines are caused by chemicals (recall Figure 5-12). Specifically, according to Kirchhoff's third law, absorption lines are seen when the spectrum of a hot, glowing object is viewed through a cool gas (recall Figure 5-13). As we have seen, a star shines with nearly a perfect blackbody spectrum. This continuous spectrum of radiation, or **continuum,** is produced deep within a star's atmosphere where the gases are hot and dense. As this light moves outward through the cooler, less dense layers of the star's upper atmosphere, atoms absorb radiation at specific wavelengths, thereby producing the spectral lines that astronomers observe.

At first glance, stellar spectra seem to come in a bewildering variety. Some prominently show the Balmer lines of hydrogen. Other spectra exhibit many absorption lines of calcium and iron. Still others are dominated by broad absorption features caused by molecules such as titanium oxide. To cope with this diversity, astronomers grouped similar-appearing stellar spectra into classes. In a popular classification scheme that emerged in the late 1800s, a star was assigned a letter from A through P according to the strength or weakness of the hydrogen Balmer lines in the star's spectrum. The A stars are those with the strongest Balmer lines, the P stars the weakest.

Nineteenth-century science could not explain how the spectral lines of a particular chemical are affected by the temperature and density of the gas. Nevertheless, a team of astronomers under the supervision of Edward C. Pickering at Harvard College Observatory forged ahead with a monumental project of examining the spectra of thousands of stars. Their goal was to develop a system of spectral classification in which all of the spectral features (not just Balmer lines) change smoothly from one spectral class to the next.

Pickering's spectral classification project was financed by the estate of Henry Draper, a wealthy physician and amateur astronomer who, in 1872, first photographed stellar absorption lines. Principal researchers on the project were Williamina P. Fleming, Antonia C. Maury, and Annie Jump Cannon. As a result of their efforts, many of the original A-through-P classes were dropped and others were consolidated. The remaining spectral classes were reordered in the sequence **OBAFGKM**. This sequence has traditionally been memorized with the sexist mnemonic: "Oh, *Be A Fine Girl, Kiss Me!*"

Annie Cannon found it useful to subdivide the original OBAFGKM sequence into finer steps. These additional steps are indicated by attaching an integer from 0 through 9 to the original letter. Thus, for example, we have . . . , F8, F9, G0, G1, G2, . . . , G9, K0, K1, K2, ... In going from one spectral type to the next, the strengths of spectral lines change in a smooth fashion. For instance, absorption lines of hydrogen become increasingly prominent as you go from spectral type B0 to A0. From A0 onward through F and G types, the hydrogen lines weaken and almost fade from view. Representative spectra of each spectral type are shown in Figure 19-11. The Sun, whose spectrum is dominated by calcium and iron, is a G2 star.

The Harvard project culminated in the *Henry Draper Catalogue*, published between 1918 and 1924 by Cannon and Pickering. It listed 225,300 stars, each of which had been personally classified by Cannon. Meanwhile, physicists had been making important discoveries about the structure of atoms. Rutherford had demonstrated that atoms have nuclei (recall Figure 5-14) and Bohr made the brilliant hypothesis that electrons move along discrete orbits around the nuclei of atoms (recall Figure 5-17). These discoveries about atoms gave scientists the conceptual and mathematical tools required to interpret and understand stellar spectra and contributed to the later development of quantum mechanics.

In the late 1920s, Harvard astronomer Cecilia Payne and physicist Meghnad Saha in India succeeded in explaining how a star's surface temperature affects its spectrum. They demonstrated that the OBAFGKM spectral sequence is actually a sequence in temperature. The hottest stars are O stars. The absorption lines seen in the spectra of O stars can occur only if these stars have surface temperatures above 25,000 K. M stars are the coolest stars. The spectral features seen in the spectrum of M stars are consistent with stellar surface temperatures of about 3000 K.

To see why the appearance of a star's spectrum is profoundly affected by the star's surface temperature, consider hydrogen. Hydrogen is by far the most abundant element in the universe, accounting for about three-quarters of the mass of a typical star. However, hydrogen lines do not necessarily show up in a star's spectrum. Recall that Balmer lines are produced when an electron in the $n = 2$ orbit of hydrogen absorbs a photon having the energy to lift it to a higher orbit. If the star is much hotter than 10,000 K, high-energy photons pouring out of the star's interior will easily knock electrons out of the hydrogen atoms in the star's outer layers.

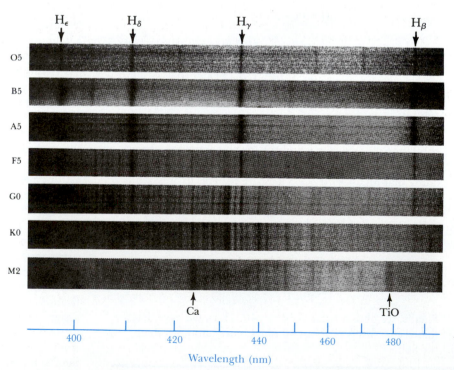

FIGURE 19-11 Principal Types of Stellar Spectra A star's spectrum is shown in the middle of each of these seven strips. The hydrogen lines are strongest in A stars, which have surface temperatures of about 10,000 K. The spectra of G and K stars exhibit numerous lines caused by metals, indicating temperatures in the range of 4000 to 6000 K. The broad, dark bands in the spectrum of an M star are caused by titanium oxide, which can exist only when the temperature is below about 3500 K. (Courtesy of N. Houk, N. Irvine, and D. Rosenbush)

TABLE 19-3
The Spectral Sequence

Spectral class	Color	Temperature (K)	Spectral lines	Examples
O	Blue-violet	28,000–50,000	Ionized atoms, especially helium	Naos (ζ Pup), Mintaka (δ Ori)
B	Blue-white	10,000–28,000	Neutral helium, some hydrogen	Spica (α Vir), Rigel (β Ori)
A	White	7,500–10,000	Strong hydrogen, some ionized metals	Sirius (α CMa), Vega (α Lyr)
F	Yellow-white	6,000–7,500	Hydrogen and ionized metals such as calcium and iron	Canopus (α Car), Procyon (α CMi)
G	Yellow	5,000–6,000	Ionized calcium and both neutral and ionized metals	Sun, Capella (α Aur)
K	Orange	3,500–5,000	Neutral metals	Arcturus (α Boo), Aldebaran (α Tau)
M	Red-orange	2,500–3,500	Strong titanium oxide and some neutral calcium	Antares (α Sco), Betelgeuse (α Ori)

This process ionizes the gas. When hydrogen's only electron is torn away, no hydrogen spectral lines can be produced.

Conversely, if the star's atmosphere is much cooler than 10,000 K, almost all of the hydrogen atoms are in the lowest ($n = 1$) energy state. The majority of the photons passing through the star's atmosphere possess too little energy to boost electrons up from the $n = 1$ to the $n = 2$ orbit of the hydrogen atoms. These unexcited atoms therefore also fail to produce hydrogen lines. In other words, a star must be hot enough to excite the electrons out of the ground state, but not so hot that the atoms become ionized.

A stellar surface temperature of about 9000 K produces the strongest hydrogen lines. Similarly, every other element also has a characteristic temperature at which it produces prominent absorption lines in the observable part of the spectrum. For example, around 25,000 K the spectral lines of helium are strong; at this temperature, photons have enough energy to excite helium atoms without tearing away the electrons altogether. In stars hotter than about 30,000 K, one of the two electrons in a helium atom is torn away. The remaining electron then produces a set of spectral lines that is recognizably different from the lines produced by un-ionized helium. When the spectral lines of singly ionized helium appear in a star's spectrum, we know that the star has a surface temperature greater than 30,000 K.

Heavy elements dominate the spectra of stars cooler than 10,000 K. Ionized metals are prominent in the spectra of stars with surface temperatures between 6000 and 8000 K. Neutral metals are strongest between approximately 5500 and 4000 K. Below 4000 K, certain atoms combine to form molecules that can survive in a star's atmosphere. As these molecules vibrate and rotate, they produce bands of spectral lines that dominate the star's spectrum. Most noticeable is titanium oxide, which is strongest in stars with surface tem-

peratures of about 3000 K. Table 19-3 summarizes the relationship between the temperature and spectra of stars. Box 19-4 discusses how astronomers measure the strengths of spectral lines and use this information to accurately determine the spectral class of a star.

19-6 Hertzsprung–Russell diagrams reveal the different kinds of stars

Around 1905 the Danish astronomer Ejnar Hertzsprung pointed out that a regular pattern appears when the absolute magnitudes of stars are plotted against their color indices. Almost a decade later, the American astronomer Henry Norris Russell independently discovered a similar regularity in a graph using spectral types instead of color indices. Plots of this kind are today known as **Hertzsprung–Russell diagrams**, or **H–R diagrams**.

Figure 19-12 is a typical Hertzsprung–Russell diagram. Each dot represents a star whose absolute magnitude and spectral type have been determined. Bright stars are near the top of the diagram, dim stars near the bottom. Hot stars (O and B stars) are toward the left side of the graph; cool stars (M stars) are toward the right side.

The most striking feature of the H–R diagram is that the data points are not scattered randomly over the graph but are grouped in several distinct regions. The band stretching diagonally across the H–R diagram includes many of the stars we see in the night sky. This band, called the **main sequence**, extends from the hot, bright, bluish stars in the upper left corner of the diagram down to the cool, dim, reddish stars in the lower right corner. A star whose properties place it in this region of an H–R diagram is called a **main sequence star**. For example, the Sun (spectral type G2, absolute magnitude +4.8) is such a star.

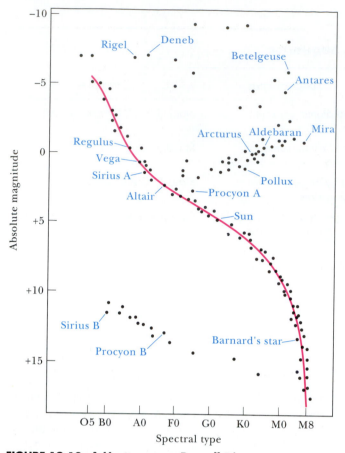

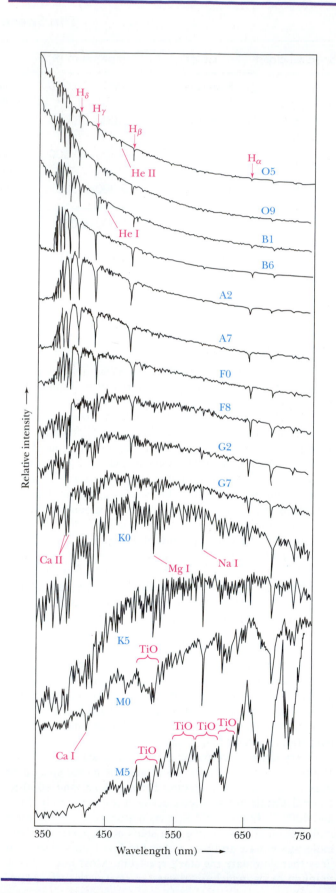

FIGURE 19-12 A Hertzsprung–Russell Diagram On an H–R diagram, the absolute magnitudes of stars are plotted against their spectral types. Each dot on this graph represents a star whose absolute magnitude and spectral class have been determined. Included on the diagram are some well-known stars. Note that the data points are grouped in specific regions on the graph. This pattern reveals the existence of different types of stars in the sky: main sequence stars, giants, supergiants, and white dwarfs. The red curve indicates the location of the main sequence.

The upper right side of the H–R diagram shows a second major grouping of data points. Stars represented by these points are both bright and cool. From the Stefan–Boltzmann law, we know that a cool object radiates much less light per unit of surface area than does a hot object. In order for these stars to be as bright as they are, they must be huge, and so they are called **giants**. As explained in Box 19-5, these stars are around 10 to 100 times larger than the Sun. Most of these stars are around 100 to 1000 times more luminous than the Sun and have surface temperatures of about 3000 to 6000 K. Cooler members of this class of stars (those with surface temperatures from about 3000 to 4000 K) are often called **red giants** because they appear reddish. Aldebaran in the constellation of Taurus and Arcturus in Boötes are two red giants that can be easily seen with the naked eye.

A few rare stars are considerably bigger and brighter than typical red giants. Appropriately enough, these superluminous stars are called **supergiants**. Betelgeuse in Orion and

---

BOX 19-4

Equivalent Widths and Line Strengths

Most astronomers today use a light-sensitive solid-state device, such as a CCD, to record the spectra of stars because it is much more sensitive than photographic film. All 14 stellar spectra in the graph on the left were obtained with such a device, which generates a plot of intensity versus wavelength. Absorption lines are seen as dips on the curve of the continuum, which has a shape quite like the blackbody curve at the temperature of the star's surface.

Astronomers designate an un-ionized atom with a roman numeral I; thus, H I is neutral hydrogen. A roman numeral II is used to identify an atom with one electron missing; thus, He II is singly ionized helium (He$^+$). Similarly, Si III is doubly ionized silicon (Si^{2+}), whose atoms are each missing two electrons. This notation is used to label some of the prominent spectral lines on the 14 spectra, which cover spectral types from O5 to M5.

By displaying a spectrum as a plot of intensity versus wavelength, an astronomer can study the shapes of individual spectral lines. The illustration on the right shows a typical spectral line consisting of a "core" flanked by "wings." The detailed shape of a spectral line, called the **line profile,** contains important information about a star. For instance, if a star is rotating, light from the approaching side is slightly blueshifted while light from the receding side is comparably redshifted. As a result, the star's spectral lines are broadened in a characteristic fashion. By measuring the shape of the spectral lines, astronomers can deduce how fast the star is rotating.

The true shape of a spectral line reflects the properties of a star's atmosphere, but the observed line profile is also broadened somewhat by the astronomer's measuring instruments. Instrumental effects do not, however, significantly alter the *total* absorption of the line, which is a measure of the energy deleted from the continuum across the entire spectral line. Thus the total absorption does not depend on the line's shape. Astronomers express the total absorption

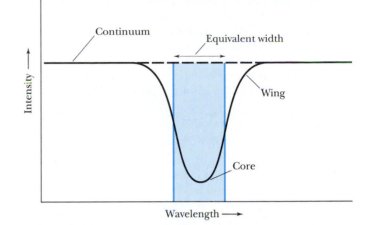

of a spectral line in terms of its **equivalent width,** which is the width of a completely dark rectangular line with the same total absorption as the observed line. For example, the equivalent width of an iron line in the Sun's spectrum is about 0.01 nm.

The equivalent width of a spectral line depends on how many atoms in the star's atmosphere are in a state in which they can absorb the wavelength in question. The more atoms there are, the stronger and broader the line is. Hence, by analyzing a star's spectrum, an astronomer can determine the star's chemical composition.

The main characteristics of the Henry Draper spectral classification scheme are summarized in the illustration below, which plots the equivalent widths of spectral lines against spectral type. This graph consolidates the information that astronomers use to deduce the surface temperature or spectral type of a star from the intensity of the lines in its spectrum. For example, a star with strong Ca II and Fe I lines in its spectrum has a surface temperature around 4500 K.

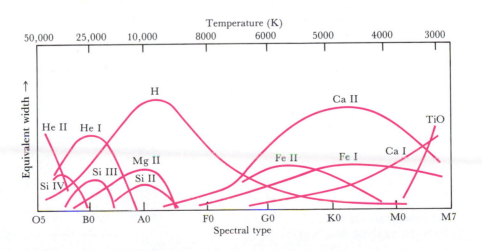

BOX 19-5

Stellar Radii

Stars behave almost exactly like blackbodies. Consequently, to find out how much energy is being radiated from a star's surface each second, we can use the Stefan–Boltzmann law, which states that the energy flux from a blackbody is proportional to the fourth power of its temperature. Specifically, in Chapter 5 we saw that the energy flux E (usually measured in joules per square meter per second) is related to the temperature T (in kelvin) by

$$E = \sigma T^4$$

where σ is a number called the Stefan–Boltzmann constant. From elementary geometry, we know that a sphere's surface area is $4\pi R^2$, where R is the sphere's radius. Because a star is spherical, we can use this expression for the surface area of a star. Multiplying the energy flux E by the star's surface area, we get the total energy output of the star per second, which is the star's luminosity L:

$$L = 4\pi R^2 \sigma T^4$$

By rearranging this equation, we can find the radius of a star in terms of its luminosity and surface temperature:

$$R = \frac{1}{T^2}\sqrt{\frac{L}{4\pi\sigma}}$$

EXAMPLE: Consider the bright reddish star Betelgeuse in the constellation of Orion. Betelgeuse has a luminosity of 10,000 $L_\odot$ and a surface temperature of 3000 K. Substituting these values (and the value of the Sun's luminosity $L_\odot$) in the equation just derived, we find that the star's radius is 2.6×10^{11} m, or nearly 2 AU. In other words, if Betelgeuse were located at the center of our solar system, this star would extend beyond the orbit of Mars.

In many calculations, it is convenient to relate everything to the Sun, which is a typical star. Specifically, for the Sun we have $L_\odot = 4\pi R_\odot^2 \sigma T_\odot^4$, where $L_\odot$ is the Sun's luminosity, $R_\odot$ is the Sun's radius, and $T_\odot$ is the Sun's surface temperature (5800 K). Dividing the general equation for L by this specific equation for the Sun, we obtain

$$\frac{L}{L_\odot} = \left(\frac{R}{R_\odot}\right)^2 \left(\frac{T}{T_\odot}\right)^4$$

This is a useful expression because the constants cancel out, making the arithmetic much easier. We can also rearrange terms to arrive at a useful alternative equation:

$$\frac{R}{R_\odot} = \left(\frac{T_\odot}{T}\right)^2 \sqrt{\frac{L}{L_\odot}}$$

EXAMPLE: Again consider Betelgeuse, for which $L = 10^4 L_\odot$ and $T = 3000$ K. Substituting these data into the

previous equation, we get

$$\frac{R}{R_\odot} = \left(\frac{5800}{3000}\right)^2 \sqrt{10^4} = 370$$

In other words, Betelgeuse's radius is 370 times larger than the Sun's.

Because the Hertzsprung–Russell diagram is a graph of luminosity versus temperature, we can use the general equation relating L, T, and R to draw lines on the H–R diagram to represent various stellar radii, as shown. Regions of the H–R diagram occupied by giants, supergiants, and white dwarfs are indicated.

Note that most of the stars on the main sequence are about the same size as the Sun. Giants have radii between 10 $R_\odot$ and 100 $R_\odot$, whereas the radii of supergiants range up to 1000 $R_\odot$. White dwarfs are roughly the same size as the Earth.

The radii of some stars have been measured with other techniques, some of which are discussed elsewhere in this text. These other methods yield values consistent with those calculated by the method described in this box.

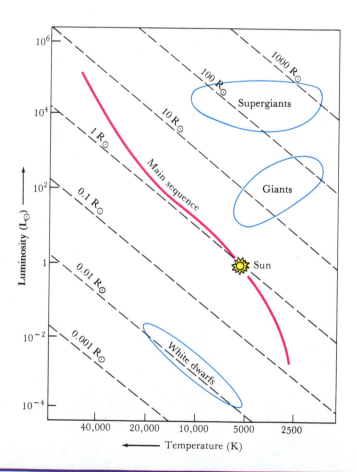

Antares in Scorpius are two supergiants you can find in the nighttime sky.

Finally, there is a third distinct grouping of data points toward the lower left corner of the Hertzsprung–Russell diagram. Because these stars are both hot and dim, they must be small. They are appropriately called **white dwarfs.** These stars, which are roughly the same size as the Earth, can be seen only with the aid of a telescope.

An H–R diagram can be constructed in another useful format. Instead of absolute magnitude, luminosity is plotted on the vertical axis of the graph; instead of spectral type, surface temperature is plotted on the horizontal axis. The resulting graph is still an H–R diagram, even though observational data (absolute magnitude and spectral type) are replaced by corresponding calculated quantities (luminosity and surface temperature).

Figure 19-13 shows this type of H–R diagram. Note that the temperature scale on the horizontal axis increases toward the left. This practice stems from the original diagrams of Hertzsprung and Russell, who placed hot O stars on the left and cool M stars on the right. This arrangement is a tradition that no one has seriously tried to change.

The existence of fundamentally different types of stars is the first important lesson to come from the H–R diagram. As we shall see in the following chapters, these different kinds of stars represent various stages of a star's life. We shall learn that the H–R diagram reflects an understanding of the life cycles of stars: how they are born and mature, and what happens when they die.

19-7   Luminosity class is a useful way of denoting a star's brightness

We have seen that a star's surface temperature largely determines which lines are prominent in its spectrum. Therefore, classifying stars by spectral type is essentially the same as categorizing them according to surface temperature.

Among stars of the same spectral type, there are subtle spectral differences caused by luminosity. For instance, for stars of spectral types B through F, the more luminous the star is, the narrower are its hydrogen lines. For example, Figure 19-14 compares the spectra of a B8 supergiant and a B8 main sequence star. Note that hydrogen lines are narrow in the spectrum of the supergiant, but quite broad in the spectrum of the main sequence star.

Hydrogen lines are good indicators of luminosity because they are affected by the density and pressure of the gas in the star's atmosphere. The higher the density and pressure, the more frequently metal ions collide and interact with hydrogen atoms. These collisions shift the energy levels in the hydrogen atoms and thus broaden the hydrogen spectral lines.

The pressure and density in the atmosphere of a luminous giant star are quite low because the star's mass is spread over a huge volume. Particles are relatively far apart and collisions between them are sufficiently infrequent that hydrogen atoms can produce narrow hydrogen lines. A main sequence star, however, is much more compact than a giant or super-

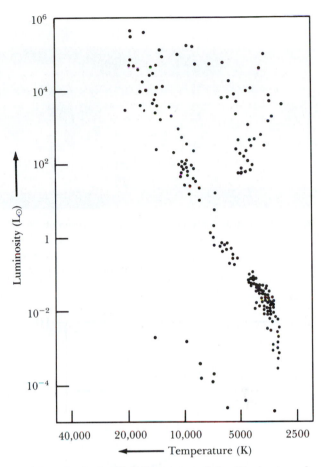

FIGURE 19-13   An H–R Diagram of the Nearest and Brightest Stars   It is often informative to draw an H–R diagram by plotting the luminosities of stars against their surface temperatures. Data for nearly 200 of the nearest and brightest stars are plotted on this diagram.

giant. In the denser atmosphere of a main sequence star, frequent interatomic collisions perturb the energy levels in the hydrogen atoms, thereby producing broader Balmer lines.

During World War II, W. W. Morgan, P. C. Keenan, and E. Kellman developed a system of **luminosity classes** based upon the subtle differences in spectral lines. When these luminosity classes are plotted on an H–R diagram (Figure 19-15), they provide a useful subdivision of the star types in the upper right half of the diagram. Luminosity class I includes all the supergiants; luminosity class V is composed of main sequence stars. The intermediate classes distinguish giant stars of various luminosities (Table 19-4).

Astronomers commonly describe a star by combining its spectral type and its luminosity class into a sort of shorthand description; for example, the Sun is said to be a G2V star. This notation supplies a great deal of information. The spectral type indicates the star's surface temperature, and the luminosity class indicates its luminosity. Thus an astronomer knows immediately that any G2V star is a main sequence star with a luminosity of about 1 $L_\odot$ and a surface tempera-

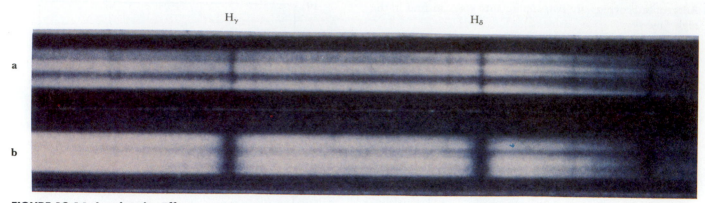

Hγ   Hδ

a

b

FIGURE 19-14 Luminosity Effects (a) The spectrum of the B8 supergiant Rigel (absolute magnitude −7.1), one of the brightest stars in Orion. (b) The spectrum of the B8 main sequence star Algol (absolute magnitude −0.2), the second brightest star in Perseus. Note the widths of the Hγ and Hδ lines in the two spectra. Both stars have a surface temperature of 13,400 K, but Rigel is about 600 times more luminous than Algol. (From *An Atlas of Stellar Spectra* by W. W. Morgan, P.C. Keenan, and E. Kellman)

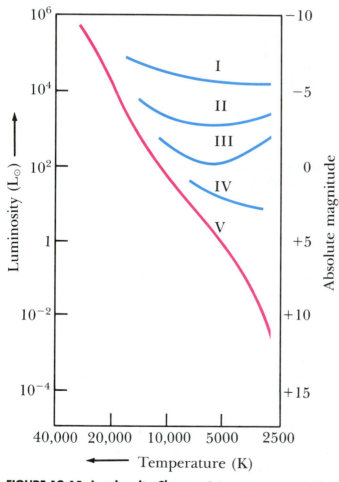

FIGURE 19-15 Luminosity Classes It is convenient to divide the H–R diagram into regions corresponding to luminosity classes. With this subdivision, finer distinctions can be made between giants and supergiants. Luminosity class V encompasses the main sequence stars, toward the lower-right side of the H–R diagram, which includes the dim red stars called red dwarfs.

ture of nearly 6000 K. Similarly, a description of Aldebaran as a K5III star tells an astronomer that it is a red giant with a luminosity of around 500 L☉ and a surface temperature of about 4000 K.

A star's spectral type and luminosity class give much of the information you need to estimate its distance from Earth. From the spectral type and the luminosity class, you can plot the star on the H–R diagram and then read off its absolute magnitude (M). If you also know the star's apparent magnitude (m), you can subtract one from the other to obtain the star's distance modulus ($m − M$). You can then get the star's distance from Table 19-1, or you can use the equation that relates m, M, and d to calculate how far away the star is. Since both the spectral type and luminosity class of a star are obtained by spectroscopy, this method of determining distance is called **spectroscopic parallax.**

As an example, we can use this method to estimate the distance to Regulus, the brightest star in Leo (the Lion). Regulus is classified as a B7V star and has an apparent magnitude of about +1.4. By plotting a B7V star on Figure 19-15, we read off an absolute magnitude of −0.6. The star's dis-

---

TABLE 19-4
Stellar Luminosity Classes

Luminosity class	Type of star
I	Supergiant
II	Bright giant
III	Giant
IV	Subgiant
V	Main sequence

tance modulus $(m - M)$ is therefore $1.4 - (-0.6) = 1.4 + 0.6 = 2.0$. From Table 19-1, we see that a distance modulus of 2.0 corresponds to 25 pc, which is the distance from Earth to Regulus.

Spectroscopic parallax is not very accurate because the locations of the luminosity classes on Figure 19-15 are somewhat uncertain. Consequently, we cannot be quite sure of the absolute magnitudes that we read off from such a graph. Nevertheless, spectroscopic parallax is often the only means that an astronomer has to estimate the distance to a remote star.

19-8   Binary stars provide crucial information about stellar masses

We now know something about the sizes, temperatures, and luminosities of stars. To complete our picture of the physical properties of stars, we need to know their masses. There is, however, no practical, direct way to measure the mass of an isolated star observed in the sky.

Fortunately for astronomers, about half of the visible stars in the night sky are not isolated individuals. Instead, they are multiple-star systems, in which two or more stars orbit each other. By carefully observing the motions of these stars, astronomers can glean important information about their masses.

A pair of stars located at nearly the same position in the night sky is called a **double star.** William Herschel made the first organized search for such pairs. Between 1782 and 1821 he published three catalogues, listing more than 800 double stars. Late in the nineteenth century, his son John Herschel discovered 10,000 more doubles. Many of these double stars are in fact true **binary stars,** or **binaries**—pairs of stars that actually orbit each other.

When astronomers can actually see the two stars orbiting each other, a binary is called a **visual binary** (Figure 19-16). After many years of patient observation, astronomers can plot the orbits of the stars in a visual binary (Figure 19-17).

Because the gravitational force between the two stars in a binary star system keeps them in orbit about each other, their orbital motions can be described by Newtonian mechanics. Specifically, their orbits obey Kepler's third law (see Box 4-3). For a binary star system, Kepler's third law can be written as

$$M_1 + M_2 = \frac{a^3}{P^2}$$

where M_1 and M_2 are the masses of the two stars expressed in solar masses, P is the orbital period in years, and a is the semimajor axis (measured in astronomical units) of the elliptical orbit of one star about the other, plotted as in Figure 19-17.

In principle, the orbital period of a visual binary is easy to determine. All you have to do is see how long it takes for the two stars to revolve once about each other. As Figure 19-17 suggests, however, more than one lifetime may be needed to complete the observations.

Determining the semimajor axis of an orbit is somewhat more difficult. The angular separation between the stars can be determined by observation, but to convert this angle into a linear distance, the distance between the binary star and the Earth must be known. In some cases this information may be obtained from parallax measurements. Once the distance between the star and Earth is known, the small-angle formula (recall Box 1-1) can be used to convert the angular separation into a distance in astronomical units. To obtain the semimajor axis, the astronomer must also take into account how the orbit is tilted to our line of sight.

Once both P and a have been determined, Kepler's third law can be used to calculate $M_1 + M_2$, the sum of the masses of the two stars in the binary system. Note that this analysis provides no information about the individual masses of the two stars. To obtain the stars' individual masses, more data about the binary orbit are needed.

Each of the two stars in a binary system actually moves in an elliptical orbit about the **center of mass** of the system

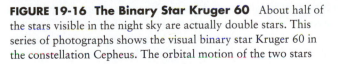

1908   1915   1920

FIGURE 19-16 The Binary Star Kruger 60 About half of the stars visible in the night sky are actually double stars. This series of photographs shows the visual binary star Kruger 60 in the constellation Cepheus. The orbital motion of the two stars about each other is apparent. This binary system has a period of $44\frac{1}{2}$ years. The maximum angular separation of the stars is about $2\frac{1}{2}$ arc sec. Their apparent magnitudes are +9.8 and +11.4. (Yerkes Observatory)

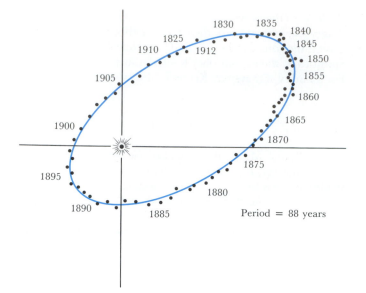

1830 1835 1840
1825 1845
1910 1912 1850
1905 1855
1860
1900 1865
1870
1895 1875
1890 1885
1880

Period = 88 years

FIGURE 19-17 The Orbit of 70 Ophiuchi After plotting observations of a visual binary star over the years, astronomers can
draw the orbit of one star with respect to the other. Once the orbit
is known, Kepler's third law can be used to deduce information
about the masses of the two stars. This illustration shows the orbit
of a faint visual double star in the constellation Ophiuchus. In plotting the orbit, either star may be regarded as the stationary one—
the shape of the orbit will be the same in either case.

(Figure 19-18). Imagine two children on a seesaw. The children must position themselves so that the seesaw balances
properly. The center of mass of this two-child system is then
located just above the support point, or fulcrum. As you no
doubt know from experience, this center of mass is offset
from the midpoint between the two children toward the
heavier child.

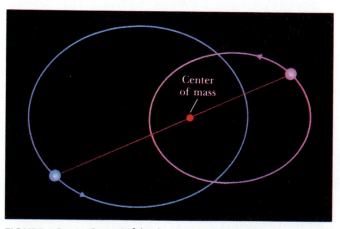

FIGURE 19-18 Star Orbits in a Binary System Both stars
in a binary system follow elliptical orbits about their common
center of mass. The center of mass is always nearer the more
massive of the two stars.

Similarly, the center of mass of a binary system can be
imagined by placing the two stars at either end of a huge
seesaw and determining where to place the fulcrum to balance the seesaw. The center of mass is always closer to the
more massive star by an amount that is proportional to the
ratio of the two masses.

In practice, of course, this experiment is impossible to
arrange. Nevertheless, the center of mass of a visual binary
can be determined by using the background stars as reference
points. The separate orbits of the two stars are plotted as in
Figure 19-18. Both elliptical orbits share a common focus
that is located at the center of mass of the two stars. This
information yields the ratio of the masses, M_1/M_2. Because
the sum $M_1 + M_2$ is already known, the individual masses of
the two stars can then be determined.

Years of careful, patient observations of binaries have
slowly yielded the masses of many stars. As the data accumulated, an important trend began to emerge: For main sequence stars, there is a direct correlation between mass and
luminosity. The more massive the star, the more luminous it
is. This **mass–luminosity relation** can be conveniently displayed as a graph (Figure 19-19). Note that the range of
stellar masses extends from about $\frac{1}{10}$ of a solar mass to about
50 solar masses. The Sun's mass lies between these extremes.

The mass–luminosity relation demonstrates that the main
sequence on an H–R diagram is a progression in mass as well
as in luminosity and surface temperature. The hot, bright,
bluish stars in the upper left corner of an H–R diagram are
the most massive main sequence stars in the sky. Likewise,
the dim, cool, reddish stars in the lower right corner of an
H–R diagram are the least massive. Main sequence stars of
intermediate temperature and luminosity also have an intermediate mass. This relationship of mass to the main sequence
will play an important role in our discussion of stellar evolution.

19-9 Binary systems that cannot be examined visually can still be detected and analyzed

Only those binary systems that are nearby or whose stars are
widely separated can be distinguished as visual binaries. The
images of the two stars in a remote binary commonly blend
together to produce the semblance of a single star. However,
spectroscopy shows that many seemingly single stars are in
fact double.

One such type of binary is discovered when the spectrum
of a star shows incongruous spectral lines. Occasionally, for
example, the spectrum of what at first appears to be a single
star may include both strong hydrogen lines (characteristic of
a type A star) and the strong absorption bands of titanium
oxide typical of a type M star. Because a single star cannot
have the differing physical properties of these two spectral
types, such a star must actually be a binary system that is too
far away for resolution of its individual stars. A binary star
detected in this way is called a **spectrum binary**.

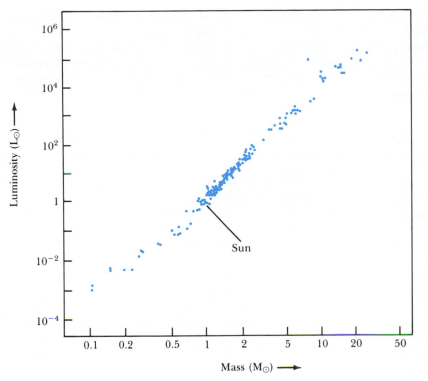

FIGURE 19-19 The Mass–Luminosity Relation   For main sequence stars, there is a direct correlation between mass and luminosity. The more massive a star, the more luminous it is.

Other binary systems can be detected from the Doppler shifts, as long as the orbital speeds of the two stars in an unresolved binary are more than a few kilometers per second. Binaries discovered and studied through periodic Doppler shifts in their spectra are called **spectroscopic binaries.** As one of the stars approaches Earth, its spectral lines are displaced toward the short-wavelength (blue) end of the spectrum. Conversely, the spectral lines of a receding star are shifted toward the long-wavelength (red) end of the spectrum. If λ_0 is the unshifted wavelength of a spectral line and λ is the wavelength of that line in a star's spectrum, then

$$\frac{\lambda - \lambda_0}{\lambda_0} = \frac{v_r}{c}$$

where v_r is the star's radial velocity toward or away from us along our line of sight and c is the speed of light. With this equation, we can convert a measurement of wavelength displacement into information about a star's motion. It is important to emphasize that the Doppler effect applies only to

motion along the line of sight. Motion perpendicular to the line of sight does not affect the observed wavelengths of spectral lines.

In many spectroscopic binaries, one of the stars is so dim that its spectral lines cannot be detected. Such a double-star system, which shows only a single set of spectral lines, is called a **single-line spectroscopic binary.** The star is obviously a binary, however, because its spectral lines shift back and forth, thereby revealing the orbital motions of two stars about their center of mass. Less frequently, an apparently single star yields spectra in which *two* complete sets of spectral lines shift back and forth. Such stars are called **double-line spectroscopic binaries.** A single-line spectroscopic binary yields less information about its two stars than does a double-line spectroscopic binary.

Figure 19-20 shows two spectra of a spectroscopic binary taken a few days apart. In Figure 19-20*a*, two sets of spectral lines are visible, offset slightly in opposite directions from the normal positions of these lines. One star, which is moving toward the Earth, has its spectral lines blueshifted; the other

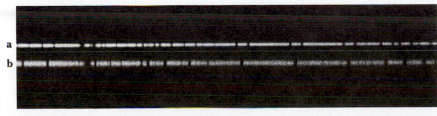

FIGURE 19-20 A Spectroscopic Binary
Spectra of κ Arietis exhibit spectral lines that shift back and forth as the two stars revolve about each other. **(a)** The stars are moving parallel to the line of sight, producing two sets of spectral lines. **(b)** Both stars are moving perpendicular to our line of sight. (Lick Observatory)

star is moving away from the Earth, so its lines are redshifted. A few days later, the stars have progressed along their orbits so that one star is moving toward the left, the other toward the right. Now neither star is moving toward or away from the Earth, so there is no Doppler shifting and both stars yield spectral lines at the same positions. That is why only one set of spectral lines appears in Figure 19-20b.

The Doppler shifts of the spectral lines reveal the star's radial velocity, including that portion of a star's orbital motion parallel to the line of sight to the star. This information is best displayed as a **radial velocity curve,** which graphs radial velocity versus time for the binary system. For example, Figure 19-21 shows the radial velocity curves that repeat every 15 days, which is the orbital period of the binary. Also note that the entire wavy pattern is displaced upward from the zero-velocity line by about 12 km/s, which is the overall motion of the binary system away from the Earth. Superimposed on this overall recessional motion are the periodic approaches and recessions of the two stars as they orbit about the center of mass.

The orbital speeds of the two stars in a binary are related to their masses by Kepler's laws and Newtonian mechanics. From a radial velocity curve, one can compare the masses of the two stars. Their individual masses can be determined only if the tilt of the binary orbits is known, because the

Doppler shifts reveal only the radial velocities of the stars rather than their true orbital speeds.

If the two stars are observed to eclipse each other periodically, then the orbit must be nearly edge-on, as viewed from the Earth. As we shall see next, individual stellar masses—and other useful data—can be determined if a spectroscopic binary also happens to be an **eclipsing binary.**

19-10   Light curves of eclipsing binaries provide detailed information about the two stars

A small fraction of all binary systems are oriented so that the two stars periodically eclipse each other as seen from the Earth. These eclipsing binaries can be detected even when the two stars cannot be resolved visually as two distinct images in the telescope. The apparent brightness of the image of the binary dims briefly each time one star blocks the light from the other.

Using a photoelectric detector at the focus of a telescope, an astronomer can measure the incoming light intensity quite accurately and draw a **light curve,** such as those shown in Figure 19-22. The shape of the light curve for an eclipsing binary reveals at a glance whether the eclipse is total or partial (compare Figures 19-22a and 19-22b).

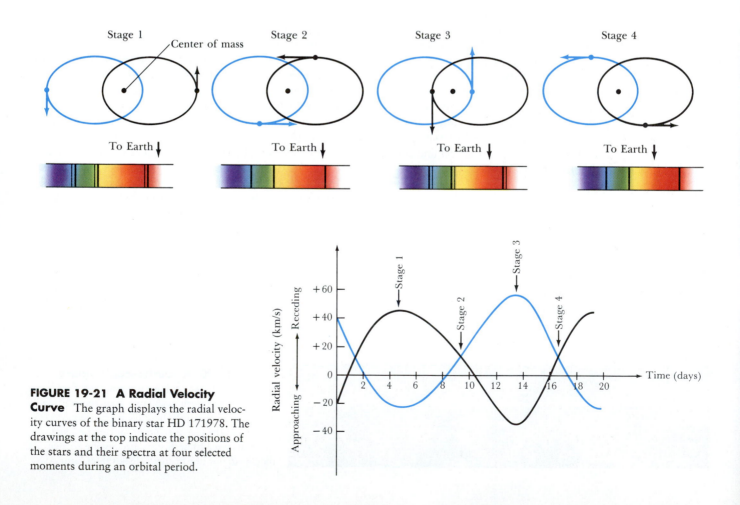

FIGURE 19-21 A Radial Velocity Curve The graph displays the radial velocity curves of the binary star HD 171978. The drawings at the top indicate the positions of the stars and their spectra at four selected moments during an orbital period.

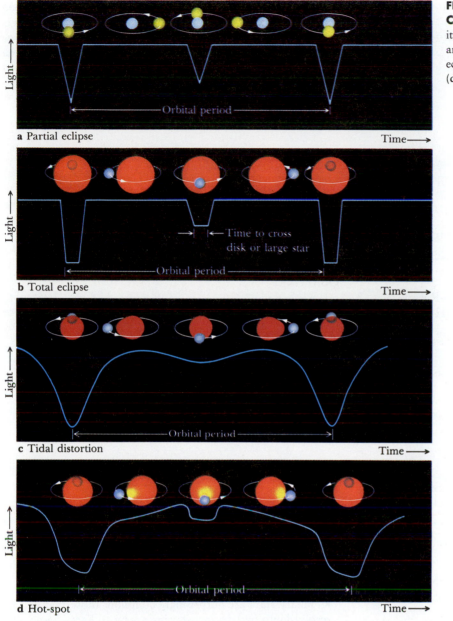

a Partial eclipse   Time ⟶

b Total eclipse   Time ⟶

c Tidal distortion   Time ⟶

d Hot-spot   Time ⟶

FIGURE 19-22 Representative Light Curves of Eclipsing Binaries The shape of its light curve usually reveals many details about an eclipsing binary. Illustrated here are (**a**) partial eclipse, (**b**) total eclipse, (**c**) tidal distortion, and (**d**) hot-spot reflection.

In fact, the light curve of an eclipsing binary can yield a surprising amount of information. For example, the ratio of the surface temperatures can be determined from how much their combined light is diminished when the stars eclipse each other. Also, the duration of mutual eclipse yields data on the relative sizes of the stars and their orbits.

If the eclipsing binary is also a double-line spectroscopic binary, an astronomer can calculate the mass and radius of each star from the light curves and the velocity curves. Unfortunately, very few binary stars are of this ideal type.

The shape of a light curve can reveal many additional details about a binary system. In some binaries, for example, the gravitational pull of one star distorts the other, much as the Moon distorts the Earth's oceans in producing tides (re-

call Figure 19-13). The effects of such tidal distortion on the light curve are shown in Figure 19-22c. Another example is "hot spot reflection," produced when one of the stars in an eclipsing binary is so hot that its radiation creates a hot spot on its cooler companion star. Every time this hot spot is exposed to our view from Earth, we receive a little extra light energy, which produces a characteristic bump on the binary's light curve (see Figure 19-22d).

Information about stellar atmospheres can also be derived from light curves. Suppose that one star of a binary is a luminous main sequence star and the other is a bloated red giant. By observing exactly how the light from the bright main sequence star is gradually cut off as it moves behind the edge of the red giant during the beginning of an eclipse, astronomers can infer the pressure and density in the upper atmosphere of the red giant.

A single star leads a straightforward, birth-to-death existence, but strange and exotic things can happen to stars in binary systems. One star in a binary might evolve rapidly, become a bloated red giant, and have its outer layers stripped away by the gravitational pull of its companion. The result is an aging star with its interior exposed to view. As we shall see in later chapters, many curious variations are possible as the two stars in a binary affect each other's evolution.

KEY WORDS

Terms preceded by an asterisk are discussed in the boxes.

* absolute bolometric magnitude
absolute magnitude
apparent brightness
apparent magnitude
binary star
* bolometric correction
* bolometric magnitude
center of mass
color index (plural indices)
continuum
distance modulus

double-line spectroscopic binary
double star
eclipsing binary
* equivalent width
giant
Hertzsprung–Russell diagram (H–R diagram)
inverse-square law
light curve
* line profile
luminosity
luminosity class

luminosity function
main sequence
main sequence star
mass–luminosity relation
moving cluster method
OBAFGKM
* open cluster
parallax
photometry
proper motion
radial velocity
radial velocity curve

red giant
single-line spectroscopic binary
spectral type
spectroscopic binary
spectroscopic parallax
spectrum binary
supergiant
tangential velocity
UBV filters
visual binary
white dwarf

KEY IDEAS

• Distances to the nearer stars can be determined by parallax, the apparent shift of a star against the background stars observed as the Earth moves along its orbit.

Distances to nearby clusters can be determined with the moving cluster method.

• The absolute magnitude of a star is the apparent magnitude it would have if viewed from a distance of 10 parsecs.

The absolute magnitude is calculated from the star's apparent magnitude and distance.

Luminosity is the amount of energy escaping from the total surface area of a star each second; energy flux is the amount of energy emitted from each square meter of a star's surface each second; apparent brightness is the energy flux (joules per square meter per second) arriving at the Earth from a star.

• Photometry measures brightness through standard filters such as the UBV filters. The color indices of a star are the differences between brightness values obtained through different filters.

The color indices of a star are a measure of its surface temperature.

• Stars are classified into spectral types (O, B, A, F, G, K, and M) based on the major patterns of spectral lines in their spectra. The spectral type of a star is directly related to its surface temperature.

• The Hertzsprung–Russell (H–R) diagram is a graph plotting the absolute magnitudes of stars against their spectral types (or, equivalently, their luminosities against surface temperatures). It reveals the existence of such different types of stars as main sequence stars, giants, supergiants, and white dwarfs.

• Binary stars are surprisingly common; those that can be resolved into two distinct star images by an Earth-based telescope are called visual binaries.

The masses of the two stars in a binary system can be computed by measuring the orbital period and orbital dimensions of the system.

Each of the two stars in a binary system moves in an elliptical orbit about the center of mass of the system.

• The mass–luminosity relation expresses a direct correlation between mass and luminosity for main sequence stars. Some binaries can be detected and analyzed, even though the system may be so distant or the two stars so close together that the two star images cannot be resolved by Earth-based telescopes.

A spectrum binary is a system detected from the presence of spectral lines for two distinctly different spectral types in the spectrum of what appears to be a single star.

A spectroscopic binary is a system detected from the periodic shift of spectral lines caused by the Doppler effect, as the orbits of the stars carry them first toward, then away from, the Earth.

An eclipsing binary is a system whose orbits are viewed nearly edge-on from the Earth, so that one star periodically eclipses the other. Detailed information about the stars in an eclipsing binary can be obtained from a study of the binary's radial velocity curve and its light curve.

REVIEW QUESTIONS

1. What is the difference between apparent magnitude and absolute magnitude?

2. Briefly describe how you would determine the absolute magnitude of a nearby star. Of what value is knowing the absolute magnitude of various stars?

3. How much dimmer does the Sun appear from Jupiter than from the Earth? (*Hint:* The average distance between Jupiter and the Sun is 5.2 AU.)

4. Suppose that a dim star were located 1 million AU from the Sun. Find (**a**) the distance to the star in parsecs and (**b**) the parallax angle of the star.

5. How far away is a star that has a proper motion of 0.005″/year and a tangential velocity of 50 km/s? What would its tangential velocity have to be in order for it to exhibit the same proper motion as Barnard's star?

6. The space velocity of a certain star is 65 km/s and its radial velocity is 35 km/s. Find the star's tangential velocity.

7. Explain why the color index of a star is related to the star's surface temperature.

8. A red star and a blue star have the same size and the same distance from Earth. Which one looks brighter in the night sky? Explain why.

9. What are the most prominent absorption lines you would expect to find in the spectrum of (**a**) a hot star, (**b**) a cool star, and (**c**) a star like the Sun? Briefly describe why these stars have such different spectra even though they have essentially the same chemical composition.

10. Suppose that you want to determine the mass, temperature, diameter, and luminosity of a star. Which of these physical quantities require you to know the distance to the star? Explain.

11. Sketch a Hertzsprung–Russell diagram. Indicate the regions on your diagram occupied by (**a**) main sequence stars, (**b**) red giants, (**c**) supergiants, (**d**) white dwarfs, and (**e**) the Sun.

12. What is the mass–luminosity relation?

13. Sketch the radial velocity curves of a binary consisting of two identical stars moving in circular orbits that are (**a**) perpendicular to and (**b**) parallel to our line of sight.

14. Sketch the light curve of an eclipsing binary consisting of two identical stars in highly elongated orbits oriented so that (**a**) their major axes are pointed toward the Earth and (**b**) their major axes are perpendicular to our line of sight.

15. Estimate the mass of a main sequence star that is 10,000 times as luminous as the Sun. What is the luminosity of a main sequence star whose mass is one-tenth that of the Sun?

ADVANCED QUESTIONS

Tools and tips . . .

Remember that a telescope's light-gathering power is proportional to the area of its objective or primary mirror. Some of the problems concerning magnitudes may require facility with logarithms, because these problems make use of the equation that relates the apparent and absolute magnitudes of a star to its distance. Make use of the H–R diagrams given in this chapter to answer the question involving spectroscopic parallax.

16. Suppose you can just barely see a twelfth-magnitude star through an amateur's 6-inch telescope. What is the magnitude of the dimmest star you could see through a 60-inch telescope?

17. Van Maanen's star, named after the Dutch astronomer who discovered it, is a nearby white dwarf whose parallax is 0.232 arc sec. How far away is the star?

18. Van Maanen's star has a proper motion of 2.95 arc sec per year and a radial velocity of +54 km/s. What is the star's actual speed relative to the Sun?

19. In the spectrum of a certain star, H_α has a wavelength of 656.50 nm. The laboratory value for the wavelength of H_α is 656.28 nm. Find (**a**) the star's radial velocity and (**b**) the wavelength at which you would expect to find H_β in the star's spectrum, given that the laboratory wavelength of H_β is 486.13 nm.

20. What would the distance between the Earth and the Sun have to be in order for the solar constant to be one watt per square meter (1 W/m^2)?

21. Explain why bolometric corrections are negative for both very hot and very cool stars.

22. A certain type of variable star is known to have an average absolute magnitude of 0.0. Such stars are observed in a particular star cluster to have an average apparent magnitude of +16.0. What is the distance to that star cluster?

23. The star Procyon in Canis Minor (the Small Dog) is a prominent star in the winter sky, because its apparent magnitude is +0.37. It is also one of the nearest stars, being only 3.51 parsecs from Earth. What is the absolute magnitude of Procyon? How many times brighter (or dimmer) than the Sun is it?

24. Barnard's star, the star with the largest known proper motion, can be seen only with a telescope because its apparent magnitude is +9.54. Its distance from Earth is only 1.81 parsecs. How much closer to Earth would Barnard's star have to be in order for it to be visible to the unaided eye?

25. Derive the equation given in the text relating proper motion and tangential velocity. (*Hint:* To do this problem, you need to understand the concept of a radian. You will also find it useful to know that 1 rad = 206,265 arc sec.)

26. Suppose two stars have the same apparent magnitude, but one star is ten times farther away than the other. What is the difference in their absolute magnitudes?

27. Suppose a star experiences an outburst in which its surface temperature doubles but its average density decreases by a factor of eight. By what factors do the star's radius and luminosity change?

28. The visual binary 70 Ophiuchi (see Figure 19-17) has a period of 87.7 years. The length of the semimajor axis is 4.5 arc sec, and the parallax of the system is 0.2 arc sec. What is the sum of the masses of the two stars?

29. An astronomer observing a binary star finds that one of the stars orbits the other once every 4 years at a distance of 2 AU. If the mass ratio of the system is 3.0, find the individual masses of the stars.

30. The bright star Rigel in the constellation of Orion has a temperature about 3 times that of the Sun and its luminosity is 64,000 times that of the Sun. What is Rigel's radius compared to the radius of the Sun?

31. The bright star Capella (α Aur) has a spectral type of G8 and an absolute magnitude of 0.7. What are the star's surface temperature, luminosity, and diameter?

32. What is the distance to Castor (α Gem), a A1V star whose apparent magnitude is 1.58?

DISCUSSION QUESTIONS

33. Why do you suppose that stars of the same spectral type but of a different luminosity class exhibit slight differences in their spectra?

34. Earth-based astronomers can measure stellar parallaxes with acceptable accuracy—10% or better—only if the angles are larger than about 0.05 arc sec. Discuss the advantages or disadvantages of making parallax measurements from a space telescope in a large solar orbit, say at the distance of Jupiter from the Sun. Assuming that this space telescope can also measure parallax angles of 0.05 arc sec, what is the distance of the most remote stars that can be accurately determined? How much bigger a volume of space would be covered compared to Earth-based observations? How many more stars would you expect to be contained in that volume?

OBSERVING PROJECTS

35. The following table lists well-known red stars and includes their coordinates (epoch 2000), apparent magnitudes, and (B – V) color indices. As indicated by the apparent magnitude column, all these stars are somewhat variable.

Star	R.A.	Decl.	Apparent magnitude	(B – V)
Betelgeuse	5^h 55.2^m	+ 7° 24′	0.4–1.3	1.85
Y CVn	12 45.1	+45 26	5.5–6.0	2.54
Antares	16 29.4	−26 26	0.9–1.8	1.83
μ Cep	21 43.5	+58 47	3.6–5.1	2.35
TX Psc	23 46.4	+ 3 29	5.3–5.8	2.60

Observe at least two of these stars both by eye and through a small telescope. Is the reddish color of the stars readily apparent, especially in contrast to neighboring stars? Incidentally, Angelo Secchi named Y Canum Venaticorum "La Superba," and μ Cephei is often called William Herschel's "Garnet Star."

36. The table of double stars (on the next page) includes vivid examples of contrasting star colors. The apparent magnitudes, spectral types, and the angular separation between the stars of each binary are given along with star names and coordinates (epoch 2000).

Observe at least four of these double stars through a telescope. Use the spectral types listed to estimate the difference in surface temperature of the stars in each pair you observe. Does the binary with the greatest difference in temperature seem to present the greatest color contrast? Based on what you see through the telescope and on what you know about the H–R diagram, explain why *all* the cool stars (spectral types K and M) listed are probably giants or supergiants.

37. Observe the eclipsing binary Algol (β Persei), using nearby stars to judge its brightness during the course of an eclipse. Algol has an orbital period of 2.87 days and, with the onset of primary eclipse, its apparent magnitude drops from 2.1 to 3.4. It remains this faint for about 2 hours. The entire eclipse, from start to finish, takes about 10 hours. Consult the "Celestial Calendar" section of the current issue of *Sky & Telescope* for the predicted dates and times of the minima of Algol. Note that the schedule is given in Universal Time (the same as Greenwich Mean Time), so you will have to convert to your own time zone. Algol is normally the second brightest star in the constellation of Perseus. Because of its northerly position (R.A. = 3^h 08.2^m, Decl. = +40° 57′), Algol is readily visible from northern latitudes during the fall and winter months.

Star	R.A.	Decl.	Apparent magnitudes	Angular separation	Spectral types
55 Psc	0^h 39.9^m	+21° 26′	5.4 and 8.7	6.5″	K0 and F3
γ And	2 03.9	+42 20	2.3 and 4.8	9.8	K3 and A0
32 Eri	3 54.3	−2 57	4.8 and 6.1	6.8	G5 and A2
ι Cnc	8 46.7	+28 46	4.2 and 6.6	30.5	G5 and A5
γ Leo	10 20.0	+19 51	2.2 and 3.5	4.4	K0 and G7
24 Com	12 35.1	+18 23	5.2 and 6.7	20.3	K0 and A3
η Boo	14 45.0	+27 04	2.5 and 4.9	2.8	K0 and A0
α Her	17 14.6	+14 23	3.5 and 5.4	4.7	M5 and G5
59 Ser	18 27.2	+ 0 12	5.3 and 7.6	3.8	G0 and A6
β Cyg	19 30.7	+27 58	3.1 and 5.1	34.3	K3 and B8
δ Cep	22 29.2	+58 25	4* and 7.5	20.4	F5 and A0

*δ Cephei is a variable star of approximately the fourth magnitude.

FOR FURTHER READING

Ashbrook, J. "Visual Double Stars for the Amateur." *Sky & Telescope,* November 1980. This informative article discusses techniques for observing double stars.

Evans, D., and others. "Measuring Diameters of Stars." *Sky & Telescope,* August 1979. This article explains how the technique of interferometry is used to determine the diameters of stars.

Davis, J. "Measuring the Stars." *Sky & Telescope,* October 1991. This article describes how astronomers use interferometry techniques to measure the diameters of stars.

Gingerich, O. "A Search for Russell's Original Diagram." *Sky & Telescope,* July 1982. The article in the "Astronomical Scrapbook" section gives entertaining insights about professional astronomy in the early 1900s.

Griffin, R. "The Radial-Velocity Revolution." *Sky & Telescope,* September 1989. This article describes recent technological advances that astronomers use to perform extremely precise Doppler shift measurements.

Hearnshaw, J. B. "Origins of the Stellar Magnitude Scale." *Sky & Telescope,* November 1992. This delightful article describes the history of an archaic magnitude scale, with which astronomers still burden themselves.

Hodge, P. "How Far Away Are the Hyades?" *Sky & Telescope,* February 1988. This excellent article looks at the trials and tribulations of determining the distance to the Hyades cluster.

Kaler, J. "Origins of the Spectral Sequence." *Sky & Telescope,* February 1986. This superb article, which traces the history of OBAFGKM, includes historical insights into the work of the astronomers who struggled to develop reliable and meaningful schemes for classifying stars.

———. *Stars.* Scientific American Library, 1992. This beautifully illustrated book gives a broad overview of our understanding of the stars.

———. *Stars and Their Spectra.* Cambridge University Press, 1989. This is an expanded version of Kaler's excellent series of articles on stellar spectroscopy that have appeared in *Sky & Telescope* over the past few years.

Labeyrie, A. "Stellar Interferometry: A Widening Frontier." *Sky & Telescope,* April 1982. This article describes how a technique called interferometry can provide extraordinarily detailed images of stars.

Lovi, G. "A Stellar Teaching Tool." *Sky & Telescope,* March 1989. In spite of a dreadfully dull title, this brief article gives a fascinating account of the motions of Sirius, including a star chart that shows Sirius's path among the constellations over the past five million years.

Mitton, J., and MacRobert, A. "Colored Stars." *Sky & Telescope,* February 1989. This brief article in the "Celestial Calendar" section gives a superb overview of star colors and includes a listing of extremely red stars and colorful double stars.

Nielsen, A. "E. Hertzsprung—Measurer of Stars." *Sky & Telescope,* January 1968. This excellent historical sketch describes the work of one of the inventors of the H–R diagram.

Phillip, A., and Green, L. "Henry N. Russell and the H–R Diagram." *Sky & Telescope,* April 1978, May 1978. These two articles give many fascinating insights into the life and times of Henry Norris Russell.

Steffey, P. C. "The Truth about Star Colors." *Sky & Telescope,* September 1992. This article dispels misconceptions that have long surrounded the colors of stars.

THE BIRTH OF STARS

REFLECTION AND EMISSION NEBULAE The two main bluish objects, NGC 6589 (top) and NGC 6590 (below), are reflection nebulae surrounding main sequence stars of spectral types B5 and B6. Interstellar dust around these two stars efficiently reflects their bluish light. Several smaller reflection nebulae are scattered around the large reddish patch of ionized hydrogen gas, IC 1283-4. Dust mixed with the gas dilutes the intense red emission with a soft blue haze. (Anglo-Australian Observatory)

OBSERVATIONS of stars, along with calculations of stellar models, have given astronomers an understanding of stellar evolution. Stars are born in cold clouds of interstellar gas and dust that are scattered abundantly throughout our Galaxy. In response to some occurrence—perhaps an encounter with one of the Galaxy's spiral arms or the detonation of a nearby supernova—an interstellar cloud begins to contract under the pull of gravity into fragments called protostars that will eventually become individual stars. The pressure within a protostar builds to resist contraction as the protostar's outer layers fall in upon the core. In time, temperatures in their cores are high enough to ignite hydrogen burning, and a full-fledged main sequence star is born. Resplendent stellar nurseries are created where clusters of young stars illuminate the surrounding interstellar gas and dust.

At a casual glance, the heavens seem eternal and unchanging; the sky that we see at night is virtually indistinguishable from the sky our ancestors saw. But this permanence is an illusion. Stars emit huge amounts of radiation: an outpouring that must produce changes and cause the stars to evolve. With careful observation and calculation, astronomers have assembled a theory of **stellar evolution**. This theory explains how stars are born in great clouds of interstellar gas and dust. They mature, grow old, and some eventually blow themselves apart in death throes that enrich interstellar space with new material for future stellar generations. The stars seem unchanging to human beings only because of the colossal time scale over which these changes occur. Major stages in the life of a star can last for millions or even billions of years.

20-1 Interstellar gas and dust pervade the Galaxy

The German-born English astronomer William Herschel, the discoverer of Uranus, was one of the greatest astronomical observers of all time. During the late 1700s, he discovered numerous double stars, nebulae, star clusters, and curious "holes in the heavens," where there seemed to be far fewer stars than might be expected. In the late 1800s, Edward E. Barnard, an American astronomer who rose from extreme poverty to become one of the greatest photographers of the heavens, turned his talents to these "holes" (Figure 20-1). His pictures suggested that the appearance of star-deficient regions is actually caused by vast clouds of interstellar matter that block our view of the stars behind them.

Astronomers were not quick to accept the idea of matter among the stars. Indeed, until the 1930s, most astronomers believed that interstellar space was essentially a perfect vacuum. Evidence for interstellar gas and dust, which together constitute the **interstellar medium,** came from various sources. (Actually, "smoke" would be a better name than "dust" because of the extremely small sizes of interstellar grains.)

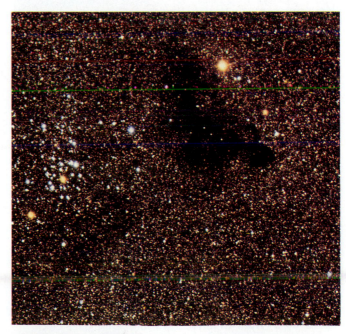

FIGURE 20-1 A Dark Nebula The dark nebula called Barnard 86 is visible in this photograph only because it blocks out light from the stars beyond it. This nebula is located in Sagittarius. The cluster of bluish stars to the left of the nebula is NGC 6520. (Anglo-Australian Observatory)

Improvements in photography helped convince many astronomers of the existence of interstellar matter. Photographs, like that of the Horsehead Nebula in Figure 20-2 and the Keyhole Nebula in Figure 20-3, clearly showed dust obscuring background nebulosity (clouds of glowing gas). Indeed, the nebulosity itself, glowing with the characteristic red light of H_α emission, is direct evidence of interstellar gas.

Some photographs also showed a bluish haze around certain stars (Figure 20-4). This haze, called a **reflection nebula**, is caused by fine grains of dust that scatter and reflect short-wavelength radiation more efficiently than long-wavelength radiation. When visible light from a star encounters interstellar dust, blue light is scattered by the dust grains, illuminating them and producing a reflection nebula. Red light manages to get through the dust without much scattering or absorption, if the cloud is not too dense.

Further evidence of interstellar atoms came from so-called stationary absorption lines discovered in the spectra of some spectroscopic binary stars. For example, certain calcium and sodium lines remain at fixed wavelengths. This was surprising because these lines appear in the spectra of double stars, whose lines shift back and forth as they orbit their common center of mass. The stationary lines are therefore not associated with the double star, but rather are caused by interstellar gas between us and the double star.

The most convincing evidence for interstellar matter came from the work of the American astronomer Robert Trumpler, who studied the brightness and distances of certain star clusters in the 1930s. Trumpler noticed that remote clusters seem to be unusually dim, dimmer than would be expected from their distance alone. His observations demonstrated that the intensity of light from remote stars is reduced as the light

FIGURE 20-2 The Horsehead Nebula Dust grains block light from the background nebulosity whose glowing gases are excited by ultraviolet radiation from young, massive stars. This nebula is located in Orion, at a distance of roughly 1600 light-years (490 parsecs) from Earth. The bright star near the top is Alnitak (ζ Ori), the easternmost star in the "belt" of Orion. (Royal Observatory, Edinburgh)

FIGURE 20-3 The Keyhole Nebula A dark notch, called the "keyhole" because of its shape, is caused by interstellar dust silhouetted against this nebulosity in the southern constellation of Carina, the keel of the Argonauts' ship. A bright variable star, called η Carinae, is located at the center of the nebula. Both the star and the nebula are about 9000 light-years from Earth. (Anglo-Australian Observatory)

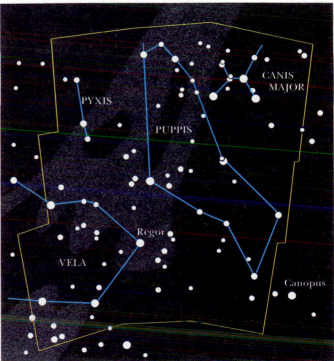

◀**FIGURE 20-4 Reflection Nebulae** Wispy bluish reflection nebulae called NGC 6726-7 surround several stars in the southern constellation of Corona Australis. This nebulosity is produced by interstellar dust grains scattering starlight. (Anglo-Australian Observatory)

passes through sparse material in interstellar space, a process called **interstellar extinction.** As mentioned above, the light from remote stars is also reddened as it passes through the interstellar medium. The appearance of remote stars is made redder because the blue component of their starlight is scattered and absorbed by interstellar dust. The effect is called **interstellar reddening.**

Measurements of both extinction and reddening in various directions in space indicate that interstellar gas and dust are largely confined to the Milky Way—that faint, hazy band of myriad stars you can see stretching across the sky on a dark, moonless night. Infrared views like Figure 20-5 clearly show clouds of gas and dust clumped along the Milky Way.

As we shall see in Chapter 25, the appearance of the Milky Way results from our inside view of the Galaxy, which is a rotating, disk-shaped collection of several hundred billion stars about 80,000 light-years in diameter. The Sun is located about 25,000 light-years from the Galaxy's center.

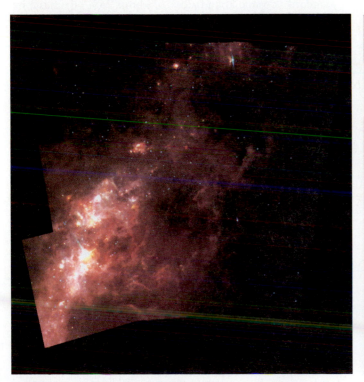

FIGURE 20-5 The Infrared Milky Way (a) This mosaic of infrared images extends over the southern constellations of Puppis and Vela, covering a region of the sky approximately 35° × 35°. Cool interstellar dust, colored red in this picture, emits radiation primarily at 60 μm. The blue dots are stars that shine brightly at 12 μm. These images were taken by the Infrared Astronomical Satellite (IRAS) in 1983. (b) This star chart shows the constellations covered by the infrared mosaic. The grayish areas indicate the visible Milky Way. (IPAC, NASA)

FIGURE 20-6 A Spiral Galaxy Spiral galaxies, like our own Milky Way Galaxy, consist of stars, gas, and dust that are largely confined to a flattened, rotating disk. Spiral arms, consisting of bright stars and glowing gas, wind outward from the galaxy's center. This galaxy, called NGC 2997, is located in the southern constellation of Antlia (the Air Pump). (Anglo-Australian Observatory)

Astronomers know these dimensions because they can map the Galaxy from the locations of bright stars and nebulae and also by using radio telescopes. As will be explained in Chapter 25, hydrogen atoms emit radio radiation that reveals the distribution of gas throughout our Galaxy. Such observations indicate that bright stars, gas, and glowing nebulae are mostly located along arching spiral arms that wind outward from our Galaxy's center. If we could view our Galaxy from a great distance, it would look somewhat like the galaxy shown in Figure 20-6.

Interstellar gas and dust are the raw material from which stars are made. The disk of our Galaxy, where most of this matter is concentrated, is therefore the site of ongoing star formation. In coming chapters, we shall see that the interstellar medium is also a dumping ground for dying stars. At the end of its life, a star can shed most of its mass in an outburst that enriches interstellar space with new chemical elements. The interstellar medium is therefore both nursery and graveyard. Because of this intimate relationship with stars, the interstellar medium evolves as successive generations of stars live out their lives. Understanding details of this cosmic symbiosis is one of the challenges of modern astronomy.

20-2 Protostars form in cold, dark nebulae and then evolve to become young main sequence stars

Stars are created by gravity acting on cold clouds of interstellar gas and dust. Denser portions of the cloud contract under their own weight into clumps called **protostars** that become new main sequence stars.

Some interstellar clouds, called **dark nebulae,** appear as dark blobs obscuring background stars or nebulosities. Many dark nebulae discovered and catalogued around 1900 by E. E. Barnard are called **Barnard objects.** There are also

some 200 relatively small, round, dark nebulae named **Bok globules,** after the Dutch-American astronomer Bart Bok, who first called attention to them in the 1940s.

A typical dark nebula contains a few thousand solar masses of gas and dust spread over a volume roughly 10 pc across. The chemical composition of this material is the standard "cosmic abundance" of about 74% (by mass) hydrogen, 25% helium, and 1% heavier elements (review Table 7-4). Emissions from molecules indicate that the cloud's internal temperature is about 10 K. As we saw in Box 7-2, the temperature of a gas is directly related to the average speed of its atoms and molecules. If an interstellar cloud is warm, its atoms move about so rapidly that there is no chance for a star to condense out of these agitated gases. At 10 K, however, the cloud is so cold and its atoms move so slowly that its gases do not provide enough internal pressure to support the cloud against its own weight. The cloud therefore contracts under the action of gravity and fragments into protostars.

In the 1950s astrophysicists such as L. Henyey in the United States and C. Hayashi in Japan performed calculations that enabled them to describe the earliest stages of a protostar. At first, a protostar is merely a cool blob of gas several times larger than our solar system. Pressure inside the protostar is still incapable of supporting all this cool gas, and so the protostar contracts. As it does so, gravitational energy is converted into thermal energy, making the gases heat up and start glowing (recall the discussion of Kelvin–Helmholtz contraction in Chapter 18). Hayashi's calculations indicate that energy from the interior of the protostar is transported outward by convection, thus warming its surface. After only a few thousand years of gravitational contraction, the surface temperature reaches 2000 to 3000 K. At this point the protostar is still quite large, so its glowing gases produce substantial luminosity. For instance, after only 1000 years of contraction, a protostar of 1 solar mass is 20 times larger in diameter and 100 times brighter than the Sun.

Astrophysicists now use high-speed computers and equations similar to those for calculating stellar structure (described in Chapter 18) to determine the conditions inside a contracting protostar. The results tell how the protostar's luminosity and surface temperature change at various stages in its contraction. This information, when plotted on a Hertzsprung–Russell diagram, gives a protostar's **evolutionary track.** This path shows us how the protostar's appearance changes because of processes in its interior.

Protostars are relatively cool when they begin to shine at visible wavelengths. Thus, the evolutionary tracks of protostars begin near the right side of the H–R diagram, the side of low temperature (Figure 20-7). However, continued gravitational contraction causes protostars to shift rapidly to the left, away from the cool region of the diagram. A protostar more massive than about five solar masses becomes hotter but shows little change in overall luminosity, because the

effect of decreasing surface area (which by itself would decrease luminosity) is counterbalanced by an increase in surface temperature (which in itself would increase luminosity). Thus, the evolutionary tracks of massive protostars traverse the H–R diagram horizontally from right to left. In contracting protostars that are less massive, the increase in surface temperature is less rapid, so it does not fully compensate for the shrinking surface area. The luminosity of low-mass protostars therefore decreases somewhat and their evolutionary tracks dip down on the H–R diagram as their surfaces warm up.

A protostar continues to shrink until the temperature at its center reaches a few million degrees, at which point hydrogen burning begins. As we saw in Chapter 18, this thermonuclear process releases enormous amounts of energy. This outpouring of energy creates enough pressure inside the protostar to stop its gravitational contraction. Once hydrostatic and thermal equilibrium are established, a stable star is born.

The evolutionary tracks of all protostars end on the main sequence, as shown in Figure 20-7. The main sequence therefore represents stars in which hydrogen burning is occurring. For most stars, this is a stable situation. For example, our Sun will remain on or very near the main sequence, quietly burning hydrogen at its core, for a total of some 10 billion years.

Note that the evolutionary tracks in Figure 20-7 end at locations along the main sequence that agree with the stellar mass–luminosity relation (recall Figure 19-19). The most massive stars are the most luminous, while the least massive stars are the least luminous. A protostar less massive than about 0.08 solar mass can never develop the necessary pressure and temperature to start hydrogen burning at its core. Instead, the failed star contracts to become a hydrogen-rich object called a **brown dwarf.** Protostars with masses greater than about 80 solar masses rapidly develop such extremely high temperatures that radiation pressure disrupts them. Main sequence stars therefore have masses between about 0.08 and 80 $M_\odot$, although the high-mass stars are extremely rare.

The evolutionary tracks of protostars begin in the red giant region of the H–R diagram, but protostars are not red giants. An H–R diagram like that in Figure 19-12 shows where stars spend *most* of their lives. Protostars, however, spend only a tiny fraction of their existence in the red giant region. A 15-$M_\odot$ protostar takes only 10,000 years to become a main sequence star, and a 1-$M_\odot$ protostar takes a few million years. By astronomical standards, these intervals are so brief that pre–main-sequence stars are quite transitory.

We are unlikely to observe the birth of a star at visible wavelengths because the surrounding interstellar cloud shields the protostar from view. The vast amount of visible light emitted by a protostar is absorbed by interstellar dust in its surrounding **cocoon nebula,** which becomes heated to a few hundred kelvin. The warmed dust then reradiates its thermal energy at infrared wavelengths, to which the dust is relatively transparent. Consequently, infrared observations reveal what is going on inside a "stellar nursery."

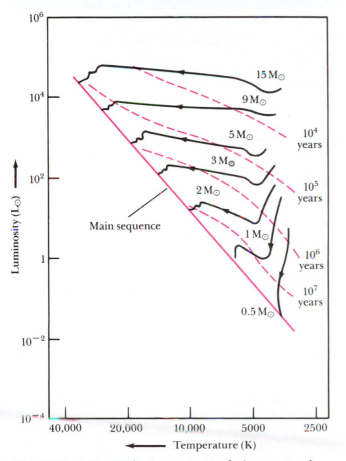

FIGURE 20-7 Pre–Main-Sequence Evolutionary Tracks
The evolutionary tracks of seven stars of different masses are shown on this H–R diagram. The dashed lines indicate the stage reached after the indicated number of years of evolution. Note that all tracks terminate on the main sequence at locations that agree with the mass–luminosity relation. (Based on stellar model calculations by I. Iben)

a

b

FIGURE 20-8 Newborn stars in the Swan Nebula
(a) This image at visible wavelengths shows an H II region called the Swan or Omega Nebula because of its characteristic shape. (b) This infrared view, constructed from images taken at wavelengths of 1.2, 1.6, and 2.2 μm, reveals hundreds of stars that do not appear in the visible view. A comparison of the visible and infrared views demonstrates that many of the stars in this star-forming region are obscured by interstellar dust. (NOAO)

Figure 20-8 shows views of a stellar nursery taken at both infrared and visible wavelengths. The visible view (Figure 20-8a) is a familiar sight to many telescope observers, but it tells only part of the story. The infrared picture (Figure 20-8b) shows hundreds of previously unseen stars, vividly demonstrating that most of the stellar nursery is hidden behind dust. Obscuring material is thickest toward the right (west) side of the nebula, where visible radiation is almost completely blocked, but numerous newborn stars shine brightly at infrared wavelengths.

20-3 A vigorous ejection of matter often accompanies the birth of a star

Spectroscopic observations demonstrate that cool, young stars often eject substantial amounts of gas as they approach the main sequence. Such gas-ejecting stars are called **T Tauri stars,** after the first example discovered in the constellation of Taurus. Some astronomers suggest that the onset of hydrogen burning is preceded by vigorous surface activity, including eruptions and flares that propel the star's outermost layers into space. In fact, an infant star going through its T Tauri stage can lose as much as 0.4 $M_\odot$ before it settles down on the main sequence.

T Tauri stars have relatively low masses of roughly 0.2 to 2.0 $M_\odot$. Young stars that are hotter and more massive than T Tauri stars lose mass through vigorous stellar winds that simply blow gas into space. Figure 20-9 is a dramatic photograph of a young, massive star experiencing mass loss driven by stellar winds.

Recent observations show that many young stars eject gas along two oppositely directed jets—a phenomenon called **bipolar outflow.** These stars are apparently surrounded by a disk of material left over from the star's formation. Gas expelled by the star cannot easily push through this matter. However, ejected gas meets little resistance in directions perpendicular to the disk. Consequently, the disk channels the ejected gas into two beams or jets that emerge at right angles to the disk's face, yielding a configuration much like an axle through a wheel.

A photograph of bipolar outflow from a young star is shown in Figure 20-10a. A jet, which is hot and luminous as it emerges from the star, travels southward approximately 0.2 light-year before cooling and becoming invisible. It travels on for another 0.5 light-year before it strikes a glowing clump called HH 34S. A second jet, which is invisible for its entire length, travels northward from the star and strikes a second glowing clump called HH 34N. These small nebulae are typical examples of **Herbig–Haro objects,** named after the astronomers George Herbig and Guillermo Haro, who first discovered them in the 1940s. Herbig–Haro objects consist of several bright knots of material that change slightly in size, shape, and brightness from year to year. They are often found near T Tauri stars. Photographs like Figure 20-10a show that a Herbig–Haro object is produced where a high-velocity jet from a young star rams into interstellar gases, causing them to glow.

FIGURE 20-9 The Mass-Loss Star HD 148937 This unusual star in the constellation of Norma is the brightest member of a triple system of stars in orbit about one another. This extremely hot, massive star is losing its outer layers continuously. Other vigorous outbursts in the past gave rise to the symmetric shells called NGC 6164 and NGC 6165 on either side of this star. (Anglo-Australian Observatory)

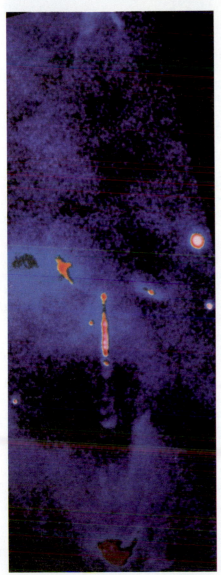

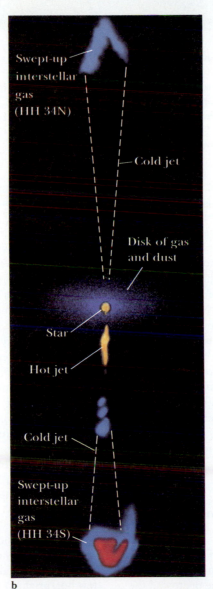

Swept-up interstellar gas (HH 34N)

Cold jet

Disk of gas and dust

Star

Hot jet

Cold jet

Swept-up interstellar gas (HH 34S)

FIGURE 20-10 Bipolar Outflow from a Young Star A young star is surrounded by material that channels ejected gases into two jets pointed in opposite directions. (a) This CCD image, taken with a 3.5-m telescope in Spain, shows a visible jet aimed southward at a Herbig–Haro object called HH 34S, a glowing cloud of swept-up gases that stop the jet's supersonic flow. A second, invisible jet traveling northward terminates at HH 34N. (b) This map identifies some of the features associated with this young star, which is about 1500 light-years from Earth. The tip-to-tip distance from HH 34N to HH 34S is about 1.6 light-years. (Courtesy of R. Mundt; Calar Alto Observatory)

a   b

a

b

c

FIGURE 20-11   Young Stars in the Orion Nebula   (a) This wide-angle view shows the entire Orion Nebula, which is an active star-forming region. It is located 1500 light-years from Earth and has a mass of about 300 solar masses. Four hot, massive stars at the center of the nebula produce the ultraviolet light that causes the gases to glow. **(b)** This image from the Hubble Space Telescope shows a plume that is probably the result of supersonic outflow from newborn stars just off the right side of the image. The plume is about $\frac{1}{4}$ light-year long and is estimated to be about 1500 years old. **(c)** This HST image shows a disk surrounding a newborn star. The disk is about 5 light-days in diameter; each square picture element (pixel) in this greatly magnified view is 50 AU across. (Anglo-Australian Observatory; Courtesy of C. R. O'Dell, NASA, ESA)

In the early 1990s astronomers using the Hubble Space Telescope discovered many examples of jets and disks around newly formed stars in the Orion Nebula. The Orion Nebula, a familiar sight to telescope observers, is one of the most prominent star-forming regions in the northern sky. Figure 20-11*a* is a wide-angle view that shows the entire nebula, which has a diameter of about 16 light-years. Figure 20-11*b*, an HST image of a small region of the nebula, shows a plume that may be a jet from young stars just off the right side of the picture. An HST image of a disk around a new star is shown in Figure 20-11*c*. A disk forms because some of the surrounding gas possesses too much angular momentum to be drawn into the contracting protostar. Instead, it spreads out into a broad, flattened disk, called a **protoplanetary disk** because it may eventually clump up to form planets. Based on the rate at which radiation from nearby hot stars are eroding material from the protoplanetary disk in Figure 20-11*c*, astronomers estimate that its initial mass was about 15 times that of Jupiter.

20-4   Young star clusters are found in H II regions

From the evolutionary tracks of protostars in Figure 20-7, we can see that high-mass stars evolve more rapidly than low-mass stars. Quite simply, the more massive the protostar, the sooner it develops the necessary central pressures and temperatures for hydrogen burning to begin. These massive main sequence stars, of spectral types O and B, are also the hottest, most luminous stars. Their surface temperatures are typically from 15,000 to 35,000 K, and thus—as indicated by Wien's law—they emit vast quantities of ultraviolet radiation. This radiation has a dramatic effect on the entire nebula in which a cluster of stars is forming. Energetic ultraviolet photons from newborn massive stars easily ionize the surrounding gas.

While some hydrogen atoms are being knocked apart by ultraviolet photons, other hydrogen atoms are reassembled, as free protons and electrons manage to get back together. During this **recombination** of hydrogen atoms, the captured electrons cascade downward through the atom's energy levels toward the ground state. These downward quantum jumps release numerous photons, causing the nebula to glow. Particularly important is the transition from $n = 3$ to $n = 2$, which produces H_α photons at 656 nm in the red portion of the visible spectrum (review Figure 5-18). Thus, the nebulosity around a newborn star cluster shines with a distinctive reddish glow. Figure 20-12 shows one of these **emission nebulae.**

Because emission nebulae are made predominantly of ionized hydrogen, they are also called **H II regions** (several famous examples are listed in Box 20-1). The collection of a few hot, bright O and B stars near the core of the nebula that

FIGURE 20-12 An H II Region Because of its shape, this emission nebula is called the Eagle Nebula. It surrounds the star cluster called M16 or NGC 6611 in the constellation of Serpens, 6500 light-years (2000 parsecs) from Earth. Several bright, hot O and B stars create the ionizing radiation that causes the nebula's gases to glow. (Anglo-Australian Observatory)

produces the ionizing ultraviolet radiation is called an **OB association.**

Individual stars in a young cluster can yield further information about stars in their infancy. Figure 20-13 shows a beautiful emission nebula surrounding the cluster NGC 2264. By measuring each star's magnitude and color index and knowing its distance, an astronomer can deduce its luminosity and surface temperature. The data for all the stars in the cluster can then be plotted on an H–R diagram, as in Figure 20-13. Note that the hottest stars, with surface temperatures around 20,000 K, are on the main sequence. These hot stars are the rapidly evolving, massive ones whose radiation causes the surrounding gases to glow. Stars cooler than about 10,000 K, however, have not yet quite arrived at the

---

BOX 20-1

Famous H II Regions

Some of the most beautiful nebulae in the sky are H II regions. The table below lists some of the most famous H II regions.

Nebula	Distance from Earth (ly)	Diameter (ly)	Mass ($M_\odot$)	Density (atoms/cm³)
M8 (Lagoon)	4000	30	1000	80
M16 (Eagle)	6000	20	500	90
M17 (Omega)	5000	30	1500	120
M20 (Trifid)	3000	12	150	100
M42 (Orion)	1500	16	300	600

Both the Lagoon and Trifid nebulae are shown in the photograph, which covers an area $2° \times 2\frac{1}{2}°$ in Sagittarius. The Lagoon Nebula is on the right, the smaller Trifid Nebula on the left.

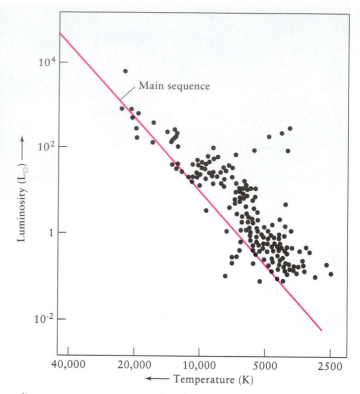

FIGURE 20-13 A Young Star Cluster and Its H–R Diagram The photograph shows an H II region and the young star cluster NGC 2264 in the constellation of Monoceros. It is located about 2600 light-years (800 parsecs) from Earth and contains numerous T Tauri stars. Each dot plotted on the H–R diagram represents a star in this cluster whose luminosity and surface temperature have been determined. Note that most of the cool, low-mass stars have not yet arrived at the main sequence. This star cluster probably started forming only 2 million years ago. (Anglo-Australian Observatory)

main sequence. These less massive stars, which are in the final stages of pre–main-sequence contraction, are just now beginning to ignite thermonuclear reactions at their centers. The locations of the data points on Figure 20-13 suggest that this particular cluster is about two million years old.

Figure 20-14 shows a young star cluster called the Pleiades that is easily visible to the unaided eye in the constellation of Taurus. In contrast with the H–R diagram for NGC 2264, nearly all the stars in the Pleiades are on the main sequence. The cluster's age is about 50 million years, which is how long it takes for the least massive stars to finally begin hydrogen burning in their cores.

A loose collection of stars such as the Pleiades or NGC 2264 is referred to as an **open cluster** or **galactic cluster**. Such clusters possess barely enough mass to hold themselves together by gravitation. Occasionally a star moving faster than average will escape, or "evaporate," from a cluster. Indeed, by the time the stars are a few billion years old, they may be so widely separated that a cluster no longer exists. If the group of stars is gravitationally unbound from the very beginning—that is, if the stars are moving away from one another so rapidly that gravitational forces cannot keep them together—then the apparent cluster that initially exists is called a **stellar association**.

20-5 Star birth can begin in giant molecular clouds

In the cold depths of interstellar space, atoms combine to form molecules. Molecules vibrate and rotate at specific frequencies that are dictated by the laws of quantum mechanics (see Box 20-2). As a molecule goes from one vibrational or rotational state to another, it either emits or absorbs a photon. This process is analogous to what happens when an atom emits or absorbs a photon as an electron jumps from one energy level to another. Many interstellar molecules emit photons with wavelengths of several millimeters. Consequently, in recent years radio telescopes tuned to wavelengths in this range have made observations that have greatly increased our knowledge of the interstellar medium.

Hydrogen is by far the most abundant element in the universe. In clouds, where starlight is absorbed by interstellar dust, much of it is in a molecular form (H_2) that is difficult to detect. The hydrogen molecule is symmetric, with two atoms of equal mass joined together, and such molecules do not emit many photons at radio frequencies. In contrast, asymmetric molecules such as carbon monoxide (CO), which consists of two atoms of unequal mass joined together, are easily detectable at radio frequencies. Carbon monoxide emits pho-

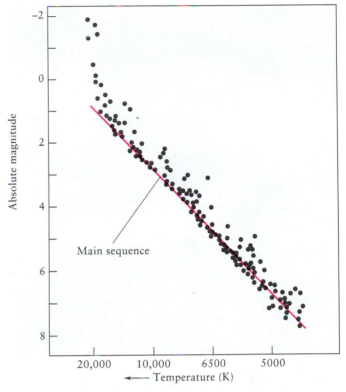

FIGURE 20-14 The Pleiades and Its H–R Diagram This open cluster called the Pleiades, which can easily be seen with the naked eye, is about 400 light-years from Earth. Each dot plotted on the H–R diagram represents a star in the Pleiades whose absolute magnitude and surface temperature have been measured.

Note that all of the cool low-mass stars have arrived at the main sequence, indicating that hydrogen burning has begun in their cores. Substantial reflection nebulosity permeates this cluster, which has a diameter of about 5 light-years and is about 50 million years old. (Anglo-Australian Observatory)

tons at wavelengths of 2.6 mm and shorter, which correspond to a transition between two rates of rotation of the molecule.

The ratio of carbon monoxide to hydrogen in interstellar space is reasonably constant: For every CO molecule, there are about 10,000 H_2 molecules. As a result, carbon monoxide is an excellent "tracer" for hydrogen gas. Wherever astronomers detect strong emission from CO, they know hydrogen gas must be abundant.

The first systematic surveys of our Galaxy looking for 2.6-mm CO radiation were undertaken in 1974 by Philip Solomon of the State University of New York at Stony Brook and Nicholas Scoville, now at Caltech. In mapping the locations of CO emission, they discovered huge clouds, now called **giant molecular clouds**, that must contain enormous amounts of hydrogen. These clouds have masses in the range of 10^5 to 2×10^6 solar masses and diameters that range from about 50 to 300 light-years. Inside one of these clouds, the density is about 200 hydrogen molecules per cubic centimeter. This is several thousand times larger than the average density of matter in the disk of our Galaxy, but 10^{17} times less dense than the air we breathe. Astronomers now estimate that our Galaxy contains about 5000 of these enormous clouds.

The constellations of Orion and Monoceros include one of the most accessible regions of the sky for studying star formation and the interaction of young stars with the interstellar medium. Figure 20-15a shows a map of this region made with a radio telescope tuned to a wavelength of 2.6 mm. Note the extensive areas of the sky covered by giant molecular clouds. Comprehensive maps of CO emission, like that shown in Figure 20-15a, help astronomers understand how the large-scale structure of the interstellar medium is related to the formation of H II regions and OB associations.

Recently U.S. astronomer Thomas M. Dame and his colleagues used CO emission to map the locations of giant molecular clouds in the inner regions of our Galaxy. The perspective drawing in Figure 20-16 shows the results of their work. They found that 17 molecular clouds clearly outline the spiral arm nearest the Sun, which is called the Sagittarius arm. These clouds lie roughly 3000 light-years apart, strung along the spiral arms like beads on a string. This arrangement is much like the spacing of H II regions in other galaxies (examine Figure 20-6).

As we shall see when we study the Galaxy in Chapter 25, spiral arms are locations where matter "piles up" temporarily as it orbits the center of the Galaxy. This enhancement of gas, dust, and stars along a spiral arm is analogous to a

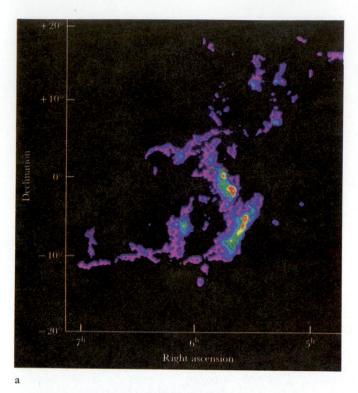

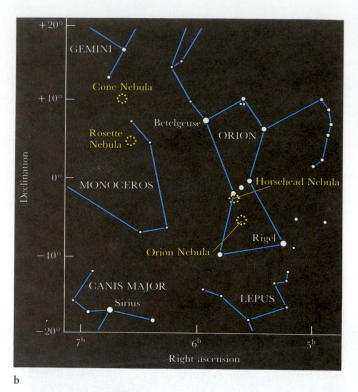

a

b

FIGURE 20-15 A Map of Carbon Monoxide Features in Orion (a) This color-coded map of a large section of the sky shows the extent of giant molecular clouds in Orion and Monoceros. The intensity of CO emission is indicated by colors in the order of the rainbow, from violet for the weakest to red for the strongest. Black indicates no detectable emission. (b) This star chart covers the same area as (a). The locations of four prominent star-forming nebulae are indicated. Note that the Orion and Horsehead nebulae are located at sites of intense CO emission. (Courtesy of R. Maddalena, M. Morris, J. Moscowitz, and P. Thaddeus)

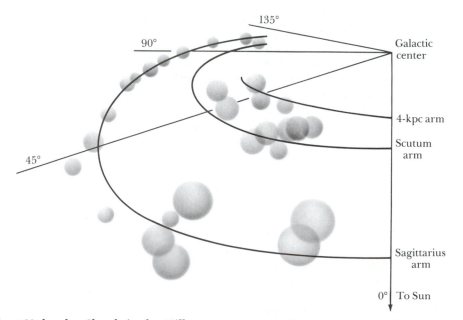

FIGURE 20-16 Giant Molecular Clouds in the Milky Way This perspective drawing shows the locations of giant molecular clouds in an inner part of our Galaxy as seen from a vantage point above the Sun. Note how the Sagittarius spiral arm is outlined by giant molecular clouds that lie along it like beads on a string. The locations of two inner spiral arms are also indicated. The distance from the Sun to the galactic center is about 25,000 light-years. (Adapted from T. M. Dame and colleagues)

traffic jam caused by construction on a freeway. Each automobile spends only a short time caught in the congestion before resuming its original speed. The traffic jam, which can be easily seen from an airplane, is simply a temporary enhancement of the number of cars in a particular location.

Spiral arms lag behind the direction in which all the stars, gas, and dust orbit the Galaxy's center. Near a galaxy's center, stars move faster than the spiral pattern and overtake it; far from a galaxy's center, the spiral pattern moves faster than the stars.

The gravitational pull of all the matter along a spiral arm compresses the interstellar medium through which it passes. When a giant molecular cloud is compressed, vigorous star formation begins in the densest regions. As soon as massive

O and B stars form, they emit ultraviolet light that ionizes the surrounding hydrogen, and an H II region is born. An H II region is thus a small, bright "hot spot" in a giant molecular cloud. The famous Orion Nebula (see Figure 20-11a) is one such H II region. Four hot, massive O and B stars at the heart of the Orion Nebula are responsible for the ionizing radiation that causes the surrounding gases to glow. The Orion Nebula is embedded in the edge of a giant molecular cloud whose mass is estimated at 500,000 solar masses.

The massive O and B stars at the core of the H II region are also responsible for the birth of stars in the rest of the giant molecular cloud. Vigorous stellar winds from the O and B stars carve out a cavity in the cloud, and the H II region, heated by the stars, expands into it. These winds travel faster

---

BOX 20-2

Interstellar Molecules

A molecule is a combination of two or more atoms—and these atoms vibrate and rotate at specific frequencies according to the laws of quantum mechanics. By either absorbing or emitting a photon, a molecule can speed up or slow down its rate of vibration or rotation. Astronomers can thus discover interstellar molecules by detecting the radiation from these vibrational or rotational transitions.

Although the first discovery of an interstellar molecule was in fact made at visual wavelengths, most molecules are strong emitters of radiation with wavelengths of around 1 to 10 mm. Consequently, observations with radio telescopes tuned to millimeter wavelengths have greatly increased the rate of discovery of interstellar molecules. Nearly 100 different kinds of interstellar molecules have been discovered so far, and the list is constantly growing. A representative partial listing is given below.

The vast majority of interstellar molecules contain carbon. Such substances are called organic molecules.

Apparently, organic chemistry is the chemistry of interstellar space as well as the chemistry of life on Earth. Astronomers have had such success in recent years in detecting interstellar molecules that they often boast of someday discovering every conceivable chemical somewhere in the universe. In fact, several organic molecules have been discovered in space that do not exist here on Earth, because they are too fragile to survive collisions with air molecules or with the walls of a container in a laboratory.

It is intriguing to speculate that interstellar molecules may have assembled somewhere else in the universe to form living organisms. After all, it took no more than a billion years for the first primitive life forms to develop on Earth from the organic molecules in the primordial oceans. The interstellar clouds in our Galaxy have had 15 billion years to accomplish the same thing. Is it likely that the interstellar clouds drifting between the stars contain extraterrestrial life?

Atoms in molecules					
2	3	4	5	6	7
CH	H_2O (water)	NH_3 (ammonia)	H_2CHN	CH_3OH (methyl alcohol)	CH_3NH_2
CO	HCO	H_2CO (formaldehyde)	H_2NCN	CH_3CN (methyl cyanide)	CH_3C_2H
H_2	HCN	HNCO	HCOOH (formic acid)	$HCONH_2$	$HCOCH_3$
SiO	SO_2	H_2CS	HC_3N		H_2CCHCN (vinyl cyanide)
CN	H_2S	HC_2H (acetylene)	H_2C_2O		
CS	OCS				
OH					
SO					

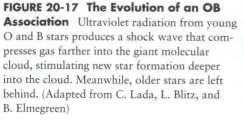

FIGURE 20-17 The Evolution of an OB Association Ultraviolet radiation from young O and B stars produces a shock wave that compresses gas farther into the giant molecular cloud, stimulating new star formation deeper into the cloud. Meanwhile, older stars are left behind. (Adapted from C. Lada, L. Blitz, and B. Elmegreen)

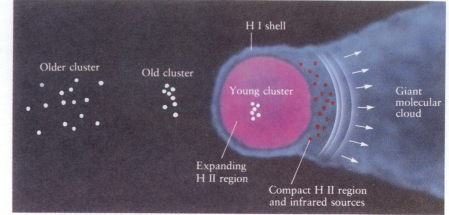

than the speed of sound in the gas, so a shock wave forms where the expanding H II region impinges on the rest of the giant molecular cloud. This shock wave compresses the hydrogen gas through which it passes, stimulating more star birth. Newborn O and B stars further expand the H II region into the giant molecular cloud. Meanwhile, the older O and B stars, which were left behind, begin to disperse (Figure 20-17). In this way, an OB association "eats into" a giant molecular cloud, leaving stars in its wake.

Infrared observations reveal many features that resemble protostars in the swept-up layer immediately behind the

shock wave from an OB association. For instance, Figure 20-18 shows both optical and infrared views of the core of the Orion Nebula. Four O and B stars, called the Trapezium, and glowing gas and dust dominate the view at visible wavelengths. Infrared observations penetrate this obscuring material to reveal dozens of infrared objects that may be cocoons of warm dust enveloping newly formed stars.

The OH and H_2O molecules in a giant molecular cloud become powerful sources of microwaves as they are excited by radiation from newborn stars. The process starts when these molecules absorb photons, whose added energy makes

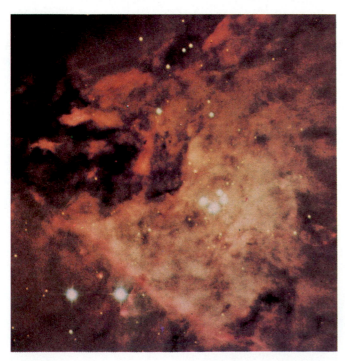

a

b

FIGURE 20-18 The Core of the Orion Nebula (a) This view at visible wavelengths shows the inner regions of the Orion Nebula (compare Figure 20-12). At the center are the four massive stars, called the Trapezium, which cause the nebula to glow. These stars are separated from each other by only 0.1 light-year. (b) This infrared composite was taken at wavelengths of 1.2 and 2.2 μm, which can penetrate interstellar dust more easily than can visible photons. Numerous infrared objects, many of which are probably new stars in early stages of formation, are seen in this view. (Anglo-Australian Observatory)

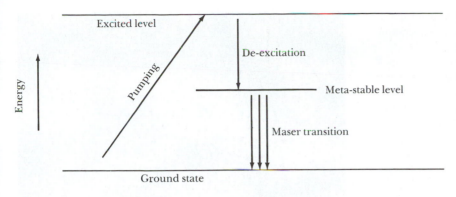

FIGURE 20-19 A Maser This energy-level diagram shows how certain molecules like OH, H_2O, and SiO can be powerful sources of microwaves. After being excited by radiation from young stars, such molecules drop down into a meta-stable level, where they remain until a passing photon stimulates de-excitation down to the ground state.

the molecules spin faster. In the language of quantum mechanics, the molecules jump from their ground state to an excited rotational energy level, a process called *pumping* (Figure 20-19). The excited molecules naturally drop back to a lower energy level by emitting photons. According to the rules of quantum mechanics, some of these molecules will drop down into an energy level that is characterized as being "meta-stable," meaning that the molecule will remain in that level for an exceptionally long time.

As a result of the pumping process, regions develop in a giant molecular cloud where vast numbers of OH or H_2O molecules are stuck in metastable states. Now imagine one of these regions being traversed by a microwave photon whose energy equals the energy difference between the meta-stable excited level and the ground level. The electromagnetic field of this photon triggers many molecules along its path to lose energy by emitting identical microwave photons traveling parallel to the original one. These new photons stimulate still more molecules to lose energy, resulting in a cascade that releases a vast amount of energy at microwave frequencies. The word **maser**, which is used to describe this phenomenon, is an acronym for *m*icrowave *a*mplification by *s*timulated

emission of *r*adiation. A single maser lasts for only a few weeks or months. Near a site of active star formation, new masers are continually being turned on while old ones, having depleted their supplies of excited molecules, simply fade away.

20-6 Processes that compress the interstellar medium can trigger star birth

Presumably any mechanism that compresses interstellar clouds can trigger the birth of stars. The most dramatic is a **supernova explosion**, caused by the violent death of a massive star. As we shall see in Chapter 22, the core of the doomed star collapses suddenly, releasing vast quantities of particles and energy that blow the star apart. The star's outer layers are blasted into space at speeds of several thousand kilometers per second.

Astronomers have found many nebulae across the sky that are the shredded funeral shrouds of these dead stars. Such nebulae, like the Cygnus Loop shown in Figure 20-20, are known as **supernova remnants**. Many supernova remnants have a distinctly arched appearance, as would be

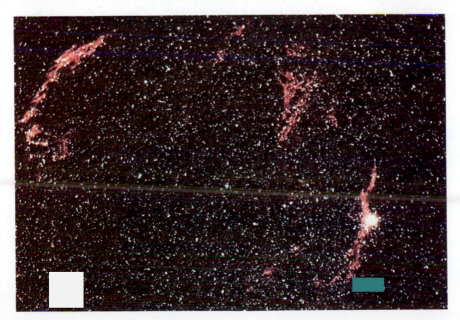

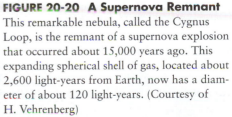

FIGURE 20-20 A Supernova Remnant This remarkable nebula, called the Cygnus Loop, is the remnant of a supernova explosion that occurred about 15,000 years ago. This expanding spherical shell of gas, located about 2,600 light-years from Earth, now has a diameter of about 120 light-years. (Courtesy of H. Vehrenberg)

expected for an expanding shell of gas. This wall of gas is typically moving away from the dead star at supersonic speeds.

Supersonic motion is always accompanied by a shock wave that abruptly compresses the medium through which it passes. Astronomers have recently studied the Cygnus Loop with the Hubble Space Telescope to examine a supernova's shock wave plowing through the interstellar medium. Figure 20-21 shows the shock wave overrunning dense clumps of gas. This collision heats and compresses the gas, causing it to glow, thereby making details of the interstellar medium visible.

When the expanding shell of a supernova remnant slams into an interstellar cloud, it squeezes the cloud, stimulating star birth. This kind of star birth can be observed in the stellar association seen in Figure 20-22. This stellar nursery is located along a luminous arc of gas about 100 light-years in length that is presumably the remnant of an ancient supernova explosion. In fact, this arc is part of an almost complete ring of glowing gas with a diameter of about 200 light-years. Spectroscopic observations of the stars along this arc reveal substantial T Tauri activity. This activity results from newborn stars that are experiencing mass loss in their final stages of contraction before they become main sequence stars.

FIGURE 20-22 The Canis Major R1 Association This luminous arc of gas, about 100 light-years long, is studded with numerous young stars. This stellar association illustrates the results of a shock wave from a supernova explosion: the triggering of star formation in the interstellar clouds through which the shock wave passes. (Courtesy of H. Vehrenberg)

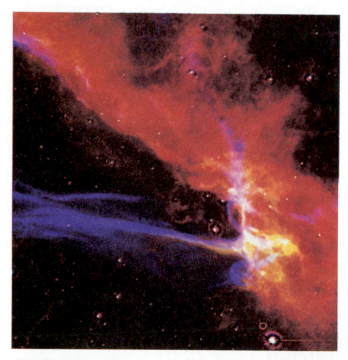

FIGURE 20-21 A Supernova's Blast Wave This image from the Hubble Space Telescope shows a small portion of the Cygnus Loop with unprecedented clarity. The bluish ribbon stretching left to right across the picture may be a knot of gas ejected by the supernova some 15,000 years ago. This knot, traveling at about 1400 km/s (3 million miles per hour), is just now catching up with the shock wave, which has been slowed by plowing though the interstellar medium. (NASA, ESA)

As we learned in Chapter 17, there is strong evidence that the Sun was once a member of one of these loose associations created by a supernova explosion. Recall that stellar associations do not remain intact for long. Individual stellar motions soon carry the stars in various directions away from their birthplaces. Nearly 5 billion years have passed since the birth of our star, so the Sun's brothers and sisters are now widely scattered across the Galaxy.

Many other processes can also trigger star formation. For instance, a collision between two interstellar clouds can create new stars. When two such clouds collide, compression occurs at the interface and vigorous star formation follows. Similarly, stellar winds from a group of O and B stars may exert strong enough pressure on interstellar clouds to cause compression, followed by star formation. The Rosette Nebula, shown in Figure 20-23, is an example of this process.

Our understanding of star birth has improved dramatically in recent years, primarily through infrared- and milli-meter-wavelength observations. Nevertheless, many puzzles and mysteries remain. For example, astronomers have gener-

FIGURE 20-23 The Core of the Rosette Nebula The Rosette Nebula is a large, circular emission nebula located near one end of a sprawling giant molecular cloud in the constellation of Monoceros. Radiation from young, hot stars has blown gas away from the center of this nebula. Some of this gas has become clumped in dark globules that appear silhouetted against the glowing background gases. (Anglo-Australian Observatory)

ally assumed that there must be a lot of interstellar dust to shield a stellar nursery from the disruptive effects of ultraviolet light from nearby massive, hot stars. However, a neighboring galaxy, called the Large Magellanic Cloud (LMC), contains young OB associations with virtually no dust. Does the process of star birth differ slightly from one galaxy to another?

Another problem is that different modes of star birth tend to produce different percentages of different kinds of stars. For example, the passage of a spiral arm through a giant molecular cloud tends to produce an abundance of massive O and B stars. In contrast, the shock wave from a supernova seems to produce fewer O and B stars, but many more of the less massive A, F, G, and K stars. We do not yet know why this is so.

In spite of these and other unanswered questions, it is now clear that star birth involves mechanisms on a colossal scale, from the deaths of massive stars to the rotation of an entire galaxy. In many respects, we have just begun to appreciate these cosmic processes. The study of cold, dark stellar nurseries will certainly be an active and exciting area of astronomical research for many years to come.

KEY WORDS

Barnard object	evolutionary track	interstellar reddening	reflection nebula
bipolar outflow	galactic cluster	maser	stellar association
Bok globule	giant molecular cloud	OB association	stellar evolution
brown dwarf	H II region	open cluster	supernova explosion
cocoon nebula	Herbig–Haro object	protoplanetary disk	supernova remnant
dark nebula	interstellar extinction	protostar	T Tauri star
emission nebula	interstellar medium	recombination	

KEY IDEAS

• Interstellar gas and dust, which make up the interstellar medium, are concentrated in the disk of the Galaxy.

Reflection nebulae are produced when starlight is reflected from dust grains in the interstellar medium, producing a characteristic bluish glow.

A cloud that is visible as a dark blot against distant stars is called a dark nebula. Some dark nebulae are named

Barnard objects and others are called Bok globules, after the astronomers who discovered them.

• Enormous cold clouds of gas called giant molecular clouds are scattered along the spiral arms of our Galaxy.

• Star formation begins when gravitational attraction causes a protostar to contract within a giant molecular cloud.

• As a protostar grows by the gravitational accretion of gases, the process known as Kelvin–Helmholtz contraction causes it to heat and begin glowing. Its relatively low temperature and high luminosity place it in the upper right region on an H–R diagram.

As a protostar evolves, it moves toward the main sequence on the H–R diagram. When its core temperatures become high enough to ignite hydrogen burning, it becomes a main sequence star.

In the final stages of pre–main-sequence contraction, when thermonuclear reactions are about to begin in the core of a protostar, the star may eject large amounts of gas into space; the low-mass stars that vigorously eject gas are called T Tauri stars.

A circumstellar disk channels the gas that is ejected by a young star into jets; clumps of glowing gas called Herbig–Haro objects are sometimes found at the ends of these jets.

The most massive protostars rapidly become main sequence O and B stars; they emit strong ultraviolet radiation that ionizes hydrogen in the surrounding cloud, thus creating the reddish emission nebulae called H II regions.

Ultraviolet radiation and stellar winds from the O and B stars at the core of an H II region create shock waves that move outward through the gas cloud, compressing the gas and triggering the formation of more protostars.

Shock waves associated with the spiral arms of our Galaxy and with supernova explosions also compress gas clouds and trigger star formation.

- A collection of newborn stars may form an open or galactic cluster, in which stars are held together by gravity; occasionally a star moving more rapidly than average will escape, or "evaporate," from such a cluster.

- A stellar association is a group of newborn stars that are moving apart so rapidly that their gravitational attraction for each other cannot pull them into orbit about each other.

REVIEW QUESTIONS

1. What evidence is there that interstellar space contains gas and dust?

2. What is a giant molecular cloud, and what role do these clouds play in the birth of stars?

3. What happens inside a protostar to slow and eventually halt its gravitational contraction?

4. Describe the energy source that causes a protostar to shine. How does this source differ from the energy source inside a true star?

5. Why do disks form around contracting protostars? What effect does a disk have on the vigorous stellar winds that accompany the birth of a star?

6. What is an H II region?

7. Why is the daytime sky blue? Why is the Sun red when seen near the horizon at sunrise or sunset? In what ways

are your answers analogous to the explanations for the bluish color of reflection nebulae and the process of interstellar reddening?

8. What is an evolutionary track, and how can one help us interpret the H–R diagram?

9. Why are low temperatures necessary in order for protostars to form inside dark nebulae?

10. Why are observations at infrared and millimeter wavelengths so much more useful in exploring interstellar clouds than are observations at visible wavelengths?

11. What is a maser?

12. Briefly describe four mechanisms that compress the interstellar medium and trigger star formation.

ADVANCED QUESTIONS

Tips and tools . . .

You may find it helpful to review Box 19-2, which describes the relationship between magnitude and brightness. Remember that the Stefan–Boltzmann law relates the temperature of a blackbody to its energy flux, as described in Box 5-2.

13. If you looked at a spectrum of a reflection nebula, would you see absorption lines, emission lines, or no lines? Explain your answer, describing how the spectrum demonstrates that the light was reflected from nearby stars.

14. Find the density (in atoms per cubic meter) of a Bok globule having a radius of 1 light-year and a mass of 100 solar masses. How does your result compare with the densities listed in Box 20-1? (Assume that the globule is made of pure hydrogen.)

15. In the direction of a particular star cluster, interstellar extinction dims starlight by 2 magnitudes per kiloparsec. If the star cluster is 1.5 kiloparsecs away, what percentage of its photons survive the trip to Earth?

16. At one stage during its birth, the protosun had a luminosity of 1000 $L_\odot$ and a surface temperature of about 1000 K. What was its radius?

17. How would you distinguish a newly formed protostar from a red giant, since they are both located in the same region on the H–R diagram?

18. The concentration or abundance of ethyl alcohol (molecular weight = 46) in a typical molecular cloud is about 1 molecule per 10^8 cubic meters. What volume of such a cloud would contain enough alcohol to make a martini (about 10 grams of alcohol)?

DISCUSSION QUESTIONS

19. What do you think would happen if our solar system were to pass through a giant molecular cloud? Do you think the Earth has ever passed through such clouds?

20. Speculate on why a shock wave from a supernova seems to produce relatively few high-mass O and B stars, compared to the lower-mass A, F, G, and K stars.

21. Speculate about the possibility of life forms and biological processes occurring in giant molecular clouds. In what ways might the conditions existing in giant molecular clouds favor or hinder biological evolution?

OBSERVING PROJECTS

22. Use a telescope to observe at least two of the following H II regions. Coordinates are for epoch 2000.

Nebula	R.A.	Decl.
M42 (Orion)	5^h 35.4^m	$-5°$ $27'$
M43	5 35.6	−5 16
M20 (Trifid)	18 02.6	−23 02
M8 (Lagoon)	18 03.8	−24 23
M17 (Omega)	18 20.8	−16 11

In each case, can you guess which stars are probably responsible for the ionizing radiation that causes the nebula to glow? Can you see any obscuration or silhouetted features that suggest the presence of interstellar dust? Draw a picture of what you see through the telescope and compare it with a photograph of the object. Take note of which portions of the nebula were not visible through your telescope.

23. On an exceptionally clear, moonless night, use a telescope to observe at least one of the following dark nebulae. These nebulae are very difficult to find because they are recognizable only by the *absence* of stars in an otherwise starry part of the sky. As usual, all coordinates are for epoch 2000.

Nebula	R.A.	Decl.
Barnard 72 (The Snake)	17^h 23.5^m	$-23°$ $38'$
Barnard 86	18 02.7	−27 50
Barnard 133	19 06.1	−6 50
Barnard 142 and 143	19 40.7	+10 57

Are you confident that you actually saw the dark nebula? Does the pattern of background stars suggest a particular shape to the nebula?

24. There are a few fine examples of objects covering such large regions of the sky that they are best seen with binoculars. If you have access to a high-quality pair of binoculars, observe the North America Nebula in Cygnus and the Pipe Nebula in Ophiuchus. Both nebulae are quite faint, so you should attempt to observe them only on an exceptionally dark, clear, moonless night.

The North America Nebula is a cloud of glowing hydrogen gas located about 3° east of Deneb, the brightest star in Cygnus. While searching for the North America Nebula, you may glimpse another diffuse H II region, the Pelican Nebula, located about 2° southeast of Deneb.

The Pipe Nebula is a 7°-long, meandering, dark nebula to the south and to the east of the star θ Ophiuchi, which is in a section of Ophiuchus that extends southward between the constellations of Scorpius and Sagittarius. Located about 12° east of the bright red star Antares, θ Ophiuchi is easily identified with the aid of star charts published during the summer months in such magazines as *Sky & Telescope* and *Astronomy*.

FOR FURTHER READING

Blitz, L. "Giant Molecular Cloud Complexes in the Galaxy." *Scientific American*, April 1982. A fine discussion of interstellar molecules is the highlight of this article about the structure and properties of giant molecular clouds.

Boss, A. "Collapse and Formation of Stars." *Scientific American*, January 1985. This article discusses how the process of star birth, although hidden from view, can nonetheless be modeled on supercomputers.

Cohen, M. *In Darkness Born: The Story of Star Formation*. Cambridge University Press, 1988. This well-written book presents an overview of our modern understanding of the early stages of stellar evolution.

Lada, C. "Energetic Outflows from Young Stars." *Scientific American*, July 1982. This article describes how radiation emitted by carbon monoxide molecules discloses bipolar outflow from newborn stars.

Reipurth, B. "Bok Globules." *Mercury*, March/April 1984. A superb collection of photographs complements this article on the relationship between Bok globules and the process of star formation.

Robinson, L. "Orion's Stellar Nursery." *Sky & Telescope*, November 1982. This article explores the Orion Nebula and its environs with the aid of high-resolution infrared observations.

Scoville, N., and Young, J. "Molecular Clouds, Star Formation, and Galactic Structure." *Scientific American*, April 1984. This article describes how the process of star formation can affect the appearance of a galaxy.

Stahler, S. W. "The Early Life of Stars." *Scientific American*, July 1991. This article examines our current understanding of protostars.

Verschuur, G. "Interstellar Molecules." *Sky & Telescope*, April 1992. This well-written article examines some of the chemical and physical processes that occur in the interstellar medium.

STELLAR MATURITY AND OLD AGE

A MASS-LOSS STAR Old stars become giants and supergiants whose bloated outer atmospheres shed matter into space. This star, HD 65750, is losing matter at a high rate and is surrounded by a reflection nebula (IC 2220) caused by starlight reflecting from dust grains. These dust grains may have condensed from material shed by the star. A typical red giant can lose 10^{-7} solar mass per year. Many red giants are surrounded by gas and dust they have ejected. (Anglo-Australian Observatory)

MATURE STARS undergo a remarkable transformation after consuming all the hydrogen in their cores. After hydrogen burning ceases, a star leaves the main sequence and expands dramatically to become a red giant. In the hot, compressed core of a red giant, helium burning ignites and becomes a new energy source. The more massive a star is, the more rapidly it consumes its thermonuclear fuels, and so the more rapidly it evolves. A cluster of stars, which naturally incorporates a range of stellar masses, therefore contains stars at various evolutionary stages, even though they are all roughly the same age. This distribution is best seen on an H–R diagram of the cluster, from which we can estimate the cluster's age. Fascinating scenarios occur when giant stars evolve in binary systems, sometimes exchanging mass or sharing a common atmosphere. An aging red giant, which is so bloated that it constantly leaks gas into space, occasionally becomes unstable and pulsates.

Thermal equilibrium is a fundamental property of all stars: Energy liberated in the interior of a star must be balanced by the energy radiated from its surface. In a main sequence star, that energy comes from a thermonuclear process, hydrogen burning, in its core.

A second fundamental property is hydrostatic equilibrium: The pressure at any depth within a star is sufficient to support the weight of the overlying layers. The outward flow of energy from **core hydrogen burning** in a main sequence star heats the star's interior so that the pressure in every layer is just enough to support the weight pressing down from above. Eventually, however, so much hydrogen in the core is used up that core hydrogen burning must cease. The termination of this energy source has a dramatic effect on the star's structure and evolution.

21-1   When core hydrogen burning ceases, a main sequence star becomes a red giant

Hydrogen has been burning in the Sun's core for the past 4.6 billion years. Initially, the Sun's chemical composition was roughly 74% hydrogen and 25% helium, with a 1% smattering of heavy elements. The ongoing fusion of hydrogen into helium in the Sun's core has significantly altered its composition, however. As shown in Figure 21-1, there is now more helium than hydrogen at the Sun's center. Nevertheless, enough hydrogen remains in the Sun's core for another

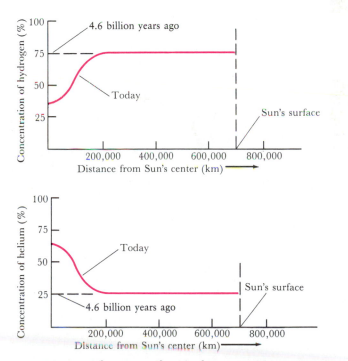

FIGURE 21-1 The Sun's Chemical Composition The Sun began with a composition of nearly 75% hydrogen and 25% helium. However, 4.6 billion years of thermonuclear reactions at the Sun's center have depleted the concentration of hydrogen and increased that of helium.

5 billion years of core hydrogen burning. The Sun's total lifetime on the main sequence will therefore be about 10 billion years.

As the supply of hydrogen at a star's center dwindles, the star begins to have difficulty supporting its outer layers. Hydrogen burning causes the number of particles in the star's core to decrease, since four hydrogen nuclei are consumed to make each helium nucleus. With fewer particles bouncing around to provide the pressure that supports the weight of the star's outer layers, the central core contracts slightly. This compression raises both the pressure and the temperature at the center of the star, increasing the rate of core hydrogen burning and making the star more luminous. This increased energy outflow also heats hydrogen around the star's core so that it ignites. By tapping this supply of fresh hydrogen, the star manages to eke out a few million more years on the main sequence.

The length of time that a star remains on the main sequence depends critically on its mass (see Box 21-1 for mathematical details). The most massive main sequence stars are the most luminous stars, and their rapid emission of energy corresponds to a rapid depletion of hydrogen in their cores. Even though a massive O or B star contains much more hydrogen fuel than a less massive main sequence star, it also consumes its hydrogen far more rapidly. Table 21-1 shows how long stars take to exhaust the supplies of hydrogen in their cores. Note that high-mass stars gobble up their hydrogen fuel in only a few million years, but low-mass stars take hundreds of billions of years to use up their hydrogen.

Finally, when all the hydrogen in the core of an aging main sequence star is used up, hydrogen burning ceases in the star's core. In this new stage, hydrogen burning still continues in the hydrogen-rich material surrounding the core. At first, this **shell hydrogen burning** occurs only in the hottest region just outside the core, where the hydrogen fuel has not yet been exhausted.

When core hydrogen burning ceases, heat flowing out of the core is no longer replaced. To maintain thermal equilibrium, the star's core again contracts, a process that converts gravitational energy into thermal energy. As a result, the tem-

---

BOX 21-1

Main Sequence Lifetimes

During hydrogen burning, a portion of a star's mass is converted into energy. We can use Einstein's famous equation relating mass and energy to calculate how long a star will remain on the main sequence.

Suppose that M is the mass of a star and f is the fraction of the star's mass that is converted into energy by hydrogen burning. The total energy E supplied by the hydrogen burning can be expressed as

$$E = fMc^2$$

where c is the speed of light.

This energy is released gradually over many years. Specifically, suppose that L is the star's luminosity and t is the total time over which the hydrogen burning occurs. Then

$$E = Lt$$

From these two equations, we see that

$$Lt = fMc^2$$

and so

$$t = \frac{fMc^2}{L}$$

Thus, a star's lifetime on the main sequence is proportional to its mass divided by its luminosity:

$$t \propto \frac{M}{L}$$

We can carry this analysis further by recalling that main sequence stars obey the mass–luminosity relation (see Figure 19-19). The distribution of data on this graph tells us that a star's luminosity is roughly proportional to the 3.5 power of its mass:

$$L \propto M^{3.5}$$

Substituting this relationship into the previous proportionality, we find that

$$t \propto \frac{1}{M^{2.5}} = \frac{1}{M^2\sqrt{M}}$$

This approximate relationship can be used to obtain rough estimates of how long a star will remain on the main sequence. It is often convenient to relate these estimates to the Sun (a typical 1-$M_\odot$ star), which will spend 10^{10} years on the main sequence.

EXAMPLE: Consider a star whose mass is 4 $M_\odot$. This star will be on the main sequence for

$$\frac{1}{4^{2.5}} = \frac{1}{4^2\sqrt{4}} = \frac{1}{32} \text{ solar lifetime}$$

Thus, a 4-$M_\odot$ star will burn hydrogen in its core for about $\frac{1}{3}$ billion years.

TABLE 21-1
Main Sequence Lifetimes

Mass ($M_\odot$)	Surface temperature (K)	Luminosity ($L_\odot$)	Time on main sequence (10^6 years)
25	35,000	80,000	3
15	30,000	10,000	15
3	11,000	60	500
1.5	7,000	5	3,000
1.0	6,000	1	10,000
0.75	5,000	0.5	15,000
0.50	4,000	0.03	200,000

perature in the core again rises and warms the surrounding gases. As the hydrogen-burning shell eats into the surrounding matter, it dumps more helium onto the core, which continues to contract and heat up as it gains mass. Over the course of hundreds of millions of years, the core of a 1-$M_\odot$ star becomes compressed to about one-tenth of its original radius and its central temperature increases from around 15 million kelvin to about 100 million kelvin.

As the hydrogen burning shell burns outward, the star's luminosity increases and the entire star expands. In other words, as the star's core contracts, its outer atmosphere expands farther and farther into space, and its surface gases begin to cool. Once the temperature of the star's bloated surface falls to about 3500 K, the gases glow with a reddish hue, in accordance with Wien's law. The star is then appropriately called a **red giant.**

Our own Sun will take about 5 billion years more to finish converting hydrogen into helium at its core. As the Sun's core contracts, its atmosphere will expand to envelop Mercury and reach almost all the way to Venus. The red giant Sun will eventually swell to a diameter of about 1 AU, and its surface temperature will decline to about 3500 K. Although the Sun's surface temperature will be much lower than it is today, the Sun will be so huge that its luminosity will be much greater than today. As a full-fledged red giant (Figure 21-2), our star will shine with the brightness of 2000 Suns. Some of the inner planets will be vaporized, and the thick atmospheres of the outer planets will boil away to reveal tiny, rocky cores. Thus, in its later years the aging Sun will destroy the planets that have accompanied it since its birth.

21-2 Helium burning begins at the center of a red giant

When a star first becomes a red giant, its hydrogen-burning shell surrounds a small, compact core of almost pure helium. In a moderately low-mass red giant, which the Sun will be 5 billion years from now, the dense helium core is about twice the size of the Earth, and the star's bloated surface has a diameter of 1 AU.

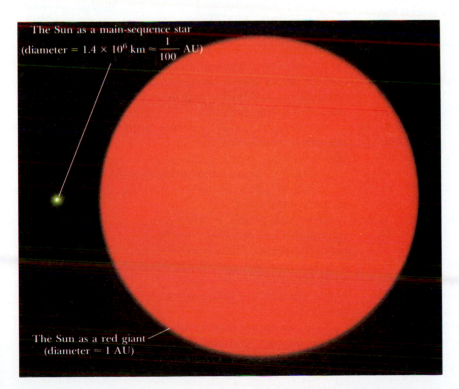

The Sun as a main-sequence star
(diameter $= 1.4 \times 10^6$ km $\approx \frac{1}{100}$ AU)

The Sun as a red giant
(diameter $\approx$ 1 AU)

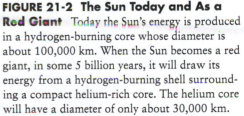

FIGURE 21-2 The Sun Today and As a Red Giant Today the Sun's energy is produced in a hydrogen-burning core whose diameter is about 100,000 km. When the Sun becomes a red giant, in some 5 billion years, it will draw its energy from a hydrogen-burning shell surrounding a compact helium-rich core. The helium core will have a diameter of only about 30,000 km.

Helium, the "ash" of hydrogen burning, is a potential fuel. At first, thermonuclear reactions do not occur in the helium core of a red giant, because the temperature is too low to fuse helium nuclei. Because each helium nucleus contains two protons, it has twice the positive electric charge of a hydrogen nucleus. Compared to hydrogen, the helium nuclei must move faster to overcome their stronger electric repulsion and get close enough to fuse together, and this requires higher temperatures (recall Box 7-2).

The hydrogen-burning shell adds mass to the helium core, forcing it to contract and further increasing the star's central temperature. When the central temperature finally reaches 100 million kelvin, **helium burning** is ignited at the star's center. As a result, the aging star has a central energy source for the first time since leaving the main sequence.

This new thermonuclear reaction occurs in two steps. First, two helium nuclei combine to form an isotope of beryllium:

$$^4He + {}^4He \rightarrow {}^8Be$$

Because this particular beryllium isotope is very unstable, it quickly breaks back down into two helium nuclei soon after it forms. However, in the star's dense core a third helium nucleus may strike the ^8Be nucleus before it has a chance to fall apart. Such a collision creates a stable, common isotope of carbon:

$$^8Be + {}^4He \rightarrow {}^{12}C + \gamma$$

In this process, a gamma-ray photon (γ) is released.

During the pioneering days of nuclear physics, helium nuclei were called **alpha particles,** and the fusion of three helium nuclei to form a carbon nucleus is still called the **triple alpha process.** Some of the carbon created in this process can fuse with an additional helium nucleus to produce oxygen:

$$^{12}C + {}^4He \rightarrow {}^{16}O + \gamma$$

Thus, both carbon and oxygen make up the "ash" of helium burning.

The second step in the triple alpha process and the process of oxygen formation releases both energy and a gamma-ray photon. This energy source, properly called **core helium burning** because of its central location, establishes thermal equilibrium, thereby preventing any further gravitational contraction of the star's core. A mature red giant burns helium in its core for about 20% as long as the time it spent burning hydrogen as a main sequence star. For example, in the distant future the Sun will consume helium in its core for about 2 billion years.

How helium burning begins at a red giant's center depends on the mass of the star. In high-mass stars (those with masses greater than about 3 $M_\odot$), helium burning begins gradually as temperatures in the star's core approach 100 million kelvin. In low-mass stars (those with masses less than about 3 $M_\odot$), helium burning begins explosively and suddenly, in what is called the **helium flash.**

The helium flash occurs because of unusual conditions that develop in the core of a low-mass star as it becomes a red giant. To appreciate these conditions we must first understand how an ordinary gas behaves; then we can explore how the densely packed electrons at the star's center alter this behavior.

When a gas is compressed, it usually becomes denser and warmer. To describe this process, scientists use the convenient concept of a **perfect gas,** which has a simple relationship between pressure, temperature, and density. Specifically, the pressure exerted by a perfect gas is directly proportional to both the density and the temperature of the gas. Many real gases actually behave like a perfect gas over a wide range of temperatures and densities.

Under most circumstances, the gases inside a star act like a perfect gas: If the gas is compressed, it heats up, and if it expands, it cools down. This behavior serves as a safety valve, ensuring that the star does not explode. For example, if energy production overheats the star's core, the core expands, cooling the gases and slowing the rate of thermonuclear reactions. Conversely, if too little energy is being created to support the star's overlying layers, the core compresses, increasing the temperature and thus speeding up the thermonuclear reactions to increase the energy output.

In a low-mass red giant, the core must be extremely compressed in order to become hot enough for helium burning to begin. At these extreme pressures and temperatures, the atoms are completely ionized; thus most of the matter in the star's core consists of detached nuclei and electrons. Eventually the free electrons become so closely crowded that a limit to further compression is reached: A law of quantum mechanics called the **Pauli exclusion principle.** According to this principle, formulated in 1925 by the Austrian physicist Wolfgang Pauli, two identical particles cannot simultaneously occupy the same "quantum state." A quantum state is a particular set of circumstances concerning locations and speeds that are available to a particle. In the submicroscopic world of atoms and particles, the Pauli exclusion principle is analogous to saying you can't have two things in the same place at the same time.

Just before the onset of helium burning, the electrons in the core of a low-mass star are so closely crowded together that any further compression would violate the Pauli exclusion principle. Because the electrons cannot be squeezed any closer together, they produce a powerful pressure that resists further core contraction.

This phenomenon, in which closely packed particles resist compression because of the Pauli exclusion principle, is called **degeneracy.** Astronomers say that the helium-rich core of a low-mass red giant is "degenerate" and is supported by **degenerate-electron pressure.** This degenerate pressure, unlike the pressure of a perfect gas, does not depend on temperature.

When the temperature in the core of a low-mass red giant reaches the high level required for the triple alpha reaction, energy begins to be released. The helium nuclei become heated, which causes the triple alpha process to proceed more rapidly. However, the pressure provided by the degenerate electrons is independent of the temperature, so the pressure does not change. Without the "safety valve" of increasing pressure, the star's core cannot expand and cool. The rising temperature causes the helium to burn at an ever-increasing rate, producing the helium flash. Eventually the temperature becomes so high that electron degeneracy is no longer a factor. The electrons then behave like a perfect gas and the star's core expands, terminating the helium flash. These events occur so rapidly that the helium flash is over in seconds.

21-3 Evolutionary tracks on the H–R diagram reveal the ages of star clusters

It is enlightening to follow the post–main-sequence evolution of mature stars on a Hertzsprung–Russell diagram (Figure 21-3). The **zero-age main sequence** (or **ZAMS**) is the location

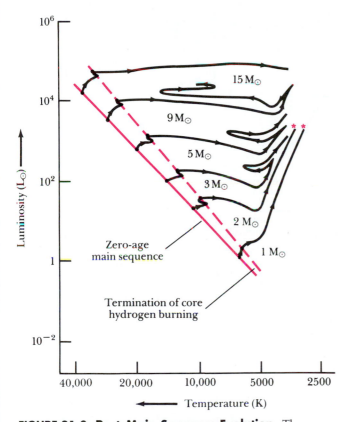

FIGURE 21-3 Post–Main-Sequence Evolution The evolutionary tracks of six stars are shown on this H–R diagram. In the high-mass stars, core helium burning ignites where the evolutionary tracks make a sharp downward turn in the red giant region of the diagram. The evolutionary tracks for low-mass stars (1 M⊙ and 2 M⊙) are shown only up to the points, indicated by the red asterisks, where the helium flash occurs at their centers. (Adapted from I. Iben)

on an H–R diagram where stars first achieve hydrostatic equilibrium, a balance between the inward force of gravity and the outward pressure produced by hydrogen burning. In subsequent years, the "evolutionary tracks" slowly inch away from the ZAMS as the hydrogen-burning core grows in mass. The dashed line on Figure 21-3 shows the locations of the stellar models when all the core hydrogen has been consumed and core hydrogen burning ceases. From there, the points representing high-mass stars move rapidly from left to right across the H–R diagram. This means that, although the star's surface temperature is decreasing, its surface area is increasing at a rate that keeps its overall luminosity roughly constant. During this transition, the star's core contracts and its outer layers expand as energy flows outward from the star's hydrogen-burning shell.

Just before core helium burning begins, the evolutionary tracks of high-mass stars turn upward in the red giant region of the H–R diagram. After the core helium burning begins, however, the evolutionary tracks back away from these temporary peak luminosities. The tracks then wander back and forth in the red giant region while the stars readjust to their new energy sources. The evolutionary tracks of two low-mass stars also appear in Figure 21-3. These two tracks are shown only to the point where the helium flash occurs in these stars.

We can summarize our understanding of stellar evolution from birth through the onset of helium burning by following the evolution of a theoretical cluster of stars. The eight H–R diagrams in Figure 21-4 show the results of a computer simulation of the evolution of a hundred stars that differ only in initial mass. All the stars, which are obliged to form at the same moment, start off as cool protostars on the right side of the H–R diagram (see Figure 21-4a), where they are spread out according to their masses. The most massive protostars are the brightest; the least massive protostars are the dimmest.

The most massive stars contract and heat up so rapidly that after only 5000 years they have already moved across the H–R diagram toward the main sequence (see Figure 21-4b). After 100,000 years, these stars have ignited hydrogen burning in their cores and settled down on the main sequence as O stars (see Figure 21-4c). After 3 million years, stars of moderate mass have also ignited core hydrogen burning and become A and B main sequence stars (see Figure 21-4d). Meanwhile, low-mass stars continue to inch their way toward the main sequence as they leisurely contract and heat up.

After a simulated 30 million years (see Figure 21-4e), the most massive stars have depleted all the hydrogen at their cores and become red giants in the upper right corner of the H–R diagram. (This simulation follows stars only to the red giant stage, after which they are simply deleted from the H–R diagram.)

After 66 million years (see Figure 21-4f), even the lowest mass protostars have finally ignited core hydrogen burning and have settled down for a long sojourn as cool, dim M

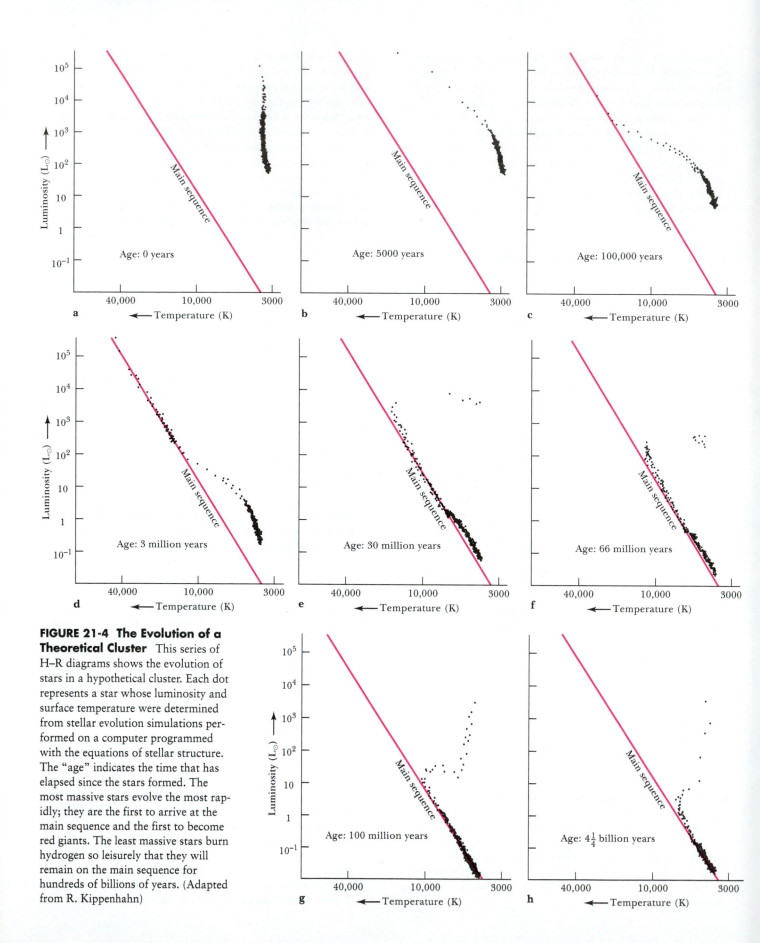

FIGURE 21-4 The Evolution of a Theoretical Cluster This series of H–R diagrams shows the evolution of stars in a hypothetical cluster. Each dot represents a star whose luminosity and surface temperature were determined from stellar evolution simulations performed on a computer programmed with the equations of stellar structure. The "age" indicates the time that has elapsed since the stars formed. The most massive stars evolve the most rapidly; they are the first to arrive at the main sequence and the first to become red giants. The least massive stars burn hydrogen so leisurely that they will remain on the main sequence for hundreds of billions of years. (Adapted from R. Kippenhahn)

FIGURE 21-5 A Globular Cluster A globular cluster is a spherical cluster that typically contains a few hundred thousand stars. This cluster, called 47 Tucanae, is located in the southern sky, roughly 19,000 light-years from Earth. (Anglo-Australian Observatory)

main sequence stars evolved long ago into red giants, leaving behind only low-mass, slowly evolving stars that still have core hydrogen burning.

The H–R diagram of a globular cluster typically shows a horizontal grouping of stars in the left-of-center portion of the diagram. As seen in Figure 21-6, these stars form a horizontal row at luminosities of about 50 L$_\odot$. These stars, called **horizontal-branch stars**, are post–helium-flash low-mass stars. In years to come, these stars will move back toward the red giant region as their fuel is devoured by core helium burning and shell hydrogen burning.

An H–R diagram can be used to determine the age of a cluster. In the H–R diagram for a very young cluster, the entire main sequence is intact. As a cluster gets older, though, stars begin to leave the main sequence. The high-mass, high-luminosity stars are the first to become red giants as the main sequence starts to burn down like a candle. Over the years, the main sequence gets shorter and shorter.

The top of the surviving portion of the main sequence, called the **turnoff point**, provides a measure of a cluster's age. The stars at the turnoff point are just now exhausting the hydrogen in their cores, and their main sequence lifetime is

stars on the main sequence. These lowest-mass stars can continue to burn hydrogen at their cores for hundreds of billions of years.

In the final two H–R diagrams, note how the main sequence gets shorter as stars exhaust their core supplies of hydrogen and evolve into red giants. These stars, whose masses range from 1 to 3 M$_\odot$, are the ones that undergo helium flash in their cores. The sudden flood of energy released by the helium flash does not disrupt a star's surface; instead it alters conditions in the star's interior.

Immediately after the helium flash, a low-mass star's superheated core expands like a perfect gas, because it is now too hot to be degenerate. Around the expanding core, temperatures fall, and so the hydrogen-burning shell reduces its energy output. This allows the star's outer layers to contract and heat up. Consequently, an old, post–helium-flash star is both smaller and hotter at the surface than a red giant.

Examples of these post–helium-flash stars are found in old star clusters called **globular clusters** because of their spherical shape. A typical globular cluster, like that shown in Figure 21-5, contains up to one million stars in a volume less than 100 parsecs across. Astronomers know that such clusters are old because they contain the remnants of massive stars (to be discussed in Chapter 23), yet no high-mass main sequence stars are seen today.

If you measure the brightness and surface temperature of many stars in a globular cluster and plot the data on an H–R diagram, as shown in Figure 21-6, you discover that the upper half of the main sequence is missing. All the high-mass

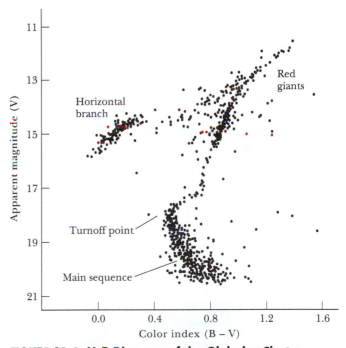

FIGURE 21-6 H–R Diagram of the Globular Cluster M55 Each dot represents a star whose V magnitude and (B – V) color index have been measured. The upper half of the main sequence is missing. The horizontal-branch stars, believed to be low-mass stars that recently experienced the helium flash, exhibit core helium burning and shell hydrogen burning. (Adapted from D. Schade, D. VandenBerg, and F. Hartwick)

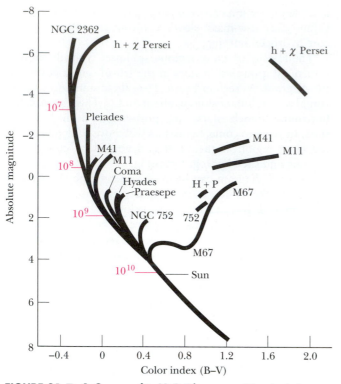

FIGURE 21-7 A Composite H–R Diagram The shaded
bands indicate where data from various open clusters fall on the
H–R diagram. The ages of turnoff points (in years) are listed in
red alongside the main sequence. The age of a cluster can be
estimated from the location of the cluster's turnoff point, where
the cluster's most massive stars are just now leaving the main
sequence. (Adapted from A. Sandage)

rich, because their spectra contain many prominent spectral
lines of heavy elements. This material originally came from
dying stars that exploded long ago, enriching the interstellar
gases with the heavy elements formed in their cores. The Sun
is a relatively young, metal-rich star.

Most of the oldest clusters are globular clusters. Such clusters are generally located outside the plane of our Galaxy,
and their spectra show only weak lines of heavy elements.
These ancient stars are thus said to be **metal poor** because
they typically contain only about 3% of the abundance of
heavy elements in the Sun. Such stars were created long ago
from interstellar gases that had not yet been substantially
enriched with heavy elements. Spectra of a metal-poor star
and of the Sun are compared in Figure 21-8.

The young, metal-rich stars, like our Sun, are commonly
called **population I** stars. The old, metal-poor stars are **population II** stars. As we shall see in later chapters, hydrogen and
helium were essentially the only two elements to emerge during the birth of the universe. Thus, the most ancient stars
should have no spectral lines of any heavy elements. Astronomers have been searching for these very old, metal-free
population III stars. So far, very few candidates have been
found.

equal to the age of the cluster (recall Table 21-1). For example, in the case of the globular cluster M55 (see Figure 21-6),
0.8-M☉ stars have just left the main sequence, indicating that
the cluster's age is roughly 15 billion years.

Data for several star clusters are plotted on Figure 21-7,
along with turnoff-point times from which the ages of the
clusters can be estimated. The youngest clusters (those with
most of their main sequences still intact) are said to be **metal**

21-4 Supergiants and red giants typically show mass loss

Both supergiant and red giant stars are so enormous that
their bloated outer layers constantly leak gases into space. At
times, this **mass loss** can be quite significant.

Mass loss can be detected spectroscopically. Gas escaping
from the tenuous outer layers of a red giant can produce
narrow absorption lines that are slightly blueshifted. According to the Doppler effect (review Figure 5-19), typical observed blueshifts correspond to a speed of about 10 km/s,
which is slightly greater than that needed to escape the star's
gravitational field. A typical mass-loss rate for a red giant is
roughly 10^{-7} M☉ per year. For comparison, the Sun's mass
loss rate is only 10^{-14} M☉ per year.

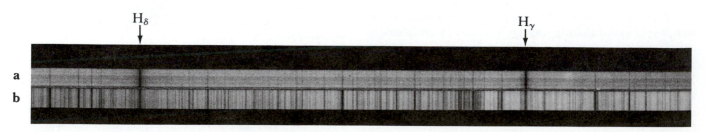

**FIGURE 21-8 Spectra of a Metal-Poor and a Metal-Rich
Star** These spectra compare (a) a metal-poor star and (b) a metal-
rich star (the Sun). Numerous spectral lines prominent in the solar
spectrum are caused by elements heavier than hydrogen and

helium. Note that corresponding lines in the metal-poor star's
spectrum are weak or absent. Both spectra cover a wavelength
range that includes H_γ and H_δ. (Lick Observatory)

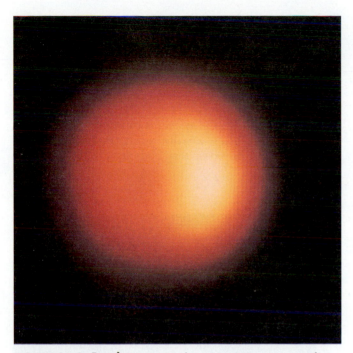

FIGURE 21-9 Betelgeuse Betelgeuse (α Orionis) is one of the largest stars, with a diameter about 800 times the Sun's. This image, taken at a deep-red wavelength of 710 μm, shows a large, bright "hot spot" on the star that may be a huge column of upwelling gas. (Courtesy of D. Buscher)

Betelgeuse, a red supergiant in the constellation of Orion, offers a good example of mass loss (Figure 21-9). Betelgeuse is 310 light-years away and has a diameter roughly equal to the diameter of Mars's orbit. Recent spectroscopic observations show that this star is losing mass at the rate of 1.7×10^{-7} $M_\odot$ per year to a huge surrounding circumstellar shell. This shell is expanding at 10 km/s and its gases have been detected out to distances of 10,000 AU from the star. Conse-

quently, the expanding circumstellar shell around Betelgeuse has an overall diameter of $\frac{1}{3}$ light-year.

Supergiant stars, which are brighter than 10^5 Suns, suffer mass loss throughout most of their existence, with mass-loss rates comparable to those of the red giants. Figure 21-10 shows a supergiant star losing mass. This particular star is a member of a class called **Wolf–Rayet stars,** named after two nineteenth-century astronomers who first drew attention to bright emission lines in the spectra of such stars. Wolf–Rayet stars have masses in the range of 30 to 50 $M_\odot$ and lie near the main sequence on the H–R diagram. They are thus fairly young stars. A large percentage of these rare and beautiful stars have been confirmed to be members of **close binary** systems—double stars whose members are separated by a distance that is comparable to the stars' sizes. In such systems, the two stars can have a significant gravitational effect on each other. Many Wolf–Rayet stars have nearby companions whose gravity plays an important role in detaching gases from the supergiants' outer atmospheres.

We shall see in the next chapter that dying stars eject vast quantities of material into space. The mass loss from supergiants and red giants accounts for roughly one-fifth of all the matter returned by the stars to the interstellar medium.

21-5 Mass transfer in close binary systems can produce unusual double stars

When one member of a double star system becomes a red giant, it can dump gas onto its companion. This process, called **mass transfer,** occurs only in close binaries, where the bloated red giant and its companion are near enough to each other that the red giant's outer layers can be gravitationally captured by the companion star.

Our modern understanding of mass transfer in close binaries is based on the work of the French mathematician Edouard Roche. In the mid-1800s, Roche studied the gravi-

FIGURE 21-10 The Wolf–Rayet Star HD 56925 This beautiful nebulosity surrounds a supergiant star that is experiencing significant mass loss. The ejected material collides and interacts with the surrounding interstellar gas and dust, thereby producing the cosmic bubble seen here. This nebulosity, referred to as NGC 2359, is located in the constellation of Canis Major. (Anglo-Australian Observatory)

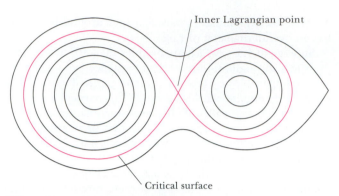

FIGURE 21-11 **Equipotential Contours** This series of curves gives the shapes that stars in binary systems can have. As a star expands and fills its Roche lobe, it become somewhat egg-shaped because of the gravitational pull of its companion and the rotation of the system as a whole. If a star overflows its Roche lobe, gas flows across the inner Lagrangian point onto the companion star.

tational field produced by two stars in a double system. His work answered the question: What shapes do stars in a binary system assume as a result of their rotation and mutual tidal interaction? This analysis assumes that the two stars in a close binary are in circular orbits about each other and keep the same side facing each other, just as our Moon keeps its same side facing the Earth. These conditions are normally produced by tidal interaction between the two stars in a close binary system.

Roche found that he could draw a series of curves, called **equipotential surfaces,** that describe the shapes of stars in a double star system (Figure 21-11). In widely separated binaries, the stars are so far apart that tidal effects are small, and so the stars are nearly perfect spheres. In close binaries, where the separation between the stars is not much greater than their sizes, tidal effects are strong, causing the stars to be somewhat egg-shaped.

Roche discovered that one of the equipotential surfaces, a figure-eight curve that encloses both stars in a binary, marks the gravitational domain of each star. This figure-eight curve, colored red in Figure 21-11, is called the **critical surface.** Each half of the curve is known as a **Roche lobe.** The more massive star is always located inside the larger Roche lobe. If gas from a star leaks over its Roche lobe, it is no longer bound by gravity to that star and is free to fall onto the companion star—or to escape from the binary system.

The point where the two Roche lobes touch, called the **inner Lagrangian point,** is a kind of balance point between the two stars in a binary. Here the effects of gravity and rotation cancel each other. When mass transfer occurs in a close binary, gases flow through the inner Lagrangian point from one star to the other.

In many binaries, the stars are so far apart that even during their red giant stage the stars' surfaces remain well inside their Roche lobes. As a result, little mass transfer can occur.

Each star thus lives out its life as if it were single and isolated. A binary system in which each star is within its Roche lobe is referred to as a **detached binary** (Figure 21-12a).

However, if the two stars are relatively close together, when one star expands to become a red giant, it may fill or overflow its Roche lobe, in which case the system is called a **semidetached binary** (Figure 21-12b). If both stars happen to fill their Roche lobes, the system is called a **contact binary,** because the two stars actually touch (Figure 21-12c). It is quite unlikely, however, that both stars exactly fill their

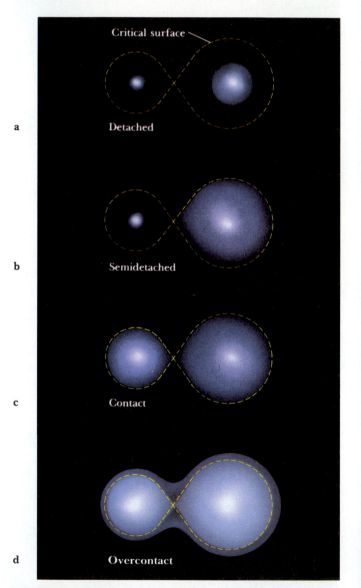

FIGURE 21-12 **Detached, Semidetached, Contact, and Overcontact Binaries** A double star is said to be a detached, semidetached, contact, or overcontact binary, depending on whether neither, either, or both of the stars fills its Roche lobe. Mass transfer is often observed in semidetached binaries. The two stars in an overcontact binary share the same outer atmosphere.

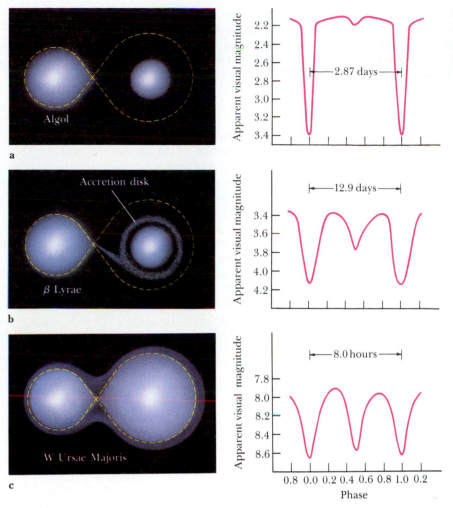

FIGURE 21-13 Three Eclipsing Binaries
Sketches of three eclipsing binaries with their light curves are shown here. (a) Algol is a semidetached binary. (b) β Lyrae is a semidetached binary in which mass transfer has produced an accretion disk around the secondary star. (c) W Ursae Majoris is an overcontact binary.

Roche lobes at the same time. It is more likely that they overflow their Roche lobes, giving rise to a common envelope of gas, which characterizes an **overcontact binary** (Figure 21-12*d*).

The eclipsing binary called Algol (from an Arabic term for "demon") gave the first clear evidence of mass transfer in close binaries. This semidetached binary can easily be seen with the naked eye in the constellation of Perseus. From Algol's light curve (Figure 21-13*a*) and spectra, astronomers calculate that the detached star is on the main sequence and its less massive companion is filling its Roche lobe. During the 1960s, astronomers labored to understand semidetached systems like Algol, which collectively came to be called Algol-type binaries.

According to stellar evolution theory, the more massive a star is, the more rapidly it should evolve. But in Algol, the more massive primary is still on the main sequence, whereas the less massive secondary has evolved to become a red giant. The apparent contradiction that the less massive star has evolved further was thus called the Algol paradox. The paradox was resolved when Zdenek Kopal at the University of Manchester and others proposed that the red giant in Algol-type binaries was originally the more massive star. As it left the main sequence to become a red giant, this star overflowed its Roche lobe, dumping gas onto the originally less massive companion. Because of the resulting mass transfer, that companion became the more massive star.

Another class of semidetached binaries, called β Lyrae variables after their prototype in the constellation of Lyra, gave evidence that matter transferred from one star can be captured into orbit about the other star to form a ring of material called an **accretion disk**. As with Algol, the less massive star in β Lyrae fills its Roche lobe, but its light curve (Figure 21-13*b*) and spectra demonstrate that the more massive detached star radiates virtually no light at all. Furthermore, the spectrum of β Lyrae is unusual, in part because of gas flowing between the stars and around the system as a whole.

The mystery of β Lyrae was solved in 1963 when Su-Shu Huang of Northwestern University published his interpretation that the less-luminous star in ß Lyrae is enveloped in a rotating disk of gas captured from its bloated companion. The disk is so thick that it completely shrouds the secondary star, making it impossible to observe at visible wavelengths.

Observations made since the 1960s have largely confirmed Huang's model, although the accretion disk is now believed to be even thicker than Huang had supposed. As the primary star overflows its Roche lobe, gases stream across the inner Lagrangian point onto the disk at the rate of 10^{-5} $M_\odot$ per year. Other gas is constantly escaping altogether from the system.

What is the fate of an Algol or β Lyrae system? If the detached star is massive enough, it will evolve rapidly, expanding to fill its Roche lobe too. The result is an overcontact binary in which both stars share the same gaseous envelope. Such binaries are sometimes called W Ursae Majoris stars, after the prototype of this class (Figure 21-13c). In Chapters 22 and 23 we shall see that mass transfer onto dead stars produces some of the most unusual objects in the sky.

21-6 Many mature stars pulsate

Astronomers have discovered several types of stars that vary in brightness because they actually change in size, alternately swelling and shrinking. The first pulsating variable star was discovered in 1596 by the Dutch astronomer David Fabricius, who noticed that the star o Ceti sometimes faded from third magnitude to invisibility. By 1640 astronomers realized these brightness variations repeated with a period of 332 days. Seventeenth century astronomers were so enthralled by this variable that they renamed the star Mira ("wonderful").

Mira is an example of a class of pulsating stars called **long-period variables.** These stars are cool red giants that vary in brightness over months or years. With surface temperatures of about 3700 K and luminosities that range from about 10 to 10,000 $L_\odot$, they occupy the right side of the H–R diagram (Figure 21-14). Some, like Mira, are quite periodic, but others are irregular. Many are known to eject large amounts of gas and dust into space.

Astronomers do not fully understand why some cool red giants become long-period variables. Because the extended, tenuous atmospheres of these huge stars make it difficult to define quantities like the star's radius, it is difficult to calculate accurate stellar models.

Astronomers have a much better understanding of a second type of pulsating star called **Cepheid variables,** or simply "Cepheids." A Cepheid variable is recognized by the characteristic way in which its light output varies: rapid brightening followed by gradual dimming. The prototype was discovered in 1784 by John Goodricke, a deaf, mute, 19-year-old English amateur astronomer. This star, δ Cephei, varies regularly in apparent magnitude from 4.3 to 3.4, with a period of 5.4 days. The surface temperatures and luminosities of these variables place them in the middle of the H–R diagram (see Figure 21-14).

After core helium burning begins, mature stars move across the middle of the H–R diagram. Figure 21-3 showed the evolutionary tracks of high-mass stars crisscrossing the H–R diagram. Post–helium-flash low-mass stars on the horizontal branch also cross the middle of the H–R diagram as they return to the red giant region.

During these transitions across the H–R diagram, a star can become unstable and pulsate. In fact, there is a region on the H–R diagram between the main sequence and the red giant branch that is called the **instability strip** (see Figure 21-14). When a star passes through this region as it evolves, the star pulsates and its brightness varies periodically. (The light curve in Figure 21-15a shows the brightness variations of δ Cephei.)

A Cepheid variable brightens and fades because of cyclic expansion and contraction of the star's outer envelope. This behavior has been deduced from spectroscopic observations. In 1894 the Russian astronomer A. A. Belopolsky noticed that spectral lines in the spectrum of δ Cephei shift back and forth with the same 5.4-day period as that of the magnitude variations. From the Doppler effect, we can translate these wavelength shifts into speeds and draw a velocity curve (Figure 21-15b). Negative speeds mean that the star's surface is expanding toward us; positive speeds mean that the star's surface is receding. Note that the light and velocity curves are mirror images of each other. The star is brighter than average as it expands and dimmer than average while it contracts.

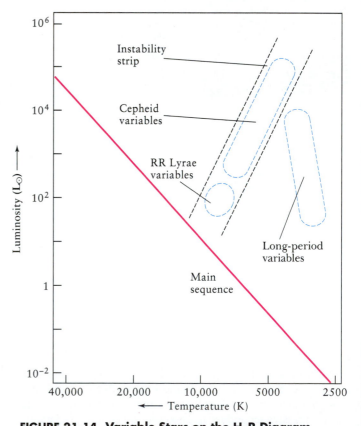

FIGURE 21-14 Variable Stars on the H–R Diagram
Long-period variables are cool red giant stars that pulsate slowly, changing their brightness in a semiregular fashion over months or years. Cepheid variables and RR Lyrae variables are located in the instability strip, which occupies a region between the main sequence and the red giant branch. A star passing through this region along its evolutionary track becomes unstable and pulsates.

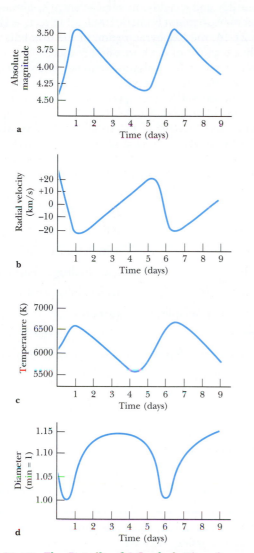

FIGURE 21-15 The Details of δ Cephei These four graphs display details about the pulsations of δ Cephei: (**a**) the star's light curve; (**b**) the velocity curve; (**c**) periodic variations in the star's surface temperature; and (**d**) periodic variations in the star's diameter.

When a Cepheid variable pulsates, the star's surface oscillates up and down like a spring. During these cyclical expansions and contractions, the star's gases alternately heat up and cool down. The temperature changes in the star's surface that result are displayed in Figure 21-15c. The periodic changes in the star's diameter are shown in Figure 21-15d.

Just as a bouncing ball eventually comes to rest, a pulsating star would soon stop pulsating without some sort of mechanism to keep its oscillations going. In 1941 the British astronomer Arthur Eddington suggested that a Cepheid pulsates because of periodic ionization and deionization of gases in its outer layers. Eddington proposed a valvelike mechanism: The star is more opaque or "light-tight" when compressed than when expanded. When the star is compressed,

trapped heat can push the star's surface outward. When the star is expanded, the heat escapes and so the star's surface, which is no longer supported, can then fall inward.

In the 1960s, the American astronomer John Cox followed up on Eddington's idea and proved that helium is the "valve" that keeps Cepheids pulsating. Normally, when a star's helium is compressed, the temperature of the gas increases and the gas becomes more transparent. In certain layers near the star's surface, however, compression may ionize helium (remove one of its electrons) instead of raising its temperature. In this zone where helium is ionized, the compressed helium is quite opaque and can trap the star's heat, which pushes the star's surface outward. Then, as the star expands, helium ions recombine with electrons, and so the gas becomes more transparent and releases the trapped energy. The star's surface then falls inward, recompressing the helium, and the whole cycle begins all over again.

Cepheid variables are very important to astronomers, because there is a direct relationship between a Cepheid's period and its average luminosity. Dim Cepheid variables pulsate rapidly, with periods of one to two days and an average brightness of a few hundred Suns. The most luminous Cepheids are the slowest variables, with periods of 100 days and average brightness equal to 10,000 Suns. This connection between period and brightness, known as the **period–luminosity relation**, is plotted in Figure 21-16. The period–luminosity relation plays an important role in measuring the overall size and structure of the universe, as we shall see in Chapter 26.

The evolutionary tracks of mature, high-mass stars pass back and forth through the upper end of the instability strip

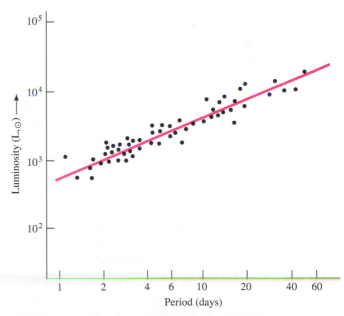

FIGURE 21-16 The Period–Luminosity Relation The period of a Cepheid variable is directly related to its average luminosity. Each dot plotted on this graph represents a Cepheid variable whose brightness and period have been measured. The line is the "best-fit" to the data. (Adapted from H. C. Arp)

on the H–R diagram. These stars become Cepheids when the ionization layers occur at just the right depth to drive the pulsations. For stars on the high-temperature side of the instability strip, helium ionization is too close to the surface and involves only an insignificant fraction of the star's mass. For stars on the cool side of the instability strip, convection in the star's outer layers prevents the storage of the energy needed to drive the pulsations. Thus Cepheids exist only in a narrow range of temperature on the H–R diagram.

Low-mass, post–helium-flash stars pass through the lower end of the instability strip as they move in the horizontal branch along their evolutionary tracks. These stars become

RR Lyrae variables, named after their prototype in the constellation of Lyra. RR Lyrae variables all have periods shorter than one day and roughly the same average brightness as the stars on the horizontal branch (100 $L_\odot$). In fact, as shown in Figure 21-14, the RR Lyrae region of the instability strip is actually a segment of the horizontal branch.

In rare cases, stellar pulsations can be quite substantial. In some cases, the expansion velocity exceeds the star's escape velocity and the star's outer layers are ejected completely. As we shall see in the next chapter, dying stars eject a significant amount of mass, renewing and enriching the interstellar medium for future generations of stars.

KEY WORDS

accretion disk	degenerate-electron pressure	mass loss	Roche lobe
alpha particle	detached binary	mass transfer	RR Lyrae variable
Cepheid variable	equipotential surface	metal-poor star	semidetached binary
circumstellar shell	globular cluster	metal-rich star	shell hydrogen burning
close binary	helium burning	overcontact binary	triple alpha process
contact binary	helium flash	Pauli exclusion principle	turnoff point
core helium burning	horizontal-branch	perfect gas	Wolf–Rayet star
core hydrogen burning	inner Lagrangian point	period–luminosity relation	zero-age main sequence (ZAMS)
critical surface	instability strip	population I, II, III	
degeneracy	long-period variable	red giant	

KEY IDEAS

• The more massive a star is, the shorter is its main sequence lifetime; the Sun has been a main sequence star for about 4.6 billion years and should remain one for about another 5 billion years.

• Core hydrogen burning ceases when the hydrogen has been exhausted in the core of a main sequence star, leaving a core of nearly pure helium surrounded by a shell through which hydrogen burning works its way outward in the star; the core shrinks and becomes hotter while the star expands to become a red giant.

  As a star becomes a red giant, its evolutionary track moves rapidly from the main sequence to the red giant region of the H–R diagram.

  When the central temperature of a red giant reaches about 100 million kelvin, the thermonuclear process of helium burning begins there; this process, also called the triple alpha process, converts helium to carbon and oxygen.

  In a more massive red giant, helium burning begins gradually; in a less massive red giant, it begins suddenly, in a process called the helium flash.

  After the helium flash, a low-mass star moves quickly from the red giant region of the H–R diagram to the horizontal branch.

• The age of a star cluster can be estimated by plotting its stars on an H–R diagram.

  The age of a star cluster is equal to the age of the main sequence stars at the turnoff point (the upper end of the remaining main sequence).

  For an older cluster, the upper portion of the main sequence is missing, because more massive main sequence stars have become red giants.

  Relatively young population I stars are metal rich; ancient population II stars are metal poor.

• Supergiants and red giants undergo extensive mass loss, sometimes producing circumstellar shells of ejected material around the stars.

• Mass transfer in a close binary system occurs when one star in a close binary overflows its Roche lobe; gas flowing from one star to the other passes across the inner Lagrangian point.

• When a star's evolutionary track carries it through a region called the instability strip in the H–R diagram, the star becomes unstable and begins to pulsate.

  Cepheid variables are high-mass pulsating variables having a direct relationship between their periods of pulsation and their luminosities.

RR Lyrae variables are low-mass pulsating variables with short periods.

REVIEW QUESTIONS

1. On what grounds are astronomers able to say that the Sun has about 5 billion years remaining in its main sequence stage?

2. Why do you suppose that the majority of the stars we see in the sky are main sequence stars?

3. What does it mean when an astronomer says that a star "moves" from one place to another on an H–R diagram?

4. What are the main sequence lifetimes of (**a**) a 25-$M_{\odot}$ star and (**b**) a 5-$M_{\odot}$ star? Compare these lifetimes with that of the Sun.

5. What will happen inside the Sun 5 billion years from now when it begins to turn into a red giant?

6. What is the helium flash?

7. How is a degenerate gas different from ordinary gases?

8. Explain how and why the turnoff point on the H–R diagram of a cluster is related to the cluster's age.

9. Why do astronomers believe that globular clusters are made of old stars?

10. What is the difference between population I and population II stars?

11. What is a Roche lobe? What is the inner Lagrangian point? Why are Roche lobes and the critical surface important in close binary star systems?

12. What is the difference between a detached binary, a semidetached binary, a contact binary and an overcontact binary?

13. What is the Algol paradox and how was it resolved?

14. Why do astronomers attribute the observed Doppler shifts of a Cepheid variable to pulsation, rather than to some other cause, such as orbital motion?

ADVANCED QUESTIONS

Tips and tools . . .

Recall that 6×10^{11} kg of hydrogen is converted into helium each second at the Sun's center, as explained in Chapter 18 (see Section 18.6). You may find it helpful to review the discussion of apparent magnitude, absolute magnitude, and luminosity found in Boxes 19-2 and 19-3. Assume that the bolometric correction for stars listed below is zero.

15. Assuming that the Sun's luminosity remains roughly constant throughout its 10 billion years on the main sequence, what fraction of the Sun's hydrogen will be converted into helium?

16. The earliest fossil records indicate that life appeared on the Earth about 2.8 billion years after the solar system was formed. If the Sun's lifetime on the main sequence is 10 billion years, what is the most mass that a star could have in order that its lifetime on the main sequence be long enough to permit life to form on one or more of its planets? Assume the evolutionary processes would be similar to those which have occurred here on the Earth.

17. When the Sun becomes a red giant, its luminosity will be 100 times greater than it is today. Assuming that this luminosity is caused *only* by the burning of the Sun's remaining hydrogen, calculate how long our star will be a red giant.

18. What observations would you make of a star to determine whether its primary source of energy is hydrogen or helium burning?

19. The star X Arietis is an RR Lyrae variable. Its apparent magnitude varies between 8.97 and 9.95 with a period of 0.65 day. Interstellar extinction dims the star by half a magnitude. Approximately how far away is the star?

20. Polaris (α CMi) is a Cepheid variable. Its apparent magnitude varies between 1.92 and 2.02 with a period of 3.97 days. Approximately how far away is the star?

21. Consult recent issues of *Sky & Telescope* or *Astronomy* to find out when Mira will next reach maximum brightness. Look up the star's location on a star chart. What insurmountable difficulty will you encounter if you want to observe Mira at maximum brightness?

DISCUSSION QUESTIONS

22. The half-life of the ^{8}Be nucleus is 2.6×10^{-16} second, which is the average time that elapses before this unstable nucleus decays into two alpha particles. How would the universe be different if the ^{8}Be half-life were zero instead? How would the universe be different if the ^{8}Be nucleus were stable?

23. Suppose that an oxygen nucleus were to be fused with a helium nucleus. What element would be formed? Look up the relative abundance of this element and comment on whether such a process is likely.

24. What observational consequences would we find in H–R diagrams for star clusters if the universe had a finite age? Could we use these consequences to establish constraints on the possible age of the universe? Explain.

OBSERVING PROJECTS

25. Observe several of the following red giants and supergiants with the naked eye and through a telescope. These stars are most easily found with the aid of star charts published every month in such magazines as *Sky & Telescope* and *Astronomy*. Epoch 2000 coordinates are given on the next page.

Star	Spectral type	R.A.	Decl.
Almach (γ And)	K3II	2ʰ 03.9ᵐ	+42° 20′
Aldebaran (α Tau)	K5III	4 35.9	+16 31
Betelgeuse (α Ori)	M2I	5 55.2	+07 24
Arcturus (α Boo)	K2III	14 15.7	+19 11
Antares (α Sco)	M1I	16 29.5	−26 26
Eltanin (γ Dra)	K5III	17 56.7	+51 29
Enif (ε Peg)	K2I	21 44.2	+09 52

Is the reddish color of these stars apparent when they are compared with neighboring stars?

26. Several of the open clusters listed in Figure 21-7 can be seen quite well with a good pair of binoculars. Observe as many of these clusters as you can, using both a telescope and a pair of binoculars. Note the overall distribution of stars in each cluster. Which clusters are seen better through binoculars than through a telescope? Which clusters can you see with the naked eye?

Star cluster	R.A.	Decl.
h Persei	2ʰ 19.0ᵐ	+57° 09′
χ Persei	2 22.4	+57 07
Pleiades	3 47.0	+24 07
Hyades	4 27	+16
Praesepe	8 40.1	+19 59
Coma	12 25	+26
M11	18 51.1	−06 16

27. There are many beautiful globular clusters scattered around the sky that can be easily seen with a small telescope. Several of the brightest and nearest globulars are listed below.

Globular cluster	R.A.	Decl.
M3 (NGC 5272)	13ʰ 42.2ᵐ	+28° 23′
M5 (NGC 5904)	15 18.6	+2 05
M4 (NGC 6121)	16 23.6	−26 32
M13 (NGC 6205)	16 41.7	+36 28
M12 (NGC 6218)	16 47.2	−1 57
M28 (NGC 6626)	18 24.5	−24 52
M22 (NGC 6656)	18 36.4	−23 54
M55 (NGC 6809)	19 40.0	−30 58
M15 (NGC 7078)	21 30.0	+12 10

Observe as many of these globular clusters as you can. How well can you distinguish individual stars toward the center of each cluster? Do you notice any differences in the overall distribution of stars between clusters?

FOR FURTHER READING

Hack, M. "Epsilon Aurigae." *Scientific American,* October 1984. This article, which focuses on a particular eclipsing binary, demonstrates how astronomers use observations and an understanding of stellar evolution to deduce details of elaborate stellar systems.

Johnson, B. "Red Giant Stars." *Astronomy,* December 1976. This well-written article surveys the properties of red giant stars.

Kafatos, M., and Michalitsianos, A. "Symbiotic Stars." *Scientific American*, July 1984. This article discusses fascinating phenomena associated with double stars that consist of a red giant and a hot, compact companion.

Kippenhahn, R. *100 Billion Suns: The Birth, Life and Death of Stars.* Basic Books, 1983. This classic book by a noted German astrophysicist describes the life cycles of stars.

MacRobert, A. "Epsilon Aurigae: Puzzle Solved?" *Sky & Telescope,* January 1988. This brief article shows how astronomers struggle to understand the complicated details of a close binary system.

Percy, J. "Observing Variable Stars for Fun and Profit." *Mercury,* May/June 1979. This entertaining article contains many important tips for a person interested in observing eclipsing binaries, Cepheids, and other variable stars.

———. "Pulsating Stars." *Scientific American,* June 1975. Although it is a few years old, this article is still the best popular description of the physics of stellar pulsations.

Tomkin, J., and Lambert, D. L. "The Strange Case of Beta Lyrae." *Sky & Telescope,* October, 1987. This fascinating article explores many of the mysteries associated with β Lyrae.

Weymann, R. "Stellar Winds." *Scientific American,* August 1978. This article discusses ongoing mass loss experienced by many stars.

Wilson, R. E. "Binary Stars—A Look at Some Interesting Developments." *Mercury,* September/October 1974. This well-written article gives a fine overview of the dynamics of binary star systems.

Wilson, O., and others. "The Activity Cycles of Stars." *Scientific American,* February 1981. This article begins with a description of the solar cycle and shows how astronomers are discovering similar cycles in other stars.

Wyckoff, S. "Red Giants: The Inside Scoop." *Mercury,* January/February 1979. This superb article, which includes many interesting details about red giants, emphasizes the effects of varying abundances of carbon and oxygen.

THE DEATHS OF STARS

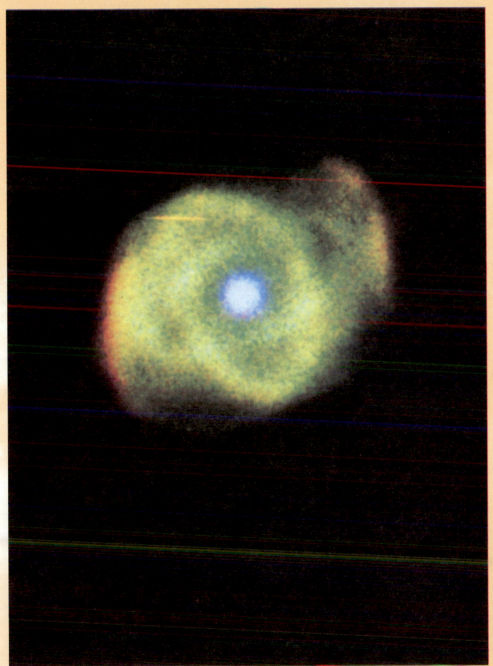

A PLANETARY NEBULA
Dying stars often eject their outer layers. A low-mass star can lose half its mass in a comparatively gentle process that produces a planetary nebula. The exposed stellar core typically has a surface temperature of about 100,000 K and is roughly one-tenth the Sun's size. UV radiation from the stellar core causes the surrounding gases to glow. The greenish color of this planetary nebula, NGC 6543, comes from doubly ionized oxygen. Its central star has exhausted all its fuel and is contracting into a white dwarf. (Lick Observatory)

AFTER ITS old age as a red giant, a star approaches the end of its life. As stars die, they eject significant amounts of matter into space. Low-mass stars may swell to become red supergiants before expelling their outer layers relatively gently, leaving planetary nebulae. The contracted, burned-out cores that remain become dead stars called white dwarfs. In contrast, the core of a high-mass star collapses suddenly, triggering a violent supernova explosion, perhaps creating a neutron star or black hole. A white dwarf accreting gas from a companion star in a close binary system can also become a supernova. Thermonuclear reactions during the final stages of stellar life produce a wide variety of heavy elements. Astronomers are developing exotic instruments, including neutrino detectors and gravitational wave antennae, to learn more about the fascinating and sometimes violent death throes of stars.

From infancy through adulthood, a star leads a fairly placid life, with hydrogen burning in its core. As the star matures, it swells slightly in response to the depletion of the star's central supply of hydrogen. These changes in an aging main sequence star—a gradual increase in luminosity accompanied by a modest decline in surface temperature—are small compared to the transformation triggered by the cessation of core hydrogen burning. As the star becomes a red giant, it takes on a seemingly schizophrenic character, with a compressed core and a bloated atmosphere. In old age, the star's behavior becomes even more dramatic as it devours its remaining nuclear fuels and begins to die.

22-1 Aging low-mass stars dredge up the ashes of thermonuclear burning from deep within their interiors

When shell hydrogen burning first ignites in a low-mass star, the outpouring of energy causes the star to expand and become a red giant. In describing the evolutionary track of the star on an H–R diagram, astronomers say that the star ascends the red giant branch for the first time. Then comes the helium flash that ignites helium burning in the star's core, at which time the star shifts over to the horizontal branch.

Carbon and oxygen are the "ash" of helium burning. After the helium flash ignites helium burning in a low-mass star, substantial amounts of carbon and oxygen begin to accumulate in the star's core.

After about a hundred million years, all the helium at the center of a low-mass star is used up, at which point helium burning ceases in the core of the star. The star's core therefore again contracts until stopped by degenerate electron pressure. This compression heats the surrounding helium-rich

gases, and a new stage of helium burning begins in a thin shell around the core. This process is called **shell helium burning.**

The renewed outpouring of energy from shell helium burning causes the star to expand again. The star ascends the red giant branch for a second and final time to become a **red supergiant.** A low-mass red supergiant consists of an inert, degenerate carbon–oxygen center and a helium-burning shell, both inside a hydrogen-burning shell, all within a volume roughly the size of the Earth. This small, dense core is surrounded by an enormous hydrogen-rich envelope roughly as big as Mars's orbit. After a while, thermonuclear reactions in the hydrogen-burning shell temporarily cease, leaving the aging star's structure as shown in Figure 22-1. These stars are also commonly called **asymptotic giant branch stars,** or **AGB stars** for short, because of the region they occupy on an H–R diagram (Figure 22-2). In addition to being as big as Mars's orbit, AGB stars typically shine with the brightness of 10,000 Suns.

In Chapter 18 we saw that energy is transported outward from a star's core by one of two processes—convection or radiative diffusion—depending on the transparency of the star's gases. In the Sun, convection dominates only the outer layers, from the photosphere down to a depth of about one-quarter solar radius (recall Figure 18-20). During the final stages of a star's life, however, the convective zone occasionally becomes so broad that it extends down to the star's core. At these times, convection can "dredge up" the ashes of thermonuclear fusion, transporting them all the way to the star's surface.

The first **dredge-up** takes place after core hydrogen burning stops, when the star becomes a red giant for the first time. Convection dips so deeply into the star that material processed by the CNO cycle is carried up to the star's surface,

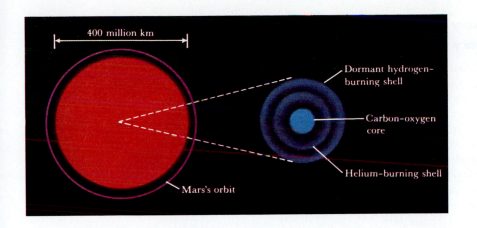

FIGURE 22-1 The Structure of an Old Low-Mass Star Near the end of its life, a low-mass star becomes a red supergiant whose diameter is almost as large as the diameter of the orbit of Mars. The star's dormant hydrogen-burning shell and active helium-burning shell are all contained within a volume roughly the size of the Earth.

changing the relative abundances of carbon, nitrogen, and oxygen. A second dredge-up occurs after core helium burning ceases, further altering the abundances of carbon, nitrogen, and oxygen. Still later, during the AGB stage, a third dredge-up is especially effective in transporting large amounts of freshly synthesized carbon to the star's surface. This third dredge-up produces a **carbon star,** a red giant or supergiant whose spectrum exhibits prominent absorption bands of carbon-rich molecules like C_2, CH, and CN.

All carbon stars undergo mass loss, some so vigorously that they are obscured in sooty cocoons of ejected matter. The Infrared Astronomical Satellite (IRAS) was very successful in finding these warm, shrouded stars, which are important because they enrich the interstellar medium with carbon and some nitrogen and oxygen. The triple alpha process is the *only* way that carbon can be made, and carbon stars are the primary avenue by which this element is dispersed into interstellar space.

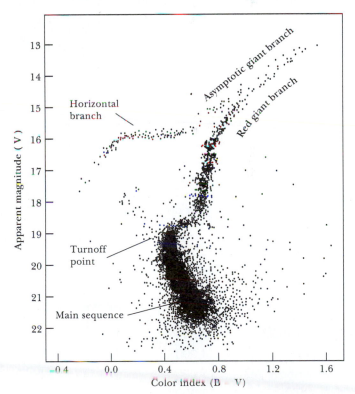

FIGURE 22-2 The Globular Cluster M3 and Its H–R Diagram The globular cluster M3 in the constellation of Canis Venatici (the Hunting Dogs) is about 30,000 light-years from Earth. Its diameter is about 100 light-years. Data for nearly 11,000 stars in M3 are plotted on the H–R diagram. Red giant branch stars are low-mass stars undergoing shell hydrogen burning; they are all destined to experience the helium flash, which will transform them into horizontal-branch stars. Asymptotic giant branch stars are low-mass stars undergoing helium shell burning; they are all post–horizontal-branch stars returning to the red giant region of the H–R diagram. (NOAO; H–R diagram adapted from R. Buonanno and colleagues).

22-2   Low-mass stars die by gently ejecting their outer layers, creating planetary nebulae

The low-mass star's impending death is signaled by instabilities, called **thermal pulses,** that develop in its helium-burning shell. As we saw in Chapter 21, the helium flash occurs because the star's core is degenerate. However, the helium-burning shell is not compressed enough to be degenerate. Instead, instabilities occur because the shell is hot and thin. A slight increase in energy output from a thin shell does little to relieve the pressure from the star's overlying layers. Temperatures rise further in the shell, increasing its energy output, which in turn drives up the temperature even more. This vicious circle is called a "thermal runaway," and it ends only when the helium-burning shell becomes thick enough to relieve the pressure of the star's outer layers.

During thermal pulses, a star's luminosity jumps from that of 1000 Suns to roughly 10,000 Suns in a series of brief bursts. As Figure 22-3 shows, the bursts are separated by relatively quiet intervals lasting about 300,000 years. During thermal pulses, thermonuclear reactions restart in the star's hydrogen-burning shell.

During these thermal pulses, the dying star's outer layers can separate completely from its carbon–oxygen core. As the ejected material expands into space, dust grains condense out of the cooling gases. Radiation pressure from the star's hot, burned-out core acts on the specks of dust, propelling them further outward, and the star sheds its outer layers altogether. A star can lose more than half of its mass in this fashion.

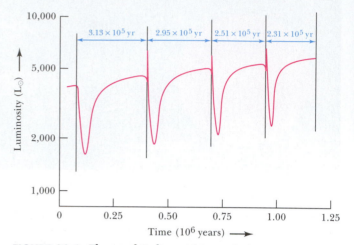

FIGURE 22-3 Thermal Pulses This graph shows how the energy output of a helium-burning shell varies in a 0.7-$M_\odot$ red supergiant. Brief periods of runaway helium burning produce thermal pulses that can eventually cause the star to eject its outer layers completely. (Adapted from I. Iben)

As a dying star ejects its outer layers, the star's hot core becomes exposed and emits ultraviolet radiation intense enough to ionize and excite the expanding shell of ejected gases. These gases therefore glow, producing a so-called **planetary nebula** (or **planetary**). Planetary nebulae have

---

Famous Planetary Nebulae

Many planetary nebulae are scattered across the sky. Some of the most famous planetaries are listed in the table below with information about their distances and diameters.

Planetary nebula	Distance from Earth (ly)	Apparent magnitude of central star	Diameter (arc min)
Dumbbell (M27)	3,500	13	8
Ring (M57)	4,000	15	1
Little Dumbbell (M76)	15,000	17	1
Owl (M97)	12,000	13	3
Saturn (NGC 7009)	3,000	12	1
Helix (NGC 7293)	400	13	15
Eskimo (NGC 2392)	10,000	10	0.5

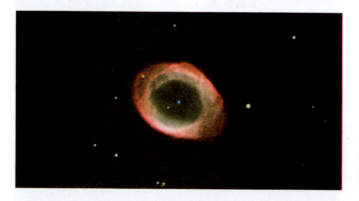

Planetary nebulae are favorite objects for amateur astronomers to observe. In many cases (such as the Ring Nebula shown in the photograph), a small telescope can reveal the nebulosity but not the central star. It is this central star, destined to become a white dwarf, that supplies the ultraviolet radiation that causes the surrounding gases to glow.

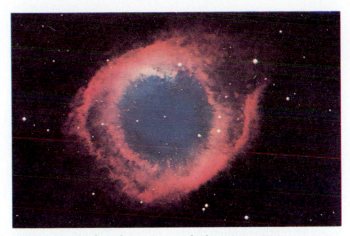

FIGURE 22-4 The Planetary Nebula NGC 7293 This beautiful object, often called the Helix Nebula, covers an area of the sky equal to half the full moon. The star that ejected these gases is at the center of the glowing shell. The bluish-green color comes from ionized oxygen, the pink and red from ionized nitrogen and hydrogen. The small radial blobs in the red shell (estimated to be about 150 AU across) give the object its alternate name, the Sunflower Nebula. It is in the constellation of Aquarius, 400 light-years from Earth. (Anglo-Australian Observatory)

FIGURE 22-5 The Planetary Nebula M27 This planetary nebula, called the Dumbbell Nebula because of its characteristic shape, is located in the constellation of Vulpecula (the Fox). Although most planetary nebulae are ring-shaped, many have a dumbbell-like appearance, possibly because a companion star caused the ejected gases to expand unevenly. This nebula, located about 1250 light-years from Earth, has an angular size equal to about one-quarter that of the full moon. (U.S. Naval Observatory)

nothing to do with planets; this misleading term was introduced in the nineteenth century because these glowing objects looked like distant planets when viewed through the small telescopes then available.

Many planetary nebulae, such as the one shown in Figure 22-4, have a distinctly spherical appearance, created by the symmetrical way in which the gases were ejected. In other cases, if the rate of expansion is not the same in all directions, the resulting nebula takes on an hourglass or dumbbell appearance, as shown in Figure 22-5.

Planetary nebulae are quite common. Astronomers estimate that there are 20,000 to 50,000 planetaries in our Galaxy alone. Several well-known examples are listed in Box 22-1. Spectroscopic observations of these nebulae show bright emission lines of ionized hydrogen, oxygen, and nitrogen. From the Doppler shifts of these lines, astronomers have concluded that the expanding shell of gas moves outward from a dying star at speeds from 10 to 30 km/s. A typical planetary nebula has a diameter of roughly 1 light-year, indicating that it must have begun expanding about 10,000 years ago.

By astronomical standards, a planetary nebula is very short-lived. After about 50,000 years, the nebula has spread out so far from the cooling central star that its nebulosity simply fades from view. The gases then mix with the surrounding interstellar medium. Astronomers estimate that all the planetary nebulae in the Galaxy return a total of about 5 $M_\odot$ to the interstellar medium each year. This amount is about 15% of all the matter expelled by all the various sorts of stars in the Galaxy each year. Because of this significant contribution, planetary nebulae play an important role in the chemical evolution of the Galaxy as a whole.

22-3 The burned-out core of a low-mass star cools and contracts until it becomes a white dwarf

Stars less massive than about 4 $M_\odot$ never develop the necessary central pressures or temperatures to ignite thermonuclear reactions that use carbon or oxygen as fuel. Instead, mass ejection just strips most outer layers from the carbon–oxygen core.

The newly exposed core of a dying star is quite hot. Figure 22-6 shows a planetary nebula surrounding a burned-out star whose surface temperature is about 200,000 K. This recently created stellar corpse is one of the hottest known stars.

The crushing weight of gases pressing inward from all sides severely compresses the stellar corpse. The electrons become so closely packed that they become degenerate throughout most of the star. At this point degenerate electron pressure alone supports the star, halting further gravitational contraction. The star is now a dense sphere roughly the same size as the Earth. Such a star is known as a **white dwarf.**

The evolutionary tracks of three burned-out stellar cores are shown in Figure 22-7. The initial red supergiants had masses of 0.8, 1.5, and 3.0 $M_\odot$. Mass ejection strips these dying stars of up to 60% of their matter. During their final spasms, the appearance of these stars changes quite rapidly. Because of the changes in luminosity and surface temperature, the points representing these stars on an H–R diagram race along their evolutionary tracks, sometimes executing loops corresponding to thermal pulses. Finally, as the ejected nebulae fade and the stellar cores cool, these dying stars'

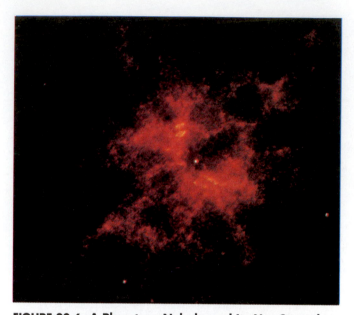

FIGURE 22-6   A Planetary Nebula and Its Hot Central Star   This photograph from the Hubble Space Telescope shows the planetary nebula NGC 2440 surrounding the burned-out core of a recently dead star. The surface temperature of this stellar corpse is about 200,000 K, making it one of the hottest stars known. Ultraviolet photons from the central star ionize the surrounding gases, causing them to glow. (NASA; ESA)

evolutionary tracks take a sharp turn downward toward the white dwarf region of the H–R diagram.

The density of matter in one of these Earth-sized stellar corpses is typically 10^9 kg/m³. In other words, a teaspoonful of white dwarf matter brought to Earth would weigh nearly 5.5 tons, as much as an elephant. In addition, as we saw in Chapter 21, degenerate gases are governed by unusual relationships between pressure, density, and temperature. Consequently, these stars have a unique **mass–radius relation:** The radius of an ideal white dwarf is inversely proportional to the cube root of its mass:

$$R \propto \frac{1}{M^{1/3}}$$

Thus, the more massive a white dwarf is, the smaller it is!

Figure 22-8 displays the mass–radius relation for white dwarfs. Note that the more degenerate matter you pile onto a white dwarf, the smaller it becomes. The electrons must be squeezed closer together to provide the greater pressure needed to support a more massive white dwarf. However, there is a limit to how much pressure degenerate electrons can produce. As a result, there is an upper limit to the mass that a white dwarf can have. This maximum mass is called the **Chandrasekhar limit** after Subrahmanyan Chandrasekhar at the University of Chicago, who pioneered theoretical studies of white dwarfs. The Chandrasekhar limit is equal to 1.4 $M_\odot$, meaning that all white dwarfs must have masses less than 1.4 $M_\odot$.

Many white dwarfs are found in the solar neighborhood, but all are too faint to be seen with the naked eye. One of the

first white dwarfs to be discovered is a companion to the bright star Sirius. The binary nature of Sirius was first deduced in 1844 by the German astronomer Friedrich Bessel, who noticed that this star was moving back and forth slightly, as if it was being orbited by an unseen object. This companion, designated Sirius B, was first glimpsed in 1863 (Figure 22-9). Recent satellite observations at ultraviolet wavelengths, where hot white dwarfs emit most of their light, demonstrate that the surface temperature of Sirius B is about 30,000 K.

The material inside a white dwarf consists mostly of ionized carbon and oxygen atoms floating in a sea of degenerate electrons. As the dead star cools, the particles in this material slow down, and electric forces between the ions begin to prevail over the random thermal motions. The ions no longer move freely. Instead, according to calculations by Hugh Van Horn and Malcolm Savedoff at the University of Rochester, they arrange themselves in orderly rows, like an immense crystal lattice. From this time on, you could say that the star is "solid." The degenerate electrons move around freely in this crystal material, just as electrons move freely through an electrically conducting metal like copper or silver. Since a

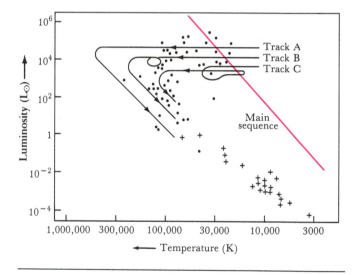

Evolutionary track	Mass ($M_\odot$)		
	Supergiant	Ejected nebula	White dwarf
A	3.0	1.8	1.2
B	1.5	0.7	0.8
C	0.8	0.2	0.6

FIGURE 22-7   Evolution from Red Supergiants to White Dwarfs   The evolutionary tracks of three low-mass red supergiants are shown as they eject planetary nebulae. The table gives the extent of mass loss in each case. The dots on the graph represent the central stars of planetary nebulae whose surface temperatures and luminosities have been determined. The crosses are white dwarfs for which similar data exist. (Adapted from B. Paczynski)

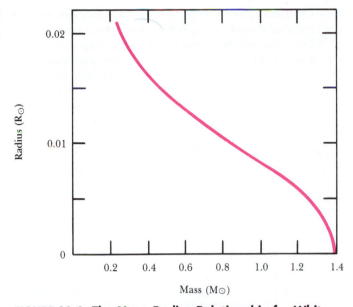

FIGURE 22-8 **The Mass–Radius Relationship for White Dwarfs** The more massive a white dwarf is, the smaller it is. This unusual relationship between mass and radius is a result of the degenerate electron pressure that supports the star. The maximum mass of a white dwarf, called the Chandrasekhar limit, is 1.4 $M_\odot$. Incidentally, 0.01 $R_\odot$ is roughly equal to the Earth's radius, which is 6960 km.

FIGURE 22-9 **Sirius and Its White Dwarf Companion** Sirius, the brightest-appearing star in the sky, is actually a double star. The secondary star is a white dwarf, seen in this photograph at the five o'clock position, almost obscured by the glare of Sirius. The spikes and rays around Sirius are the result of optical effects within the telescope. (Courtesy of R. B. Minton)

diamond is crystallized carbon, a cool carbon–oxygen white dwarf resembles an immense spherical diamond!

As a white dwarf cools off, its size remains constant because the degenerate electron pressure supporting the dead star does not depend on its temperature. However, the white dwarf's luminosity and its surface temperature decline. Its energy now comes only from thermal energy (heat) as it cools while remaining at a constant size. Consequently, the evolutionary tracks of aging white dwarfs point toward the lower right corner of the H–R diagram (Figure 22-10). As billions of years pass, white dwarfs grow dimmer and dimmer as their surface temperatures drop toward absolute zero. Our Sun will evolve into a cold, dark, diamond sphere of carbon and oxygen, about the size of the Earth.

22-4 High-mass stars create heavy elements in their cores

High-mass stars end their lives quite differently from low-mass stars. A high-mass star is capable of igniting a host of thermonuclear reactions that create heavy elements in its core.

After core helium burning ceases, gravitational compression drives the star's central temperature up to 600 million kelvin, at which time the first of the new thermonuclear reactions, **carbon burning**, begins. This thermonuclear process produces neon, magnesium, oxygen, and helium. Only stars with masses greater than about 4 $M_\odot$ develop central temperatures high enough to ignite carbon burning.

The more massive a star is, the greater the temperatures that can be reached in its core by gravitational compression. Nuclear reactions involving heavy nuclei require high temperatures because the strong electric charges of these nuclei exert enough force to tend to keep the nuclei apart. Only at

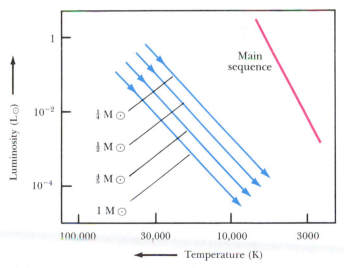

FIGURE 22-10 **White-Dwarf "Cooling Curves"** The evolutionary tracks of four white dwarfs of different masses are shown here. As these dead stars lose heat at a constant size, they become dimmer and cooler, moving toward the lower right on the H–R diagram. The more massive a white dwarf is, the smaller and hence fainter it is.

the great speeds associated with very high temperatures can the nuclei travel fast enough to penetrate each other's repulsive electric fields and fuse together.

If a star has a main-sequence mass of at least 9 $M_\odot$ (before mass ejection), its central temperature can rise to 1 billion kelvin, at which temperature the process of **neon burning** begins. This process uses up the neon accumulated from carbon burning and further increases the concentrations of oxygen and magnesium in the star's core. If the central temperature of the star reaches about 1.5 billion K, **oxygen burning** begins. The principal product of oxygen burning is sulfur.

As the star consumes increasingly heavier nuclei, the thermonuclear reactions produce a wider variety of products. For example, oxygen burning also produces silicon, phosphorus, and magnesium. This is typical of how thermonuclear reactions in the cores of massive stars create many different elements.

Some thermonuclear reactions that create heavy elements also release neutrons. Recall that a neutron is very much like a proton except that it does not have an electric charge. Therefore, a neutron is not repelled by the positively charged nuclei. Consequently, neutrons can easily collide and combine with nuclei, creating new isotopes in the process. As described in Box 22-2, this absorption of neutrons, called **neutron capture**, creates many elements and isotopes that are not produced directly in the other nuclear fusion reactions.

When a nuclear fuel is exhausted in the core of a massive star, gravitational contraction to ever-higher densities drives up the star's central temperature, thereby igniting the "ash" of the previous fusion stage and possibly the outlying shell of unburned fuel. Each successive thermonuclear reaction occurs with increasing rapidity. This speed up occurs because fusion reactions, which combine low-mass nuclei to make higher-mass ones, reduce the number of nuclei. Fewer nuclei means less fuel, which translates into shorter burning time. For example, calculations by Stanford E. Woosley at the University of California and Thomas A. Weaver at Lawrence Livermore Laboratory demonstrate that carbon burning in a 25-$M_\odot$ star occurs for 600 years, neon burning for 1 year, and oxygen burning for only 6 months.

After half a year of core oxygen burning in a 25-$M_\odot$ star, a final gravitational compression forces the central temperature up toward 3 billion kelvin, at which point **silicon burning** begins. This thermonuclear process happens so furiously that the entire core supply of silicon in a 25-$M_\odot$ star is used up in one day.

In order for an element to serve as a thermonuclear fuel, energy must be given off when its nuclei collide and fuse. This energy comes from packing together the neutrons and pro-

---

BOX 22-2

The s and r Processes

A nucleus is composed of protons and neutrons. Generally, there are a few more neutrons than protons in a nucleus. This happens because neutrons, which have no electric charge, can be packed together more easily than protons. In fact, neutrons act as buffers between the protons, which have mutually repulsive electric fields. A nucleus with too many protons or too many neutrons is unstable and will eject surplus particles until a stable balance of neutrons and protons has been restored.

There are many reactions in stellar cores that produce free neutrons. These neutrons can then be absorbed by nuclei, thus transmuting one isotope into another and creating new elements as the unstable isotopes decay. For example, successive neutron captures by ^{110}Cd can produce four stable isotopes of cadmium:

$$^{110}\text{Cd} + {}^1\text{n} \rightarrow {}^{111}\text{Cd}$$

$$^{111}\text{Cd} + {}^1\text{n} \rightarrow {}^{112}\text{Cd}$$

$$^{112}\text{Cd} + {}^1\text{n} \rightarrow {}^{113}\text{Cd}$$

$$^{113}\text{Cd} + {}^1\text{n} \rightarrow {}^{114}\text{Cd}$$

The specific isotopes produced by a nucleus that absorbs neutrons depend on how many neutrons are bombarding the nucleus. If the rate of neutron bombardment is low, then unstable nuclei have a chance to decay radioactively between neutron absorptions. For example, the absorption of a neutron by ^{114}Cd forms the unstable nucleus ^{115}Cd, which decays into an isotope of indium:

$$^{114}\text{Cd} + {}^1\text{n} \rightarrow {}^{115}\text{Cd}$$

$$^{115}\text{Cd} \rightarrow {}^{115}\text{In} + \text{e}^- + \nu$$

The ^{115}In nucleus can absorb another neutron, creating an unstable isotope (^{116}In), which decays into a stable isotope of tin (^{116}Sn):

$$^{115}\text{In} + {}^1\text{n} \rightarrow {}^{116}\text{In}$$

$$^{116}\text{In} \rightarrow {}^{116}\text{Sn} + \text{e}^- + \nu$$

The production of ^{115}In and ^{116}Sn occurs only if the rate of neutron absorption is low. This slow reaction is termed an **s process**, and these two isotopes are called **s-process isotopes.**

Occasionally a flood of neutrons is released in a star's core, and so neutrons can be absorbed rapidly. In fact, a

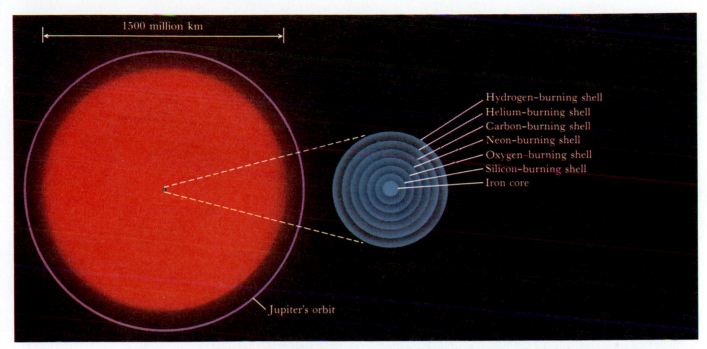

FIGURE 22-11   The Structure of an Old High-Mass Star
Near the end of its life, a high-mass star becomes a red supergiant almost as big as the orbit of Jupiter. The star's energy comes from six concentric burning shells, all contained within a volume roughly the same size as the Earth.

nucleus can soak up a large number of neutrons without having the chance to decay radioactively between absorptions. Such a rapid reaction, termed an **r process**, produces **r-process isotopes** that are often different from those created by the corresponding s process.

EXAMPLE: Consider again ^{114}Cd, the stable isotope of cadmium. A deluge of neutrons can produce a highly unstable isotope:

$$^{114}\text{Cd} + 8\ ^{1}\text{n} \rightarrow\ ^{122}\text{Cd}$$

This isotope decays into an unstable isotope of indium:

$$^{122}\text{Cd} \rightarrow\ ^{122}\text{In} + e^- + \nu$$

The indium then decays into a stable isotope of tin:

$$^{122}\text{In} \rightarrow\ ^{122}\text{Sn} + e^- + \nu$$

This particular isotope of tin can be produced only by the r process, just as ^{116}Sn can be produced only by the s process. Both isotopes are stable and are found on Earth.

By measuring the relative abundances of various isotopes in the world around us, astronomers can thus deduce conditions that once existed inside the stars that manufactured the elements of which our planet is composed.

tons more tightly in the ash nuclei than in the fuel nuclei. Silicon burning involves many hundreds of nuclear reactions. The major final product, or "ash," of this process is a stable isotope of iron, ^{56}Fe. The 56 protons and neutrons inside the iron nuclei are so tightly bound together that no further energy can be extracted by fusing still more nuclei with iron. Because ^{56}Fe does not act as a fuel for any other thermonuclear reactions, the sequence of burning stages ends here.

The buildup of an inert, iron-rich core signals the impending violent death of a massive star. Successive layers of shell burning surrounding this iron core consume the star's remaining reserves of fuel (Figure 22-11). The entire energy-producing region of the star is contained in a volume only as big as the Earth, whereas the star's enormously bloated atmosphere is nearly as big as the orbit of Jupiter.

22-5   High-mass stars die violently by blowing themselves apart in supernova explosions

The core of an aging, massive star gets progressively hotter as it contracts to ignite successive stages of thermonuclear burning. As we saw in Chapter 5, temperature governs the properties of radiation emitted by matter. Specifically, Wien's law and Planck's law together tell us that as the temperature of a blackbody increases, so does the energy of the photons it emits. When the temperature in the core of a massive star reaches a few hundred million kelvin, the photons are energetic enough to initiate a host of nuclear reactions that create

---

TABLE 22-1
Evolutionary Stages of a 25-M$_\odot$ Star

Stage	Temperature (K)	Density (kg/m^3)	Duration of stage
Hydrogen burning	4×10^7	5×10^3	7×10^6 years
Helium burning	2×10^8	7×10^5	7×10^5 years
Carbon burning	6×10^8	2×10^8	600 years
Neon burning	1.2×10^9	4×10^9	1 year
Oxygen burning	1.5×10^9	10^{10}	6 months
Silicon burning	2.7×10^9	3×10^{10}	1 day
Core collapse	5.4×10^9	3×10^{12}	$\frac{1}{4}$ second
Core bounce	2.3×10^{10}	4×10^{15}	milliseconds
Explosive	about 10^9	varies	10 seconds

---

neutrinos. These neutrinos, which carry off energy, escape from the star's core, just as solar neutrinos flow freely out of the Sun. To compensate for the energy drained by the neutrinos, the star must provide energy either by burning more thermonuclear fuel, or by contracting, or both. When the star's core is converted into iron, the only source of energy is contraction and rapid heating. In about a tenth of a second, the core temperature skyrockets to 5 billion K. The gamma-ray photons associated with this intense heat have so much energy that they begin to break down the iron nuclei into alpha particles, or helium nuclei, a process called **photodisintegration.**

Within another tenth of a second, as the densities continue to climb rapidly, the electrons are forced to combine with protons to produce neutrons. This process also releases a flood of neutrinos (ν):

$$e^- + p \rightarrow n + \nu$$

At about 0.25 second after the collapse begins, the density in the core reaches 4×10^{17} kg/m^3, which is the density with which neutrons and protons are packed together inside nuclei. At nuclear density matter is virtually incompressible. Thus, when the neutron-rich material of the core reaches nuclear density, it suddenly becomes very stiff and the core collapse comes to a sudden halt.

During this critical stage, the unsupported regions surrounding the core are plunging inward at speeds up to 15% of the speed of light. As this material crashes onto the now-rigid core, enormous temperatures and pressures develop, causing the falling material to bounce back. In just a fraction of a second, a wave of matter begins to move back outward toward the star's surface, propelled in part by the flood of neutrinos trying to escape from the star's core. This wave rapidly accelerates as it encounters less and less resistance and soon becomes a shock wave. After a few hours, this shock wave reaches the star's surface, by which time the star's outer layers have begun to lift away from the core. In this way the star becomes a **supernova.**

This particular description of the death of a 25-M$_\odot$ star is based on detailed computer calculations of an evolving stellar model. The key stages in the star's evolution are summarized in Table 22-1. The final stage in the evolution of other massive stars probably follow similar scenarios, though details may differ.

This particular 25-M$_\odot$ star ejects 24 M$_\odot$ of material, leaving behind a 1-M$_\odot$ corpse called a **neutron star.** Under slightly different conditions, a massive star might blow itself completely apart, leaving no corpse at all. For example, the bounce and subsequent shock wave might occur at the center of the star rather than at the surface of its neutronized core. Alternatively, under a different set of initial conditions inside the star, a massive burned-out core might collapse to form a **black hole.**

Supercomputer simulations give many insights into the complex and rapidly changing conditions deep inside a star as it is torn apart by a supernova explosion. For example, Figure 22-12 shows the first 20 milliseconds after the stiffening of the inner core, which becomes a neutron star. Color represents temperature, red being the hottest areas (20 billion kelvin) and blue the coolest (600 million kelvin). Note the turbulent swirls and eddies that grow behind the shock wave that moves outward from the star's core. Ongoing research suggests that this turbulence may be an important feature of a supernova explosion. Without the turbulence, the shock wave might not be able to plow through the bulk of the doomed star, which continues to fall inward. Turbulence, which involves plumes of matter moving at speeds up to half the speed of light, can revitalize the explosion and insure that the star blows up.

22-6 In 1987 a nearby supernova gave us a close-up look at the death of a massive star

As the outer layers of a massive dying star are blasted into space, the star's luminosity suddenly increases by a factor of approximately 10^8, which is equivalent to a jump of 20 mag-

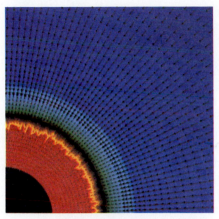

5 milliseconds

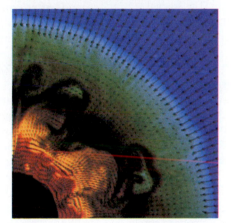

10 milliseconds

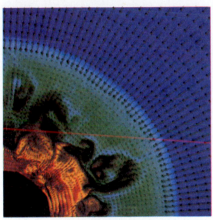

15 milliseconds

20 milliseconds

FIGURE 22-12 The Core of a Supernova These four images show the core of a massive star during the first 20 milliseconds of a supernova explosion. Color indicates temperatures that range from 20 billion kelvin (red) to 600 million kelvin (blue). Matter that bounces off the stiffened core of the star (colored black) produces a shock wave. Instabilities generate turbulent eddies that grow as the shock wave moves outward. This simulation, which describes only the first $\frac{1}{50}$ second in the birth of a neutron star, took six hours on a supercomputer. (Courtesy of A. Burrows and B. Fryxell)

nitudes in brightness. For a few days, a supernova can shine as brightly as an entire galaxy.

On February 23, 1987, a supernova was discovered in the Large Magellanic Cloud (LMC), which is a companion galaxy to our own Milky Way. The supernova (designated SN 1987A because it was the first to be discovered that year) occurred near an enormous H II region in the LMC called 30 Doradus or the Tarantula Nebula because of its spiderlike appearance (Figure 22-13). The supernova was so bright that it could be seen with the naked eye.

A bright supernova is a rare event. In 1885 a supernova in the Andromeda Galaxy was just barely visible to the naked eye. We have to go back to 1604 to find another supernova bright enough to have been seen without a telescope. SN 1987A gave astronomers the unique opportunity to study the death of a massive, nearby star using modern equipment.

SN 1987A was unusual because it did not rise to its expected maximum brightness; instead it stopped at only a tenth of the luminosity typical for an exploding massive star (like the one described in Table 22-1). For 85 days after the detonation, the brightness of SN 1987A gradually increased; then it settled into a slow decline characteristic of an ordinary supernova (Figure 22-14).

Fortunately, the doomed star had been observed prior to becoming a supernova, and these observations helped explain why SN 1987A was not altogether typical. The Large

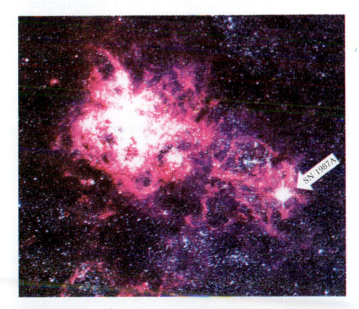

FIGURE 22-13 The Supernova SN 1987A In 1987 a supernova was discovered in a nearby galaxy called the Large Magellanic Cloud (LMC). This photograph shows a portion of the LMC that includes the supernova and a huge H II region called the Tarantula Nebula. At maximum brightness the supernova could be seen without a telescope by observers in the southern hemisphere. (European Southern Observatory)

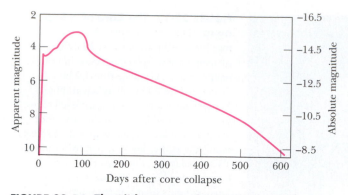

FIGURE 22-14   The Light Curve of SN 1987A   This graph shows how the brightness of SN 1987A changed during its first 600 days. The sudden rise in luminosity following detonation stalled at only a tenth of the anticipated maximum. The gradual decline that began after day 100 is nevertheless typical of massive dying stars.

Magellanic Cloud is about 160,000 light-years from Earth—near enough to us that many of its stars have been individually observed and catalogued. The doomed star was identified as a B3I supergiant (see Figure 22-15). When this star was on the main sequence, its mass was about 20 $M_\odot$, although by the time it exploded it probably had shed a few solar masses.

SN 1987A was initially less luminous than expected because the doomed star was a blue supergiant when it exploded rather than a red supergiant. The evolutionary track for an aging 20-$M_\odot$ star wanders back and forth across the top of the H–R diagram, and so the star alternates between being a hot (blue) supergiant and a cool (red) supergiant. The star's size changes significantly as the surface temperature changes. A blue supergiant is only 10 times larger in diameter than the Sun, but a red supergiant of the same luminosity would be 1000 times larger. Because the doomed star was

relatively small when it exploded, it reached only a tenth of the brightness that an exploding red supergiant would have.

Calculations by Stanford Woosley, Thomas Weaver, and their colleagues at Lick Observatory suggest that the interior of the doomed star contained 6 $M_\odot$ of helium and 2 $M_\odot$ of elements heavier than helium. At the time of detonation, the star's iron core had a mass of 1.5 $M_\odot$ at a temperature of 10 billion kelvin. Energy released by the decay of radioactive isotopes created during the explosion contributed significantly to the supernova's brightness as it began to fade. Some astronomers suspect that a neutron star may be uncovered in a few years.

Most of the energy of a supernova explosion is carried away by the neutrinos, which are produced in great profusion during the collapse of the star's core. During the first ten seconds after core collapse, the supernova emits a burst of neutrinos with a luminosity of 10^{46} watts. For comparison, the total optical luminosity of all the stars and galaxies in the observable universe is 10^{45} watts. Thus, for a fraction of a second, SN 1987A shone ten times brighter than the entire rest of the universe!

About three and a half years after SN 1987A exploded, astronomers using the Hubble Space Telescope obtained a picture of the supernova that showed a ring of glowing gas around the exploded star (Figure 22-16). This gas is a relic of a hydrogen-rich outer atmosphere that was ejected by gentle stellar winds from the doomed star when it was a red supergiant, about 100,000 years ago. The diffuse gas was subsequently compressed into a narrow shell by high-speed stellar winds that flowed from the star when it became a blue supergiant. The gaseous ring seen in Figure 22-16 is glowing because it was ionized by the initial flash of ultraviolet radiation from the supernova. Astronomers predict that material ejected from the supernova itself will strike the circumstellar shell sometime between 1997 and 2004. This collision will cause the shell of gas to brighten considerably and emit copious radiation at X-ray, ultraviolet, and visible wavelengths.

FIGURE 22-15   The Doomed Star and SN 1987A
The photograph on the left shows a small section of the Large Magellanic Cloud before the outburst, with the doomed star

—a B3 supergiant—identified by an arrow. The view on the right shows the supernova a few days after the explosion. (Anglo-Australian Observatory)

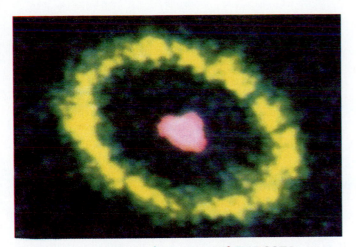

FIGURE 22-16 A Shell of Gas Around SN 1987A Intense radiation from the supernova explosion caused a circumstellar shell of gas around SN 1987A to glow. The luminescent ring seen in this view from the Hubble Space Telescope is about 1.3 light-years in diameter. Between 1997 and 2004, high-speed gases ejected by the supernova will reach the ring and cause it to brighten significantly, but not enough to be visible to the naked eye. (NASA; ESA)

Astronomers have been lucky with SN 1987A. Its progenitor had been studied and its distance from Earth (160,000 ly) was known. Observations were facilitated by the LMC's location in an unobscured part of the heavens, as well as the supernova's timing in the middle of the southern hemisphere summer with its dry, clear skies. Furthermore, neutrino detectors in Japan and the United States detected a burst of neutrinos three hours ahead of the light emitted from the exploding star's surface. Finally, the supernova was unusual, which advanced our understanding of supernovae in general. Because SN 1987A is such an important object, astronomers will be monitoring its progress for many years to come.

As this book was going to press, an amateur astronomer in Spain discovered a supernova, called SN 1993J, in the galaxy M81 in Ursa Major (the Great Bear). Although this galaxy is 10 million light-years from Earth, astronomers were able to identify the supernova's progenitor on photographs taken during the 1980s. Box 22-3 summarizes some of the early observations of SN 1993J.

22-7 White dwarfs in close binary systems can also become supernovae

Astronomers often discover supernovae in distant galaxies (Figure 22-17). The frequency of these discoveries suggests that a typical galaxy, like our own Milky Way, has one supernova explosion every 20 to 50 years. Unfortunately, interstellar gas and dust obscure much of the Milky Way, so the majority of supernovae in our galaxy go unnoticed.

When astronomers began discovering supernovae in distant galaxies, they classified them into two categories, designated Type I and Type II, depending on their spectra. Hydrogen lines are prominent in Type II supernovae but

a

b

FIGURE 22-17 A Distant Supernova Sometime during 1940, a supernova exploded in the galaxy NGC 4725 in the constellation of Coma Berenices. (a) The galaxy before the outburst.

(b) By the time this photograph was taken in 1941, the supernova (at arrow) had faded from its maximum brightness. (The Carnegie Observatories)

absent in Type I. This classification tells us if the doomed star had hydrogen in its outer layers.

Type II supernovae are caused by the death of massive stars that still have ample hydrogen in their atmospheres when they explode. The lack of hydrogen lines in the spectrum of a Type I supernova tells us that its progenitor lost its hydrogen-rich outer layers before exploding.

Astronomers have recently subdivided the Type I classification into two subclasses: Type Ia, which show no helium lines in their spectra, and Type Ib, which are helium-rich. Many astronomers suspect that Type Ib supernovae are caused by dying massive stars that have been stripped of their outer layers before they explode. A star can lose its outer layers to a strong wind, as in Wolf-Rayet stars (recall Figure 21-10), or to a companion star in a close binary. Either mechanism can strip material all the way down to the level of the hydrogen-burning shell, thereby exposing a star's helium-rich interior. If such a star is still massive enough to undergo core collapse, it dies as a supernova that exhibits no hydrogen but many helium lines in its spectrum.

All supernovae begin with a sudden rise in brightness that occurs in less than a day (Figure 22-18). A Type Ia supernova typically reaches an absolute magnitude of −19 at peak brightness, but a Type II supernova is usually about 2 magnitudes fainter. Type Ia supernovae then settle into a gradual decline that lasts for over a year, whereas the Type II light curve has a steplike appearance, caused by alternating periods of steep and gradual declines in brightness.

A Type Ia supernova can begin with a carbon–oxygen-rich white dwarf in a close, semidetached binary system. To trigger the outburst, a swollen red giant companion overflows its Roche lobe and dumps gas onto the white dwarf. When the white dwarf's mass gets close to the Chandrasekhar limit, carbon burning begins. Because the white dwarf is composed of degenerate matter, the usual "safety valve" between pressure and temperature does not operate. In a catastrophic runaway process reminiscent of the helium flash, the rate of carbon burning skyrockets, because increasing temperatures inside the white dwarf are not relieved by corresponding increases in pressure. Soon the temperature gets so high that the electrons are no longer degenerate and the star blows up.

If the carbon burning moves through the white dwarf faster than the speed of sound, a **detonation** rips the star

---

BOX 22-3

The Supernova 1993J

On March 28, 1993, F. Garcia Diez, a Spanish amateur astronomer, discovered the brightest supernova to be seen in the northern hemisphere in 21 years. Called SN 1993J, the supernova is located in the spiral galaxy about 10 million light-years from Earth. Two of the photographs below (left and middle) show pre- and post-explosion views of a small portion of the galaxy M81. The area covered in these views is indicated by a yellow square in the wide-angle photograph of the galaxy on the right. The presence of hydrogen lines in the spectra of SN 1993J indicated that it is a Type II supernova, caused by the core collapse of a massive, evolved star.

Because M81 is a regularly observed galaxy, astronomers soon found stars at the supernova's location on photographs and CCD images taken during the preceding ten years. Apparently two stars make up the progenitor's image, one a red supergiant, the other a blue supergiant.

Because the supernova's peak luminosity (10^9 $L_{\odot}$) was too bright to have been produced by the explosion of a blue supergiant, it is probable that the red supergiant exploded. Thus SN 1993J may be a more "normal" Type II supernova than was SN 1987A. Astronomers will follow the behavior of SN 1993 during the coming years with great interest.

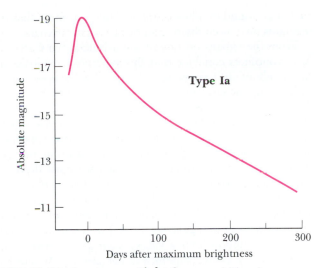

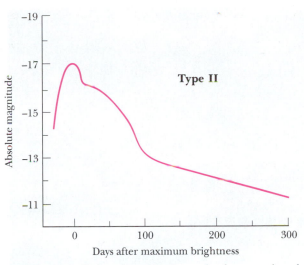

FIGURE 22-18 Supernova Light Curves A Type Ia super-nova, which exhibits a gradual decline in brightness, is caused by an exploding white dwarf in a close binary system. A Type II supernova usually has alternating intervals of steep and gradual decline. Type II supernovae are caused by the death of massive stars.

apart. However, if the carbon burning propagates slower than the speed of sound, the white dwarf is consumed in a process called **deflagration.** Both processes supply about the same total amount of energy output.

A Type Ia supernova is powered by nuclear energy, and the resulting spectacle in the sky is simply the fallout from a thermonuclear explosion. A wide array of unstable isotopes are produced during the outburst, most notably ^{56}Ni, which decays into ^{56}Co. The smooth decline of the light curve of a Type Ia supernova directly results from the gamma rays emitted by radioactive decay of ^{56}Co, which turns into a stable isotope of iron.

Type II supernovae are powered by gravitational energy, because detonation is triggered by the collapse of the star's iron-rich core. The fact that Type Ia and Type II reach roughly the same absolute magnitude at maximum brightness is a coincidence.

22-8 A supernova remnant can be detected at many wavelengths for many years

Astronomers find the debris of supernova explosions, called **supernova remnants,** scattered across the sky. A beautiful example is the Veil Nebula, seen in Figure 22-19. The doomed star's outer layers were blasted into space with such violence that they are still traveling through the interstellar medium at supersonic speeds. As this expanding shell of gas plows through space, it collides with atoms in the interstellar medium, exciting this sparse gas and causing it to glow.

A few nearby supernova remnants cover sizable areas of the sky. The largest is the Gum Nebula, named after Colin Gum, who first noticed its faint glowing wisps on photographs of the southern sky (Figure 22-20). It is also called the Vela Supernova Remnant, because its 60° diameter is centered on the constellation of Vela.

The Gum Nebula looks big simply because it is quite nearby. Its near side is only about 330 ly (100 pc) from Earth, and the center of the nebulosity is just 1300 ly (460 pc) away. Studies of the nebula's expansion rate suggest that this supernova exploded around 9000 BC. It could have been witnessed by people living in places such as Egypt and India. At maximum brilliance, the exploding star probably reached an apparent magnitude of −10, equal to the brightness of the Moon at first quarter.

Many supernova remnants are virtually invisible at optical wavelengths. However, when the expanding gases collide

FIGURE 22-19 The Veil Nebula This nebulosity is a portion of the Cygnus Loop (see Figure 20-20), which is the remnant of a supernova that exploded about 15,000 years ago. The distance to the nebula is about 2,600 light-years, and the overall diameter of the loop is about 120 light-years. (Palomar Observatory)

FIGURE 22-20 The Gum Nebula The Gum Nebula, which spans 60° of the sky, has the largest angular size of any known supernova remnant. Only the central regions of the nebula are shown here. The nearest portions of this expanding nebula are only 100 pc (330 ly) from the Earth. The supernova explosion occurred about 11,000 years ago, and the supernova remnant now has a diameter of about 700 pc (2300 ly). (Royal Observatory, Edinburgh)

have been found on photographic plates, but more than 100 remnants have been discovered by radio astronomers.

From the expansion rate of the nebulosity in Cassiopeia A, astronomers conclude that this supernova explosion occurred about 300 years ago. Although telescopes were in wide use by the late 1600s, no one saw the outburst. In fact, the last supernova seen in our Galaxy, which occurred in 1604, was observed by Johannes Kepler. A few years earlier, in 1572, Tycho Brahe also recorded the sudden appearance of an exceptionally bright star in the sky. To find any other accounts of nearby bright supernovae, we must delve into astronomical records that are almost 1000 years old.

At first glance, this apparent lack of nearby supernovae may seem puzzling. Astronomers have seen more than 600 supernovae in distant galaxies. From this frequency, it is reasonable to suppose that a galaxy such as our own should have as many as five supernovae per century. Where have they been?

As we shall see when we study galaxies in Chapters 25 and 26, the plane of our Galaxy is where massive stars are born and supernovae explode. This region is so rich in interstellar dust, however, that we simply cannot see very far into space in the directions occupied by the Milky Way. In other words, supernovae probably do in fact erupt every few decades in remote parts of our Galaxy, but their detonations are hidden from our view by intervening interstellar matter.

with the interstellar medium, they radiate energy at a wide range of wavelengths, from X rays through radio waves. For example, Figure 22-21 shows both X-ray and radio images of the supernova remnant Cassiopeia A. Optical photographs of this part of the sky reveal only a few small, faint wisps. In fact, radio searches for supernova remnants are more fruitful than optical searches. Only two dozen supernova remnants

22-9 Neutrinos and gravity waves emanate from supernovae

In recent years astronomers have devised ingenious ways of detecting supernovae based on two of nature's most elusive phenomena, neutrinos and gravity waves. We have seen that

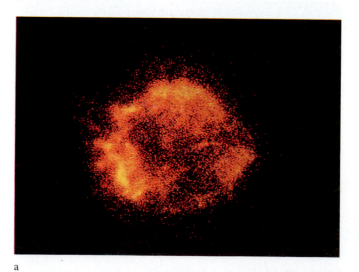

a

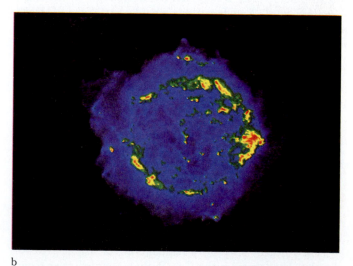

b

FIGURE 22-21 Cassiopeia A Supernova remnants, such as Cassiopeia A, are sometime strong sources of X rays and radio waves. (a) An X-ray picture of "Cas A" taken by the Einstein Observatory. (b) A corresponding radio image produced by the

Very Large Array. The supernova explosion that produced this nebula occurred 300 years ago, 10,000 light-years from Earth. (Smithsonian Institution and the Very Large Array)

a Type II supernova explosion is triggered by the collapse of a massive star's core, producing a flood of neutrinos. These neutrinos hardly interact with matter at all. In fact, under most conditions, matter is transparent to neutrinos. Consequently, when a supernova explodes, vast quantities of neutrinos pour out into space and stream across the Galaxy, passing easily through any gas, dust, or nebulae they happen to encounter. This deluge of neutrinos eventually rains down on—and through—the Earth. By detecting the arrival of these particles, astronomers could learn about the supernova.

Detecting neutrinos is a difficult, tricky business. Some neutrino "telescopes" consist of large drums or tanks of water (Figure 22-22) buried deep in the Earth. Water (H_2O) contains lots of protons, the nuclei of hydrogen atoms. Astronomers detect neutrinos by observing brief flashes of light when protons are struck by the neutrinos.

The neutrinos from a supernova explosion travel at or very near the speed of light and carry a lot of energy, typically 20 MeV or more. (The energy unit of electron volt, eV, was introduced in Box 5-3; 1 MeV = 10^6 electron volts.) Most neutrinos pass right through the Earth. On rare occasions, however, a neutrino hits a proton in the water-filled drum. This collision produces a positron, which then recoils at such a high speed that it emits a brief flash of light known as **Cerenkov radiation.** This type of radiation, first observed in 1934 by the Russian physicist Pavel A. Cerenkov, occurs whenever a particle moves through a medium such as water at a speed that is faster than the speed of light in that medium. Such motion does not violate the tenet that the speed of light in a vacuum (3×10^8 m/s) is the ultimate speed limit in the universe. The speed of light through a substance like water or glass is much less than its speed in a vacuum. The recoiling electrons in the water-filled drums can easily exceed this reduced light speed, thus producing flashes that reveal the existence of an erupting supernova somewhere in our Galaxy.

On February 23, 1987, a supernova exploded in the Large Magellanic Cloud (recall Figure 22-13). Teams of physicists working with neutrino detectors excitedly reported Cerenkov flashes from bursts of neutrinos three hours before astronomers saw the exploding star in the sky. At the Kamiokande II detector in Kamioka, Japan, scientists reported seeing flashes from 12 neutrinos at about the same time that 8 were detected by a team at the IMB (Irvine–Michigan–Brookhaven) detector in a salt mine under Lake Erie.

As we saw in Chapter 18, the mass of the neutrino is a controversial topic. The 1987 supernova gave scientists the opportunity to estimate that mass from the time during which the neutrino arrived. If neutrinos were massless, they would travel at the speed of light, and all the supernova's neutrinos would arrive at Earth at essentially the same moment. The more massive the neutrino is, however, the greater the spread in arrival times. The interval over which the bursts were observed suggests that the mass of a neutrino is probably between 0 and 16 eV. For comparison, the mass of an electron is 511,000 eV.

The ability to detect neutrinos from a supernova offers exciting opportunities for astronomers. Ordinary telescopic observations of a supernova explosion can show us only the expanding outer layers of the dying star. Even X-ray or radio observations fail to see through the hot gases being blasted into space. Thus, ordinary techniques do not allow us to observe the extraordinary events occurring in and around the doomed star's core. However, neutrinos easily penetrate a star's outer layers. These particles carry information about the conditions under which they were created. By measuring the energy and momentum carried by the escaping neutrinos, we should be able to learn many facts about the star's collapsing core.

Another exotic technique being developed to probe details of supernova explosions involves a phenomenon called gravitational radiation. During and immediately after core collapse, vast amounts of matter crushed to nuclear density are hurled about at enormous speeds. This dense mass of material is surrounded by a strong gravitational field. Because this

FIGURE 22-22 Inside a Neutrino Detector A diver is shown servicing one of the photomultiplier tubes in the Irvine–Michigan–Brookhaven neutrino detector. Neutrinos from SN 1987A were detected by this apparatus when they struck protons in the water, producing brief flashes of light that were picked up by the photomultipliers. (Courtesy of K. S. Luttrell)

matter moves so rapidly during the explosion, the gravitational field around the star changes quickly.

As we saw in Chapter 4, Einstein's general theory of relativity is the most accurate description of gravity we have. According to general relativity, gravity actually curves the fabric of space. During a supernova explosion, then, the sudden vigorous changes in gravity around the star can create ripples in the geometry of space. These ripples, which move outward from the supernova at the speed of light, are called **gravitational radiation**, or **gravitational waves**.

Like neutrinos, gravitational waves should bear the imprint of conditions in the star's collapsing core. Unfortunately, gravitational radiation carries very little energy, making the waves very difficult to detect. Nevertheless, in the 1960s, Joseph Weber at the University of Maryland pioneered the construction of gravitational-wave antennas by gluing sensitive crystals onto a large aluminum cylinder. Because gravitational waves are ripples in the geometry of space, objects vibrate slightly when a gravitational wave passes through them. If the cylinder oscillates, the crystals produce an electrical voltage that can be amplified and recorded. In this way, Weber tried to detect the passage of a gravitational wave through his apparatus.

Teams of physicists around the world are working hard to perfect new techniques for detecting gravitational waves. Scientists, such as Kip Thorne at Caltech, are optimistic that highly sensitive antennas using lasers will be operational by the late 1990s. Work has begun on two such antennas, one in Louisiana and the other in Washington. Each will consist of an L-shaped vacuum pipe with arms 4 km long for the laser beams. Mirrors mounted at the ends of the pipes will reflect the laser beams back and forth. If a gravitational wave passes by, the mirrors will move, changing the length of the light path. These changes will be detected by combining the laser beams from the two arms of the antenna.

These antennas, called LIGO for *Laser Interferometer Gravitational-wave Observatory*, should be able to detect bursts of gravitational waves from the collapsing cores of supernovae as far away as 30 million parsecs (about 100 million light-years). This volume of space is so huge and contains so many galaxies that astronomers might be able to detect supernova explosions as frequently as once a month. By analyzing a burst of gravitational waves from a supernova, astronomers should be able to figure out many details concerning the creation of such objects as neutron stars and black holes.

KEY WORDS

Terms preceded by an asterisk are discussed in the boxes.

asymptotic giant branch star (AGB star)	detonation	oxygen burning	shell helium burning
black hole	dredge-up	photodisintegration	silicon burning
carbon burning	gravitational radiation	planetary nebula	supernova
carbon star	gravitational waves	*r process	supernova remnant
Cerenkov radiation	mass–radius relation	*r-process isotope	thermal pulses
Chandrasekhar limit	neon burning	red supergiant	white dwarf
deflagration	neutron capture	*s process	
	neutron star	*s-process isotope	

KEY IDEAS

• A low-mass star becomes a red giant when shell hydrogen burning begins, a horizontal-branch star when core helium burning begins, and a red supergiant when the helium in the core is exhausted and shell helium burning begins.

Thermal runaways in the helium-burning shell produce thermal pulses during which more than half a star's mass may be ejected into space.

Ultraviolet radiation from a hot carbon–oxygen core ionizes and excites ejected gases, producing a planetary nebula.

• The burned-out core of a low-mass star becomes a degenerate, dense sphere about the size of the Earth called a white dwarf, which glows from thermal radiation; as the sphere cools, it becomes dimmer and eventually cold.

• A high-mass star undergoes a sequence of different thermonuclear reactions in its core and shells: carbon burning, neon burning, oxygen burning, and silicon burning.

In the last stages of its life, a high-mass star has an iron-rich core surrounded by concentric shells hosting the various thermonuclear reactions.

A high-mass star dies in a violent cataclysm in which its core collapses and most of its matter is ejected into space at high speeds; the luminosity of the star increases suddenly by a factor of around 10^8 during this explosion, producing a supernova.

The ejected matter, moving at supersonic speeds through interstellar gases and dust, glows as a nebula called a supernova remnant.

• An accreting white dwarf in a close binary system can also become a supernova when carbon burning ignites explosively throughout such a degenerate star.

Type Ia supernovae are produced by accreting white dwarfs in close binaries; Type II supernovae are created by the deaths of massive stars.

Most supernovae occurring in our Galaxy are hidden from our view by interstellar dust and gases.

• Bursts of neutrinos and gravitational waves are emitted during a supernova explosion; the detection and study of these emissions will give important insight into phenomena occurring deep inside an exploding star.

REVIEW QUESTIONS

1. What is the difference between a red giant and a red supergiant?

2. What is the asymptotic red giant branch?

3. How is a planetary nebula formed?

4. What is a white dwarf?

5. What is the significance of the Chandrasekhar limit?

6. A 1-$M_\odot$ white dwarf has a radius of about 17,000 km. What is the radius of a white dwarf whose mass is equal to the Chandrasekhar limit?

7. On an H–R diagram, sketch the evolutionary track that the Sun will follow as it leaves the main sequence to become a white dwarf. Approximately how much mass will the Sun have when it becomes a white dwarf? Where will the rest of the mass have gone?

8. Why do you suppose that all the white dwarfs known to astronomers are relatively close to the Sun?

9. What prevents thermonuclear reactions from occurring at the center of a white dwarf? If no thermonuclear reactions are occurring in its core, why doesn't the star collapse?

10. Why is the temperature in a star's core so important in determining which nuclear reactions can occur there?

11. Suppose that the brightness of a star becoming a supernova increases by 20 magnitudes. Show that this corresponds to an increase of 10^8 in luminosity.

12. What is the difference between Type Ia, Type Ib, and Type II supernovae?

13. Why have radio searches for supernovae remnants been more fruitful than optical searches?

ADVANCED QUESTIONS

Tips and tools

You may find it useful to review Box 19-5, which discusses stellar radii and their relationship to temperature and luminosity. Important data and a photograph of the Ring Nebula are found in Box 22-1. The formula for escape velocity was given in Box 7-2. The all-important relationship between absolute magnitude, apparent magnitude, and distance was discussed in Box 19-2.

14. The central star in a newly formed planetary nebula has a luminosity of 1000 $L_\odot$ and a surface temperature of 100,000 K. How big is the star?

15. You want to determine the age of a planetary nebula. What observations should you make, and how would you use the resulting data?

16. The Ring Nebula is expanding at the rate of about 20 km/s. How long ago did the central star shed its outer layers?

17. Find the average density of a 1-$M_\odot$ white dwarf having the same size as the Earth. What speed is required to eject gas from the white dwarf's surface? Suppose some interstellar gas fell onto the white dwarf. With what speed would it strike the star's surface?

18. What kinds of stars would you monitor if you wished to observe a supernova explosion from its very beginning? Look up lists of the brightest and nearest stars. Which, if any, of these stars are possible supernova candidates? Explain.

19. Suppose that the red supergiant star Betelgeuse became a supernova. At the height of the outburst, how bright would it appear in the sky? How would it compare with the brightness of the full moon?

20. In mid-April of 1979, a supernova exploded in a galaxy called M100 (NGC 4321) in the constellation of Coma Berenices. It reached magnitude of +12 at maximum brilliance and its spectrum showed hydrogen lines. How far away is the galaxy?

21. Consult recent issues of *Sky & Telescope* or *Science News* to learn about the latest observations of SN 1993J. Has the shape of the supernova's light curve been adequately explained? Has the supernova produced any surprises?

DISCUSSION QUESTIONS

22. Suppose that you discover a small, glowing disk of light while searching the sky with a telescope. How would you decide if this object is a planetary nebula? Could your object be something else? Explain.

23. Why are astronomers interested in building neutrino "telescopes" and instruments to detect gravitational waves?

24. Speculate about the connection between carbon stars and life on Earth. How might the origin and evolution of life been affected if the convective zone in AGB stars did not reach all the way down into their carbon-rich cores?

OBSERVING PROJECTS

25. Although they represent a fleeting stage at the end of a star's life, planetary nebulae are found all across the sky. Some of the brightest, with epoch 2000 coordinates, are listed in the table on the next page.

Planetary nebula	R.A.	Decl.
Little Dumbbell (M76)	1^h 42.4^m	+51° 34′
NGC 1535	4 14.2	−12 44
Eskimo	7 29.2	+20 55
Ghost of Jupiter	10 24.8	−18 38
Owl (M97)	11 14.8	+55 01
Ring (M57)	18 53.6	+33 02
Blinking Planetary	19 44.8	+50 31
Dumbbell (M27)	19 59.6	+22 43
Saturn	21 04.2	−11 22
NGC 7662	23 25.9	+42 33

Observe as many of these planetary nebulae as you can on a clear, moonless night using the largest telescope at your disposal. Note and compare the various shapes of the different nebulae. In how many cases can you see the central star? The so-called Blinking Planetary (NGC 6826) affords an excellent demonstration of central versus peripheral vision: The nebula seems to disappear when you look straight at it, but it reappears as soon as you look toward the side of your field of view. This effect occurs because the central region of the human retina, on which an image is focused when you look straight at an object, is less sensitive than surrounding regions of the retina.

26. Two supernova remnants can be seen through modest telescopes, one in the winter sky and the other in the summer sky. Both are quite faint, however, so you should schedule your observations for a moonless night. The winter sky contains the Crab Nebula, which is discussed in detail in the next chapter. The coordinates are R.A. = 5^h 34.5^m and Decl. = 22° 00′, which places the object near the star marking the eastern horn of Taurus (the Bull). Whereas the entire Crab Nebula easily fits in the field of view of an eyepiece, the Veil or Cirrus Nebula in the summer sky is so vast that you can see only a small fraction of it at a time. The easiest way to find the Veil Nebula is to aim the telescope at the star 52 Cygni (R.A. = 20^h 45.7^m and Decl. = +30° 43′), which lies on one of the brightest portions of the nebula. If you then move the telescope slightly north or south until 52 Cygni is just out of the field of view, you should see faint wisps of glowing gas. (52 Cygni is the bright star toward the lower right corner of Figure 20-20, which shows the entire Cygnus Loop.)

27. Use a telescope to observe the remarkable triple star 40 Eridani, whose coordinates are R.A. = 4^h 15.3^m and Decl. = −7° 39′. The primary, a 4.4-magnitude yellowish star like the Sun, has a 9.6-magnitude white dwarf companion, the most easily seen white dwarf in the sky. On a clear, dark night with a moderately large telescope, you should also see that the white dwarf has an 11th-magnitude companion, which completes this most interesting trio.

FOR FURTHER READING

Bethe, H., and Brown, G. "How a Supernova Explodes." *Scientific American,* May 1985. This enlightening article describes many fascinating details about the detonation of a Type II supernova.

Kahn, R. "Desperately Seeking Supernovae." *Sky & Telescope,* June 1987. This article describes the trials and tribulations of developing an automated telescope to search for supernovae in galaxies.

Malin, D., and Allen, D. "Echoes of the Supernova." *Sky & Telescope,* January 1990. Two noted Australian astronomers discuss their observations and photographs of "light echoes" caused by radiation from SN 1987A reflecting off dust clouds in the vicinity of the supernova.

Marschall, L. *The Supernova Story.* Plenum, 1988. This superbly written book is an excellent choice for someone who wishes to learn more about supernovae, including SN 1987A.

Murdin, P., and Murdin, L. *Supernovae.* Cambridge University Press, 1985. This book tells the story of our growing understanding of supernovae, pieced together from historical records, modern observations, and theoretical studies.

Reddy, F. "Supernovae: Still a Challenge." *Sky & Telescope,* December 1983. This article explains how Type I and Type II supernovae explode.

Schorn, R. "Supernova Shines On." *Sky & Telescope,* May 1987. This article contains many outstanding pictures of the early stages of SN 1987A.

Seward, F., Gorenstein, P., and Tucker, W. "Young Supernova Remnants." *Scientific American,* August 1985. This article focuses on X-ray observations of the remnants of supernovae whose detonations were seen within the past thousand years.

Soker, N., "Planetary Nebulae." *Scientific American,* May 1992. This article describes the structure and evolution of fluorescent clouds of gas that represent the last gasp of dying low-mass stars.

Tierney, J. "Quest for Order: Profile of S. Chandrasekhar." *Science 82,* September 1982. This biographical sketch introduces the Indian astrophysicist who gave us our basic understanding of white dwarfs.

Wheeler, C., and Harkness, R. "Helium-rich Supernovas." *Scientific American,* November 1987. This article discusses Type Ib supernovae, whose progenitors are massive stars that have shed all their hydrogen prior to detonating.

Woosley, S., and Weaver, T. "The Great Supernova of 1987." *Scientific American,* August 1989. Two preeminent authorities on supernovae teamed up to write this superb article, which describes many fascinating details of SN 1987A.

NEUTRON STARS

THE CREATION OF A NEUTRON STAR? The bright star near the center of this photograph is SN 1987A one week after detonation. Many astronomers suspect that the supernova's core may have collapsed to form a neutron star. In 1990 the supernova's light curve had leveled off, indicating the presence of an underlying, continual source of energy, possibly caused by gases falling back down onto a rapidly rotating neutron star. Astronomers are keeping a careful watch on the supernova's expanding layers in the hope of seeing pulsed radiation. This would confirm the presence of a spinning neutron star. (NOAO)

A SUPERNOVA EXPLOSION can produce a stellar corpse called a neutron star, which is an extremely dense sphere of degenerate neutrons. Because of their extraordinary properties, neutron stars give rise to a variety of fascinating phenomena. For example, a pulsar is a rapidly spinning neutron star with a powerful magnetic field that sweeps beams of radiation around the sky. A pulsating X-ray source can result when a neutron star in a binary system accretes gas from its companion. Isolated neutron stars spin more slowly as they radiate energy, but the fastest pulsars are in close binary systems, where they have been spun up by accretion from their companion stars. Models of neutron stars devised by astronomers to explain these phenomena invoke many bizarre attributes of neutron stars, such as the superconductivity and superfluidity that dominate their interiors. These models also demonstrate that there is an upper limit to the mass of a neutron star, just as the Chandrasekhar limit sets the maximum mass of a white dwarf. In close binary systems, each of these two kinds of stellar corpses—white dwarfs and neutron stars—can undergo explosive thermonuclear reactions on their surfaces. Hydrogen burning on a white dwarf produces a nova, whereas helium burning on a neutron star yields a burster.

The brightening eastern sky was aglow with pink and red as Yang Wei-T'e waited patiently for the sunrise. As imperial astronomer to the Chinese court during the Sung dynasty, Yang was thoroughly familiar with the constellations. He knew at once that something quite extraordinary had happened that night. Preceding the Sun by only a few minutes, a dazzling object ascended above the eastern horizon. It was far brighter than Venus and more resplendent than any star he had ever seen. Yang dutifully recorded his observations and thoughts on this unusual celestial event:

> I make my kowtow. I observed the phenomenon of a guest star. Its color was slightly iridescent. Following an order of the Emperor, I respectfully make the prediction that the guest star does not disturb Aldebaran. This indicates that . . . the Empire will gain great power. I beg to store this prediction in the Department of Historiography.

The date would be officially recorded as "the day of Ch'ih Ch'iu in the fifth moon of the first year of the Shih-huo period"; we would call it July 4, 1054.

Yang's "guest star" was so brilliant that it could easily be seen during broad daylight for the rest of July. After a year, however, "it faded and became invisible." Actually, the su-pernova had occurred some 6500 years earlier, and its light had taken this long to reach the Earth. Today we realize that Yang Wei-T'e had witnessed the creation of a neutron star.

23-1 The discovery of pulsars stimulated interest in neutron stars

The neutron was discovered in 1932 during laboratory experiments. Within a year, two astronomers predicted the existence of neutron stars. Inspired by the realization that white dwarfs are supported by degenerate electron pressure, Fritz Zwicky at the California Institute of Technology (Caltech) and his colleague Walter Baade at Mount Wilson Observatory proposed that a highly compact ball of neutrons could similarly produce a powerful pressure. This **degenerate neutron pressure** could also support a stellar corpse, perhaps one even more massive than is permitted by the Chandrasekhar limit. "With all reserve," Zwicky and Baade theorized, "we advance the view that supernovae represent the transition from ordinary stars into **neutron stars,** which in their final stages consist of extremely closely packed neutrons." In other words, there could be at least two types of stellar corpses: white dwarfs and neutron stars.

This prophetic proposal was politely ignored by most scientists for years. After all, a neutron star must be a rather weird object. As we saw in Chapter 22, in order to transform

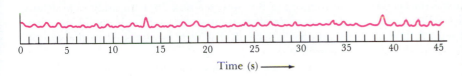

Time (s) ——➤

FIGURE 23-1   A Recording of the First Pulsar This chart recording shows the intensity of radio emission from the first known pulsar, CP1919 (this designation means "Cambridge pulsar at a right ascension of $19^h 19^m$"). Note that some pulses are weak, others strong. Nevertheless, the spacing between pulses is exactly 1.3373011 seconds. (Adapted from A. Hewish)

protons and electrons into neutrons, the density in the star must be equal to nuclear density, about 4×10^{17} kg/m^3. If a single thimbleful of neutron star matter were brought back to Earth, it would weigh 100 million tons. A star compacted to nuclear density must be very small. A 1-M$_\odot$ neutron star would have a diameter of only 30 km, about the same size as San Francisco or Manhattan. The surface gravity on one of these neutron stars would be so strong that an object would have to travel at one-half the speed of light to escape into space. All of these conditions seemed so outrageous that few astronomers paid serious attention to the subject of neutron stars—until 1968, when astronomers discovered pulsating radio sources.

As a young graduate student at Cambridge University, Jocelyn Bell spent many months helping construct an array of radio antennas covering $4\frac{1}{2}$ acres of the English countryside. The instrument was completed by fall of 1967, and Bell and her colleagues began detecting and scrutinizing radio emissions from various celestial sources. Their goal was to see if radio sources exhibit a phenomenon similar to the twinkling of stars, namely tiny, random fluctuations in brightness caused by gas in between the source and the observer. While searching for random flickering, Bell noticed that the antennas had detected regular pulses of radio noise from one particular location in the sky. Repeated observations demonstrated that the radio pulses were arriving with a regular period of 1.3373011 seconds (Figure 23-1).

The regularity of this pulsating radio source was so striking that the Cambridge team suspected they might be detecting signals from an advanced alien civilization. This possibility was soon discarded, however, when several more of these pulsating radio sources, which came to be called **pulsars**, were discovered across the sky. In all cases, the periods were extremely regular, ranging from about $\frac{1}{4}$ second for the fastest to about $1\frac{1}{2}$ seconds for the slowest.

The discovery of pulsars was officially announced in early 1968, after which many astronomers around the world began proposing all sorts of explanations for these regular radio pulsations. Many of these theories were quite bizarre, and arguments raged among astronomers for months. However, by late 1968, all this controversy was laid to rest with the discovery of a pulsar, now called the Crab pulsar, in the middle of the Crab Nebula.

Yang Wei-T'e and his colleagues had been quite precise about the location of the supernova of 1054. When we turn a telescope toward this location in the constellation of Taurus, we find the Crab Nebula, shown in Figure 23-2. This object, which looks like an exploded star, is a supernova remnant.

The presence of a pulsar in a supernova remnant tells us that pulsars are probably associated with dead stars. Before this discovery, most astronomers believed that all stellar corpses are white dwarfs. The number of white dwarfs in the sky seemed to account for all the stars that must have died since our Galaxy was formed. It was generally assumed that all dying stars—even the most massive ones, which produce supernovae—somehow manage to eject enough matter so that their corpses do not exceed the Chandrasekhar limit.

FIGURE 23-2   The Crab Nebula This beautiful nebula, named for the armlike appearance of its filamentary structure, is the remnant of the supernova of AD 1054. The distance to the nebula is about 6500 ly (2000 pc), so its present angular size (4 by 6 arc min) corresponds to linear dimensions of about 7 by 10 ly. (Palomar Observatory)

The discovery of the Crab pulsar showed these conservative opinions to be in error.

At the time of its discovery, the Crab pulsar was the fastest pulsar known to astronomers. Its period is 0.033 second, which means that it flashes 30 times each second. It was immediately apparent to astrophysicists that white dwarfs are too big and bulky to produce 30 signals per second. Calculations demonstrated that a white dwarf can neither rotate nor vibrate that fast. The existence of the Crab pulsar clearly indicates that the stellar corpse at the center of the Crab Nebula is much smaller and more compact than a white dwarf. As Thomas Gold of Cornell University emphasized, astronomers would now have to seriously consider the existence of neutron stars.

23-2 Pulsars are rapidly rotating neutron stars with intense magnetic fields

As we have seen, a neutron star is small and dense. It should also rotate rapidly. All stars rotate, but most of them do so in a leisurely fashion. For example, our Sun takes nearly a full month to rotate once about its axis. However, just as an ice skater doing a pirouette speeds up when she pulls in her arms, a collapsing star also speeds up as its size shrinks. (This

phenomenon is a direct consequence of a law of physics called the conservation of angular momentum, which was introduced in Box 4-4; see Box 23-1 for additional details.) In fact, a star that rotates once a month would spin faster than once a second, if it were compressed to the size of a neutron star.

Neutron stars also have intense magnetic fields. It seems safe to say that every star possesses some magnetic field, but the strength is typically quite low. For a star like the Sun, the magnetic field is spread out over millions upon millions of square kilometers of its surface. However, if this star collapses down to a neutron star, its surface area (which is proportional to the square of its radius) shrinks by a factor of 10^9. The magnetic field, which is bonded to the star's gases, becomes concentrated onto one-billionth of its original area, and so its strength increases by a factor of one billion.

The strength of a magnetic field is usually measured in the unit called the gauss (G), named after the famous German mathematician and astronomer Karl Friedrich Gauss. As we saw in Box 18-2, a magnetic field splits spectral lines into two or more lines, whose spacing reveals the field's strength. For example, magnetically split lines in the solar spectrum reveal that the Sun's overall magnetic field has a strength in the range of 1 to 2 G, although fields thousands of times

---

BOX 23-1

The Conservation of Angular Momentum

Angular momentum, introduced briefly in Box 4-4, is a measure of the momentum carried by an object because of its rotation or revolution about an axis. A mass m moving with a speed v as it revolves about an axis has an angular momentum L given by

$$L = mvr$$

where r is the distance from the mass to the axis of revolution.

The **conservation of angular momentum** requires that neither the magnitude nor the direction of the angular momentum ever change. For a mass revolving about an axis, this requires the value of L to remain constant and the direction of the axis of rotation to remain fixed in space.

When speaking about angular motion, it is often convenient to define the **angular velocity** ω (omega) as

$$\omega = \frac{v}{r}$$

With this definition, we can write the angular momentum of a mass m as

$$L = mr^2\omega$$

For a rotating object of some size, not all the mass is located at the same distance from the axis of rotation. In

that case, the object's angular momentum is given by

$$L = I\omega$$

where I is the object's moment of inertia, an expression of the way matter is distributed throughout the object. For instance, the moment of inertia of a solid wheel of mass M and radius R is $\frac{1}{2}MR^2$. For a bicycle wheel, which has all its mass in a ring a distance R from the axis of rotation, the angular momentum is MR^2. That is twice as great as for the solid wheel, much of whose mass is closer in.

These examples with wheels demonstrate a general property of moments of inertia: The more concentrated the mass is around the axis of rotation, the smaller the moment of inertia is. This trait helps us understand why an ice skater doing a pirouette speeds up as she pulls in her arms. When she does so, her moment of inertia (I) decreases, and so her angular velocity (ω) must increase in order for her angular momentum ($I\omega$) to remain unchanged.

For a sphere of uniform density with mass M and radius R, the moment of inertia is given by

$$I_{\text{sphere}} = \frac{2}{5}MR^2$$

Thus, the angular momentum of a uniform sphere rotating with an angular velocity ω can be written as

$$L = \frac{2}{5}MR^2\omega$$

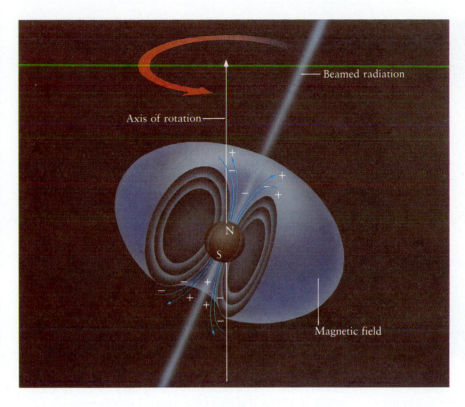

FIGURE 23-3 A Rotating, Magnetized Neutron Star It is reasonable to suppose that a neutron star is rotating rapidly and possesses a powerful magnetic field. Charged particles that have been accelerated near the star's magnetic poles produce two oppositely directed beams of radiation. As the star rotates, the beams sweep around the sky. If the Earth happens to lie in the path of the beams, we detect a pulsar.

Labels in figure: Beamed radiation; Axis of rotation; N; S; Magnetic field

EXAMPLE: Suppose that a star collapses from solar dimensions ($R = R_\odot = 7 \times 10^5$ km) down to neutron star dimensions ($R = R_* = 16$ km) with no loss of mass (so that M is constant). The conservation of angular momentum tells us that the original angular velocity (ω_0) is related to the final angular velocity (ω_f) by

$$L = \tfrac{2}{5} M R_\odot^2 \, \omega_0 = \tfrac{2}{5} M R_*^2 \, \omega_f$$

$$R_\odot^2 \, \omega_0 = R_*^2 \, \omega_f$$

$$\omega_0 (7 \times 10^5)^2 = \omega_f (16)^2$$

or

$$\frac{\omega_f}{\omega_0} = \left(\frac{7 \times 10^5}{16} \right)^2 = 2 \times 10^9$$

In other words, the star is rotating two billion times faster after the collapse than it was before. For example, suppose the star rotates about once a month, as the Sun does. Because one month equals 3×10^6 s, when the star becomes a neutron star it is rotating nearly 1000 times per second.

In reality, material escaping from the collapsing star would carry away a significant amount of its angular momentum. Consequently, the resulting neutron star would probably not rotate quite as fast as this calculation suggests.

stronger have been measured in sunspots. The spectra of certain white dwarfs reveal splitting that corresponds to field strengths of 1 million gauss or more, considerably stronger than anything ever produced in a laboratory. By our usual standards, the trillion-gauss (10^{12} G) fields surrounding neutron stars are extremely powerful.

The axis of rotation of a typical neutron star is inclined at an angle to the magnetic axis, the line connecting the north and south magnetic poles (Figure 23-3). After all, there is no reason for these two axes to coincide. In 1969 Peter Goldreich at Caltech pointed out that the combination of such a powerful magnetic field and rapid rotation would act like a giant electric generator, creating extremely intense electric fields near the neutron star's surface. At its surface, there are plenty of protons and electrons, because the pressures there are too low to combine them into neutrons. The powerful electric fields push these charged particles out from the neutron star's polar regions along its curved magnetic field, as sketched in Figure 23-3. As the particles stream along the curved field, they become accelerated and emit energy. The result is two thin beams of radiation pouring out of the neutron star's north and south magnetic polar regions.

A rotating, magnetized neutron star is somewhat like a lighthouse beacon. As the star rotates, the beams of radiation sweep around the sky. If the Earth happens to be located in the right direction, a brief flash is observed each time the beam sweeps through our line of sight.

When this scenario was first proposed in the late 1960s, a team of astronomers at the University of Arizona began wondering if pulsars might also emit flashes at wavelengths other than radio frequencies. After aiming a telescope at the center

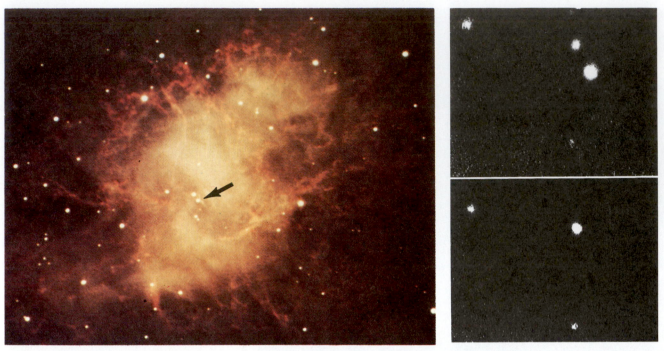

a

b

FIGURE 23-4 The Crab Pulsar A pulsar is located at the center of the Crab Nebula. (a) The Crab pulsar is identified by the arrow. (b) These two photographs show the Crab pulsar in its on and off states. Like the radio pulses, these visual flashes have a period of 0.033 s. (Palomar Observatory; Lick Observatory)

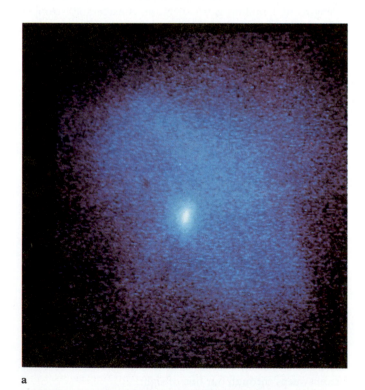

a

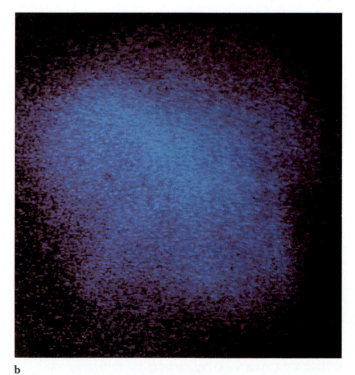

b

FIGURE 23-5 X-Ray Views of the Crab Nebula and Pulsar These views of the Crab Nebula were obtained with an X-ray telescope on the Earth-orbiting Einstein Observatory. The pulsar is shown in its (a) on and (b) off states. (Harvard–Smithsonian Center for Astrophysics)

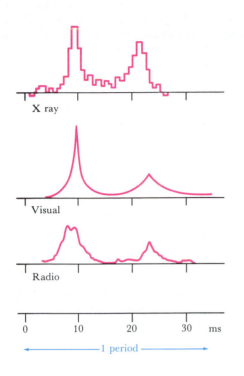

of the Crab Nebula, they used a spinning disk with a slit in it to "chop," or interrupt, incoming light at rapid intervals. To their surprise, they found that one of the stars at the center of the nebula is actually flashing on and off 30 times each second (Figure 23-4).

Since those pioneering days, the Crab pulsar has been observed emitting flashes over a wide range of wavelengths. For example, Figure 23-5 shows X-ray pictures from the Einstein Observatory with the pulsar in its on and off states; the light curves of the Crab pulsar at optical, X-ray, and radio wavelengths are displayed in Figure 23-6. Note that, in addition to the main pulse, there is a second pulse about halfway through the pulsar's period. Many pulsars exhibit both main and secondary pulses, simply because neutron stars have both north and south magnetic poles. When one of the magnetic poles is more directly aligned toward the Earth than the other, we detect one strong pulse and one weak one during each rotation of the neutron star.

Since 1968 radio astronomers have discovered about 450 pulsars scattered across the sky. Each one is presumed to be the neutron star corpse of a massive extinct star. The Vela pulsar, located at the core of the Gum Nebula (Figure 23-7), is visibly flashing like the Crab pulsar, but with a period of 0.089 s, making it the slowest pulsar ever detected at visible wavelengths.

Several supernovae have been seen throughout history, but they did not necessarily produce pulsars. For instance, those observed by Tycho Brahe in 1572 and by Johannes Kepler in 1604 were probably Type Ia supernovae, and it is difficult to imagine how an exploding white dwarf might become a neutron star. The primary source of pulsars is Type II supernovae, in which the cores of massive stars collapse.

FIGURE 23-6 Light Curves of the Crab Pulsar These three graphs show the intensity of radiation emitted by the Crab pulsar at X-ray, optical, and radio wavelengths. One millisecond equals 10^{-3} s. Thus, the pulsar's period is 33 ms.

a

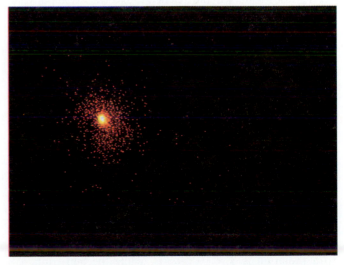

b

FIGURE 23-7 The Vela Pulsar in X Rays The Vela pulsar, named for its location in the southern constellation of Vela, is at the middle of a supernova remnant called the Gum Nebula. Like the Crab pulsar, the Vela pulsar can be detected at X-ray, optical, and radio wavelengths. (a) The location of the pulsar is identified by the arrow. (b) This view was obtained from one of the X-ray telescopes on the orbiting Einstein Observatory. (Harvard–Smithsonian Center for Astrophysics)

Because the great supernova of 1987 was a Type II, astronomers have been carefully observing its remains for signs of a pulsar. In 1990 SN 1987A's light curve had leveled off to a steady luminosity, just about what would be expected from gas falling onto a neutron star's surface. Meanwhile, NASA scientists are designing a satellite called the *X-ray Timing Explorer*, to be launched in 1995. Its X-ray sensors will be capable of seeing through much of the gas and dust surrounding the supernova to search for rapid X-ray variations indicative of a rapidly spinning neutron star. Obviously, SN 1987A will occupy astronomers' attention for many years to come.

23-3   Pulsars gradually slow down as they radiate energy into space

The Crab pulsar is one of the youngest pulsars, its creation having been observed by Yang Wei-T'e some 900 years ago. The Vela pulsar is also quite young. From the size of its supernova remnant and the rate at which it is expanding, astronomers estimate that the Vela supernova occurred roughly 11,000 years ago. Only the very youngest pulsars seem to be energetic enough to emit both optical flashes and radio pulses. As a pulsar ages, it gradually slows down because it loses energy.

Many of the details of how pulsars emit energy are still poorly understood. Nevertheless, without the concept of a rapidly rotating, magnetized neutron star, it is virtually impossible to explain why the Crab Nebula shines the way it does.

The diffuse part of the Crab Nebula—not the reddish filaments—shines with an eerie light. This light has been identified by the Russian astronomer Iosif Shklovskii as the same type of radiation first observed in 1947 in a particle accelerator built at the General Electric Company in Schenectady, New York. This machine, called a synchrotron, was designed to accelerate electrons up to speeds close to the speed of light for experiments in high-energy nuclear physics. The electrons were whirled along a circular path and held in orbit by powerful magnets. Through a small window in the side of the doughnut-shaped machine, scientists noticed a strange light being emitted by the circulating beam of electrons. It is now known that this light, called **synchrotron radiation,** is emitted whenever high-speed electrons move along curved paths through a magnetic field. These electrons are said to be *relativistic*, because their velocities are near the speed of light and Einstein's theory of relativity must be applied to understand their motions. Consequently, the Crab Nebula must contain quite a large number of relativistic electrons spiraling in an extensive magnetic field.

The total energy output of the Crab Nebula in synchrotron radiation is 3×10^{31} watts, while the Sun emits 4×10^{26} watts. Thus, the Crab Nebula is 75,000 times more luminous than the Sun. In 1966, two years before the discovery of pulsars, John A. Wheeler at Princeton University and Franco Pacini from Italy had already speculated that the ultimate source of this prodigious energy output might be a

spinning neutron star. Their prophetic idea was confirmed by the discovery that the Crab pulsar is slowing down.

Although pulsars were first noted for their regular periods, careful measurements by radio telescopes soon revealed that many pulsars are indeed gradually slowing down. The period of a typical pulsar increases by a few billionths of a second each day. The Crab pulsar, which has one of the shortest periods, is slowing more quickly than most other pulsars: Its period increases by 3×10^{-8} second each day. Thirty-billionths of a second may sound like a trivial amount, but a rapidly spinning neutron star possesses so much rotational energy that even the slightest slowdown corresponds to a tremendous loss of energy. In fact, the observed rate of slowing for the Crab pulsar corresponds to an energy loss equal to the entire luminosity of the Crab Nebula. In other words, although the details are still poorly understood, the pulsar's rotational energy is constantly being converted into synchrotron radiation, which is why the Crab Nebula shines.

Astronomers have also noticed that pulsars sometimes exhibit a sudden, unexpected speedup, called a **glitch**. For example, Figure 23-8 shows accurate period measurements of the Vela pulsar during 1975 and 1976. On this graph, the pulsar's gradual slowdown is shown as a steady increase in its period. In September 1975, however, there was an abrupt speedup, after which the pulsar continued to slow down at its usual rate. This glitch was the third one observed for the Vela pulsar; similar glitches have been observed for the Crab pulsar.

Astronomers speculate that a glitch is caused by a sudden movement of the crust of a neutron star, called a **starquake**. According to the law of the conservation of angular momentum, a rotating object speeds up when it contracts, as we have seen. A neutron star is so dense that a typical glitch (which involves a sudden speedup, causing the period to decrease by 10^{-7} s) corresponds to a settling of the star's crust

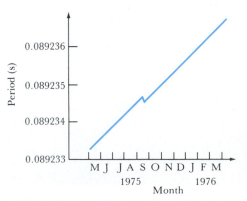

FIGURE 23-8   A Glitch of the Vela Pulsar This graph shows how the period of the Vela pulsar decreased suddenly during September 1975. The speedup, called a glitch, is probably caused by a starquake, during which the star's crust cracks and settles.

by less than 1 mm. Because the speedup was sudden, the starquake must have been equally abrupt. The starquakes that caused glitches of the Vela and Crab pulsars, were ten billion times more energetic that the most powerful Earthquakes.

23-4 Superfluidity and superconductivity are among the strange properties of neutron stars

Starquakes tell us about the crust of a pulsar, but astrophysicists must use elaborate calculations to surmise conditions inside a neutron star. As we have seen, a neutron star consists of closely packed degenerate neutrons. Detailed models strongly suggest that this sea of densely packed neutrons can flow without any friction whatsoever, a phenomenon called **superfluidity**.

Superfluidity is observed in laboratory experiments with liquid helium at temperatures near absolute zero. Because it is frictionless, a superfluid exhibits such strange properties as being able to creep up the walls of a container in apparent defiance of gravity. Within a neutron star, rapid friction-free whirlpools of superfluid neutrons may develop and interact with the crust above, perhaps causing a starquake.

Although a neutron star is made up predominantly of neutrons, some protons and electrons are also scattered throughout the star's interior. Indeed, a pulsar's magnetic field must be anchored to the neutron star by charged particles. A neutron, of course, is electrically neutral. Thus, without the protons and electrons in its interior, a neutron star would rapidly lose its magnetic field.

Models of the internal structure of a neutron star strongly suggest that the protons in a neutron star's core can move around without experiencing any electrical resistance whatsoever. This phenomenon, called **superconductivity**, is observed in the laboratory with certain substances at very low temperatures.

As shown in Figure 23-9, the structure of a neutron star probably consists of a superfluid and superconducting core surrounded by a superfluid mantle, which in turn is surrounded by a brittle crust only 1 km thick.

23-5 The fastest pulsars were probably created by mass transfer in close binary systems

In 1982 astronomers at the Arecibo Observatory discovered a pulsar whose period is only 1.557 ms (one millisecond equals a thousandth of a second). This remarkable pulsar, called PSR 1937+21 (a name that gives its position in the sky), must consist of a neutron star spinning 642 times per second.

Like all pulsars, PSR 1937+21 is slowing down, but at a very gradual rate. A pulsar slows down because its spinning magnetic field radiates energy into space; the stronger the field, the greater the energy loss, and thus the greater is the

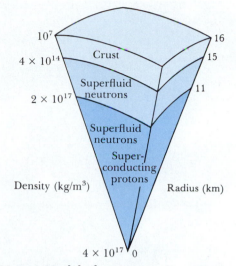

FIGURE 23-9 A Model of a Neutron Star This model for a 1.3-$M_\odot$ neutron star has a superconducting, superfluid core 32 km in diameter. The core is surrounded by a superfluid mantle of neutrons 4 km thick. The star's crust is probably composed of heavy nuclei (such as iron) and free electrons.

slowdown. Because PSR 1937+21 is slowing down so gradually, its magnetic field must be quite weak, perhaps only about 10^8 gauss. Presumably, its field is weak because the pulsar is very old, having radiated away its magnetism over billions of years. Why, then, is the pulsar spinning so rapidly?

In the decade since 1982, astronomers have discovered about two dozen of these very fast pulsars, which are now called **millisecond pulsars**. All have periods between 1 and 10 ms, which means that their neutron stars are spinning at rates of 100 to 1000 rotations per second. The majority are in binary systems with orbital periods between 10 and 100 days. Orbital periods this short tell us that the separation between the two stars in these binaries is quite small—they are all close binary systems. This fact suggests a scenario for the origin and evolution of millisecond pulsars.

Imagine a binary system consisting of a high-mass star and a low-mass star. The high-mass star evolves more rapidly than the low mass star and soon becomes a Type II supernova that creates a neutron star. Like most newborn neutron stars, it spins a few times per second, and so initially we see a pulsar rather like the Crab or Vela pulsar. Over the next few billion years, the pulsar slows down as it radiates energy into space. Meanwhile, the slowly evolving low-mass star begins to expand as it evolves away from the main sequence to become a red giant. When it gets big enough to fill its Roche lobe, it starts to spill gas over the inner Lagrangian point onto the neutron star. The in-falling gas strikes the neutron star's surface at high speed and at an angle that causes the star to spin faster. In this way, a slow, aging pulsar is "spun up" by mass transfer from its bloated companion.

What about the few millisecond pulsars, like PSR 1937+21, that are not members of close binaries? It may be that the companions of solitary millisecond pulsars have

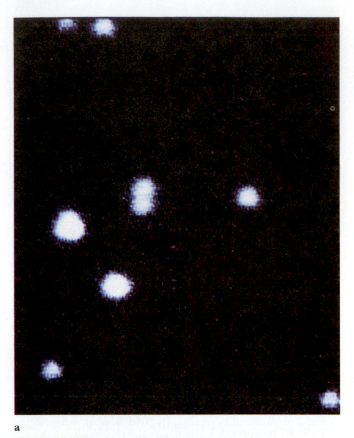

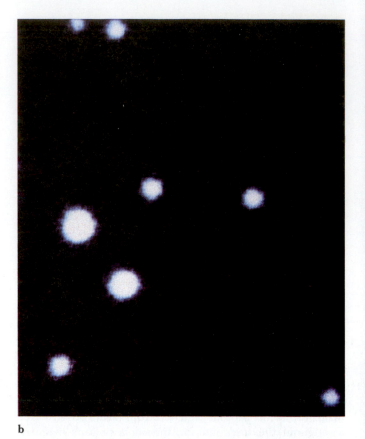

a

b

FIGURE 23-10 The Black Widow Pulsar The millisecond pulsar PSR 1957+20 is a rapidly spinning neutron star in a close binary system. The neutron star and its companion revolve about each other with an orbital period of 9.16 hours. Most of the visible light from this system comes from the side of the companion star facing the neutron star. That hemisphere is exposed to our view in (**a**) but turned away from us in (**b**). An unrelated background star less than 1 arc sec to the northeast shines constantly. (Courtesy of Jan van Paradijs)

been worn away by the high-energy particles emitted by the pulsar after it was spun up. The recently discovered "Black Widow Pulsar" may in fact be caught in the act of destroying its companion in just such a process (Figure 23-10). This system, an eclipsing binary, consists of a neutron star spinning 622 times per second and an ordinary companion revolving about each other with a period of 9.16 hours. The companion blocks out the pulsar's radio signals for 45 minutes during eclipse. Just before and after eclipse, however, the pulsar's signals are delayed substantially, as if the radio waves were being slowed as they pass through a cloud of ionized gas surrounding the companion star. This circumstellar material is probably the star's outer layers, dislodged by particles and radiation from the pulsar. In a few hundred million years or so, the companion of the pulsar will have completely disintegrated, leaving behind only a spun-up, solitary millisecond pulsar.

23-6 Pulsating X-ray sources are neutron stars in close binary systems

During the 1960s, astronomers obtained tantalizing views of the X-ray sky during brief rocket and balloon flights that momentarily lifted X-ray detectors above the Earth's atmo-sphere. A number of strong X-ray sources were discovered, each being named after the constellation in which it is located. For example, Scorpius X-1 is the first X-ray source found in the constellation of Scorpius.

Astronomers were so intrigued by these preliminary discoveries that they began designing an Earth-orbiting, X-ray–detecting satellite that could make observations 24 hours a day. Their hopes and dreams were realized with the launch of *Explorer 42* on December 12, 1970 (Figure 23-11). Because it was to be placed in an equatorial orbit, the satellite was launched from Kenya. In recognition of the hospitality of the Kenyan people, *Explorer 42* was christened *Uhuru*, which means "freedom" in Swahili.

Uhuru gave us our first comprehensive look at the X-ray sky. As the satellite slowly rotated, its X-ray detectors swept across the heavens. Each time an X-ray source came into view, signals were transmitted to receiving stations on the ground. Before its battery and transmitter failed in early 1973, *Uhuru* had succeeded in locating 339 X-ray sources.

The discovery of pulsars was still fresh in everyone's mind, so the *Uhuru* team was very excited to detect X-ray pulses coming from Centaurus X-3 in early 1971. Figure 23-12 shows data from one sweep of *Uhuru's* detectors across Centaurus X-3. The X-ray pulses have a regular period of 4.84 s.

FIGURE 23-11 *Uhuru* *Uhuru* was a small satellite designed to detect astronomical sources of X rays. During three years of flawless operation, it located more than 300 different X-ray objects across the sky. (NASA)

A few months later, similar pulses were discovered coming from a source designated Hercules X-1, which had a period of 1.24 s. Because the periods of these two X-ray sources are so short, astronomers began to suspect that they had found some rapidly rotating neutron stars.

It soon became clear, however, that these systems are not ordinary pulsars like the Crab or Vela pulsars. For instance, Centaurus X-3 turns on and off periodically. Every 2.087 days, it turns off for almost 12 hours. Apparently Centaurus X-3 is an eclipsing binary, and it takes nearly 12 hours for the X-ray source to pass behind its companion star.

The binary nature of Hercules X-1 is even more compelling. It too has an off state, corresponding to a 6-hour eclipse every 1.7 days. Moreover, careful timing of its X-ray pulses shows a periodic Doppler shifting every 1.7 days, which is direct evidence of orbital motion about a companion star: When the X-ray source is approaching us, its pulses are separated by slightly less than 1.24 seconds, and when the source is receding from us, slightly more than 1.24 seconds elapse between the pulses.

Careful optical searches of the location of Hercules X-1 soon led astronomers to a variable dim star that had already been catalogued as HZ Herculis. The apparent magnitude of this star varies between 13 and 15, with a period of 1.7 days. Because this period is exactly the same as the orbital period of the X-ray source, astronomers concluded that HZ Herculis is the companion star around which Hercules X-1 orbits.

Putting all the pieces together, astronomers now realize that pulsating X-ray sources, like Centaurus X-3 and Hercules X-1, are double stars in which one of the stars is a neutron star. All these double stars have very short orbital periods, which means that the distance between the two stars is quite small. The neutron star can therefore capture gases escaping from its ordinary companion.

Pulsating X-ray sources are close binaries consisting of a neutron star and an ordinary giant star that fills (or nearly fills) its Roche lobe. In Hercules X-1, the ordinary star fills its Roche lobe to overflowing, thereby transferring matter across the inner Lagrangian point onto the neutron star. Although the ordinary star in Centaurus X-3 does not quite fill its Roche lobe, mass transfer still occurs, because stellar winds propel the star's outer layers over the critical surface (Figure 23-13). A typical rate of mass transfer from the ordinary star to the neutron star is roughly 10^{-9} solar masses per year.

Because of its strong gravity, a neutron star in a pulsating X-ray source easily captures much of the gas escaping from its companion. But like an ordinary pulsar, the neutron star is rotating rapidly and has a powerful magnetic field inclined to the axis of rotation (recall Figure 23-3). As the gas falls toward the neutron star, its magnetic field funnels the incoming matter down onto the star's north and south magnetic polar regions. The neutron star's gravity is so strong that the gas is traveling at nearly half the speed of light by the time it

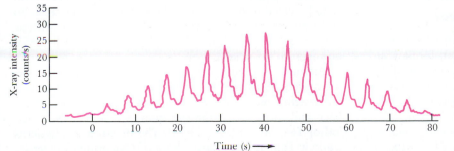

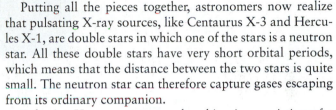

FIGURE 23-12 **X-Ray Pulses from Centaurus X-3** This graph shows the intensity of X rays detected by *Uhuru* as Centaurus X-3 moved across the satellite's field of view. Successive pulses are separated by 4.84 s. The gradual variation in the height of the pulses from left to right is the result of the changing orientation of *Uhuru*'s X-ray detectors toward the source as the satellite rotated. (Adapted from R. Giacconi and colleagues)

FIGURE 23-13 A Model of a Pulsating X-Ray Source Gas escaping from the ordinary star is captured by the neutron star. The in-falling gas is funneled down onto the neutron star's magnetic poles. It strikes the star with enough energy to create two X-ray–emitting hot spots. As the neutron star spins, beams of X rays from the hot spots sweep around the sky.

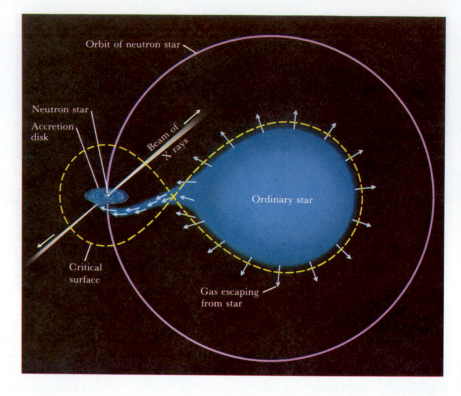

crashes onto the star's surface. This violent impact creates hot spots at both poles having temperatures of about 10^8 K; these hot spots therefore emit abundant X rays. In fact, their X-ray luminosity is roughly 10^{31} watts, nearly 100,000 times brighter than the Sun.

As the neutron star rotates, the beams of X rays from its polar caps sweep around the sky. If the Earth happens to be in the path of one of the two beams, we can observe a pulsating X-ray source. The neutron star's pulse period is thus equal to its rotation period. For example, the neutron star in Hercules X-1 spins at the rate of once every 1.24 seconds. The X rays from the neutron star can heat the exposed hemisphere of the companion star. This side of the companion star therefore becomes hot and bright, whereas the other side is cooler and dimmer. We can now understand why the brightness of HZ Herculis varies periodically: As the system rotates, the two hemispheres are alternately exposed to our view.

23-7 High-speed jets of matter can be ejected from an accreting neutron star

In a binary system, gases captured by the gravity of a neutron star may go into orbit about the neutron star, forming an accretion disk, as shown in Figure 23-13. Accretion disks have been detected in many close binaries where mass transfer is occurring between the two stars (recall Figure 21-13b).

With ordinary pulsating X-ray sources, like Hercules X-1 and Centaurus X-3, the gas is falling onto the neutron star at a rate low enough to allow the resulting X rays to escape. If

the companion star is dumping vast amounts of material onto the neutron star, however, the resulting energy cannot escape easily. Instead, tremendous pressures build up as newly arrived gases crowd down onto the neutron star. These pressures can be relieved only if gas is ejected. Ejection cannot occur in the plane of the accretion disk, where more gas is continually spiraling in toward the neutron star. Matter can more easily escape perpendicularly to the accretion disk, along its rotation axis. The result can be two powerful beams of high-velocity hot gases.

Such jets of gas apparently explain the weird behavior of a star called SS433. (SS433 is so named because it is the 433rd star on a list of similar objects published in 1977 by C. Bruce Stephenson and Nicholas Sanduleak of Case Western Reserve in Ohio.) In the autumn of 1978, Bruce Margon and some of his colleagues at UCLA began taking a series of spectrograms of the star SS433, which has been noted for its strong emission lines. To everyone's surprise, the spectrum of SS433 contained several complete sets of spectral lines. One set was greatly redshifted away from its usual wavelengths and another set was comparably blueshifted. Somehow, SS433 seemed to be coming and going at the same time. To make matters even more puzzling, the wavelengths of these redshifted and blueshifted lines change dramatically from one night to the next.

Astronomers had never seen anything like this, and soon many were observing SS433. By mid-1979, it was clear that the system's redshifted and blueshifted lines actually move back and forth across the spectrum of SS433, with a period of 164 days (Figure 23-14). The British astrophysicists Andrew Fabian and Martin Rees of Cambridge University

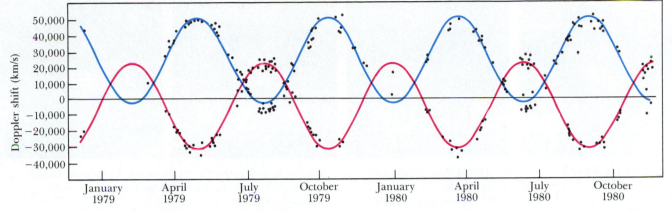

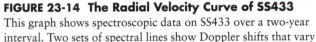

FIGURE 23-14 The Radial Velocity Curve of SS433
This graph shows spectroscopic data on SS433 over a two-year interval. Two sets of spectral lines show Doppler shifts that vary with a period of 164 days. These spectral lines are probably caused by material ejected along jets that point alternately toward and away from the Earth. (Adapted from B. Margon)

were quick to point out that the two sets of spectral lines could be caused by two oppositely directed jets of gas, one angled toward us, the other angled away. Their model is sketched in Figure 23-15. Furthermore, the 164-day variation could be explained by a precession, or "wobble," of the accretion disk and its two jets. As the two jets circle about the sky every 164 days, we see a periodic variation in the Doppler shift.

To account for the large redshifts and blueshifts Margon discovered, the gas in the two oppositely directed jets must have a speed of 78,000 km/s, or roughly one-quarter the speed of light. In addition, the accretion disk must be tilted with respect to the orbital plane of the two stars in the binary system. Just as the tilt of the Earth's axis with respect to the plane of the ecliptic causes the Earth's rotation axis to precess, the changing tilt of the accretion disk results in the 164-day precession of the two jets.

Figure 23-16 is a high-resolution radio view of SS433. Note the two oppositely directed appendages emerging from the central source. As we shall see in Chapter 28, many peculiar galaxies and faraway objects called quasars have similar radio structures, though on a much larger scale. Quasars are so incredibly remote that they are difficult to study. The real significance of SS433 may be that it gives us a miniature quasarlike object right in our own celestial backyard.

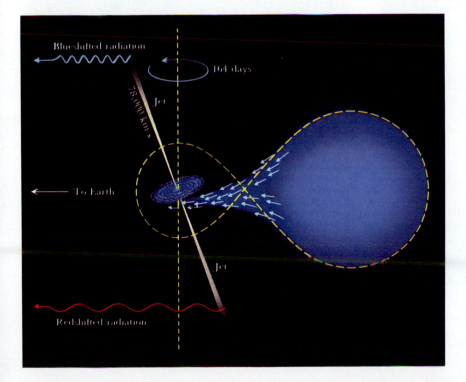

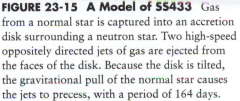

FIGURE 23-15 A Model of SS433 Gas from a normal star is captured into an accretion disk surrounding a neutron star. Two high-speed oppositely directed jets of gas are ejected from the faces of the disk. Because the disk is tilted, the gravitational pull of the normal star causes the jets to precess, with a period of 164 days.

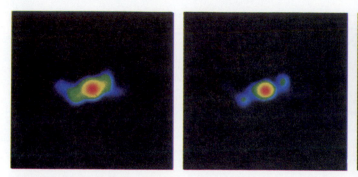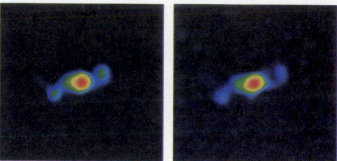

FIGURE 23-16 SS433 These four views of SS433, made with the Very Large Array in early 1981, show jets extending out to one-sixth light-year on either side of SS433. Three-quarters of the radio emission comes from SS433 itself (the red central blob).

SS433 is located at the center of a supernova remnant roughly 16,000 ly (5000 pc) from Earth in the constellation of Aquila (the Eagle). (VLA)

23-8 Explosive thermonuclear processes on white dwarfs and neutron stars produce novae and bursters

Occasionally a star suddenly brightens by a factor of 10^6. This phenomenon is called a **nova** (not to be confused with a supernova, which involves a 10^8 increase in brightness). Novae are fairly common. Their abrupt rise in brightness is followed by a gradual decline that may stretch out over several months or more (Figures 23-17 and 23-18).

Painstaking observations of numerous novae by Robert Kraft, Merle Walker, and their colleagues at Lick Observa-

tory strongly suggest that all novae are members of close binary systems containing a white dwarf. Gradual mass transfer from the ordinary companion star (which presumably fills its Roche lobe) deposits fresh hydrogen onto the white dwarf. Because of the white dwarf's strong gravity, this hydrogen is compressed into a dense layer covering the hot surface of the white dwarf. As more gas is deposited and compressed, the temperature in the hydrogen layer increases. Finally, when the temperature reaches about 10^7 K, hydrogen burning ignites throughout the gas layer, embroiling the white dwarf's surface in a thermonuclear holocaust that we see as a nova.

a

b

FIGURE 23-17 Nova Herculis 1934 These two pictures show a nova (**a**) shortly after peak brightness as a magnitude +3 star, and (**b**) two months later, when it had faded to magnitude +12. Novae are named after the constellation and year in which they appear. (Lick Observatory)

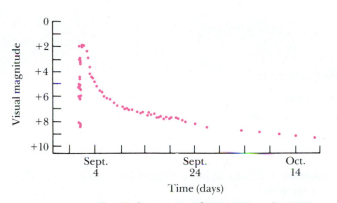

FIGURE 23-18 The Light Curve of Nova Cygni 1975
This graph shows the history of a nova that blazed forth in the constellation of Cygnus in September 1975. Its rapid rise and gradual decline in magnitude are characteristic of all novae. This nova, officially designated V1500 Cyg, was easily visible to the naked eye for nearly a week.

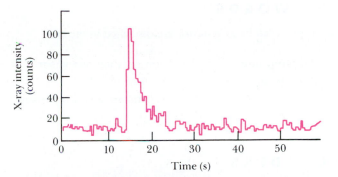

FIGURE 23-19 X Rays from a Burster A burster emits a constant low intensity of X rays interspersed with occasional powerful bursts of X rays. This particular burst is typical. It was recorded on September 28, 1975, by an Earth-orbiting X-ray telescope pointed toward the globular cluster NGC 6624. About one-third of all known bursters are located in globular clusters. (Adapted from W. Lewin)

A similar phenomenon occurs with neutron stars. Beginning in late 1975, astronomers analyzing data from X-ray satellites realized that their instruments had detected sudden, powerful bursts of X rays from objects in the sky. The record of a typical burst is shown in Figure 23-19. The source emits X rays at a constant low level until suddenly, without warning, there is an abrupt increase in X rays, followed by a gradual decline. An entire burst typically lasts for only 20 s. Sources that behave in this fashion are known as **bursters.** Several dozen bursters have been discovered, most being located toward the center of our Galaxy.

Bursters, like novae, are believed to involve close binaries whose stars are engaged in mass transfer. With a burster, however, the stellar corpse is a neutron star rather than a white dwarf. Gases escaping from the ordinary companion star fall onto the neutron star. The burster's magnetic field is probably not strong enough to funnel the falling material toward the magnetic poles, so the gases are distributed more evenly over the surface of the neutron star. The energy released as these gases crash down onto the neutron star's surface produces the low-level X rays that are continuously emitted by the burster.

Most of the gas falling onto the neutron star is hydrogen, which the star's powerful gravity compresses against its hot surface. In fact, temperatures and pressures in this accreting layer become so high that the arriving hydrogen is promptly converted into helium by hydrogen burning. This constant hydrogen burning soon produces a layer of helium that covers the entire neutron star.

Finally, when the helium layer is about 1 m thick, helium burning ignites explosively and we observe a sudden burst of X rays. In other words, whereas explosive hydrogen burning on a white dwarf produces a nova, explosive helium burning on a neutron star produces a burster. In both cases, the process is explosive, because the fuel is compressed so tightly

against the star's surface that it becomes degenerate, like the star itself. As we saw with the helium flash inside red giants, the ignition of a degenerate thermonuclear fuel involves a sudden thermal runaway because an increase in temperature does not produce a corresponding increase in pressure that would otherwise relieve compression of the gases and slow the nuclear reactions.

Just as there is an upper limit to the mass of a white dwarf, there is an upper limit to the mass of a neutron star. Above this limit, degenerate neutron pressure cannot support the overpowering weight of the star's matter pressing inward from all sides. The Chandrasekhar limit for a white dwarf is 1.4 $M_\odot$. The corresponding upper limit for a neutron star is probably 2.5 to 3 $M_\odot$.

Before pulsars were discovered, most astronomers believed all dead stars to be white dwarfs. Dying stars were thought to eject enough material somehow so that their corpses could be below the Chandrasekhar limit. Obviously, this idea proved incorrect. Inspired by this lesson, astronomers soon began wondering what might happen if a dying massive star failed to eject enough matter to get below the upper limit for a neutron star. For example, what might a 5-$M_\odot$ stellar corpse be like?

The gravity associated with a neutron star is so strong that the escape velocity from it is roughly one-half the speed of light. With a stellar corpse greater than 3 $M_\odot$, there is so much matter crushed into such a small volume that the escape velocity actually exceeds the speed of light. Because nothing can travel faster than light, nothing—not even light—can leave this dead star. It therefore disappears from the universe, its powerful gravity leaving a hole in the fabric of space and time. In this way, the discovery of neutron stars inspired astrophysicists to examine seriously one of the most bizarre and fantastic objects ever predicted by modern science, the black hole.

KEY WORDS

Terms preceded by an asterisk are discussed in the box.

*angular momentum

*angular velocity

burster

*conservation of angular
  momentum

degenerate neutron pressure

glitch

millisecond pulsar

neutron star

nova (*plural* novae)

pulsar

starquake

superconductivity

KEY IDEAS

• A neutron star is a dense stellar corpse consisting of closely packed degenerate neutrons.

A neutron star typically has a diameter of about 30 km, a mass less than 3 $M_\odot$, a magnetic field a trillion times stronger than that of the Sun, and a rotation period of roughly 1 second.

A neutron star consists of a superfluid, superconducting core surrounded by a superfluid mantle and a thin, brittle crust.

Energy pours out of the north and south polar regions of a neutron star in intense beams produced by streams of charged particles moving in the star's intense magnetic field.

• A pulsar is a source of periodic pulses of radio radiation. These pulses are produced as beams of radio waves from a neutron star's polar regions sweep past the Earth.

The steady slowing of a pulsar's pulse rate presumably reflects the gradual slowing of the rotation rate of the neutron star; glitches (sudden speedups of the pulse rate) correspond to starquakes.

• Mass transfer from an ordinary star to a neutron star in a close binary system can significantly speed up the neutron star's rotation rate, giving rise to a millisecond pulsar.

• Some X-ray sources exhibit regular pulses as well as the effects of orbital motion; these objects are neutron stars in close binary systems with ordinary stars.

Gases from the ordinary star in a close binary system fall onto the dense neutron star at great speeds, creating hot spots near its poles; these hot spots then radiate intense beams of X rays.

• Gases falling onto the neutron star can form an accretion disk around the neutron star; beams of high-velocity gases may shoot out from opposite faces of the accretion disk.

• Material from the ordinary star in a close binary can fall onto the surface of the companion white dwarf or neutron star to produce a surface layer in which thermonuclear reactions can explosively ignite.

Explosive hydrogen burning may occur in the surface layer of a companion white dwarf, producing the sudden increase in luminosity that we call a nova. The brightness increase in a nova is only a hundredth of that observed in a supernova.

Explosive helium burning may occur in the surface layer of a companion neutron star, producing the sudden increase in X-ray radiation that we call a burster.

REVIEW QUESTIONS

1. What is a neutron star?

2. Why do astronomers believe that pulsars are rapidly rotating neutron stars?

3. During the weeks immediately following the discovery of the first pulsar, one suggested explanation was that the pulses might be signals from an extraterrestrial civilization. Why did astronomers soon discard this idea?

4. Compare a white dwarf and a neutron star. Which of these two types of stellar corpse is more common? Explain.

5. How does the law of the conservation of angular momentum help explain why neutron stars rotate so much more rapidly than ordinary stars?

6. What is the difference between superconductivity and superfluidity?

7. How does a starquake affect a pulsar's period?

8. How do you think astronomers have deduced that the Vela pulsar is about 11,000 years old?

9. Why do astronomers think that millisecond pulsars are very old?

10. Describe a pulsating X-ray source like Hercules X-1 or Centaurus X-3. What produces the pulsation?

11. What is the difference between a nova and a Type Ia supernova?

12. What is SS433? In what ways does it resemble X-ray pulsars like Hercules X-1? In what ways is it unlike an X-ray pulsar?

ADVANCED QUESTIONS

Tips and tools . . .

Recall that the Doppler effect was introduced in Chapter 5 (see Figure 5-20). The volume of a sphere of radius r is $\frac{4}{3}\pi r^3$. Recall that the small-angle formula was given in Box 1-1. In

Chapter 19 we saw that the tangential velocity (v_t) of an object is related to its proper motion (μ) by $v_t = 4.74\mu d$, where the tangential velocity is measured in kilometers per second, the proper motion is in arc seconds per year, and d is the object's distance in parsecs.

13. The distance to the Crab Nebula is about 2000 pc. When did the star actually explode?

14. Why do you suppose that most of the angular momentum contained in the solar system resides with the planets rather than with the far more massive Sun?

15. How do we know that the Crab pulsar is really embedded in the Crab Nebula and not simply located at a different distance along the same line of sight?

16. To determine accurately the period of a pulsar, astronomers must take into account the Earth's orbital motion about the Sun. Explain why. Knowing that the Earth's orbital velocity is 30 km/s, calculate the maximum correction to a pulsar's period because of the Earth's motion. Explain why the size of the correction is greatest for pulsars located near the ecliptic.

17. The mass of a neutron is about 1.7×10^{-27} kg and its radius is about 10^{-15} m. Compare the density of matter in a neutron with the density of a neutron star.

18. Propose an explanation for the fact that X-ray pulsars are speeding up but ordinary (radio) pulsars are slowing down.

19. From the data given in the caption to Figure 23-2, calculate the rate of expansion of the Crab Nebula. Assuming that your telescope can distinguish features as small as 1 arc sec, how long would you have to wait to see a change in the size of the Crab Nebula?

20. The apparent expansion rate of the Crab Nebula is 0.23 arc sec per year, and emission lines in its spectra exhibit a Doppler shift indicating an expansion velocity of about 1200 km/s. The apparent size of the Crab Nebula is about 4 by 6 arc min. From these data, calculate the distance to the Crab Nebula and estimate the date on which it exploded. Do your answers agree with figures quoted in this chapter? If not, can you point to assumptions you made in your computations that lead to the discrepancies? Or do you think your calculations suggest additional physical effects are at work in the Crab Nebula, over and above a constant rate of expansion?

21. Consult recent issues of such magazines as *Sky & Telescope* and *Science News* for information about the latest observations of the stellar remnant at the center of SN 1987A. Has a pulsar been detected? Has the supernova's debris thinned out enough to give a clear view of the neutron star?

22. If the model for Hercules X-1 discussed in the text is correct, at what orientation of the binary system do we see its maximum optical brightness? Explain your answer.

DISCUSSION QUESTIONS

23. Compare novae and bursters. What do they have in common? In what ways are they different?

24. How might astronomers be able to detect the presence of an accretion disk in a close binary system?

OBSERVING PROJECTS

25. If you did not take the opportunity to observe the Crab Nebula as part of Chapter 22's exercises, do so now. The Crab Nebula is visible from October through March. Its epoch 2000 coordinates are R.A. = 5^h 34.5^m and Decl. = $22°00'$, which is near the star marking the eastern horn of Taurus (the Bull). Be sure to schedule your observations for a moonless night. The larger the telescope you use, the better, because the Crab Nebula is quite dim.

26. Consult such publications as the current issue of *Sky & Telescope* or recent International Astronomical Union (IAU) *Circulars* to see if any novae or supernovae have been sighted recently. If by good fortune one has been sighted, what is its magnitude? Is it within reach of a telescope at your disposal? If so, arrange to observe it. Draw what you see through the eyepiece, noting the object's brightness in comparison with other stars in the field of view. If possible, observe the same object a few weeks or months later to see how its brightness has changed.

FOR FURTHER READING

Clark, D. *The Quest for SS433*. Dutton, 1985. This excellent book describes the discovery and developing understanding of SS433 in the format of a detective story.

Feinberg, R. T. "Pulsars, Planets, and Pathos." *Sky & Telescope*, May 1992. The article discusses the controversial discovery of a planet orbiting a millisecond pulsar.

Lewin, W. "The Sources of Celestial X-ray Bursts." *Scientific American*, May 1981. This informative article describes the ways in which matter falling on neutron stars can produce powerful bursts of X rays.

Margon, B. "The Bizarre Spectrum of SS433." *Scientific American*, October 1980. The astronomer who discovered the unusual properties of SS433 discusses the star's spectrum and its interpretation.

Seward, F. "Neutron Stars in Supernova Remnants." *Sky & Telescope*, January 1986. The author of this well-written article proposed that differences in the structure of supernova remnants may result from differences in the way neutron stars form.

Shaham, J. "The Oldest Pulsars in the Universe." *Scientific American*, February 1987. This article describes the evolution of millisecond pulsars and concludes that they are extremely old neutron stars.

BLACK HOLES

A BLACK HOLE IN A DOUBLE STAR SYSTEM This artist's rendition shows the close binary system that contains Cygnus X-1. Cygnus X-1 is a strong source of X rays and is widely believed by astronomers to be a black hole. Gas from the companion star is captured into orbit about the black hole, forming an accretion disk. As gases spiral in toward the black hole, friction heats them to high temperatures. At the inner edge of the accretion disk, the gases are so hot that they emit X rays. (Courtesy of D. Norton, Science Graphics)

A DYING high-mass star can give rise to a stellar corpse too massive to be supported by degenerate pressure, and so it is doomed to collapse to a single point of infinite density. Such an object, called a black hole, is predicted by Einstein's general theory of relativity. A black hole is strange but simple. It is a place of inconceivably intense gravity from which nothing—not even light—can escape. Matter that falls into a black hole literally disappears forever from the universe. Nevertheless, a black hole is an uncomplicated object because its structure is completely specified by only three quantities: its mass, electric charge, and angular momentum. In recent years astronomers have found evidence that certain binary systems may contain black holes. Each of these binaries is a powerful source of X rays presumably produced by hot gases in an accretion disk surrounding the black hole.

As we saw in the previous chapter, a neutron star is an astonishing compact stellar corpse. A typical neutron star consists of roughly a solar mass compressed to nuclear density within a sphere barely 30 km in diameter. The dead star's gravity is so strong that the speed needed to escape from its surface equals half the speed of light.

Suppose, however, that the mass of a dying star's burned-out core exceeds 3 solar masses (3 $M_\odot$). This mass is well above the Chandrasekhar limit, so degenerate electron pressure cannot support the resulting stellar corpse. The star therefore cannot become a white dwarf. But because 3 $M_\odot$ is also above the mass limit for neutron stars, degenerate neutron pressure cannot support the star's crushing weight either. Instead, burned-out matter presses inexorably inward toward the dead star's center, compressing its matter to densities even greater than nuclear density. If this massive stellar corpse can be neither a white dwarf nor a neutron star, what might it become?

It does not take very much further compression to cause the escape speed from this stellar corpse to exceed the speed of light. Because nothing can travel faster than the speed of light, nothing—not even light—can manage to escape from the dead star: It becomes a **black hole**. Once 3 $M_\odot$ of matter is squeezed into a sphere 18 km in diameter, it literally disappears from the observable universe. Yet some of its effects can still be detected, because its gravity alters the very fabric of space and time.

24-1 The general theory of relativity describes gravity in terms of the geometry of space and time

To appreciate fully the nature of a massive stellar corpse, we must use the best theory of gravity at our disposal. The gravitational field around one of these massive dead stars is so

strong that Isaac Newton's theory of gravity is not valid. Instead, we must turn to Albert Einstein's general theory of relativity.

According to the classical physics of Newton, space is perfectly uniform and fills the universe like a rigid framework. Similarly, time passes at a monotonous, unchanging rate. It is always possible to know exactly how fast you are moving through this rigid fabric of space and time, and the results of your observations depend on your state of motion.

Albert Einstein began a revolution in physics with his **special theory of relativity** in 1905. Einstein was guided in his thinking by one lofty idea: that neither our location in space and time nor our motion through space and time shall prejudice our description of physical reality. A surprising fact emerged when Einstein applied this idea to his studies of electricity and magnetism: Everyone who measures the speed of light gets the same answer (3×10^8 m/s), regardless of the person's state of motion. This conflicts with the Newtonian view that a stationary person and a moving person should measure different speeds. Einstein's efforts to incorporate light's constant speed into physics gave us a new understanding of the nature of space and time.

The basic equations of the special theory of relativity logically follow from the fact that the speed of light is the same to all observers (see Box 24-1). These equations relate measurements by observers moving at different speeds and ensure, for instance, that both you on Earth and a friend in a rocketship traveling near the speed of light have the same complete description of electricity and magnetism, unaffected by any pitfalls or paradoxes caused by your relative motion.

In developing the special theory of relativity, Einstein found that he had to abandon old-fashioned, rigid notions of space and time. For example, imagine a friend whizzing across our solar system in a rocketship while you remain here on Earth. Einstein proved that your friend's clocks will seem to tick more slowly than your own. In addition, your friend's rulers held parallel to the direction of motion will seem shorter than yours. In brief, moving clocks are slowed and moving rulers are shortened in the direction of motion. These details of the special theory of relativity, which are discussed in Box 24-1, are a direct consequence of the speed of light being an absolute constant.

After developing the special theory of relativity, Einstein turned his attention to gravity. He began by demonstrating that it is not necessary to think of gravity as a force. According to Newton's theory, an apple falls to the floor because the force of gravity pulls the apple down. Einstein pointed out that the apple would appear to behave in exactly the same way in space, far from Earth's gravity, if the floor were to accelerate upward. In other words, the floor comes up to meet the apple.

In Figure 24-1, the two famous gentlemen watching an apple fall toward the floor of their closed compartments have no way of telling who is at rest on the Earth and who is in the hypothetical elevator moving upward at a constantly increasing speed. This is an example of Einstein's **principle of equivalence:** In a small volume of space, the downward pull

---

BOX 24-1

Some Comments on Special Relativity

The special theory of relativity describes how motion affects measurements of time, distance, and mass. Einstein proved that these measurements must depend on the speed of the observer in order for all people, whether moving or stationary, to agree on certain basic physical phenomena, especially those involving the behavior of light.

Imagine that you are standing on the Earth while a friend is traveling across our solar system at a high speed, as shown in the accompanying sketch. You set off a flashbulb that emits a sudden bright flash of light. The radiation moves away from you equally in all directions, and thus you see an expanding spherical shell of light.

What does your high-speed friend see? Einstein argued that this person must also see an expanding spherical shell of light. She does not see, for example, an expanding football-shaped shell.

By requiring that both people observe a spherical shell, Einstein derived a series of equations to relate specific measurements of time and distance between two people. These equations are named the **Lorentz transformations,** after the famous Dutch physicist Hendrik Antoon Lorentz, a contemporary of Einstein's, who developed these equations independently. These equations tell us exactly how a moving person's clocks slow down and how rulers shrink.

To appreciate the Lorentz transformations, again imagine that you are on the Earth while a friend is moving at a speed v with respect to you. Suppose that you both observe the same phenomenon on Earth, which appears to occur over an interval of time. According to your (stationary) clock, the phenomenon lasts for T_0 seconds; according to your friend's (moving) clock, the same phenomenon lasts for T seconds. The Lorentz transformation for time tells us that these two time intervals are related by

$$T = \frac{T_0}{\sqrt{1 - (v/c)^2}}$$

where c is the speed of light.

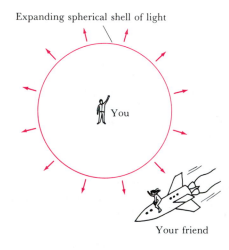

Expanding spherical shell of light

You

Your friend

EXAMPLE: Suppose that your friend is moving at 98% of the speed of light. Then

$$\frac{v}{c} = 0.98$$

so that

$$T = \frac{T_0}{\sqrt{1 - (0.98)^2}} \approx 5T_0$$

Thus, a phenomenon that lasts for 1 second on a stationary clock is stretched out to 5 seconds on a clock moving at 98% of the speed of light. This phenomenon is often called the **dilation of time.**

The Lorentz transformation for time is plotted in the accompanying graph. This graph shows how 1 s measured on a stationary clock is stretched out when measured by a moving clock. Note that significant differences between the recordings of the stationary and moving clocks occur only at speeds near the speed of light. For speeds less than about half the speed of light, the mathematical factor of

of gravity can be accurately and completely duplicated by an upward acceleration of the observer.

The principle of equivalence allowed Einstein to focus entirely on motion, rather than force, in discussing gravity. From his special theory of relativity he already knew how rulers and clocks are affected by motion, and he used this information to understand the effects of gravity. In other words, he described gravity entirely in terms of its effects on space and time. Far from a source of gravity the acceleration is small, so the effect on clocks and rulers is small. Likewise, nearer a source of gravity the acceleration is larger, so the

distortion of clocks and rulers is larger. In this way, Einstein "generalized" his special theory to arrive at his **general theory of relativity.**

The general theory of relativity describes gravity entirely in terms of the geometry of space and time. Far from a source of gravity, like a planet or a star, space is "flat" and clocks tick at their normal rate. Closer to a source of gravity, however, clocks slow down and space is curved.

Where gravity is weak, Einstein's general theory of relativity gives the same results as the classical theory of Newton. In stronger gravity, however, such as near the Sun's surface, the its mass M, would be given by

$$M = \frac{M_0}{\sqrt{1 - (v/c)^2}}$$

Thus, a 1-kg brick traveling at 98% of the speed of light would behave as though it had a mass of 5 kg, according to a stationary observer.

The mathematical expression $\sqrt{1 - (v/c)^2}$ becomes significantly different from 1 only when v is nearly as large as c. This explains why the effects of relativity on time, distance, and mass become noticeable only at extremely high speeds. Particle accelerators like cyclotrons and synchrotrons that propel electrons and protons to velocities near the speed of light have tested these predictions of special relativity to a high degree of accuracy. For instance, when a high-speed particle smashes into a stationary target, the resulting debris is scattered in a way that would be consistent with the fact that the impacting particle had a mass higher than its proper mass by precisely the amount determined by the above equation.

Finally, special relativity explains why it is impossible to travel faster than the speed of light. Suppose that a friend climbs aboard a rocketship containing an infinite supply of fuel. With rocket engines blazing, your friend attempts to "break the light barrier." You keep in touch by radio and, as she gets closer and closer to the speed of light, you begin to notice the effects of the dilation of time. For instance, her speech becomes drawn out to a leisurely drawl as her clock slows down relative to yours. In the same way, the rocket engines appear to you to be shutting down. The gallons-per-minute rate at which fuel pours into the rocket engines is directly associated with the passage of time and is therefore subject to time dilation. As the rocket's velocity approaches the speed of light, the factor $\sqrt{1 - (v/c)^2}$ approaches zero. Thus your friend never gets the chance to burn that last drop of fuel that would put her past the speed of light.

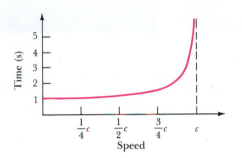

$\sqrt{1 - (v/c)^2}$ is almost exactly equal to 1, so stationary and slowly moving clocks tick at almost exactly the same rate.

In the language of relativity, we say that a clock at rest measures **proper time** (T_0) and a ruler at rest measures **proper distance** (L_0). According to the Lorentz transformations, distances perpendicular to the direction of motion are unaffected. However, a ruler of proper length L_0 held parallel to the direction of motion shrinks to a length L, given by

$$L = L_0 \sqrt{1 - (v/c)^2}$$

EXAMPLE: If your friend is traveling at 98% of the speed of light relative to you, you have concluded that her clocks are ticking only one-fifth as fast as yours. You will also conclude that her 1-foot ruler is only about 2.4 inches long when held parallel to the direction of motion:

$$L = 12 \sqrt{1 - (0.98)^2} = 2.4$$

This shrinkage of length is often called **Fitzgerald–Lorentz contraction.**

Albert Einstein also demonstrated that measurements of mass are affected by the relative velocity of the observer. Specifically, suppose that an object has a **proper mass** M_0 when at rest. If this same object is moving with a velocity v,

FIGURE 24-1 The Equivalence Principle
The equivalence principle asserts that you cannot distinguish between being at rest in a gravitational field and being accelerated upward in a gravity-free environment. This idea was an important step in Einstein's quest to develop the general theory of relativity.

two theories give different predictions. Chapter 4 already discussed two examples, the precession of Mercury's perihelion and the deflection of a light ray grazing the Sun (review Figures 4-21 and 4-24). General relativity has withstood these and many other tests. It is by far the most elegant—and accurate—description of gravity yet devised.

24-2   A black hole is a simple object that has only a "center" and a "surface"

Imagine a dying star too massive to become either a white dwarf or a neutron star. The overpowering weight of the star's burned-out matter pressing inward from all sides causes the star to contract rapidly. As the star's matter becomes compressed to enormous densities, the strength of gravity at the surface of this rapidly shrinking sphere also increases dramatically. According to the general theory of relativity, distortions of space and time around the star be-

come increasingly pronounced. Finally, the escape velocity from the dying star's surface equals the speed of light, and thus the star disappears. At this stage, space becomes so severely curved that a hole is punched in the fabric of the universe. The dying star disappears into this cavity, leaving behind only a black hole.

The geometry of space around a black hole is sketched in Figure 24-2. Note that space far from the hole is flat, because gravity is weak there. Near the hole, however, gravity is strong and the curvature of space is severe.

Surrounding a black hole, where the escape speed from the hole just equals the speed of light, is the **event horizon**. This sphere is also sometimes thought of as the "surface" of the black hole. Once a massive dying star collapses to within its event horizon, it disappears permanently from the universe. The term *event horizon* is in fact quite appropriate, because this surface it is like a horizon beyond which we cannot see any events.

In addition to making space curve, gravity slows time. In fact, as seen by an outside observer, time seems to stop entirely at the event horizon. If you stood at a safe distance and watched a friend fall toward a black hole, you would note that her clocks begin to tick more and more slowly. From your point of view, she never has enough time to reach the event horizon, where her clocks would stop ticking. However, your friend would not notice this slowing of time as she glances at her wristwatch. From her point of view, she continues her fall through the event horizon into the black hole.

Once a dying star has contracted inside its event horizon, no known forces in the universe can prevent the complete collapse of the star down to a single point. The star's entire mass is crushed to infinite density at this point, known as the **singularity**, at the center of the black hole.

We now can see that the structure of a nonrotating black hole is quite simple. As sketched in Figure 24-3, it has only two parts: a singularity, or center, and the event horizon, or

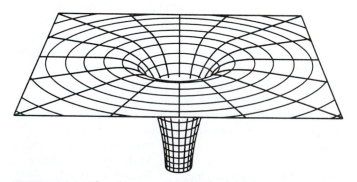

FIGURE 24-2 The Geometry of a Black Hole This diagram shows how the shape of space is distorted by the gravitational field of a black hole. Far from the hole, gravity is weak and space is thus "flat." Near the hole, gravity is strong and space is highly curved.

surface, that surrounds it. The distance between the singularity and the event horizon is called the **Schwarzschild radius** (abbreviated R_{Sch}), after the German astronomer Karl Schwarzschild, who in 1916 was the first to solve Einstein's equations of general relativity. The Schwarzschild radius is related to the mass M of the black hole by

$$R_{Sch} = \frac{GM}{c^2}$$

where c is the speed of light and G the universal constant of gravitation. For example, for a 10-$M_\odot$ black hole, the Schwarzschild radius is 30 km.

To understand why the complete collapse of such a doomed star is inevitable, think about your own life here on Earth, far from any black holes. You have the freedom to move as you wish through the three dimensions of space: up and down, left and right, or forward and back. But you do not have the freedom to move at will through the dimension of time. Whether we like it or not, we are all carried inexorably from the cradle to the grave.

Inside a black hole, a powerful gravity distorts the structure of space and time so severely that the directions of space and time become interchanged. In a sense, inside a black hole you acquire a limited ability to affect the passage of time. This seeming gain does you no good, however, because you lose a corresponding amount of freedom to move through space. Whether you like it or not, you will be dragged inexorably from the event horizon toward the singularity. Just as no force in the universe can prevent the forward march of time from past to future outside a black hole, no force in the universe can prevent the inward march of space from event horizon to singularity inside a black hole.

At a black hole's singularity, the strength of gravity is infinite, so the curvature of space and time there is infinite. Space and time at the singularity are thus all jumbled up. They do not exist as separate, distinctive entities.

This confusion of space and time has profound implications for what goes on inside a black hole. All the laws of physics require a clear, distinct background of space and time. Without this identifiable background, we can not speak rationally about the arrangement of objects in space or the ordering of events in time. Because space and time are all jumbled up at the center of a black hole, the singularity there does not obey the laws of physics. The singularity behaves in a random, capricious fashion, totally devoid of rhyme or reason.

Fortunately, we are shielded from the singularity by the event horizon. In other words, although random things do happen at the singularity, none of their effects manage to escape beyond the event horizon. Consequently, the outside universe remains understandable and predictable.

The chaotic, random behavior of the singularity has been so disturbing to physicists that in 1969 the British mathematician Roger Penrose and his colleagues proposed the **law of cosmic censorship:** "Thou shalt not have naked singularities." In other words, every singularity must be completely surrounded by an event horizon, because an exposed singularity could affect the universe in an unpredictable and random way.

Incidentally, science fiction abounds with nasty rumors that black holes are evil things that go around gobbling up everything in the universe. Not so! The bizarre effects created by highly warped space and time are limited to a region quite near the hole. For example, relativistic effects predominate only within a thousand kilometers of a 10-$M_\odot$ black hole. Beyond a thousand kilometers, gravity is weak enough that Newtonian physics can adequately describe everything.

24-3   The structure of a black hole can be completely described with only three numbers

In addition to shielding us from singularities, the event horizon prevents us from ever knowing much about anything that falls into a black hole. For example, there is no way we could ever discover the chemical composition of a massive star whose collapse has produced a particular black hole. Even if someone were to go into a black hole and make a measurement or conduct a chemical test, there is no way the observer could get any of this information back to the outside world. A black hole is in fact an "information sink," because in-falling matter carries with it many properties, such as its chemical composition, texture, color, shape, and size, that are then forever removed from the universe.

Because this information has completely vanished, it cannot affect the structure or properties of the hole. For example, consider two hypothetical black holes—one made from

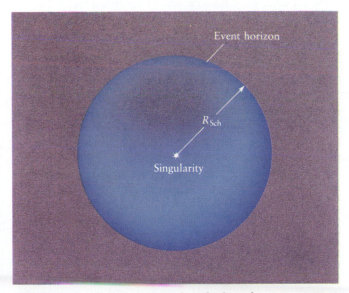

FIGURE 24-3 The Structure of a Black Hole A nonrotating black hole has only two parts: a singularity and a surrounding event horizon. The distance from the singularity to the event horizon is called the Schwarzschild radius (R_{Sch}). Inside the event horizon, the escape velocity exceeds the speed of light, so the event horizon is a one-way surface. Things can fall in, but nothing can get out.

the gravitational collapse of 10 $M_\odot$ of iron and the other made from the gravitational collapse of 10 $M_\odot$ of peanut butter. Obviously quite different substances went into the creation of the two holes. Once the event horizons of these two black holes have formed, however, both the iron and the peanut butter will have permanently disappeared from the universe. As seen from the outside, the two holes are absolutely identical, making it impossible for us to tell which ate the peanut butter and which ate the iron. A black hole is thus unaffected by the information it destroys.

Because a black hole is indeed an information sink, it is reasonable to wonder whether we can determine anything at all about a black hole. In other words, what properties characterize a black hole?

First, we can measure the *mass* of a black hole. One way to do this would be by placing a satellite into orbit about the hole. After measuring the size and period of the satellite's orbit, we could use Kepler's third law (recall Box 4-3) to determine the mass of the black hole. This mass is equal to the total mass of all the material that has gone into the hole.

Second, we can also measure the total *electric charge* possessed by a black hole. Like gravity, the electric force acts over long distances—it is a long-range interaction that is felt in the space around the hole. Appropriate equipment on a space probe passing near the hole could measure the intensity of the electric field, and the electric charge could thus be determined.

In actuality, we would not expect a black hole to possess any appreciable electric charge. For instance, if a hole did happen to start off with a sizable positive charge, it would vigorously attract vast numbers of negatively charged electrons from the interstellar medium, which would soon neutralize the hole's charge. For this reason, astronomers neglect electric charge when discussing real black holes.

BOX 24-2

Gravitational Radiation

A gravitational wave is a ripple in the overall geometry of space and time. As an example of how these ripples are produced, think of a man whose mass is 80 kg. All matter is a source of gravity and, according to general relativity, gravity curves space and slows down time. So the 80-kg man is surrounded by a slight warping of space and time commensurate with his mass.

Now suppose that this man begins waving his arms. Although his total mass does not change, how his mass is distributed does change. The geometry of space and time must adapt to these changes, because the gravitational field of the man with his hands over his head is slightly different from that of the man when he has his hands at his sides. These minor readjustments appear as tiny ripples in the overall geometry of space and time surrounding the man. In the same way, a bouncing ball, the Moon going around the Earth, or binary stars all produce gravitational waves. From the equations of general relativity, it is possible to prove that gravitational radiation moves outward from its source at the speed of light.

Gravitational waves are difficult to detect, because they carry very little energy. To appreciate how weak gravitational waves are, imagine two electrons separated by a short distance. Because they each possess mass and charge, these electrons exert both gravitational and electric forces on each other. The gravitational force is about 10^{42} times weaker than the electric force. If these two electrons are made to wiggle back and forth, they will radiate both gravitational and electromagnetic waves. Because gravity is so much weaker than electromagnetism, the resulting gravitational waves are subdued by a factor of 10^{-42} compared to the electromagnetic waves.

Processes involving dramatic changes in intense gravitational fields produce the strongest bursts of gravitational radiation. For example, the collapse of a massive star's core during a supernova explosion emits substantial gravitational radiation. Of course, we cannot observe the actual collapse of the core with ordinary telescopes, because the outer layers of the supernova emit such an overpowering amount of light. However, gravitational waves from the collapsing core carry detailed information about how this dense matter is being rearranged. With a gravitational wave antenna, we should be able to observe directly the creation of a neutron star or black hole.

Although an actual burst of gravitational waves has not yet been conclusively detected, many astronomers believe that the effects of gravitational radiation have been observed. In 1974 Joseph Taylor and his colleagues at the University of Massachusetts discovered a pulsar in a binary system. The system apparently consists of two neutron stars separated by only 2.8 solar radii. One of the two stars emits radio pulses every 0.059 second, and the orbital period of the two stars about each other is only 7.75 hours. The average orbital velocity of these stars is thus enormous—about 0.1% of the speed of light.

Because these two stars have strong gravitational fields and are moving so rapidly, this entire binary system should be a substantial source of gravitational waves. As gravitational radiation carries energy away from the system, the two stars should gradually spiral in closer and closer to each other, causing the orbital period of the two stars to decrease. Because one of the stars is a pulsar, radio astronomers have been able to measure its orbital period with extreme accuracy. These observations prove that the two stars are indeed spiraling in toward each other at exactly the rate required by the emission of gravitational waves.

Although a black hole might theoretically have a tiny electric charge, it can have no magnetic field of its own whatsoever. When a black hole is created, the collapsing star may possess an appreciable magnetic field. This magnetic field must be radiated away in the form of electromagnetic and gravitational waves before the dead star can settle down inside its event horizon. As we saw in Chapter 22, gravitational waves are ripples in the overall geometry of space. Some physicists are exploring the possibility of observing the creation of black holes by detecting bursts of gravitational radiation emitted by collapsing massive stars. A further discussion of gravitational radiation appears in Box 24-2.

Third and finally, we can detect the effects of a black hole's rotation. Specifically, we can measure a black hole's *angular momentum*. Because of the conservation of angular momentum (recall Box 23-1), we expect a black hole to be spinning rapidly. Einstein's theory makes the startling prediction that this rotation causes space and time to be dragged around the hole. A spinning black hole is thus surrounded by space that rotates with the hole. In fact, around the event horizon of every rotating black hole, a region exists where this dragging of space and time is so severe that it is impossible to stay in the same place. No matter what you do, you get pulled around the hole, along with the rotating geometry of space and time. This region, where it is impossible to be at rest, is called the **ergosphere** (Figure 24-4).

To measure a black hole's angular momentum, we could hypothetically place two satellites in orbit about the hole. Suppose that one satellite circles the hole in the same direction the hole rotates, and the other in the opposite direction. One satellite is thus carried along by the geometry of space and time, but the other is constantly fighting its way "upstream." The two satellites will thus have different orbital periods. By comparing of these two periods, astronomers can deduce the total angular momentum of the hole.

And that is all. A black hole possesses no qualities other than mass, charge, and angular momentum. This simplicity is the essence of the famous **no-hair theorem** first formulated in the early 1970s: "Black holes have no hair." Any and all additional properties carried by the matter that has fallen into the hole have disappeared from the universe and thus can have no effect on the structure of the hole.

24-4 A black hole distorts the images of background stars and galaxies

Are there any black holes out there? Many astronomers think so, but finding black holes in the sky is a difficult business. Since light cannot escape from inside the event horizon, you obviously cannot observe a black hole directly in the way that you can observe a star or a planet. The best you can hope for is to detect the effects of a black hole's powerful gravity.

One option to pursue is the distortion that a black hole creates in the appearance of background objects. For exam-

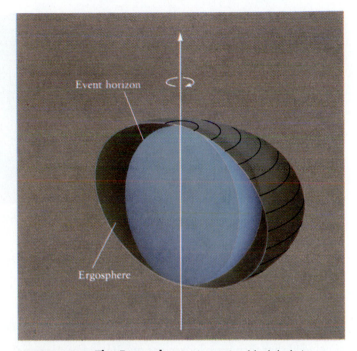

FIGURE 24-4 The Ergosphere A rotating black hole is surrounded by a region called the ergosphere, where the dragging of space and time around the hole is so severe that nothing can remain at a fixed location. Because the ergosphere (the shaded area in this cross-sectional diagram) is outside the event horizon, this bizarre region is accessible to us, and astronauts or asteroids could travel through it without disappearing into the black hole. According to detailed calculations, objects grazing the ergosphere could be catapulted back out into space at tremendous speeds. In other words, the ejected object could leave the ergosphere with more energy than it had initially, having extracted added energy from the hole's rotation. This is called the *Penrose process* after Roger Penrose, the British mathematician who proposed it.

ple, suppose that the Earth, a black hole, and a background star are in nearly perfect alignment, as sketched in Figure 24-5. Because of the warped space around the black hole, light from the star curves around the black hole as it heads toward us. As a result, there are two paths along which light rays can travel from the background star to us here on Earth. Thus, we should see two images of the star.

A powerful source of gravity that distorts background images is called a **gravitational lens**. This distortion is virtually the only way we can hope to find an isolated black hole in our Galaxy. Unfortunately, the alignment between the Earth, the black hole, and a remote star must be almost perfect. Without nearly perfect alignment, the second image of the background star is too faint to be noticeable.

A number of gravitational lenses have in fact been discovered, involving images of remote, luminous objects called quasars. As we shall see in Chapter 27, a quasar is a very bright, starlike object. Typical quasars shine as brightly as

FIGURE 24-5 A Gravitational Lens A black hole can deflect light rays from a distant star so that an observer sees two images of the star. A number of so-called gravitational lenses have been discovered in which light from a remote quasar is deflected by an intervening galaxy. No gravitational lenses caused by black holes have yet been discovered, however.

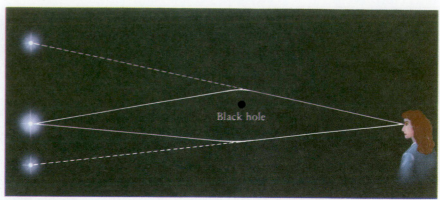

Black hole

hundreds of galaxies and are located billions of light-years from Earth. More than 4000 quasars have been discovered across the sky.

In 1979 astronomers Dennis Walsh, Robert Carswell, and Ray Weymann were surprised to find two faint quasars separated by only 6 arc sec. They took spectra of both quasars and discovered that they were nearly identical. They thus concluded that they were looking at two images of the same quasar. Further observations revealed a galaxy between the quasar images (Figure 24-6). The gravitational field of this galaxy bends the light from the remote quasar and thus acts like a gravitational lens.

By 1990 nearly two dozen candidates for gravitational lenses had been reported, of which six cases are quite convincing. Each of the six involves a remote quasar whose starlike image is split by a galaxy located between us and the

quasar. When such a galaxy deflects the light from a remote quasar, three or four images can be produced, because the galaxy's mass is spread over a volume rather than being concentrated at a point, as in the case of a black hole. A fine example of four images is the "Einstein cross" in Figure 24-7. Searches for more gravitational lenses are under way.

Computers can be used to simulate the "gravitational lensing," or distortion of the image of an extended object like a galaxy, caused by a black hole (Figure 24-8). The resulting images are elongated arcs that partly encircle the location of the deflecting black hole. These arcs connect to form a ring-like image when the galaxy, the black hole, and the observer are in nearly perfect alignment.

Several fine examples of arched gravitational images were discovered in the late 1980s. In 1987, Roger Lynds at Kitt Peak National Observatory and Vahe Petrosian of Stanford

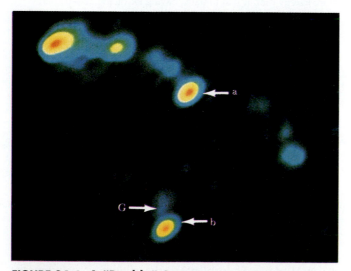

FIGURE 24-6 A "Double" Quasar Two images, labeled **a** and **b**, of the same quasar are seen in this radio view made by the Very Large Array. Light from the distant quasar is deflected to either side of a massive galaxy located between us and the quasar. A faint image of the deflecting galaxy **G** is seen directly above image **b**. The jetlike feature protruding from the upper image **a** does not appear alongside the lower image because the jet is too far away from the required quasar–galaxy–Earth alignment. (VLA; NRAO)

FIGURE 24-7 The Einstein Cross This photograph from the Hubble Space Telescope shows the gravitational lensing of a quasar in the constellation of Pegasus. The quasar, about 8 billion light-years from Earth, is seen as four separate images surrounding a galaxy that is only 400 million light-years away. The diffuse image at the center of this Einstein cross is the core of the intervening galaxy. (NASA and ESO)

a

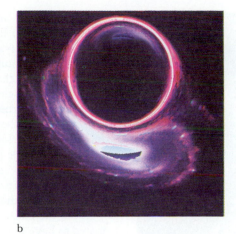

b

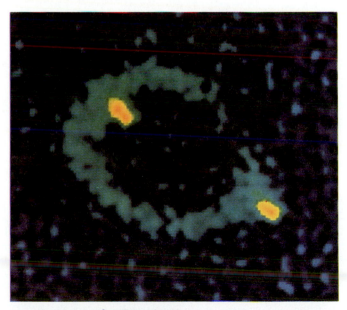

c

FIGURE 24-8 The Gravitational Lensing of a Galaxy
A computer was used to calculate these views of a galaxy seen through a gravitational lens consisting of a massive black hole $(8 \times 10^{12}\ M_\odot)$ located halfway between us and the galaxy. (a) An undistorted view of the galaxy. (b) A view of the galaxy seen with a nearly perfect alignment between the galaxy, the black hole, and the Earth. (c) A view of the galaxy seen with the black hole displaced slightly from perfect alignment. (Courtesy of E. Falco, M. Kurtz, and M. Schneps; Smithsonian Astrophysical Observatory)

University found a huge luminous arc more than 300,000 light-years long in a remote cluster of galaxies (Figure 24-9). This arc is now known to be the light from an extremely remote galaxy or quasar that has been stretched out into a curved image by the gravitational field of an intervening galaxy. In the early 1990s Anthony Tyson at Bell Telephone Laboratories showed that detailed observations of these distorted, arc-shaped background images can be used to map the mass of the foreground cluster of galaxies. (You may wish to glance ahead to Figure 26-26.)

In 1988 Jacqueline Hewitt of MIT discovered a ringlike image of a remote radio galaxy. Albert Einstein was the first to point out that a ring-shaped image would be seen if a massive body were located directly between us and a remote source of light, and so the object shown in Figure 24-10 is called an **Einstein ring.** Several additional examples of huge arcs and Einstein rings have recently been identified. Once its optical problems are solved, the Hubble Space Telescope may uncover many more cases of gravitational lensing as it examines remote galaxies and quasars.

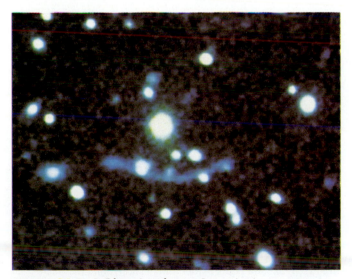

FIGURE 24-9 A Giant Luminous Arc This luminous arc is about 300,000 light-years long and is located in a cluster of galaxies about 5 billion light-years from Earth. Almost every fuzzy spot on this picture is a galaxy. Spectroscopic observations strongly suggest that this arc is the image of an extremely remote galaxy or quasar that has been stretched out into an arc by the gravitational field of an intervening galaxy. (NOAO)

FIGURE 24-10 The Einstein Ring A gravitational lens should produce a ringlike image if the background light source, the observer, and the deflecting galaxy or black hole are perfectly aligned. This radio object in the constellation of Leo may be the first known example of a so-called Einstein ring. (VLA; NRAO)

FIGURE 24-11 HDE 226868 This star is the optical companion of the X-ray source Cygnus X-1. The star is a B0 supergiant located 8000 ly from Earth. Some astronomers suspect that Cygnus X-1 might be a black hole. This photograph was taken with the 200-in. telescope at Palomar. (Courtesy of J. Kristian, Carnegie Observatories)

226868 is a B0 supergiant with a surface temperature of about 31,000 K. Because such stars do not emit significant amounts of X rays, HDE 226868 alone cannot be Cygnus X-1. Because double stars are very common, astronomers began to suspect that the visible star and the X-ray source are in orbit about each other.

Further spectroscopic observations soon showed that the spectral lines in the spectrum of HDE 226868 shift back and forth with a period of 5.6 days. This behavior is characteristic of a single-line spectroscopic binary; the companion of HDE 226868 is just too dim to produce its own set of spectral lines. The clear implication is that HDE 226868 and Cygnus X-1 are the two components of a double-star system.

From the mass–luminosity relation, HDE 226868 is estimated to have a mass of roughly 30 $M_\odot$. As a result, Cygnus X-1 would have a mass of about 7 $M_\odot$; otherwise, it would not exert enough gravitational pull to make the B0 star wobble by the amount deduced from the periodic Doppler shifting of its spectral lines. Because seven solar masses is too large for either a white dwarf or a neutron star, Cygnus X-1 might be a black hole.

The case for Cygnus X-1 being a black hole is not a strong one. HDE 226868 might be undermassive for its spectral type, which would imply a somewhat lower mass for Cygnus X-1. In addition, uncertainties in the distance to the binary system could further reduce estimates of the mass of Cygnus X-1. If all these uncertainties combined in just the right way, the estimated mass of Cygnus X-1 could be pushed down to about 3 $M_\odot$. Thus there is a slim chance that Cygnus X-1 might contain the most massive possible neutron star rather than a black hole.

24-5 Black holes have been discovered in binary star systems

Binary stars offer the best chance of finding black holes in our Galaxy. For instance, if a black hole were to capture gas from its companion star, the fate of this material might reveal the existence of the hole. In fact, since the early 1970s several good black hole candidates have been discovered in just this way.

Shortly after the launch of *Uhuru*, astronomers became intrigued with an X-ray source designated Cygnus X-1. This source is highly variable and irregular. Its X-ray emission flickers on time scales that are as short as a hundredth of a second. One of the fundamental concepts in physics is that nothing can travel faster than the speed of light (recall Box 24-1). Because of this limitation, an object cannot flicker faster than the time required for light to travel across the object. Because light travels 3000 kilometers in a hundredth of a second, Cygnus X-1 must be smaller than the Earth.

Cygnus X-1 occasionally emits radio radiation, and in 1971 radio astronomers used these outbursts to pinpoint the source, coincident with the star HDE 226868 (Figure 24-11). Spectroscopic observations promptly revealed that HDE

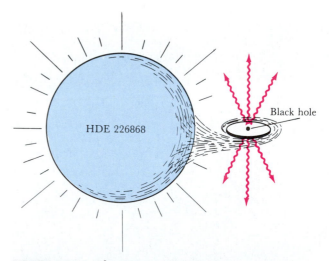

FIGURE 24-12 The Cygnus X-1 System A stellar wind from HDE 226868 pours matter onto an accretion disk surrounding a possible black hole. The infalling gases from the disk are heated to high temperatures as they spiral in toward the hole. At the inner edge of the disk, just above the black hole, the gases become so hot that they emit vast quantities of X rays. An artist's rendition of this system is seen at the beginning of this chapter.

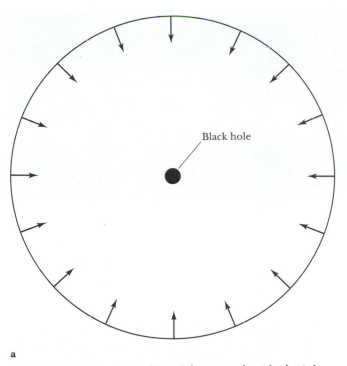

a

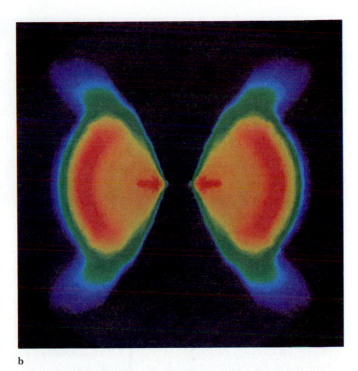

b

FIGURE 24-13   An Accretion Disk Around a Black Hole
A supercomputer can be used to solve the complicated relativistic equations that govern the in-falling of matter toward the black hole. This supercomputer simulation begins with matter possessing angular momentum falling toward a black hole, as shown schematically in (a). After a while the in-falling gases become captured into a doughnut-shaped accretion disk shown in cross section in (b). False color is used here to display density, from purple and blue for the least dense regions through the colors of the rainbow to red, which indicates the densest regions. (Courtesy of L. L. Smarr and J. F. Hawley)

If Cygnus X-1 does contain a black hole, the X rays do not come from the black hole itself. Gas captured from HDE 226868 goes into orbit about the hole, forming an accretion disk about four million kilometers in diameter (Figure 24-12). As material in the disk gradually spirals in toward the hole, friction heats the gas to temperatures approaching 2 million kelvin. In the final 200 kilometers above the hole, these extremely hot gases emit the X rays that our satellites detect. Presumably the X-ray flickering is caused by small hot spots on the rapidly rotating inner edge of the accretion disk. In this way, the black hole's existence is announced by doomed gases just before they plunge to oblivion.

Details about the structure of an accretion disk around a black hole have been elucidated by supercomputer simulations. For instance, in the 1980s Larry L. Smarr at the University of Illinois and John F. Hawley, now at the University of Virginia, used a supercomputer to solve the equations that describe how accreting gas is distributed around a black hole. As shown in Figure 24-13a, the simulation begins with gas falling toward the black hole. The in-falling gas is endowed with angular momentum, which causes the gas to orbit about the hole. Eventually the orbiting gas settles into a thick disc centered about the black hole, as shown in cross section in Figure 24-13b. The inner edge of an accretion disk occurs where the outward-directed "centrifugal force" on the gas just balances the inward pull of gravity. Along this inner edge, the flow of gas is unstable and hot bubbles can develop (Figure 24-14).

In the early 1980s, another candidate for a black hole was found—a binary system similar to that of Cygnus X-1 in a nearby galaxy called the Large Magellanic Cloud. The X-ray source, called LMC X-3, exhibits rapid fluctuations just like those of Cygnus X-1. LMC X-3 circles a B3 main sequence star every 1.7 days. From the orbital data, astronomers conclude that the mass of the compact X-ray source is probably about 6 $M_\odot$. X rays are emitted from the hot, inner regions of an accretion disk that surrounds the 6-$M_\odot$ object. Once again, however, the case for a black hole is not firm. Some astronomers argue that the accretion disk filters light from the visible star, misleading us to an overestimate of the object's mass.

Another black hole candidate is a spectroscopic binary in the constellation of Monoceros that contains the flickering X ray source A0620-00 (the A refers to the British satellite *Ariel 5* that discovered the source; the numbers refer to its position in the sky). The visible companion of A0620-00 is an orange dwarf star of spectral type K called V616 Monocerotis, which orbits the X-ray source every 7.75 hours. From orbital data, astronomers estimate the mass of A0620-00 to be about 9 $M_\odot$.

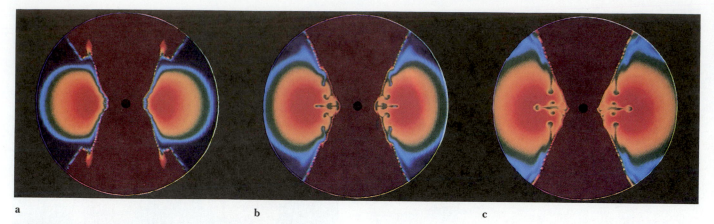

a   b   c

FIGURE 24-14 The Inner Edge of an Accretion Disk
Three stages in the evolution of the inner edge of an accretion disk are shown here in cross section using false color. The black hole is the dot in the center of each view. View (**a**) shows the undisturbed accretion disk. In (**b**), bubbles and fingers of hot gas develop because of instabilities along the inner edge of the disk. In (**c**), the bubbles have extended into the disk. (Courtesy of L. L. Smarr and J. F. Hawley)

Perhaps the most convincing black hole candidate is the spectroscopic binary called V404 Cygni, which consists of an X-ray source orbited by a K0 IV star whose mass is about 1 $M_\odot$. Doppler shift measurements reveal that the line-of-sight velocity of the visible star varies by more than 400 km/s as it orbits its unseen companion every 6.47 days. These data imply that the mass of the companion star is between 8 and 12 $M_\odot$. Eight solar masses is much higher than the lower

---

BOX 24-3

Wormholes and Time Machines

The mathematics of general relativity is so rich, complex, and fascinating that theoretical physicists have spent years investigating the geometry of black holes. These efforts have yielded some surprising results. For instance, in the 1930s Einstein and his colleague, Nathan Rosen, discovered that the full geometry of a black hole can connect our universe with a second domain of space and time that is separate from ours. The first diagram shows the geometry of this connection, called an **Einstein–Rosen bridge.** You could think of the upper surface as our universe and the lower surface as a "parallel universe." An alternative interpretation of this geometrical oddity is that the upper and lower surfaces are different regions of our own universe. The Einstein–Rosen bridge therefore connects our universe with itself, forming a **wormhole,** shown in the second diagram.

These mathematical curiosities have inspired some scientists to speculate about using a wormhole as a shortcut to get from one place in our universe to another, or to a parallel universe. But detailed calculations reveal a major obstacle: The powerful gravity of a black hole causes the wormhole to collapse almost as soon as it forms. As a result, to get from one side of a wormhole to the other, you would have to travel faster than the speed of light, which is not possible.

Recently Caltech physicists Kip Thorne, Michael Morris, and Ulvi Yurtsever have proposed a scheme that might get around the difficulty of a collapsing wormhole. They point out that, according to general relativity,

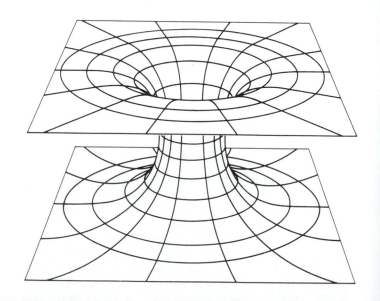

limits for the black hole candidates in Cygnus X-1, LMC X-3, or A0620-00, all of which just might contain 3-$M_\odot$ neutron stars rather than black holes. In comparison, the argument for a black hole in V404 Cygni seems to be quite persuasive.

There are several other rapidly flickering X-ray sources in binary systems; all are excellent candidates for black holes. They include Circinus X-1 and GX 339-4, the latter in Scorpius. Some astronomers also argue that SS433 may contain a black hole rather than a neutron star. With these and similar objects, a great deal of observational effort is necessary to rule out all non–black-hole explanations of the data. Only then can we feel confident that additional black holes have been discovered.

Although finding black holes is a tedious and tricky business, it is becoming clear that they should be fairly common. Although a black hole is created by the gravitational collapse of a burned-out star whose mass exceeds 3 $M_\odot$, this is not the only way a black hole can form. A white dwarf or a neutron star in a binary system can be transformed into a black hole by accreting enough matter from its companion star. This transformation can occur when the companion star becomes a red giant and dumps a significant part of its mass over its Roche lobe.

Another possibility is two dead stars coalescing to form a black hole. For example, imagine a binary system consisting of two neutron stars, such as the binary pulsar discussed in Box 24-2. Because of the emission of gravitational radiation, the two stars gradually spiral in toward each other and eventually merge. If their total mass exceeds 3 $M_\odot$, the entire system may become a black hole.

A black hole is one of the most bizarre and fantastic concepts ever to emerge from modern physical science. Indeed, some physicists have argued that the highly warped geometry of black holes allows for the possibility of wormholes to "other universes" and time machines (see Box 24-3). Although the idea of black holes initially met with skepticism, it is now clear that many of the stars we see in the sky are doomed to disappear from the universe someday, leaving only black holes behind. Even more astounding is the idea that enormous black holes, containing millions or even billions of solar masses, are located at the centers of many galaxies and quasars. As we shall see in the next chapter, one of these monstrosities may even be lurking at the center of our Milky Way, only 25,000 light-years from the Earth.

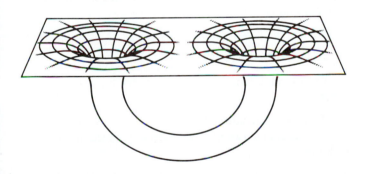

pressure as well as mass can be a source of gravity. Normally we don't see the gravitational effects of pressure because they are so small. Thorne and his colleagues speculate that a technologically advanced civilization might someday develop a means of using pressure to produce antigravity strong enough to keep the wormhole open.

If a wormhole could be held open, it could be converted into a "time machine." To see how, imagine you take one end of a wormhole and move it around for a while at speeds very near the speed of light. As we saw in Box 24-1, such motion causes clocks to slow down. Thus, when you stop moving that end of the wormhole, you find that it has not aged as much as the stationary end. In other words,

one side of the wormhole has a different time than the other. As a result, you could go into one end of the wormhole at a late time and come out at an early time. For example, you might go in at 10 AM and come out at 9 AM.

Time machines are illogical. If you could get back from a trip an hour before you left, you could meet yourself and tell yourself what a nice journey you had. Then both of you could take the trip. If you and your twin return just before you both left, there would be four of you. And all four of you could take the trip again. And then all eight of you. Then all sixteen of you . . . .

Making copies of yourself is an example of how time machines violate **causality,** the notion that effects must follow their causes. If time machines are possible, then phenomena are possible that are fundamentally illogical, irrational, and unpredictable.

We have never seen a phenomenon that violates causality. To the contrary, the universe seems remarkably rational. Scientists are therefore motivated to explain why time machines cannot exist. Indeed, British astrophysicist Stephen Hawking astutely points to strong observational evidence against time machines: We are *not* being visited by hordes of tourists from the future. If we could discover why nature precludes time machines, we would have a much deeper understanding of the nature of space and time.

KEY WORDS

Terms preceded by an asterisk are discussed in the boxes.

black hole

ergosphere

law of cosmic censorship

*proper time

*causality

event horizon

*Lorentz transformations

Schwarzschild radius (R_{Sch})

*dilation of time

*Fitzgerald–Lorentz contraction

no-hair theorem

singularity

Einstein ring

general theory of relativity

principle of equivalence

special theory of relativity

*Einstein–Rosen bridge

gravitational lens

*proper distance

*wormhole

*proper mass

KEY IDEAS

• The special theory of relativity asserts that an observer will note a slowing of clocks and a shortening of rulers that are moving with respect to the observer; this effect becomes significant only if the clock or ruler is moving with a speed near the speed of light.

• The general theory of relativity asserts that gravity causes space to become curved and time to slow down; these effects are significant only in the vicinity of large masses or compact objects.

• If a stellar corpse has a mass greater than about 3 $M_\odot$, gravitational compression will make the object so dense that its escape speed exceeds the speed of light; the corpse then contracts rapidly to a single point called a singularity.

The singularity is surrounded by a surface called the event horizon, where the escape velocity equals the speed of light; nothing—not even light—can escape from inside the event horizon.

• A black hole—a singularity surrounded by an event horizon—has only three physical properties: mass, electric charge, and angular momentum.

In the region called the ergosphere, around the outside of the event horizon, space and time themselves are dragged along with the rotation of the black hole.

• The general theory of relativity predicts the existence of gravitational radiation; gravitational waves are ripples in the overall geometry of space and time that are produced by moving masses.

It may be possible to study how stars collapse by measuring the gravitational waves emitted during these events; antennas for detecting gravitational radiation are being developed.

• An isolated black hole might be detected by how it works as a gravitational lens, distorting the image of a star or galaxy behind it.

• Some binary star systems are thought to contain a black hole; in such a system, gases captured from the companion star by the black hole emit detectable X rays.

REVIEW QUESTIONS

1. Under what circumstances are degenerate electron pressure and degenerate neutron pressure incapable of preventing the complete gravitational collapse of a dead star?

2. Find the Schwarzschild radius for an object having a mass equal to that of the planet Jupiter.

3. In what way is a black hole blacker than black ink or a black piece of paper?

4. If the Sun suddenly became a black hole, how would the Earth's orbit be affected?

5. According to the general theory of relativity, why can't some sort of yet-undiscovered degenerate pressure prevent the matter inside a black hole from collapsing all the way down to a singularity?

6. What is the law of cosmic censorship?

7. What is the no-hair theorem?

8. What kind of black hole is surrounded by an ergosphere?

9. What is a gravitational lens? Why do you suppose that no gravitational lenses have yet been discovered in our Galaxy?

10. Why do you suppose that all the black hole candidates mentioned in the text are members of very short-period binary systems?

11. As a binary system loses energy by emitting gravitational waves, why do its members speed up and why does the period become shorter?

ADVANCED QUESTIONS

Tips and tools. . .

Remember that the density of an object is its mass divided by its volume. The volume of a sphere of radius r is $\frac{4}{3}\pi r^3$. Recall that Box 4-3 contains Newton's formulation of Kepler's third law, which explicitly includes masses.

12. To what density must the matter of a dead $10\text{-}M_\odot$ star be compressed in order for the star to disappear inside its event horizon?

13. Prove that the density of matter needed to produce a black hole is inversely proportional to the square of the mass of the hole. If you wanted to make a black hole from matter compressed to a density of water (1000 kg/m^3), how much mass would you need?

14. What is the Schwarzschild radius of a black hole whose mass is (**a**) the mass of the Earth, (**b**) the mass of the Sun, and (**c**) the mass of the Milky Way Galaxy, which is about $1.1 \times 10^{11} \ M_\odot$? In each case, also calculate what the density would be if the matter were spread uniformly throughout the volume of the event horizon.

15. A clock on board a moving starship shows a total elapsed time of 2 minutes, while an identical clock on Earth shows a total elapsed time of 10 minutes. What is the speed of the starship relative to the Earth?

16. How fast should a meter stick be moving in order to appear to be a "centimeter stick"?

17. To what speed should a particle be accelerated in order for its mass to appear to double to an at-rest observer?

18. Find the total mass of the neutron star binary system described in Box 24-2, for which the orbital period is 7.75 hours and the average distance between the neutron stars is 2.8 solar radii. Is your result reasonable for a pair of neutron stars? Explain.

19. The orbital period of the single-line spectroscopic binary containing A0620-00 is 0.32 day, and Doppler shift measurements reveal that its radial velocity peaks at 457 km/s (about 1 million miles per hour). By scaling this system to the Sun–Earth system and using Kepler's third law, prove that the mass of the X-ray source must be at least 3.1 times the mass of the Sun. (*Hint*: Assume that the mass of the K5V visible star—about $0.5 \ M_\odot$ from the mass–luminosity relationship—is negligible compared to that of the invisible companion.)

DISCUSSION QUESTIONS

20. Discuss why the fact that the speed of light is the same for all observers, regardless of their motion, requires us to abandon the Newtonian view of space and time.

21. Describe the kinds of observations you might make in order to locate and identify black holes.

22. Speculate on the effects you might encounter on a trip to the center of a black hole.

OBSERVING PROJECT

23. As you well know, you cannot see a black hole with a telescope. Nevertheless, you might want to observe the visible companion of Cygnus X-1. The epoch 2000 coordinates of this ninth-magnitude star are R.A. = $19^h \ 58.4^m$ and Decl. = $+35°12'$, which is quite near the bright star η Cygni. Compare what you see with the photograph in Figure 24-11.

FOR FURTHER READING

Abramowicz, M. A. "Black Holes and the Centrifugal Force Paradox." *Scientific American*, March 1993. This fascinating article discusses the paradoxical prediction that an object orbiting near a black hole feels a centrifugal force pushing *inward* rather than outward.

Chaffee, F. "The Discovery of a Gravitational Lens." *Scientific American*, November 1980. This article, written soon after the first gravitational lens was discovered, describes the lens phenomenon with the aid of excellent diagrams.

Davies, P. "Wormholes and Time Machines." *Sky & Telescope*, January 1992. This intriguing article by a noted scientist discusses some of the fanciful aspects of black holes.

Jeffries, A., and others. "Gravitational Wave Observatories." *Scientific American*, June 1987. This article, which includes an excellent description of gravitational radiation, explains how lasers can be used to detect gravitational waves.

Kaufmann, W. *Black Holes and Warped Spacetime*. W. H. Freeman and Company, 1979. This slim book gives a nontechnical overview of black holes and general relativity.

———. *Cosmic Frontiers of General Relativity*. Little, Brown, 1977. This book describes many fascinating aspects of black holes, including views that might greet an astronaut who plunges through a wormhole connecting our universe with another.

McClintock, J. "Do Black Holes Exist?" *Sky & Telescope*, January 1988. This well-written article presents evidence for black holes in certain X-ray binaries, especially A0620-00.

Price, R., and Thorne, K. "The Membrane Paradigm for Black Holes." *Scientific American*, April 1988. Two renowned physicists present a new way of picturing the interaction of a black hole with its environment.

Stokes, G., and Michalsky, J. "Cygnus X-1." *Mercury*, May/June 1979. This brief article describes evidence that Cygnus X-1 is a black hole.

Thorne, K. "The Search for Black Holes." *Scientific American*, December 1974. This superb article describes various aspects of black holes that may be useful in searches for these elusive objects.

Turner, E. "Gravitational Lenses." *Scientific American*, July 1988. Beautiful illustrations make this article on gravitational lensing especially clear.

OUR GALAXY

OUR GALAXY This wide-angle photograph, taken from Australia, spans 180° of the Milky Way, from Cygnus at the far right to the Southern Cross at the left. The center of our Galaxy is in the constellation of Sagittarius, in the middle of this photograph. Figure 25-1 is a comparable photograph showing the northern Milky Way. (Courtesy of D. di Cicco)

THE MILKY WAY, that hazy band of myriad faint stars stretching across the night sky, is our edge-on view of the disk of our own Galaxy. Throughout the twentieth century, astronomers have struggled to determine the size, shape, and rotation of our Galaxy. We now realize that we live in a vast disk-shaped assemblage of stars, gas, and dust roughly 25 kpc (80,000 ly) in diameter. Huge spiral arms gracefully arching outward from the Galaxy's nucleus pose many puzzles for astronomers. Apparently gravitational interactions with a nearby galaxy or random bursts of star formation within our own Galaxy can compress the interstellar medium and give rise to a spiral structure. Most mysterious is the remarkable source of energy at the very center of the Galaxy. Vast amounts of radiation pour from this compact source, which may be a supermassive black hole. By exploring our Galaxy, we gain important insights into the properties of galaxies in general, thereby preparing ourselves to widen our perspective on the universe and ask fundamental questions about the cosmos.

On a clear, moonless night, far from city lights, you can often see a hazy, luminous band stretching across the sky. In the Northern Hemisphere, summer nights provide the best view, with conspicuous patches of the band in Cygnus and Sagittarius. Ancient peoples devised fanciful myths to account for this "milky way" among the constellations. In the southern sky, portions are so bright that South American Indians named them, much as ancient astronomers named the constellations.

Galileo was the first person to look at the Milky Way through a telescope. He immediately discovered it to be composed of countless dim stars. Today we realize that this hazy band is actually our view from inside a vast disk of several hundred billion stars that includes the Sun. The Milky Way stretches all the way around the sky in a continuous band that is almost perpendicular to the plane of the ecliptic. Figure 25-1 is a wide-angle photograph showing roughly one-third of the Milky Way.

FIGURE 25-1 The Milky Way This wide-angle photograph shows the northern Milky Way, from Cassiopeia on the left to Sagittarius on the right. Note the dark lanes and blotches. This mottling is caused by interstellar gas and dust obscuring the light from the background stars. (Steward Observatory)

In studying the Milky Way, we explore the universe on a grand scale. Instead of examining individual stars, we look at an entire system of stars. And instead of focusing on the location and life of an isolated star, we look at the overall arrangement and history of a huge stellar community of which the Sun is a member.

25-1 Interstellar dust hides much of the Milky Way Galaxy from view

Because the Milky Way completely encircles us, astronomers in the eighteenth century began to suspect that the Sun and all the stars in the sky were part of an enormous disk of stars—the Milky Way Galaxy. Today astronomers simply call it "the Galaxy" (with a capital G). We now know that the Milky Way also contains copious interstellar dust that hides so many stars from view that astronomers were long misled into thinking that we lie at the Galaxy's center.

In the 1780s, William Herschel attempted to deduce the Sun's location in the Galaxy by counting the number of stars in each of 683 regions of the sky. Herschel reasoned that he should see the greatest number of stars toward the Galaxy's center and a lesser numbers toward the Galaxy's edge.

Herschel found roughly the same density of stars all along the Milky Way. Therefore, he concluded that we are at the center of our Galaxy (Figure 25-2). In the early 1900s the Dutch astronomer J. C. Kapteyn, analyzing the brightness and proper motion of a large number of stars, came to essentially the same conclusion. According to Kapteyn, the Milky Way is about 10 kpc (33,000 ly) in diameter and about 2 kpc (6500 ly) thick, with the Sun near its center.

Both Herschel and Kapteyn were wrong about the Sun's being at the center of our Galaxy. The reason for their mistake was finally discerned in the 1930s by R. J. Trumpler. While studying star clusters, Trumpler discovered that the more remote clusters appear unusually dim—more so than would be expected from their distances alone. As a result, Trumpler concluded that interstellar space must not be a perfect vacuum: It must contain dust that absorbs or scatters light from distant stars. Like the stars themselves, this obscuring material is concentrated in the plane of the Galaxy. Great patches of this interstellar dust are clearly visible in such wide-angle photographs as Figure 25-1. Because Herschel and Kapteyn were actually seeing only the nearest stars in the Galaxy, they had no true idea of either the enormous size of the Galaxy or of the vast number of stars concentrated around the galactic center.

As we saw in Chapter 20, the dimming of light by the interstellar medium is called **interstellar extinction.** In the plane of our Galaxy, the average interstellar extinction at visible wavelengths is about 1 magnitude per kiloparsec. For example, a star in the Milky Way that is 5 kpc (16,300 ly) from Earth appears 5 magnitudes dimmer than it should from its distance alone. And in regions where there are dense interstellar clouds, such as toward the galactic center, the extinction is even greater. In fact, at optical wavelengths the center of our Galaxy is totally obscured from view.

The amount of interstellar extinction that light suffers is roughly inversely proportional to its wavelength. In other words, the longer the wavelength, the farther the radiation can travel through interstellar dust without being scattered or absorbed. As a result, we can see farther into the plane of the Milky Way at infrared wavelengths than at visible wavelengths, and radio waves can traverse the Galaxy freely.

Starlight warms interstellar dust grains to temperatures in the range of 10 to 90 K. According to Wien's law, this dust emits infrared radiation at wavelengths predominantly from 30 to 300 μm. In 1983 the Infrared Astronomical Satellite (IRAS) scanned the sky with a 60-cm reflector at wavelengths covering this far-infrared range, giving the panoramic view of the dust along the Milky Way shown in Figure 25-3a.

In 1990 an instrument on the COBE satellite scanned the sky at shorter infrared wavelengths, producing a near-infrared view of the plane of the Milky Way (Figure 25-3b). Because interstellar dust does not emit very much light at near-infrared wavelengths, it does not interfere significantly with this view, which reveals the distribution of stars deep in the Milky Way.

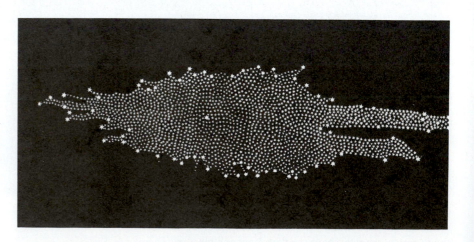

FIGURE 25-2 Herschel's Map of Our Galaxy William Herschel attempted to map the Milky Way Galaxy by counting the numbers of stars in various parts of the sky. Because of interstellar dust that blocked his view of distant stars, Herschel erroneously concluded that the Sun is at the center of the Galaxy. (Yerkes Observatory)

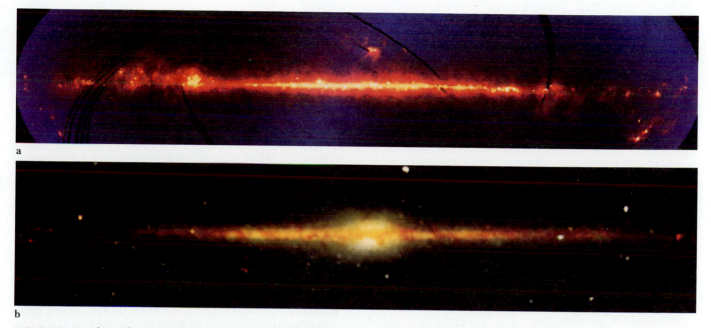

a

b

FIGURE 25-3 The Infrared Milky Way (a) This far-infrared view, constructed from IRAS observations at 25, 60, and 100 μm, shows the distribution of interstellar dust along the Milky Way.

(b) In this near-infrared view, constructed from COBE observations at 1.2, 2.2, and 3.4 μm, we see much farther through the dust in the Milky Way than we can at visible wavelengths. (NASA)

25-2 The Sun is located in the disk of our Galaxy, about 25,000 light-years from the galactic center

Although interstellar dust in the plane of our Galaxy hides the sky covered by the Milky Way, our view to either side of the Milky Way is relatively unobscured. Knowledge of our position in the Galaxy came from studies of globular clusters in these unobscured regions of the sky. As we saw in Chapter 21 (recall Figure 21-5), these spherical clusters typically contain a million stars packed in a volume only a few hundred light-years across. Variable stars in these clusters provided the key astronomers needed to gauge the size of our Galaxy.

In 1912 the American astronomer Henrietta Leavitt reported her important discovery of the period–luminosity relation for Cepheid variables. As we saw in Chapter 21 (recall Figure 21-15), Cepheid variables are pulsating stars that periodically vary in brightness. Leavitt studied numerous Cepheids in the Small Magellanic Cloud (a small galaxy near the Milky Way) and found their periods to be directly related to their average luminosities. As Figure 25-4 shows, the longer the period, the higher the luminosity.

The period–luminosity law is an important tool in astronomy, because it can be used to determine distances. For instance, suppose you find a Cepheid variable in the sky. By measuring its period and using a graph like Figure 25-4, you can determine the star's average luminosity. Knowing the star's average luminosity, you can find out how far away the star must be in order to give the observed brightness (recall Box 19-2).

Shortly after Leavitt's discovery of the period–luminosity law, Harlow Shapley, a young astronomer at the Mount Wilson Observatory in California, began studying a family of pulsating stars closely related to Cepheid variables called the **RR Lyrae variables**, which we first discussed in Chapter 21. The light curve of an RR Lyrae variable (Figure 25-5) is

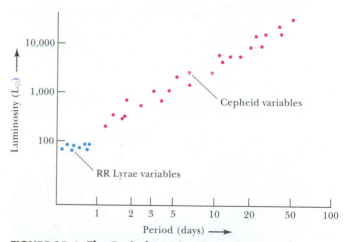

FIGURE 25-4 The Period–Luminosity Relation This graph shows the relationship between the periods and luminosities of Cepheid variables. The brighter the Cepheid, the longer is its pulsation period. Data for RR Lyrae variables are also shown. RR Lyrae variables are horizontal branch stars with average luminosities of about 100 $L_\odot$ and pulsation periods of less than a day.

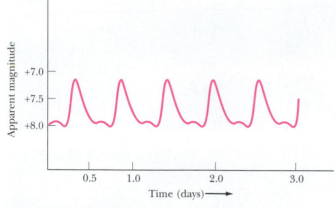

FIGURE 25-5 The Light Curve of RR Lyrae An RR Lyrae variable can be recognized by its characteristic variations in brightness. All RR Lyrae variables have periods of less than a day and have brightnesses that vary by about 1 magnitude. The example shown here is the curve of RR Lyrae, the prototype of this class of stars.

similar to that of a Cepheid (see Figure 21-13), and Shapley believed RR Lyrae variables to be short-period Cepheids like those Leavitt had studied. RR Lyrae variables are commonly found in globular clusters (Figure 25-6).

By 1915 Shapley had noticed a peculiar property of globular clusters. Ordinary stars and open star clusters are rather uniformly spread along the Milky Way. However, the majority of the 93 globular clusters that Shapley studied are located in one half of the sky, widely scattered around the portion of the Milky Way that is in the constellation of Sagittarius. Figure 25-7 shows two globular clusters in this part of the sky.

Shapley used the period–luminosity relation to determine the distances to the then-known 93 globular clusters in the sky. From their directions and distances, he mapped out the three-dimensional distribution of these clusters in space. By 1917 Shapley had discovered that the globular clusters form a huge spherical distribution. Shapley then made the bold conjecture, which was subsequently confirmed, that the arrangement of the globular clusters indicates the true size and extent of the Galaxy. The distribution of globular clusters is centered not on the Earth, as Herschel and Kapteyn would have believed, but rather about a point in the Milky Way toward the constellation of Sagittarius.

Since Shapley's pioneering observations, many astronomers have measured the distance from the Sun to the **galactic nucleus,** the center of our Galaxy. Shapley's estimate was about three times larger than the now generally accepted distance of about 8.6 kpc (28,000 ly). In fact, recent measurements based on radio observations of gas clouds containing powerful water masers (recall Figure 20-19) around the galactic center suggest a still shorter distance of about 7 kpc (23,000 ly). In this book, we adopt a compromise distance of 7.7 kpc (25,000 ly) from the Sun to the galactic nucleus.

The distance to the center of the Galaxy establishes a scale with which the dimensions of other features can be determined. The **disk** of our Galaxy is about 25 kpc (80,000 ly) in diameter and about 0.6 kpc (2000 ly) thick, as shown in Figure 25-8. The galactic nucleus is surrounded by a distribution of stars, called the **central bulge,** that is about 4.6 kpc

FIGURE 25-6 The Globular Cluster M55 Three RR Lyrae variables are identified with arrows in this globular cluster, located in the constellation of Sagittarius. From their average apparent brightness (as seen in this photograph) and average true brightness (known to be roughly 100 Suns), astronomers have deduced that the distance to this cluster is 20,000 light-years. (Harvard Observatory)

FIGURE 25-7 A View Toward the Galactic Center More than one million stars are visible in this photograph. This view looks toward a relatively clear "window" just 4° south of the galactic nucleus in Sagittarius. There is surprisingly little matter to obscure the view in this tiny section of the sky. The two globular clusters are NGC 6522 and NGC 6528. (NOAO)

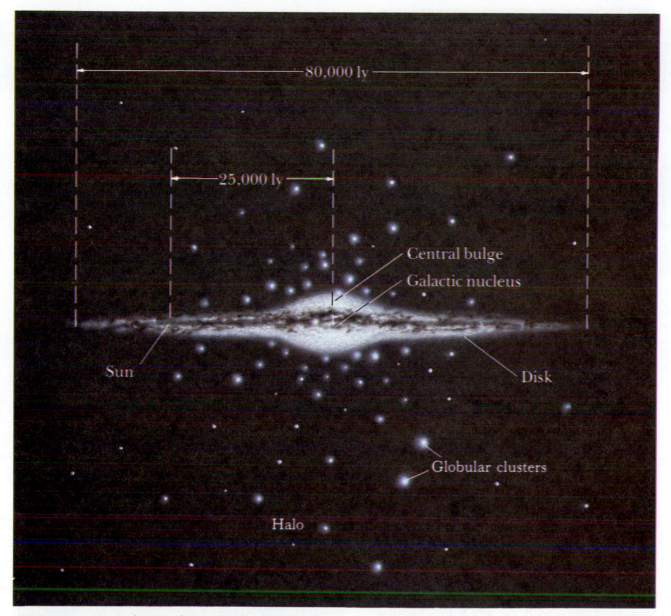

FIGURE 25-8 Our Galaxy (Schematic Edge-On View)
There are three major components of our Galaxy: a thin disk, a
central bulge, and a halo. The disk contains gas and dust along
with metal-rich (population I) stars. The halo is composed almost
exclusively of old, metal-poor (population II) stars. The central
bulge is a mixture of population I and population II stars.

(15,000 ly) in diameter. The near-infrared observations show
that stars around the Galaxy's center are arranged along a
"bar," seen nearly end-on in Figure 25-3*b*. The spherical
distribution of globular clusters traces the **halo** of the Galaxy.
Our Galaxy, seen edge-on from a great distance, would probably look somewhat like NGC 4565, shown in Figure 25-9.

Different kinds of stars are found in the various components of our Galaxy. The globular clusters in the halo are
composed of old, metal-poor, population II stars. Although
these clusters are conspicuous, their stars constitute only
about 1% of the total number in the halo. The halo also
contains individual population II stars, called **high-velocity**
stars because of their high speed relative to the Sun. These
ancient stars orbit the Galaxy along paths tilted at random
angles to the plane of the Milky Way.

The stars in the disk are mostly young, metal-rich, population I stars like the Sun. The disk of a galaxy appears
bluish, as in Figure 25-9, because its light is dominated by
radiation from young, hot O and B stars. The central bulge
contains both population I and II stars, suggesting that some
stars there are quite ancient whereas others were created
more recently. The central bulge looks reddish because of
many red giants and red supergiants clustered around the
center of our Galaxy.

FIGURE 25-9   Edge-On View of the Galaxy NGC 4565
If we could view our Galaxy edge-on from a great distance, it would probably look like this galaxy in the constellation of Coma Berenices. A layer of dust and gas is clearly visible in the plane of the galaxy. Also note the reddish color of the bulge that surrounds the galaxy's nucleus. (U.S. Naval Observatory)

25-3   The spiral structure of our Galaxy has been plotted from radio and optical observations of star-forming regions

Because interstellar dust obscures our visual view in the plane of our Galaxy, a detailed understanding of the structure of the galactic disk had to wait until the development of radio astronomy. Because of their long wavelengths, radio waves easily penetrate the interstellar medium without being scattered or absorbed. As we shall see in this section, radio and optical observations reveal that our Galaxy has **spiral arms**, spiral-shaped concentrations of gas and dust unwinding from the center in a shape reminiscent of a pinwheel.

Hydrogen is by far the most abundant element in the universe. Hence, by looking for concentrations of hydrogen gas we should be able to detect important clues about the structure of the disk of our Galaxy. Unfortunately, the major electron transitions in the hydrogen atom (review Box 5-5) produce photons at ultraviolet and visible wavelengths, which do not penetrate the interstellar medium. How can we detect all this gas? Fortunately, hydrogen atoms also emit radio waves.

In addition to having mass and charge, particles such as protons and electrons possess a tiny amount of angular momentum commonly called **spin**. An electron or a proton can in fact be crudely visualized as a tiny spinning sphere. According to the laws of quantum mechanics, the electron and proton in a hydrogen atom can be spinning in either exactly the same direction or opposite directions (Figure 25-10), but

not at random angles. The electron can also flip over from one configuration to the other. When this happens, and the hydrogen atom gains or loses a tiny amount of energy. For instance, in going from "parallel" to "antiparallel" spins, the atom emits a low-energy photon whose wavelength is 21 cm—a radio wavelength.

In 1944 the Dutch astronomer H. C. van de Hulst predicted that 21-cm radiation from this **spin–flip transition** in interstellar hydrogen would be detected as soon as appropriate radio telescopes were constructed. In 1951 a team of astronomers at Harvard did succeed in detecting the faint **21-cm radio emission** from interstellar hydrogen. Figure 25-11 shows the results of a more recent 21-cm survey of the entire sky; hydrogen gas in the plane of the Milky Way stands out prominently as a bright band across the middle of this image. The distribution of gas in the Milky Way is quite frothy, as described in Box 25-1.

The detection of 21-cm radio radiation was a major breakthrough that permitted astronomers to probe the galactic disk. To see how they did so, suppose that you aim a radio telescope along a particular line of sight across the Galaxy, as sketched in Figure 25-12. Your radio receiver picks up 21-cm emission from hydrogen clouds at points 1, 2, 3, and 4 (point S is the location of the Sun). However, the radio waves from these various clouds are Doppler shifted by slightly different amounts, because the clouds are moving at different speeds as they travel with the rotating Galaxy.

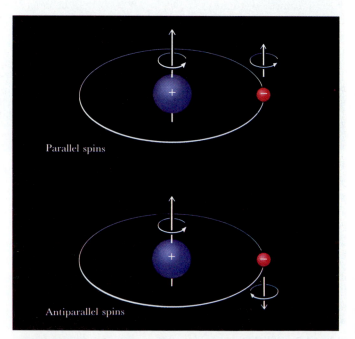

FIGURE 25-10   The Hyperfine Structure of the Hydrogen Atom   In the lowest orbit of the hydrogen atom, the electron and the proton can be spinning in either the same or opposite directions. When the electron flips over, the atom either gains or loses a tiny amount of energy. This energy is either absorbed or emitted as a radio photon having a wavelength of 21 cm.

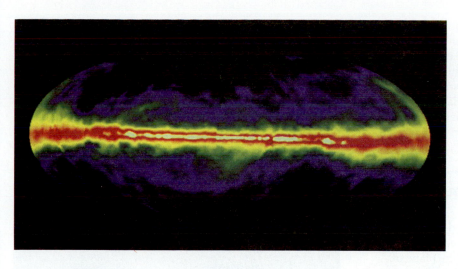

FIGURE 25-11 The Entire Sky at 21 Centimeters This false-color image shows the results of an all-sky survey of 21-cm emission from interstellar hydrogen. Colors indicate the density of hydrogen gas, ranging from black and blue for the lowest to red and white for the highest. Note how hydrogen gas is concentrated along the Milky Way, which extends horizontally across the image. (Courtesy of C. Jones and W. Forman; Harvard–Smithsonian Center for Astrophysics)

It is important to remember that the Doppler shift reveals only motion parallel to the line of sight (review Figure 5-19). Note that cloud 2 has the highest speed along our line of sight, because it is moving directly toward us. Consequently, the radio waves from cloud 2 exhibit a larger Doppler shift than those from the other three clouds along our line of sight. Because clouds 1 and 3 are at the same distance from the galactic center, they have the same orbital speed. The fraction of their velocity parallel to our line of sight is also the same, so their radio waves exhibit the same Doppler shift, which is less than the Doppler shift of cloud 2. Finally, cloud 4 is the same distance from the galactic center as is the Sun. This cloud is thus orbiting the Galaxy at the same speed as the Sun, resulting in no net motion along the line of sight. Radio waves from cloud 4, as well as from hydrogen gas near the Sun, are not Doppler-shifted at all.

These various Doppler shifts cause the 21-cm radiation to be smeared out over a range of wavelengths. That is, radio waves from gases in different parts of the Galaxy arrive at our radio telescopes with slightly different wavelengths. It is therefore possible to sort out the various gas clouds and thus produce a map of the Galaxy, such as that shown in Figure 25-13.

FIGURE 25-13 A Map of Neutral Hydrogen in Our Galaxy This map, from radio-telescope surveys of 21-cm radiation, shows arched lanes of gas that suggest a spiral structure for our Galaxy. The Sun's location is indicated by a yellow arrow. Details in the blank, wedge-shaped region of the map cannot be determined, because gas in this part of the sky is moving perpendicular to our line of sight and thus does not exhibit a detectable Doppler shift. (Courtesy of G. Westerhout)

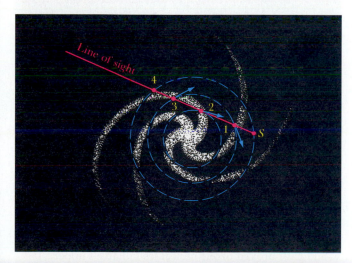

FIGURE 25-12 A Technique for Mapping Our Galaxy Hydrogen clouds at different locations along our line of sight are moving at slightly different speeds. As a result, radio waves from various gas clouds are subjected to slightly different Doppler shifts, permitting radio astronomers to sort out the gas clouds and map the Galaxy.

A 21-cm map of our Galaxy reveals numerous arched lanes of neutral hydrogen gas. Photographs of other galaxies show spiral arms outlined by bright stars and emission nebulae, which indicate active star formation (Figure 25-14). We can obtain additional clues about the spiral structure of our own Galaxy by mapping the locations of similar star-forming complexes, which are marked by OB associations, H II regions, and molecular clouds.

Interstellar absorption limits the range of visual observations in the plane of the Galaxy to less than 3 kpc (10,000 ly) from the Earth. Nevertheless, there are enough OB associations and H II regions visible in the sky to plot the spiral arms in the vicinity of the Sun. As we saw in Chapter 20, carbon monoxide (CO) is a good "tracer" of molecular clouds. Radio observations of this molecule have recently been used to plot remote regions of the Galaxy.

BOX 25-1

The Local Bubble and the Geminga Pulsar

As we saw in Chapter 20, interstellar dust and gas (mostly hydrogen) together make up the interstellar medium. This material is not uniformly spread throughout the plane of the Milky Way, but rather is quite lumpy and frothy, churned up over billions of years by passing stars and exploding supernovae. Indeed, astronomers think a recent, nearby supernova may have blown an enormous hole in the interstellar medium in the vicinity of the Sun.

Evidence for such holes comes from observations in the so-called extreme-ultraviolet, which covers wavelengths from 10 to 100 nm. Until recently most astronomers believed that this wavelength range was unobservable because high-energy ultraviolet photons are readily absorbed as they ionize the abundant hydrogen. This opinion was disproved in 1975, when a team of astronomers used a telescope aboard a spacecraft to observe several hot stars at extreme-ultraviolet wavelengths. Their discovery proved there are holes in the interstellar gas through which distant objects can be viewed in this part of the spectrum.

Later space missions showed that the Sun lies inside a huge cavity roughly 300 light-years across, dubbed the **Local Bubble.** Although its boundaries are quite uncertain (see illustration), the Local Bubble may extend as far as 1000 light-years in the direction of the star β Canis Majoris. In 1992 NASA launched a satellite called the Extreme-Ultraviolet Explorer (EUVE) that is now probing the Local Bubble more thoroughly than any previous spacecraft.

Some astronomers think they can identify the supernova that blew the Local Bubble. In 1992 the ROSAT X-ray satellite detected pulses from a mysterious gamma-ray source in the constellation of Gemini that had defied explanation for nearly 20 years. Indeed, the source was so illusive that is was nicknamed *Geminga* from a word in an Italian dialect meaning "it's not there" or "does not exist." The ROSAT discovery of pulsed radiation with a period of 0.237 s proved that the source is a pulsar.

A team of Italian astronomers recently discovered the optical counterpart of the Geminga Pulsar—a very faint star with a large proper motion. Because of the high speed with which this pulsar is moving against the background stars, it must be quite nearby, probably less than 300 light-

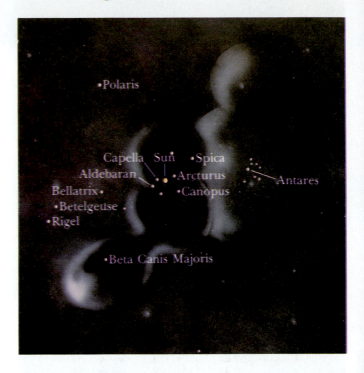

years from Earth. Furthermore, the pulsar's period is lengthening at a rate suggesting that the pulsar is quite young, perhaps only 300,000 years old. Calculations demonstrate a Type II supernova would have had enough energy to evacuate a cavity in the interstellar medium that we now call the Local Bubble. Quite possibly, the Geminga Pulsar is the remnant of that supernova.

An amusing postscript to this story is the effect of the Geminga supernova explosion on Earth. This supernova could have been so nearby that it reached an apparent magnitude of nearly −13, which is slightly brighter than the full moon. High-energy radiation from the supernova would have broken ozone molecules in Earth's upper atmosphere, possibly depleting 10% of the world's ozone. Because this loss would have reduced protection against the Sun's ultraviolet light, our Neanderthal ancestors may have had quite a sunburn!

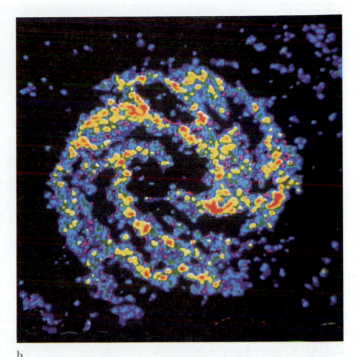

a

b

FIGURE 25-14   A Spiral Galaxy   This galaxy, called M83, is in the southern constellation of Centaurus about 12 million light-years from Earth. (**a**) This photograph at visual wavelengths clearly shows the spiral arms illuminated by young stars and glowing H II regions. (**b**) This radio view at a wavelength of 21 cm shows the emission from neutral hydrogen gas (H I). Note that the spiral arms are better distinguished by emission from ionized hydrogen than from neutral hydrogen. (Anglo-Australian Observatory; VLA, NRAO)

Taken together, all these observations demonstrate that our Galaxy has four major spiral arms and several short arm segments (Figure 25-15). The Sun is located on a relatively short arm segment called the Orion arm, which includes the Orion Nebula and neighboring sites of vigorous star formation in that constellation. Two major spiral arms border either side of the Sun's position. The Sagittarius arm is on the side toward the galactic center. This is the arm you see during the summer months when you look at the portion of the Milky Way stretching across Scorpius and Sagittarius (see the photograph on the first page of this chapter). During winter, when our nighttime view is directed away from the galactic center, we see the Perseus arm. The remaining two major spiral arms are usually referred to as the Centaurus arm and the Cygnus arm.

25-4   Moving at half a million miles per hour, the Sun takes about 200 million years to complete one orbit of our Galaxy

The pinwheel shape of our Galaxy suggests that it rotates. Indeed, if the stars in our Galaxy were not orbiting the galactic center, they would all fall into the galactic center. However, measuring the rotation of our Galaxy accurately is a difficult business.

Radio observations of 21-cm radiation from hydrogen gas give important clues about our Galaxy's rotation. Doppler shift measurements of this radiation indicate that our Galaxy does not rotate like a rigid body. Instead, the orbital speed of stars and gas about the galactic center is fairly constant throughout much of the Galaxy's disk. As a result, stars inside the Sun's orbit complete a trip around the galactic center more quickly than the Sun does because they have a shorter distance to travel. Conversely, stars outside the Sun's orbit take longer to go once around the galactic center because they have farther to travel. Consequently, as seen by Earth-based astronomers moving along with the Sun, stars inside Sun's orbit overtake and pass us, while we overtake and pass stars outside the Sun's orbit (Figure 25-16).

The 21-cm observations reveal only how fast things are moving relative to the Sun. Of course, the Sun itself is also moving. To get a complete picture of our Galaxy's rotation, we must therefore find out how fast the Sun is traveling. In other words, we need to find a kind of background that is not rotating along with the rest of the Galaxy.

A method of doing this was proposed by the Swedish astronomer Bertil Lindblad. Not all the stars in the sky move in the orderly pattern sketched in Figure 25-16. Distant galaxies and some components of our own Galaxy, like globular clusters and high-velocity stars, do not participate in the general rotation of the Galaxy but have instead more or less random motions. Lindblad took the average of these random

FIGURE 25-15 Our Galaxy (Face-On View) Our Galaxy has four major spiral arms and several shorter arm segments. The Sun is located on the Orion arm, between two major spiral arms. The Galaxy's diameter is about 80,000 light-years, and the Sun is about 25,000 light-years from the galactic center.

motions to establish a stationary background. He then used the Doppler shifts of the stars in this background to deduce that the Sun's speed along its orbit about the galactic center is 230 km/s. That's about a half million miles per hour!

We know that we are 25,000 light-years from the galactic center, so we can use the Sun's speed v to calculate the time required for one complete trip around the Galaxy, a time called the Sun's orbital period T. For the sake of simplicity, we assume that the Sun travels along a circular orbit about the center of the Galaxy. The distance traveled by the Sun is the circumference of this circle, which is given by $2\pi r$ (with r

being the radius of the circle). The time required for one orbit is equal to the distance traveled divided by the Sun's speed:

$$T = \frac{2\pi r}{v} = \frac{2\pi \times 25,000 \text{ ly}}{230 \text{ km/s}} \times \frac{9.46 \times 10^{12} \text{km}}{1 \text{ ly}}$$

$$= 6.5 \times 10^{15} \text{ s} = 2.0 \times 10^8 \text{ years}$$

Traveling at a half million miles per hour, it takes the Sun about 200 million years to complete one trip around the Galaxy. These results demonstrate how vast our Galaxy is.

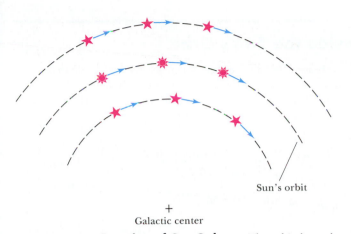

FIGURE 25-16 Rotation of Our Galaxy The orbital speed of stars and gas around the galactic center is fairly constant throughout most of our Galaxy. This schematic diagram shows the Sun and two stars traveling at the same speed but at different distances from the galactic center. Although they start off lined up, the Sun and the stars become increasingly separated as they move along their orbits. As seen by an observer traveling with the Sun, stars inside the Sun's orbit overtake and move ahead of the Sun while stars far from the galactic center lag behind.

Once we know the basic features of the Sun's orbit around the Galaxy, we can use Kepler's third law to estimate the mass of all the matter (for example, stars, gas, dust) *inside* the Sun's orbit. As shown in Box 25-2, calculations yield $9.4 \times 10^{10} \, M_\odot$. The mass of the entire Galaxy must be larger than this, because this figure does not include any matter *outside* the Sun's orbit.

In recent years, astronomers have been astonished to discover how much matter actually lies beyond the orbit of the Sun. The clues come from 21-cm radiation emitted by hydrogen in spiral arms that extend far beyond the Sun's orbit. Because we know the true speed of the Sun, we can convert the Doppler shifts of this radiation into actual speeds for the spiral arms. This calculation gives us a **rotation curve**, a graph of the velocity of galactic rotation measured outward from the galactic center (Figure 25-17). Note that the rotation curve keeps rising, even out to a distance of 60,000 light-years from the galactic nucleus.

According to Kepler's third law, the orbital speeds of stars or gas clouds beyond the confines of most of the Galaxy's mass should decrease with increasing distance from the Galaxy's center, just as the orbital speeds of the planets decrease with increasing distance from the Sun. Yet the Galaxy's rotation curve is quite flat, indicating uniform orbital speeds well beyond the visible edge of the galactic disk. To explain this constant rotation speed of the outer parts of the Galaxy, astronomers postulate the existence of a large amount of matter beyond the $10^{11} \, M_\odot$ lying inside the Sun's orbit. Because of this matter, the total mass of our Galaxy could be at least $6 \times 10^{11} \, M_\odot$.

To make matters even more puzzling, this matter is dark. No one has detected any radiation coming from it, and it does not show up on photographs. Various explanations of this **dark matter** have been proposed. It might, for instance, be composed of numerous Jupiterlike planets and dim, low-mass stars. Or it might be made of something much more exotic, such as black holes, neutrinos possessing mass, or a variety of hypothetical particles predicted by speculative theories at the forefront of physics. Dark matter is one of the most baffling mysteries in modern astronomy.

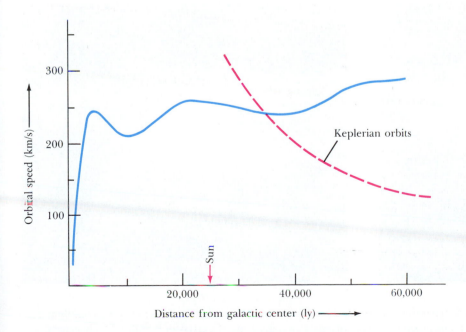

FIGURE 25-17 The Galaxy's Rotation Curve The blue curve shows the orbital speeds of stars and gas out to a distance of 60,000 light-years from the galactic center. The dashed red curve indicates how this orbital speed should decline beyond the confines of most of the Galaxy's mass. Since the data (blue curve) do not show any such decline, astronomers still have not detected the actual edge of the Galaxy. Indeed, these data indicate that an abundance of dark matter must extend far beyond the visible disk of our Galaxy.

BOX 25-2

Estimating the Mass Inside the Sun's Orbit

We can use Kepler's third law to estimate the mass interior to the Sun's orbit about the galactic center. As we saw in Box 4-3, Kepler's law is:

$$P^2 = \frac{4\pi^2 a^3}{GM}$$

where P is the orbital period of the Sun, a is the semimajor axis of the Sun's orbit, M is the mass of the Galaxy inside the Sun's orbit, and G is the constant of gravitation.

Rearranging terms and assuming the Sun's orbit is a circle of radius r, it can prove that:

$$M = \frac{rv^2}{G}$$

where v is the Sun's orbital speed.

Now we can insert known values to obtain the mass within the Sun's orbit. Since $r = 25{,}000$ ly $= 2.4 \times 10^{20}$ m, $v = 230$ km/s $= 2.3 \times 10^5$ m/s, and $G = 6.67 \times 10^{-11}$ m^3 kg^{-1} s^{-2}, we get:

$$M = \frac{2.4 \times 10^{20} \times (2.3 \times 10^5)^2}{6.67 \times 10^{-11}} = 1.9 \times 10^{41} \text{ kg}$$

$$= 9.4 \times 10^{10} \text{ M}_\odot$$

This estimate involves only mass interior to the Sun's orbit, because matter outside the Sun's orbit does not affect the Sun's motion and thus does not enter into Kepler's third law.

25-5   Self-sustaining star formation can produce spiral arms

That spiral arms should exist at all was another puzzle that confounded astronomers for many years. One explanation involves the effects of ongoing stellar evolution in the disk of a galaxy.

Many galaxies exhibit the beautiful arching arms outlined by brilliant H II regions and OB associations (recall Figure 25-14a). As we think through the effects of a galaxy's rotation, a dilemma arises. All spiral galaxies have rotation curves similar to our own (glance ahead at Figure 26-25). As Figure 25-17 demonstrated, the velocity of stars and gases is fairly constant over a large portion of a galaxy's disk. However, the farther that stars are from a galaxy's center, the farther they must travel to complete one orbit of the galaxy. Thus, stars and gases on the outskirts of a galaxy take much longer to complete an orbit than does material near the galaxy's center. Consequently, the spiral arms should eventually "wind up," wrapping themselves tightly around the galaxy's nucleus. After a few galactic rotations, the spiral structure should disappear altogether.

a

b

FIGURE 25-18 Variety in Spiral Arms The differences from one spiral galaxy to another suggest that more than one process can create spiral arms. (a) The galaxy NGC 7793 has fuzzy, poorly defined spiral arms. (b) The galaxy NGC 628 has thin, well-defined spiral arms. (Courtesy of P. Seiden, D. Elmegreen, B. Elmegreen, and A. Mobarak; IBM)

The appearance of spiral arms varies widely. In some galaxies, called **flocculent spiral galaxies** (Figure 25-18a), the spiral arms are broad, fuzzy, chaotic, and poorly defined. In other galaxies, called **grand-design spiral galaxies** (Figure 25-18b), the spiral arms are thin, delicate, graceful, and well defined. This variation in shape suggests that more than one mechanism gives rise to the spiral structure of a galaxy.

The theory of **self-propagating star formation** explains how flocculent spiral arms can result from a galaxy's rotation. Imagine that star formation begins in a dense interstellar cloud somewhere in the disk of a galaxy that does not yet have spiral arms. As soon as hot, massive stars form, their radiation compresses nearby matter, triggering the formation of additional stars in that gas. When massive stars become supernovae, they produce shock waves that further compress the surrounding interstellar medium, thus encouraging still more star formation. Although all parts of this broad, star-forming region have roughly the same orbital speed about the galaxy's center, the inner regions have a shorter distance to travel to complete one orbit than the outer regions. As a result, the inner edges of the star-forming region move ahead of the outer edges as the Galaxy rotates. The bright O and B stars and their nearby glowing nebulae soon become stretched out in the form of a spiral arm.

Spiral arms produced by bursts of star formation come and go essentially at random across a galaxy. Bits and pieces of spiral arms appear where star formation has recently begun but fade and disappear at other locations where all the massive stars have died off. Self-propagating star formation therefore tends to produce flocculent spiral galaxies that have a chaotic appearance with poorly defined spiral arms, like NGC 7793 in Figure 25-18a. To account for the orderly appearance of grand-design spirals, we must turn to another explanation for spiral structure.

25-6 Spiral arms are caused by density waves that sweep around a galaxy

In the 1920s Bertil Lindblad proposed that the spiral arms of a galaxy are a persistent pattern that moves among the stars, much like ripples on a pond. As waves on the surface of a pond move across the water, the individual water molecules simply bob up and down in little ellipses: A cork in the water merely bobs up and down as the waves ripple by. The waves are simply a pattern that moves across the water; no water actually travels along with the wave pattern. Lindblad spoke similarly of **density waves** that sweep around a galaxy to cause its spiral structure.

In the mid-1960s American astronomers C. C. Lin and Frank Shu enhanced and embellished this density-wave theory. Lin and Shu argued that density waves passing through the disk of a galaxy cause stars and the interstellar medium to "pile up" temporarily. A spiral arm is thus a temporary compression of the material in a galaxy.

Imagine a traffic jam created by workers painting a line down a busy freeway. The cars normally cruise along the freeway at 55 miles per hour, but the crew of painters causes a temporary bottleneck. The cars must slow down temporarily to avoid hitting anyone. As seen from the air, there is a noticeable congestion of cars around the painters. An individual car spends only a few moments in the traffic jam, though, before resuming its usual speed. But the traffic jam itself lasts all day, inching its way along the road as the painters advance. The traffic jam, seen clearly from an airplane, is simply a temporary crowding of the number of cars in a particular location.

To understand better how a density wave operates in a galaxy, think once again about a pond. If the water molecules were left completely undisturbed, the surface of the pond would be perfectly smooth. In reality, however, the molecules are constantly buffeted by small disturbances, or perturbations, such as the wind. A perturbation pushes one molecule, which pushes the next one, which pushes the next, and so on. The result is a water wave. Individual molecules on the surface of the pond move in tiny elliptical paths as the wave pattern moves across the water (Figure 25-19a).

In a galaxy, stars are separated by vast distances. Nevertheless, stars do interact, because they are affected by each other's gravity. In water waves and sound waves, molecular forces are responsible for orchestrating the motions of molecules. In a galaxy, the force of gravity controls the interactions between stars.

a Water wave

b Kinematic wave

FIGURE 25-19 Water Waves and Star Orbits (a) In a water wave, each molecule rotates about a point on the undisturbed water level in a tiny ellipse. (b) Similarly, even a small disturbance in the orbit of a star can cause the star to oscillate in tiny ellipses about its original, nearly circular orbit (dots). The resulting path of the star is an ellipselike curve that precesses.

Seen from above, the undisturbed orbit of a star would be a nearly perfect circle about the center of a galaxy. However, other matter in the galaxy produces small gravitational perturbations that cause the star to deviate from its undisturbed orbit and to oscillate back and forth. To describe these oscillations, Lindblad thought of the star as being attached to a tiny "epicycle." As seen in Figure 25-19*b*, the star rotates counterclockwise around the epicycle, while the epicycle itself moves clockwise along the undisturbed path (shown as dots). The final path of the star is nearly an ellipse that slowly rotates, or precesses. Meanwhile, the gravity of this star affects the motions of its neighbors, creating a wavelike disturbance that propagates from one stellar orbit to the next.

In 1973 Agris J. Kalnajs in Australia argued that the precessing elliptical orbits of stars are not randomly oriented, as depicted in Figure 25-20*a*. Instead, there is a strict correlation between orbits: Each precessing elliptical orbit is tilted with respect to its neighbor at a specific angle. A spiral pattern resembling a grand-design spiral galaxy results where the ellipses are bunched closest together, as shown in Figure 25-20*b*. Although stars in a galaxy are scattered randomly along their orbits, they tend to crowd together along spiral arms, because that is where their orbits are closest together for the longest stretches.

An increased density of stars in a spiral arm increases the gravitational attraction all along the spiral. This force has almost no effect on the stars, which simply continue to lumber along their orbits. However, the atoms and molecules in the interstellar medium are readily pulled into the spiral arm, forming the crest of the density wave.

A density wave moves through the material of a galaxy at a speed roughly 30 km/s slower than the speed of the stars and the interstellar medium. On its own, the interstellar gas can transport a disturbance, such as a slight compression, at the speed of sound in the interstellar medium, or about 10 km/s. The density wave is faster than that, or supersonic. As happens with a supersonic airplane traveling through the air, a shock wave builds up along the leading edge of the density wave. Shock waves are characterized by a sudden,

abrupt compression of the medium through which they move, which is why you hear a sonic boom from a supersonic airplane. Similarly, the interstellar medium is violently compressed by the cosmic sonic boom of the density wave.

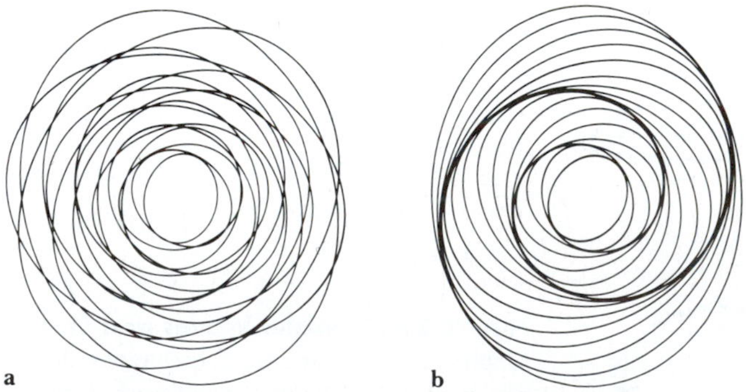

As density waves sweep through the plane of a galaxy, they recycle the interstellar medium. Old gases and dust left behind from ancient, dead stars are compressed into new nebulae in which new stars form. The sprawling dust lanes alongside the string of emission nebulae that outlines a spiral arm attest to the recent passage of a compressional shock wave (examine Figure 25-18*b*). Because the material left over from the death of ancient stars is enriched in heavy elements, new generations of stars are likely to be more metal-rich than their ancestors.

There must be a driving mechanism that keeps density waves going in spiral galaxies. After all, density waves expend an enormous amount of energy to compress the interstellar gas and dust. Debra and Bruce Elmegreen at IBM have pointed out how gravity can supply that needed energy. Grand-design spirals invariably have either a central bar extending across their nuclei or else a nearby companion galaxy. The asymmetric gravitational field of such a bar or companion pulls on the gas, stars, and dust of a galaxy to generate density waves. In the next chapter we shall further explore how gravitational interactions between galaxies induce spiral structure.

25-7 Infrared and radio observations are used to probe the galactic nucleus, whose nature is poorly understood

The nucleus of our Galaxy is an active, crowded place. The number of stars in Figure 25-7 gives some hint of the stellar congestion there. If you lived on a planet near the galactic center, you could see a million stars as bright as Sirius, the brightest single star in our own night sky. The total intensity of starlight from all those nearby stars would be equivalent to 200 of our full moons. In effect, night would never really fall on a planet near the center of the Milky Way.

Because of the severe interstellar absorption at visual wavelengths, some of our most important information about the galactic center comes from infrared and radio observations. Figure 25-21 shows two views of the center of our Galaxy. Figure 25-21*a* is a wide-angle photograph at visible wavelengths of the Milky Way, centered on the galactic nucleus in Sagittarius. Figure 25-21*b*, an IRAS view of the galactic center, covers the area indicated by the white rectangle in Figure 25-21*a*. The prominent band across this infrared image is interstellar dust in the plane of the Galaxy. The numerous knots and blobs along the layer of dust are interstellar clouds heated by young O and B stars. Numerous arches and streamers of dust surround the galactic nucleus. The strongest infrared emission comes from Sagittarius A, which is also a strong source of synchrotron radiation. As we saw in Chapter 23, this type of radio radiation is produced by high-speed electrons spiraling around a magnetic field. In spite of its compact size, Sagittarius A is one of the brightest radio sources in the entire sky.

FIGURE 25-20 The Origin of Spiral Density Waves
Both drawings have exactly the same number of ellipses, each one representing the orbit of a star. (**a**) Randomly oriented ellipses. (**b**) Ellipses with a correlation between the orientations of adjacent ellipses.

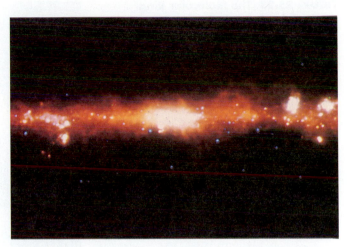

FIGURE 25-21 The Galactic Center (a) This wide-angle photograph of the Milky Way at visible wavelengths is centered on the nucleus of our Galaxy in Sagittarius. (b) This infrared view from IRAS covers the area outlined by the white rectangle in (a). Black represents the dimmest regions of infrared emission, with blue the next dimmest, followed by yellow, red, and then white for the strongest emission. The prominent band across this photograph is interstellar dust in the plane of the Galaxy. Streamers of dust extending above and below the dust layer can be seen. (Courtesy of D. di Cicco; NASA)

Doppler shift measurements of 21-cm radiation reveal the motions of hydrogen gas around the center of our Galaxy. Using a radio telescope in 1960, Jan H. Oort and G. W. Rougoor discovered two enormous expanding arms of hydrogen. One arm, located between us and the galactic center, is approaching us at a speed of 53 km/s. The other arm, on the other side of the galactic nucleus, is receding from us at a rate of 135 km/s. These expanding arms contain several million solar masses of hydrogen. Something quite extraordinary must have happened about 10 million years ago to expel such an enormous amount of gas from the center of our Galaxy.

Some of the most detailed radio images of the galactic center were produced by the VLA. Figure 25-22a is a wide-angle view of Sagittarius A covering an area about 250 light-years across. Huge filaments perpendicular to the plane of

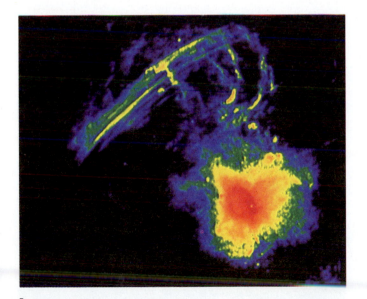

a

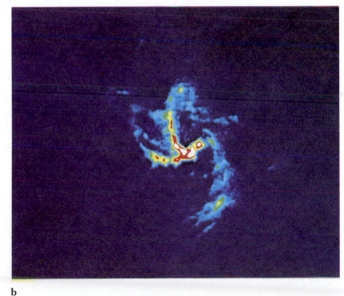

b

FIGURE 25-22 Two Radio Views of the Galactic Nucleus These two pictures show the appearance of the center of our Galaxy at radio wavelengths. The strongest radio emission is shown in red, weaker emission being colored green through blue. (a) This view covers an area of the sky about the same size as the full moon, corresponding to a distance of 250 ly across. The parallel filaments may be associated with a magnetic field. The galactic nucleus is toward the lower right, at the center of the strongest emission. (b) This high-resolution view shows details of the galactic center covering an area 20 ly across. The pinwheel-like structure is centered on Sagittarius A*. (VLA, NRAO; Courtesy of K. Y. Lo and N. Killeen, VLA, NCSA)

FIGURE 25-23 The Molecular Ring Around the Galactic Center This high-resolution map of HCN emission reveals a ring of neutral atomic and molecular gas surrounding the galactic nucleus. The ring is about 15 ly in diameter and consists of gas at a temperature of about 300 K, which is much warmer than the dust (80 K) in that region. The pinwheel feature in Figure 25-22b is outlined in white. Note that one of the pinwheel's arms (the western arc) coincides with the ring, but the other two arms do not. (Courtesy of R. Güsten and M.C.H. Wright)

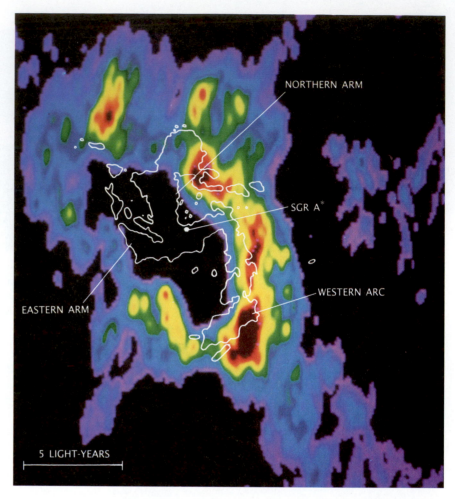

NORTHERN ARM

SGR A*

WESTERN ARC

EASTERN ARM

5 LIGHT-YEARS

the Galaxy stretch 200 light-years northward of the galactic center, then abruptly arch southward toward Sagittarius A, at the core of the strongest emission. The orderly arrangement of these filaments, reminiscent of quiescent prominences on the Sun, suggests that a magnetic field may be controlling the flow of ionized gas.

The inner core of Sagittarius A, shown in Figure 25-22b, covers an area about 20 light-years across. Here a pinwheel-like feature surrounds a source called Sagittarius A* (pronounced "Sagittarius A star") at the center of this view. One of the arms of this pinwheel is part of a ring of gas and dust orbiting the galactic center.

FIGURE 25-24 The Galactic Center This infrared image ▶ shows about 300 stars less than 1 light-year from Sagittarius A*, which is at the center of the picture. Details as small as $\frac{1}{50}$th of a light-year can be seen in this remarkable image, obtained with the New Technology Telescope at the European Southern Observatory in Chile. The distribution of stars around the galactic center implies a very high stellar density (about a million solar masses per cubic light-year) there. (Courtesy of A. Eckart, R. Genzel, R. Hofmann, B. J. Sams, and L. E. Tacconi-Garman; ESO)

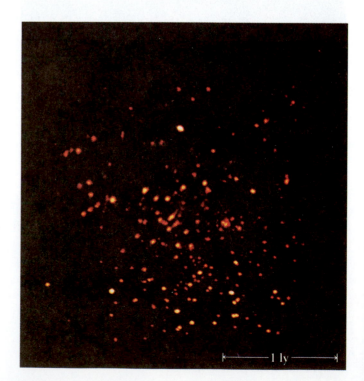

1 ly

Radio radiation from HCN molecules in the ring surrounding Sagittarius A* yields the view in Figure 25-23. Inclined by about 70° to our line of sight, this ring is quite lumpy and consists of turbulent clouds of gas. The total amount of gas, both atomic and molecular, orbiting the galactic nucleus along with this ringlike structure is estimated at 3×10^4 $M_\odot$. All the material less than about 6 light-years from the galactic nucleus seems to have been ionized by a source that coincides with Sagittarius A*.

Sagittarius A* is believed to be the galactic nucleus. It is the brightest radio source within several degrees of the galactic center. Its position, pinpointed with simultaneous observations by radio telescopes scattered around the world, seems to be very near the dynamical and gravitational center of the Galaxy. The immediate vicinity of Sagittarius A* is shown Figure 25-24, a high-resolution infrared image that covers a region at the galactic center about 2 light-years across.

Sagittarius A* is neither a star nor a pulsar, based on its high luminosity and radio spectrum. It also cannot be a supernova remnant because it is not expanding rapidly. Some evidence suggests that it might be a very massive black hole. The strongest evidence comes from Doppler shifts of streamers of ionized gas in the vicinity of Sagittarius A*. Their observed speeds, as high as 300 km/s, can be explained as orbital motion around the galactic center. A source of gravity must therefore be holding this high-speed gas in orbit about the galactic center. Using Kepler's third law, astronomers estimate that one million solar masses is needed to prevent this gas from flying off into interstellar space. Many astronomers argue that an object this massive and this compact could be only a black hole. Because of its enormous mass, it is called a **supermassive black hole.** As we shall see in Chapter 27, astronomers have also found extraordinary activity in the nuclei of many other galaxies, which indicates the possibility of supermassive black holes at their centers. The idea of an extremely massive black hole at the center of our own Galaxy would thus have many precedents.

Many astronomers disagree, however, with this idea of a supermassive black hole at the center of our Galaxy. First of all, the gravity of a supermassive black hole should cause stars to crowd around the galactic center much more than is actually observed. Second, accretion by the black hole should produce an infrared source, but Sagittarius A* is not associated with any infrared object. Finally, infrared sources near Sagittarius A* do not seem to be tidally disrupted, as would be expected from the gravity of a supermassive black hole.

Astronomers are still groping for a better understanding of the galactic center. During the coming years, observations from Earth-orbiting satellites as well as from radio and infrared telescopes on the ground will certainly add to our knowledge of the core of the Milky Way.

KEY WORDS

Terms preceded by an asterisk are discussed in the boxes.

central bulge (of a galaxy)	galactic nucleus	*Local Bubble	spin
dark matter	grand-design spiral galaxy	Milky Way Galaxy	spin–flip transition
density wave	halo (of a galaxy)	rotation curve	spiral arm
disk (of a galaxy)	high-velocity star	RR Lyrae variable	supermassive black hole
flocculent spiral galaxy	interstellar extinction	self-propagating star formation	21-cm radio emission

KEY IDEAS

• Our Galaxy has a disk about 25 kpc (80,000 ly) in diameter and about 600 pc (2000 ly) thick, with a high concentration of interstellar dust and gas in the disk.

The galactic nucleus is surrounded by a spherical distribution of stars called the central bulge; the entire Galaxy is surrounded by a spherical distribution of globular clusters and old stars called the halo of the Galaxy.

OB associations, H II regions, and molecular clouds in the galactic disk outline huge spiral arms.

• Interstellar dust obscures our view at visual wavelengths along lines of sight that lie in the plane of the galactic disk.

Hydrogen clouds can be detected despite the intervening interstellar dust by the 21-cm radio waves emitted by the spin–flip transition.

The galactic nucleus has been studied through its radio and infrared emissions, which also pass readily through the intervening interstellar dust.

• The Sun is located about 7.7 kpc (25,000 ly) from the galactic nucleus, between two major spiral arms.

The Sun moves in its orbit at a speed of about one-half million miles per hour and takes about 200 million years to complete one orbit about the Galaxy.

From studies of the rotation of the Galaxy, astronomers estimate that the total mass of the Galaxy is about 6×10^{11} $M_\odot$, with much of this mass being in some nonvisible, unknown form spread beyond the edge of the luminous material in the Galaxy.

• According to the theory of self-propagating star formation, spiral arms are caused by the birth of stars over an extended region in a galaxy. Differential rotation of the galaxy stretches the star-forming region into an elongated arch of stars and nebulae.

- According to the density-wave theory, spiral arms are created by density waves that sweep around the Galaxy.

    Each star moves in a precessing ellipse about the galactic nucleus; gravity correlates their orbits, creating a spiral pattern.

    The gravitational field of this spiral pattern compresses the interstellar clouds through which it passes, thereby triggering the formation of the OB associations and H II regions that illuminate the spiral arms.

- Infrared and radio observations have revealed many details of the galactic nucleus, but astronomers are still puzzled by the processes occurring there.

    A strong radio source called Sagittarius A* is located at the galactic center.

    A supermassive black hole with a mass of about 10^6 $M_\odot$ may exist at the galactic center.

REVIEW QUESTIONS

1. Why do you suppose the Milky Way is far more prominent in July than in December?

2. How did interstellar extinction mislead astronomers into believing that we are at the center of our Galaxy?

3. How would the Milky Way appear if the Sun were relocated to the edge of the Galaxy?

4. Why don't astronomers detect 21-cm radiation from the hydrogen in giant molecular clouds?

5. Why do astronomers believe that vast quantities of dark matter surround our Galaxy?

6. Describe the rotation curve you would obtain if the Galaxy were rotating like a rigid body.

7. Explain why globular clusters spend most of their time in the galactic halo, even though their eccentric orbits take them close to the galactic center.

8. A gas cloud located in the spiral arm of a distant galaxy is observed to have an orbital velocity of 400 km/s. If the cloud is 20,000 pc from the center of the galaxy and is moving in a circular orbit, find (a) the orbital period of the cloud and (b) the mass of the galaxy contained within the cloud's orbit.

9. Approximately how many times has our solar system orbited the center of our Galaxy since the Sun and planets were formed approximately five billion years ago?

10. How would you estimate the total number of stars in our Galaxy?

11. Compare the kinds of spiral arms produced by density waves with those produced by self-propagating

star formation. Examine a map of our Galaxy and cite evidence that both processes probably occur in our Galaxy.

ADVANCED QUESTIONS

Tips and tools . . .

Several of the following questions make extensive use of Kepler's third law, and so you might find it helpful to review Box 4-3. According to the Pythagorean theorem, an isosceles right triangle has a hypotenuse that is longer than its sides by a factor of $\sqrt{2}$.

12. What can you surmise about galactic evolution from the fact that the galactic halo is dominated by population II stars, whereas population I stars are predominantly found in the galactic disk?

13. The mass of our Galaxy interior to the Sun's orbit is calculated from the radius of the Sun's orbit and its orbital speed. By how much would this estimate be in error if the calculated distance to the galactic center were off by 10%? By how much would this estimate be in error if the calculated orbital velocity were off by 10%?

14. Speculate on the reasons for the rapid rise in the Galaxy's rotation curve (see Figure 25-17) at distances close to the galactic center.

15. According to the Galaxy's rotation curve in Figure 25-17, a star 60,000 ly from the galactic center has an orbital velocity of about 300 km/s. What is the mass within that star's orbit?

16. Suppose you were to use a radio telescope to measure the Doppler shift of 21-cm radiation in the plane of the Galaxy. At an angle of 45° from the galactic center, you detect 21-cm radiation from a cloud of atomic hydrogen that has the highest Doppler shift along your line of sight. How far is that cloud from you and from the galactic center?

17. The Galaxy is about 25 kpc in diameter and 600 pc thick. If supernovae occur randomly in the Galaxy at the rate of about three each century, how often, on average, should we expect to see a supernova within 300 pc (1000 ly) of the Sun?

18. Show that the form of Kepler's third law stated in Box 4-3 ($P^2 = 4\pi^2 a^3/GM$) is equivalent to $M = rv^2/G$, which is the form of the law used in this chapter, provided the orbit is a circle.

19. The photographs on the next page show the spiral galaxy M74, located about 55 million light-years from Earth in the constellation of Pisces. The image on the left is an optical photograph taken at visible wavelengths. On the right is an ultraviolet photograph taken with NASA's Ultraviolet Imaging Telescope carried aloft by the Space Shuttle during the *Astro-1* mission in 1990. Compare these two images and, based on what you know about stellar evolution and spiral structure, explain the differences you see.

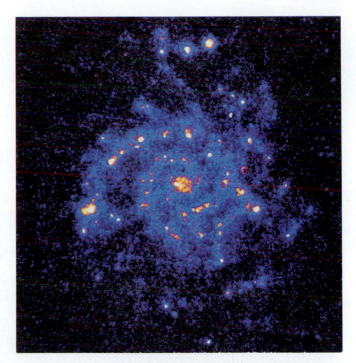

DISCUSSION QUESTIONS

20. From what you know about stellar evolution, the interstellar medium, and the density-wave theory, explain the appearance and structure of the spiral arms of grand-design spiral galaxies.

21. What observations would you make to determine the nature of the hidden mass in our Galaxy's halo?

OBSERVING PROJECT

22. Examine the Milky Way on a clear, moonless night with a high-quality pair of binoculars or a telescope with a wide field of view. During the summer months, look toward the galactic center in Sagittarius and follow the Milky Way arching across the sky through the constellations of Aquila and Cygnus. In the winter, follow the Milky Way as it sweeps across such recognizable constellations as Canis Major, Auriga, and Perseus. Take your time to note details in the structure of the Milky Way. For instance, toward the galactic center you should be able to see a few H II regions along with extensive star clouds. Other stretches of the Milky Way clearly display dark patches and mottling where clouds of interstellar gas and dust have blocked the light from background stars. Enjoy the spectacle.

FOR FURTHER READING

Bok, B. "A Bigger and Better Milky Way." *Astronomy,* January 1984. This entertaining and informative article was written by a renowned Dutch-American astronomer who made significant contributions to our modern understanding of the Milky Way.

———. "The Milky Way Galaxy." *Scientific American,* March 1981. This article clearly shows how astronomers have struggled to understand the structure and dynamics of our Galaxy.

Bok, B., and Bok, P. *The Milky Way.* 5th ed. Harvard University Press, 1981. This classic book presents an up-to-date, in-depth survey of the Milky Way.

Chaisson, E. "Journey to the Center of the Galaxy." *Astronomy,* August 1980. This well-written article describes many of the mysteries associated with the galactic center.

Chien, P. "EUVE Probes the Local Bubble." *Sky & Telescope,* February 1992. This article explains how the Extreme-Ultraviolet Explorer is giving us views of the heavens at wavelengths previously thought to be unobservable.

Geballe, T. "The Central Parsec of the Galaxy." *Scientific American,* July 1979. This article explains how astronomers explore the swirling mass of stars, gas, and dust that envelops the galactic center.

Kraus, J. "The Center of Our Galaxy." *Sky & Telescope,* January 1983. This brief article explains how radio astronomers investigate the galactic center.

Palmer, E. S. "Unveiling the Hidden Milky Way." *Astronomy,* November 1989. Beautiful illustrations grace this article, which explains how radio astronomers trace the spiral structure of our Galaxy.

Scoville, N., and Young, J. "Molecular Clouds, Star Formation and Galactic Structure." *Scientific American,* April 1984. This article shows how star formation and giant molecular clouds can affect the structure of the Galaxy.

CHAPTER

26

GALAXIES

A SPIRAL GALAXY This galaxy, called NGC 1365, is the largest and most impressive member of a cluster of galaxies about 60 million light-years from Earth. NGC 1365 is classified as a barred spiral because of the bar crossing its nucleus. From the ends of the bar two distinct and quite open spiral arms branch off. The yellow color of the nucleus and the bar shows that old, relatively cool stars occupy these parts of the galaxy. Light from young, hot stars causes the blue color of the arms. (European Southern Observatory)

THE UNIVERSE IS populated with galaxies, which come in many shapes and sizes. Some are disk shaped, like our own Milky Way, with arching spiral arms that are active sites of star formation. Others are featureless, ellipse-shaped agglomerations of stars, virtually devoid of interstellar gas and dust. Some galaxies are only a hundredth of the size and one ten-thousandth of the mass of the Milky Way. Others are giants, typically five times the size and fifty times the mass of the Milky Way. Most galaxies are located in groups and clusters that stretch across the universe forming huge, lacy patterns. Galaxies occasionally collide, producing such spectacular phenomena as a violent burst of star formation and enormous trails of stars and gas cast outward into intergalactic space. Most important, remote galaxies are receding from us with speeds that are proportional to their distances from our Galaxy. This relationship between distance and recessional velocity, called the Hubble law, reveals that we live in an expanding universe.

William Parsons was the third Earl of Rosse in Ireland. He was rich, he liked machines, and he was fascinated with astronomy. Accordingly, he set about building gigantic telescopes. In February 1845 his pièce de résistance was finished. The telescope's massive mirror measured 6 ft in diameter and was mounted at one end of a 60-ft tube controlled by cables, straps, pulleys, and cranes. For many years, this triumph of nineteenth-century engineering was the largest telescope in the world.

With this new telescope, Lord Rosse examined many of the nebulae that had been discovered and catalogued by Wil-liam Herschel. Lord Rosse observed that some of these nebulae have a distinct spiral structure. One of the best examples is M51 (also called NGC 5194).

Lacking photographic equipment, the earl had to make drawings of what he saw. Figure 26-1 shows a drawing he made of M51, and Figure 26-2 shows a modern photograph of it. Views such as this inspired Lord Rosse to echo the famous German philosopher Immanuel Kant, who in 1755 had suggested that these objects might be "island universes"—vast collections of stars far beyond the confines of the Milky Way.

FIGURE 26-1 Lord Rosse's Sketch of M51 Using a large telescope of his own design, Lord Rosse was able to distinguish spiral structure in this "spiral nebula." This galaxy is today called the Whirlpool Galaxy because of its distinctive appearance. (Courtesy of Lund Humphries)

FIGURE 26-2 The Spiral Galaxy M51 This spiral galaxy (also called NGC 5194) in the constellation of Canes Venatici is about 15 million light-years from Earth. The "blob" at the end of one spiral arm is the companion galaxy NGC 5195. (Courtesy of D. Elmegreen, B. Elmegreen, and P. Seiden; IBM)

Many astronomers did not subscribe to this notion of island universes. A considerable number of the objects listed in the NGC are in fact scattered throughout the Milky Way, and it seemed likely that the intriguing "spiral nebulae" could also be components of our Galaxy.

The astronomical community became increasingly divided over the nature of the spiral nebulae. Finally, in April 1920, a debate was held at the National Academy of Sciences in Washington, D.C. On one side was Harlow Shapley, a young, brilliant astronomer renowned for his recent determination of the size of the Milky Way Galaxy. Shapley believed the spiral nebulae to be relatively small, nearby objects scattered around our Galaxy like the globular clusters he had studied. Opposing Shapley was Heber D. Curtis of the Lick Observatory near San Jose, California. Curtis championed the island universe theory, arguing that each of these spiral nebulae is a rotating system of stars much like our own Galaxy.

The Shapley–Curtis debate generated much heat but little light. Nothing was decided, because no one could present conclusive evidence to demonstrate exactly how far away the spiral nebulae were. Astronomy desperately needed a defini-tive determination of the distance to a spiral nebula. Such a measurement became the first great achievement of a young man who studied astronomy at the Yerkes Observatory, near Chicago. His name was Edwin Hubble.

26-1   Hubble proved that the spiral nebulae are far beyond the Milky Way

After completing his studies, Edwin Hubble joined the staff of the Mount Wilson Observatory in Pasadena, California, and on October 6, 1923, took a historic photograph of the Andromeda "Nebula," one of the spiral nebulae around which controversy raged. A modern photograph of this object appears in Figure 26-3. Hubble carefully examined his photographic plate and discovered what he at first thought to be a nova. Referring to previous plates of that region, he soon realized that the object was actually a Cepheid variable star. Further scrutiny of additional plates over the next several months revealed several more Cepheids, two of which are identified in Figure 26-4.

FIGURE 26-3 The Andromeda Galaxy This nearby galaxy (also called M31 or NGC 224) covers an area of the sky roughly five times as large as the full moon. Under good observing conditions, the galaxy's bright central bulge can be glimpsed with the naked eye in the constellation of Andromeda. The distance to this galaxy is $2\frac{1}{4}$ million light-years. The white rectangle outlines the area shown in Figure 26-4. (Palomar Observatory)

FIGURE 26-4 Cepheid Variables in the Andromeda Galaxy Two Cepheid variables are identified (see arrows) in this view showing the outskirts of the Andromeda Galaxy. Because these stars appear so faint, even though they are intrinsically luminous, Edwin Hubble successfully demonstrated that the Andromeda "Nebula" is extremely far away. (The Carnegie Observatories)

As we saw in Chapter 21, Cepheid variables help astronomers determine distances. An astronomer begins by carefully measuring the variations in brightness of a Cepheid variable, recording the results in the form of a light curve (a plot of apparent magnitude versus time). This graph gives the period and average apparent magnitude of the variable. The astronomer then uses the period–luminosity relation (recall Figure 21-16) to find the average absolute magnitude of the Cepheid variable. Knowing both the absolute and apparent magnitudes, the astronomer calculates the distance to the star (Box 26-1).

Cepheid variables are intrinsically bright: They typically have luminosities of a few thousand Suns. Hubble realized that for these luminous stars to appear as dim as they were

---

BOX 26-1

Cepheid Variables as Indicators of Distance

Because their periods are directly linked to their intrinsic brightness, Cepheid variables are one of the most reliable tools astronomers have for determining the distances to galaxies. To this day, astronomers use this link—much as Hubble did back in the 1920s—to measure intergalactic distances. In 1992, for instance, a team of astronomers used the Hubble Space Telescope to find Cepheid variables in a galaxy called IC 4182 and deduce its distance from Earth.

The team of astronomers used the Hubble Space Telescope on 20 separate occasions to take pictures of the stars in IC 4182. By comparing these pictures, the astronomers could pick out which stars vary in brightness. The photographs below show one of 27 Cepheids they discovered. From these images, the astronomers estimated the apparent magnitudes of the Cepheid variables and plotted their light curves.

The accompanying graph shows the light curve of one such Cepheid, whose period is 42.0 days and whose average apparent magnitude (m) is +22.0. According to the period–luminosity law, a Cepheid variable with a period of 42.0 days should have an average absolute magnitude (M) of −6.5. Therefore the distance modulus ($m − M$) for this star is

$$(m - M) = (22.0 + 6.5) = 28.5$$

In Chapter 19, we saw that the distance modulus of a star is related to its distance in parsecs (d) by

$$d = 10^{(m - M + 5)/5}$$

Inserting the value for the distance modulus in this equation, we get the distance to the galaxy:

$$d = 10^{(28.5 + 5)/5} = 10^{6.7} = 5 \times 10^6$$

The galaxy is therefore 5 Mpc (16 million light-years) from Earth.

Astronomers are interested in IC 4182 because a Type Ia supernova was observed there in 1937. All Type Ia supernovae are exploding white dwarfs that reach the same maximum brightness at the peak of their outburst. If astronomers knew the peak absolute magnitude of Type Ia supernovae, they could use this value to determine the distances to galaxies so dim and remote that only an occasional supernova becomes bright enough to be seen clearly.

At maximum brightness, the 1937 supernova reached an apparent magnitude of +8.6. Since the distance modulus of the galaxy ($m − M$) is 28.5, we see that the maximum brightness of a Type Ia supernova is

$$M = 8.6 - 28.5 = -19.9$$

Whenever astronomers find a Type Ia supernova in a remote galaxy, they can combine this absolute magnitude with the observed maximum apparent magnitude to get the galaxy's distance modulus, from which the galaxy's distance can be easily calculated.

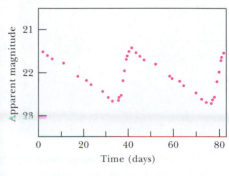

on his photographs of the Andromeda "Nebula," they must be extremely far away. Straightforward calculations using modern data reveal that M31 is $2\frac{1}{4}$ million light-years from Earth. This result proves that the Andromeda "Nebula" is not a nearby nebula but an enormous stellar system, far beyond the confines of the Milky Way. Today, this system is properly called the Andromeda Galaxy.

Hubble's results, which were presented at a meeting of the American Astronomical Society on December 30, 1924, settled the Shapley–Curtis debate once and for all. The universe was recognized to be far larger and populated with far bigger objects than anyone had thus far seriously imagined. Hubble had discovered the realm of the galaxies.

26-2   Hubble devised a system for classifying galaxies according to their appearance

Millions of galaxies spread across the sky. They can be seen in every unobscured direction. A typical spiral galaxy contains roughly 100 billion stars and measures about 100,000 light-years in diameter.

Hubble found that galaxies can be classified by their appearance into four broad categories. These categories form the basis for what is now referred to as the **Hubble classification scheme.** The four types of galaxies are the spirals, barred spirals, ellipticals, and irregulars.

Both M51 and M31, seen in Figures 26-2 and 26-3, respectively, are examples of **spiral galaxies,** characterized by arched lanes of stars and glowing nebulae. While studying spiral galaxies, Hubble noted that this class can be further subdivided, according to the relative size of the central bulge, the winding of the spiral arms, and the knottiness of the arms (Figure 26-5). Spirals with tightly wound spiral arms and a prominent, fat central bulge are called Sa galaxies. Those with moderately wound spiral arms and a moderate-sized central bulge, such as M51 and M31, are Sb galaxies. And

galaxies with loosely wound spirals and a tiny central bulge are Sc galaxies. Generally, the arms become more "knotty" in appearance as the spiral pattern becomes more open.

Fortunately, the size of the central bulge and the degree of winding of the spiral arms go hand in hand, so we can even classify spiral galaxies that are viewed nearly edge-on. Note, for example, the large central bulge of the Sa galaxy in Figure 26-6. If we could see this galaxy face-on, we would find that its spiral arms are tightly wound around its voluminous bulge. An Sb galaxy has a smaller central bulge, whereas the tiny central bulge of an Sc is barely noticeable in an edge-on view.

In **barred spiral galaxies,** the spiral arms originate at the ends of a bar running through the galaxy's nucleus rather than from the nucleus itself (Figure 26-7). As with ordinary spirals, Hubble subdivided barred spirals according to the relative size of their central bulge, the winding of their spiral arms and their knottiness. An SBa galaxy has a large central bulge and thin, tightly wound spiral arms. Likewise, an SBb galaxy is a barred spiral with a moderate central bulge and moderately wound spiral arms, while an SBc galaxy has lumpy, loosely wound spiral arms and a tiny central bulge. As we saw in the previous chapter, some astronomers think that the Milky Way Galaxy is a barred spiral with its bar pointing roughly toward Earth so that we see it almost end-on (recall Figure 25-3b).

Elliptical galaxies, so named because of their distinctly elliptical shapes, have no spiral arms. Hubble subdivided these galaxies according to how round or flattened they look. The roundest elliptical galaxies he called E0 galaxies; the flattest, E7 galaxies. Elliptical galaxies with intermediate amounts of flattening receive intermediate designations (Figure 26-8). Of course, an E1 or E2 galaxy might actually be a very flattened disk of stars that we just happen to view face-on, and a cigar-shaped E7 galaxy might look spherical if seen end-on. The Hubble scheme classifies galaxies entirely by how they appear from Earth.

Sa   Sb   Sc

FIGURE 26-5 Various Spiral Galaxies Viewed Face-On Edwin Hubble classified spiral galaxies according to the winding of their spiral arms and the relative size of their central bulges.

Three examples are shown here. (Courtesy of J. D. Wray; McDonald Observatory)

Sa

Sb

Sc

FIGURE 26-6 Various Spiral Galaxies Viewed Edge-On
An Sa galaxy has a large central bulge. An Sb galaxy has a more modest central bulge. An Sc galaxy has a small central bulge. (Courtesy of J. D. Wray; McDonald Observatory)

SBa

SBb

SBc

FIGURE 26-7 Various Barred Spiral Galaxies As with spiral galaxies, Edwin Hubble classified barred spirals according to the winding of their spiral arms and the relative size of their central bulges. Three examples are shown here. (Courtesy of J. D. Wray; McDonald Observatory)

E0

E3

E6

FIGURE 26-8 Various Elliptical Galaxies Edwin Hubble classified elliptical galaxies according to how round or flattened they look. An E0 galaxy is round. The flattest elliptical galaxies are E7s. Three examples are shown here. (Courtesy of J. D. Wray; McDonald Observatory)

Elliptical galaxies look far less dramatic than their spiral and barred spiral cousins because they are virtually devoid of interstellar gas and dust. Consequently, there is little material from which stars could have recently formed, and so there is no evidence of young stars in most elliptical galaxies. For the most part, star formation in elliptical galaxies ended long ago.

Elliptical galaxies have a wide range of sizes and masses. Both the biggest and the smallest galaxies in the universe are elliptical. Figure 26-9 shows an example of a **giant elliptical galaxy** that is about 20 times larger than a normal galaxy. This giant elliptical is located near the middle of a large cluster of galaxies in the constellation of Coma Berenices.

Giant ellipticals are rather rare, but **dwarf elliptical galaxies** are quite common. Dwarf ellipticals are only a fraction the size of their normal counterparts and contain so few stars—only a few million—that these galaxies are completely transparent. You can actually see straight through the center of a dwarf galaxy and out the other side, as shown in Figure 26-10.

Because of the Doppler effect, the average motions of stars in a galaxy can be deduced from the shape of the galaxy's spectral lines. Elliptical galaxies exhibit spectral lines whose shapes reveal that star motions are quite random. In a very round elliptical galaxy, this randomness is **isotropic,** meaning "equal in all directions." Because the stars are whizzing around equally in all directions, the galaxy is genuinely

spherical. In a flattened elliptical galaxy, the randomness of the stellar motions is **anisotropic,** which means that the range of star speeds is different in different directions. This anisotropy explains why E5s, E6s, and E7s are fatter in one direction than in another.

Edwin Hubble connected the three main types of galaxies—spirals, barred spirals, and ellipticals—in his now-famous "tuning fork diagram" (Figure 26-11). According to this scheme, the S0 and SB0 galaxies, also called **lenticular galaxies,** are midway in appearance between ellipticals and the two kinds of spirals. Although they look somewhat elliptical, lenticular galaxies—literally "lens-shaped" galaxies—have both a central bulge and a disk like spiral galaxies, but no discernible spiral arms. They are therefore sometimes referred to as "armless spirals." Figure 26-12 shows typical S0 and SB0 galaxies.

The tightness of a galaxy's spiral windings is related to the dust and gas content of the galaxy. Sa and SBa galaxies tend to have less gas and dust than do Sc and SBc galaxies. As stars form, galaxies gradually exhaust their gas and dust content. Although supernovae and planetary nebulae return some of this material to the interstellar medium, as a galaxy ages, more and more of its gas and dust become locked inside stars.

Galaxies that do not fit into Hubble's scheme of spirals, barred spirals, and ellipticals are usually referred to as **irregular galaxies.** Astronomers today recognize two types of

FIGURE 26-9 A Giant Elliptical Galaxy This huge galaxy sits near the center of a rich cluster of galaxies located about 70 Mpc (230 million light-years) from Earth. Many normal-sized galaxies surround this giant elliptical, which is about 2 million light-years in diameter. (Courtesy of R. Schild)

FIGURE 26-10 A Dwarf Elliptical Galaxy This nearby dwarf elliptical, called Leo I, is about 1 million light-years from Earth. Note how difficult it is to make out the galaxy, which fills the picture. Leo I contains so few stars that you can see through the galaxy's center. (Anglo-Australian Observatory)

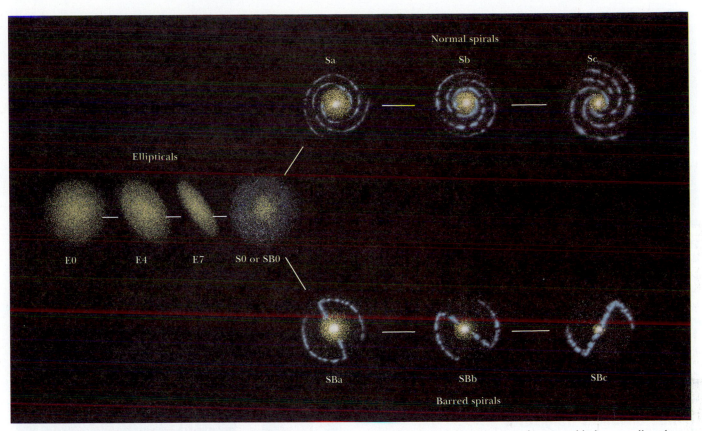

FIGURE 26-11 Hubble's Tuning Fork Diagram Hubble summarized his classification scheme for regular galaxies with this tuning fork diagram. An elliptical galaxy is classified according to how flattened it appears. A spiral or barred spiral galaxy is classified according to the size of its central bulge as well as the winding and lumpiness of its spiral arms. An S0 or SB0 galaxy is midway in appearance between ellipticals and spirals.

irregulars. Irr I galaxies, like underdeveloped spiral galaxies, contain many OB associations and H II regions. Irr II galaxies have asymmetrical, distorted shapes that seem to have been caused by collisions with other galaxies or by violent activity in their nuclei.

The best-known examples of Irr I galaxies are the Large Magellanic Cloud (LMC) and the Small Magellanic Cloud (SMC), which are nearby companions of our Milky Way that can be seen with the naked eye from southern latitudes. Telescopic views of the Magellanic clouds can easily resolve individual stars (Figures 26-13 and 26-14). The SMC does not exhibit any of the geometric symmetry characteristic of spirals or ellipticals, but the LMC does have a faint barlike structure.

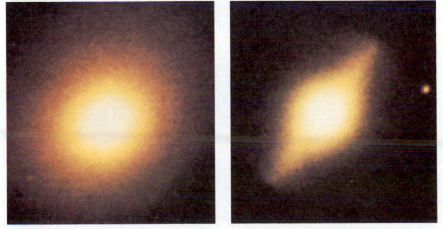

FIGURE 26-12 Lenticular Galaxies
A lenticular galaxy has a disk but no spiral arms. An SB0 galaxy seems to have a barlike structure across its nucleus, whereas an S0 galaxy does not. (Courtesy of J. D. Wray; McDonald Observatory)

FIGURE 26-13 The Large Magellanic Cloud (LMC)   At a distance of only 160,000 light-years, this galaxy is the nearest companion of our Milky Way Galaxy. Note the huge H II region (called the Tarantula Nebula, or 30 Doradus) toward the left side of the photograph. Its diameter of 800 light-years and mass of 5 million Suns make it the largest known H II region. (Anglo-Australian Observatory)

FIGURE 26-14 The Small Magellanic Cloud (SMC)   The SMC is only slightly farther away from us than the LMC. Because of its sprawling, asymmetrical shape, the SMC is classified as an irregular galaxy. Note that the SMC is rich in young, blue stars. (Anglo-Australian Observatory)

26-3   The Hubble law states that the redshifts of remote galaxies are proportional to their distances from Earth

Whenever an astronomer finds an object in the sky that can be seen or photographed, the natural inclination is to attach a spectrograph to a telescope and record the spectrum. As long ago as 1914, V. M. Slipher, working at the Lowell Observatory in Arizona, began taking spectra of "spiral nebulae." He was surprised to discover that of the 15 spiral nebulae he studied, the spectral lines of 11 were shifted toward the red end of the spectrum, indicating that they were all moving away from Earth. This marked dominance of redshifts was presented by Curtis in the Shapley–Curtis debate as evidence that these spiral nebulae could not be ordinary nebulae in our Milky Way Galaxy.

During the 1920s, Edwin Hubble and Milton Humason photographed the spectra of many galaxies with the 100-inch telescope on Mount Wilson. Five representative elliptical galaxies and their spectra are shown in Figure 26-15. As

FIGURE 26-15 Five Galaxies and Their Spectra   The ▶ photographs of these five elliptical galaxies all have the same magnification. They are arranged, from top to bottom, in order of increasing distance from Earth. The spectrum of each galaxy is the hazy band between the comparison spectra. In all five cases, the so-called H and K lines of singly ionized calcium can be seen. The recessional velocity, calculated from the Doppler shifts of the H and K lines, is given below each spectrum. Note that the more distant a galaxy is, the greater its redshift. (The Carnegie Observatories)

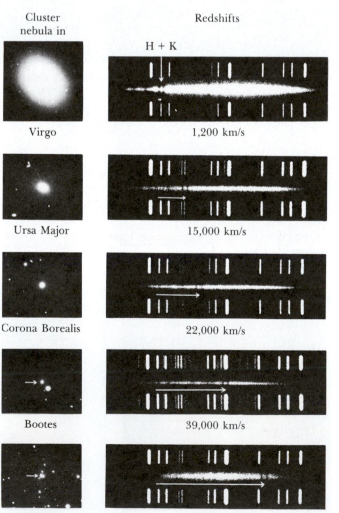

Cluster nebula in   Redshifts

H + K

Virgo   1,200 km/s

Ursa Major   15,000 km/s

Corona Borealis   22,000 km/s

Bootes   39,000 km/s

Hydra   61,000 km/s

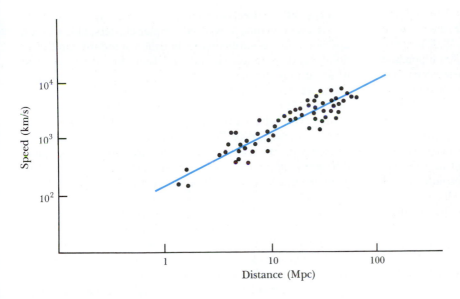

FIGURE 26-16 The Hubble Law The distances and recessional velocities of 60 Sc spiral galaxies are plotted on this graph. The straight line is the best fit for the data. This linear relationship between distance and speed is called the Hubble law. (Adapted from Sandage and Tammann)

indicated by this illustration, Hubble and Humason found a direct correlation between the distance to a galaxy and its redshift. In other words, nearby galaxies are moving away from us slowly, and more distant galaxies are rushing away from us much more rapidly. This universal recessional movement is sometimes referred to as **Hubble flow.**

Hubble estimated the distances to a number of galaxies and then used the Doppler formula (review Figure 5-19) to calculate the speed at which these galaxies are receding from us. Plotting the data on a graph of distance versus speed, Hubble found that the points lie near a straight line. Figure 26-16 is a modern version of Hubble's graph based on recent data.

This relationship between the distances to galaxies and their redshifts is one of the most important astronomical discoveries of the twentieth century. As we shall see in Chapter 28, this relationship tells us that we are living in an expanding universe. In 1929 Hubble published this discovery, which is now known as the **Hubble law.**

The Hubble law is most easily stated as a formula:

$$v = H_0 r$$

where v is the recessional velocity, r is the distance, and H_0 is a constant commonly called the **Hubble constant.** This formula is the equation for the straight line displayed in Figure 26-16. The Hubble constant H_0 is the slope of the line in Figure 26-16. From the data plotted on this graph we find that

$$H_0 = 50 \text{ km s}^{-1} \text{ Mpc}^{-1}$$

(say "fifty kilometers per second per megaparsec"). In other words, for each million parsecs to a galaxy, the galaxy's speed away from us increases by 50 km/s. For example, a galaxy located 100 million parsecs from Earth should be rushing away from us with a speed of 500 km/s. Incidentally, some people prefer to write the units of the Hubble constant with slashes rather than exponents: 50 km/s/Mpc.

The exact value of the Hubble constant is a topic of heated debate among astronomers today. The data plotted in Figure 26-16, as well as the values of H_0 given above, are based on work by Allan Sandage, Gustav Tammann, and their colleagues. However, other prominent astronomers, such as Sidney van den Bergh in Canada, Gerard de Vaucouleurs of the University of Texas, and John Huchra at Harvard, argue strongly that the Sandage–Tammann value for H_0 is too low. Independent, meticulous observations by these three astronomers and their colleagues have led them to infer that the true value for H_0 is about 90 km/s/Mpc. Many astronomers simply use a number between the two extremes, such as 75 km/s/Mpc.

The Hubble constant is one of the most important numbers in all astronomy. It expresses the rate at which the universe is expanding and even gives the age of the universe. Furthermore, the Hubble law can be used to determine the distances to extremely remote galaxies and quasars. If the redshift of a galaxy is known, the Hubble law can be used to convert that redshift into a distance from Earth, as described in Box 26-2.

26-4   Astronomers use "standard candles" to determine the distance to a remote galaxy

Teams of astronomers are making concerted efforts to determine an accurate value for the Hubble constant. The main challenge they face is accurately measuring distances to galaxies.

The distances to nearby galaxies can be determined fairly reliably. For example, Cepheid variables can be seen out to 20 Mly from Earth. The distances to galaxies in this nearby volume of space can thus be determined from the period–luminosity law.

Beyond 20 Mly even the brightest Cepheid variables, which have absolute magnitudes of about −6, fade from view.

Astronomers then turn to more luminous stars. The brightest red supergiants have absolute magnitudes of −8; the brightest blue supergiants have absolute magnitudes of −9. These two types of stars can be seen out to distances of 50 million and 80 million light-years, respectively. Out to these limits, you can determine the distances to galaxies from the apparent magnitudes of these luminous supergiants.

Beyond 80 Mly individual stars are no longer discernible. Astronomers therefore turn to entire star clusters and nebulae. The brightest globular clusters, which have a total absolute magnitude of about −10, can be seen out to 130 Mly from Earth. The brightest H II regions have absolute magnitudes of −12 and can be detected out to 300 Mly. From the measured apparent magnitudes of these clusters and nebulae, distances to remote galaxies can be estimated.

Finally, to get beyond 300 Mly, astronomers must wait for supernova explosions. Some supernovae reach an absolute magnitude of about -20 at the peak of their outbursts (Figure 26-17). These brilliant outbursts can, in theory, be seen out to distances of 8 billion light-years from Earth.

These varied objects—Cepheid variables and the most luminous supergiants, as well as globular clusters, H II regions, and supernovae—are commonly called **standard candles**. A good standard candle should have four characteristics. First of all, it must be bright, so that you can see it out to great distances. Second, it must have a well-defined luminosity, so that you know its intrinsic brightness. Third, it should be easily identifiable—for example, by the shape of its light curve or unique color. Finally, it must be relatively common, so that you can use it to determine the distances to many different galaxies. As you might suspect, astronomers go to great lengths to check the accuracy and reliability of their standard candles.

The major obstacle in determining the Hubble constant is that the farther we look into space, the fewer standard candles we have. For example, the distance to a nearby galaxy can be cross-checked in many ways. The distance computed from the period–luminosity relation can be compared to the distance determined from the magnitudes of the most luminous supergiants. Then these results can be compared with

---

BOX 26-2

The Hubble Law as a Distance Indicator

Suppose that you aim a telescope at an extremely distant galaxy. You take a spectrum of the galaxy and find that the spectral lines are shifted toward the red end of the spectrum. For instance, you find a spectral line whose normal wavelength is λ_0 at a longer wavelength λ. The spectral line has thus been shifted by an amount $\Delta\lambda = \lambda - \lambda_0$.

The redshift of the galaxy (usually denoted by z) is given by

$$z = \frac{\Delta\lambda}{\lambda_0} = \frac{\lambda - \lambda_0}{\lambda_0}$$

According to the Doppler effect (review Figure 5-19), this wavelength shift corresponds to a speed v, where

$$z = \frac{v}{c}$$

and c is the speed of light. (This equation is valid only if the speed v is much less than the speed of light.)

According to the Hubble law, the recessional velocity v of a galaxy is related to its distance r from Earth by

$$v = H_0 r$$

where H_0 is the Hubble constant.

Combining the equation for the Doppler shift with the equation for the Hubble law, we see that the distance to a galaxy is related to its redshift by

$$r = \frac{zc}{H_0}$$

EXAMPLE: Suppose that you observe the giant elliptical galaxy NGC 4889. The so-called K line of singly ionized calcium normally has a wavelength of 393.3 nm. In the spectrum of NGC 4889, you find this spectral line at 401.8 nm. Thus, the redshift of the galaxy is

$$z = \frac{401.8 - 393.3}{393.3} = 0.0216$$

The galaxy is therefore moving away from us with a speed of

$$v = zc = (0.0216)(3 \times 10^5 \text{ km/s}) = 6500 \text{ km/s}$$

With $H_0 = 50$ km/s/Mpc, the Hubble law gives the distance to the galaxy:

$$r = \frac{zc}{H_0} = \frac{6500}{50} = 130 \text{ Mpc} = 420 \text{ Mly}$$

Note that if you had used $H_0 = 100$ km/s/Mpc, you would have concluded that the distance is

$$r = \frac{zc}{H_0} = \frac{6500}{100} = 65 \text{ Mpc} = 210 \text{ Mly}$$

Unfortunately, astronomers must deal with this kind of uncertainty. Lacking an accurate value for the Hubble constant, most astronomers use a number between 50 and 90 km/s/Mpc. Thus, the distance to NGC 4889 is probably somewhere between 200 and 400 million light-years. A compromise distance of about 300 million light-years has been adopted for this text.

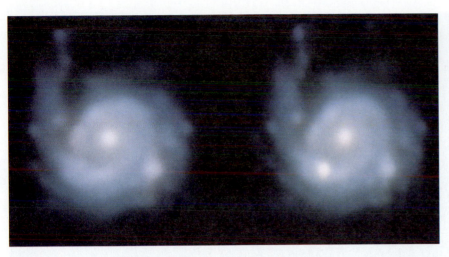

FIGURE 26-17 A Supernova in a Spiral Galaxy In 1991 a supernova erupted in the spiral galaxy NGC 3310. The left view shows the galaxy before the outburst. The right view shows the supernova near maximum brightness in one of the spiral arms of the galaxy. Supernovae, which can be seen even in extremely remote galaxies, are important "standard candles" used to determine the distances to these faraway galaxies. (NRAO)

the magnitudes of the galaxy's globular clusters and the angular sizes of its H II regions. The results from all these methods can then be averaged to obtain a distance to the galaxy in which astronomers have confidence.

As we turn to more distant galaxies, however, we can see fewer and fewer standard candles. With fewer standard candles, fewer cross-checks can be made. The distance to these remote galaxies thus becomes less certain. It is this uncertainty in determining distances that is responsible for our uncertainty in the value of H_0.

In the 1970s astronomers Brent Tully and Richard Fisher discovered that the width of the hydrogen 21-cm emission line of a spiral galaxy is related to the galaxy's absolute magnitude: The broader the line, the brighter the galaxy. This correlation, called the **Tully–Fisher relation,** is one of the best distance indicators currently available.

That such a relationship should exist can be seen as follows. Radiation from the approaching side of a rotating galaxy is blueshifted while that from the galaxy's receding side is redshifted. Thus the 21-cm line is Doppler broadened by an amount directly related to how fast a galaxy is rotating, which is related to the galaxy's mass by Kepler's third law. Furthermore, the more massive a galaxy is, the more stars it contains, and so the brighter it is. Consequently, the width of a galaxy's 21-cm line is directly related to the galaxy's luminosity. Since line widths can be measured quite accurately, astronomers can use the Tully–Fisher relation to determine the luminosities of spiral galaxies. Astronomers are hopeful that techniques, such as the Tully–Fisher relation, along with instruments like the Hubble Space Telescope and the Keck Telescope, will solve many of the problems in determining the precise value for H_0.

26-5 Galaxies are grouped in clusters that are members of superclusters

Galaxies are not scattered randomly throughout the universe but are found in **clusters.** A typical cluster, called the Fornax cluster because it is located in the southern constellation of Fornax (the Furnace), is seen in Figure 26-18.

A cluster is said to be either **poor** or **rich,** depending on how many galaxies it contains. Poor clusters, which far outnumber rich ones, are often called **groups.** For example, the Milky Way Galaxy, the Andromeda Galaxy, and the Large and Small Magellanic clouds belong to a poor cluster familiarly known as the **Local Group.** The Local Group contains nearly 30 galaxies, most of which are dwarf ellipticals. Box 26-3 contains a map of the Local Group.

The nearest fairly rich cluster is the Virgo cluster. It is a sprawling collection of over 1000 galaxies covering a $10° \times 12°$ area of the sky. The distance to the Virgo cluster is about 50 million light-years, too far away for Cepheid variables to be seen from our Galaxy. Instead, the distance to the Virgo cluster has been determined by the apparent faintness of O and B supergiant stars and by the brightness of globular clusters surrounding some of the galaxies. The overall diameter of the Virgo cluster is about 7 million light-years.

The center of the Virgo cluster is dominated by three giant elliptical galaxies. Two of them appear in Figure 26-19. These enormous galaxies are 2 million light-years in diameter, 20 times as large as an ordinary elliptical or spiral. In other words, one giant elliptical is roughly the same size as the entire Local Group!

Astronomers also categorize clusters of galaxies as regular or irregular, depending on the overall shape of the cluster. The Virgo cluster, for example, is called **irregular,** because its galaxies are scattered throughout a sprawling region of the sky. Our own Local Group is also an irregular cluster. In contrast, a **regular** cluster has a distinctly spherical appearance, with a marked concentration of galaxies at its center.

The nearest example of a rich, regular cluster is the Coma cluster, located about 300 million light-years from us toward the constellation of Coma Berenices. Despite the great distance to this cluster, more than 1000 bright galaxies within it are easily visible on photographic plates. Certainly, many thousands of dwarf ellipticals are too faint to be detected from our distance. The membership of the Coma cluster may total as many as 10,000 galaxies. The core of the Coma cluster is dominated by two giant ellipticals surrounded by many normal-sized galaxies (Figure 26-20).

FIGURE 26-18 A Cluster of Galaxies This cluster of galaxies, called the Fornax cluster, is about 60 million light-years from Earth. Both elliptical and spiral galaxies are easily identified. The barred spiral galaxy at the lower left is NGC 1365, the largest and most impressive member of the cluster. (Anglo-Australian Observatory)

FIGURE 26-19 The Center of the Virgo Cluster The Virgo cluster is a rich, sprawling collection of galaxies about 50 million light-years from Earth. Only the center of this huge cluster appears in this photograph. Note the two giant elliptical galaxies (M84 and M86). (Royal Observatory, Edinburgh)

The overall shape of a cluster is related to the dominant types of galaxies it contains. Rich, regular clusters contain mostly elliptical and S0 galaxies. For example, about 80% of the brightest galaxies in the Coma cluster are ellipticals; only a few spiral galaxies are scattered around the cluster's outer regions. Irregular clusters, such as the Virgo cluster and the Hercules cluster (Figure 26-21), have a more even mixture of galaxy types.

Clusters of galaxies are themselves grouped together in huge associations called **superclusters**. A typical supercluster contains dozens of individual clusters spread over a region of space up to 100 million light-years across. Patterns in the distribution of galaxies can be seen in maps such as the one in Figure 26-22, which covers many hundreds of square degrees. Note the delicate, lacy pattern spread across the sky.

In the 1980s several teams of astronomers undertook the monumental task of mapping the locations of galaxies in space. They used redshifts of galaxies in unobscured regions of the sky to construct maps that display the actual distribution of the galaxies in three dimensions. Each map, which typically covers a narrow strip 6° wide and 153° long, encompasses a pie-shaped slice of the universe that reaches out

FIGURE 26-20 The Center of the Coma Cluster This rich, regular cluster is about 300 million light-years from Earth. Only the cluster's center, dominated by two giant galaxies, appears in this view. Regular clusters are composed mostly of elliptical and S0 galaxies and are common sources of X rays. (Courtesy of R. Schild)

FIGURE 26-21 The Hercules Cluster This irregular cluster, which is about 500 million light-years from Earth, is much less dense than the Coma cluster. The Hercules cluster contains many spiral galaxies, often associated in pairs and small groups. (Courtesy of R. Schild)

FIGURE 26-22 Two Million Galaxies This map shows the distribution of roughly two million galaxies over about 10% of the sky. Each dot on the map is shaded according to the number of galaxies it contains. Dots are black where there are no galaxies, white where there are more than 20, and gray for a number between 1 and 19. Note that the clusters are not smoothly distributed across the sky; instead, they form a lacy, filamentary structure. The small, bright patches are individual galaxy clusters. The larger, elongated bright areas are superclusters and filaments, which generally surround darker voids containing few galaxies. Statistical analysis of this map shows that galaxies clump together on distance scales up to about 150 million light-years. Over distances greater than this, the distribution of galaxies in the universe appears to be roughly uniform. (S. J. Maddox, W. J. Sutherland, G. P. Efstathiou, and J. Loveday; Oxford Astrophysics)

BOX 26-3
The Local Group

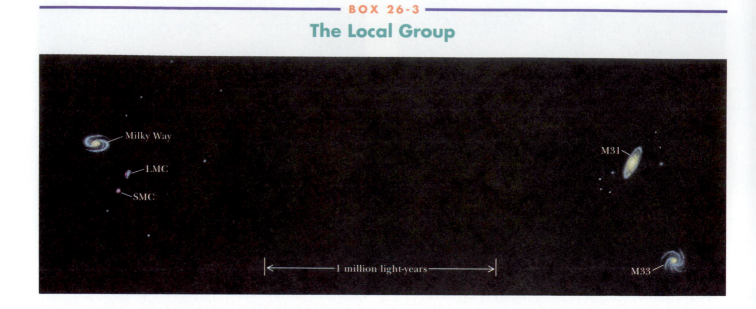

The Andromeda Galaxy, M31, is the biggest and most massive galaxy in the Local Group. Our Milky Way Galaxy takes second place. Scattered around these two primary galaxies are at least 25 smaller ones, most of which are dwarf ellipticals. The illustration shows an artist's rendition of the Local Group; the table lists these galaxies in order of increasing distance from Earth (the letter *d* preceding a Hubble type indicates a dwarf galaxy). The ten galaxies closest to us are satellites of the Milky Way Galaxy. Similarly, the Andromeda Galaxy has eight satellites.

Astronomers occasionally find additional nearby dwarf galaxies that are also members of the Local Group.

We may never know the total number of galaxies in the Local Group because dust in the plane of the Milky Way obscures our view over a considerable region of the sky. Nevertheless, we can be certain that no additional large spiral galaxies are hidden by the Milky Way because radio astronomers would have detected 21-cm radiation from them, even though their visible light is completely absorbed by interstellar dust.

Galaxy	Type	Distance (10^3 ly)	Diameter (10^3 ly)	Galaxy	Type	Distance (10^3 ly)	Diameter (10^3 ly)
Milky Way	Sb	—	80	NGC 147	E5	2200	10
LMC	Irr	160	20	NGC 185	E2	2200	6
SMC	Irr	190	15	IC 1613	Irr	2500	12
Draco	dE3	220	0.5	NGC 205	E5	2200	10
Ursa Minor	dE5	220	1	M32	E2	2250	5
Sculptor	dE3	240	1	M31	Sb	2250	130
Carina	dE3	300	0.5	Andromeda I	dE0	2250	2
Sextans I	dE	300	3	Andromeda II	dE2	2250	2
Fornax	dE3	500	3	Andromeda III	dE0	2250	2
Leo I	dE4	600	1	M33	Sc	2500	50
Leo II	dE0	600	0.5	DDO 210	Irr	3000	4
NGC 6822	Irr	1700	8	Pisces	Irr	3000	0.5
IC 5152	Irr	2000	5	Tucana	dE5	3000	2
WLM	Irr	2000	7				

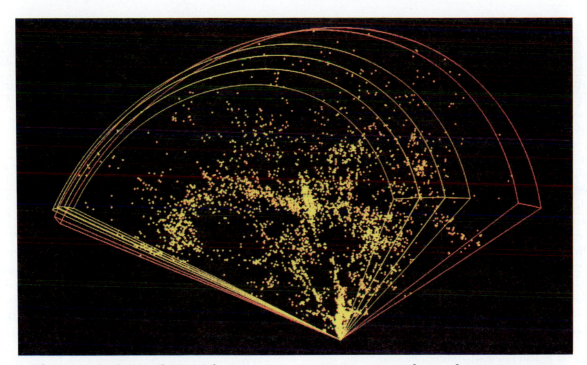

FIGURE 26-23   The Large-Scale Distribution of Galaxies   This map shows the distribution of nearly 4000 galaxies in four slices extending out to a redshift of $z = 15,000$ km/s. The Earth is located at the apex of the pie-shaped slices. Yellow dots indicate galaxies in the right ascension range 8^h to 17^h and declination range $26.5°$ to $44.5°$; orange dots cover the declination range $8.5°$ to $14.5°$. The "Great Wall" crosses the survey nearly parallel to the outer boundary. The vertical clumping of dots near the center of the map is the Coma cluster (see Figure 26-20), which forms part of the Great Wall. Note the prominent voids along the inner edge of the Great Wall. (Courtesy of M. J. Geller; Smithsonian Astrophysical Observatory)

to a redshift of 15,000 km/s. If the Hubble constant is 75 km/s/Mpc, this redshift corresponds to a distance of 200 Mpc (650 million light-years) from Earth. Figure 26-23 shows four such slices. The Earth is at the apex of the wedge-shaped map; each dot represents a galaxy.

These maps reveal several important characteristics of the large-scale distribution of galaxies in the universe. For instance, the maps show enormous **voids** where exceptionally few galaxies are found. These voids, which seem to be somewhat elongated or tube-shaped, measure 100 million to 400 million light-years across. Clusters of galaxies are concentrated on the surfaces of these voids. As we shall see in Chapter 29, many astronomers suspect that this large-scale pattern contains important clues about conditions shortly after the Big Bang that led to the formation of clusters of galaxies.

The maps also reveal large, coherent structures extending vast distances. For example, Figure 26-23 shows a long band of galaxies, dubbed the "Great Wall," roughly parallel to the curved outer boundary of the map. If the Hubble constant equals 75 km/s/Mpc, then this wall covers an area at least 80 by 230 Mpc (about $\frac{1}{4}$ by $\frac{3}{4}$ billion light-years).

In recent years astronomers have time and again been startled to find vast structures whose sizes seem to be limited only by the area of the sky covered in their surveys. Indeed, when larger areas are explored, larger features are found.

Furthermore, some of these features exhibit significant movement, indicating large-scale motions in the universe (see the essay by Alan Dressler that follows this chapter).

At present astrophysicists are working to explain how these enormous structures came into existence. Ongoing redshift surveys to explore even larger volumes of the universe may well uncover even more surprises during the next few years.

26-6   Most of the matter in the universe has yet to be discovered

A cluster of galaxies must be held together by gravity. In other words, there must be enough matter in the cluster to prevent the galaxies from wandering away. Nevertheless, careful examination of a rich cluster, like the Coma cluster, reveals that the mass of the visually luminous matter is not at all sufficient to bind the cluster gravitationally. The observed line-of-sight speeds of the galaxies, measured by Doppler shifts, are so large that the cluster should have broken apart long ago. Considerably more mass than has been observed is needed to keep the galaxies bound in orbit about the center of the cluster.

This dilemma is called the **dark-matter problem.** A lot of nonluminous matter must be contained within each cluster, or else the galaxies would have long ago dispersed in random directions and the cluster would no longer exist today. Analyses demonstrate that the total mass needed to bind a typical rich cluster is about 10 times greater than the mass of material that shows up on photographs.

Some of this mystery has been solved recently by X-ray astronomers. Satellite observations of rich clusters have revealed that X rays pour from the space between their galaxies (Figure 26-24). This flow is evidence of substantial amounts of hot intergalactic gas at temperatures between 10 and 100 million kelvin. This gas is typically as massive as all the visible galaxies in the cluster together.

However, the discovery of hot intergalactic gas in rich clusters solves only a small part of the dark-matter problem. Most astronomers agree that a great deal of dark matter still remains to be discovered in rich clusters. One popular speculation is that these clusters may contain a lot of undetected dim stars. These faint stars could be located in extended halos surrounding individual galaxies or scattered throughout the spaces between the galaxies of a cluster.

Evidence for massive extended halos comes from the rotation curves of galaxies. As mentioned in Chapter 25, many galaxies have rotation curves similar to that of our Milky Way Galaxy (recall Figure 25-17). For example, Figure 26-25 shows the rotation curves of four spiral galaxies. These rotation curves remain remarkably flat out to surprisingly great distances from a galaxy's center. In other words, the orbital

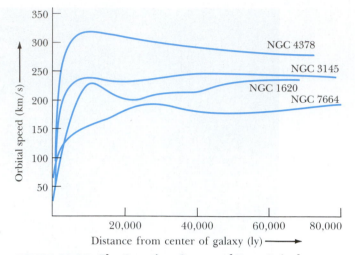

FIGURE 26-25 The Rotation Curves of Four Spiral Galaxies This graph shows the orbital speed of material in the disks of four spiral galaxies. Many galaxies have flat rotation curves, indicating the presence of extended halos of dark matter. (Adapted from V. Rubin and K. Ford)

speed of the stars remains roughly constant out to distances beyond which the H II regions are so dim and widely scattered that reliable measurements are not possible. This tells us that we still have not detected the true edges of these and many similar galaxies. In the outer portions of a galaxy we should see a decline in orbital speed, in accordance with Kepler's third law. Because this decline has not been observed, astronomers conclude that there must be a considerable amount of dark material extending well beyond the visible portion of a galaxy's disk.

Although we do not know what dark matter is made of, it is possible to investigate its mass and location in a cluster of galaxies. As we saw in Chapter 24, the gravitational deflection of light by a foreground galaxy can produce an arc-shaped image of a background galaxy (recall Figure 24-9). In 1992 J. Anthony Tyson of AT&T Bell Laboratories and his colleagues demonstrated that these distorted images can be used to learn about the mass of foreground galaxies. Figure 26-26 shows a fine example of the gravitational lensing of a distant cluster by a nearer one. The faint blue arcs circling a nearby cluster of yellowish galaxies are elongated images of more distant blue galaxies. By measuring the distortion of the images of the background galaxies, Tyson and his colleagues determined that dark matter, which constitutes about 90% of the cluster's mass, is distributed much like the visible matter in the cluster. In other words, the overall arrangement of visible galaxies seems to trace the location of dark matter.

Many proposals have been made to account for this hidden matter. Some of the suggestions made so far include black holes and various types of massive particles predicted by speculative theories. The more mundane suggestions include dim stars and Jupiterlike planets. The nature of this unseen matter is one of the greatest mysteries in modern astronomy.

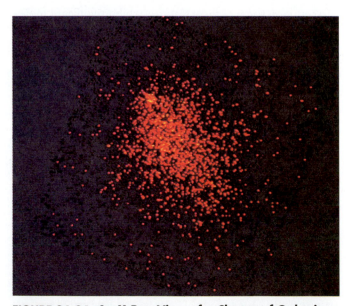

FIGURE 26-24 An X-Ray View of a Cluster of Galaxies This view from the Einstein Observatory shows the rich cluster Abell 1367 (that is, the 1367th cluster in a list of rich clusters catalogued by UCLA astronomer George O. Abell). The X-ray emission shown here comes from hot gas between the galaxies. (Harvard–Smithsonian Center for Astrophysics)

FIGURE 26-26 Gravitational Lensing of a Cluster of Galaxies This cluster of galaxies, called Abell 2218, is about 2 million light-years away in the constellation of Draco (the Dragon). This CCD image records not only the cluster's galaxies (yellowish blobs) but also hundreds of faint bluish galaxies in the distant background. Gravitational lensing by the matter in the foreground cluster makes many of these background galaxies appear as short arcs. (Courtesy of J. A. Tyson; AT&T Bell Laboratories)

26-7 Colliding galaxies produce starbursts, spiral arms, and other spectacular phenomena

Occasionally two galaxies in a cluster collide. When they do, their stars pass by each other. There is so much space between the stars that the probability of two stars crashing into each other is extremely small. However, the galaxies' huge clouds of interstellar gas and dust do slam into each other and are stopped in their tracks. In this way, two colliding galaxies can be stripped of their interstellar gas and dust. The violence of the collision between the interstellar clouds heats the gas to extremely high temperatures. This process may be a major source of the hot intergalactic gas often observed in rich, regular clusters.

In a less violent collision or a near-miss between two galaxies, the compressed interstellar gas may have more time to cool, allowing many protostars to form. Such collisions may thus account for **starburst galaxies**, which blaze with the light of numerous newborn stars. These galaxies are characterized by bright centers surrounded by clouds of warm interstellar dust, indicating recent, vigorous star birth (Figure 26-27). The warm dust is so abundant that starburst galaxies

FIGURE 26-27 A Starburst Galaxy Prolific star formation is occurring at the center of this galaxy, called M82 or NGC 3034, located about 12 million light-years from Earth. This activity was probably triggered by gravitational interactions with neighboring galaxies. Note the turbulent appearance of the interstellar gas and dust around the galaxy's center. (Lick Observatory)

a

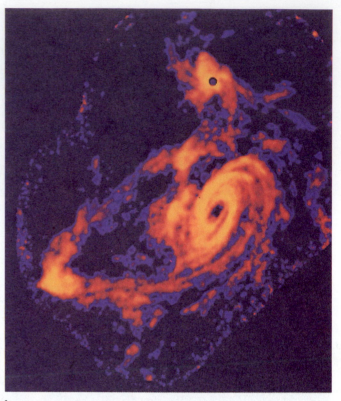

b

FIGURE 26-28 The M81–M82–NGC 3077 Cluster The
starburst galaxy M82 is in a nearby cluster whose three members
are connected by streamers of hydrogen gas. (**a**) This wide-angle
photograph shows the three galaxies at visual wavelengths.

(**b**) This mosaic of 13 fields observed with radio telescopes of the
Very Large Array shows the streams of hydrogen gas that connect
the three galaxies. (Palomar Sky Survey; M. S. Yun, VLA, and
Harvard)

FIGURE 26-29 Merging Galaxies This
contorted object in the constellation of Ophiuchus
consists of two spiral galaxies in the process of
merging. The collision between the two galaxies has
triggered an immense burst of star formation.
(Courtesy of W. C. Keel)

are among the most luminous objects in the universe at infrared wavelengths.

The starburst galaxy M82 shown in Figure 26-27 is one member of a nearby cluster that includes the beautiful spiral galaxy M81 and a fainter companion called NGC 3077 (Figure 26-28a). Radio surveys of that region of the sky reveal enormous streams of hydrogen gas connecting the three galaxies (Figure 26-28b). The loops and twists in these streamers suggest that the three galaxies have had several close encounters over the ages. A similar stream of hydrogen gas connects our Galaxy with its nearest neighbor, the Large Magellanic Cloud.

Supercomputer simulations show that colliding galaxies can hurl thousands of stars out into intergalactic space along huge, arching streams. While some of the stars are flung far and wide, other stars lose energy and momentum. As these stars slow down, the galaxies may merge. Several dramatic examples of such **galaxy mergers** have recently been discovered (Figure 26-29).

The behavior of merging galaxies is dramatically illustrated in supercomputer simulations by Joshua Barnes, now at the Institute for Astronomy on Hawaii. One of his simulations is shown in Figure 26-30, which displays the collision and eventual merging of two galaxies. As the two galaxies pass through each other, gravitational interactions severely distort the distribution of their stars, throwing out a pair of extended tails. The interaction also prevents the galaxies from continuing on their original paths. They instead fall

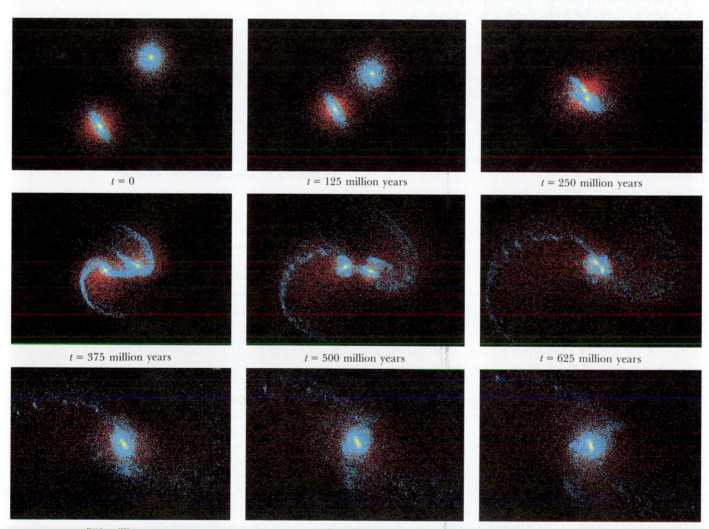

FIGURE 26-30 A Simulated Collision Between Two Galaxies These frames from a supercomputer simulation show the collision and merger of two disk-shaped galaxies. Stars in the disk of each galaxy are colored blue, while stars in their central bulges are yellow. Red indicates dark matter that surrounds each galaxy. The pictures display progress at 125-million-year intervals. Compare the final frames with the photograph of NGC 2623 in Figure 26-31. (Courtesy of J. Barnes; CITA)

FIGURE 26-31 Colliding Galaxies with "Antennae"
Many pairs of colliding galaxies exhibit long "antennae" of stars ejected by the collision. This particular system, called NGC 2623, is also a significant source of radio emission. Supercomputer simulations, like the one shown in Figure 26-30, give important insights into possible histories of such systems. (Palomar Observatory)

galaxy. Although much material is stripped from the satellite, most of its stars plunge into the nucleus of the large galaxy.

Close encounters between galaxies provide a third way of forming spiral arms (in addition to density waves and self-propagating star formation, discussed in Sections 25-5 and 25-6). Supercomputer simulations clearly demonstrate that spiral arms can be created during a collision, either by drawing out long streamers of stars or by compressing clouds of interstellar gas. For instance, the spiral arms of M51 (examine Figure 26-2) may have been produced by a close encounter with a second galaxy. The disruptive galaxy is now located at the end of one of the spiral arms created by the collision. Some astronomers argue that the spiral arms of our Milky Way Galaxy were similarly produced by a close encounter with the Large Magellanic Cloud.

In a rich cluster there must be many near-misses between galaxies. If galaxies are surrounded by extended halos of dim stars, these near-misses could strip the galaxies of their outlying stars. In this way, a loosely dispersed sea of dim stars might come to populate the space between galaxies in a cluster. Searching for these dim stars in extended halos and intergalactic space is one of the projects for the Hubble Space Telescope.

back together for a second encounter (at 625 million years), when they look remarkably similar to the actual colliding galaxies in Figure 26-31. The simulated galaxies merge shortly thereafter, leaving a single object.

When two galaxies merge, the result is a bigger galaxy. If this new galaxy is located in a rich cluster, it may capture and devour additional galaxies, growing to enormous dimensions by **galactic cannibalism.** Cannibalism differs from mergers in that the dining galaxy is bigger than its dinner, whereas merging galaxies are about the same size.

Many astronomers suspect that galactic cannibalism is the reason that giant ellipticals are so huge. As we have seen, giant galaxies typically occupy the centers of rich clusters. In many cases, smaller galaxies are located around these giants (examine Figure 26-9). As they pass through the extended halo of a giant elliptical, these smaller galaxies slow down and are eventually devoured by the larger galaxy.

Figure 26-32 shows a supercomputer simulation in which a large, disk-shaped galaxy devours a small satellite galaxy. The large galaxy consists, by mass, of 90% stars (in blue) and 10% gas (in white). It is surrounded by a halo of dark matter having a mass about 3.3 times that of the disk. The satellite galaxy, which has a tenth of the mass of the large galaxy, contains only stars (in orange). Initially the satellite is in circular orbit about the large galaxy. Spiral arms appear in the large galaxy as the collision proceeds. Two billion years elapse as the satellite spirals in toward the core of the large

26-8 Spiral galaxies were more common in the past than they are today

Astronomers can probe the past to gain important clues about galactic evolution simply by looking deep into space. The more distant a galaxy is, the longer its light takes to reach us. Consequently, as we examine galaxies that are at increasing distances from Earth, we are actually looking farther and farther back in time, seeing galaxies at increasingly earlier stages of their lives.

By observing remote galaxies, astronomers have discovered that galaxies were bluer in the past than they are today. This trait was noticed in 1978 by astronomers Harvey Butcher and Augustus Oemler, who drew attention to large numbers of blue galaxies in the remote, rich clusters they were studying. Spectroscopic studies of these galaxies in the 1980s by James Gunn and Alan Dressler demonstrated that most owe their blue color to vigorous star formation, often occurring in intense, episodic bursts.

In 1992 Dressler, Oemler, Gunn, and Butcher teamed up to use the Hubble Space Telescope to examine two remote, rich clusters of galaxies with redshifts of 0.4. This redshift corresponds to a look-back in time of about 4 billion years; thus the HST photograph in Figure 26-33 shows us how galaxies looked 4 billion years ago.

Careful examination of the HST images revealed a surprisingly large number of spiral galaxies. About 30% of the galaxies in these remote clusters are spirals, whereas only 5% of the galaxies in nearby rich clusters are spirals. Spiral galaxies were therefore much more common 4 billion years ago

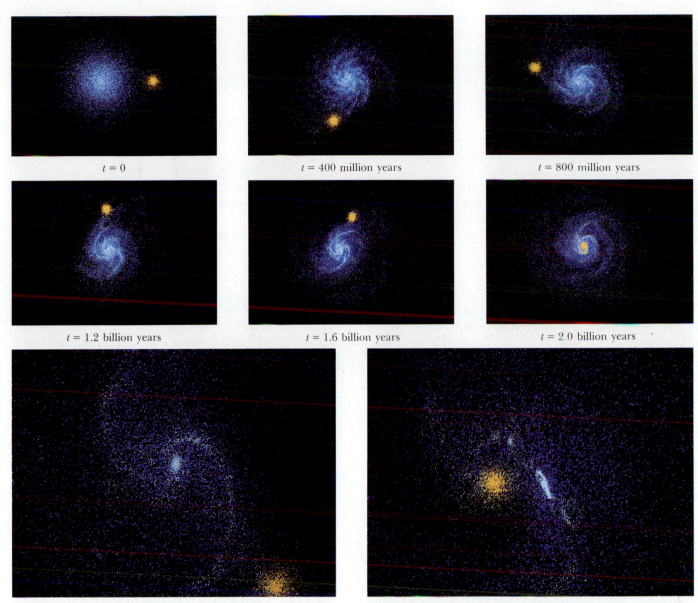

$t = 0$ $t = 400$ million years $t = 800$ million years

$t = 1.2$ billion years $t = 1.6$ billion years $t = 2.0$ billion years

Close-up view: $t = 1.8$ billion years

Close-up view: $t = 1.9$ billion years

FIGURE 26-32 Simulated Galactic Cannibalism This simulation, performed at the Pittsburgh Supercomputing Center, shows a small galaxy (stars in orange) being devoured by a larger, disk-shaped galaxy (stars in blue, gas in white). The upper six pictures display progress at 400-million-year intervals. Note how interaction with the satellite galaxy induces spiral arms in the disk galaxy. The lower two pictures, which cover the interval from 1.8 to 1.9 billion years, are close-up views of the satellite galaxy plunging into the core of the disk galaxy. As the satellite galaxy sweeps through the inner regions of the disk galaxy, a significant amount of gas becomes concentrated along one of the spiral arms. Vigorous star formation would be expected in these gas clouds. (Courtesy of L. Hernquist)

than they are today. Moreover, the HST images enabled the astronomers to match up color and spectral information about these galaxies with their shapes. They found that most of the blue, star-forming galaxies are spirals, many of which show signs of collisions or mergers.

Collisions and mergers are probably responsible for depleting rich clusters of their spiral galaxies. During a collision, interstellar gas in the colliding galaxies is vigorously compressed, which triggers a burst of star formation. A succession of collisions produces a series of star-forming episodes that create numerous bright, hot O and B stars that become disbursed along arching spiral arms by the galaxy's rotation. Eventually, however, the gas is used up; star formation then ceases and the spiral arms fade away.

Although collisions and mergers deplete the population of spiral galaxies, these events do not produce elliptical or S0 galaxies. The HST images show that elliptical and S0 galaxies were already well-developed 4 billion years ago. Indeed, the preponderance of old population II stars in elliptical galaxies demonstrates that their stars formed in a burst of activity 10 to 15 billion years ago. In contrast, spiral galaxies have been forming stars continually over the past 10 to 15 billion years, although at a gradually decreasing rate as their interstellar gas gets used up. Figure 26-34 compares the rates at which spiral and elliptical galaxies form stars.

The different types of spiral galaxies can be understood in terms of the speed with which their gas was used up to create stars. Sa and SBa galaxies presumably used up their gas more quickly than Sc or SBc galaxies. In irregular galaxies, star formation has taken place much more slowly and at a roughly constant rate. These galaxies still contain a significant supply of primordial gas.

One of the challenges that faces astronomers today is to understand how galaxies were created. One theory, proposed in the 1960s, argues that galaxies formed from the gravitational contraction of huge clouds of primordial gas. The rate of star formation in a contracting gas cloud determines whether it becomes a spiral or an elliptical galaxy. If the rate of star formation is low, then the gas has plenty of time to settle in a flattened disk. A flattened disk is the natural con-

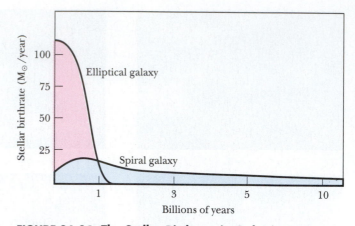

FIGURE 26-34 The Stellar Birthrate in Galaxies Most of the stars in an elliptical galaxy are created in a brief burst of star formation when the galaxy is very young. In spiral galaxies, star formation occurs at a more leisurely pace and may extend over billions of years. (Adapted from J. Silk)

sequence of the overall rotation of the original gas cloud. Star formation continues in the disk because it contains an abundance of hydrogen, and thus a spiral galaxy is created. But if the stellar birthrate is high, then virtually all of the gas is used up in the formation of stars before a disk has time to form. In this case, an elliptical galaxy is created.

An alternative theory, proposed in 1977, suggests that galaxies formed by the merging of several gas clouds rather than from the gravitational contraction of huge, isolated clouds. A third theory also contends that galaxies formed from mergers of gas clouds, except that the ancestral fragments were rather small and quite numerous. These three main theories of galaxy formation are illustrated in Figure 26-35.

Recent pictures from the Hubble Space Telescope seem to favor the idea that galaxies formed from the merging of gas clouds. Figure 26-36 shows galaxylike objects estimated to be between 3 and 10 billion light-years away. They all have unusual shapes, quite unlike the familiar spiral or elliptical shapes possessed by galaxies in the nearby universe. Because some of these images show apparent mergers, these objects may be the building blocks of today's large galaxies.

Galactic evolution is a difficult and controversial subject full of unresolved questions. Where did the pregalactic clouds of gas come from? And what happened in the early universe to cause the primordial hydrogen and helium to clump up in clouds destined to evolve into galaxies, instead of becoming objects a million times bigger or smaller? Even more troublesome is the issue of the dark matter. We know that the observable stars, gas, and dust in a galaxy constitute only 10% of the mass associated with the galaxy. We have no idea what the remaining 90% looks like or what it's made of. These questions are among the most challenging issues that face astronomers today.

FIGURE 26-33 A Remote Cluster of Galaxies This image from the Hubble Space Telescope shows a rich cluster of galaxies, called Abell 851, located about 4 billion light-years from Earth in the constellation of Ursa Major. Astronomers combined ten 36-minute exposures—one per orbit as HST circled the Earth—to construct this image, which is equivalent to one 6-hour exposure. (NASA, ESA)

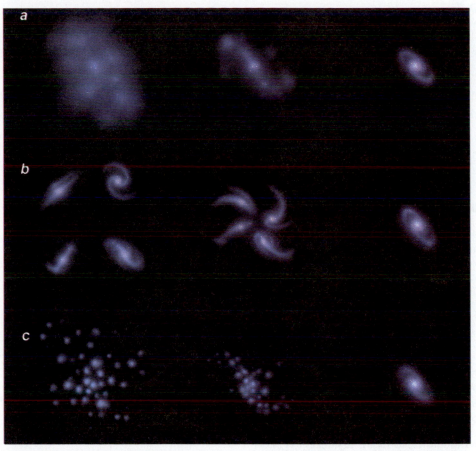

FIGURE 26-35 Theories of Galaxy Formation Theories of galaxy formation fall into three categories. (**a**) A single huge primordial gas cloud contracts to form a galaxy. (**b**) Several moderate-sized gas clouds merge to form a galaxy. (**c**) Numerous small gas clouds coalesce to form a galaxy. (Adapted from van den Bergh and Hesser)

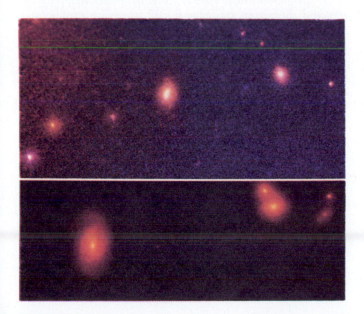

FIGURE 26-36 Embryonic Galaxies? These images from the Hubble Space Telescope show galaxylike objects that are 3 to 10 billion light-years from Earth. These objects may be galaxy fragments in the process of merging to form large galaxies like those we see in the nearby universe today. (NASA, ESA)

KEY WORDS

anisotropic	galaxy merger	irregular cluster (of galaxies)	rich cluster (of galaxies)
barred spiral galaxy	giant elliptical galaxy	irregular galaxy	spiral galaxy
cluster (of galaxies)	group (of galaxies)	isotropic	standard candle
dark-matter problem	Hubble classification scheme	lenticular galaxy	starburst galaxy
dwarf elliptical galaxy	Hubble constant	Local Group	supercluster
elliptical galaxy	Hubble flow	poor cluster (of galaxies)	Tully–Fisher relation
galactic cannibalism	Hubble law	regular cluster (of galaxies)	void

KEY IDEAS

• Galaxies can be grouped under the Hubble classification into four major categories: spirals, barred spirals, ellipticals, and irregulars.

The disks of spiral and barred spiral galaxies are sites of active star formation.

Elliptical galaxies are nearly devoid of interstellar gas and dust, a deficiency that severely inhibits star formation.

• There is a simple linear relationship between the distance from the Earth to a galaxy and the redshift of that galaxy (which is a measure of the speed with which it is receding from us); this relationship is the Hubble law, $v = H_0 r$.

Because of difficulties in measuring the distances to galaxies, the value of the Hubble constant, H_0, is not known with certainty. Some astronomers believe that H_0 is about 50 km/s/Mpc, whereas others argue that the value is closer to 90 km/s/Mpc. As a result, distances to remote galaxies calculated from the Hubble law are uncertain by a factor of 2.

Standard candles, such as Cepheid variables and the most luminous supergiants, globular clusters, H II regions, and supernovae in a galaxy, are used in estimating intergalactic distances.

The Tully–Fisher relation, which correlates the width of the 21-cm line of hydrogen in a spiral galaxy with its absolute magnitude, is one of the best available methods for determining distance.

• Galaxies are grouped into clusters rather than being scattered randomly through the universe.

A rich cluster contains hundreds or even thousands of galaxies; a poor cluster, often called a group, may contain only a few dozen.

A regular cluster has a nearly spherical shape with a central concentration of galaxies; in an irregular cluster, galaxies are distributed asymmetrically.

Our Galaxy is a member of a poor, irregular cluster called the Local Group.

Rich, regular clusters contain mostly elliptical and S0 galaxies; irregular clusters contain spiral and irregular galaxies along with ellipticals.

Giant elliptical galaxies are often found near the centers of rich clusters.

• The observable mass of a cluster of galaxies is not large enough to account for the observed motions of the galaxies; a large amount of unobserved mass must be present between the galaxies. This situation is called the dark-matter problem.

Hot intergalactic gases emit X rays in rich clusters; massive extended halos probably surround all the galaxies.

Gravitational lensing of remote galaxies by a foreground cluster enables astronomers to glean information about the distribution of dark matter in the foreground cluster.

• When two galaxies collide, their stars pass each other, but their interstellar media collide violently, either stripping the gas and dust from the galaxies or triggering prolific star formation.

The gravitational effects during a galactic collision can throw stars out of their galaxies into intergalactic space.

Galactic mergers may occur; a large galaxy in a rich cluster may tend to grow steadily through galactic cannibalism, perhaps producing in the process a giant elliptical galaxy.

• Observations of remote clusters of galaxies reveal that spiral galaxies were more common billions of years ago than they are today. This excess of spiral galaxies probably resulted from collisions and mergers when galaxies had an abundance of interstellar gas with which to form stars.

• By studying galaxies that are 5 to 10 billion light-years away, astronomers hope to learn more about how galaxies formed.

Most astronomers believe that galaxies probably formed either by the gravitational contraction of huge, individual clouds of gas or by the merging of several smaller gas clouds.

A repaired Hubble Space Telescope and a new generation of large, Earth-based telescopes will provide important clues about how galaxies formed billions of years ago.

REVIEW QUESTIONS

1. What was the Shapley–Curtis debate all about? Was a winner declared at the end of the debate? Whose ideas turned out to be correct?

2. How did Edwin Hubble prove that the Andromeda "Nebula" is not a nebula in our Milky Way Galaxy?

3. What is the Hubble classification scheme? Which category includes the biggest galaxies? Which includes the smallest? Which type of galaxy is the most common?

4. Are there any galaxies besides our own that can be seen with the naked eye? If so, which one(s)?

5. What is the Hubble law?

6. Some galaxies in the Local Group exhibit blueshifted spectral lines. Why aren't these blueshifts violations of the Hubble law?

7. How is it possible that galaxies in our Local Group still remain to be discovered? In what part of the sky would these galaxies be located? What sorts of observations might reveal these galaxies?

8. How would you distinguish star images from unresolved images of remote galaxies on a photographic plate?

9. Explain why the dark matter in galaxy clusters could not be neutral hydrogen.

10. What types of galaxies are most likely to have new stars forming? Describe the observational evidence that supports your answer.

11. On what grounds do astronomers believe that spiral galaxies were more numerous in the past than they are today? What could account for this excess of spiral galaxies in the distant past?

12. Why do you suppose there are many discordant determinations of H_0?

13. What kinds of stars would you expect to find populating space between galaxies in a cluster?

ADVANCED QUESTIONS

Tips and tools . . .

The relationship between apparent magnitude, absolute magnitude, and distance is discussed in Box 19-2. As explained in Box 25-2, a useful form of Kepler's third law is $M = rv^2/G$, where M is the mass within an orbit of radius r, v is the orbital speed, and G is the gravitational constant.

14. A certain galaxy is observed to be receding from the Sun at a rate of 75,000 km/s. The distance to this galaxy is measured independently and found to be 1.4×10^9 pc. What is the value of the Hubble constant for these data?

15. Suppose you discover a Type Ia supernova in a distant galaxy. At maximum brilliance, the supernova reaches an apparent magnitude of +10. How far away is the galaxy?

16. Two galaxies separated by 600 kpc are orbiting each other with a period of 40 billion years. What is the total mass of the two galaxies?

17. The average radial velocity of galaxies in the Hercules cluster pictured in Figure 26-21 is 10,800 km/s. How far away is the cluster? How does your answer depend on the value of the Hubble constant?

18. The rotation curve of the Sa galaxy NGC 4378 is shown in Figure 26-25. Using data from that graph, calculate the orbital period of stars 20 kpc from the galaxy's center. What is the mass of the galaxy out to 20 kpc from its center?

19. How might you determine what part of a galaxy's redshift is caused by the galaxy's orbital motion about the center of mass of its cluster?

DISCUSSION QUESTIONS

20. Discuss the advantages and disadvantages of using the various "standard candle" distance indicators to obtain extragalactic distances.

21. Discuss whether the various Hubble types of galaxies actually represent some sort of evolutionary sequence.

OBSERVING PROJECTS

22. Using a telescope with an aperture of at least 30 cm (12 in.), observe as many of the spiral galaxies listed in the table on the following page as you can. Many of these galaxies are members of the Virgo cluster, which can best be seen during the spring. Since all galaxies are quite faint, be sure to schedule your observations for a moonless night. The best view is obtained when a galaxy is near the meridian.

Spiral galaxy	R.A.	Decl.	Hubble type
M31 (NGC 224)	0^h 42.7^m	+41° 16′	Sb
M58 (NGC 4579)	12 37.7	+11 49	Sb
M61 (NGC 4303)	12 21.9	+4 28	Sc
M63 (NGC 5055)	13 15.8	+42 02	Sb
M64 (NGC 4826)	12 56.7	+21 41	Sb
M74 (NGC 628)	1 36.7	+15 47	Sc
M83 (NGC 5236)	13 37.0	−29 52	Sc
M88 (NGC 4501)	12 32.0	+14 25	Sb
M90 (NGC 4569)	12 36.8	+13 10	Sb

Spiral galaxy	R.A.	Decl.	Hubble type
M91 (NGC 4548)	12^h 35.4^m	+14° 30′	SBb
M94 (NGC 4736)	12 50.9	+41 07	Sb
M98 (NGC 4192)	12 13.8	+14 54	Sb
M99 (NGC 4254)	12 18.8	+14 25	Sc
M100 (NGC 4321)	12 22.9	+15 49	Sc
M101 (NGC 5457)	14 03.2	+54 21	Sc
M104 (NGC 4594)	12 40.0	−11 37	Sa
M108 (NGC 3556)	11 11.5	+55 40	Sc

While at the eyepiece, make a sketch of what you see. Can you distinguish any spiral structure? After completing your observations, compare your sketches with photographs found in such popular books as *Galaxies* by Timothy Ferris (Sierra Club Books, 1980) and *The Color Atlas of the Galaxies* by James Wray (Cambridge University Press, 1988).

23. Using a telescope with an aperture of at least 30 cm (12 in.), observe as many of the following elliptical galaxies as you can. Six of these galaxies are in the Virgo cluster, which is conveniently located in the evening sky from March through June. The photograph below, which covers a 2° × 5° swath of the cluster, may help you find some of these galaxies. As in the previous exercise, be sure to schedule your observations for a moonless night, when the galax-ies you wish to observe will be near the meridian. Do these elliptical galaxies differ in appearance from spiral galaxies?

Elliptical galaxy	R.A.	Decl.	Hubble type
M58 (NGC 4472)	12^h 29.8^m	+8° 00′	E4
M59 (NGC 4621)	12 42.0	+11 39	E3
M60 (NGC 4649)	12 43.7	+11 33	E1
M84 (NGC 4374)	12 25.1	+12 53	E1
M86 (NGC 4406)	12 26.2	+12 57	E3
M89 (NGC 4552)	12 35.7	+12 33	E0
M110 (NGC 205)	00 40.4	+41 41	E6

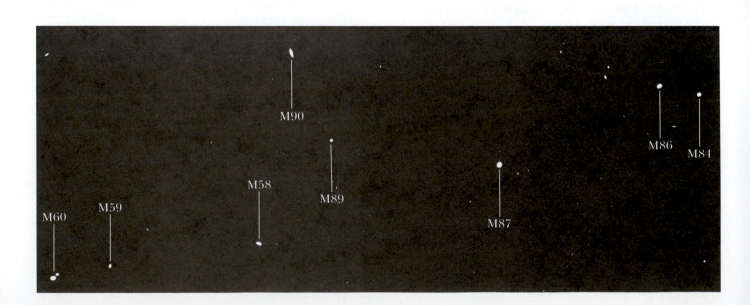

24. Using a telescope with an aperture of at least 30 cm (12 in.), observe as many of the following interacting galaxies as you can. As in the previous exercises, be sure to schedule your observations for a moonless night when the galaxies you wish to observe will be near the meridian.

While at the eyepiece, make a sketch of what you see. Can you distinguish hints of interplay among the galaxies? After completing your observations, compare your sketches with photographs found in such popular books as *Galaxies* by Timothy Ferris (Sierra Club Books, 1980).

Interacting galaxies	R.A.	Decl.
M51 (NGC 5194)	13^h 29.9^m	$+47°$ $12'$
NGC 5195	13 30.0	+47 16
M65 (NGC 3623)	11 18.9	+13 05
M66 (NGC 3627)	11 20.2	+12 59
NGC 3628	11 20.3	+13 36
M81 (NGC 3031)	9 55.6	+69 04
M82 (NGC 3034)	9 55.8	+69 41
M95 (NGC 3351)	10 44.0	+11 42
M96 (NGC 3368)	10 46.8	+11 49
M105 (NGC 3379)	10 47.8	+12 35

FOR FURTHER READING

de Vaucouleurs, G. "The Distance Scale of the Universe." *Sky & Telescope,* December 1983. This article gives fascinating historical insights into the trials and tribulations of determining distances to galaxies.

Dressler, A. "Galaxies Far Away & Long Ago." *Sky & Telescope,* April 1993. This article discusses recent observations by the Hubble Space Telescope of remote spiral-rich clusters of galaxies.

————. "Observing Galaxies Through Time." *Sky & Telescope,* August 1991. This article explains how observations of remote galaxies can give clues about how galaxies evolve.

Elmegreen, D. M., and Elmegreen, B. "What Puts the Spiral in Spiral Galaxies?" *Astronomy,* September 1993. This excellent overview of spiral structure includes an assortment of beautiful computer-enhanced images of galaxies.

Field, G. "The Hidden Mass in Galaxies." *Mercury,* May/June 1982. This well-written article describes evidence for dark matter in the universe.

Gorenstein, P., and Tucker, W. "Rich Clusters of Galaxies." *Scientific American,* November 1978. This article summarizes the main properties of rich clusters of galaxies.

Hartley, K. "Elliptical Galaxies Forged by Collision." *Astronomy,* May 1989. This brief article summarizes arguments for the idea that colliding and merging spiral galaxies may give birth to elliptical galaxies.

Hodge, P. *Galaxies.* Harvard University Press, 1986. Written in a friendly, informal style, this superb book covers galaxies, galactic evolution, and closely related topics.

Keel, W. "Crashing Galaxies, Cosmic Fireworks." *Sky & Telescope,* January 1989. This article discusses the many galactic forms and phenomena that result from collisions between galaxies.

Lake, G. "Cosmology of the Local Group." *Sky & Telescope,* December 1992. This article explains how astronomers can learn about galaxy dynamics and dark matter by studying the Local Group.

————. "Understanding the Hubble Sequence." *Sky & Telescope,* May 1992. This article describes current thinking about how galaxies get their shapes.

Osterbrock, D., Brashear, R., and Gwinn, J. "Young Edwin Hubble." *Mercury,* January/February 1990. This fascinating article chronicles the adolescence and early adulthood of one of the greatest astronomers of all time.

Rubin, V. "Dark Matter in Spiral Galaxies." *Scientific American,* June 1983. This fine article on dark matter gives many insights into how astronomers practice their profession.

Schroeder, M., and Comins, N. "Galactic Collisions on Your Computer." *Astronomy,* December 1988. This article includes a short program that simulates collisions between galaxies. If you have access to a personal computer, you can watch the interaction that produced the Whirlpool Galaxy.

Silk, J. "Formation of the Galaxies." *Sky & Telescope,* December 1986. Taking clues from the many properties of galaxies, the author of this article presents an overview of how galaxies probably form.

Smith, R. *The Expanding Universe: Astronomy's Great Debate.* Cambridge University Press, 1982. This book gives a detailed, meticulously documented history of the birth and development of extragalactic astronomy.

Tully, R. "Unscrambling the Local Supercluster." *Sky & Telescope,* June 1982. This article explains how astronomers discover and try to understand large-scale structures like the Local Supercluster, which includes the Local Group and all nearby clusters.

van den Bergh, S., and Hesser, J. E. "How the Milky Way Formed." *Scientific American,* January 1993. Although this article focuses on the Milky Way, it contains an excellent summary of theories of galaxy formation in general.

Wray, J. *The Color Atlas of Galaxies.* Cambridge University Press, 1988. This book contains a magnificent collection of color photographs of more than 600 galaxies.

ALAN DRESSLER

The Great Attractor

ALAN DRESSLER is an astronomer at the Observatories of the Carnegie Institution in Pasadena, California. As a teenager he polished mirrors for 4- and 8-inch telescopes; he later studied physics at the University of California at Berkeley and received his Ph.D. in astronomy from the University of California at Santa Cruz.

In addition to his research in large-scale structure of the universe, Dressler has focused his studies on the nature of galaxies—their present-day structure and stellar populations and how these have evolved. Dressler and colleague James E. Gunn of Princeton University have pursued the difficult task of obtaining electronic pictures and spectra of very faint, high-redshift galaxies. Looking far into space is a view to earlier cosmic times, so Dressler and Gunn have been able to study the evolution of galaxies directly, finding that many galaxies formed stars at a much higher rate 5 to 10 billion years ago.

Dressler was one of the first to find observational evidence for massive black holes in the centers of galaxies. He found that the speeds of stars increase rapidly as they cross the very center of the Andromeda Galaxy (M31), indicating a central mass concentration of tens of millions of solar masses in a region only a few light-years across. He is now leading a team of astronomers who are taking pictures of high-redshift galaxies with the Hubble Space Telescope to see for the first time whether galaxies looked noticeably different 5 billion years ago.

Our gaze into a starry sky scarcely reveals the magnificent forms that nature has sculpted into the distribution of matter and galaxies over vast intergalactic distances. To see these features we must look beyond our Milky Way to the billions of other galaxies that, interwoven with mysterious invisible matter, make up our universe.

A true three-dimensional view of where the galaxies are in space requires some knowledge of their distances, which can be estimated by measuring galaxy redshifts and applying Hubble's discovery that redshifts are approximately proportional to distance. Advances in telescopes have greatly increased the number of galaxy redshifts that have been measured, and a new picture is emerging of the distribution of galaxies in space. More like a sculpture than spatter, galaxies are found linked in long chains and giant walls surrounding nearly empty regions called voids.

Such regularity probably descends from the Big Bang itself, when the physical properties of unimaginably hot matter carved these patterns into the distribution of primeval matter. As cosmic time passed, these once-small fluctuations grew. Gravity drew more and more matter into denser regions and emptied out the less dense areas. Thus, today's distribution of galaxies should be a contrast-enhanced picture of the universe's structure at birth. These patterns engraved by the Big Bang may be our best opportunity to test theories of particle physics that describe the nature of matter in its high-density, high-temperature form.

However, the theories predict the distribution of *all* matter, and most of the matter in the universe might not be in galaxies but *in between* them in a form that gives off little or no light. Evidence for this view has built steadily since the 1930s, when physicist Fritz Zwicky noted that galaxies in great clusters move very rapidly—with typical speeds of about 1000 km/s. If gravity is to prevent their escape, there must be much more matter in clusters of galaxies than can be seen in the galaxies themselves. This "dark matter" is probably in a very different form than the "ordinary matter" of protons, neutrons, and electrons from which the stars, our world, and we ourselves are fashioned. A popular model suggests that dark matter is in the form of elementary particles like the neutrino, which interact only through the relatively weak force of gravity. Such particles, if they exist, are immune to electromagnetism, nature's strongest force, and therefore would not clump together to form such things as stars, planets, or puppy dogs. In fact, these mysterious particles would pass right through them!

To test these theories, we must map the distribution of mass in at least a representative chunk of the universe. Galaxies may mark where the dark matter has collected due to gravity, but we will not know this for certain until we make a map of the *mass* distribution in some region of space and compare it with the *galaxy* distribution. Fortunately, the pull of gravity from dark matter allows us to map its distribution even though we cannot see it directly. This is because galaxies, which we can see, are attracted to regions of great mass density and evacuate the low-density zones. Thus, by measuring the *motions* of galaxies as well as their positions in space, we can infer the distribution of dark matter simply by finding out which way galaxies are moving.

Such observations are not simple because the dominant motion of a galaxy is its flight in the Hubble expansion of the universe. However, if the distance to a galaxy can be estimated, the expansion velocity can be removed, thus uncovering any remaining "peculiar velocity" due to the pull of unseen matter.

In 1986 I and six of my colleagues extended the work done by others before us and made the largest such map of peculiar motions of galaxies. We discovered that our Milky Way Galaxy and its neighbors are streaming toward a distant point, roughly in the direction marked by the Centaurus constellation. Galaxies over nearly 100 million light-years are cruising at speeds of 400 to 1000

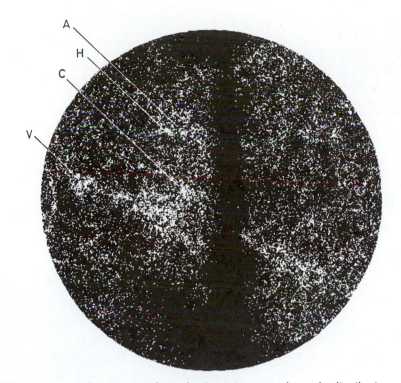

This map, centered approximately on the Great Attractor, shows the distribution of galaxies over half of the sky. The dark vertical band is obscured by the Milky Way. The Virgo (V), Centaurus (C), Hydra (H), and Antlia (A) clusters are indicated. The great concentration of galaxies just below the Centaurus cluster is obvious, which suggests that some of the matter associated with the Great Attractor might be visible.

km/s and are converging toward a region roughly 150 million light-years from our Galaxy. We identified the cause of this galaxy streaming as the gravitational pull of a *Great Attractor,* an extensive continent of dark matter and galaxies whose higher-than-average density is tugging on galaxies over a huge volume of space.

Though the Great Attractor was discovered from the *motions* of galaxies, subsequent maps showed that the number of galaxies in this region is also far enhanced. Our observations verified that dark matter accounts for at least 10 times as much mass as is contained in the visible galaxies, which supports the idea that most of the mass in the Great Attractor is in a form other than ordinary matter. The total mass in the

Great Attractor is approximately 10^{17} times the mass of our Sun and equivalent to the most massive superclusters of galaxies, with an extent of roughly 500 million light-years.

Our discovery of the Great Attractor confirms the general picture that the universe is very lumpy, as full of large, coherent piles of dark matter as of highly clustered galaxies. This lumpiness probably originated during the very earliest moments of the universe. If so, there must be many structures like the Great Attractor that we have yet to find. Further exploration of the large-scale structure of the universe promises to teach us more about how the universe was born and how our destinies were forged in the first moments of creation.

QUASARS, BLAZARS, AND ACTIVE GALAXIES

THE CORE OF A RADIO GALAXY
This artist's rendition shows a scenario that many astronomers believe is responsible for the enormous energy output from the centers of active galaxies. A supermassive black hole at the center of the galaxy is surrounded by an accretion disk. In the inner regions of the accretion disk, matter crowding toward the hole is diverted outward along two oppositely directed beams, which deposit energy into two huge, radio-emitting lobes on either side of the galaxy. (*Astronomy*)

SINCE THE 1960s astronomers have discovered thousands of incredibly luminous objects in the sky. Many of these quasars, blazars, and active galaxies shine with the brilliance of 1000 normal galaxies. Furthermore, this enormous energy output seems to come from a very small region, probably less than a light-year across. Many active galaxies also exhibit highly energetic features, such as jets of relativistic particles streaming outward from their centers. Many astronomers now believe that accretion onto supermassive black holes is responsible for the energy output of these ultraluminous objects. Recent observations from both Earth-based and orbiting observatories suggest that supermassive black holes reside at the centers of certain otherwise "normal" galaxies.

The development of radio astronomy in the late 1940s ranks among the most important scientific accomplishments of the twentieth century. Before this technology, everything known about the distant universe had to be gleaned from visual observations. Radio telescopes provided a view of the universe in a wavelength range far removed from that of visible light. Many unexpected and surprising discoveries emerged with this new ability to examine the previously invisible universe.

As we saw in Chapter 6, the first radio telescope was built in 1936 by an amateur astronomer, Grote Reber, in his backyard in Illinois. By 1944 Reber had detected strong radio emissions from Sagittarius, Cassiopeia, and Cygnus. Two of these sources, nicknamed Sgr A and Cas A, happen to be in our own Galaxy—they are the galactic nucleus and a supernova remnant. The location of the third source, called Cygnus A (Cyg A), was finally established in 1951, using a newly constructed radio interferometer. Walter Baade and Rudolph Minkowski then used the 200-inch optical telescope on Palomar Mountain to discover a strange-looking galaxy at that position. A photograph of the visible counterpart of Cygnus A is shown in Figure 27-1.

When Baade and Minkowski photographed the spectrum of Cygnus A, they found a number of bright emission lines, all shifted by 5.7% toward the red end of the spectrum. As we saw in Box 26-2, astronomers denote redshift by the letter z, and so this object has a redshift of $z = 0.057$. This redshift corresponds to a speed of 17,000 km/s. If the Hubble constant equals 75 km/s/Mpc, this speed corresponds to a distance of about 700 million light-years from Earth, farther away than any galaxy known to astronomers at that time.

Because of its enormous distance, Cygnus A must be one of the most luminous radio sources in the sky. Although it is hundreds of millions of light-years from Earth, its radio waves can be picked up by amateur astronomers with backyard equipment. In fact, Cygnus A shines with a radio luminosity that is 10^7 times as great as that of an ordinary galaxy like the Milky Way. Obviously, the object that comprises Cygnus A must be something quite extraordinary.

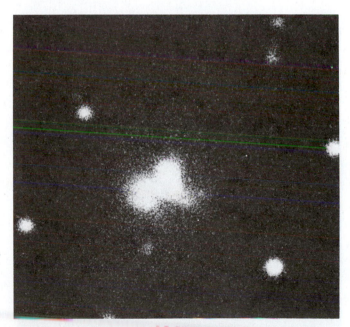

FIGURE 27-1 Cygnus A (3C 405) This strange-looking galaxy was discovered at the location of the radio source Cygnus A. This galaxy has a substantial redshift, which means that it must be extremely far away (about 700 million light-years from Earth). Because Cyg A is one of the brightest radio sources in the sky, the energy output of this remote galaxy must be enormous. (Palomar Observatory)

27-1 Quasars look like stars but have huge redshifts

During the late 1950s and early 1960s, radio astronomers were busy making long lists of all the radio sources they kept finding across the sky. One of the most famous of these lists, titled the *Third Cambridge Catalogue,* was published in 1959 and lists 471 radio sources (the first two versions of this catalogue were incomplete and had some inaccuracies). Even today astronomers often refer to sources by their "3C numbers." With the discovery of the extraordinary luminosity of Cygnus A (designated "3C 405" because it is the 405th source on the Cambridge list), astronomers were eager to learn whether any other sources in the 3C catalogue had similarly extraordinary properties.

One interesting case was 3C 48. In 1960 Allan Sandage used the 200-inch telescope to discover a "star" at the location of this radio source (Figure 27-2). Because ordinary stars are not strong sources of radio emission, astronomers knew 3C 48 must be something unusual. Indeed, its spectrum showed a series of bright spectral lines that no one could identify. Although 3C 48 was clearly an oddball, many astronomers thought it was just a strange star in our own Galaxy.

Another such "star" was discovered in 1962 when astronomers at the Australian National Radio Observatory observed 3C 273 as the Moon passed in front of it. By noting when radio emission from 3C 273 was blocked by the edge of the Moon, the Australian astronomers determined the exact location of 3C 273. Armed with a precise position, astronomers were then able locate the optical counterpart of 3C 273. As Figure 27-3 shows, it looks like a star with a luminous jet protruding from one side. And, as was the case with 3C 48, its visible spectrum contains a series of bright spectral lines that no one could identify.

The stumbling block that prevented astronomers from deciphering these spectra was the idea that 3C 48 and 3C 273 are nearby stars. All stars in our Galaxy have comparatively small Doppler shifts, because any star with an extremely high speed would have long ago escaped from the Galaxy. Astronomers therefore erroneously assumed that the mysterious spectral lines exhibited by 3C 48 and 3C 273 were not shifted far from their normal wavelengths.

A breakthrough occurred in 1963, when Maarten Schmidt at the California Institute of Technology realized that the spectrum of 3C 273 in fact exhibits an enormous redshift. Schmidt identified four of its brightest spectral lines as Balmer lines of hydrogen that have suffered a redshift corresponding to a speed of 45,000 km/s, which is 15% of the speed of light. According to the Hubble law, this huge redshift implies that the distance to 3C 273 is roughly 2 billion light-years. Obviously, 3C 273 is not a star, but rather a very remote, very powerful source of radio radiation.

Upon learning how Schmidt deciphered the spectrum of 3C 273, two other Caltech astronomers, Jesse Greenstein and T. A. Matthews, found they could identify the spectral lines of 3C 48 as having suffered a redshift of $z = 0.367$. That shift corresponds to a velocity of nearly one-third the speed

FIGURE 27-2 The Quasar 3C 48 For several years astronomers erroneously believed that this object was simply a peculiar nearby star that happened to emit radio waves. Actually, the redshift of this starlike object is so great that, according to the Hubble law, it must be roughly 4 billion light-years away. (Palomar Observatory)

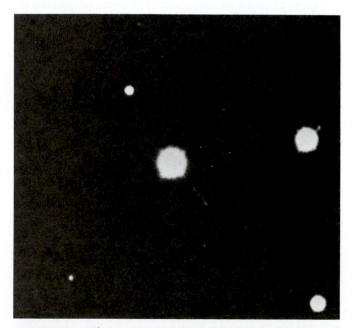

FIGURE 27-3 The Quasar 3C 273 This greatly enlarged view shows the starlike object associated with the radio source 3C 273. Note the luminous jet projecting from the object's core and pointing toward the lower right. The redshift of this object is so great that its distance from Earth, according to the Hubble law, is about 2 billion light-years. (NOAO)

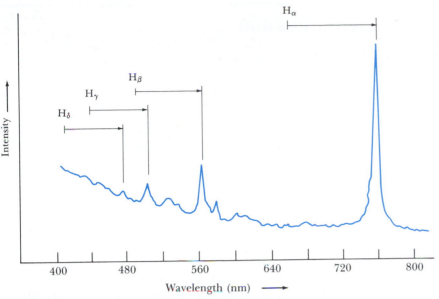

FIGURE 27-4 The Spectrum of 3C 273
Four bright emission lines of hydrogen dominate the spectrum of 3C 273. The arrows indicate how far these spectral lines are redshifted from their usual wavelengths.

of light, which places 3C 48 twice as far away as 3C 273, or approximately 4 billion light-years from Earth.

Because of their strong radio emission and starlike appearance, 3C 48 and 3C 273 were dubbed **quasi-stellar radio sources**, a term soon shortened to **quasars**. After the first quasars were discovered by their radio emission, many similar, high-redshift, starlike objects were found that emit little or no radio radiation. These "radio-quiet" quasars were originally called **quasi-stellar objects** to distinguish them from radio emitters. Today, however, the term quasar is often used to include both types. Only about 10% of quasars are "radio-loud."

Figure 27-4 shows the spectrum of 3C 273 displayed as obtained with a charge-coupled device (CCD). As we saw in Chapter 6 (review Figure 6-20), a spectrum is a graph of intensity versus wavelength on which emission lines appear as peaks that rise above the overall background intensity. Emission lines are caused by excited atoms that emit radiation at specific wavelengths. Spectra of ordinary galaxies (recall Figure 26-15) are dominated by dark absorption lines. Most quasars and many peculiar galaxies exhibit strong emission lines in their spectra, a sign that something unusual is going on.

Thousands of quasars have been discovered since the pioneering days of the early 1960s. Quasars look rather like stars, and all have large redshifts, ranging from 0.06 up to the current maximum of 4.9. Most quasars are more than 3 billion light-years (1000 Mpc) from Earth. For example, the quasar OH 471, shown in Figure 27-5, has a redshift of $z = 3.4$, which corresponds to a speed slightly greater than 90% of the speed of light and a distance of roughly 10 billion light-years from Earth.

A value of z greater than 1 does not mean that a quasar is receding from us faster than the speed of light. At high speeds, the formula expressing the redshift must be modified by the special theory of relativity, as explained in Box 27-1.

According to the relativistic formula, as the speed of a receding source approaches the speed of light, its redshift (z) increases without bound. Indeed, $z = \infty$ corresponds to a speed equal to the speed of light.

For very remote objects, the relationship between redshift and distance from Earth depends on how the universe evolved. As we shall see in Chapter 28, the Hubble law reveals that the universe is expanding. In other words, if you could watch the motions of widely separated clusters of

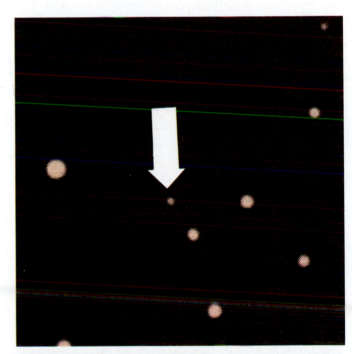

FIGURE 27-5 The Quasar OH 471 This quasar has a redshift that corresponds to a speed slightly greater than 90% of the speed of light. (Palomar Observatory)

BOX 27-1

The Relativistic Redshift

As we saw in Box 26-2, redshift (z) is defined as

$$z = \frac{\lambda - \lambda_0}{\lambda_0}$$

where λ is the observed wavelength of a spectral line in the spectrum of a star, galaxy, or quasar, and λ_0 is the unshifted wavelength of that same spectral line, as deduced from laboratory experiments.

EXAMPLE: The strongest emission line in the spectrum of a quasar called PKS 2000-330 is the Lyman-alpha line of hydrogen, observed at a wavelength of 582.5 nm. From laboratory measurements, we know that this spectral line is normally seen in the ultraviolet at a wavelength of 121.6 nm. Thus, this quasar has a redshift of

$$z = \frac{582.5 - 121.6}{121.6} = 3.79$$

The average redshift for this quasar, determined from measurements of a number of spectral lines, is $z = 3.78$.

For low velocities, we ignore the effects of relativity and use the following equation for the Doppler shift:

$$z = \frac{v}{c}$$

where c is the speed of light. Thus, for example, a 5% shift in wavelength ($z = 0.05$) corresponds to a velocity of 5% of the speed of light ($v = 0.05c$). For high velocities, however, we must use the relativistic equation for the Doppler shift:

$$z = \sqrt{\frac{c + v}{c - v}} - 1$$

We can also write this relationship in the useful form

$$\frac{v}{c} = \frac{(z + 1)^2 - 1}{(z + 1)^2 + 1}$$

This relativistic relationship between z and v is displayed in the graph. Note that z approaches infinity as v approaches the speed of light.

EXAMPLE: As stated earlier, quasar PKS 2000-330 has a redshift of $z = 3.78$. Using this value and applying the full, relativistic equation to find the radial velocity for the quasar, we obtain

$$\frac{v}{c} = \frac{(4.78)^2 - 1}{(4.78)^2 + 1} = \frac{21.85}{23.85} = 0.92$$

---

galaxies over millions of years, you would see them gradually moving away from each other. The intergalactic distance is affected by this expansion, as explained in Box 27-2. Table 27-1 relates the redshift (z) to recessional speed (v/c) and distance for a reasonable choice of cosmological quantities (a Hubble constant of 75 km/s/Mpc and a "deceleration parameter," described in Box 27-2, of $\frac{1}{2}$). It is useful to remember that the distance to a remote object expressed in light-years tells you the "look-back time"—how far back into the past you are seeing when you view that object.

The distance to a remote object depends on the value of the Hubble constant, which is somewhat uncertain. Some astronomers prefer to talk about the factor by which the universe has expanded since some ancient time, because such a figure does not depend on H_0. As we shall see in Chapter 28, an object at a redshift z emitted its light when the universe was more compact that it is today by a factor of $1/(1 + z)$. For example, a quasar with a redshift of $z = 4$ emitted its light when the distances between clusters of galaxies were $1/(1 + 4)$, or one-fifth of what they are today. At that time, galaxies were more densely crowded together than now. Since density is inversely proportional to the cube of a volume's dimensions, the density of galaxies in space back then was 5^3, or 125 times the value today.

TABLE 27-1

Redshift and Distance

z	v/c	Distance (10^9 ly)	Distance (Mpc)
0.00	0.00	0.0	0
0.10	0.10	1.2	400
0.25	0.22	2.9	900
0.50	0.38	5.0	1500
0.75	0.51	6.6	2000
1.00	0.60	7.8	2400
1.50	0.72	9.4	2900
2.00	0.80	10.4	3200
3.00	0.88	11.5	3500
4.00	0.92	12.0	3700
5.00	0.95	12.3	3800
∞	1.00	13.0	4000

In other words, this quasar appears to be receding from us with a velocity of 92% of the speed of light.

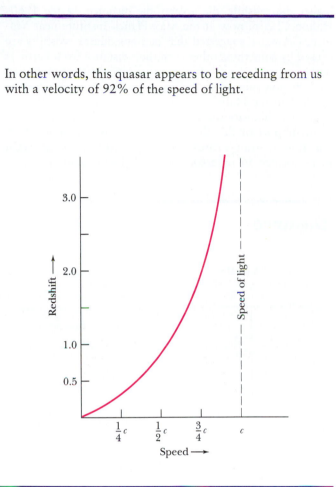

27-2 Quasars are the ultraluminous centers of distant galaxies

Quasars must be extraordinarily luminous because they can be seen from Earth in spite of their enormous distances. Indeed, aside from a handful of peculiar galaxies, quasars are the only objects that can be detected at redshifts greater than $z = 1$.

A quasar's luminosity can be calculated from its apparent brightness and distance. For example, 3C 273 has a luminosity of about 10^{40} watts, which is equivalent to 25 trillion Suns. Generally, quasar luminosities range from about 10^{38} watts up to nearly 10^{42} watts. For comparison, a typical large galaxy, like our own Milky Way, shines with a luminosity of 10^{37} watts, which equals 25 billion Suns. Thus an average quasar is a thousand times more luminous than the Milky Way Galaxy.

Some quasars emit radiation that is not like the radiation from stars. As we saw in Chapter 5, a star behaves like a blackbody, which means that a plot of the intensity of its radiation against wavelength (or frequency) follows a blackbody curve, like those in Figure 5-8. This type of radiation, which depends only on an object's temperature, is called

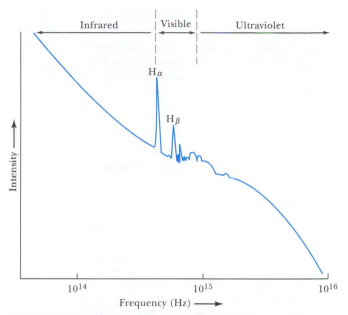

FIGURE 27-6 The Spectrum of a Quasar The intensity of radiation from the quasar PKS 0405-123 is plotted against frequency. The prominent Balmer emission lines H_α and H_β are identified. Note the overall decline of intensity with increasing frequency.

thermal radiation. Figure 27-6 shows the intensity of radiation from a quasar plotted against frequency. Note that the intensity of the radiation decreases as frequency increases. This behavior is characteristic of **nonthermal radiation.**

Synchrotron radiation is a type of nonthermal radiation (Figure 27-7). As we saw in the discussion of the Crab Nebula in Chapter 23, synchrotron radiation is produced by relativistic electrons traveling in a strong magnetic field. (Recall that "relativistic" means "traveling near the speed of light.")

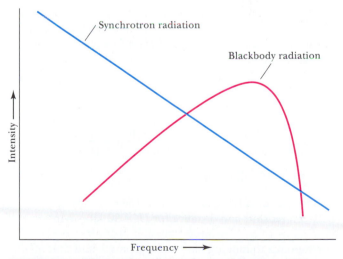

FIGURE 27-7 Thermal and Nonthermal Spectra This schematic graph compares the spectra of synchrotron radiation with blackbody radiation. A blackbody spectrum always has a hump, while the nonthermal spectrum of synchrotron radiation shows a steady decline with frequency.

As the electrons spiral around the magnetic field, they emit electromagnetic radiation. This radiation is also **polarized**, which means that the electric fields of the waves are oriented in a specific direction. (Unpolarized light, like light from the Sun, consists of waves whose electric fields are oriented at random angles.)

When quasars were first discovered, their energy output seemed so absurdly huge that a few astronomers began to question long-held beliefs, such as the Hubble law. Traditionally, astronomers had used the Hubble law to calculate the distance to a galaxy or quasar from its redshift. The higher the redshift, the greater the distance. In the 1960s Halton C. Arp, now at the Max Planck Institute near Munich, Germany, suggested that perhaps quasar redshifts are caused by something other than their distance from Earth. If part of a quasar's redshift were caused by some yet-undiscovered phenomenon, then the quasar could be much closer to Earth than the Hubble law would have us believe. If so, a quasar's luminosity would not be so incredibly large.

In support of this theory, Arp and his colleagues drew attention to strange cases where high-redshift quasars seem to be located in or associated with low-redshift galaxies.

BOX 27-2

Redshift and Distance

As we shall see in Chapter 28, the Hubble law reveals that the universe is expanding. In other words, if you could watch the motions of widely separated clusters of galaxies over millions of years, you would see them gradually migrating away from each other. The rate of recession is given by the Hubble constant. For instance, if H_0 is 75 km/s/Mpc, then a galaxy 100 Mpc away should exhibit a recessional speed of 7500 km/s due to the expansion of the universe. This expansion is noticeable only over distances larger than about 100 Mpc; over smaller distances it is masked by the random motions of galaxies and clusters.

The gravity of all the matter in the universe is slowing the expansion of the universe. The rate of this slowing is described by the so-called **deceleration parameter**, q_0. The distance to a high-redshift object depends on both H_0 and q_0, neither of which is accurately known. We have seen that H_0 probably lies between 50 and 100 km/s/Mpc, and most astronomers suspect that q_0 is probably between 0 and 1. If q_0 is larger than $\frac{1}{2}$, the expansion of the universe will eventually halt and reverse itself, while if q_0 is smaller than $\frac{1}{2}$, the expansion will continue forever.

If $q_0 = 0$, there is no deceleration and the universe expands forever at a constant rate. In this case, the age of the universe is $1/H_0$. For example, if H_0 is 50 km/s/Mpc, then the age of the universe is

$$\frac{1}{H_0} = \frac{1}{50 \text{ km/s/Mpc}}$$

$$= \frac{1 \text{ s Mpc}}{50 \text{ km}} \left(\frac{3.09 \times 10^{19} \text{ km}}{1 \text{ Mpc}} \right) \left(\frac{1 \text{ yr}}{3.16 \times 10^7 \text{ s}} \right)$$

$$= 20 \text{ billion years}$$

The accompanying figure shows the relationship between redshift and distance for $H_0 = 50$ km/s/Mpc and two values of q_0. The largest possible redshift ($z = \infty$) corresponds to looking back through space and time all the way back to the Big Bang, the creation event during which the universe came into existence. Further discussion of the deceleration parameter and the Big Bang is found in Chapter 28.

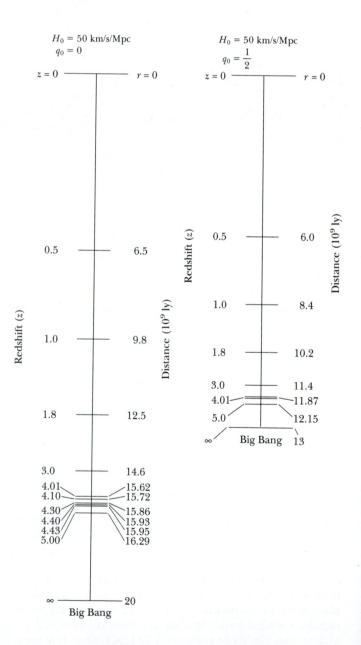

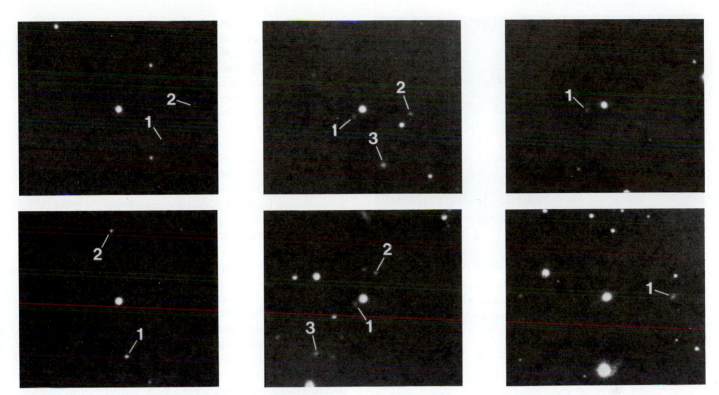

FIGURE 27-8 Quasars in Remote Clusters of Galaxies
A quasar is located at the center of each of these photographs.
Very distant galaxies, identified by the numbers, are faintly visible
near each of the quasars. In each case, the redshift of the quasar is
virtually the same as the redshifts of the galaxies surrounding it.
(Courtesy of A. Stockton)

These examples of "discordant redshifts" fueled a heated debate during the 1970s reminiscent of the Shapley–Curtis Debate 50 years earlier. By the 1980s, however, the preponderance of evidence clearly favored the standard interpretation of redshifts according to the Hubble law. Most astronomers today believe that Arp's discordant redshifts are simply a "projection effect," whereby a distant quasar just happens to be in the same part of the sky as a nearby galaxy.

The redshift debate of the 1970s was largely put to rest by observations showing that quasars are associated with remote galaxies. By taking long-exposure photographs, astronomers discovered that quasars are often found in groups or clusters of galaxies. Several examples are shown in Figure 27-8. In all these cases, the redshift of the quasar is essentially the same as the redshifts of the galaxies that surround it. Thus the quasar's redshift indicates its distance from Earth, just as the redshifts of galaxies do.

A second important link between quasars and galaxies was established in the 1980s by astronomers who examined the "fuzz" seen around the images of some quasars. The spectrum of this fuzz shows stellar absorption lines, indicating that each of these quasars is embedded in a galaxy. In each case, the star's absorption lines have the same redshift as the quasar's emission lines, further supporting the belief that quasars are at distances indicated by their redshifts and the Hubble law.

It is very difficult to observe the "host galaxy" in which a quasar is located because the quasar's light overwhelms light from the stars. Nevertheless, painstaking observations have revealed some basic properties of these host galaxies. Radio-quiet quasars seem to be located in spiral galaxies, whereas radio-loud quasars seem to be located in ellipticals. However, a large percentage of these host galaxies have distorted shapes or are otherwise peculiar. Many have nearby companion galaxies, suggesting a link between collisions or mergers and the quasar itself. Astronomers look forward to using the repaired Hubble Space Telescope to examine quasars and the host galaxies in greater detail.

27-3 Seyfert galaxies and radio galaxies bridge the gap between normal galaxies and quasars

We have seen that some scientists in the 1970s preferred to challenge the Hubble law rather than accept the existence of such highly luminous objects. One reason for their skepticism was the huge gap in energy output between normal galaxies and quasars. The gap was soon bridged, however, when astronomers realized that the centers of certain peculiar galaxies look like low-luminosity quasars.

Examples of peculiar galaxies had been known since the early 1900s. It was not until 1943, however, that Carl Seyfert at the Mount Wilson Observatory made the first systematic study of these strange objects. Seyfert focused his attention on spiral galaxies with bright, compact nuclei that seem to

FIGURE 27-9 A Seyfert Galaxy This Sc spiral galaxy, called NGC 1566, is a Seyfert galaxy in the southern constellation of Dorado. The nucleus of this galaxy is a strong source of radiation whose spectrum shows emission lines of highly ionized atoms. (Anglo-Australian Observatory)

show signs of intense and violent activity. An example is shown in Figure 27-9. The nuclei of these galaxies, which are intense sources of quasarlike radiation, have strong emission lines in their spectra. These galaxies are now referred to as **Seyfert galaxies.**

About 10% of the most luminous spiral galaxies are Seyfert galaxies. When this book was published, about 700 Seyferts were known and more continue to be found. Seyfert galaxies range in luminosity from about 10^{36} to 10^{38} watts, which makes the brightest Seyferts as luminous as faint quasars. Like radio-quiet quasars, they also tend to be weak

sources of radio radiation. Some Seyferts are members of interacting pairs or exhibit the vestiges of mergers and collisions. For example, the Seyfert galaxy NGC 1275 (Figure 27-10) is actually two colliding galaxies.

While Seyfert galaxies resemble dim, radio-quiet quasars, certain elliptical galaxies, called **radio galaxies** because of their strong radio emission, are like dim, radio-loud quasars. The first of these peculiar galaxies was discovered in 1918 by H. D. Curtis. His short-exposure photograph of M87 revealed a bright, starlike nucleus with a protruding jet. Figure 27-11*a* shows an overall view of the entire giant elliptical. Figure 27-11*b*, from the Hubble Space Telescope, shows the jet extending outward 5000 light-years from the galaxy's nucleus. Light from the galaxy's central regions is thermal radiation, suggesting a profusion of stars crowded around the galaxy's nucleus. Light from the jet is synchrotron radiation, signifying that relativistic particles are being ejected from the nucleus.

During the 1960s and 1970s, radio astronomers systematically examined extragalactic radio sources. They found that a radio galaxy generally consists of two **radio lobes** on either side of a parent galaxy. The radio lobes generally span a distance that is 5 to 10 times the size of the parent galaxy, although smaller and much larger examples are known. The parent galaxy—almost always a giant elliptical—sits midway between the radio lobes. Because of their overall shape, radio galaxies are sometimes called **double radio sources.** Cygnus A, shown in Figure 27-12, is a fine example.

The spectrum and polarization of the radiation from a radio galaxy bear all the characteristics of synchrotron emission, which suggests that radio galaxies should have jets of relativistic particles. In fact, some radio galaxies appear to have a "head" of concentrated radio emission, with a weaker "tail" trailing behind it. A good example of the **head–tail sources** is the active elliptical galaxy NGC 1265 in the Perseus cluster of galaxies. This galaxy is known to be mov-

a

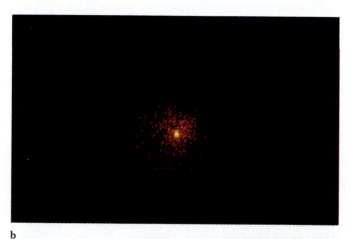

b

FIGURE 27-10 The Seyfert Galaxy NGC 1275 This Seyfert galaxy, also called 3C 84, located in the Perseus cluster, is a strong source of X rays and radio radiation. Spectroscopic studies indicate that NGC 1275 is actually two galaxies in collision. (**a**) A photograph of the galaxy at visible wavelengths shows streamers

of gas, which are probably the vestiges of the collision. (**b**) An X-ray image from the Einstein Observatory shows that most of the galaxy's X-ray emission comes from its nucleus. (NOAO; Harvard–Smithsonian Center for Astrophysics)

FIGURE 27-11 The Radio Galaxy M87 This giant elliptical galaxy is in the Virgo cluster, about 60 million light-years from Earth. (a) This photograph shows the full extent of the galaxy, which measures about 300,000 light-years across. The numerous fuzzy dots that surround the galaxy are globular clusters. (b) This HST image shows the galaxy's bright nucleus from which a jet extends outward some 5000 light-years. (Anglo-Australian Observatory; NASA, ESA)

ing at a high speed (2500 km/s) relative to the cluster as a whole. Figure 27-13 is a radio image of NGC 1265. Note that its radio emission has a distinctly windswept appearance. Just as smoke pouring from a steam locomotive trails behind a rapidly moving train, particles ejected along two jets from this galaxy are deflected by the galaxy's passage through the sparse intergalactic medium.

Like most giant ellipticals, many radio galaxies are found near the centers of rich clusters of galaxies and thus are probably subjected to collisions and mergers. A spectacular

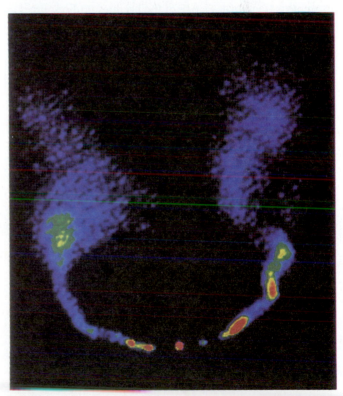

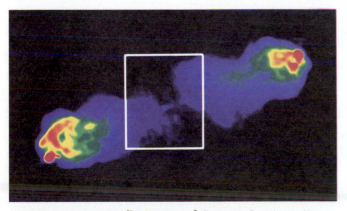

FIGURE 27-12 A Radio Image of Cygnus A This radio image shows that most of the radio emission from Cygnus A comes from the radio lobes located on either side of the visible peculiar galaxy. These two radio lobes are each about 160,000 light years from the optical galaxy. Each lobe contains a brilliant, condensed region of radio emission. The white rectangle indicates the area shown in Figure 27-1. (VLA; NRAO)

FIGURE 27-13 The Head–Tail Source NGC 1265 The elliptical galaxy NGC 1265 would probably be an ordinary double radio source except that this galaxy is moving at a high speed through the intergalactic medium. Because of this motion, its two jets trail behind the galaxy, giving this radio source its distinctly windswept appearance. (NRAO)

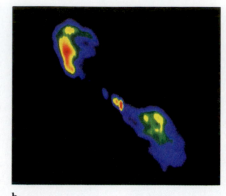
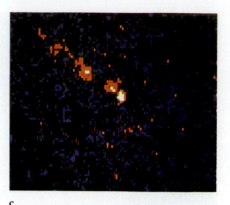

a   b   c

FIGURE 27-14 The Radio Galaxy NGC 5128 This extraordinary galaxy (also called Cen A), located in the southern constellation of Centaurus, is roughly 13 million light-years from Earth. These three views show visible, radio, and X-ray images of this galaxy. (a) This photograph at visible wavelengths shows a broad dust lane across the face of the galaxy. A bluish glow along the edges of the dust lane disclose the presence of numerous young stars. (b) Vast quantities of radio radiation pour from extended regions of the sky on either side of the dust lane. Oppositely directed radio jets emanating from the core of NGC 5128 are nearly perpendicular to the galaxy's dust lane. The structure shown here is usually referred to as the inner lobes, to differentiate them from the much larger lobes that extend much farther out from the galaxy. (c) This X-ray image from the Einstein Observatory shows that NGC 5128 has a bright X-ray nucleus. An X-ray jet protrudes from this nucleus along a direction perpendicular to the galaxy's dust lane. Diffuse X-ray emission comes from the regions surrounding the galaxy's center. (Cerro Tololo Inter-American Observatory, NOAO; VLA, NRAO; Harvard–Smithsonian Center for Astrophysics)

example is Centaurus A, one of the brightest radio sources in the southern sky. The peculiar parent galaxy of Centaurus A, shown in Figure 27-14a, has a broad dust lane studded with young, hot, massive stars. This galaxy, called NGC 5128, is thought to be an elliptical galaxy that has collided with a spiral galaxy, seen edge-on by us. As Figure 27-14b shows, radio waves pour from two lobes on either side of the galaxy. A second set of radio lobes farther from the galaxy spans a volume 2 million light-years across. The entire system is about 13 million light-years from Earth. X-ray images of NGC 5128 reveal an X-ray jet (Figure 27-14c) sticking out of the galaxy's nucleus. This jet, which is perpendicular to the galaxy's dust lane, is aimed toward one of the radio lobes.

The energy output of radio galaxies covers roughly the same range as that of Seyferts. As Table 27-2 shows, these galaxies bridge the gap between quasars and normal galaxies. Because they share so many properties of their remote, ultraluminous cousins, Seyfert galaxies are probably the nearby relics or analogs of radio-quiet quasars, whereas radio galaxies are the remnants of radio-loud quasars. The connection between radio galaxies and radio-loud quasars is reinforced by the discovery that some high-redshift double radio sources have quasars midway between their radio lobes.

There are no nearby quasars. The nearest one is 800 million light-years from Earth; the majority are at least 3 billion light-years away. Since distance corresponds to look-back time, the absence of nearby quasars means that quasar activity tapered off about a billion years ago.

27-4 Quasars, blazars, Seyferts, and radio galaxies are active galaxies

Inspired by the discovery of quasars and luminous peculiar galaxies, astronomers during the 1960s and 1970s searched for unusual objects that might offer clues about these powerful energy sources. One of the objects they focused on was BL Lacertae, a starlike object in the constellation of Lacerta (the Lizard).

BL Lacertae (Figure 27-15) was discovered in 1929, when it was at first mistaken for a variable star, largely because its brightness varies by a factor of 15 within only a few months. BL Lac's most intriguing characteristic was its featureless spectrum, exhibiting neither absorption nor emission lines. Careful examination, however, also revealed some fuzz (visible in Figure 27-15) around its bright, starlike core. In the early 1970s, Joseph Miller at the Lick Observatory blocked out the light from the bright center of BL Lac and managed to obtain a spectrum of the fuzz. This spectrum, which contains stellar absorption lines, strongly resembles the spectrum

---

TABLE 27-2

Galaxy and Quasar Luminosities

Object	Luminosity (W)
Sun	4×10^{26}
Milky Way Galaxy	10^{37}
Seyfert galaxies	10^{36}–10^{38}
Radio galaxies	10^{36}–10^{38}
Quasars	10^{38}–10^{42}

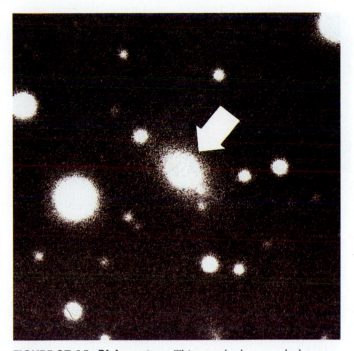

FIGURE 27-15 BL Lacertae This superb photograph shows fuzz around BL Lacertae itself. BL Lacertae objects are giant elliptical galaxies with bright, starlike nuclei that have many quasarlike properties. BL Lacertae objects contain much less gas and dust than do Seyfert galaxies. (Courtesy of T. D. Kinman; Kitt Peak National Observatory)

of an elliptical galaxy. In other words, BL Lacertae is an elliptical galaxy with a bright, starlike center.

BL Lacertae is the prototype of a class of galaxies called **BL Lacertae objects,** or, more simply, **blazars.** Their light is polarized and has a nonthermal spectrum typical of synchrotron radiation. Another important characteristic is their variability. For example, Figure 27-16 shows brightness fluctuations of the blazar 3C 279, determined by carefully examining old photographic plates on which it had been photographed inadvertently in the past. Note the prominent outbursts that occurred around 1937 and 1943. During these outbursts, the luminosity of 3C 279 increased by a factor of at least 25. Because of the enormous distance to 3C 279, at the peak of each of these outbursts, this blazar must have been shining with a brilliance at least 10,000 times as great as that of the entire Milky Way Galaxy.

Detailed radio observations show that blazars are probably double radio sources seen end-on. For instance, high-resolution studies of blazars with the Very Large Array have revealed diffuse radio emission around a bright core in almost all cases. A faint radio halo around a bright radio core is exactly what you would expect if you were looking straight down the jet from a radio galaxy: As seen from this angle, the galaxy's bright nucleus is surrounded by weaker emission from its radio lobes.

The idea that blazars are radio galaxies with their jets aimed toward Earth was supported by the surprising discovery of movement that appeared to be faster-than-light. Such **superluminal motion** is also observed in some quasars, where very-long-baseline interferometry reveals a lumpy structure that changes with time. For example, Figure 27-17 shows four high-resolution images of the quasar 3C 273 spanning three years. During this interval, the two "blobs" that compose 3C 273 moved away from each other at a rate of almost 0.001 arc sec per year. Taking into account the distance to the source, this rate of angular separation corresponds to a speed ten times that of light!

Superluminal motion was puzzling when it was first discovered because it seemed to violate one of the basic tenets of the special theory of relativity: Nothing can move faster than light. Astronomers soon realized, however, that superluminal motion can be explained as movement slower than light—once we take into account the angle at which we view the radio source. If a relativistic beam of material is aimed close to your line of sight, it can *appear* to be moving faster than the speed of light. Quasars generally exhibit lower superluminal speeds (1 to 5 c) than blazars (5 to 10 c), probably because the relativistic jets from quasars are not aimed as close to our line of sight as blazar jets are.

Because of the many properties they share, quasars, blazars, Seyfert galaxies, and radio galaxies are now collectively called **active galaxies.** Since the bright energy source that powers these objects is located at the centers of galaxies, astronomers say that these galaxies possess **active galactic nuclei.**

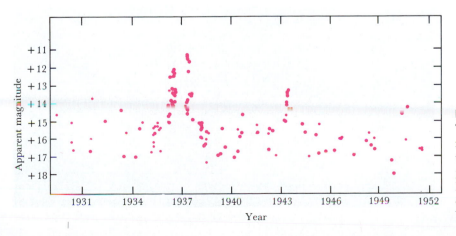

FIGURE 27-16 The Brightness of 3C 279 This graph shows variations in apparent magnitude of the blazar 3C 279. The data were obtained by carefully examining old photographic plates at Harvard College Observatory. Note the large outburst in 1937 and the somewhat smaller one in 1943. (Adapted from L. Eachus and W. Liller)

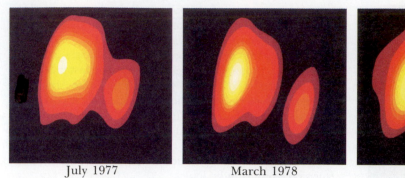
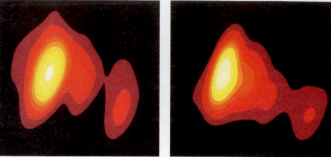

| July 1977 | March 1978 | July 1979 | July 1980 |

FIGURE 27-17 Superluminal Motion in 3C 273 These four high-resolution radio maps of the quasar 3C 273 show "blobs" that seem to be moving apart at ten times the speed of light. In fact, a beam of relativistic particles from 3C 273 is aimed almost directly at the Earth, giving the illusion of faster-than-light motion.

An important clue about the energy sources that power active galactic nuclei comes from their variability. Some quasars vary in brightness over a few weeks or months. Some blazars actually fluctuate from night to night. X-ray observations reveal that some blazars vary in brightness over as little as three hours.

These fluctuations in brightness allow astronomers to place strict limits on the maximum size of a light source, because an object cannot vary in brightness faster than light's travel time across that object. For example, an object that is one light-year in diameter cannot vary significantly in brightness over a period of less than one year.

To understand this limitation, imagine an object that measures one light-year across, as shown in Figure 27-18. Suppose the entire object emits a brief flash of light. Photons from that part of the object nearest the Earth arrive at our telescopes first. Photons from the middle of the object arrive at Earth six months later. Finally, light from the far side of the object arrives a year after the first photons. Although the object emitted a sudden flash of light, we observe a gradual variation in brightness that lasts a full year. In other words, the flash is stretched out over an interval equal to the difference in the light travel time between the nearest and most remote observable regions of the object.

The rapid flickering exhibited by active galactic nuclei means that they draw their energy from a small volume, possibly less than one light-day across. Astrophysicists therefore face the challenge of explaining how the energy output of a thousand galaxies can be produced in such a very small volume.

27-5 Supermassive black holes may be the "central engines" that power active galaxies

As long ago as 1968, the British astronomer Donald Lynden-Bell, working at Caltech, pointed out that an extremely massive black hole could produce the energy output of a quasar or active galactic nucleus from a volume roughly the size of our solar system. Lynden-Bell theorized that gravitational energy released by gases falling onto a black hole would be converted into enough radiation to account for the luminosity of a quasar.

There is a natural limit, called the **Eddington limit** after the famous British astrophysicist Sir Arthur Eddington, to the luminosity that can be radiated by accretion onto a compact object like a black hole. If the luminosity exceeds the Eddington limit, the surrounding gas no longer plunges in-

FIGURE 27-18 A Limit on the Speed of Variations in Brightness The rapidity with which the brightness of an object can vary significantly is limited by the time it takes light to travel across the object. Even if the object, shown here to be 1 light-year in size, emits a sudden flash of light, photons from point A arrive at Earth one year before photons from point C.

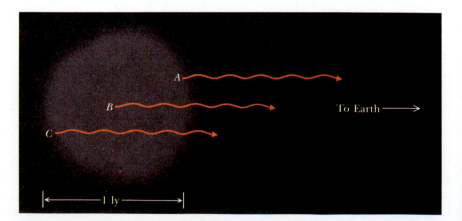

FIGURE 27-19 The Central Bulge of M31
The central bulge of the Andromeda Galaxy can be seen with the naked eye on a clear, moonless night. A photograph of the entire galaxy is shown in Figure 26-3. Spectroscopic observations suggest that a supermassive black hole is located at the center of this large, nearby galaxy. (NOAO)

ward, but instead is blown outward by radiation pressure. This limit allows us to calculate the mass of a quasar.

Numerically, if the mass of the black hole (M) is expressed in solar masses, the Eddington limit (L_{Edd}) in solar units is given by

$$L_{Edd} = 30,000 \left(\frac{M}{M_\odot} \right) L_\odot$$

Anything as luminous as a quasar must have a very high Eddington limit, and this equation tells us that the mass of the black hole must also be quite large. For example, consider the quasar 3C 273, whose luminosity is about 3×10^{13} $L_\odot$. To calculate the minimum mass of a black hole that could continue to attract gas to power the quasar, assume that the quasar's luminosity equals the Eddington limit. Inserting $L_{Edd} = 3 \times 10^{13}$ $L_\odot$ into the above equation, we find that $M = 10^9$ $M_\odot$. Therefore, if a black hole is responsible for the energy output of 3C 273, its mass must be greater than a billion Suns!

Astronomers use the term **supermassive black hole** when referring to a black hole that contains millions or billions of solar masses. As Box 27-3 explains, extremely massive black holes are not as far-fetched as they might first seem, because extreme conditions of pressure and density are not needed to produce them. In fact, evidence for supermassive black holes has been found at the centers of several nearby normal galaxies. For example, the Andromeda Galaxy (M31) is the largest, most massive galaxy in the Local Group. At a distance of only 2.2 million light-years from Earth, M31 is so close to us that details in its core as small as one parsec across can be resolved under the best seeing conditions. A wide-field view of M31 is seen in Figure 26-3. A photograph of the central bulge of M31 is shown in Figure 27-19.

In the mid-1980s, several astronomers made high-resolution spectroscopic observations of the core of M31. By measuring the Doppler shifts of spectral lines at various locations in the core, they determined the orbital speeds of the stars about the galaxy's nucleus. Results are plotted in Figure 27-20 for the innermost 850 light-years of the galaxy.

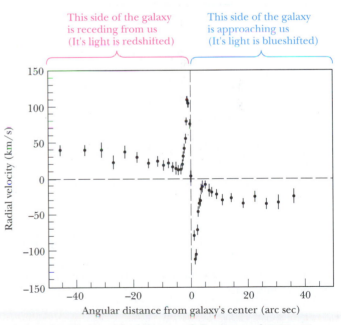

FIGURE 27-20 Rotation Curve of the Core of M31 The radial velocity of matter in the core of M31 is plotted against the angular distance from the galaxy's center. Note the sharp peaks, one blueshifted and one redshifted, within 5 arc sec of the galaxy's center. At the distance of M31, 1 arc sec corresponds to 10.7 ly. (Adapted from J. Kormendy)

(At M31's distance, 850 light-years equals 80 arc sec.) Note that the rotation curve in the galaxy's nucleus does not follow the trend set in the outer core. Rather, there are sharp peaks—one on the approaching side of the galaxy and the other on the receding side—within 5 arc sec of the galaxy's center.

The most straightforward interpretation of these remarkably symmetrical peaks in M31's velocity curve is that they are caused by the rotation of a disk of stars orbiting M31's center. One side of the disk is approaching us while the other side is receding from us. The highest observed radial velocity (at 1.1 arc sec from the galaxy's center; see Figure 27-20) is 110 km/s. This is surely an underestimate, because unsteadiness of the Earth's atmosphere prohibits the detection of features smaller than about 0.5 arc sec across.

The high-speed stars orbiting close to M31's center indicate the presence of a massive central object. Kepler's third law can be used to demonstrate that there must be about 10 million solar masses within 10 light-years of the galaxy's center. That much matter confined to such a small volume strongly suggests the presence of a supermassive black hole.

Located near M31 is a small satellite galaxy called M32, an elliptical with a bright, starlike nucleus (Figure 27-21a). High-resolution spectroscopy of this galaxy also indicates that stars quite close to M32's center are orbiting its nucleus at exceptionally high speeds. These orbital motions suggest the presence of a black hole of roughly 8 million solar masses.

In addition, a picture taken by the Hubble Space Telescope (Figure 27-21b) shows a remarkable concentration of stars at the core of M32. The density of stars there is more than a hundred million times greater than the density of stars in the Sun's neighborhood. This strong concentration of stars further suggests the presence of a supermassive black hole whose powerful gravity causes the stars to crowd around it.

Astronomers have uncovered more evidence of supermassive black holes in distant galaxies. For instance, high-resolution spectroscopy of M104 (Figure 27-22) reveals high-speed orbital motions inside a bright, starlike nucleus. These motions suggest that a billion solar masses lie within 3.5 arc sec of the galaxy's center. Assuming that M104 is 60 million light-years from Earth, all this material must be jammed into a region only 2000 light-years across. Once again the observations imply the presence of a supermassive black hole.

There is no conclusive proof for supermassive black holes at the centers of galaxies. All of the observations we have discussed give only circumstantial indications of black holes. Nevertheless, the possibility that supermassive black holes might be quite commonplace has inspired astrophysicists to examine processes that could generate the energy output of active galactic nuclei and account for their basic properties. The scenario that has emerged is called the "unified model," because it explains all active galaxies as just being different views of one type of object—a supermassive black hole surrounded by an accretion disk.

BOX 27-3

The Plausibility of Extremely Massive Black Holes

Despite their exotic properties, massive black holes do not necessarily require exotic circumstances for their creation. To see why, we shall examine the density of material needed to produce a massive black hole. However, first a word of caution. As we saw in Chapter 24, space around a black hole is highly curved. Simple geometric equations (such as a volume $V = \frac{4}{3}\pi R^3$ for a sphere of radius R) are thus not exactly correct. Nevertheless, they are accurate enough to ensure that the following arguments remain valid.

We want to know how tightly we must compress a mass M inside a sphere of radius R in order to create a black hole. Just before the creation of the hole, the average density (ρ) of the compressed matter is the mass M divided by the volume $\frac{4}{3}\pi R^3$:

$$\rho = \frac{3M}{4\pi R^3}$$

As explained in Chapter 24, however, the radius of a black hole is related to its mass by Schwarzschild's equation:

$$R = \frac{2GM}{c^2}$$

where c is the speed of light and G is the gravitational constant. Substituting this expression for R into the previous equation, we obtain

$$\rho = \frac{3c^6}{32\pi G^3 M^2}$$

This formula tells us that the density required to create a black hole is inversely proportional to the square of the mass of the hole. In other words, as the mass increases, the density needed to make a black hole drops dramatically.

EXAMPLE: To make a 1-$M_\odot$ black hole, the equation tells us that we must compress this mass to a density of roughly 10^{19} kg/m³, which is about 20 times the typical density inside a neutron star.

To make a 10^9-$M_\odot$ black hole, however, the required average density is lowered by a factor of 10^{18}. Thus, we need to squeeze the matter to a density of only 10 kg/m³, only one-hundredth the density of water.

a

b

FIGURE 27-21 The Elliptical Galaxy M32 (a) This small galaxy is a satellite of M31, a portion of which is seen at the left of this wide-angle photograph. Both galaxies are about 2.2 million light-years from Earth. (b) This high-resolution image from the Hubble Space Telescope shows the center of M32. Note the concentration of stars at the nucleus of the galaxy. The area covered in this view is 175 light-years across. (Palomar Observatory; NASA, ESA)

Imagine a billion-solar-mass black hole sitting at the center of a galaxy, surrounded by a huge rotating accretion disk of matter captured by the hole's gravity. According to Kepler's third law, the inner regions of this accretion disk would orbit the hole more rapidly than would the outer parts. Thus, the rapidly spinning inner regions would constantly rub against the slower moving gases in the outer re-gions. This friction would heat up the gases and cause them to spiral inward toward the hole.

Supercomputer simulations give important insights into the structure of an accretion disk surrounding a black hole. These simulations, which combine general relativity with equations describing the dynamics of gas flow, demonstrate that matter is accelerated to supersonic speeds as it plunges

FIGURE 27-22 The Sombrero Galaxy This spiral galaxy in Virgo is nearly edge-on to our Earth-based view. Spectroscopic observa-tions suggest that a billion-solar-mass black hole is located at the galaxy's center. (ESO)

FIGURE 27-23 Flow Patterns along the Inner Edge of an Accretion Disk ▶ This computer-generated picture shows the flow pattern of gas in an accretion disk surrounding a black hole. The black hole itself is located at the center of this cross-sectional diagram, where the broad-angled inner edges of the accretion disk point. The colors display gas pressure, from red for high pressure across the spectrum to blue for low pressure. The flow pattern, indicated by white arrows, shows how the in-falling gas is channeled into two oppositely directed jets. (Courtesy of J. F. Hawley and L. L. Smarr)

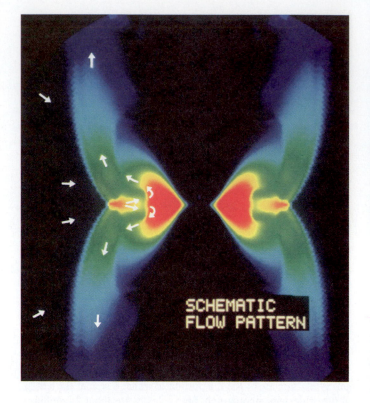

toward a black hole. Near the hole, however, this inward rush is abruptly stopped because of the angular momentum that the matter possesses. Because of the conservation of angular momentum, the inward-spiraling gases do not necessarily reach the black hole, but rather can become concentrated in high-speed orbits quite close to the hole. These orbits mark the location of the "centrifugal barrier," where the supersonic inflow is halted so abruptly that a shock wave is created. This shock wave defines the inner edge of the accretion disk, as shown in cross-section in Figure 27-23. Funnel-shaped cavities surrounding the axis of rotation are kept empty by the centrifugal effect.

FIGURE 27-24 A Supermassive Black Hole as the "Central Engine" The energy output of an active galaxy may involve a supermassive black hole that captures matter from its surroundings. In the scenario depicted here, the inflow of matter through an accretion disk is redirected to produce powerful jets of particles traveling near the speed of light. (Also see artist's rendition on the first page of this chapter.)

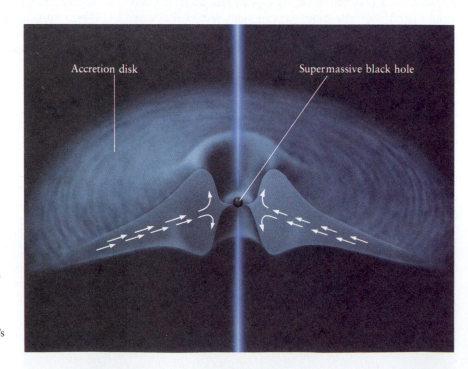

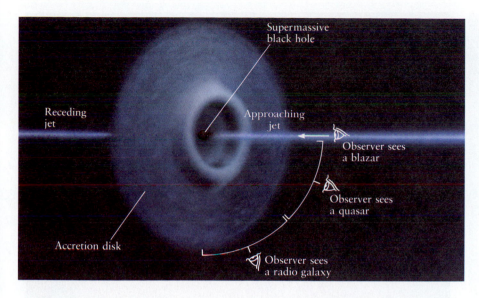

FIGURE 27-25 The "Unified Model" of Active Galaxies Double radio sources, quasars, and blazars may be the same type of object viewed at different angles. If one of the jets is aimed almost exactly at the Earth, we see a blazar. If the jet is somewhat tilted to our line of sight, we see a quasar. If the jets are nearly perpendicular to our line of sight, we see a double radio source.

Because of the constant inward crowding of hot gases, pressures rapidly climb in the inner accretion disk. To relieve this congestion, matter is expelled from the hole along the inner edges of the empty funnel-shaped cavity. The result is two beams of relativistic particles, oppositely directed at right angles to the accretion disk, like the axle through a wheel. The scenario is sketched in Figure 27-24, and an artist's rendition is seen on the opening page of this chapter.

Why is this ejected matter confined in narrow jets? Note that the jets must penetrate a galaxy's interstellar gas. Recent studies demonstrate that there is a natural focusing of high-speed matter as it bores through matter rather than empty space. For instance, think of water squirting out of an ordinary garden hose. As water squirts out of the nozzle, the unconfined stream broadens and the spray fans out through a wide angle. However, if the nozzle is placed in a swimming pool, the stream of water does not broaden nearly as much in the water as it did in air. Similarly, as the two jets of hot gas leave the vicinity of a black hole, they must blast their way through the interstellar gas of the galaxy in which it resides. Passage through this material causes the jets to become narrow beams. This focusing helps explain not only double radio sources but also the jets and beams we see protruding from active galaxies and some quasars.

This model of a supermassive black hole surrounded by an accretion disk with oppositely directed high-speed jets has received considerable attention in recent years, because it apparently offers a single explanation for the wide variety of behavior exhibited by active galaxies. Many astronomer today believe that the main difference between double radio galaxies, quasars, and blazars is only the angle at which the central engine is viewed. As Figure 27-25 shows, an observer sees a double radio source when the accretion disk is viewed nearly edge-on, so that the jets are nearly in the plane of the sky. At a steeper angle, the observer sees a quasar. If one of the jets is aimed almost directly at Earth, a blazar is observed.

In addition to accounting for the overall properties of active galaxies, this unified model can also explain some important details. For example, astronomers recognize two types of Seyfert galaxies according to their spectra. Type I Seyfert galaxies have both broad and narrow emission lines, usually with a narrow, weak emission peak superimposed on each of the broad lines. The spectra of Type II Seyfert galaxies have only narrow emission lines. These spectral differences may simply be the result of viewing the "central engine" from different angles. A Type II Seyfert galaxy may be a mostly edge-on view of the accretion disk around a supermassive black hole, whereas a Type I Seyfert provides a more pole-on view, allowing us to see the turbulent region where the broad spectral lines are formed.

Recent observations of active galactic nuclei provide additional circumstantial evidence for this unified model. In 1992 astronomers used the Hubble Space Telescope to examine the nucleus of the radio galaxy NGC 4261 in the Virgo cluster. Their image of the galaxy's center, shown in Figure 27-26b, reveals a cold, dark, dusty disk that could represent the outer regions of an accretion disk surrounding a suspected supermassive black hole.

The possible existence of supermassive black holes in quasars and active galaxies is one of the most exciting topics in modern astronomy. In the coming years, instruments like the Hubble Space Telescope and the twin Keck Telescopes will be used to probe the nuclei of galaxies with unprecedented resolution. These observations will undoubtedly give us a much better understanding of the powerful engines at the cores of active galaxies and quasars.

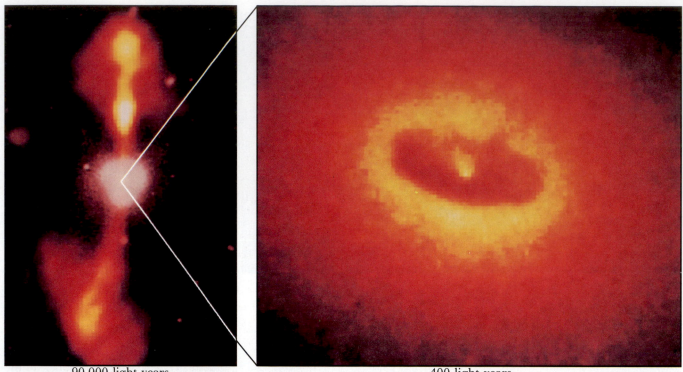

90,000 light-years   400 light-years

FIGURE 27-26 The Core of an Active Galaxy The giant elliptical galaxy NGC 4261 is a double radio source located in the Virgo cluster, about 45 million light-years from Earth. (a) An optical photograph of the galaxy (white) is combined with a radio image (orange) to show both the visible galaxy and its jets. (b) This HST image of the nucleus of NGC 4261 shows a disk of gas and dust about 400 light-years in diameter. (NASA, ESA)

KEY WORDS

Terms preceded by an asterisk are discussed in the boxes.

active galactic nucleus	double radio source	quasar	Seyfert galaxy
active galaxy	Eddington limit	quasi-stellar object	superluminal motion
BL Lacertae object	head–tail source	quasi-stellar radio source	supermassive black hole
blazar	nonthermal radiation	radio galaxy	thermal radiation
*deceleration parameter	polarized radiation	radio lobes	

KEY IDEAS

• A quasar is an object in the sky that looks like a star but has a huge redshift. This corresponds to an extreme distance from the Earth, according to the Hubble law.

To be seen at such large distances, quasars must be very luminous, typically about 1000 times brighter than an ordinary galaxy.

About 10% of all quasars are strong sources of radio emission and are therefore called "radio-loud"; the remaining 90% are radio-quiet, because they are weak radio sources at best.

Some of the energy emitted by quasars is produced by high-speed particles traveling in a strong magnetic field.

• Seyfert galaxies are spiral galaxies with bright nuclei that are strong sources of radiation. Seyfert galaxies seem to be nearby, low-luminosity, radio-quiet quasars.

• Radio galaxies are elliptical galaxies located midway between the lobes of a double radio source. Radio galaxies seem to be nearby, low-luminosity, radio-loud quasars.

Relativistic particles are ejected from the nucleus of a radio galaxy along two oppositely directed beams.

• Blazars are probably radio galaxies seen end-on, with a jet of relativistic particles aimed toward the Earth.

Rapid fluctuations in the brightness of blazars indicate that the energy-producing region is quite small.

- Quasars, blazars, and Seyfert and radio galaxies are examples of active galaxies. The energy source at the center of an active galaxy is called an active galactic nucleus.

- The preponderance of evidence suggests that an active galactic nucleus consists of an accretion disk surrounding a supermassive black hole.

  As gases spiral in toward the supermassive black hole, they are redirected to become two jets of high-speed particles that are aligned perpendicularly to the accretion disk.

  An observer sees a double radio source when the accretion disk is viewed nearly edge-on, so that the jets are nearly in the plane of the sky. At a steeper angle, the observer sees a quasar. If one of the jets is aimed almost directly at Earth, a blazar is observed.

REVIEW QUESTIONS

1. Suppose you saw an object in the sky that you suspected might be a quasar. What sort of observations might you perform to find out if it was indeed a quasar?

2. Explain why astronomers cannot use any of the standard candles described in Chapter 26 to determine the distances to quasars.

3. How would you distinguish between thermal and nonthermal radiation?

4. It was suggested in the 1960s that quasars might be compact objects ejected at high speeds from the centers of nearby ordinary galaxies. Why does the absence of blueshifted quasars disprove this hypothesis?

5. Why do you suppose there are no quasars relatively near our Galaxy?

6. What is a Seyfert galaxy? Why do astronomers think that Seyfert galaxies may be related to radio-quiet quasars?

7. What is a radio galaxy? What is a double radio source? Why do astronomers think these objects may be related to radio-loud quasars?

8. Compare and contrast SS433 (review Figure 23-15) with a typical double radio source.

9. How could a supermassive black hole, from which nothing—not even light—can escape, be responsible for the extraordinary luminosity of a quasar?

10. Why do some astronomers suspect that the centers of certain nearby galaxies contain supermassive black holes?

11. Explain how the unified model of active galaxies accounts for the basic properties of quasars, blazars, and double radio sources.

ADVANCED QUESTIONS

Tips and tools . . .

Relativistic redshift is discussed in Box 27-1. Relationships between the density of matter needed to form a black hole and the Schwarzschild radius are given in Section 24-2. Relationships between apparent magnitude, absolute magnitude, and distance are discussed in Boxes 19-2 and 19-3. You may find it useful to know that 1 ly = 63,240 AU and 1 km/s = 0.211 AU/yr.

12. What redshift would be observed for an object receding from the Sun at a rate of 99% of the speed of light?

13. The quasar PC 1158+4635 discovered in 1989 in the constellation of Ursa Major has a redshift of $z = 4.73$. At what speed does this quasar seem to be receding from us?

14. Suppose that someday an astronomer discovers a quasar with a redshift of 6.0. With what velocity would this quasar seem to be receding from us?

15. How much water at a density of 1 g/cm^3 would it take to make a black hole?

16. What density is required to make a black hole whose mass is equal to the Earth's?

17. Calculate the Schwarzschild radius of a 10^9-$M_\odot$ black hole. How does your answer compare with the size of our solar system?

18. By direct calculation verify that the brightest normal galaxies are too faint to be detected beyond 10 billion light-years.

19. Explain how the existence of gravitational lenses involving quasars (review Figures 24-6 and 24-7) constitutes evidence that quasars are located at the great distances from Earth inferred from the Hubble law.

20. Verify by direct calculation the statement in the text that Kepler's third law applied to the observed orbital velocity (110 km/s) at 1.1 arc sec from the center of M31 demonstrates the presence of about 10 million solar masses within 3.6 pc of the galaxy's center.

21. Calculate the maximum luminosity that could be generated by accretion onto a 10-million-solar-mass black hole.

22. Imagine driving down a street toward a traffic light. How fast would you have to go so that the red light would appear green?

DISCUSSION QUESTIONS

23. Some quasars show several sets of absorption lines whose redshifts are less than the redshifts of the quasars' emission lines. For example, the quasar PKS 0237-23 has five sets of absorption lines with redshifts in the range of 1.364 to 2.202, whereas the quasar's emission lines have a redshift of 2.223. Propose an explanation for these sets of absorption lines.

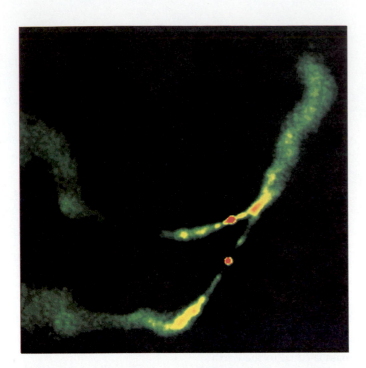

24. The accompanying image from the VLA shows the radio galaxy 3C 75, which has several jets. High-resolution optical photographs reveal that the galaxy has two nuclei, which are the two reddish spots near the center of the VLA image. Propose a scenario that might explain the appearance of 3C 75.

OBSERVING PROJECTS

25. Use a telescope with an aperture of at least 20 cm (8 in.) to observe the Seyfert galaxy NGC 1068. Located in the constellation of Cetus (the Whale), this galaxy is most easily seen during autumn and winter (September through January). The epoch 2000 coordinates are: R.A. = $2^h 42.7^m$ and Decl. = $-0°01'$. Sketch what you see. Is the galaxy's nucleus diffuse or starlike? How does this compare with other galaxies you have observed?

26. Use a telescope with an aperture of at least 20 cm (8 in.) to observe the two companions of the Andromeda Galaxy, M32 and NGC 205. Both are small elliptical galaxies, but only one is suspected of harboring a supermassive black hole. They are located on opposite sides of the Andromeda Galaxy and are most easily seen during autumn and winter (September through January). The epoch 2000 coordinates are:

Galaxy	R.A.	Decl.
M32 (NGC 221)	$0^h 42.7^m$	+40° 52'
NGC 205	0 40.4	+41 41

Make a sketch of each galaxy. Can you see any obvious difference in the appearance of these galaxies? How does this difference correlate with what you know about these galaxies?

27. If you have access to a telescope with an aperture of at least 30 cm (12 in.), you should definitely observe the Sombrero Galaxy, M104. It is located in Virgo and can most easily be seen during the spring and early summer (March through July). The epoch 2000 coordinates are: R.A. = $12^h 40.0^m$ and Decl. = $-11° 37'$. Make a sketch of the galaxy. Can you see the dust lane? How does the nucleus of M104 compare with the centers of other galaxies you have observed?

28. If you have access to a telescope with an aperture of at least 30 cm (12 in.), observe M87 and compare it with the other two giant elliptical galaxies, M84 and M86, that dominate the central regions of the Virgo cluster. The epoch 2000 coordinates are:

Galaxy	R.A.	Decl.
M84 (NGC 4374)	$12^h 25.1^m$	+12°53'
M86 (NGC 4406)	12 26.2	+12 57
M87 (NGC 4486)	12 30.8	+12 24

The photograph from the Palomar Sky Survey shown below may be helpful in identifying the galaxies. This photograph covers an area 3° × 3°; north is at the top.

29. If you have access to a telescope with an aperture of at least 40 cm (16 in.), you might try to observe the brightest-appearing quasar, 3C 273, which has an apparent magnitude of nearly +13. It is located in Virgo at coordinates R.A. = $12^h 29^m 07^s$ and Decl. = $+2° 03' 07''$. You may find it helpful to consult Mood's article in *Astronomy*, listed in the readings.

FOR FURTHER READING

Arp, H. *Quasars, Redshifts and Controversies.* Interstellar Media, 1987. In this book, which has both technical and nontechnical sections, Arp sets out his controversial theories with considerable conviction.

———. "Related Galaxies with Different Redshifts." *Sky & Telescope*, April 1983. This brief article includes puzzling photographs that seem to show high-redshift, galaxylike objects connected to low-redshift galaxies.

Blandford, R., and others. "Cosmic Jets." *Scientific American*, May 1982. This article, which describes jets emanating from the centers of active galactic nuclei, includes an explanation of superluminal (i.e., faster-than-light) motions sometimes observed.

Burns, J., and Price, R. "Centaurus A: The Nearest Active Galaxy." *Scientific American*, November 1983. By examining the nearby active galaxy NGC 5128, the authors piece together a comprehensive model of active galactic nuclei and double radio sources.

Courvoisier, T. J.-L., and Robson, E. I. "The Quasar 3C 273." *Scientific American*, June 1991. This article takes a close look at the brightest-appearing quasar in the sky and shows how its energy output could be derived from a supermassive black hole.

Downes, A. "Radio Galaxies." *Mercury*, March/April 1986. This superb article on double radio sources includes an interesting overview of the techniques used by radio astronomers.

Ferris, T. "The Spectral Messenger: The Redshift Controversy." *Science 81*, October 1981. This well-written article describes the major issues involved in the redshift controversy.

Finkbeiner, A. "Active Galactic Nuclei: Sorting Out the Mess." *Sky & Telescope*, August 1992. This article describes the unified model, which explains how quasars and active galactic nuclei may derive their energy from an accretion disk surrounding a supermassive black hole.

McCarthy, P. "Measuring Distances to Remote Galaxies and Quasars." *Mercury*, January/February 1988. This brief article presents an exceptionally clear summary of how the distances to remote objects depend on such factors as the Hubble constant.

Miley, G. K., and Chambers, K. C. "The Most Distant Radio Galaxies." *Scientific American*, June 1993. This fascinating article explains how observations of remote radio galaxies give a glimpse of the early evolution of giant galaxies.

Mood, J. "Star Hopping to a Quasar." *Astronomy*, May 1987. If you have access to a fairly large telescope and want to hunt for the brightest-appearing quasar in the sky, you might consult this article on 3C 273.

Preston, R. *First Light.* Atlantic Monthly Press, 1987. This book takes the reader behind the scenes at the Palomar Observatory. It includes an exciting section on Maarten Schmidt's pioneering research on quasars.

Rees, M. J. "Black Holes in Galactic Centers." *Scientific American*, November 1990. This article explores the idea that supermassive black holes are located at the centers of active galaxies and quasars.

Shipman, H. *Black Holes, Quasars, and the Universe.* 2nd ed. Houghton Mifflin, 1980. This clear, cogent text has an excellent section on quasars.

Smith, D. "Mysteries of Cosmic Jets." *Sky & Telescope*, March 1985. This one-page article describes recent observations of the jet in M87 and contrasts it with the jet in 3C 273.

Sulentic, J. "Are Quasars Far Away?" *Astronomy*, October 1984. This article by one of Arp's co-workers raises some interesting questions about the distances to quasars.

Tananbaum, H., and Lightman, A. "Cosmic Powerhouses: Quasars and Black Holes." In Cornell, J., and Lightman, A., eds., *Revealing the Universe.* MIT Press, 1982. This chapter describes observations and theoretical work that lead astronomers to believe that quasars are powered by supermassive black holes.

Verschuur, G. *The Invisible Universe Revealed.* Springer Verlag, 1987. This fascinating book has four excellent chapters summarizing our knowledge of radio galaxies and quasars.

Weedman, D. *Quasar Astronomy.* Cambridge University Press, 1986. This somewhat technical monograph summarizes our current understanding of quasars.

COSMOLOGY:
THE CREATION AND FATE
OF THE UNIVERSE

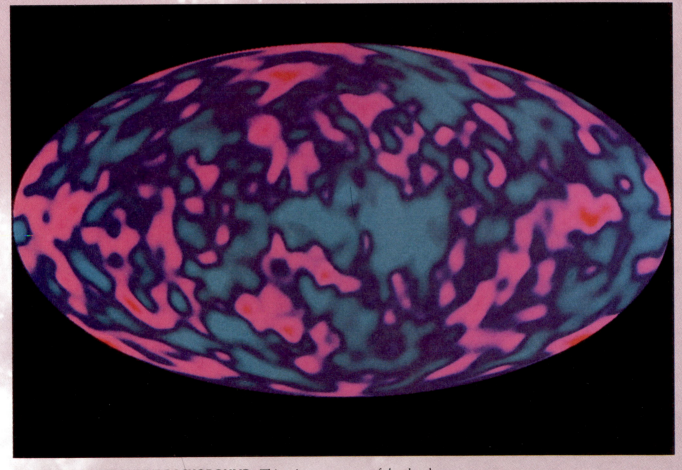

THE COSMIC MICROWAVE BACKGROUND This microwave map of the sky shows temperature variations in the cosmic microwave background, a profusion of photons left over from the Big Bang. Pink regions are about 0.0003 K warmer than the average temperature of 2.73 K; blue regions are about 0.0003 K cooler than the average. These tiny fluctuations in temperature date back to the earliest moments of the universe and may be directly related to the large-scale structure of the universe today. (Goddard Space Flight Center; NASA)

WE LIVE in an expanding universe. This expansion of space began with an explosive event at the beginning of time called the Big Bang. The universe is now full of microwave photons, ghostly relics of the primordial fireball that filled all space shortly after the Big Bang. Today barely detectable, this radiation dominated the universe for nearly a million years after the Big Bang. Whether the universe will expand forever or collapse in a Big Crunch depends on the density of matter throughout space. Clues to the ultimate fate of the universe can be gleaned from the motions of extremely remote galaxies. If the universe expands forever, bizarre phenomena such as the evaporation of black holes may occur in the extremely distant future.

As foolish as it may seem, one of the most profound questions you can ask is, "Why is the sky dark at night?" This question apparently haunted Johannes Kepler as long ago as 1610. It was brought to public attention in the early 1800s by the German amateur astronomer Heinrich Olbers.

To appreciate the problem, consider the universe as Olbers and his contemporaries saw it—an infinite expanse with stars scattered more or less randomly throughout space. Isaac Newton himself argued that this model of the universe is the only concept that makes sense. If the universe were not infinite, or if stars were grouped in only one part of the universe, the gravitational forces acting between the stars would soon cause all this matter to fall together into a compact blob. Obviously, this has not happened. Thus, as classical Newtonian mechanics would have it, we must be living in a universe that is both infinite and static. According to this model, the universe is infinitely old and will continue to exist forever without major changes in its structure. Such an infinite expanse, however, presents us with a puzzle.

Imagine looking out into space in a static, infinite universe. Because space goes on forever, with stars scattered throughout it, your line of sight must eventually hit a star. No matter where you look in the sky, you should ultimately see a star. The entire sky should thus be as bright as an average star. Even at night, the entire sky should be blazing like the surface of the Sun. That this is not so is the dilemma called **Olbers's paradox.**

28-1 We live in an expanding universe

Olbers's paradox foretold that there is something wrong with Newton's idea of an infinite, static universe. According to the classical, Newtonian picture of reality, space is laid out in all directions like a great, flat sheet of inflexible, rectangular graph paper. This rigid, flat space stretches on and on, totally independent of stars or galaxies or anything else. Similarly, a Newtonian clock ticks steadily and monotonously forever, never slowing down or speeding up. Furthermore, Newtonian space and time are unrelated; measurements made with rulers are independent of measurements made with clocks.

Albert Einstein demonstrated that this view of space and time is wrong. In his special theory of relativity (recall Box 24-1), he proved that measurements with clocks and rulers depend on the motion of the observer. As explained in Chapter 24, Einstein's general theory of relativity tells us that gravity curves the fabric of space. As a result, the matter that occupies the universe influences the overall shape of space throughout the universe.

Shortly after formulating his general theory of relativity in 1915, Einstein applied his ideas to the structure of the universe. At that time, the Newtonian view that the universe is static prevailed. Einstein was therefore dismayed to find that his calculations could not produce a truly static universe; according to general relativity, the universe must be either expanding or contracting. In desperation, he added a term called the **cosmological constant** (Λ) to his equations to ensure that his calculations would predict a static universe. This cosmological constant was supposed to represent a pressure that balances gravitational attraction in the static universe and prevents it from collapsing. It was reported that Einstein, in his later years, called the cosmological constant "the greatest blunder of my life." Because he doubted his original equations, he missed the opportunity of postulating that we live in an expanding universe. He could have beat Hubble to the punch by at least ten years.

Edwin Hubble is usually credited with discovering that we live in an expanding universe. As we saw in Chapter 26 (review Figure 26-16), Hubble found a simple linear relationship between the distances to remote galaxies and the redshifts of those galaxies' spectral lines. This relationship, now called the Hubble law, states that the greater the distance to a galaxy, the greater is the galaxy's redshift. Thus, remote galaxies are moving away from us with speeds proportional to their distances. Specifically, the recessional velocity v of a galaxy is related to its distance r from Earth by the equation

$$v = H_0 r$$

where H_0 is the Hubble constant. Because the clusters of galaxies are getting farther and farther apart as time goes on, astronomers say that the universe is expanding.

What does it actually mean to say that the universe is expanding? According to general relativity, space itself is not rigid. The amount of space between widely separated locations changes over time. A good analogy is that of a person blowing up a balloon, as sketched in Figure 28-1. Small coins, each representing a galaxy, are glued onto the surface of the balloon. As the balloon expands, the amount of space between the coins gets larger and larger. In the same way, as the universe expands, the amount of space between widely separated galaxies increases. The expansion of the universe *is* the expansion of space.

The expanding-balloon analogy includes several important characteristics about the expanding universe. For example, imagine that you are stationed on one of the coins in Figure 28-1. As the balloon expands, you see all the other coins moving away from you. Specifically, you observe that the nearby coins are moving away from you slowly, whereas the more distant coins are moving away more rapidly. You find the same relationship, which is just like the Hubble law, no matter which coin you decide to call home. Your coin is not at the center of the balloon, of course. Indeed, the surface of the balloon does not actually have a center: You could move around the surface indefinitely without ever finding a center. Likewise, the surface of the balloon has no edge; you could explore every inch of it and never find an edge.

Just as the surface of the balloon has neither center nor edge, our universe has no center or edge. No matter which galaxy you call home, all the other galaxies are receding from you. No one is ever at the "center" of the universe. Questions such as "What is beyond the edge of the universe?" or "What is the universe expanding into?" are as meaningless as asking "What is north of the Earth's north pole?"

The ongoing expansion of space explains why photons from remote galaxies are redshifted. Imagine a photon coming toward us from a distant galaxy. As the photon travels through space, the space is expanding, so the photon's wavelength becomes stretched. When the photon reaches our eyes, we see a longer wavelength than usual, which is the redshift. The longer the photon's journey, the more its wavelength will have been stretched. Thus, photons from distant galaxies have larger redshifts than those of photons from nearby galaxies, as expressed by the Hubble law.

A redshift caused by the expansion of the universe is properly called a **cosmological redshift**, to distinguish it from a Doppler shift. Doppler shifts are caused by an object's *motion through space*, whereas a cosmological redshift is caused by the *expansion of space*.

We can calculate the factor by which the universe has expanded since some ancient time from the redshift of light emitted by objects at that time. As we have seen before, redshift (z) is defined as

$$z = \frac{\lambda - \lambda_0}{\lambda_0}$$

where λ_0 is the unshifted wavelength of a photon, and λ is the wavelength we observe. For instance, λ_0 could be the wavelength of a particular spectral line in the spectrum of light leaving the surface of a remote quasar. As the quasar's light travels through space, its wavelength is stretched by the expansion of the universe. Thus, at our telescopes we observe the spectral line to have a wavelength λ. The ratio λ/λ_0 is a measure of the amount of stretching. By rearranging terms in the preceding equation, we can solve for this ratio to obtain

$$\frac{\lambda}{\lambda_0} = 1 + z$$

For example, consider a quasar with a redshift of $z = 3$. Since the time that light left that quasar, the universe has expanded by a factor of $1 + z = 1 + 3 = 4$. In other words, when the light left that quasar, representative distances between widely separated galaxies were only one-quarter as large as they are today. A representative volume of space, which is proportional to the cube of its dimensions, was only $1/4^3$, or $\frac{1}{64}$th as large as it is today. Thus the density of matter in such a volume was 64 times greater than it is today.

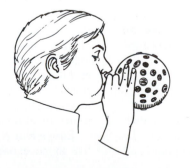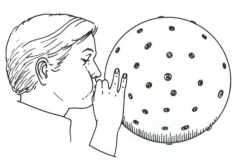

FIGURE 28-1 The Expanding-Balloon Analogy The expanding universe can be compared to the expanding surface of an inflating balloon. All the coins on the balloon recede from one another as the balloon expands, just as all the galaxies recede from one another as the universe expands. (Adapted from C. Misner, K. Thorne, and J. Wheeler)

Finally, it is important to realize that the expansion of space occurs primarily in the intergalactic space that separates the various clusters of galaxies. Just as the coins in Figure 28-1 do not expand as the balloon inflates, galaxies themselves do not expand. Einstein and others have established that an object that is held together by its own gravity, such as a galaxy, is always contained within a patch of nonexpanding space. A galaxy's gravitational field produces this nonexpanding region, which is indistinguishable from the flat, rigid space described by Newton. Thus, the Earth and your body, for instance, are not getting any bigger. Only the distance between widely separated galaxies increases with time.

28-2 The expanding universe probably emerged from an explosive event called the Big Bang

Cosmology is unique among the sciences because it depends on some philosophical assumptions. Since there is only one observable universe, we cannot carry out controlled experiments or even make comparisons: We must accept certain assumptions or abandon hope of making progress in understanding the nature of the universe. The Hubble law provides a classic example; it could be interpreted as implying that we are at the center of the universe. We reject this interpretation, however, because it violates a cosmological extension of Copernicus's belief that we do not occupy a special location in space.

When Einstein began applying his general theory of relativity to cosmology, he made a daring assumption—one that gives precise meaning to the idea that we do not occupy a special location in space. Einstein assumed that over very large distances the universe is *homogeneous* (that is, every region is the same as every other region) and *isotropic* (that is, the universe looks the same in every direction). In other words, if you could stand back and look at a very large region of space, you could not tell one part of the universe from another. The assumption that the universe is homogeneous and isotropic constitutes the **cosmological principle.**

Models of the universe based on the cosmological principle have proven to be surprisingly successful in describing the structure and evolution of the universe and in interpreting observational data. All of our discussion about the universe in this chapter and the next assumes that the universe is homogeneous and isotropic on the largest scale.

Because the universe has been expanding for billions of years, there must have been a time in the very distant past when all the matter in the universe was concentrated in a state of infinite density. Presumably some sort of colossal explosion initiated the expansion of the universe. This explosion, commonly called the **Big Bang,** marks the creation of the universe.

To calculate the time elapsed since the Big Bang, imagine watching a movie of any two galaxies separated by a distance r and receding from each other with a velocity v. Now run the film backward, and observe the two galaxies approaching each other as time runs in reverse. We can calculate the time T_0 it will take for the galaxies to collide by using the simple equation

$$T_0 = \frac{r}{v}$$

Using the Hubble law, $v = H_0 r$, to replace the velocity v in this equation, we get

$$T_0 = \frac{r}{H_0\,r} = \frac{1}{H_0}$$

Note that the distance of separation, r, has cancelled out and does not appear in this equation. As a result, T_0 is the same for all galaxies. This is the time in the past when all galaxies were crushed together, the time back to the Big Bang. In other words, H_0 gives us an estimate of the age of the universe:

$$T_0 = \frac{1}{75 \text{ km/s/Mpc}} = 13 \text{ billion years}$$

where we have made use of the fact that 1 Mpc equals 3.09×10^{19} km to convert T_0 into units of time.

As we saw in Chapter 26, the value of the Hubble constant is uncertain by roughly a factor of two. Some astronomers think that H_0 is as low as 50 km/s/Mpc whereas others think it may be as high as 90 km/s/Mpc. This range of H_0 corresponds to a range in T_0 from about 11 to 20 billion years. Many astronomers suspect that the true age of the universe is probably about 15 billion years, which is the age of the oldest stars.

The finite age of the universe offers a resolution of Olbers's paradox. The entire sky is not as bright as the surface of the Sun because we cannot see any objects that are more than 15 billion light-years away. Because the universe is about 15 billion years old, the light from stars more than 15 billion light-years away has just not had enough time to get here. This is true even if the universe is infinite, with galaxies scattered throughout its limitless expanse.

You can think of the Earth as being at the center of an enormous sphere having a radius of roughly 15 billion light-years (Figure 28-2). The surface of this sphere is called the **cosmic particle horizon.** Our entire **observable universe** is located inside this sphere. We cannot see anything beyond the cosmic particle horizon because the travel time for light coming from these incredibly remote distances is greater than the age of the universe. Throughout the observable universe, galaxies are distributed sparsely enough that most of our lines of sight hit no stars, which explains why the night sky is dark.

The finite age of the universe gives us one way out of Olbers's paradox. However, there is a second effect that also contributes significantly to the darkness of the night sky: the redshift. The Hubble law tells us that the redshift of a galaxy is directly related to its distance from Earth—the greater the distance, the greater the redshift. Recall from Chapter 5 that

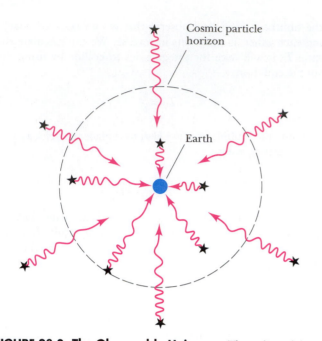

FIGURE 28-2 The Observable Universe The radius of the cosmic particle horizon is equal to the distance that light has traveled since the Big Bang. This event occurred about 15 billion years ago, so the cosmic particle horizon is about 15 billion light-years away. We cannot see objects beyond the cosmic particle horizon, because their light has not had enough time to reach us.

the energy E carried by a photon is related to its wavelength λ by the expression

$$E = \frac{hc}{\lambda}$$

where h is Planck's constant and c is the speed of light.

When a photon is redshifted, its wavelength is lengthened and its energy is decreased. The greater the redshift, the less the energy. Consequently, even though there are many galaxies far from the Earth, they have large redshifts and their light does not carry much energy. A galaxy nearly at the cosmic particle horizon has a nearly infinite redshift, meaning that its light carries practically no energy at all. This decrease in photon energy because of the expansion of the universe decreases the brilliance of remote galaxies, thus also contributing to the darkness of the night sky.

There are many misconceptions about the Big Bang. For one thing, it was not at all like an exploding bomb. When a bomb explodes, pieces of debris fly off *into space* from a central location. If you could trace all the pieces back to their origin, you could find out exactly where the bomb had been. This process is not possible with the universe, though, because the universe itself always has and always will consist of all space. There is genuinely nothing—not even space—beyond the "edge" of the universe. As we have seen, the universe logically cannot have an edge.

The concept of a Big Bang origin for the universe is a straightforward, logical consequence of having an expanding universe. If you can just imagine far enough back into the

past, you can arrive at a time roughly 15 billion years ago when the density throughout the universe was infinite. The entire universe was in effect like the center of a black hole.

Comparing the Big Bang to the center of a black hole can help us appreciate certain aspects of the creation of the universe. As we saw in Chapter 24, matter at the center of a black hole is crushed to infinite density. This location, called the singularity, is characterized by infinite curvature in which space and time are all tangled up. Without a clear background of space and time, such concepts as "past," "future," "here," and "now" cease to have meaning. For this reason, a better name for the Big Bang is the **cosmic singularity**.

At the moment of the Big Bang, a state of infinite density filled the universe. Throughout the universe, space and time were completely jumbled up in a condition of infinite curvature like that at the center of a black hole. Thus, we cannot use the laws of physics to tell us exactly what happened at the moment of the Big Bang. And we certainly cannot use science to tell us what existed before the Big Bang. These things are fundamentally unknowable. The phrases "*before* the Big Bang" or "at the *moment* of the Big Bang" are meaningless, because time did not really exist until *after* that moment.

A very short time after the Big Bang, space and time did begin to behave in the way we think of them today. This interval, called the **Planck time** (t_P), is given by the expression

$$t_P = \frac{Gh}{c^5} = 1.35 \times 10^{-43} \text{ s}$$

where G is the constant of gravitation, h is Planck's constant, and c is the speed of light (Box 28-1).

From the Big Bang, at time $t = 0$, to the Planck time 10^{-43} second later, all known science fails us. We do not know how space, time, and matter behaved in that brief interval. Nevertheless, some physicists have speculated that space and time as we know them today burst forth from a seething, foam-like, space–time mishmash during the Planck time. We can think of the Big Bang as an explosion of space at the beginning of time.

28-3   The microwave radiation that fills all space is evidence of a hot Big Bang

One of the major successes in modern astronomy involves discoveries about the origin of the heavy elements. We know today that all the heavy elements are created in the infernos at the centers of stars (review Box 22-2 for details) and in supernovae. As astronomers began to understand the details of thermonuclear synthesis in the 1960s, a new problem arose: There is too much helium around. For example, the Sun consists of about 74% hydrogen and 25% helium by mass, leaving only 1% for all the remaining heavier elements combined. This 1% can be understood as material produced inside earlier generations of massive stars that long ago cast these heavy elements out into space when they became supernovae. Some freshly made helium certainly accompanied these heavy elements, but this amount was not nearly enough to account for one-quarter of the Sun's mass.

Shortly after World War II, Ralph Alpher and Robert Hermann proposed that the universe immediately following the Big Bang must have been so incredibly hot that thermonuclear reactions occurred everywhere throughout space. Following up this idea in 1960, the Princeton physicists Robert Dicke and P. J. E. Peebles discovered that they could indeed account for today's high abundance of helium by assuming that the early universe had been at least as hot as the Sun's center, where helium is currently being produced. The early universe must therefore have been filled with many high-energy, short-wavelength photons, which formed a radiation field whose temperature can be given by Planck's blackbody law (review Figure 5-8).

The universe has expanded so much since those ancient times that all those short-wavelength photons now have wavelengths that are so stretched that they have become low-

---

BOX 28-1

The Planck Time: A Limit of Knowledge

In the general theory of relativity, space and time are treated together as a continuum. Because of quantum effects, however, this smooth picture of space and time breaks down at extremely small distances or short time intervals. To see why this breakdown occurs, we must first examine some basic ideas in quantum mechanics.

Quantum concepts were introduced around 1900 to explain fundamental properties of light, particularly details of blackbody radiation and the photoelectric effect. As we saw in Chapter 5, this new understanding revealed that light has *both* wavelike and particlelike properties. If you perform an experiment to study the wavelike properties of light (for example, Young's double-slit experiment, shown in Figure 5-4), you find that light behaves like waves. If you perform an experiment to study the particlelike properties of light (for example, the photoelectric effect shown in Figure 5-10), you find that light behaves like particles. This chameleonlike behavior is called **wave–particle duality.**

As experiment and theory continued to progress through the 1920s, physicists discovered that particles, such as electrons, can behave like waves. In other words, wave–particle duality applies to particles as well as photons. Experiments performed by American physicist Arthur H. Compton with X-ray photons striking matter helped confirm the wave–particle duality of nature. Because massive particles have wavelike properties, they typically cannot be precisely located over a distance λ_C given by

$$\lambda_C = \frac{h}{mc}$$

where h is Planck's constant, m is the particle's mass, c is the speed of light, and λ_C is now called the **Compton wavelength.** For example, an electron has a Compton wavelength of 2.43×10^{-3} nm, which is roughly a thousandth the size of an atom.

The Compton wavelength is a measure of how fuzzy the physical world is at the quantum level. Quantum effects limit our ability to locate a particle on scales smaller than its Compton wavelength, which is a characteristic distance over which the particle exists.

General relativity, the second great revolution in physics to emerge in the early twentieth century, tells us that events around a mass m that are within the Schwarzschild radius are hidden from the outside world (recall Figure 24-3). This limiting distance (D) is

$$D = \frac{Gm}{c^2}$$

where G is the constant of gravitation.

To see why general relativity breaks down, imagine a black hole whose mass is so tiny that its Schwarzschild radius is smaller than its Compton wavelength. Because of quantum-mechanical uncertainty, we cannot be sure that the singularity is inside the event horizon. Indeed, there is a probability that the singularity spends some time outside the event horizon.

The limiting case of the smallest black hole that does not suffer from this breakdown is obtained by equating D and λ_C. The resulting equation can be solved for the **Planck mass** (m_P):

$$m_P = \sqrt{\frac{hc}{G}} = 5.46 \times 10^{-8} \text{ kg}$$

The Planck mass must not be confused with the mass of a proton, which is 10^{19} times smaller. A black hole whose mass is less than the Planck mass is not correctly described by general relativity.

The limiting distance corresponding to $m = m_P$ is called the **Planck length** (λ_P):

$$\lambda_P = \sqrt{\frac{Gh}{c^3}} = 4.05 \times 10^{-35} \text{ m}$$

Note that the size of a proton (about 10^{-15} m) is enormous compared to the Planck length.

The light-travel time across this length is called the Planck time (t_P):

$$t_P = \frac{\lambda_C}{c} = \frac{Gh}{c^5} = 1.35 \times 10^{-43} \text{ s}$$

Because general relativity breaks down for times earlier than the Planck time, no one knows how to describe the universe between $t = 0$ and $t = t_P$. At the moment $t = t_P$, the mass contained within the cosmic particle horizon at each point in space was of the order of the Planck mass.

FIGURE 28-3 The Bell Labs Horn Antenna Using this horn antenna in New Jersey, Arno Penzias and Robert Wilson detected the microwave radiation coming from all parts of the sky. (Bell Labs)

energy, long-wavelength photons. The temperature of this cosmic radiation field is now quite low, only a few degrees above absolute zero. By Wien's law, radiation at such a low temperature should have its peak intensity at microwave wavelengths of roughly one millimeter. In the early 1960s, Dicke and his colleagues began designing an antenna to detect this microwave radiation.

Meanwhile, just a few miles from Princeton University, Arno Penzias and Robert Wilson of Bell Telephone Laboratories were working on a new microwave horn antenna designed to relay telephone calls to Earth-orbiting communications satellites (Figure 28-3). Penzias and Wilson were deeply puzzled when, no matter where in the sky they pointed their antenna, they detected faint background noise. Thanks to a colleague, they happened to learn about the work of Dicke and Peebles, and came to realize that they had discovered the cooled-down cosmic background radiation left over from the hot Big Bang.

Since those pioneering days, scientists have made many measurements of the intensity of this background radiation at a variety of wavelengths. The most accurate measurements come from the Cosmic Background Explorer satellite (COBE), which was placed in orbit about the Earth in 1989 (Figure 28-4). Data from COBE's spectrometer shown in Figure 28-5 demonstrate that this ancient radiation has the spectrum of a blackbody with a temperature of 2.73 K. This radiation field, which fills all of space, is commonly called the **cosmic microwave background.**

An important feature of the microwave background is its intensity, which is almost perfectly isotropic (that is, the same

FIGURE 28-4 The Cosmic Background Explorer (COBE) This satellite, launched in 1989, is measuring the spectrum and angular distribution of the cosmic microwave background over a wavelength range of 1 μm to 1 cm. COBE (pronounced CO-bee) is designed to detect deviations from a perfect blackbody spectrum and from perfect isotropy. (Courtesy of J. Mather; NASA)

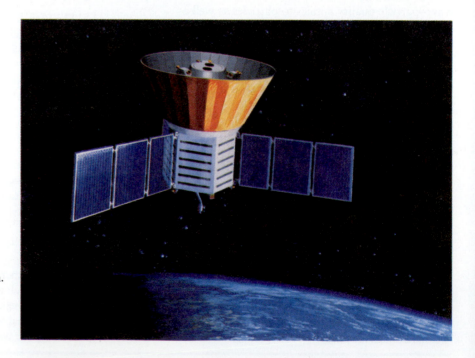

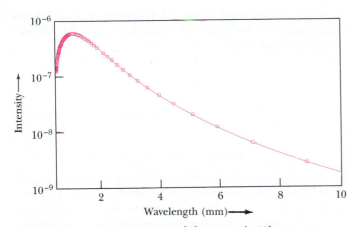

FIGURE 28-5 The Spectrum of the Cosmic Microwave Background The little squares on this graph are COBE's measurements of the brightness of the cosmic microwave background plotted against wavelength. The data fall along a blackbody curve for 2.73 K to a remarkably high degree of accuracy. Note that the peak of the curve, at a wavelength of 1.1 mm, is in accordance with Wien's law. (Courtesy of E. Cheng; NASA COBE Science Team)

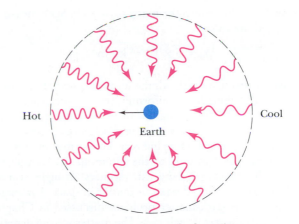

FIGURE 28-7 Our Motion Through the Microwave Background Because of the Doppler effect, the microwave background is slightly warmer in that part of the sky toward which we are moving. Recent measurements indicate that our Galaxy, along with the rest of the Local Group, is moving in the general direction of the Virgo and Hydra clusters.

in all directions). In other words, we detect nearly the same background intensity from all parts of the sky. However, extremely accurate measurements first made from high-flying airplanes and more recently from COBE reveal a very slight variation in temperature across the sky. The microwave background is slightly warmer than average toward the constellation of Leo and slightly cooler than average in the opposite direction toward Aquarius. Between the warm spot in Leo and the cool spot in Aquarius, the background temperature declines smoothly across the sky. A map of the microwave sky showing this anisotropy is seen in Figure 28-6.

This variation in temperature can be explained as a result of the Earth's overall motion through the cosmos. If we were at rest with respect to the microwave background, the radiation would be truly isotropic. Because we are moving through this radiation field, however, we see a Doppler shift. Specifically, we see shorter-than-average wavelengths in the direction toward which we are moving, as sketched in Figure 28-7. A decrease in wavelength corresponds to an increase in photon energy and thus in temperature. The temperature excess observed corresponds to a speed of 390 km/s. Conversely, we see longer-than-average wavelengths in that part of the sky from which we are receding. An increase in wavelength corresponds to a decline in photon energy and hence a decline in temperature.

We are thus traveling away from Aquarius toward Leo at a speed of 390 km/s. Taking into account the known velocity of the Sun around the center of our Galaxy, we find that the entire Milky Way Galaxy is moving at 600 km/s toward the Hydra–Centaurus supercluster, possibly because of the gravitational pull of an enormous mass, dubbed the **Great Attractor**, lying in that direction.

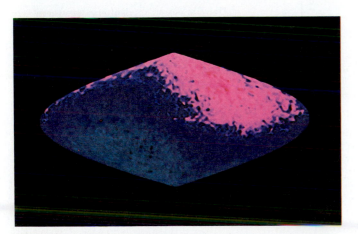

FIGURE 28-6 The Microwave Sky This map of the microwave sky was produced from data taken by instruments on board COBE. The galactic center is in the middle of the map, and the plane of the Milky Way runs horizontally across the map. Color indicates temperature: Magenta is warm and blue is cool. The temperature variation across the sky is caused by the Earth's motion through the microwave background. (NASA)

28-4 Cosmic background radiation dominated the universe during the first million years

Everything in the universe falls into one of two categories: matter or energy. The matter is contained in such objects as stars, planets, and galaxies, and is composed of particles such as electrons, protons, and neutrons. If we take a large volume V of space and add up the total mass M of all the stars and galaxies we can see in it, we can calculate the **average density of matter** (ρ_m) in the universe at the present time simply by dividing the volume into the mass: $\rho_m = M/V$. Such measurements yield the value

$$\rho_m = 5 \times 10^{-28} \text{ kg/m}^3$$

In other words, if we took all the luminous matter we could find and distributed it uniformly throughout space, we would have 5×10^{-28} kilogram in every cubic meter. The mass of a hydrogen atom is 1.7×10^{-27} kg, so ρ_m is equivalent to about one hydrogen atom in every 3 cubic meters of space.

This density refers only to the luminous matter that we can actually observe. Many astronomers strongly suspect that there is a lot of nonluminous matter in space. For example, the famous dark-matter problem discussed in Chapter 26 applies to clusters of galaxies: The matter we can actually detect is not sufficient to hold clusters together gravitationally. The true density of the matter in the universe, including both the luminous and the nonluminous forms, may be as high as 10^{-26} kg/m³. This value is equivalent to the mass of six hydrogen atoms per cubic meter of space. As we shall see in the next section and the final chapter, a density this great has profound implications for the ultimate fate of the entire universe.

The radiation energy in the universe consists of photons. There are, of course, many starlight photons traveling across space, but the vast majority of photons in the universe belong to the 3-K microwave background.

To compare this radiation field with the matter in the universe, recall Einstein's famous equation $E = mc^2$, which tells us that we can think of the energy in the universe as being equal to mass multiplied by the square of the speed of light. This connection between matter and energy allows us

to speak of the **mass density of radiation** (ρ_{rad}). Einstein's equation ($E = mc^2$) can be combined with the Stefan–Boltzmann law to give

$$\rho_{rad} = \frac{4\sigma T^4}{c^3}$$

where σ is the Stefan–Boltzmann constant. For $T = 3$ K, this equation yields

$$\rho_{rad} = 7 \times 10^{-31} \text{ kg/m}^3$$

Note that ρ_m is about a thousand times larger than ρ_{rad}. In other words, the density of matter (which we have probably underestimated) is clearly much greater than the mass density of radiation. For this reason we say we are living in a **matter-dominated universe.**

Although matter dominates the universe today, this was not always the case. Matter prevails over radiation today only because the energy now carried by microwave photons is so small. Nevertheless, the number of photons in the microwave background is astounding. From the physics of blackbody radiation it can be demonstrated that there are today 550 million photons in every cubic meter of space. In other words, the photons in space outnumber atoms by roughly a billion to one. In terms of total number of particles, the universe thus consists almost entirely of microwave photons. This radiation field no longer has much "clout," though, because its photons have been redshifted to long wavelengths and low energies after nearly 15 billion years of being stretched by the expansion of the universe.

In contrast, think back toward the Big Bang. The universe becomes increasingly compressed, and so the density of mat-

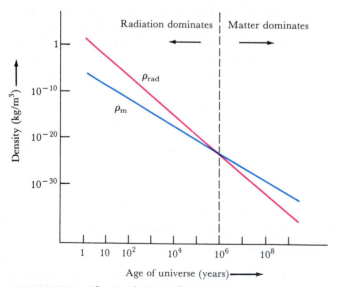

FIGURE 28-8 The Evolution of Density During roughly the first million years, the mass density of radiation (ρ_{rad}) exceeded the matter density (ρ_m), and the universe was radiation-dominated. Later, however, continued expansion of the universe caused ρ_{rad} to become less than ρ_m, at which point the universe became matter-dominated.

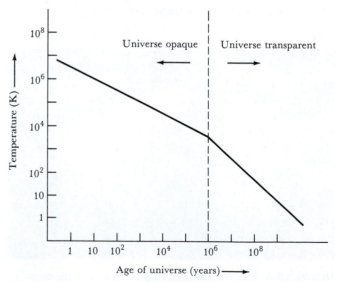

FIGURE 28-9 The Evolution of Temperature As the universe expanded, the photons in the radiation background became increasingly redshifted and the temperature of the radiation fell. Roughly 1 million years after the Big Bang, when the temperature fell below 3000 K, hydrogen atoms formed and the radiation field "decoupled" from the matter in the universe.

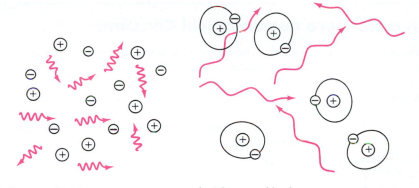

FIGURE 28-10 The Era of Recombination
(a) Before recombination, photons in the cosmic background prevented protons and electrons from forming hydrogen atoms. (b) As soon as hydrogen atoms could survive, the universe became transparent. This transition from opaque to transparent occurred roughly 1 million years after the Big Bang.

a Before recombination   b After recombination

ter increases as we go back in time. The photons in the background radiation also become more crowded together, but an additional effect must be considered. As we go back in time, the photons become less redshifted and thus have shorter wavelengths and higher energy than they do today. Because of this added energy, the mass density of radiation (ρ_{rad}) increases more quickly than the density of matter (ρ_m). In fact, as shown in Figure 28-8, there was a time in the ancient past when ρ_{rad} equaled ρ_m. Before this time, ρ_{rad} was greater than ρ_m, and so radiation held sway over matter. Astronomers call this state a **radiation-dominated universe.**

This transition from a radiation-dominated universe to a matter-dominated universe occurred nearly a million years after the Big Bang, at a time that corresponds to a redshift of about $z = 1000$. In other words, since that time the wavelengths of photons have been stretched by a factor of 1000. Today these microwave photons typically have wavelengths of about 1 mm, but when the universe was about a million years old, they had wavelengths of about 0.001 mm.

We can use Wien's law to calculate the temperature of the cosmic background radiation at the time of this transition from a radiation-dominated universe to a matter-dominated one. Just as a peak wavelength λ_{max} of 1 mm corresponds to a blackbody temperature of 3 K, a peak wavelength of 0.001 mm corresponds to 3000 K. In other words, the temperature of the radiation background is proportional to the redshift and has been declining over the ages, as shown in Figure 28-9. It just so happens that $T = 3000$ K when $z = 1000$ and $\rho_m = \rho_{rad}$.

The nature of the universe changed in a fundamental way when z was equal to 1000. To see how, recall that hydrogen is by far the most abundant element in the universe—hydrogen atoms outnumber helium atoms about 12 to 1. Of course, a hydrogen atom consists of a single proton orbited by a single electron. It takes only a little energy to knock the proton and electron apart. In fact, a radiation field warmer than about 3000 K easily ionizes hydrogen. Thus, hydrogen atoms could not survive in the radiation-dominated universe that existed before $z = 1000$ (that is, in the first million years after the Big Bang). As sketched in Figure 28-10a, the background photons prior to $t = 1$ million years had energies

great enough to prevent electrons and protons from binding to form hydrogen atoms. Only since $t = 1$ million years have these photons been redshifted enough to permit hydrogen atoms to exist (Figure 28-10b).

Prior to $t = 1$ million years, the universe was completely filled with a shimmering expanse of high-energy photons colliding vigorously with protons and electrons. This state of matter, called a **plasma**, is opaque, just like the glowing gases inside a discharge tube (like a neon advertising sign) or a fluorescent light bulb. Princeton University physicist P. J. E. Peebles coined the term **primordial fireball** to describe the universe during this time.

After $t = 1$ million years, the photons no longer had enough energy to keep the protons and electrons apart. As soon as the temperature of the radiation field fell below 3000 K, protons and electrons began combining to form hydrogen atoms. Because hydrogen gas is transparent, the universe suddenly became transparent! All the photons that had been vigorously colliding with charged particles just a few seconds earlier could now stream unimpeded across space. Today we see these same photons in the microwave background.

This dramatic moment, when the universe went from being opaque to being transparent, is referred to as the **era of recombination.** Because the universe was opaque prior to $t = 1$ million years, we cannot see any farther into the past than the era of recombination. The microwave background, whose photons have suffered a redshift of $z = 1000$, contains the most ancient photons we shall ever be able to observe. This microwave background is today only a ghostly relic of its former dazzling splendor.

28-5 The future of the universe will be determined by the average density of matter in it

During the 1920s general relativity was applied to cosmology by Alexandre Friedmann in Russia, Georges Lemaître in Belgium, Willem de Sitter in the Netherlands, and, of course, Einstein himself. The resulting picture of the structure and evolution of the universe, called **relativistic cosmology,** agrees surprisingly well with our intuitive notion that gravity

---

BOX 28-2

Cosmology with a Nonzero Cosmological Constant

Some interesting effects result when the cosmological constant Λ is added to the equations of general relativity that describe cosmological models. The cosmological constant behaves like a constant energy density, present even if the universe is totally devoid of matter and radiation. Λ can therefore be thought of as the energy density of the vacuum.

Einstein introduced Λ to provide a cosmological repulsion that can balance the gravity of the universe, thereby making the universe static. The critical value of the cosmological constant for which this balance occurs is called Λ_c.

If Λ is only slightly larger than Λ_c, the universe starts with a Big Bang but then has a long "quasi-stationary" phase with gravity and cosmological repulsion almost in balance before cosmological repulsion wins and the expansion continues. This scenario is sometimes called a Lemaître model, after the French cosmologist Georges Lemaître, who discovered it. If Λ is much larger than Λ_c, then the universe will expand forever.

If Λ is less than Λ_c, there are two possibilities: a bounce model or an oscillating model. A **bounce universe** contracts from an infinitely extended state (no Big Bang) until cosmological repulsion halts the contraction and the universe rebounds out to infinity again. An **oscillating universe** expands from a Big Bang, reaches a maximum size, and then collapses, only to erupt again in another Big Bang. Whether the universe bounces or oscillates depends on factors such as the overall curvature of space. If the cosmological constant is negative (Λ), then only oscillating models are

possible. These various scenarios are sketched in the accompanying graph.

Astronomers who have attempted to estimate Λ typically suggest that its value lies somewhere between 2×10^{-51} m^2 and -2×10^{-51} m^2, which is a range extremely close to zero. Thus most researchers find it convenient to assume that $\Lambda = 0$.

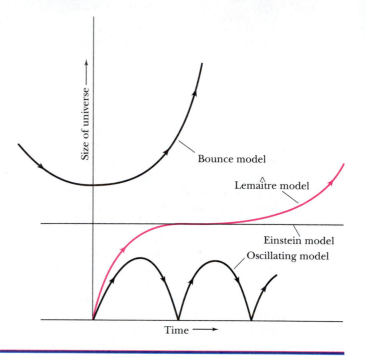

should be slowing the cosmological expansion. In the discussion that follows, we shall assume that Einstein's infamous cosmological constant is precisely zero ($\Lambda = 0$).

To envision the effect of gravity on the expansion of the universe, consider a cannonball shot upward from the surface of the Earth. If the cannonball's speed is less than the escape velocity from the Earth of 11.2 km/s (review Box 7-2), the ball will fall back to Earth. If the cannonball's speed equals the escape velocity, it will just barely escape falling back to Earth. And if the ball's speed exceeds the escape velocity, it can easily leave the Earth and never fall back or stop, despite the relentless pull of gravity. Gravity acts similarly to rein in the expansion of the universe and to determine its ultimate fate. Instead of talking about some sort of "mutual escape velocity" between galaxies, however, astronomers speak of the average density of matter in the universe.

If the average density of matter throughout space is low, the gravity associated with this matter is weak and the expansion of the universe will continue forever. Even infinitely far into the future, galaxies will continue to rush away from each other. In such a case, we say that the universe is **unbounded**, or **open**.

Conversely, if the density of matter throughout space is high, the resulting gravity will be strong enough to eventually halt the expansion of the universe. At some point, the universe will reach a maximum size and then begin contracting as gravity starts to pull the galaxies back toward each other. In such a case, we say that the universe is **bounded**, or **closed**.

Separating these two scenarios is the situation in which we say that the universe is **marginally bounded.** In this case, the density of matter across space exactly equals the **critical density** (ρ_c), so that the galaxies just barely manage to keep moving away from each other. This situation is analogous to that of a cannonball leaving the Earth with a speed exactly equal to the escape velocity. The critical density is given by the expression

$$\rho_c = \frac{3H_0^2}{8\pi G}$$

where H_0 is the Hubble constant and G is the gravitational constant. Using a Hubble constant of 75 km/s/Mpc, we get

$$\rho_c = 1.1 \times 10^{-26} \text{ kg/m}^3$$

This density is equivalent to about six hydrogen atoms per cubic meter of space.

As we saw earlier in this chapter, the average density of the luminous matter we see in the sky seems to be about 5×10^{-28} kg/m^3, which is only about one-twentieth of this critical density. However, the dark-matter problem suggests that the true density of matter across space may be equal to ρ_c. Because of these uncertainties, present-day estimates of density are not accurate enough to tell us whether the universe will continue to expand forever.

Cosmological models of the universe become somewhat more complicated if the cosmological constant, Λ, is not equal to zero. Like Einstein, most theoreticians suspect that Λ is zero, or at least extremely small. Box 28-2 is supplied for the reader who wishes to get a taste of some of the complexities introduced by a nonzero cosmological constant.

28-6 The rate of deceleration of the universe can be determined by observing extremely distant galaxies

Another way of determining the future of the universe is to measure the rate at which the cosmological expansion is gradually slowing down. This slowing can be measured because it causes the relationship between redshift and distance for extremely remote galaxies to deviate from the direct pro-

portion specified by the Hubble law. To see why, reexamine the Hubble law displayed in Figure 26-16. This straight line reflects a simple relationship between the distance to a galaxy and its recessional velocity from the Earth. Figure 26-16 was based, however, on data extending to only a billion light-years from Earth. All we can see is a straight line, which tells us that we live in an expanding universe.

Suppose, however, that you were to measure the redshifts of galaxies several billion light-years from Earth. The light from these galaxies has taken billions of years to arrive at your telescope, so what your measurements will reveal is how fast the universe was expanding billions of years ago. Because the universe was then expanding faster than it is today, your data will deviate slightly from the straight-line Hubble law.

The curves in Figure 28-11a display the deviation from the straight-line Hubble law for different rates of deceleration of the universe. Astronomers denote the amount of deceleration with the **deceleration parameter** (q_0). Appropriately, $q_0 = 0$ corresponds to no deceleration at all. This is possible if the universe is completely empty and thus has no gravity to slow down the expansion. As sketched in Figure 28-11b, a $q_0 = 0$ universe expands forever at a constant rate. As explained earlier in this chapter, the present age of this empty universe, with no deceleration, is exactly $1/H_0$.

The case $q_0 = \frac{1}{2}$ corresponds to a marginally bounded universe. Such a universe just barely manages to continue

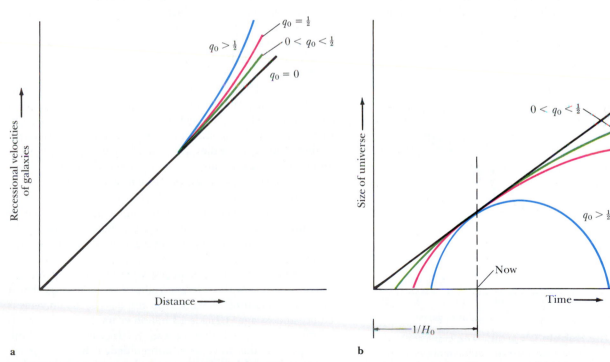

a

b

FIGURE 28-11 Deceleration and the Hubble Diagram
These two graphs compare the appearance of the Hubble diagram with the evolution of the universe. (a) This diagram shows the relationship between recessional velocity and distance for various values of q_0. (b) This diagram shows the size of the universe for various values of q_0. The case $q_0 = 0$ is an empty universe ($\rho = 0$). The case $q_0 = \frac{1}{2}$ is a marginally bounded universe ($\rho = \rho_c$). If $0 < q_0 < \frac{1}{2}$, then the universe is unbounded and will expand forever. If $q_0 > \frac{1}{2}$, the universe is bounded and will someday collapse.

TABLE 28-1

Deceleration, Density, and the Age of the Universe ($\Lambda = 0$)

Deceleration parameter (q_0)	Density parameter (Ω_0)	Average density (ρ)	Age of universe (T_0)
$q_0 > \frac{1}{2}$	$\Omega_0 > 1$	$\rho > \rho_c$	$T_0 < \frac{2}{3}(1/H_0)$
$q_0 = \frac{1}{2}$	$\Omega_0 = 1$	$\rho = \rho_c$	$T_0 = \frac{2}{3}(1/H_0)$
$0 < q_0 < \frac{1}{2}$	$0 < \Omega_0 < 1$	$\rho < \rho_c$	$\frac{2}{3}(1/H_0) < T_0 < 1/H_0$
$q_0 = 0$	$\Omega_0 = 0$	$\rho = 0$	$T_0 = 1/H_0$

expanding forever. If the cosmological constant, Λ, is zero (so there is no pressure propelling the expansion of the universe), then the universe contains matter at the critical density ρ_c. In this case, the present age of the universe equals $\frac{2}{3}(1/H_0)$. For simplicity, we shall continue to assume that $\Lambda = 0$.

If q_0 is between 0 and $\frac{1}{2}$, the universe is unbounded and will continue to expand forever. Such a universe contains matter at less than the critical density and has a present age between $\frac{2}{3}(1/H_0)$ and $1/H_0$.

If q_0 is greater than $\frac{1}{2}$, the universe is bounded and is filled with matter having a density greater than ρ_c. The present age of such a universe is less than $\frac{2}{3}(1/H_0)$. The gravitational pull of all matter through all space is strong enough to eventually halt the expansion of such a universe.

Some astronomers prefer to characterize the behavior of the universe in terms of the **density parameter** (Ω_0), which is the density of matter in the universe divided by the critical density. If $\Lambda = 0$, then $\Omega_0 = 2q_0$. For instance, the marginally bounded case, $q_0 = \frac{1}{2}$, corresponds to $\Omega_0 = 1$, so that the universe is filled with matter at the critical density. Table 28-1 shows how the deceleration parameter, the density parameter, and the age of the universe are related.

In principle, it should be possible to determine the nature of the universe by measuring the redshifts and distances of many remote galaxies and then plotting the data on a Hubble diagram like the one in Figure 28-11a. If the data points fall above the $q_0 = \frac{1}{2}$ line, the universe is bounded. If the data points fall between the $q_0 = 0$ and $q_0 = \frac{1}{2}$ lines, the universe is unbounded.

Unfortunately, such observations are extremely difficult to make. Galaxies nearer than a billion light years are of no help in determining q_0. And beyond a billion light-years, uncertainties cloud determinations of distance. As we saw in Chapter 26, the main problem is that the objects used as standard candles cease to be reliable distance indicators, primarily because of galaxy evolution, which is poorly understood. Nearby galaxies and their components, which are used as standard candles, have evolved more than remote galaxies, which we observe at an earlier stage of development simply because of the travel time of light to Earth. Such evolutionary effects, as well as the ravages of galaxy collisions, alter the chemical compositions, luminosities, and sizes of these standard candles, reducing their usefulness. For example, Figure 28-12 shows data obtained by Allan Sandage of the Carnegie Observatories. Note how the data points are scattered about the $q_0 = \frac{1}{2}$ line. This scatter prevents us from knowing conclusively whether the universe is bounded or unbounded. Nevertheless, because the data lie close to the $q_0 = \frac{1}{2}$ line, we suspect that the universe might be marginally bounded.

A troublesome dilemma arises when we calculate the age of the universe with different values for H_0 and q_0. For a Hubble constant of 75 km/s/Mpc and $q_0 = 0$, the universe is empty and its age ($1/H_0$) is 13 billion years. If the universe is marginally bounded, then its age, $\frac{2}{3}(1/H_0)$, is about 8.7 billion years. If the universe is unbounded, it has an age between 8.7 and 13 billion years. If the universe is bounded, then its age is less than 8.7 billion years.

All of these estimates for the age of the universe are uncomfortably low. It is obviously impossible for the universe to be younger than the stars it contains, some of which seem

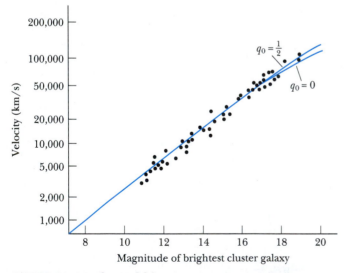

FIGURE 28-12 The Hubble Diagram This graph shows the Hubble diagram extended to include extremely remote galaxies. The magnitude of the brightest galaxy in a cluster is directly correlated with the distance to the cluster. If the data points fall between the curves marked $q_0 = 0$ and $q_0 = \frac{1}{2}$, the universe is unbounded. If the data points fall above the curve marked $q_0 = \frac{1}{2}$, the universe is bounded. (Adapted from A. Sandage)

to be about 15 billion years old. Either astronomers' understanding of the ages of stars is not correct, or our straightforward relativistic cosmology involving only two constants (H_0 and q_0) does not adequately describe the universe.

The cosmological models discussed above, which give a maximum age of the universe of 13 billion years, assume that $\Lambda = 0$. Inspired by such discrepancies as the ages of the oldest stars, some astronomers have resurrected the idea of a nonzero cosmological constant. The presence of such a term in the equations of general relativity greatly complicates the behavior of the universe. One possibility is that the universe started with an explosive Big Bang and vigorous expansion, but then slowed down and entered a quasi-stationary state during which it did not expand very much. This quasi-stationary state ended when the repulsive pressure associated with a nonzero Λ took over and propelled the universe to the expansion rate we see today. The age of this "loitering universe" could be much greater than the ages of the oldest stars, yet still give us a Hubble constant of 75 km/s/Mpc.

28-7   The shape of the universe is another indicator of its ultimate fate

There is another way of tackling the issue of the fate of the universe. Our present understanding of the universe is based on Einstein's general theory of relativity, which explains that gravity curves the fabric of space. The gravity of all matter scattered across space should thus give the universe some overall shape (or geometry). Gravity also determines the fate of the universe. Thus, by measuring the shape of space, we should be able to discover whether the universe is bounded or unbounded.

To see what astronomers mean by the geometry of the universe, imagine shining two powerful laser beams out into space. Suppose that we can align these two beams so that they are perfectly parallel as they leave the Earth. Finally, suppose that nothing gets in the way of these two beams, so that we can follow them for billions of light-years across the universe, across the space whose curvature we wish to detect.

With this arrangement there are only three possibilities. First, we might find that our two beams of light remain perfectly parallel, even after traversing billions of light-years. In this case, space would not be curved: The universe would have **zero curvature**, and space would be **flat**.

Alternatively, we might find that our two beams of light gradually converge. In such a case, space would not be flat. Recall that the lines of longitude on the Earth's surface are parallel at the equator but intersect at the poles. Thus, in this case the three-dimensional geometry of the universe would be analogous to the two-dimensional geometry of a spherical surface. We would then say that space is **spherical** and that the universe has **positive curvature**.

The third and final possibility is that the two initially parallel beams of light would gradually diverge, becoming farther and farther apart as they moved across the universe. In this case the universe would still have to be curved, but in the opposite sense from the spherical model. We would then say that the universe had **negative curvature**. In the same way that a sphere is a positively curved two-dimensional surface, a saddle is a good example of a negatively curved two-dimensional surface. Parallel lines drawn on a sphere always converge, but parallel lines drawn on a saddle always diverge. Mathematicians say that saddle-shaped surfaces are hyperbolic. Thus, in a negatively curved universe, we would describe space as **hyperbolic**.

These three situations are sketched in Figure 28-13. To illustrate these cases, three surfaces are drawn: a sphere, a plane, and a saddle. Of course, real space is three-dimensional, but it is much easier to visualize an analogous two-dimensional surface. Thus, as you examine the drawings in Figure 28-13, remember that the real universe has one more dimension. For example, if the universe is in fact hyperbolic, then the geometry of space must be the three-dimensional analogue of the two-dimensional surface of a saddle.

There is no easy way of measuring the curvature of space across the universe. In theory, if you drew an enormous triangle whose sides were each a billion light-years long (see Figure 28-13), you could determine the curvature of space by measuring the three angles of the triangle. If their sum equaled $180°$, space would be flat. If the sum was greater than $180°$, space would be spherical. And if the sum of the three angles was less than $180°$, space would be hyperbolic. Unfortunately, this direct method for measuring the curvature of space is not practical.

Each of the three possible geometries corresponds to a different behavior and fate of the universe. Flat space corresponds to the marginally bounded case ($q_0 = \frac{1}{2}$) in which the galaxies just barely manage to keep receding from each other. This flat-space scenario divides the positive-curvature cases from the negative-curvature cases. If the density across space is greater than the critical density, then $q_0 > \frac{1}{2}$, and space is positively curved. Conversely, if the density across space is less than the critical density, then $0 < q_0 < \frac{1}{2}$, and space is negatively curved. These relationships are summarized in Table 28-2.

According to the cosmological principle, the universe does not have an "edge" or a "center." This is clearly the case for both the flat and hyperbolic universes, because they are infinite and extend forever in all directions. In contrast, a spherical universe is finite, but it also lacks a center and an edge. To see why, think about the Earth. Even though the Earth has a finite surface area (511 million square kilometers), you could walk forever around it without finding a center or an edge. Likewise, relativistic cosmology strictly rules out any possibility of a center or an edge to the universe.

The curvature of space should affect the density of galaxies we see at great distances from Earth. To understand why this is so, imagine grains of sand randomly sprinkled on three surfaces—a sphere, a plane, and a saddle—as shown along the top of Figure 28-14. For each surface, suppose you make a map on ordinary (flat) paper of the locations of the sand grains. Your approach might be to plot the distance of each grain from some reference point (the cross).

FIGURE 28-13 The Geometry of the Universe The shape of space is determined by the matter contained in the universe. The curvature is either positive (**a**), zero (**b**), or negative (**c**), depending on whether the density is greater than, equal to, or less than the critical density. In theory, the curvature of the universe could be determined by measuring the angles of a huge triangle in space.

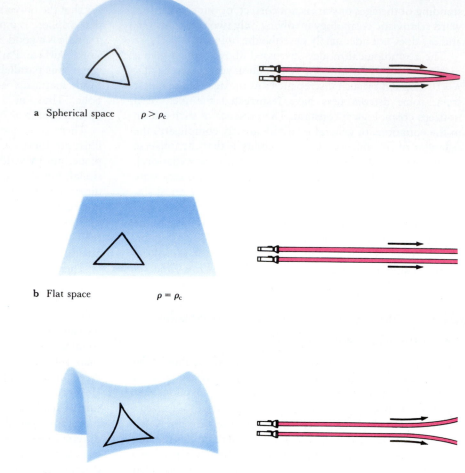

a Spherical space $\rho > \rho_c$

b Flat space $\rho = \rho_c$

c Hyperbolic space $\rho < \rho_c$

By making a flat map, you are essentially constraining the shape of the universe to a flat surface, as sketched in the middle row of Figure 28-14. For example, making a flat map of a positively curved universe is like squashing a hemisphere onto a plane, as if you were to crush half an orange on a kitchen table; cracks and gaps appear around the edges. Conversely, if you squash a saddle-shaped surface onto a plane, excess surface piles up around the edges.

As shown at the bottom of Figure 28-14, the resulting maps exhibit noticeable differences. Only the map of the flat plane still shows a uniform distribution of dots just like the random distribution of sand grains. The dots on the map of the spherical surface appear to thin out, because making a flat map of a sphere amounts to stretching the spherical surface until it flattens out. Conversely, the density of dots on the map of the saddle surface increases near the edges, because the surface in effect piles up there when mapped onto a flat piece of paper.

Instead of making maps, astronomers count the number of galaxies per cubic megaparsec at various distances from

---

TABLE 28-2

The Geometry and Fate of the Universe ($\Lambda = 0$)

Geometry of space	Curvature of space	Average density (ρ)	Deceleration parameter (q_0)	Type of universe	Ultimate future of universe
Spherical	Positive	$\rho > \rho_c$	$q_0 > \frac{1}{2}$	Closed	Eventual collapse
Flat	Zero	$\rho = \rho_c$	$q_0 = \frac{1}{2}$	Flat	Perpetual expansion (just barely)
Hyperbolic	Negative	$\rho < \rho_c$	$0 < q_0 < \frac{1}{2}$	Open	Perpetual expansion

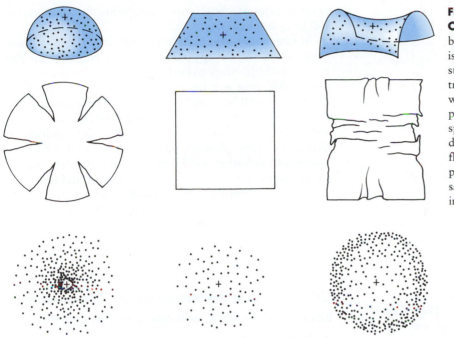

FIGURE 28-14 Galaxy Counts and the Curvature of Space This illustration shows, by analogy, why the number density of galaxies is affected by the geometry of space. Three surfaces representing the three possible geometries are sprinkled with sand grains. In each case, we measure the distances d from a reference point (the cross) to the grains of sand. On the spherical surface, the number of grains within distance d increases more slowly than d^2. On the flat surface, the number of grains increases precisely as the square of the distance. On the saddle-shaped surface, the number of grains increases more rapidly than d^2.

Earth. The rate at which this number density declines with increasing distance depends on the shape of space and hence on Ω_0, as shown in Figure 28-15. Also shown are the recent data of Princeton physicists Edwin Loh and Earl Spillar, who examined 1000 galaxies in five small patches of the sky. Although it's difficult to choose a value of Ω_0 from the data plotted in Figure 28-15, the results are consistent with our universe being nearly flat and very close to marginally bounded.

There is yet another method of determining the geometry of the universe. Recall that light rays from distant galaxies are bent slightly by the curvature of space. This deflection of light rays thus distorts the appearance of remote galaxies. Normally, the more distant an object is, the smaller it appears. However, for extremely remote galaxies the curvature of the universe can bend light so that images are actually magnified. An extremely distant galaxy can thus appear bigger than would normally be expected.

The graph in Figure 28-16 shows how the diameters of the images of galaxies depend on Ω_0. We must examine galaxies with redshifts greater than $z = 1$ to decide among various values of Ω_0. These galaxies are so far away that they appear only as faint, hazy blobs through Earth-based telescopes. The data shown in Figure 28-16 are derived from recent observations of jets in compact radio galaxies and quasars by Kenneth Kellermann of the National Radio Astronomy Observatory. By assuming that all such jets have roughly the same length, Kellermann used them as distance indicators. A plot of the angular sizes of these "standard rods" against their redshifts seems to favor the case $\Omega_0 = 1$.

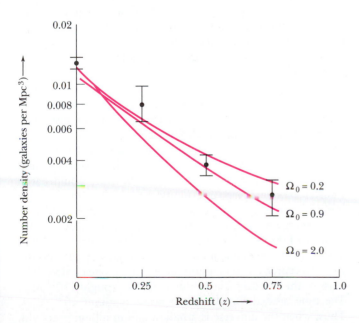

◀ **FIGURE 28-15 The Number Density of Galaxies** The number density of galaxies, expressed in galaxies per cubic megaparsec, is plotted here against redshift. Curves for three different values of Ω_0 are shown. The data imply that the density of matter in the universe is nearly equal to the critical density and thus that the universe is nearly flat. (Adapted from E. Loh and E. Spillar)

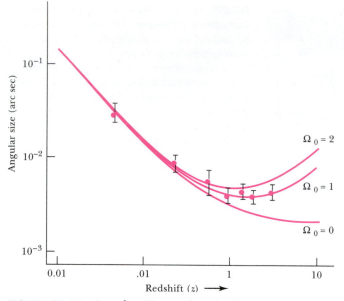

FIGURE 28-16 Angular Size and Redshift The curvature of the universe can magnify the images of remote galaxies. The data, which seem to follow the curve $\Omega_0 = 1$, show how the average angular sizes of radio jets in 82 active galaxies and quasars depend on redshift. (Adapted from K. Kellermann)

28-8   If the universe is bounded, we are living inside a universe that will ultimately collapse in a Big Crunch

If the average density of matter across space is greater than the critical density, the universe will someday stop expanding. Eventually the mutual gravitational attraction between galaxies will cause them to move back toward each other. As the universe begins to contract, the redshifts we see today in galactic spectra will be replaced by blueshifts.

The specific events leading up to the death of a contracting universe mimic those of the birth of the universe, but in reverse. For many billions of years, this contraction will be noticeable only to astronomers. At first, only the nearest clusters of galaxies will exhibit blueshifts in their spectra, because these are the galaxies we see in the most recent past. Further into the future, however, increasing numbers of more remote galaxy clusters will have their redshifts replaced by blueshifts.

The contraction of the universe will also cause the temperature of the radiation background to increase. Just as the expansion of the universe is ultimately the expansion of space, so the contraction of the universe is actually the contraction of space. Photons traveling across this contracting space will be subjected to a shortening of their wavelengths. Their energies will then increase, and thus the temperature of the radiation field will increase.

Just as the expanding universe began with a Big Bang, the contracting universe will end with a **Big Crunch.** About 70 million years before the Big Crunch, the temperature of the background radiation field will be up to 300 K, at which

point galaxies will be so crowded together that the night sky will be as bright as day.

At one million years before the Big Crunch, the radiation temperature will be up to 3000 K. The photons will then possess enough energy to begin dissociating atoms and molecules. With three minutes to go before the Big Crunch, the temperature will be 10 million kelvin, and the stars and planets will dissolve. With about a second to go, the temperature will climb above 10 billion kelvin, and the resulting extremely energetic photons will break up nuclei. Finally, all matter and radiation will be crushed out of existence, as a cosmic singularity comes to envelop all space and time.

Just as we cannot use physics to tell us what existed before the Big Bang, so science cannot reveal what might happen after the Big Crunch. The infinite warping of space and time in a cosmic singularity prevents us from extrapolating any further into the future.

28-9   If the universe expands forever, black hole evaporations will occur in the extremely distant future

If the average density of matter across space is less than or equal to the critical density, the universe will expand forever. Extrapolating into the extremely distant future is a risky business. However, we can still reasonably predict a few significant events that will occur in a universe that exists forever.

Although the universe today consists mostly of hydrogen and helium, these gases will eventually be used up in the thermonuclear fires of generation after generation of stars. Astronomers estimate that this depletion will occur in about 10^{12} years. No more stars will form, because the hydrogen and helium required will no longer exist. Galaxies will grow dim as the final generation of stars dies off. Eventually, roughly a trillion years after the Big Bang, all the matter in the universe will consist of either dead stars (white dwarfs, neutron stars, and black holes) or cold lumps (meteorites, used spaceships, rocks, and so forth).

Still further into the future, the effects of extremely rare events will become important. For example, as noted in an earlier chapter, the probability of two stars colliding somewhere in our Galaxy (or any other galaxy) is extremely small. Indeed, the probability of two stars even passing near each other is almost infinitesimal. Over trillions upon trillions of years, however, these events will sooner or later occur.

A close encounter between two stars in a galaxy would likely be significant because one of the stars could easily transfer enough energy and momentum to the other to kick it out of its galaxy. The remaining star would then fall into a lower-energy orbit closer to the galaxy's nucleus. Eventually, because of the emission of gravitational radiation (recall Box 24-2), these remaining stars would gradually spiral into the galaxy's center, coalescing into an enormous black hole. The mass of the resulting black hole would be roughly 10^{11} M$_\odot$. The time needed for all this to occur is about 10^{27} years. Thus, when the universe is a billion billion billion years old, galaxies will consist of enormous black holes surrounded by dead stars (Figure 28-17).

The orbital motions of galaxies in clusters also emit gravitational radiation. As a result of this energy loss, entire galaxies spiral toward each other, eventually coalescing into even more gigantic black holes. It will take roughly 10^{31} years for these colossal objects to form, each of which might contain 10^{15} $M_{\odot}$.

Even further into the future, an extraordinary process called **black hole evaporation** will become important. As discussed in Box 28-3, the laws of quantum mechanics allow a black hole to emit particles. Although this quantum-mechanical phenomenon is significant only for low-mass black holes at normal time scales, even the most massive black holes will eventually evaporate completely if the universe exists forever. From the equations given in Box 28-3, we find that ordinary black holes (such as Cygnus X-1) evaporate after roughly 10^{67} years. The gigantic black holes discussed above will evaporate after 10^{97} to 10^{106} years. All of these evaporations end the same way—in a burst of particles and radiation equivalent to the detonation of a billion megaton hydrogen bombs, releasing a total of 10^{24} joules.

Astronomers usually think of a black hole as an "information sink": Material falls in, but nothing gets out. Specifically, as explained in Chapter 24, all of the information carried by in-falling matter (such as its shape, color, and chemical composition) is removed from the universe. This conception is valid only for ordinary, massive black holes where the quantum mechanical emission of particles is negligible. However, during the final stages of evaporation, vast quantities of particles pour out of a black hole. The black hole in effect behaves like a **white hole,** dumping new material into the universe. Because this material carries information (such as color, shape, size, and so on), a white hole is an "information source." The British physicist Stephen W. Hawking has proven that, during the final moments of evaporation, a black hole is actually a white hole that randomly pumps new matter and information into the universe in an unpredictable fashion.

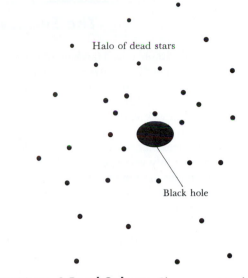

FIGURE 28-17 A Dead Galaxy Close encounters between stars can eject a star from its galaxy. Over a span of 10^{27} years, a galaxy might lose up to 99% of its stars in this manner. Meanwhile, the remaining stars would coalesce into an enormous black hole that might have a mass of about 10^{11} $M_{\odot}$.

To emphasize the totally unpredictable way in which evaporating black holes can randomly dump new information into the universe, Hawking has humorously—though improbably—proposed that one of these holes "could emit a television set or the works of Proust in ten leather volumes." Thus, as we stand at the frontiers of human knowledge, groping dimly into the remote future for an understanding that might not even exist, it is perhaps amusing to recall the words of Mark Twain:

There is something fascinating about science. One gets such a wholesale return of conjecture out of such a trifling investment of fact.

—*Life on the Mississippi* (1883)

KEY WORDS

Terms preceded by an asterisk are discussed in the boxes.

average density of matter	cosmological principle	marginally bounded universe	*primordial black hole
Big Bang	cosmological redshift	mass density of radiation	primordial fireball
Big Crunch	critical density	matter-dominated universe	radiation-dominated universe
black-hole evaporation	deceleration parameter	negative curvature	
bounded (closed) universe	density parameter	observable universe	relativistic cosmology
*Compton wavelength	era of recombination	Olbers's paradox	spherical space
cosmic microwave background	flat space	*Planck length	unbounded (open) universe
cosmic particle horizon	Great Attractor	*Planck mass	*virtual pair
cosmic singularity	*Heisenberg uncertainty principle	Planck time	*wave–particle duality
cosmological constant	hyperbolic space	plasma	white hole
		positive curvature	zero curvature

BOX 28-3
The Evaporation of Black Holes

During the early 1970s Stephen W. Hawking of Cambridge University proposed that numerous small black holes, called **primordial black holes,** could have been created during the Big Bang. Theoretically, these black holes could have had masses as small as the Planck mass: 5×10^{-8} kg.

Most astronomers feel that primordial black holes probably do not exist. Only very special kinds of density fluctuations during the earliest moments of the universe could have led to the creation of numerous small black holes. We have no reason to suspect that the early universe obliged us with the necessary special conditions. Nevertheless, theoretical investigations of the properties of small black holes have produced some surprising and significant discoveries.

First, it is instructive to realize how tiny low-mass black holes indeed are. As we saw in Chapter 24, the Schwarzschild radius of a black hole is given by

$$R_{Sch} = \frac{2GM}{c^2}$$

where G is the gravitational constant, M is the mass of the hole, and c is the speed of light. For example, if the Earth, with its mass of 6×10^{24} kg, were crushed to become a black hole, it would be about the size of a Ping-Pong ball. A typical large asteroid with a mass of 10^{17} kg crushed to become a black hole would be about the size of an atom. A black hole made from a billion metric tons of matter (roughly 10^{12} kg) would be about the size of a proton.

Because low-mass black holes are submicroscopic, we must use quantum mechanics to understand their properties. The **Heisenberg uncertainty principle** is a basic tenet of submicroscopic physics. This principle states that you cannot determine precisely both the position and the speed of a subatomic particle. Over extremely short distances or times, a certain amount of "fuzziness" is built into the nature of reality.

The Heisenberg uncertainty principle leads logically to the concept of **virtual pairs,** which explains that at every point in space, pairs of particles and antiparticles are constantly being created and destroyed. As we shall see in greater detail in the next chapter, an antiparticle is quite like an ordinary particle except that it has an opposite electric charge and can annihilate an ordinary particle so that both disappear, usually leaving a pair of photons in their place. In the case of virtual pairs, the process of creation and annihilation occurs over such incredibly brief time intervals that these virtual particles and antiparticles cannot be directly observed.

Think about a tiny black hole. Furthermore, think about the momentary creation of a virtual proton–antiproton pair just outside the hole's event horizon. It may happen that one of the particles falls into the black hole. Its partner is

then deprived of a counterpart with which to annihilate and must therefore become a real particle. To accomplish this conversion, some of the energy of the black hole's gravity must be converted into matter, according to $E = mc^2$. This decreases the mass of the black hole by a corresponding amount, and the particle is free to escape from the hole. In this way, particles can quantum-mechanically "leak" out of a black hole, carrying some of the hole's mass with them.

The smaller a black hole is, the more easily particles can leak out through its event horizon. Jacob Bekenstein and Stephen Hawking proved mathematically that you can speak of the *temperature* of a black hole as a way of describing the amounts of energy carried away by particles leaking out of it. This temperature is given by

$$T = \frac{hc^3}{16\pi^2 kGM}$$

where h is Planck's constant, k is the Boltzmann constant, and M is the mass of the hole. For example, a 1-billion-ton black hole emits particles and energy as if its temperature were nearly 10^{12} K. This relationship between the mass of a black hole and its temperature is graphed in the diagram below.

For ordinary black holes, such as Cygnus X-1, this effect is negligible over time spans of billions of years. For example, the temperature of a 10-$M_\odot$ black hole is about 10^{-7} K, barely above absolute zero. In other words, particles hardly ever manage to escape from ordinary black holes.

As particles escape from a small black hole, the mass of the black hole decreases, making its temperature go up. As its temperature rises, still more particles escape, further decreasing the hole's mass and forcing the temperature still

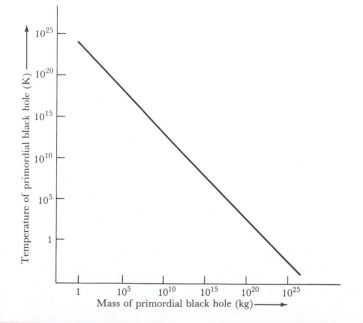

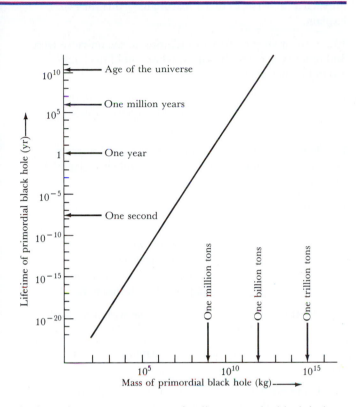

higher. This runaway process finally causes the black hole to evaporate completely. During its final seconds of evaporation, the hole gives up the last of its mass in a violent burst of energy equal to the detonation of a billion megaton hydrogen bombs.

The smaller a black hole is, the hotter it is and the faster it evaporates. Hawking proved mathematically that the lifetime t of a black hole is

$$t = 10{,}240\,\pi^2\left(\frac{G^2 M^3}{hc^4}\right)$$

This value tells us that the very low mass primordial black holes evaporated fairly soon after the Big Bang. For example, a one-million-ton black hole would last for only 3000 years. The relationship between the mass of a black hole and its lifetime is graphed in the diagram above.

Because the universe is about 20 billion years old, it is possible to calculate the minimum mass needed for a primordial black hole to have survived to the present. The answer is roughly 2×10^{11} kg (about 200 million tons). In other words, only those primordial black holes with masses greater than 200 million tons are cool enough and thus evaporating slowly enough to have survived to the present time.

Although few people think that such small black holes actually exist, the ideas and formulas developed by Stephen Hawking reveal that there are fundamental interrelations among thermodynamics, quantum mechanics, and general relativity.

KEY IDEAS

- The universe began as an infinitely dense cosmic singularity that expanded explosively in the event called the Big Bang, which can be described as an explosion of space at the beginning of time.

  The Hubble law describes the continuing expansion of space; it is meaningless to speak of an edge or center to the universe.

  The observable universe extends about 15 billion light-years in every direction from the Earth; we cannot see objects beyond the cosmic particle horizon at the distance of 15 billion years because light from these objects has not had enough time to reach us.

  Before the Planck time (about 10^{-43} second after the calculated time of the Big Bang), the universe was so dense that the known laws of physics do not properly describe the behavior of space, time, and matter.

- The cosmic microwave background radiation, corresponding to radiation from a blackbody at a temperature of nearly 3 K, is the greatly redshifted remnant of the hot universe that existed about 1 million years after the Big Bang.

  The universe today is matter-dominated, but during the first million years, before the background radiation became greatly redshifted, the universe was radiation-dominated.

  During the first million years of the universe, matter and energy formed an opaque plasma called the primordial fireball.

  By 1 million years after the Big Bang, the expansion of the universe had caused the temperature of the primordial fireball to fall below 3000 K so that protons and electrons could combine to form hydrogen atoms; this event is called the era of recombination.

  The universe became transparent during the era of recombination, meaning that the 3-K microwave background radiation contains the oldest photons in the universe.

- If the average density of matter in the universe is greater than the critical density ρ_c, space is spherical (with positive curvature), the density parameter Ω_0 has a value greater than 1, and the universe is closed (bounded) and will ultimately collapse in a Big Crunch.

- If the average density of matter in the universe is less than ρ_c, space is hyperbolic (with negative curvature), Ω_0 has a value less than 1, and the universe is open (unbounded) and will continue to expand forever.

- If the average density of matter in the universe is exactly equal to ρ_c, space is flat (with zero curvature), Ω_0 is exactly equal to 1, the universe is marginally bounded, and its expansion will just barely continue forever (decreasing toward a zero rate of expansion at an infinite time in the future).

• Using present techniques, we cannot measure ρ_c or Ω_0 precisely enough to determine which of the preceding cases exists; the best evidence now available suggests that the average density is very near ρ_c and thus Ω_0 is near 1.

If the universe is closed (bounded), it will collapse in a Big Crunch during which the events following the Big Bang will recur, but in a reverse sequence.

If the universe is open (unbounded) or marginally bounded (flat), black holes will eventually evaporate, pouring matter and energy randomly back into the universe.

REVIEW QUESTIONS

1. What does it mean when astronomers say that we live in an expanding universe? What is actually expanding?

2. What is Olbers's paradox, and how can it be resolved?

3. How does modern cosmology preclude the possibility of either a center or an edge to the universe?

4. Explain the difference between a Doppler shift and a cosmological redshift.

5. Suppose that the universe will expand forever. What will eventually become of the microwave background radiation?

6. What does it mean to say that the universe is matter-dominated? When was the universe radiation-dominated?

7. What was the era of recombination? What significant events occurred in the universe during this era?

8. What is meant by the critical density of the universe? Why is this quantity important to cosmologists?

9. Describe how the critical density, the deceleration parameter, and the geometry of space are related.

10. Under what circumstances does a black hole behave like a white hole?

ADVANCED QUESTIONS

Tips and tools . . .

You may find it useful to know that 1 parsec equals 3.26 light-years and a year contains 3.16×10^7 seconds. The lifetimes of black holes are discussed in Box 28-3.

11. Calculate the maximum age of the universe for a Hubble constant of 50 km/s/Mpc, 75 km/s/Mpc, and 100 km/s/Mpc. On the basis of your answers, explain how the ages of globular clusters could be used to place a limit on the maximum value of the Hubble constant.

12. The so-called creationists claim that the universe came into being about 6000 years ago. Find the Hubble constant for such a universe. Is this a reasonable value for H_0? Explain.

13. If the matter density of radiation in the universe were 300 times larger than it is now, what would the background temperature be?

14. What would the critical density of matter in the universe be if the value of the Hubble constant was 100 km/s/Mpc?

15. Find the Compton wavelength of a one-kilogram mass. Do you think that this is a detectable wavelength? Explain.

16. Using the values for h, c, and G given in the text, verify that the Planck mass is about 5×10^{-8} kg.

17. With a diagram, show how you would expect the observed number of galaxies to depend on the distance from Earth for various values of q_0. Discuss some of the problems of using such a diagram to determine q_0.

18. What is the lifetime of a black hole with a mass of $10 \ M_\odot$?

19. Find the mass of a black hole having a temperature of 3 K.

20. Using the values of h, c, and G given in the text, calculate the value of the Planck time.

21. What is the mass of a black hole that has a lifetime equal to the age of the universe?

DISCUSSION QUESTIONS

22. Suppose we were living in a radiation-dominated universe. Discuss how such a universe would be different from what we now observe.

23. Prior to the discovery of the 3-K cosmic microwave background, it seemed possible that we might be living in a "steady-state universe" whose overall properties do not change with time. The steady-state model, like the Big Bang model, assumes an expanding universe but not a "creation event." Instead, matter is assumed to be created continuously everywhere in space to ensure that the average density of the universe remains constant. Explain why the existence of the cosmic microwave background is a fatal blow to the steady-state theory.

24. How can astronomers be certain that the 3-K cosmic microwave background is not some sort of localized phenomenon?

25. Discuss the implications of the idea that we cannot use science to tell us what existed before the Big Bang.

26. Suppose that numerous primordial black holes had been produced during the Big Bang. What kinds of observations might you perform to search for these black holes? Is it absurd to suggest that one of these black holes might be relatively nearby (perhaps in our own solar system!) and have escaped being discovered by astronomers? How might you use data from a gamma-ray survey of the sky to place an upper limit on the number of primordial black holes created during the Big Bang?

27. Do you think there can be "other universes," regions of space and time that are not connected to our universe? Should astronomers be concerned with such possibilities? Why or why not?

FOR FURTHER READING

Brush, S. G. "How Cosmology Became a Science." *Scientific American*, August 1992. This fascinating historical article shows how the discovery of the cosmic microwave background established the Big Bang theory and made cosmology an empirical science.

Cohen, N. *Gravity's Lens: Views of the New Cosmology*. Wiley, 1988. This appealing introduction to cosmology includes glimpses of recent research on the Big Bang, the cosmic microwave background, gravitational lenses, quasars, active galaxies, and dark matter.

Davies, P. "Everyone's Guide to Cosmology." *Sky & Telescope*, March 1991. This article by a noted scientist gives an excellent overview of modern cosmology with special emphasis on misconceptions about the Big Bang.

Dicus, D., and others. "The Future of the Universe." *Scientific American*, March 1983. This fascinating article, which speculates about the future of the universe through the year 10^{100}, includes descriptions of the decay of protons and the evaporation of black holes.

Disney, M. *The Hidden Universe*. Macmillan, 1984. This book on dark matter in the universe illustrates how astronomers develop and investigate new ideas.

Dressler, A. "The Large-Scale Streaming of Galaxies." *Scientific American*, September 1987. This article describes the strategy that astronomers used to piece together observations that led to the discovery of the Great Attractor.

Freedman, W. L. "The Expansion Rate and Size of the Universe." *Scientific American*, November 1992. This article describes the challenges that astronomers face in determining the Hubble constant.

Gott, J., and others. "Will the Universe Expand Forever?" *Scientific American*, March 1976. This article nicely summarizes the theoretical and observational consequences of the expansion of the universe.

Gribbin, J. *In Search of the Big Bang*. Bantam, 1986. This very readable introduction to modern cosmology describes the history, observations, and idea[s] modern understanding of the creation and s[tructure of the] universe.

Harrison, E. *Darkness at Night*. Harvard University Press, 1987. This definitive book on Olbers's paradox uses eloquent, nontechnical language to explain why the night sky is dark.

Hawking, S. "The Quantum Mechanics of Black Holes." *Scientific American*, January 1977. This lucid article explains how matter can quantum-mechanically escape from black holes, causing them to evaporate.

Islam, J. "The Ultimate Fate of the Universe." *Sky & Telescope*, January 1979. This fascinating article uses clear explanations of quantum-mechanical phenomena to speculate about the extremely distant future.

Kippenhahn, R. *Light from the Depths of Time*. Springer Verlag, 1987. This delightful introduction to cosmology by a noted German astronomer features the space-travel daydreams of an "Everyman" named Herr Meyer and good analogies to explain such phenomena as curved space and the Big Bang.

Osterbrock, D. E., and others. "Edwin Hubble and the Expanding Universe." *Scientific American*, July 1993. This article traces the life and career of the astronomer who shaped our modern understanding of galaxies and cosmology.

Shu, F. "The Expanding Universe and the Large-Scale Geometry of Space–Time." *Mercury*, November/December 1983. This lucid article describes the relationships between gravitation, relativity, and the expansion of the universe.

Trefil, J. *The Dark Side of the Universe*. Scribner, 1988. This book presents an excellent summary of the current status of research on such cosmological issues as dark matter, the ultimate fate of the universe, and cosmic strings.

Wagoner, R., and Goldsmith, D. *Cosmic Horizons: Understanding the Universe*. W. H. Freeman and Company, 1983. This superb introduction to cosmology is a clear, concise guide to modern ideas about the origin and evolution of the universe.

Webster, A. "The Cosmic Background Radiation." *Scientific American*, August 1974. This article describes the radiation, ranging from radio waves to gamma rays, that fills the universe.

Weinberg, S. *The First Three Minutes*. 2nd ed. Basic Books, 1988. This updating of a classic 1977 introduction to cosmology lucidly describes the events immediately following the Big Bang.

STEPHEN W. HAWKING

The Edge of Spacetime

STEPHEN HAWKING is a theoretical physicist at the University of Cambridge. Born in Oxford in 1942, he obtained his B.A. at Oxford in 1962 and went on to earn his doctorate working on gravity and cosmology under Dr. Dennis W. Sciama at Cambridge. He continued his work on these subjects at Cambridge and, in 1974 and 1975, at the California Institute of Technology. A fellow of the Royal Society since 1974, he has received many honors and awards, including the Eddington Medal of the Royal Astronomical Society, the Dannie Heinemann Prize of the American Institute of Physics and the American Physical Society, the Maxwell Medal and Prize of the Institute of Physics, and the Einstein Medal. In 1979 he was elected to Newton's old chair, Lucasian Professor at Cambridge.

His contributions have been in the areas of general relativity, gravitation, and quantum theory as it relates to black holes and their thermodynamics. Though afflicted by a progressive disease of the nervous system since 1961, which has confined him to a wheelchair, he continues to be phenomenally productive as both a writer and a researcher.

From the dawn of civilization, people have asked questions like: Did the universe have a beginning in time? Will it have an end? Is the universe bounded or infinite in spatial extent? In this essay I shall outline some answers to these questions that are suggested by modern developments in science. Most of what I describe is now generally accepted, though some of it was controversial not long ago. My final conclusion, however, is based on some very recent work on which there has not yet been time to reach a consensus.

In most early mythological and religious accounts, the universe, or at least its human inhabitants, was created by a Divine Being at some date in the fairly recent past like 4004 BC. Indeed, the necessity of a "first cause" to account for the creation of the universe was used as an argument to prove the existence of God. Conversely, Greek philosophers such as Plato and Aristotle did not like the thought of such direct divine intervention in the affairs of the world and preferred to believe that the universe had existed and would exist forever. Most people in the ancient world believed that the universe is spatially bounded. In the earliest cosmologies the world was a flat plate with the sky as a pudding basin overhead. The Greeks, however, realized that the world is round. They constructed an elaborate model in which the Earth was a sphere surrounded by a number of spheres that carried the Sun, Moon, and the planets. The outermost sphere carried the so-called fixed stars, which maintain the same relative positions but appear to rotate across the sky.

This model with the Earth at the center was adopted by the Christian church. This view had the great attraction that it left plenty of room outside the sphere of the stars for Heaven and Hell, though how these were situated was never quite clear. The Earth-centered model of the universe remained in favor until the seventeenth century, when the observations of Galileo showed that it had to be replaced by the Copernican model, in which the Earth and the other planets orbit the Sun. Not only did this model get rid of the spheres, but it also showed that the fixed stars must be at a very great distance because they did not show any apparent movement as the Earth went round the Sun (apart from that caused by the Earth's rotation about its axis).

With the Earth no longer at the center of the universe, it was fairly natural to postulate that the stars are other Suns like our own and that they are distributed roughly uniformly throughout an infinite universe. This uniform distribution, however, raised a problem: According to Newton's theory of gravity, published in 1687, each star would be attracted toward every other star in the universe. Why, then, did not the stars all fall together to a single point? Newton himself tried to argue that this would indeed happen for a bounded collection of stars; however, if the universe were infinite, the gravitational force on a star caused by the attraction of stars on one side of the star would be balanced by the force arising from stars on the other side. The net force on any star would therefore be zero and so the stars could remain motionless.

This argument is in fact an example of the fallacy one can fall into when one adds up an infinite number of quantities: By adding them up in different orders one can get different results. We now know that an infinite distribution of stars cannot remain motionless if they are all attracting each other; they will start to fall toward each other. The only way that one can have a static, infinite universe is if the force of gravity becomes repulsive at large distances. Even then the universe would be unstable, because if the stars got slightly nearer to each other, the attraction would win out over the repulsion and the stars would fall together. On the other hand, if they got slightly farther away from each other, the repulsion would win and they would move away from each other.

Despite these and other difficulties, nearly everyone in the eighteenth and nineteenth centuries believed that the universe was essentially unchanging in time. For such a universe the question of whether it had a beginning was metaphysical: One could equally well believe that it had existed forever or that it had been created in its present form a finite time ago. The belief in a static universe still persisted in 1915 when Einstein formulated his general theory of relativity, which modified Newton's theory of gravity to make it compatible with discoveries about the propagation of light. Einstein added a so-called cosmological constant, which produced a repulsive force between particles at a great distance. This repulsive force could balance the normal gravitational attraction and allow a static, uniform universe. This solution was unstable, but it had the interesting property that in it space was finite but unbounded, just as the surface of the

Part of the Vela supernova remnant. (Anglo-Australian Telescope board)

Earth is finite in area but does not have any boundary or edge. Time in this solution, however, could be infinite.

Einstein's static model of the universe was one of the great missed opportunities of theoretical physics: If he had stuck to his original version of general relativity without the cosmological constant, he could have predicted that the universe ought to be either expanding or collapsing. As it happened, however, it was not realized that the universe was changing with time until astronomers like Slipher and Hubble began to observe the light from other galaxies.

Visible light is made up of waves—like radio waves but with a much shorter wavelength, or distance, between wave crests. If one passes the light through a prism, it is decomposed into its constituent wavelengths—colors like a rainbow. Slipher and Hubble found the same characteristic patterns of wavelengths or colors as for the light from stars in our own Galaxy, but the patterns were all shifted toward the red, or longer-wavelength end of the rainbow or spectrum. The only reasonable explanation of this was that the galaxies are moving away from us. This effect, known as the Doppler

shift, is used by the police to measure the speed of cars: As the cars speed toward the police, the distance between the wave crests is crowded up and the wavelength is reduced.

During the 1920s Hubble observed the remarkable fact that the red shift is greater the further the galaxy is from our own. This meant that other galaxies are moving away from us at rates that are roughly proportional to their distance from us. The universe is not static, as had been previously thought, but is expanding. The rate of expansion is very low: It will take something like 20 billion years for the separation of two galaxies to double, but this expansion completely changes the nature of the discussion about whether the universe has a beginning or an end. This is not just a metaphysical question as in the case of a static universe: As I shall describe, there may be a very real physical beginning or end of the universe.

The first model of an expanding universe that was consistent with Einstein's general theory of relativity and Hubble's observations of red shifts was proposed by the Russian physicist and mathematician Alexander Friedmann in 1922. However, it received very little attention until

(continued)

HAWKING (continued)

similar models were formulated by other people toward the end of the 1920s. In the Friedmann model and its later generalizations, the universe was assumed to be the same at every point in space and in every direction. This is obviously not a good approximation in our immediate neighborhood; there are local irregularities like the Earth and the Sun, and there are many more visible stars in the direction of the center of our Galaxy than in other directions. However, if we look at distant galaxies, we find that they are distributed roughly uniformly throughout the universe, the same in every direction. Thus it does seem to be a good approximation on a large scale. Even better evidence comes from observations of the background of microwave radiation that was discovered in 1965 by two scientists at the Bell Telephone Laboratories. The universe is very transparent to radio waves with wavelengths of a few inches, so this radiation must have traveled to us from very great distances. Any large-scale irregularities in the universe would cause the radiation reaching us from different directions to have different intensities. Yet—to a very high degree of accuracy—the observed intensity is the same in every direction.

There are three kinds of generalized Friedmann models. In one of them, the galaxies are moving apart sufficiently slowly that the gravitational attraction between them will eventually stop their moving apart and start their approaching each other. The universe will expand to a maximum size and then recollapse. In the second model, the galaxies are moving apart so fast that gravity can never stop them and the universe expands forever. Finally, there is a third model in which the galaxies are moving apart at just the critical rate to avoid recollapse. In principle we could determine which model corresponds to our universe by comparing the present rate of expansion with the present average mass density. The

A spiral galaxy. (Anglo-Australian Telescope Board)

mass of the matter in the universe that we can observe directly is not enough to stop the expansion. However, we have indirect evidence that there is more mass that we cannot see. Whether this "invisible" mass could be enough to stop the expansion eventually remains an open question.

In the Friedmann model that recollapses eventually, space is finite but unbounded, as in the Einstein static model. In the other two Friedmann models, which expand forever, space is infinite. Time, on the other hand, has a boundary or edge. In all the models the expansion starts from a state of infinite density called the Big Bang singularity. In the model that recollapses there is another singularity, called the Big Crunch, at the end of the recollapse.

Singularities are places where the curvature of spacetime is infinite and the concepts of space and time cease to have any meaning. Scientific theories assume a background of space and time, so they all break down at a singularity. If there were events before the Big Bang, they would not enable one to predict the present state of the universe because predictability would break down at the Big Bang. Similarly, there is no way that one can determine what happened before the Big Bang from a knowledge of events after the Big Bang. This means that the existence or non-

existence of events before the Big Bang is purely metaphysical; they have no consequences for the present state of the universe. One might as well apply the principle of economy, known as Occam's razor, to cut such events out of the theory and say that time began at the Big Bang. Similarly, there is no way that we can predict or influence any events after the Big Crunch, so one might as well regard it as the end of time.

This beginning and possible end of time predicted by the Friedmann models are very different from earlier ideas. Prior to the Friedmann solutions, the beginning or end of time was something that had to be imposed from outside the universe; there was no necessity for a beginning or an end. In the Friedmann models, on the other hand, the beginning and end of time occur for dynamical reasons. One could still imagine the universe being created by an external agent in a state corresponding to some time after the Big Bang, but it would not have any meaning to say that the universe was created *before* the Big Bang. From the present rate of expansion of the universe, we can estimate that the Big Bang should have occurred between 10 and 20 billion years ago.

Many people disliked the idea that time has a beginning or an end because it smacks of divine intervention. There were therefore a number of attempts to avoid this conclusion. One of these was the steady-state model of the universe proposed in 1948 by Herman Bondi, Thomas Gold, and Fred Hoyle. In this model it was proposed that, as the galaxies move farther away from each other, new galaxies are formed in between out of matter that is being "constantly created." The universe would therefore look more or less the same at all times and the density would be roughly constant. This model had the great virtue that it made definite predictions that could be tested by observations. Unfortunately, observations of radio sources by Martin Ryle and his collaborators at Cambridge in the 1950s and early 1960s showed that the number of radio sources must have been greater in the past, contradicting the steady-state model. The final nail in the coffin of the steady-state theory was the discovery of microwave background radiation in 1965. There was no way this radiation could be accounted for in the model.

Another attempt to avoid a beginning of time was the suggestion that the singularity was simply a consequence of the high degree of symmetry of the Friedmann solutions: The relative motion of any two galaxies is restricted to the line joining them. It would therefore not be surprising if all galaxies collided with each other at some time. However, in the real universe, the galaxies also have some random velocities perpendicular to the line joining them These "transverse velocities" could cause the galaxies to miss each other and allow the universe to pass from a contracting phase to an expanding one without the density ever becoming infinite. Indeed, in 1963 two Russian scientists claimed that this would happen in nearly every solution of the equations of general relativity. They based this claim on the fact that all the solutions with a singularity that they constructed had to satisfy some constraint or symmetry. They later realized, however, that there was a more general class of solutions with singularities that did not have to obey any constraint or symmetry.

This realization showed that singularities *could* occur in general solutions of general relativity, but it did not answer the question of whether they necessarily *would* occur. However, between 1965 and 1970, a number of theorems were proved that showed that any model of the universe must have a Big Bang singularity if the model obeys general relativity, satisfies a few other reasonable assumptions, and assumes as much matter as we observe in the universe, The same theorems predict that there will be a singularity that will be an end of time if the whole universe recollapses. Even if the universe is expanding too fast to collapse in its entirety, we nevertheless expect some localized regions, such as massive burnt-out stars, to collapse and form black holes. The theorems predict that the black holes will contain singularities that will be an end of time for anyone unfortunate or foolhardy enough to fall in.

Einstein's general theory of relativity is probably one of the two greatest intellectual achievements of the twentieth century. It is, however, incomplete because it is what is called a classical theory; that is, it does not incorporate the uncertainty principle of the other great discovery of this century, quantum mechanics. The uncertainty principle states that certain pairs of quantities, such as the position and velocity of a particle, cannot be predicted simultaneously with an arbitrarily high degree of accuracy. The more accurately one predicts the position of the particle, the less accurately one will be able to predict its velocity, and vice versa.

Quantum mechanics was developed in the early years of this century to describe the behavior of very small systems such as atoms or individual elementary particles. In particular, there was a problem with the structure of the atom, which was supposed to consist of a number of electrically charged particles called electrons orbiting around a central nucleus, as the planets orbit the Sun. The previous classical theory predicted that electrons would radiate light waves because of their motion. The waves would carry away energy and so would cause the electrons to spiral inwards until they collided with the nucleus. However, such behavior is not allowed by quantum mechanics because it would violate the uncertainty principle: If an electron were to sit on the nucleus, it would have both a definite position and a definite velocity. Instead, quantum mechanics predicts that the electron does not have a definite position; rather, the probability of finding it is spread out over some region around the nucleus, remaining finite even at the nucleus.

The prediction of classical theory of an infinite probability of finding the electron at the nucleus is rather similar to the prediction of classical general relativity that there should be a Big Bang singularity of infinite density. Thus one might hope that if one were able to combine general relativity and quantum mechanics into a theory of quantum gravity, one would find that the singularities of gravitational collapse or expansion were smeared out as in the case of the collapse of an atom.

The first indication that singularities might be smeared out came with the discovery that black holes, formed by the collapse of localized regions such as stars, would not be completely black if one took into account the uncertainty principle of quantum mechanics. Instead, a black hole would radiate particles and radiation like a hot body. The radiation would carry away energy and so reduce the mass of the black hole. This in turn would increase the rate of emissions. Eventually the black hole would disappear completely in a tremendous burst of emissions. All the matter that collapsed to form the black hole—and any astronaut who was unlucky enough to fall into the black hole—would disappear, at least from our region of the universe. However, the energy that corre-

(continued)

HAWKING (continued)

sponded to this mass (by Einstein's famous equation $E = mc^2$) would be emitted by the black hole in the form of radiation. Thus the astronaut's mass–energy would be recycled to the universe. However, this would be rather a poor sort of immortality, because the astronaut's subjective concept of time would almost certainly come to an end, and the particles out of which he had been composed would not in general be the same as the particles that would be re-emitted by the black hole. Still, black hole evaporation did indicate that gravitational collapse might not lead to a complete end of time.

The real problem with spacetime having an edge or boundary at a singularity is that the laws of science do not determine the initial state of the universe at the singularity but only how it evolves thereafter. This problem would remain even if there were no singularity and time continued back indefinitely: The laws of science would not fix what the state of the universe was in the infinite past. In order to pick out one particular state for the universe from among all possible states that are allowed by the laws of science, one has to supplement the laws by boundary conditions that say what the state of the universe was at an initial singularity or in the infinite past. Many scientists are embarrassed at talking about the boundary conditions of the universe because they feel that it verges on metaphysics or religion. After all, they might say, the universe could have started off in a completely arbitrary state. That may be so, but in that case it could also have evolved in a completely arbitrary manner. Yet all the evidence that we have suggests it evolves in a well-determined way according to certain laws. It is therefore not unreasonable to suppose that there may also be simple laws that govern the boundary conditions and determine the state of the universe.

In the classical general theory of relativity, which does not incorpo-

rate the uncertainty principle, the initial state of the universe is a point of infinite density. It is very difficult to define what the boundary conditions of the universe should be at such a singularity. However, when quantum mechanics is taken into account, there is the possibility that the singularity may be smeared out and that space and time together may form a closed four-dimensional surface without boundary or edge, like the surface of the Earth but with two extra dimensions. This would mean that the universe is completely self-contained and does not require boundary conditions. One would not have to specify the state in the infinite past, and there would not be any singularities at which the laws of physics break down. One could say that the boundary conditions of the universe are that it has no boundary.

It should be emphasized that this is simply a *proposal* for the boundary conditions of the universe. One cannot deduce them from some other principle: One can merely pick a reasonable set of boundary conditions, calculate what they predict for the present state of the universe, and see if they agree with observations. The calculations are very difficult and have been carried out so far only in simple models with a high degree of symmetry. However, the results are very encouraging. They predict that the universe must have started out in a fairly smooth and uniform state. It would have undergone a period of what is called exponential or "inflationary" expansion, during which its size would have increased by a very large factor but the density would have remained the same. The universe would then have become very hot and would have expanded to the state that we see today, cooling as it expanded. It would be uniform and the same in every direction on very large scales but would contain local irregularities that would develop into stars and galaxies.

What happened at the beginning of the expansion of the universe? Did spacetime have an edge at the Big

Bang? The answer is that if the boundary conditions of the universe are that it has no boundary, time ceases to be well-defined in the very early universe just as the direction north ceases to be well-defined at the North Pole of the Earth. Asking what happened before the Big Bang is like asking for a point one mile north of the North Pole. The quantity that we measure as time had a beginning, but that does not mean spacetime has an edge, just as the surface of the Earth does not have an edge at the North Pole, or at least, so I am told: I have not been there myself.

If spacetime is indeed finite but without boundary or edge, this would have important philosophical implications. It would mean that we could describe the universe by a mathematical model that was completely determined by the laws of science alone; they would not have to be supplemented by boundary conditions. We do not yet know the precise form of the laws: At the moment we have a number of partial laws governing the behavior of the universe under all but the most extreme conditions. However, it seems likely that these laws are all part of some unified theory that we have yet to discover. We are making progress and there is a reasonable chance that we will discover it by the end of the century.

At first sight it might appear that such a theory would enable us to predict everything in the universe. However, our powers of prediction would be severely limited, first by the uncertainty principle that states that certain quantities cannot be exactly predicted but only their probability distribution, and second, and even more importantly, by the complexity of the equations, which makes them impossible to solve in any but very simple situations. Thus we would still be a long way from omniscience.

EXPLORING
THE EARLY UNIVERSE

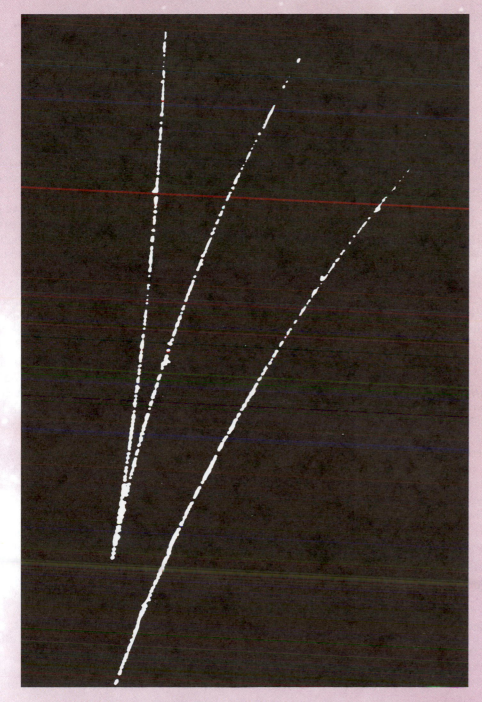

THE CREATION OF MATTER FROM ENERGY A gamma ray (from below) enters a bubble chamber, a device filled with liquid hydrogen and designed to make the path of a charged particle visible. Near the bottom of the photograph, the energy carried by the gamma ray is converted into an electron and an antielectron. Because of a magnetic field surrounding the bubble chamber, the electron is deflected to the right and the antielectron veers toward the left. The path of a stray electron can also be seen. The path of the gamma ray is not visible, because photons are electrically neutral. (Courtesy of Lawrence Berkeley Laboratory)

THE SMOOTHNESS of the cosmic microwave background suggests that the newborn universe may have suddenly "inflated" shortly after the beginning of time. This brief episode of incredibly vigorous expansion explains many attributes of the Big Bang. For instance, using the laws of physics to explore these earliest moments of the universe, astrophysicists now understand how all the matter and energy in the universe may have burst forth from a vacuum. Astrophysicists are also using supercomputer simulations to try to understand how the large-scale structure of the universe could have arisen from tiny density fluctuations left over when inflation stopped. As the universe continued to expand, the four physical forces became separate and distinct. To understand the origins of these forces, researchers have developed speculative theories that predict such oddities as "cosmic strings" and ten dimensions of space.

In recent years there has been a productive marriage of two apparently diverse fields of science. Cosmology, the study of the universe on the largest scale, has become intimately tied to particle physics, the study of matter on the smallest scale. This union has already yielded extra-ordinary insights into the creation of the universe and the fundamental nature of matter and physical forces.

This union of cosmology and particle physics stems from the extreme conditions that existed throughout space immediately after the Big Bang. The universe then was so dense and hot that particles were everywhere constantly colliding at speeds nearly equal to the speed of light. These collisions were violent enough to shatter particles, breaking them into their most fundamental constituents, in much the same way that physicists explore the structure of matter by using particle accelerators. Thus, particle physics gives us important clues about the nature of the early universe.

Similarly, advances in cosmology benefit particle physics. Accelerators are limited to energies that are quite low compared to those prevailing during the Big Bang. Intriguing phenomena have been predicted at energies that are much higher than any we can achieve on Earth. Because of the extreme energies with which particles interacted during the Big Bang, the earliest moments of the universe are a physicist's laboratory in which the most deeply hidden secrets of matter and forces are explored.

29-1 The flatness of the universe and the isotropy of the microwave background suggest that the newborn universe had a brief period of vigorous expansion

Ever since Edwin Hubble discovered that the universe is expanding, astronomers have struggled to determine its rate of deceleration. Since the 1960s, various teams of astronomers have determined different values for the deceleration parameter (q_0), some slightly larger than $\frac{1}{2}$ and some smaller. Because $q_0 = \frac{1}{2}$ is the special case of a flat universe, which separates a bounded cosmological model from an unbounded one, the predicted fate of the universe has swung back and forth. According to some data, the universe seems to be just barely open and infinite, whereas other data indicate that it is just barely closed and is ultimately doomed to collapse.

As we saw in Chapter 28, if the cosmological constant (Λ) is zero, then $q_0 = \frac{1}{2}$ means that the average density of matter throughout space (ρ) equals the critical density (ρ_c). The density parameter (Ω_0) is the ratio ρ/ρ_c, and so $q_0 = \frac{1}{2}$ is the same as $\Omega_0 = 1$.

Motivated by the fact that Ω_0 is nearly equal to 1, physicists began looking at the special conditions associated with a flat universe, in which $\rho = \rho_c$. Specifically, suppose that the average density of matter during the Big Bang was slightly larger, or smaller, than the critical density. How would this deviation have grown, or decreased, as the universe evolved?

Calculations demonstrate that, immediately after the Big Bang, the slightest deviation from the precise critical density would mushroom very rapidly, multiplying itself every 10^{-35} s. If the density had been slightly less than ρ_c, the universe would soon have become wide open and virtually empty. If, on the other hand, the density had been slightly greater than ρ_c, the universe would soon have become tightly closed and so packed with matter that the entire cosmos would have rapidly collapsed in a Big Crunch. In other words, immediately after the Big Bang the fate of the universe hung in the balance, like a pencil teetering on its point, so that the tiniest deviation from the precise equality $\rho = \rho_c$ would have rapidly propelled the universe away from the special case of $\Omega_0 = 1$.

Observations reveal that q_0 is today approximately $\frac{1}{2}$ and Ω_0 is approximately 1. Consequently, the density of the universe immediately after the Big Bang must have been equal to the critical density to an incredibly high order of precision. Calculations demonstrate that, in order for space to be as flat as it is today, the value of ρ right after the Big Bang must have been equal to ρ_c to more than 50 decimal places!

What could have happened during the first few moments of the universe to ensure that $\rho = \rho_c$ to such an astounding degree of accuracy? Because $\rho = \rho_c$ means that space is flat, this enigma is called the **flatness problem.**

A second enigma, closely related to the flatness problem, is the isotropy of the cosmic microwave background. As we saw in Chapter 28, the microwave background is so incredibly uniform across the sky that sensitive temperature measurements can reveal our own motion through this radiation field (recall Figure 28-5). Subtracting the effects of our own motion, we find that the temperature of the microwave background is the same in all parts of the sky to an accuracy of 1 part in 10,000.

To appreciate this so-called **isotropy problem,** think about microwave radiation coming toward us from two opposite parts of the sky. This radiation left over from the primordial fireball has been traveling for about 15 billion years. The total distance between the opposite sides of the observable universe is roughly 30 billion light-years, so these widely separated regions have absolutely no connection with each other. Why, then, do these unrelated parts of the universe have the same temperature?

In the early 1980s Alan Guth at Stanford University offered a remarkable solution, called **inflation,** to the problems of the flatness of the universe and the isotropy of the microwave background. Guth analyzed the suggestion that the universe might have experienced a brief period of extremely rapid expansion shortly after the Planck time (Figure 29-1). During this **inflationary epoch,** which lasted for only about 10^{-24} s, the universe ballooned outward in all directions to become 10^{50} times its original size. This inflation moved much of the material that was originally near our location far beyond the edge of the observable universe. But today, as the universe continues to expand, the cosmic particle horizon recedes into these inaccessible regions, enabling us to see matter and radiation that was in close thermal contact with our location before inflation began.

In Guth's mathematical analysis, the violent expansion of the universe during the inflationary epoch occurred because, for a brief interval, the cosmological constant Λ was huge. Recall that the cosmological constant was introduced by Einstein to hold up a static universe. Today Λ is very tiny and may, in fact, be zero (see Box 28-2 for details). During the inflationary epoch, however, Λ was so large that it provided repulsive forces 10^{120} times larger than Einstein needed in his static cosmology to prop up the universe. This cosmic repulsion drove a violent expansion of space in all directions, literally producing the event we today call the Big Bang.

Inflation accounts for the flatness of the universe. To see why, think about a small portion of the Earth's surface, such as your backyard. For all practical purposes, it is impossible to detect the Earth's curvature over such a small area, and so your backyard looks flat. Similarly, the observable universe is such a tiny fraction of the entire inflated universe that any overall curvature in it is virtually undetectable (Figure 29-2). Like your backyard, the segment of space we can observe looks flat.

The existence of an inflationary epoch would also account for the isotropy of the microwave background. In examining microwaves that are from opposite parts of the sky, what we are seeing is radiation from parts of the universe that were originally in intimate contact with each other. This common origin is why they have the same temperature.

Finally, it is important to note that the concept of inflation does not violate Einstein's dictum that nothing can travel faster than the speed of light. Remember that the expansion of the universe is the expansion of space and does not involve the motion of objects through space. The inflationary epoch was a brief moment during which the distances between particles suddenly increased by an enormous amount. The important point is that this expansion was accomplished entirely by a sudden vigorous *expansion of space.* Particles did not move through space; space itself inflated.

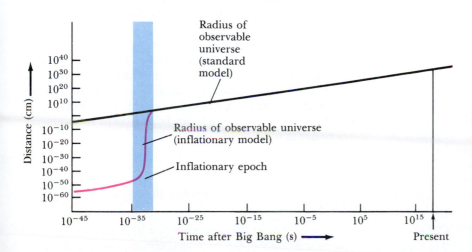

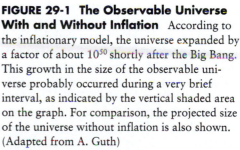

FIGURE 29-1 The Observable Universe With and Without Inflation According to the inflationary model, the universe expanded by a factor of about 10^{50} shortly after the Big Bang. This growth in the size of the observable universe probably occurred during a very brief interval, as indicated by the vertical shaded area on the graph. For comparison, the projected size of the universe without inflation is also shown. (Adapted from A. Guth)

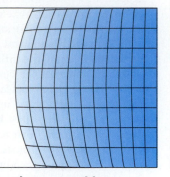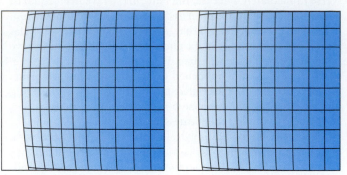

FIGURE 29-2 Inflation Solves the Flatness Problem
This sequence of drawings shows how inflation can produce a locally flat geometry. In each successive frame, the sphere is inflated by a factor of three (while the number of grid lines on the sphere is increased by the same factor). Note how the curvature of the surface quickly becomes undetectable on the scale of the illustration. (Adapted from A. Guth and P. Steinhardt)

29-2   During inflation, all the mass and energy in the universe burst forth from the vacuum of space

The inflationary epoch was a brief but truly stupendous expansion of the universe that occurred almost immediately after the beginning of time. Physicists now realize that inflation helps explain where all the matter and radiation in the universe came from. To see how material particles could be created from a violent expansion of space, we must first understand what quantum mechanics tells us about space.

Quantum mechanics is the branch of physics that explains the behavior of subatomic particles and photons. Because it is able to deal with even the tiniest phenomena, quantum mechanics tells us, for example, how to calculate the structure of atoms and the interactions between nuclei.

There are significant differences between the ordinary world around us and the submicroscopic world of quantum mechanics. In the ordinary world we have no trouble knowing where things are. You know where your house is; you know where your car is; you know where this book is. In the subatomic world of electrons and nuclei, however, you can no longer speak with this same confidence. As we saw in Box 28-3, a certain amount of fuzziness, or uncertainty, enters into the description of reality. At the incredibly small dimensions of the quantum world, things are no longer sharp and distinct.

To appreciate the reasons for this uncertainty, imagine trying to measure the position of a single electron. To find out where it is located, you must observe it. And to observe it, you must shine a light on it. However, the electron is so tiny and has such a small mass that the photons in your beam of light possess enough energy to give the electron a mighty kick. As soon as a photon strikes the electron, it imparts momentum to the electron, which recoils in some unpredictable direction. Consequently, no matter how carefully you try to measure the precise location of an electron, you necessarily introduce some uncertainty into the speed of that electron.

These ideas are at the heart of the famous **Heisenberg uncertainty principle**, first formulated in 1927 by Werner Heisenberg, one of the founders of quantum mechanics. This principle states that there is a reciprocal uncertainty between position and momentum (or mass times velocity). The more precisely you try to measure the position of a particle, the more unsure you become of how the particle is moving. Conversely, the more accurately you determine the momentum of a particle, the less sure you are of its location.

There is an analogous uncertainty involving energy and time. You cannot know with infinite precision the exact energy of a quantum-mechanical system at every moment in time. Over short time intervals, there can be great uncertainty about the amounts of energy in the subatomic world. Specifically, let ΔE be the uncertainty in energy measured over a short interval of time Δt. Then

$$\Delta E \times \Delta t \geq \frac{h}{2\pi}$$

where h is Planck's constant.

We might look upon the Heisenberg uncertainty principle as being merely an unfortunate limitation on our ability to know everything with infinite precision. Alternatively, we can explore how this principle may be helpful in our quest to understand the universe.

We have seen that one of the important conclusions of Einstein's special theory of relativity is the equivalence of mass and energy: $E = mc^2$. There is nothing uncertain about the speed of light (c), which is an absolute constant. Therefore, according to Einstein's equation, any uncertainty in the energy of a physical system can be attributed to an uncertainty Δm in the mass. Thus

$$\Delta E = c^2 \Delta m$$

Combining this expression with the previous equation, we obtain

$$\Delta m \times \Delta t \geq \frac{h}{2\pi c^2}$$

This result is astonishing. It means that, in a very brief interval Δt of time, we cannot be sure how much matter there is in a particular location, even in "empty space." During this brief moment, matter can spontaneously appear and then disappear. Whenever this happens, for each particle created, so is an *antiparticle*.

There is nothing mysterious about antimatter. A particle and an antiparticle are identical in almost every respect; their main distinction is that they carry opposite electric charges. An antielectron (or positron, e^+) has a positive charge, for example, whereas an ordinary electron (e^-) carries a negative charge. Particles and antiparticles are always created or destroyed in equal numbers, thus ensuring that the total electric charge in the universe remains constant, regardless of how many pairs of particles and antiparticles come and go. Physicists often refer to this kind of balance between matter and antimatter as a **symmetry.**

The time interval over which a particle–antiparticle pair can exist is incredibly brief. For example, consider an electron and an antielectron, each with a mass 9.1×10^{-31} kg. To see how long a newborn electron–antielectron pair might exist without violating the principle of conservation of mass, we substitute their combined mass in the previous equation for Δm and we find that Δt equals 6.5×10^{-22} s. In other words, during a time interval shorter than 6.5×10^{-22} s, an electron and an antielectron can spontaneously appear and then disappear without violating any laws of physics.

This process can happen absolutely anywhere and at any time, not just under the unusual conditions of the early universe. It can also occur with more massive particles, but because of the reciprocal nature of the uncertainty principle, the more massive the particle, the shorter the time interval it can exist. For example, the proton is about 2000 times heavier than the electron. Pairs of protons and antiprotons can appear and disappear spontaneously, but they exist for only $\frac{1}{2000}$ as long as pairs of electrons and antielectrons do.

One of the fundamental doctrines of subatomic physics, sometimes called the **principle of totalitarianism,** is that if a quantum-mechanical process is not strictly forbidden, then it must occur. Pairs of every conceivable particle and antiparticle are constantly being created and destroyed at every location across the universe. We have no way of observing these pairs of particles and antiparticles directly without violating the uncertainty principle. The particle–antiparticle pairs exist for such short intervals that it is impossible to observe them. For this reason, these pairs of particles are called **virtual pairs.** They don't "really" exist; they "virtually" exist.

Although these virtual pairs of particles and antiparticles cannot be directly observed, their effects have been detected. Imagine, for instance, an electron in orbit about the nucleus of an atom, such as a hydrogen atom. Ideally, the electron should follow its orbit in a smooth, unhampered fashion. However, because of the constant brief appearance and disappearance of pairs of particles and antiparticles, minuscule electric fields exist for extremely short intervals of time. These tiny, fleeting electric fields cause the electron to jiggle slightly in its orbit. The effect of this jiggling was first detected in 1947 by W. E. Lamb and R. C. Retherford, who noticed a tiny shift in the spectrum of light from the hydrogen atom. This phenomenon, known as the **Lamb shift,** provides powerful support for the idea that every point in space, all across the universe, is seething with virtual pairs of particles and antiparticles. This constant appearance and disappearance of particles and antiparticles is sketched in Figure 29-3.

The concept of virtual pairs also readily explains another phenomenon called **pair production.** It has been known for years that highly energetic gamma rays (photons) can convert their energy into pairs of particles and antiparticles. Quite simply, the gamma ray disappears upon colliding with a second photon, and a particle and an antiparticle appear in its place. This process is routinely observed in high-energy nuclear experiments (see illustration on the opening page of this chapter). Indeed, it is one of the ways in which physicists can manufacture many of the more exotic species of particles and antiparticles. This process of pair production and its inverse process of **annihilation** are sketched in Figure 29-4.

These particles and antiparticles come from nature's ample supply of virtual pairs. The gamma rays provide a virtual pair with so much energy that the virtual particles can materialize and appear as real particles in the real world. The only requirement is that nature's balance sheet be satisfied. To create a particle and an antiparticle having a total mass M, the incoming gamma-ray photons must possess an amount of energy E that satisfies the condition $E \geq Mc^2$. If the photons carry too little energy, pair production will not occur. Likewise, the more energetic the photons, the more massive the particles and antiparticles that can be manufactured.

Around 1980 physicists began applying these ideas to their thinking about the creation of the universe. During the inflationary epoch, space was expanding with explosive vigor. Yet, as we have seen, all space is seething with virtual pairs of particles and antiparticles. Normally a particle and

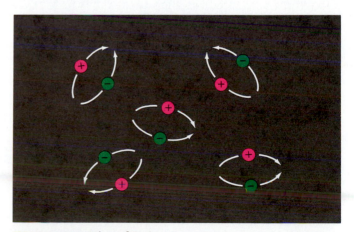

FIGURE 29-3 Virtual Pairs Pairs of particles and antiparticles can appear and then disappear anywhere in space provided that each pair exists only for a very short time interval, as dictated by the uncertainty principle.

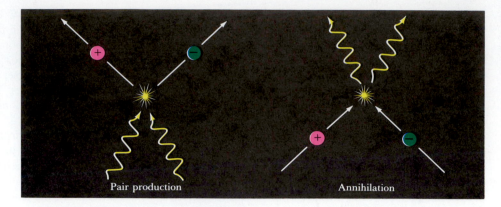

FIGURE 29-4 Pair Production and Annihilation Pairs of virtual particles can be converted into real particles by high-energy gamma-ray photons. Conversely, a particle and an antiparticle can annihilate each other and be transformed into energy in the form of gamma rays.

an antiparticle have no trouble getting back together in a time interval (Δt) short enough as required by the uncertainty principle. During inflation, however, the universe expanded so fast that particles were rapidly separated from their corresponding antiparticles. Deprived of the opportunity to recombine, these virtual particles had to become *real* particles in the real world. In this way, the universe was flooded with particles and antiparticles created by the violent expansion of space.

29-3   As the primordial fireball cooled off, most of the matter and antimatter in the early universe annihilated each other

As soon as the flood of matter and antimatter appeared in the universe, collisions between particles and antiparticles began to produce numerous high-energy gamma rays. As these gamma rays collided, they promptly gave back the particle and antiparticles from which they came. As a result, the rate of pair production soon equaled the rate of annihilation. For example, for every proton (p) and antiproton ($\overline{p}$) that annihilated each other to create gamma rays, two gamma rays collided elsewhere to produce a proton and an antiproton. In other words, reactions such as

$$p + \overline{p} \rightarrow \gamma + \gamma$$

and

$$\gamma + \gamma \rightarrow p + \overline{p}$$

proceeded with equal vigor. In this state, called **thermal equilibrium**, precisely as many particles and antiparticles are being created as are being destroyed.

Now let us think about what happened soon after the Big Bang. As the universe continued to expand, all the gamma-ray photons became increasingly redshifted; as a result, the temperature of the radiation field fell. The temperature eventually became so low that the gamma rays no longer had enough energy to create particular kinds of particles and antiparticles. We say that the temperature dropped below the particles' **threshold temperature** (Box 29-1). Collisions between particles and antiparticles did add photons to the cos-

mic-radiation background, but collisions between photons could no longer replenish the supply of particles and antiparticles.

When the universe was about 0.0001 second old (that is, at time $t = 10^{-4}$ s), the temperature of the radiation fell below 10^{13} K. At that moment, the universe became cooler than the threshold temperatures of both protons and neutrons. No new particles appeared, but the annihilation of protons by antiprotons and of neutrons by antineutrons continued vigorously everywhere throughout space. This wholesale annihilation dramatically lowered the matter content (particles and antiparticles) of the universe, while simultaneously increasing the radiation (photon) content.

A little later, when the universe was about 1 second old, its temperature fell below 6×10^9 K, the threshold temperature for electrons and antielectrons. A similar annihilation of pairs of electrons and antielectrons further decreased the matter content of the universe while raising its radiation content. This radiation field, which fills all space, *is* the primordial fireball we discussed in Section 28-4. The primordial fireball, which dominated the universe for the first million years, therefore derived its energy from the annihilation of particles and antiparticles during the first second after the Big Bang.

Now we have a dilemma. If the symmetry between particles and antiparticles were truly valid, then for every proton there should have been an antiproton. For every electron, there should likewise have been an antielectron. Consequently, by the time the universe was 1 second old, every particle would have been annihilated by an antiparticle, leaving no matter at all in the universe.

Obviously, this did not happen. The planets, stars, and galaxies we see in the sky are made of matter, not antimatter. Thus there must have been an excess of matter over antimatter immediately after the Big Bang. Physicists therefore say that somehow a **symmetry breaking** occurred during the earliest moments of the universe, meaning that the particles emerging from the Big Bang outnumbered the antiparticles.

We can estimate the extent of this symmetry breaking because, as noted in Chapter 28, there are roughly a billion photons today in the microwave background for each proton and neutron in the universe. Thus, for every billion antiprotons, there must have been a billion plus one ordinary protons, leaving one surviving proton after annihilation. Similarly, for every billion anti-electrons, there must have been a billion plus one ordinary electrons.

29-4 A background of neutrinos and most of the helium in the universe are relics of the primordial fireball

The early universe must have been populated with vast numbers of neutrinos (ν) and antineutrinos (ν̄). These particles take part in the nuclear reaction that transforms neutrons into protons and vice versa. For example, a neutron can decay into a proton by emitting an electron and an antineutrino:

$$n \rightarrow p + e^- + \bar{\nu}$$

This radioactive decay happens quickly (its half-life is about 10.5 minutes), which is why we do not find free neutrons floating around in the universe today.

In the early universe, collisions between particles kept the number of protons approximately equal to the number of neutrons. This balance was maintained only as long as the density of matter in the universe remained high. By the time the universe was about 2 seconds old, however, matter was thinned out enough that the natural tendency for neutrons to decay into protons then took over, and the number of neutrons began to decline.

Before all the neutrons could decay into protons, the primordial fireball had cooled enough that the first heavier element, helium, could survive. A helium nucleus consists of two protons and two neutrons. To create a helium nucleus, first a single proton and a single neutron must combine to form deuterium (²H), sometimes called "heavy hydrogen." We can write this reaction as

$$p + n \rightarrow {}^2H$$

Deuterium is easily destroyed because a proton and a neutron do not stick together very well. In the early universe, high-energy gamma rays easily broke deuterium nuclei back down into independent protons and neutrons. As a result, the synthesis of helium could not get beyond the first step. This block to the creation of helium is called the **deuterium bottleneck**.

BOX 29-1

Elementary Particles and Threshold Temperatures

The **rest energy** of a particle equals its mass times the square of the speed of light. It is the energy that would be released if all the mass of the particle were converted into energy. For example, consider protons and antiprotons, each having a mass of 1.7×10^{-27} kg. Substituting this mass into the equation $E = mc^2$, we find that the rest energy of each of these particles is 938 MeV (where MeV stands for "million electron volts," a unit of energy commonly used by physicists and discussed in Box 5-4). This tells us how much energy colliding gamma-ray photons must have in order to create a proton and an antiproton. If the combined energy of the two photons is less than 2×938 MeV, the reaction $\gamma + \gamma \rightarrow p + \bar{p}$ will not occur.

A photon energy of 938 MeV corresponds to a particular temperature, called the threshold temperature for the creation of protons and antiprotons, according to the equation

$$E = kT$$

where h is the Boltzmann constant ($k = 1.38 \times 10^{-23}$ J/K = 8.6×10^{-5} eV/K). For gamma-ray photons having an energy of 938 MeV, the temperature is 10.9 trillion kelvin. Thus, as soon as the temperature of the radiation field falls below 10.9×10^{12} K, proton–antiproton pairs can no longer be created.

Each kind of particle has its own threshold temperature. For example, electrons and antielectrons have masses about 1800 times smaller than protons and antiprotons. Thus, the threshold temperature for the creation of electron–antielectron pairs is $\frac{1}{1800}$ that for proton–antiproton pairs. Consequently, when the radiation temperature falls below 5.9 billion kelvin, the reaction $\gamma + \gamma \rightarrow e^+ + e^-$ that creates electron–antielectron pairs can no longer take place. We thus say that the threshold temperature for electrons and antielectrons is 5.9×10^9 K. Several common types of particles, with their masses and threshold temperatures, are listed in the table below.

Particle	Symbol	Rest energy (MeV)	Threshold temperature (10⁹ K)
Neutrino	ν, ν̄	0.00001 (?)	0.0001
Electron	e⁻, e⁺	0.5110	5.930
Muon	μ⁻, μ⁺	105.66	1,226.2
Pi meson	π⁰	134.96	1,556.2
	π⁺, π⁻	139.57	1,619.7
Proton	p, p̄	938.26	10,888
Neutron	n, n̄	939.55	10,903

When the universe was about 3 minutes old, the photons in the background radiation became so redshifted that they could no longer break up the deuterium. By this time, most of the neutrons had decayed into protons, and protons outnumbered neutrons by six to one. Because deuterons could now survive, the remaining neutrons combined with protons and rapidly produced helium (review Box 18-3). The result was what we find in the universe today: one helium atom for every ten hydrogen atoms.

Where are all those primordial neutrinos and antineutrinos that had interacted so vigorously with the protons and neutrons before the universe was 2 seconds old? As we saw in our discussion of the Sun (recall Section 18-8), neutrinos and antineutrinos are very difficult to detect because they do not interact strongly with matter. The Earth itself is virtually transparent to the neutrinos from the Sun.

By the time the universe was about 2 seconds old, matter was sufficiently spread out so that the universe had became transparent to neutrinos and antineutrinos. From that time on, neutrinos and antineutrinos could travel across the universe unimpeded. Consequently, a neutrino–antineutrino background should fill the universe much as the cosmic microwave background does. Indeed, these ancient neutrinos and antineutrinos may be about as populous today as the photons in the microwave background ($5.5 \times 10^8 \text{ m}^{-3}$). The neutrino–antineutrino background should be slightly cooler than the photon background, which received extra energy from the electron–antielectron annihilations. Physicists estimate that the current temperature of the neutrino–antineutrino background is about 2 K, as opposed to 3 K for the microwave background.

Until recently no one paid much attention to this elusive neutrino–antineutrino background. Because these particles, like photons, are generally believed to be massless, they just did not seem very important. However, around 1980, some experiments suggested that neutrinos and antineutrinos might have mass after all. If they do, it is probably less than 10^{-4} of the mass of the electron. More recent experiments have failed to confirm the existence of the neutrino mass, however. As we saw in Chapter 22, analyses of the burst of neutrinos detected from the 1987 supernova constrain the neutrino mass to be less than about 10 eV, or about $\frac{1}{50,000}$ the mass of an electron.

So many neutrinos and antineutrinos are spread throughout space that these particles could account for most of the matter in the universe if this tiny mass does exist. Indeed, the matter contained in neutrinos and antineutrinos may be ten times greater than all the matter in the stars, planets, and galaxies combined. Consequently, the neutrino–antineutrino background is a candidate for the dark matter that seems to pervade the universe.

29-5 Galaxies were formed from density fluctuations in the early universe

The distribution of matter in the universe today is quite lumpy. Stars are grouped together in galaxies, galaxies into

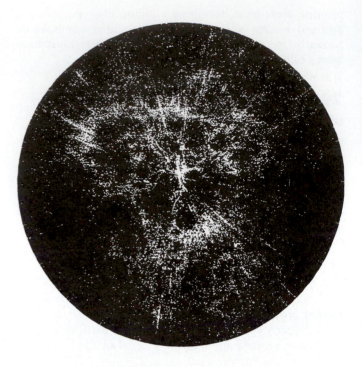

FIGURE 29-5 The Large-Scale Structure of the Universe This map shows the distribution of 14,000 galaxies in a wedge-shaped slice that sweeps all the way around the sky. The Earth is at the center of this map, which covers an expanse about 1 billion light-years in diameter. Note that the galaxies are mostly clustered around voids in which few galaxies are found. (Courtesy of J. P. Huchra and M. J. Geller, Smithsonian Astrophysical Observatory)

clusters, and clusters into **superclusters** that stretch across 100 million light-years.

Since 1985, several teams of astronomers plotted the positions of clusters and superclusters of galaxies in space. As we saw in Chapter 26 (recall Figure 26-23), these maps show that galaxies are concentrated along enormous sheets that surround voids measuring 100 million to 400 million light-years across. These features, which characterize the large-scale structure of the universe, are seen in Figure 29-5.

Although there is a lot of lumpiness in the universe today, the early universe must have been exceedingly smooth. To see why, think back to the era of recombination that occurred roughly 1 million years after the Big Bang. Before recombination, high-energy photons were constantly colliding vigorously with charged particles throughout all of space. After recombination, the universe became transparent, and these photons stopped interacting with the matter in the universe. Astronomers say that matter "decoupled" from radiation during the era of recombination. Because, as we have seen, the 3-K microwave background is extremely isotropic, we can conclude that the matter with which these photons once collided so frequently must also have been spread smoothly across space.

Although matter must have been distributed quite smoothly across space during the early universe, it could not have been perfectly uniform. If it had been, it would still have to be absolutely uniform today; there would now be neither stars nor galaxies, only a few atoms per cubic meter throughout space. Consequently, there must have been slight lumpiness, or **density fluctuations,** in the distribution of matter in the early universe. Through the action of gravity, these fluctuations eventually grew to become the galaxies and clusters of galaxies that we see today throughout the universe.

The COBE satellite has given us a map of the density fluctuations from which the large-scale structure of the universe eventually emerged (Figure 29-6). Recall that, according to the general theory of relativity, gravity causes time to slow down (recall Figure 4-25). As a result, photons in the primordial fireball experienced small gravitational redshifts that varied from place to place depending on the distribution of matter. The slight variations in temperature that COBE found across the sky mimic fluctuations in density that existed when the universe became transparent a million years after the Big Bang. From that moment on, photons had negligible effect on matter, and the density fluctuations were free to grow and eventually yield the large-scale structure we see today.

Our understanding of how gravity can amplify density fluctuations dates back to 1902, when the British physicist James Jeans solved a problem first proposed by Isaac Newton. Suppose that you have a gas with only very tiny fluctuations in density, as shown in Figure 29-7. These regions of higher density will then gravitationally attract nearby material and thus gain mass. As this happens, however, the pressure of the gas inside these regions will also increase, which can make these regions expand and disperse. The question then becomes: Under what conditions does gravity overwhelm gas pressure so that a permanent object can form?

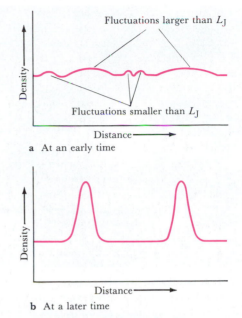

a At an early time

b At a later time

FIGURE 29-7 The Growth of Density Fluctuations
(a) Small density fluctuations in the distribution of matter shortly after the era of recombination. (b) If the size of a fluctuation is greater than the Jeans length (L_J), it becomes gravitationally unstable and can grow in amplitude.

James Jeans proved that an object will grow from a density fluctuation provided that the fluctuation extends over a distance that exceeds the so-called **Jeans length** L_J, given by

$$L_J = \frac{\pi k T}{m G \rho}$$

where k is the Boltzmann constant, T the temperature of the gas, G the gravitational constant, m the mass of a single particle in the medium, and ρ the average density of the gas.

We can apply the Jeans formula to the conditions prevalent during the era of recombination, when $T = 3000$ K and $\rho = 10^{-18}$ kg/m^3. Taking m to be the mass of the hydrogen atom ($m = 2 \times 10^{-27}$ kg), we find that $L_J = 100$ ly, the diameter of a typical globular cluster. Furthermore, the mass contained in this volume ($\rho L_J^3 = 5 \times 10^5$ M$_\odot$), equals the mass of a typical globular cluster. As we saw in Chapter 21, globular clusters are composed of the most ancient stars we can find in the sky. For these reasons, Robert Dicke and P. J. E. Peebles at Princeton have proposed that globular clusters (Figure 29-8) were among the first objects to form after matter decoupled from radiation. But, interesting as these calculations may be, they do not shed light on how galaxies and clusters of galaxies came to be formed.

Astronomers are now working on a variety of other ideas to explain this large-scale structure of the universe. One theory is that giant explosions dominated the universe shortly after the first galaxies and stars formed. These explosions, which could have involved colossal supernovae or some sort of quasarlike activity, would have been strong enough to

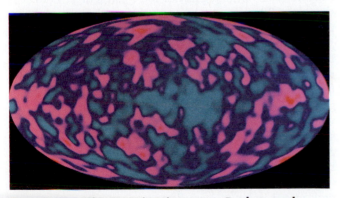

FIGURE 29-6 The Cosmic Microwave Background
This map of the sky shows temperature variations in the cosmic microwave background. Pink regions are about 0.0003 K warmer than the average temperature of 2.73 K; blue regions are about 0.0003 K cooler than the average. These tiny temperature fluctuations date back to the earliest moments of the universe and may be directly related to the large-scale structure of the universe today. (Goddard Space Flight Center; NASA)

FIGURE 29-8   A Globular Cluster A typical globular cluster contains 10^5 to 10^6 stars. Cluster diameters range from about 20 to 400 ly. Because these parameters are comparable to the Jeans length (L_J) during the era of recombination, astronomers believe that globular clusters must have been among the first objects to form in the universe. (NASA)

clear out bubblelike voids, while sweeping aside and compressing material for the next generation of stars and galaxies. Unfortunately, these explosions would have produced only small voids, one-tenth as large as those that have been observed.

Most astronomers therefore think that large-scale structure arose from the gravitational amplification of density fluctuations revealed by COBE. We have seen, however, that about 90% of the matter in the universe is dark. Since the nature of this dark matter is not known, researchers have the freedom to hypothesize different types of dark matter in their studies of the formation of large-scale structure. Neutrinos are an example of **hot dark matter**, so named because it consists of lightweight particles traveling at high speeds. **Cold dark matter**, on the other hand, consists of massive particles traveling at slow speeds. Examples include WIMPs (weakly interacting massive particles, discussed in Section 18-8), and exotic, hypothetical particles predicted by speculative theory, discussed in the next section.

Researchers favoring hot dark matter have used neutrinos to account for the observed large-scale structure. Today there should be roughly 100 million neutrinos, on the average, per cubic meter. If each neutrino has a mass of about 10 eV, then the cosmic neutrino background alone has an average density very nearly equal to the critical density.

If a neutrino has no mass, it will always travel at the speed of light, just as a photon does. If, however, it has a little mass, it must slow down as the universe expands and cools. In this case, neutrinos would accumulate over time in density fluctuations. The gravitational pull of these neutrinos on surrounding matter would eventually lead to the formation of clusters of galaxies.

Typical effects of cold dark matter are shown in Figure 29-9, which presents the results of a supercomputer simulation of the development of large-scale structure. The simulation follows the motions of more than two million particles of cold dark matter in a box that expands as the universe expands. The box representing the present time is about 50 Mpc on a side.

The simulation begins with an almost perfectly uniform distribution of particles, mimicking the tiny density fluctuations that must have been present just after inflation. A supercomputer programmed with Newton's laws in an expanding universe is then used to calculate the motions of particles. At selected intervals, a slice is taken through the box and the distribution of particles is examined. The slice through today's universe is about 6 Mpc thick.

Nine representative slices through the computational box are shown in Figure 29-9. Each slice is designated by an "expansion factor" a, which is scaled so that the final slice at the end of the simulation is set at $a = 1$. Thus, for instance, the first frame ($a = 0.04$) shows a slice of the universe when the computational box was only 4% of its size in the last frame. The first four slices ($a = 0.04$ through $a = 0.10$) show the appearance of the universe at very early times, while the last five views ($a = 0.20$ through $a = 1.00$) are from much later times. The slice labeled $a = 0.60$ seems to resemble the universe today, so the simulation extends from the past into the future.

Note how the tiny fluctuations barely visible in the first slice gradually become more pronounced as time passes. These features merge to form larger structures as gravity continues to draw matter together, overcoming the expansion of the universe. The development of both voids and filaments is clearly seen. Ordinary matter, which follows the course set by the cold dark matter, accumulates along density enhancements forming galaxies and clusters of galaxies. The slice $a = 0.20$ seems to represent the era during which our Galaxy formed.

Simulations using cold dark matter are not entirely successful because they do not produce enough large-scale structure like that seen in Figures 26-23 and 29-5. The density fluctuations deduced from COBE observations taken together with cold dark matter give too few superclusters and voids. Simulations using hot dark matter are not entirely successful either because high-speed particles tend to erase galaxy-sized density fluctuations. A primary difference between simulations based on cold and hot dark matter is the stage at which galaxies form. In calculations based on cold dark matter, galaxies form first and then group together in clusters and then superclusters. In calculations based on hot dark matter, huge supercluster-sized sheets of matter form first and then fragment into galaxies. Many astronomers suspect that simulations involving both cold and hot dark matter may yield the best agreement between simulations and the observed large-scale structure.

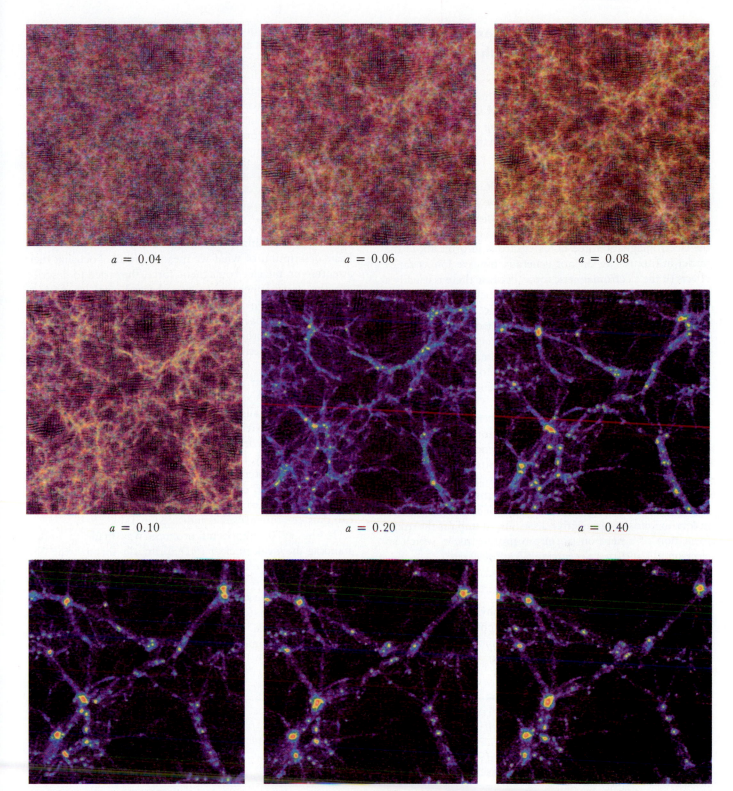

$a = 0.04$ $a = 0.06$ $a = 0.08$

$a = 0.10$ $a = 0.20$ $a = 0.40$

$a = 0.60$ $a = 0.80$ $a = 1.00$

FIGURE 29-9 A Cold-Dark-Matter Simulation These nine views show slices through a large, box-shaped volume of space as the universe expands. The "expansion factor" a indicates the relative size of the box at the time when the slice was taken. The first four views represent early epochs within the first few billion years after the Big Bang. The last five views are from much later times, with the $a = 0.60$ view possibly representing the present time. Galaxy formation probably occurs around the time of the $a = 0.20$ slice. (Courtesy of J. M. Gelb and E. Bertschinger; MIT)

29-6 Grand unified theories explain that all the forces had the same strength immediately after the Big Bang

The behavior and interactions of everything in the universe can be understood as the result of just *four* physical forces: gravity, electromagnetism, and the strong and weak nuclear forces. We are all intimately familiar with the force of gravity, the long-range force that dominates the universe over astronomical distances.

The electromagnetic force is also a long-range force (in principle, its influence extends to infinity, as gravity's does), but it is much stronger than the gravitational force. Just as the force of gravity holds the Moon in orbit about the Earth, the electromagnetic force holds electrons in orbit about the nuclei in atoms. We do not generally observe longer-range effects of the electromagnetic force, because there is usually a negative electric charge for every positive charge and a south magnetic pole for every north magnetic pole. Thus, over great volumes of space the effects of electromagnetism effectively cancel out. No similar canceling occurs with gravity because there is no equivalent "negative mass."

Both the strong and the weak nuclear forces are said to be short-range because their influence extends only over distances that are less than about 10^{-15} m. The **strong nuclear force** holds protons and neutrons together inside the nuclei of atoms. Without the strong nuclear force, nuclei would disintegrate because of the electromagnetic repulsion of the positively charged protons. In fact, the strong nuclear force overpowers the electromagnetic forces inside nuclei.

The weak nuclear force is so weak that it cannot hold anything together. Instead, the **weak nuclear force** is at work in certain kinds of radioactive decay. An example is the transformation of a neutron (n) into a proton (p), in which an electron (e^-) and an antineutrino ($\bar{\nu}$) are released:

$$n \rightarrow p + e^- + \bar{\nu}$$

Numerous experiments in nuclear physics strongly suggest that protons and neutrons are composed of more basic particles called **quarks**, the most common varieties being "up" (u) quarks and "down" (d) quarks. A proton is composed of two up quarks and one down quark, a neutron of two down quarks and one up quark.

In the 1970s the concept of quarks led to a breakthrough in our understanding of the strong and weak nuclear forces. The strong nuclear force holds quarks together, while the weak nuclear force is at work whenever a quark changes from one variety to another. For example, when a neutron decays into a proton, one of the neutron's down quarks changes into an up quark. Thus, the weak nuclear force is responsible for transformations such as

$$d \rightarrow u + e^- + \bar{\nu}$$

In the 1940s, the physicists Richard P. Feynman, Julian S. Schwinger, and Sinitiro Tomonaga succeeded in developing a basic description of what we mean by force. Focusing their attention on the electromagnetic force, they tried to describe exactly what happens when two charged particles interact. According to their theory, now called **quantum electrodynamics,** charged particles interact by exchanging *virtual photons.* Virtual photons, like virtual particles, cannot be observed directly, because they exist for extremely short time intervals.

Quantum electrodynamics has proven to be one of the most successful theories in modern physics. It accurately describes many details of the electromagnetic interaction between charged particles. Inspired by these successes, physicists have tried to develop similar theories for the other three forces. Specifically, the weak nuclear force occurs when particles exchange **intermediate-vector bosons,** the gravitational force occurs when particles exchange **gravitons,** and quarks stick together by exchanging **gluons.** The four physical forces can thus be summarized as indicated in Table 29-1.

In the 1970s important progress was made in understanding the weak nuclear force, primarily through the Nobel-prizewinning efforts of physicists Steven Weinberg, Sheldon Glashow, and Abdus Salaam. They proposed a theory that predicted the existence of three types of intermediate-vector bosons, that are exchanged in various manifestations of the weak force. These three particles were actually discovered in experiments in the 1980s, thus providing strong support for this theory.

TABLE 29-1

The Four Forces

Force	Relative strength	Particles exchanged	Particles acted upon	Range	Example
Strong	1	Gluons	Quarks	10^{-15} m	Holding nuclei together
Electromagnetic	1/137	Photons	Charged particles	Infinite	Holding atoms together
Weak	1/10,000	Intermediate-vector bosons	Quarks, electrons, neutrinos	10^{-16} m	Radioactive decay
Gravity	6×10^{-39}	Gravitons	Everything	Infinite	Holding the solar system together

One of the startling predictions of the Weinberg–Glashow–Salaam theory is that the weak force and the electromagnetic force should be identical to each other at energies greater than 100 GeV. In other words, if particles are slammed together with a total energy greater than 100 billion electron volts, then electromagnetic interactions become indistinguishable from weak interactions. We can thus say that above 100 GeV the electromagnetic force and the weak force are "unified."

This unification occurs because the three types of intermediate-vector bosons behave just like photons above 100 GeV. Physicists describe this similarity by saying that "symmetry is restored" above 100 GeV. In the world around us, however, particles interact with very much less energy than this. Below 100 GeV, intermediate-vector bosons behave like massive particles, but photons are always massless. Because intermediate-vector bosons and photons are not similar at low energies, we say that "symmetry is broken" below 100 GeV, which is why the electromagnetic and the weak forces behave so differently in the world around us.

In the 1970s Sheldon Glashow and Howard Georgi proposed a **grand unified theory** (or **GUT**), which predicts that the strong, weak, and electromagnetic forces are unified at energies above 10^{14} GeV. In other words, if particles were to collide at energies greater than 10^{14} GeV, the strong, weak, and electromagnetic interactions would be indistinguishable from each other.

Many physicists suspect that all four forces may be unified at energies greater than 10^{19} GeV (Figure 29-10); that is, if particles were to collide at these colossal energies, there would be no difference between the gravitational, electromagnetic, and nuclear forces. However, no one has yet succeeded in working out the details of such a **supergrand unified theory**, which is sometimes also called a **theory of everything** (or **TOE**).

Physicists use particle accelerators to examine the unification of the weak and electromagnetic forces by slamming particles together at energies around 100 GeV. There is no hope, though, of ever constructing machines capable of making particles collide with energies in the trillions of GeV. It may thus be impossible to test directly the grand and supergrand unified theories in a laboratory. However, the universe immediately after the Big Bang was so hot and its particles were moving so fast that they did indeed collide with energies in the trillions of GeV. The earliest moments of the universe have thus become a laboratory for testing some of the most elegant, sophisticated theories in physics.

Many of the ideas connecting particle physics and cosmology are controversial and still quite speculative. Nevertheless, it is possible to summarize our understanding with the aid of Figure 29-11. During the Planck time (from $t = 0$ s to $t = 10^{-43}$ s), particles collided with energies greater than 10^{19} GeV, and all four forces were unified. Because we do not yet have a TOE that properly describes the behavior of gravity, we remain ignorant of what was going on during the first 10^{-43} second of the universe's existence. We know, however, that by the end of the Planck time the energy of particles in

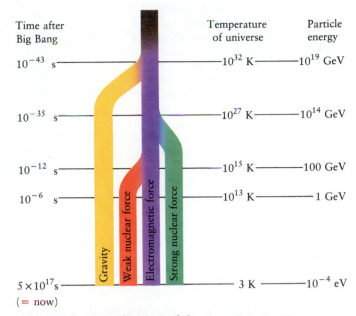

FIGURE 29-10 Unification of the Four Forces The strength of the four physical forces depends on the speed or energy with which particles interact. As shown in this schematic diagram, the higher the energy, the more the forces resemble each other. Also included here are the temperature of the universe and the time after the Big Bang when the strengths of the forces are thought to have been equal.

the universe had fallen to 10^{19} GeV, below which gravity is no longer unified with the other three forces. We can thus say that, at $t = 10^{-43}$ s, there was a **spontaneous symmetry breaking** in which gravity was "frozen out" of the otherwise unified hot soup that filled all space. As we saw earlier, energy (E) is related to temperature (T) by $E = kT$, where k is the Boltzmann constant (roughly 10^{-4} eV/K). Thus, the temperature of the universe was 10^{32} K when gravity emerged as a separate force.

At $t = 10^{-35}$ s, the energy of particles in the universe had fallen to 10^{14} GeV, equivalent to a temperature of 10^{27} K, below which the strong nuclear force is no longer unified with the electromagnetic and weak nuclear forces. Thus, at $t = 10^{-35}$ s, there was a second spontaneous symmetry breaking, at which time the strong nuclear force made its appearance, freezing out of the otherwise unified hot soup that had already given us gravity. Calculations suggest that the inflationary epoch lasted from $t = 10^{-35}$ s to about $t = 10^{-24}$ s, during which the universe increased its size by a factor of roughly 10^{50}.

At $t = 10^{-12}$ s, the temperature of the universe had dropped to 10^{15} K, the energy of the particles had fallen to 100 GeV, and there was a final spontaneous symmetry breaking and "freeze-out" that separated the electromagnetic force from the weak nuclear force. From that moment on, all four forces have interacted with particles essentially as they do today.

FIGURE 29-11 The Early History of the Universe As the universe cooled, the four forces "froze out" of their unified state as a result of spontaneous symmetry breaking. The inflationary epoch lasted from 10^{-35} to 10^{-24} s after the Big Bang. Neutrons and protons froze out of the hot "quark soup" during a stage called confinement one millionth of a second after the Big Bang. The universe became transparent to light (that is, photons froze out) when the universe was a million years old.

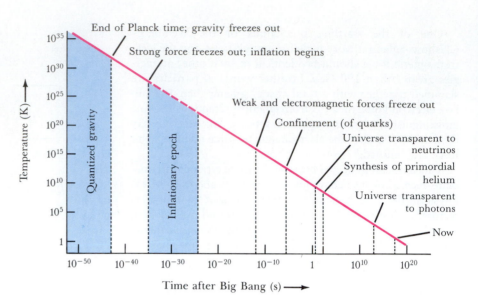

The next significant event occurred at $t = 10^{-6}$ s, when the temperature was 10^{13} K and particles were colliding with energies of roughly 1 GeV. Prior to this moment, particles collided so violently that individual protons and neutrons could not exist, being constantly fragmented into quarks. After this time, appropriately called the period of **confinement,** quarks were finally able to stick together to form individual protons and neutrons.

As previously outlined, the universe became transparent to neutrinos at $t = 2$ s, and all the primordial helium was produced by $t = 3$ min. Finally, the universe became transparent to photons at $t = 10^6$ yr.

29-7 Cosmic strings and other oddities may be relics of the early universe

We have seen that, according to quantum mechanics, the vacuum of empty space is in fact seething with activity. Just as an atom can be excited into a higher energy level, GUTs explain that the vacuum of space can have an excited state. This excited or **false vacuum** would look identical to what we normally regard as empty space (a true vacuum) except that it would contain a significant amount of energy.

Alan Guth has proposed that inflation occurred when the universe went from the high-energy symmetric state of a false vacuum to the low-energy symmetry-broken state of a true vacuum. The energy released during this transition gave rise to a powerful "negative pressure" that caused space to balloon outward in all directions. This is the source of the large value of Einstein's cosmological constant Λ during the inflationary epoch.

An analogy using pencils may help to illustrate the nature of the transition from a symmetric false vacuum to an asymmetric true vacuum. When a pencil is balanced on its point, no particular horizontal direction is singled out over any other. When the pencil falls over, however, this symmetry is broken, because the pencil then points toward a specific di-

rection of the compass. One direction has become singled out over all others.

Now imagine thousands of pencils precariously balanced on their points, as shown in Figure 29-12a. Of course, the pencils soon fall over, with the result shown in Figure 29-12b. Symmetry has been broken everywhere, because every pencil points in a particular horizontal direction.

As soon as the universe cooled to 10^{27} K, when the strong nuclear force froze out, the false vacuum that then filled space became unstable, like thousands of pencils balanced on their points. The rapid transition to a true vacuum is analogous to the pencils falling over. But note that something odd might occur. As thousands upon thousands of pencils knock each other over, it could happen that waves of falling pencils might fall together in such a way that a single pencil remains upright, as shown in Figure 29-12c. For that pencil, symmetry would not have been broken. An analogous relic left over from the early universe, called a **cosmic string,** is a long, thin, massive wire of unbroken symmetry along which the strong, weak, and electromagnetic forces remain unified.

Some physicists suspect that cosmic strings might exist in the universe today. Such structures would be incredibly massive (about a trillion tons per millimeter) and detectable only by their gravitational effects. Thus, cosmic strings are an ideal candidate for the dark matter in the universe. Perhaps more importantly, though, they may have served as the gravitational focus along which clusters of primordial galaxies could have condensed billions of years ago. This suspicion arises partly from the way in which clusters of galaxies seem to be scattered across the sky.

Figure 29-13 shows the locations of 400,000 galaxies, covering approximately one-quarter of the sky. On this map, the sky is divided into tiny squares, each measuring about 10 arc min across; colors indicate the number of galaxies within each square. White indicates high densities of galaxies, whereas the darker colors denote low-density areas. The brown and black areas are voids that are nearly free of gal-

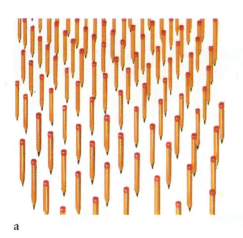
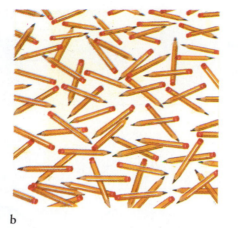

a   b   c

FIGURE 29-12 Symmetry Breaking and Cosmic Strings
The symmetry that existed in the early universe can be likened to orderly rows of pencils balanced on their points, as sketched in (**a**). The state of broken symmetry that exists today is analogous to all the pencils having fallen over, as sketched in (**b**). A situation could arise, however, in which the falling pencils might topple toward each other so that one pencil is supported upright, as in (**c**). Some physicists theorize that comparable structures, called cosmic strings, might be left over from the symmetry breaking that occurred during the earliest moments of the universe.

axies. Note that many of the white regions seem to be connected to each other by the green squares. These large-scale structures are typically 100 Mly long and contain up to one million galaxies with a mass of roughly 10^{16} $M_\odot$. Some physicists suspect that cosmic strings may be responsible for these structures.

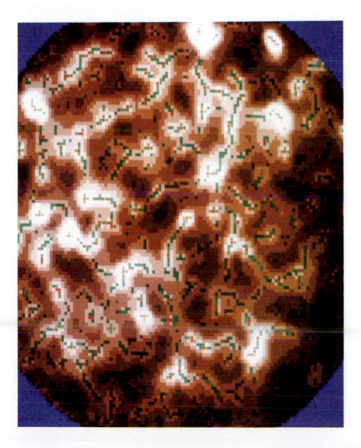

29-8 Kaluza–Klein theories and supergravity predict that the universe may have 11 dimensions

In relativity theory, it is customary to combine time with the three dimensions of ordinary space, resulting in a four-dimensional combination called **space–time**. In 1919 a virtually unknown mathematician, Theodor Franz Éduard Kaluza, proposed that the four dimensions of space–time should be supplemented with a fifth dimension. Kaluza's aim in introducing a fifth dimension was to give a unified geometric description of the only forces of nature known at that time: gravity and electromagnetism. Kaluza had discovered that many of the effects of both these forces could be fully described as the curvature of five-dimensional space–time, just as Einstein had proven that the effects of gravity alone are manifested in the curvature of four-dimensional space–time.

Kaluza's hypothetical fifth dimension exists at every point in ordinary space but, like a very tiny loop, is curled up so tightly that it is not directly observable. A particle always follows the straightest possible path in five dimensions, but

◄ **FIGURE 29-13 The Distribution of 400,000 Galaxies**
This map shows the locations of nearly half a million galaxies spanning 100° of the sky. The north pole of our Galaxy (in the constellation of Coma Berenices) is at the center of the map. Each small square covers 10 × 10 arc min. The color indicates the number of galaxies found in each square. White indicates a high density of galaxies, and darker reddish regions indicate few galaxies. Green squares emphasize filaments along which galaxies are concentrated, and the single red squares (near the centers of the white regions) indicate the points where the galaxy counts reach a maximum. (Courtesy of E. L. Turner, J. E. Moody, and J. R. Gott)

FIGURE 29-14 Hidden Dimensions of Space Hidden dimensions of space might exist provided they are curled up so tightly that we cannot observe them. This drawing shows how an ordinary two-dimensional plane might contain two additional dimensions. At every point on the plane, there is a very tiny sphere so small that it cannot be seen. To pinpoint a particular location, you need to give not only a position on the plane but also a position on the sphere that is tangent to the plane at that point, as indicated by the red lines. (Adapted from D. Freedman and P. Van Nieuwenhuizen)

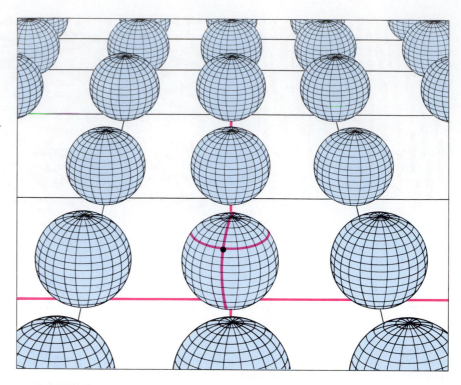

when viewed in the three dimensions of ordinary space, the path appears curved, exactly as if the particle had been deflected by gravitational and electromagnetic forces.

In 1926 the Swedish physicist Oskar Klein expanded upon Kaluza's five-dimensional theory to make it compatible with quantum mechanics. In doing so, Klein discovered that particles could be identified with particular vibrations of the compact loop of the fifth dimension. Today, any quantum-mechanical theory that uses more than four dimensions to provide a unified description of the forces of nature is called a **Kaluza–Klein theory.**

Today we know of four physical forces, so modern Kaluza–Klein theories must have even more than five dimensions. The best agreement between theory and experiment is in fact obtained if space–time has eleven dimensions. In other words, at every point in ordinary space and at every moment of time there must be a very compact seven-dimensional structure far too tiny for us to detect (Figure 29-14). Particles travel along the straightest possible paths in this eleven-dimensional space, but their paths in ordinary three-dimensional space appear curved, as if the particles were being deflected by the four forces.

The choice of eleven dimensions is supported by a super-grand unified theory called **supergravity.** Edward Witten at Princeton has proven that at least seven hidden dimensions must be added to ordinary space–time to account for the four forces. But if more than eleven dimensions are used, the partnership between the various types of particles breaks down. Although eleven dimensions would seem to be the best choice, some theorists have recently argued in favor of as many as twenty-six dimensions.

A surprising prediction of supergravity theories is that every type of ordinary particle has a "supersymmetric partner," meaning that there may be many particles scattered throughout the universe that have never been detected. These hypothetical particles are named in Table 29-2.

Some of these supersymmetric particles might be quite massive. They may be the WIMPs that were proposed to account for the observed deficiency of neutrinos from the Sun. Or they could be the material of the dark matter that pervades the universe and holds clusters of galaxies together. Nevertheless, most scientists are cautiously skeptical about these recent theoretical developments. These speculative multidimensional theories could well have absolutely nothing to do with reality.

TABLE 29-2
Some Particles and Their Supersymmetric Partners

	Ordinary particle	Supersymmetric partner
Basic particles of matter	Electron	Selectron
	Quark	Squark
	Neutrino	Sneutrino
Particles that mediate forces	Graviton	Gravitino
	Photon	Photino
	Gluon	Gluino

It may be that the universe is composed mostly of bizarre, undetected particles whose existence suggests hidden dimensions of space. If this turns out to be the case, the subject of astronomy may soon become more fantastic than the most fanciful science fiction.

We will first understand
How simple the universe is
When we realize
How strange it is.
—*Anonymous*

KEY WORDS

Terms preceded by an asterisk are discussed in the boxes.

annihilation

cold dark matter

confinement

cosmic string

density fluctuation

deuterium bottleneck

false vacuum

flatness problem

gluon

grand unified theory (GUT)

graviton

Heisenberg uncertainty principle

hot dark matter

inflation

inflationary epoch

intermediate-vector boson

isotropy problem

Jeans length

Kaluza–Klein theory

Lamb shift

pair production

principle of totalitarianism

quantum electrodynamics

quark

*rest energy

space–time

spontaneous symmetry breaking

strong nuclear force

supercluster

supergrand unified theory

supergravity

symmetry

symmetry breaking

theory of everything (TOE)

thermal equilibrium

threshold temperature

KEY IDEAS

• A brief period of rapid expansion, called the inflation, explains why the universe is nearly flat and the 3-K microwave background is almost perfectly isotropic.

Inflation occurred immediately after the Big Bang; during this tiny fraction of a second, the universe expanded to a size many times larger than it would have reached through its normal expansion rate.

During the inflationary period, much of the material originally near our location moved far beyond the limits of our observable universe; the observable universe today is therefore expanding into space containing matter and radiation that was originally in close contact with our matter and radiation immediately after the Big Bang.

• Heisenberg's uncertainty principle states that the amount of uncertainty in the momentum of a subatomic particle increases as its position is known more precisely, and vice versa; a similar relationship exists between mass and time.

Because of the uncertainty principle, particle–antiparticle pairs can spontaneously form and disappear within a fraction of a second; these pairs, which can never be detected directly, are called virtual pairs.

A virtual pair can become a real particle–antiparticle pair when photons collide in the process of pair production; the photons disappear, and their energy is replaced by the mass of the particle–antiparticle pair.

A corresponding process of annihilation involves the disappearance of a colliding particle–antiparticle pair and the appearance of photons.

• Just after inflation, the universe was filled with particles and antiparticles formed by pair production and with numerous high-energy photons formed by annihilation; a state of thermal equilibrium existed in this hot plasma.

As the universe expanded, its temperature decreased; as the temperature fell below the threshold temperature required to produce each kind of particle, annihilation dominated production of that kind of particle.

The present domination of matter over antimatter results because particles and antiparticles were not created in exactly equal numbers just after the Planck time; this imbalance is an example of the process called symmetry breaking.

• Helium could not have been produced until the cosmological redshift eliminated most of the high-energy photons. These photons created a deuterium bottleneck by breaking down deuterons before they could combine to form helium.

• Galaxies are generally located on the surfaces of roughly spherical voids; thus, the arrangement of clusters and superclusters of galaxies, which characterizes the large-scale structure of the universe, has a sudsy appearance.

Astronomers use supercomputers to simulate how the large-scale structure of the universe arose from primordial density fluctuations. Neither cold nor hot dark matter models fully account for details of the large-scale structure.

• Four basic forces—gravity, electromagnetism, the strong nuclear force, and the weak nuclear force—explain all of the interactions observed in the universe.

Grand unified theories (GUTs) are attempts to explain two or more forces in terms of a single consistent set of physical laws; a supergrand unified theory would explain all four forces.

GUTs suggest that all four physical forces were equivalent just after the Big Bang; however, because we have no satisfactory supergrand unified theory, we can say nothing about the nature of the universe during this period before the Planck time.

- At the Planck time ($t = 10^{-43}$ s after the Big Bang), gravity froze out to become a distinctive force in a spontaneous symmetry breaking.

During a second spontaneous symmetry breaking, the strong nuclear force became a distinct force.

A final spontaneous symmetry breaking separated the electromagnetic force from the weak nuclear force; from that moment on, the universe behaved as it does today.

- Cosmic strings—long, massive remnants of the early universe—may yet be found to exist.

- Kaluza–Klein theories predict that there may be hidden dimensions of space.

REVIEW QUESTIONS

1. What is the Heisenberg uncertainty principle, and how does it lead to the idea that all space is filled with virtual particle–antiparticle pairs?

2. What is the difference between an electron and an antielectron?

3. In what way was the universe was matter-dominated immediately after the Big Bang? How did this matter-dominated period differ from the matter-dominated universe in which we live today?

4. Where did most of the photons in the 3-K microwave background come from?

5. What is meant by the threshold temperature of a particle?

6. Why is it reasonable to suppose that all space is filled with a neutrino background analogous to the 3-K microwave background?

7. What is the deuterium bottleneck, and why was it important during the formation of heavy nuclei in the early universe?

8. Describe an example of each of the four basic types of interactions in the physical universe. Do you think it possible that a fifth force might be discovered someday? Explain your answer.

9. What is the observational evidence for (**a**) the Big Bang, (**b**) the inflationary epoch, (**c**) the era of recombination, and (**d**) the confinement of quarks?

10. Describe the large-scale structure of the universe as revealed by the distribution of clusters and superclusters of galaxies.

11. Why is it reasonable to say that the inflationary epoch was in fact the phenomenon called the Big Bang?

ADVANCED QUESTIONS

Tips and tools . . .

The mass of a proton is 1.67×10^{-27} kg. The mass of an electron is 9.11×10^{-31} kg. As explained in Chapter 28, the critical density ρ_C is proportional to the square of the Hubble constant H_0 and is equal to about 1.1×10^{-26} kg/m^3 if H_0 is 75 km/s/Mpc. It is also useful to know that 1 m^3 = 10^6 cm^3.

12. An electron has a lifetime of 1.0×10^{-8} s in a given energy state before it makes a transition to a lower state. What is the uncertainty in the energy of the photon emitted in this process?

13. How long can a proton–antiproton pair exist without violating the principle of the conservation of mass?

14. Calculate the threshold temperature for electrons and antielectrons.

15. Using the physical conditions present in the universe during the era of recombination ($T = 3000$ K and $\rho = 10^{-18}$ kg/m^3), verify that the Jeans length for the universe at that time was about 100 ly and that the total mass contained in a sphere with this diameter is about 4×10^5 M$_\odot$.

16. Suppose that the average density of neutrinos throughout space is 100 neutrinos per cubic centimeter and that these neutrinos are responsible for giving the universe a density equal to the critical density. What range of neutrino masses corresponds to the often-quoted range of the Hubble constant, namely 50–100 km/s/Mpc?

DISCUSSION QUESTIONS

17. Some GUTs predict that the proton is unstable. What would it be like to live at a time when protons were decaying in large numbers?

18. Suppose we were now living in a radiation-dominated universe. How would such a universe be different from what we currently observe?

FOR FURTHER READING

Barrow, J., and Silk, J. *The Left Hand of Creation: Origin and Evolution of the Universe.* Basic Books, 1983. This authoritative book tells the exciting story of how space, time, and matter may have been created.

Burns, J. "Very Large Structures in the Universe." *Scientific American,* July 1986. This article describes observations that led to the discovery of enormous superclusters and immense voids in space.

Carringan, R., and Trower, W. *Particle Physics in the Cosmos* and *Particles and Forces: At the Heart of Matter*. W. H. Freeman and Company, 1989. These two superb collections of articles from *Scientific American* cover many exciting discoveries and developments in particle physics and cosmology during the 1980s.

Close, F. *The Cosmic Onion*. American Institute of Physics, 1983. This well-written book uses clever drawings to explain some details of particle physics.

Davies, P. *The Forces of Nature*. 2nd ed. Cambridge University Press, 1986. This is perhaps one of the best introductions to particle physics and grand unified field theory.

———. *Superforce*. Simon & Schuster, 1984. This lucid and entertaining treatise examines the quest for a grand unified theory of physical reality.

Freedman, D., and Van Nieuwenhuizen, P. "The Hidden Dimensions of Spacetime." *Scientific American*, March 1985. This challenging article describes speculative work by theoretical physicists who think there might be more than three dimensions of space.

Gregory, S., and Thompson, L. "Superclusters and Voids in the Distribution of Galaxies." *Scientific American*, March 1982. This article shows how redshift surveys establish the existence of superclusters and voids.

Guth, A., and Steinhardt, P. "The Inflationary Universe." *Scientific American*, May 1984. This interesting article explains why the universe probably experienced a sudden but brief period of vigorous expansion very early in its history.

Haber, H., and Kane, G. "Is Nature Supersymmetric?" *Scientific American*, June 1986. This article explains how theoretical physicists use geometry in their quest for a deeper understanding of physical reality.

Halliwell, J. J. "Quantum Cosmology and the Creation of the Universe." *Scientific American*, December 1991. This article describes recent efforts in applying quantum mechanics to the universe as a whole.

Kaufmann, W., ed. *Particles and Fields*. W. H. Freeman and Company, 1980. This collection of readings on particle physics from *Scientific American* explores the fundamental nature of matter.

Krauss, L. "Dark Matter in the Universe." *Scientific American*, December 1986. This article surveys evidence for dark matter and speculates about its nature.

Peat, F. *Superstrings and the Search for the Theory of Everything*. Contemporary Books, 1988. This interesting book explores the idea that all elementary particles—including protons, neutrons, and electrons—are actually tiny strings vibrating in ten spacial dimensions.

Quigg, C. "Elementary Particles and Forces." *Scientific American*, April 1985. This article surveys the quest for a grand unified field theory from an experimental viewpoint.

Silk, J. *The Big Bang*. W. H. Freeman and Company, 1989. This updated edition of a fascinating book describes many of the processes that made the universe the way it is.

Silk, J., Szalay, A., and Zeldovich, Ya. B. "The Large-Scale Structure of the Universe." *Scientific American*, October 1983. The article shows how the large-scale structure of the universe may have resulted from perturbations in the matter density that emerged from the Big Bang.

THE SEARCH FOR EXTRATERRESTRIAL LIFE

THE PREVALENCE OF LIFE This painting, entitled *DNA Embraces the Planets*, artistically expresses the suspicion of many scientists that carbon-based life may be a common phenomenon in the universe. Other scientists argue, however, that we may be unique and no intelligent alien civilizations exist. In either case, humanity has the clear mandate to preserve and protect the abundance of life forms with which we share our planet. (Courtesy of J. Lomberg)

The heavens inspire us to contemplate profound questions, from the creation of the universe to the nature of the stars and the formation of the Earth. Of all the fascinating subjects we might explore, perhaps none is as compelling as extraterrestrial life. Are we alone? Does life exist elsewhere in the universe? What are the chances that we might someday make contact with an alien civilization?

There are no certain answers to questions about extraterrestrial life. Earth is the only planet on which life is known to exist, and our spacecraft have so far failed to detect any life forms on other worlds. These explorations do not, however, eliminate the possibility of extraterrestrial biology.

As you have seen throughout this book, one of the great lessons of modern astronomy is that our circumstances are quite ordinary. Contrary to the beliefs of our ancestors, we do not occupy a special location, like the "center of the universe." Over the past four centuries it has become clear that we inhabit one of nine planets orbiting an unremarkable star—just one of billions in an undistinguished galaxy. Is it possible that we are also biologically commonplace? The answer to this question is sought by scientists involved in the search for extraterrestrial intelligence, or SETI.

All terrestrial life is based on the unique properties of the carbon atom. Although the possibility cannot be ruled out that other forms of biochemistry exist, scientists confine their search to life as we now know it. Carbon is an extremely versatile element; its atoms can form chemical bonds to create especially long and complex molecules. Among these carbon-based compounds, called *organic molecules*, are the molecules of which living organisms are made.

Organic molecules can be linked together to form elaborate structures, such as chains, lattices, and fibers. Some of these structures are capable of complex, self-regulating chemical reactions. Furthermore, the primary constituents of organic molecules—carbon, hydrogen, nitrogen, oxygen, sulfur, and phosphorus—are among the most abundant elements in the universe. Indeed, the versatility and abundance of carbon suggest that extraterrestrial biology, also called *exobiology*, may also be based on organic chemistry.

Organic molecules are liberally scattered throughout the Galaxy. In interstellar clouds, carbon atoms have combined with other elements to produce an impressive variety of organic compounds. Since the 1960s, radio astronomers have detected telltale microwave emission lines from interstellar clouds that help identify dozens of these carbon-based chemicals (recall Box 20-2). Examples include ethyl alcohol (CH_3CH_2OH), formaldehyde (H_2CO), methyl cyanoacetylene (CH_3C_3N), and acetaldehyde (CH_3CHO), to name a few.

Further evidence of extraterrestrial organic molecules comes from newly fallen meteorites called *carbonaceous chondrites* (Figure A-1), which are often found to contain a variety of organic substances. As noted in Chapter 17, carbonaceous chondrites are ancient meteorites dating from the formation of the solar system. So it seems reasonable to conclude that, even from their earliest days, the planets have been continually bombarded with organic compounds.

Interstellar space is not the only source of organic material. In a classic experiment performed in 1952, American chemists Stanley Miller and Harold Urey demonstrated that simple chemicals can combine to form prebiological compounds under conditions that supposedly prevailed on the primitive Earth. In a closed container, they subjected a mixture of hydrogen, ammonia, methane, and water vapor to an electric arc (to simulate lightning bolts) for a week. At the end of this period, the inside of the container had become coated with a reddish-brown substance rich in compounds essential to life.

Scientists today tend to believe that Earth's primordial atmosphere was probably composed of carbon dioxide, nitrogen, and water vapor outgassed from volcanoes, along with some hydrogen. Modern versions of the Miller–Urey experiment (Figure A-2) using these common gases have also succeeded in producing a wide variety of organic compounds.

FIGURE A-1 A Carbonaceous Chondrite Carbonaceous chondrites are ancient meteorites that date back to the formation of the solar system. Chemical analyses of newly fallen specimens disclose that they are rich in organic molecules, many of which are the chemical building blocks of life. This sample is a piece of the Allende meteorite, a large carbonaceous chondrite that fell in Mexico in 1969. (From the collection of Ronald A. Oriti)

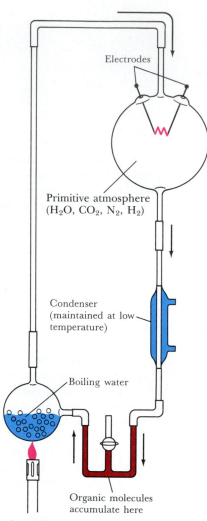

Electrodes

Primitive atmosphere
(H_2O, CO_2, N_2, H_2)

Condenser
(maintained at low
temperature)

Boiling water

Organic molecules
accumulate here

FIGURE A-2 The Miller–Urey Experiment (Updated)
Modern versions of this classic experiment prove that numerous organic compounds important to life can be synthesized from gases that were present in Earth's primordial atmosphere. This experiment supports the hypothesis that life on Earth arose as a result of ordinary chemical reactions.

It is important to emphasize that scientists have not created life in a test tube. Biologists have yet to figure out how these organic molecules gathered themselves into cells and developed systems for self-replication. Nevertheless, since so many chemical components of life are so easily synthesized under conditions that simulate the primordial Earth, it seems reasonable to suppose that life could have originated as the result of chemical processes.

An abundance of organic building blocks does not guarantee that life is commonplace throughout the universe. If a planet's environment is hostile, life may never get started or may quickly become extinct. It seems quite likely, however, that there are planets orbiting other stars. Perhaps conditions on some of these worlds are suitable for life as we know it.

The development of life on Earth seems to suggest that extraterrestrial life, including intelligent species, might evolve on habitable planets, given sufficient time and hospitable conditions. How might we ascertain whether such worlds exist, given the tremendous distances that separate us from them? Many astronomers hope to learn about extraterrestrial civilizations by detecting radio transmissions from them. As we have seen, radio waves can travel immense distances without being significantly degraded by the gas and dust through which they pass. Because of this ability to penetrate the interstellar medium, radio waves are a logical choice for interstellar communication.

Over the past several decades, astronomers have proposed various ways to search for alien radio transmissions, and several limited searches have been undertaken. In 1960 Frank Drake used a radio telescope at the National Radio Astronomy Observatory in West Virginia to listen to two Sunlike stars, τ Ceti and ε Eridani, without success. About 40 similar unsuccessful searches using radio telescopes have taken place since then in both the United States and the Soviet Union. In 1973, for instance, astronomers listened to 600 nearby solar-type stars for half an hour each, but no unusual signals were detected. Since 1983 a radio telescope belonging to Harvard University has been used along with a sophisticated computer program to scan a wide range of frequencies over a large portion of the sky. So far, nothing has turned up.

Should we be discouraged by this lack of success? What are the chances that a radio astronomer might someday detect radio signals from an extraterrestrial civilization? The first person to tackle this issue was Frank Drake, now at the University of California at Santa Cruz. Drake proposed that the number of technologically advanced civilizations in the Galaxy (designated by the letter N) could be estimated by the equation

$$N = R_* f_p n_e f_l f_i f_c L$$

where

R_* = the rate at which solar-type stars form in the Galaxy

f_p = the fraction of stars that have planets

n_e = the number of planets per solar system suitable for life

f_l = the fraction of those habitable planets on which life actually arises

f_i = the fraction of those life forms that evolve into intelligent species

f_c = the fraction of those species that develop adequate technology and then choose to send messages out into space

L = the lifetime of that technologically advanced civilization

The Drake equation is enlightening because it expresses the number of extraterrestrial civilizations in a series of terms, some of which can be estimated from what we know about stars and stellar evolution. For instance, the first two factors, R_* and f_p, can be determined by observation. In estimating R_*, we should probably exclude massive stars (those larger than about 1.5 $M_\odot$), because they have main sequence lifetimes shorter than the time it took to develop intelligent life here on Earth. Life on Earth originated some 3.5 to 4.0 billion years ago. If that is typical of the time needed to evolve higher life forms, then a massive star probably becomes a red giant or a supernova before intelligent creatures appear on any of its planets.

Although low-mass stars have much longer lifetimes, they, too, seem unsuited for life because they are so cool. Only planets very near a low-mass star would be sufficiently warm for life as we know it, and a planet that close can become tidally coupled to the star, with one side continually facing the star, while the other is in perpetual, frigid darkness. This leaves us with main sequence stars like the Sun, those with spectral types between F5 and M0. Based on statistical studies of star formation in the Milky Way, some astronomers estimate that roughly one of these Sunlike stars forms in the Galaxy each year, thus setting R_* at 1 per year.

We learned in Chapter 7 that the planets in our solar system formed as a natural consequence of the birth of the Sun, and we have seen evidence suggesting that similar processes of planetary formation may be commonplace around single stars. Yet so far no planet outside our solar system has been observed directly. Nevertheless, many astronomers believe that most Sunlike stars probably do have planets, and so they give f_p a value of 1.

Unfortunately, the rest of the terms in the Drake equation are very uncertain. Let's play with some hypothetical values. The chances that a planetary system has an Earthlike world suitable for life are not known. Were we to consider our own solar system as representative, we could put n_e at 1. Let's be more conservative, however, and suppose that one in ten solar-type stars is orbited by a habitable planet, making $n_e = 0.1$. From what we know about the evolution of life on Earth, we might assume that, given appropriate conditions,

the development of life is a certainty, which would make $f_l = 1$. This is an area of intense interest to biologists.

For the sake of argument, we might also assume that evolution might naturally lead to the development of intelligence (a conjecture that is hotly debated) and also make $f_i = 1$. It's anyone's guess as to whether these intelligent extraterrestrial beings would attempt communication with other civilizations in the Galaxy, but were we to assume they would, f_c would be put at 1 also.

The last variable, L, involving the longevity of civilization, is the most uncertain of all, and certainly cannot be subjected to testing! Looking at our own example, we see a planet whose atmosphere and oceans are increasingly polluted by creatures that possess nuclear weapons. If we are typical, perhaps L is as short as 100 years. Putting all these numbers together, we arrive at

$$N = \frac{1}{yr} \times 1 \times 0.1 \times 1 \times 1 \times 1 \times 100 \text{ yr} = 10$$

In other words, out of the hundreds of billions of stars in the Galaxy, we would estimate that there are only ten technologically advanced civilizations from which we might receive communications.

A wide range of values have been proposed for the terms in the Drake equation, and these various guesses produce vastly different estimates of N. Some scientists argue that there is exactly one advanced civilization in the Galaxy and that we are it. Others speculate that there may be hundreds or thousands of planets inhabited by intelligent creatures.

If extraterrestrial beings were purposefully sending messages into space, it seems reasonable that they might choose a frequency that is fairly free of interference from extraneous sources. SETI pioneer Bernard Oliver has drawn attention to a range of relatively noise-free frequencies in the neighborhood of the microwave emission lines of hydrogen (H) and hydroxide (OH) (Figure A-3). This region of the microwave

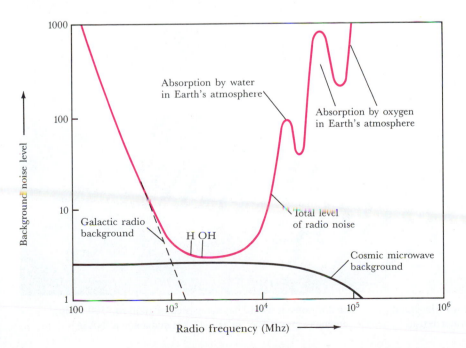

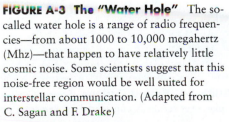

FIGURE A-3 The "Water Hole" The so-called water hole is a range of radio frequencies—from about 1000 to 10,000 megahertz (Mhz)—that happen to have relatively little cosmic noise. Some scientists suggest that this noise-free region would be well suited for interstellar communication. (Adapted from C. Sagan and F. Drake)

spectrum is called the "water hole," a humorous reference to the H and OH lines being so close together. Or perhaps such beings would choose to transmit at a wavelength of 21 cm, because astronomers studying the distribution of hydrogen around the Galaxy would already have their radio telescopes tuned to that wavelength (recall Figure 25-10).

Even if there are only a few alien civilizations scattered across the Galaxy, we have the technology to detect radio transmissions from them. The most ambitious plan ever proposed is Project Cyclops, which would consist of from 1000 to 2500 radio antennas, each 100 ft in diameter (Figure A-4). This colossal array would be so sensitive that it could detect signals from virtually anywhere in the Galaxy. Unfortunately, it seems unlikely that such a project will ever be funded.

On a more modest scale, NASA has funded a SETI program called the Microwave Observing Project (MOP), which began in October 1992. NASA scientists are using radio telescopes along with sophisticated electronic equipment and powerful computers to conduct both a targeted search and an all-sky survey.

The targeted search will examine some 800 nearby solar-type stars over a frequency range that covers the long-wavelength side of the water hole. The largest radio telescopes are being used in order to achieve the highest possible sensitivity (Figure A-5). Specialized equipment for processing digital signals enables NASA scientists to study tens of millions of individual frequency channels simultaneously. This equipment can automatically detect continuous waves or pulses, whether the frequency remains constant or drifts slowly because of relative motion between the transmitter and the receiver.

The all-sky survey uses several dish-shaped antennas of NASA's Deep Space Network to extend the search over the entire sky. Some sensitivity is sacrificed, but the survey does cover the entire water hole. Digital signal-processing equipment automatically examines ten million frequency channels in an effort to detect signals from fixed positions in the sky. New technology developed for the Microwave Observing Project permits recognition of a wide variety of signals during the observations. The project is almost completely automated, and incoming data can be sifted at a rate far beyond human capabilities.

In addition to NASA's efforts, three teams of scientists are actively involved in SETI programs of their own. At Ohio State University, a huge radio telescope has been upgraded with a new receiver from Stanford University capable of monitoring ten million separate microwave frequencies in an all-sky search. In 1992 scientists from the University of California at Berkeley attached a new four-million channel receiver, dubbed SERENDIP III, to the 1000-ft Arecibo antenna in Puerto Rico. (The acronym stands for search for extraterrestrial radio emissions from nearby, developed, intelligent populations.) During the first six months of operation, a fifth of the sky visible to Arecibo's fixed antenna was examined over a frequency range of 425 to 435 Mhz. In 1994 the Berkeley team will install a receiver capable of monitoring 106 million channels, at which time the SERENDIP III hardware will be shipped to Russia.

A SETI program that has used a Harvard radio telescope since the 1980s was recently revitalized and now involves twin 8.4-million channel receivers, one in rural Massachusetts and the other nerar Buenos Aires in Argentina. This

FIGURE A-4 Project Cyclops Project Cyclops was the grandest plan ever seriously considered for detecting extraterrestrial civilizations. It would have consisted of 1000 to 2500 radio antennas spread over an area 16 km in diameter. (NASA)

FIGURE A-5 A Deep-Space Tracking Antenna This dish-shaped antenna, located in the Mojave Desert in California, was originally built by NASA to track interplanetary spacecraft. It is now being used along with an antenna at the Arecibo Observatory in Puerto Rico to search for extraterrestrial intelligence. In 1996 an antenna in Canberra, Australia, will join the network. (Jet Propulsion Laboratory)

expanded effort, called Project META (for *m*egachannel *ex*tra*terrestrial *a*ssay), is sponsored by the Planetary Society. Like the Ohio State and Berkeley projects, META is conducting an all-sky search.

SETI researchers are deeply concerned that increasing pollution of the radio spectrum by Earth-based radio transmissions will soon severely hamper their efforts. These scientists have therefore proposed building a SETI antenna on the far side of the Moon or placing a shielded radio dish in Earth orbit (Figure A-6). With a budgetary Sword of Damocles hanging over all NASA projects, however, it may take an actual signal detection to inspire the necessary funding.

Detecting a message from an alien civilization would be one of the greatest events in human history. Such a message could dramatically change the course of civilization through the sharing of scientific information or an awakening of social or humanistic enlightenment. In only a few years our technology, industry, and social structure might advance the equivalent of centuries into the future. Such changes would touch every person on Earth. Mindful of these profound implications, scientists push ahead with the search for extraterrestrial communication.

FIGURE A-6 A SETI Antenna in Space This artist's conception shows a 300-m radio dish in orbit. The dish can be very thin because it does not need to be supported. Two satellites at its focus relay transmission back to Earth. A large disk-shaped shield protects the antenna from human radio transmissions. (NASA)

FOR FURTHER READING

Baugher, J. *On Civilized Stars: The Search for Intelligent Life in Outer Space.* Prentice-Hall, 1985. This nontechnical book, written by a physicist, speculates on extraterrestrial intelligence and the possibility of our eventually communicating with it.

Beatty, J. K. "The New, Improved SETI." *Sky & Telescope,* May 1983. This article discusses SETI research that emerged in the early 1980s.

Davies, P. *The Cosmic Blueprint.* Simon & Schuster, 1988. This extraordinary book examines connections between physics and biology in order to understand the process of life more fully.

Dawkins, R. *The Blind Watchmaker.* Norton, 1987. This well-written book surveys our modern understanding of evolution.

Feinberg, G., and Shapiro, R. *Life beyond Earth: The Intelligent Earthling's Guide to Life in the Universe.* This excellent introduction to exobiology includes some fascinating speculation about possible forms of life.

Goldsmith, D., ed. *The Quest for Extraterrestrial Life: A Book of Readings.* University Science Books, 1980. This fascinating compilation includes articles by Carl Sagan, Frank Drake, Fred Hoyle, and Bernard Oliver, to name just a few.

Goldsmith, D., and Owen, T. *The Search for Life in the Universe.* Benjamin/Cummings, 1980. This book sets the stage with long sections on astronomy, biology, and planetary evolution before dealing with the search for extraterrestrial life.

Gould, S. *Wonderful Life.* Scribner, 1990. This fascinating book explains how a recent reinterpretation of fossils from the "Cambrian explosion"—a sudden proliferation of life forms that occurred about 600 million years ago—is changing our ideas about evolution.

Hart, M., and Zuckerman, B., eds. *Extraterrestrials—Where Are They?* Pergamon Press, 1982. This thought-provoking volume examines the implications of our failure to observe extraterrestrials.

Horowitz, N. *To Utopia and Back: The Search for Life in the Solar System.* W. H. Freeman and Company, 1986. This concise, nontechnical introduction to the processes and origin of life pays particular attention to the results from the Viking landers on Mars.

Kutter, G. S. *The Universe and Life.* Jones and Bartlett, 1987. The first half of this well-written textbook covers astronomy while the second half is devoted to biology and evolution.

Marx, G., ed. *Bioastronomy—The Next Steps.* Kluwer, 1988. Although this book consists of the proceedings of a technical conference on SETI, many of its papers can easily be understood by someone who has read this textbook.

Naeye, R. "SETI at the Crossroads." *Sky & Telescope,* November 1992. This article eloquently describes ongoing SETI programs.

Papagiannis, M. D. "Bioastronomy: The Search for Extraterrestrial Life." *Sky & Telescope,* June 1984. This article eloquently but briefly examines various issues involved in SETI research.

Regis, E. *Extraterrestrials: Science and Alien Intelligence.* Cambridge University Press, 1985. This engrossing collection of essays explores philosophical and moral issues surrounding the search for alien intelligence.

Rood, R. and Trefil, J. *Are We Alone?* Scribner, 1981. This book takes an informal, friendly, occasionally skeptical look at modern ideas about extraterrestrial life.

Sagan, C., and Drake, F. "The Search for Extraterrestrial Intelligence." *Scientific American,* May 1975. This classic article explores the possibility and methodology of communicating with extraterrestrials.

Schorn, R. A. "Extraterrestrial Beings Don't Exist." *Sky & Telescope,* September 1981. This one-page article looks at the possibility that intelligent life might be extremely rare.

Shklovskii, I. S., and Sagan, C. *Intelligent Life in the Universe,* Holden-Day, 1966. This somewhat dated but nevertheless classic book was one of the first to examine the question of extraterrestrial life in light of our modern understanding of astronomy and biology.

Wilson, A. C. "The Molecular Basis of Evolution." *Scientific American,* October 1985. This interesting article surveys recent developments in our understanding of evolution at the molecular level.

APPENDIXES

EARTH
MARS
MERCURY
MOON
IO
EUROPA
GANYMEDE
CALLISTO
VENUS
TITAN
TRITON

The Terrestrial Worlds This montage of photographs taken by various spacecraft shows the terrestrial planets and the seven largest moons of the solar system at the same scale. (Prepared for NASA by S. P. Meszaros)

1 The Planets: Orbital Data

Planet	Semimajor axis (AU)	Semimajor axis (10⁶ km)	Sidereal period (yr)	Sidereal period (d)	Synodic period (d)	Mean orbital speed (km/s)	Orbital eccentricity	Inclination of orbit to ecliptic (°)
Mercury	0.3871	57.9	0.2408	87.97	115.88	47.9	0.206	7.00
Venus	0.7233	108.2	0.6152	224.70	583.92	35.0	0.007	3.39
Earth	1.0000	149.6	1.0000	365.26	——	29.8	0.017	0.00
Mars	1.5237	227.9	1.8809	686.98	779.94	24.1	0.093	1.85
(Ceres)	2.7656	413.7	4.603		466.6	17.9	0.097	10.61
Jupiter	5.2028	778.3	11.862		398.9	13.1	0.048	1.31
Saturn	9.5388	1427.0	29.458		378.1	9.6	0.056	2.49
Uranus	19.1914	2871.0	84.01		369.7	6.8	0.046	0.77
Neptune	30.0611	4497.1	164.79		367.5	5.4	0.010	1.77
Pluto	39.5294	5913.5	248.54		366.7	4.7	0.248	17.15

2 The Planets: Physical Data

Planet	Equatorial diameter (km)	Equatorial diameter (Earth = 1)	Mass (Earth = 1)	Mean density (kg/m³)	Rotation period* (d)	Inclination of equator to orbit (°)	Surface gravity (Earth = 1)	Albedo	Brightest visual magnitude	Escape velocity (km/s)
Mercury	4,878	0.38	0.055	5430	58.65	2 (?)	0.39	0.106	−1.9	4.3
Venus	12,102	0.95	0.815	5250	−243.01	177.3	0.88	0.65	−4.4	10.4
Earth	12,756	1.00	1.000	5520	0.997	23.4	1.00	0.37	——	11.2
Mars	6,786	0.53	0.107	3950	1.026	25.2	0.38	0.15	−2.0	5.0
Jupiter	142,984	11.21	317.94	1330	0.410	3.1	2.34	0.52	−2.7	59.6
Saturn	120,536	9.45	95.18	690	0.426	26.7	0.93	0.47	+0.7	35.5
Uranus	51,118	4.01	14.53	1290	−0.746	97.9	0.79	0.50	+5.5	21.3
Neptune	49,528	3.88	17.14	1640	0.800	29.6	1.12	0.5	+7.8	23.3
Pluto	2,300	0.18	0.002	2030	−6.387	122.5	0.04	0.5	+15.1	1.1

*Negative values indicate retrograde rotation.

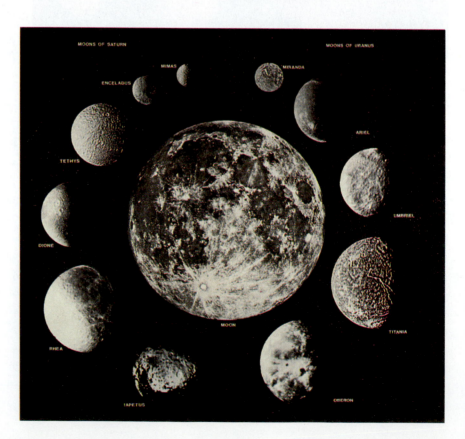

Moons of Saturn and Uranus This montage of photographs taken by the Voyager spacecraft shows the medium-sized satellites of Saturn and Uranus along with our Moon, all at the same scale. (Prepared for NASA by S. P. Meszaros)

3 Satellites of the Planets

Planet	Satellite	Discoverer(s)	Mean distance from planet (km)	Sidereal period (d)	Orbital eccentricity	Diameter of satellite* (km)	Approximate magnitude at opposition
Earth	Moon	——	384,400	27.322	0.05	3476	−13
Mars	Phobos	Hall (1877)	9,380	0.319	0.01	$28 \times 23 \times 20$	+11
	Deimos	Hall (1877)	23,460	1.263	0.00	$16 \times 12 \times 10$	12
Jupiter	Metis	Synnott (1979)	127,960	0.295	0.00	(40)	+18
	Adrastea	Jewitt et al. (1979)	128,980	0.298	0 (?)	$24 \times 16 \times 20$	19
	Amalthea	Barnard (1892)	181,300	0.498	0.00	$270 \times 200 \times 155$	14
	Thebe	Synnott (1979)	221,900	0.675	0.01	(100)	16
	Io	Galileo (1610)	421,600	1.769	0.00	3630	5
	Europa	Galileo (1610)	670,900	3.551	0.01	3138	5
	Ganymede	Galileo (1610)	1,070,000	7.155	0.00	5262	5
	Callisto	Galileo (1610)	1,883,000	16.689	0.01	4800	6
	Leda	Kowal (1974)	11,094,000	238.72	0.15	(16)	20
	Himalia	Perrine (1904)	11,480,000	250.57	0.16	(180)	15
	Lysithea	Nicholson (1938)	11,720,000	259.22	0.11	(40)	18
	Elara	Perrine (1905)	11,737,000	259.65	0.21	(80)	17
	Ananke	Nicholson (1951)	21,200,000	631	0.17	(30)	19
	Carme	Nicholson (1938)	22,600,000	692	0.21	(44)	18
	Pasiphae	Melotte (1908)	23,500,000	735	0.38	(70)	17
	Sinope	Nicholson (1914)	23,700,000	758	0.28	(40)	18
Saturn	Pan	Showalter (1990)	133,570	0.573	0.00	20	+19
	Atlas	Terrile (1980)	137,640	0.602	0 (?)	$40 \times 30 \times 30$	18
	Prometheus	Collins et al. (1980)	139,350	0.613	0.00	$140 \times 80 \times 100$	16
	Pandora	Collins et al. (1980)	141,700	0.629	0.00	$110 \times 70 \times 100$	16
	Epimethius	Walker (1966)	151,422	0.694	0.01	$140 \times 100 \times 100$	16
	Janus	Dolfus (1966)	151,472	0.695	0.01	$220 \times 160 \times 200$	14
	Mimas	Herschel (1789)	185,520	0.942	0.02	390	13
	Enceladus	Herschel (1789)	238,020	1.370	0.00	500	12
	Tethys	Cassini (1684)	294,660	1.888	0.00	1050	10
	Telesto	Smith et al. (1980)	294,660	1.888	0 (?)	(24)	19
	Calypso	Smith et al. (1980)	294,660	1.888	0 (?)	$30 \times 20 \times 25$	19
	Dione	Cassini (1684)	377,400	2.737	0.00	1120	10
	Helene	Laques et al. (1980)	377,400	2.737	0.01	$40 \times 30 \times 30$	18
	Rhea	Cassini (1672)	527,040	4.518	0.00	1530	10
	Titan	Huygens (1655)	1,221,850	15.945	0.03	5150	8
	Hyperion	Bond (1848)	1,481,000	21.277	0.10	$410 \times 260 \times 220$	14
	Iapetus	Cassini (1671)	3,561,000	79.331	0.03	1440	11
	Phoebe	Pickering (1898)	12,952,000	550.48	0.16	220	16

(continued)

─────────────────────── **3 Satellites of the Planets** *(continued)* ───────────────────────

Planet	Satellite	Discoverer(s)	Mean distance from planet (km)	Sidereal period (d)	Orbital eccentricity	Diameter of satellite* (km)	Approximate magnitude at opposition
Uranus	Cordelia	*Voyager 2* (1986)	49,750	0.335	0 (?)	(30)	+24
	Ophelia	*Voyager 2* (1986)	53,760	0.376	0 (?)	(30)	24
	Bianca	*Voyager 2* (1986)	59,160	0.435	0 (?)	(40)	23
	Cressida	*Voyager 2* (1986)	61,770	0.464	0 (?)	(70)	22
	Desdemona	*Voyager 2* (1986)	62,660	0.474	0 (?)	(60)	22
	Juliet	*Voyager 2* (1986)	64,360	0.493	0 (?)	(80)	22
	Portia	*Voyager 2* (1986)	66,100	0.513	0 (?)	(110)	21
	Rosalind	*Voyager 2* (1986)	69,930	0.558	0 (?)	(60)	22
	Belinda	*Voyager 2* (1986)	75,260	0.624	0 (?)	(70)	22
	Puck	*Voyager 2* (1986)	86,010	0.762	0 (?)	150	20
	Miranda	Kuiper (1948)	129,780	1.414	0.00	470	16
	Ariel	Lassell (1851)	191,240	2.520	0.00	1160	14
	Umbriel	Lassell (1851)	265,970	4.144	0.00	1170	15
	Titania	Herschel (1787)	435,840	8.706	0.00	1580	14
	Oberon	Herschel (1787)	582,600	13.463	0.00	1520	14
Neptune	Naiad	*Voyager 2* (1989)	48,230	0.296	0 (?)	(50)	+25
	Thalassa	*Voyager 2* (1989)	50,070	0.312	0 (?)	60	24
	Despoina	*Voyager 2* (1989)	52,530	0.333	0 (?)	80	23
	Galatea	*Voyager 2* (1989)	61,950	0.429	0 (?)	180	23
	Larissa	*Voyager 2* (1989)	73,550	0.554	0 (?)	150	21
	Proteus	*Voyager 2* (1989)	117,640	1.121	0 (?)	415	20
	Triton	Lassell (1846)	354,800	5.877	0.00	2700	14
	Nereid	Kuiper (1949)	5,513,400	359.16	0.75	(340)	19
Pluto	Charon	Christy (1978)	19,640	6.387	0.00	1190	+17

*A diameter given in parentheses is estimated from the amount of sunlight it reflects.

4 The Nearest Stars

Name	Parallax (arc sec)	Distance (ly)	Spectral type*	Radial velocity (km/s)	Proper motion (arc sec/yr)	Apparent visual magnitude	Luminosity (Sun = 1.0)
Sun			G2 V			−26.7	1.0
Proxima Centauri	0.772	4.2	M5e	−16	3.85	11.05	0.00006
α Centauri A	0.750	4.3	G2 V	−22	3.68	−0.01	1.6
α Centauri B			K0 V			1.33	0.45
Barnard's star	0.545	5.9	M5 V	−108	10.31	9.54	0.00045
Wolf 359	0.421	7.6	M8e	+13	4.70	13.53	0.00002
BD +36°2147	0.397	8.1	M2 V	−84	4.78	7.50	0.0055
Luyten 726-8A	0.387	8.4	M6e	+29	3.36	12.52	0.00006
Luyten 726-8B (UV Ceti)			M6e	+32		13.02	0.00004
Sirius A	0.377	8.6	A1 V	−8	1.33	−1.46	23.5
Sirius B			wd			8.3	0.003
Ross 154	0.345	9.4	M5e	−4	0.72	10.45	0.00048
Ross 248	0.314	10.3	M6e	−81	1.60	12.29	0.00011
ε Eridani	0.303	10.7	K2 V	+16	0.98	3.73	0.30
Ross 128	0.298	10.8	M5	−13	1.38	11.10	0.00036
61 Cygni A	0.294	11.2	K5 V	−64	5.22	5.22	0.082
61 Cygni B			K7 V			6.03	0.039
ε Indi	0.291	11.2	K5 V	−40	4.70	4.68	0.14
BD +43°44A	0.290	11.2	M1 V	+13	2.90	8.08	0.0061
BD +43°44B			M6 V	+20		11.06	0.00039
Luyten 789-6	0.290	11.2	M7e	−60	3.26	12.18	0.00014
Procyon A	0.285	11.4	F5 IV–V	−3	1.25	0.37	7.65
Procyon B			wd			10.7	0.00055
BD +59°1915A	0.282	11.5	M4	0	2.29	8.90	0.0030
BD +59°1915B			M5	+10	2.27	9.69	0.0015
CD −36°1668	0.279	11.7	M2 V	+10	6.90	7.35	0.0013
G51-15	0.278	11.7			1.27	14.81	0.00001
τ Ceti	0.277	11.9	G8 V	−16	1.92	3.50	0.45
BD +5°1668	0.266	12.2	M5	+26	3.77	9.82	0.0015
Luyten 725-32 (YZ Ceti)	0.261	12.4	M5e	+28	1.32	12.04	0.0002
CD −39°14192	0.260	12.5	M0 V	+21	3.46	6.66	0.028
Kapteyn's star	0.256	12.7	M0 V	+245	8.72	8.84	0.0039
Kruger 60 A	0.253	12.8	M3	−26	0.86	9.85	0.0016
Kruger 60 B			M5e			11.3	0.0004

Note: In the case of double stars, like Sirius A and B or Procyon A and B, both members have the same parallax, distance, etc. These redundant values are not repeated for the second (dimmer) member.

* "wd" stands for white dwarf; "e" means that the star's spectrum contains emission lines.

5 The Visually Brightest Stars

Star	Name	Apparent visual magnitude	Spectral type	Absolute magnitude	Distance (ly)	Radial velocity (km/s)	Proper motion (arc sec/yr)
α CMa A	Sirius	−1.46	A1 V	+1.4	9	−8	1.324
α Car	Canopus	−0.72	F0 I	−8.5	1200	+21	0.025
α Boo	Arcturus	−0.04	K2 III	−0.2	36	−5	2.284
α Cen A	Rigel Kents	0.00	G2 V	+4.4	4	−25	3.676
α Lyr	Vega	0.03	A0 V	+0.5	26	−14	0.345
α Aur	Capella	0.08	G8 III	−0.5	42	+30	0.435
β Ori A	Rigel	0.12	B8 Ia	−7.1	910	+21	0.001
α CMi A	Procyon	0.38	F5 IV	+2.6	11	−3	1.250
α Eri	Achernar	0.46	B3 IV	−1.6	85	+19	0.098
α Ori	Betelgeuse	0.50	M2 Iab	−5.6	310	+21	0.028
β Cen	Hadar	0.61	B1 II	−5.1	460	−12	0.035
α Aql	Altair	0.77	A7 IV–V	+2.2	17	−26	0.658
α Tau A	Aldebaran	0.85	K5 III	−0.3	68	+54	0.202
α Sco A	Antares	0.96	M1 Ib	−4.7	330	−3	0.029
α Vir	Spica	0.98	B1 V	−3.5	260	+1	0.054
β Gem	Pollux	1.14	K0 III	+0.2	36	+3	0.625
α PsA	Fomalhaut	1.16	A3 V	+2.0	22	+7	0.367
α Cyg	Deneb	1.25	A2 Ia	−7.5	1800	−5	0.003
β Cru	Mimosa	1.25	B0.5 III	−5.0	420	+20	0.049
α Leo A	Regulus	1.35	B7 V	−0.6	85	+4	0.248

Note: Acrux, the brightest star in Crux (the Southern Cross), appears to the naked eye
as a star of magnitude +0.87, which suggests that it should be in this table. A small telescope,
however, reveals that Acrux is actually a double star whose blue-white components have
visual magnitudes of 1.4 and 1.9, and so they are dimmer than any of the stars listed here.

6 Some Important Astronomical Quantities

Astronomical unit:	$1 \text{ AU} = 1.496 \times 10^{11} \text{ m}$
Parsec:	$\text{pc} = 3.086 \times 10^{16} \text{ m} = 3.262 \text{ ly}$
Light year:	$1 \text{ ly} = 9.460 \times 10^{15} \text{ m} = 63{,}240 \text{ AU}$
Solar mass:	$1 \text{ M}_\odot = 1.989 \times 10^{30} \text{ kg}$
Solar radius:	$1 \text{ R}_\odot = 6.960 \times 10^8 \text{ m}$
Solar luminosity:	$1 \text{ L}_\odot = 3.90 \times 10^{26} \text{ W}$

7 Some Important Physical Constants

Speed of light:	$c = 2.998 \times 10^8 \text{ m/s}$
Gravitational constant:	$G = 6.668 \times 10^{-11} \text{ N m}^2 \text{ kg}^{-2}$
Planck constant:	$h = 6.625 \times 10^{-34} \text{ J s}$
	$= 4.136 \times 10^{-15} \text{ eV s}$
Boltzmann constant:	$k = 1.380 \times 10^{-23} \text{ J K}^{-1}$
	$= 8.617 \times 10^{-5} \text{ eV K}^{-1}$
Stefan–Boltzmann constant:	$\sigma = 5.669 \times 10^{-8} \text{ W m}^{-2} \text{ K}^{-4}$
Mass of electron:	$m_e = 9.108 \times 10^{-31} \text{ kg}$
Mass of ^{1}H atom:	$m_H = 1.673 \times 10^{-27} \text{ kg}$

GLOSSARY

A ring One of three prominent rings encircling Saturn. (pp. 256–257)

absolute bolometric magnitude The absolute magnitude of a star measured above the Earth's atmosphere over all wavelengths. (pp. 346–347)

absolute magnitude The apparent magnitude that a star would have if it were at a distance of 10 parsecs. (pp. 341–343)

absolute zero A temperature of $-273°$C (or 0 K) where all molecular motion stops; the lowest possible temperature. (p. 84)

absorption line spectrum Dark lines superimposed on a continuous spectrum. (p. 91)

acceleration A change in velocity. (p. 66)

accretion The gradual accumulation of matter in one location, typically due to the action of gravity. (p. 138)

accretion disk A disk of gas orbiting a star or black hole. (p. 395)

accretion hypothesis The theory that planetary atmospheres came from ices and gas trapped in planetesimals.

active galactic nucleus (AGN) The center of an active galaxy. (p. 515)

active galaxy A galaxy that is emitting exceptionally large amounts of energy; a Seyfert galaxy or a quasar. (p. 515)

active Sun The Sun during times of frequent solar activity such as sunspots, flares, and associated phenomena.

adaptive optics A technique for improving a telescopic image by altering the telescope's optics in a way that compensates for distortion caused by the Earth's atmosphere. (p. 110)

aerosol Tiny droplets of liquid dispersed in a gas. (p. 265)

AGB star An asymptotic giant branch star. (pp. 402–403)

albedo The fraction of sunlight that a planet, asteroid, or satellite reflects. (p. 180)

alpha decay The decay of a radioactive isotope by the emission of alpha particles.

alpha particle The nucleus of a helium atom, consisting of two protons and two neutrons. (pp. 92, 388)

amino acids The chemical building blocks of proteins. (p. 299)

Amor asteroid An asteroid whose orbit brings it closer to the Sun than the distance of Mars's orbit. (p. 294)

angle The opening between two lines that meet at a point. (p. 7)

angstrom (Å) A unit of length equal to 10^{-10} meter.

angular diameter The angle subtended by the diameter of an object. (p. 7)

angular measure The size of an angle, usually expressed in degrees, minutes of arc, and seconds of arc. (p. 7)

angular momentum A measure of the momentum associated with rotation. (pp. 71, 137, 421–422, 424–425)

angular resolution The angular size of the smallest feature that can be distinguished with a telescope. (p. 108)

angular size The angle subtended by an object. (p. 7)

angular velocity The speed with which an object revolves about an axis. (p. 424)

anisotropic Possessing unequal properties in different directions; not isotropic. (p. 480)

annihilation The process by which the masses of a particle and antiparticle are converted into energy. (pp. 557–558)

annular eclipse An eclipse of the Sun in which the Moon is too distant to cover the Sun completely, so that a ring of sunlight is seen around the Moon at mid-eclipse. (pp. 45–46)

anorthosite Rock commonly found in ancient, cratered highlands on the Moon. (pp. 170–171)

Antarctic Circle A circle of latitude $23\frac{1}{2}°$ north of the Earth's South Pole. (p. 27)

anticyclone The circulation of winds around a high-pressure system in a planet's atmosphere. (p. 235)

anticyclonic flow Wind-flow pattern around an anticyclone. (p. 279)

antielectron A positron. (p. 557)

antimatter Matter consisting of antiparticles such as antiprotons, antielectrons (positrons), and antineutrons. (p. 557)

aperture The diameter of an opening; the diameter of the primary lens or mirror of a telescope.

aphelion The point in its orbit where a planet is farthest from the Sun. (p. 62)

apogee The point in its orbit where a satellite or the Moon is farthest from the Earth. (p. 45)

Apollo asteroid An asteroid whose orbit brings it closer to the Sun than the distance of the Earth's orbit. (p. 294)

apparent brightness The flux of a star's light arriving at the Earth. (p. 344)

apparent magnitude A measure of the brightness of light from a star or other object as measured from Earth. (pp. 47, 341–343)

apparent solar day The interval between two successive transits of the Sun's center across the local meridian. (p. 29)

apparent solar time Time reckoned by the position of the Sun in the sky. (p. 29)

Arctic Circle A circle of latitude $23\frac{1}{2}°$ south of the Earth's North Pole. (p. 27)

ascending node The point along an orbit where an object crosses a reference plane (usually the ecliptic or the celestial equator) from south to north.

association A loose cluster of stars whose spectra, motions, or positions suggest that they had a common origin. (p. 374)

asteroid One of tens of thousands of small, rocky, planetlike objects in orbit about the Sun. (pp. 132, 291)

asteroid belt A region between the orbits of Mars and Jupiter that encompasses the orbits of many asteroids. (pp. 132, 292)

asthenosphere A warm, plastic layer of the mantle beneath the lithosphere of the Earth. (p. 152)

astigmatism An optical defect of a lens or mirror whereby light rays in different planes do not focus at the same point.

astrometry The branch of astronomy dealing with the precise determination of the positions and motions of celestial objects.

Astronomical Almanac A book published annually that gives information about a wide variety of astronomical phenomena including eclipses, phases of the moon, positions of the planets, and so forth. (pp. 31, 46)

astronomical unit (**AU**) The semimajor axis of the Earth's orbit; the average distance between the Earth and the Sun. (p. 10)

astronomy The branch of science dealing with objects and phenomena that lie beyond the Earth's atmosphere.

astrophysics That part of astronomy dealing with the physics of astronomical objects and phenomena.

asymptotic giant branch star A red supergiant star that has ascended the giant branch for the second and final time; an AGB star. (pp. 402–403)

atmosphere (**atm**) A unit of atmospheric pressure. (pp. 153, 155)

atmospheric pressure The force per unit area exerted by a planet's atmosphere. (pp. 153, 155)

atom The smallest particle of an element that has the properties characterizing that element. (pp. 91, 131)

atomic mass unit (**amu**) A unit of mass (1.67×10^{-27} kg) nearly equal to the mass of a hydrogen atom.

atomic number The number of protons in the nucleus of an atom of a particular element. (p. 129)

atomic weight The mass of an atom in atomic mass units.

aurora Light radiated by atoms and ions in the Earth's upper atmosphere, mostly in the polar regions. (p. 156)

aurora australis Aurorae seen from southern latitudes; the southern lights. (p. 156)

aurora borealis Aurorae seen from northern latitudes; the northern lights. (p. 156)

autumnal equinox The intersection of the ecliptic and the celestial equator where the Sun crosses the equator from north to south. (p. 24)

average density The mass of an object divided by its volume. (p. 127)

B ring One of three prominent rings encircling Saturn. (pp. 256–257)

Balmer line An emission or absorption line in the spectrum of hydrogen caused by an electron transition between the second and higher energy levels. (p. 93)

Balmer series The entire pattern of Balmer lines. (pp. 93–95)

bar A unit of pressure. (p. 155)

Barnard object A dark nebula discovered by E. E. Barnard. (p. 368)

barred spiral galaxy A spiral galaxy in which the spiral arms begin from the ends of a "bar" running through the nucleus rather than from the nucleus itself. (pp. 478–479)

basin A large, shallow, circular depression on the surface of the Moon or a terrestrial planet.

belt A dark band in Jupiter's atmosphere. (p. 230)

belt asteroid An asteroid whose orbit lies in the asteroid belt, between the orbits of Mars and Jupiter. (p. 292)

Big Bang An explosion of all space, roughly 20 billion years ago, from which the universe emerged. (pp. 6, 529ff)

Big Crunch The fate of the universe if it is bounded and ultimately collapses in upon itself. (p. 542)

binary asteroid Two asteroids that revolve about each other as they orbit the Sun.

binary star Two stars revolving about each other. (pp. 355–360)

biosphere The layer of soil, water, and air surrounding the Earth in which living organisms thrive. (p. 157)

bipolar outflow Oppositely directed jets of gas expelled from a young star. (pp. 370–371)

bipolar sunspot group A sunspot group that has roughly comparable areas covered by north and south magnetic polarity. (p. 319)

BL Lacertae object A type of active galaxy whose nucleus does not exhibit spectral lines. (p. 515)

black hole An object whose gravity is so strong that the escape velocity exceeds the speed of light. (pp. 5, 410, 438ff)

black hole evaporation The process by which black holes emit particles. (pp. 543–544)

blackbody A hypothetical perfect radiator that absorbs and re-emits all radiation falling upon it. (pp. 87–88)

blackbody curve The intensity of radiation emitted by a blackbody plotted as a function of wavelength or frequency. (p. 88)

blackbody radiation The radiation emitted by a blackbody; thermal radiation. (p. 88)

blazar A BL Lacertae object. (p. 515)

blueshift A decrease in the wavelength of photons emitted by an approaching source of light. (p. 96)

Bode's law A numerical sequence that gives the approximate average distances of the planets from the Sun in astronomical units. (p. 290)

Bohr atom A model of the atom, described by Niels Bohr, in which electrons revolve about the nucleus in certain allowed circular orbits. (p. 93)

Bok globule A small, roundish, dark nebula. (p. 368)

bolometric correction The difference between the visual and bolometric magnitudes of a star. (pp. 346–347)

bolometric magnitude A measure of the brightness of a star or other object as detected by a device above the Earth's atmosphere and sensitive to all wavelengths of radiation. (pp. 346–347)

Bonner Durchmusterung A famous nineteenth-century star catalogue. (p. 23)

bounded universe A universe throughout which the average density exceeds the critical density. (p. 536)

breccia A rock formed by the sudden amalgamation of various rock fragments under pressure.

brown oval An elongated brownish feature usually seen in Jupiter's northern hemisphere. (p. 233)

brown dwarf A starlike object that is not massive enough to ignite hydrogen burning in its core. (p. 369)

burster A nonperiodic X-ray source that emits powerful bursts of X rays. (pp. 5, 435)

butterfly diagram A plot of sunspot latitude versus time. (p. 318)

C ring One of three prominent rings encircling Saturn. (pp. 256–257)

caldera The crater at the summit of a volcano. (p. 213)

capture theory The hypothesis that the Moon was gravitationally captured by the Earth. (p. 173)

carbon burning The thermonuclear fusion of carbon to produce heavier nuclei. (p. 407)

carbon cycle A series of nuclear reactions, involving carbon as a catalyst, by which hydrogen is transformed into helium; the CNO cycle. (p. 324)

carbon star A peculiar red giant star whose spectrum shows strong absorption by carbon and carbon compounds. (p. 403)

carbonaceous chondrite A type of meteorite that has a high abundance of carbon and volatile compounds. (p. 299)

Cassegrain focus An optical arrangement in a reflecting telescope in which light rays are reflected by a secondary mirror to a focus behind the primary mirror. (p. 106)

Cassini division An apparent gap between Saturn's A and B rings. (pp. 256–257, 260)

causality The notion that effects follow their causes. (p. 451)

celestial equator A great circle on the celestial sphere 90° from the celestial poles. (p. 24)

celestial mechanics The branch of astronomy dealing with the motions and gravitational interactions of objects in the solar system.

celestial poles Points about which the celestial sphere appears to rotate. (p. 24)

celestial sphere A sphere of very large radius centered on the observer; the apparent sphere of the sky. (p. 24)

center of mass That point in an isolated system that moves at a constant velocity in accordance with Newton's first law. (pp. 68, 355–356)

central bulge A spherical distribution of stars around the nucleus of a spiral galaxy. (p. 458–459)

centrifugal force A fictitious force directed outward from the axis of rotation in a rotating system. (p. 236)

Cepheid variable A type of yellow, supergiant, pulsating star. (pp. 396–397, 477)

Cerenkov radiation   The radiation emitted by a particle traveling through a medium at a speed greater than the speed of light in that medium. (p. 417)

Ceres   The largest asteroid and the first to be discovered. (pp. 291–292)

Chandrasekhar limit   The maximum mass of a white dwarf. (p. 406)

charge-coupled device (CCD)   A type of solid-state silicon wafer designed to detect photons. (p. 112)

chemical differentiation   The process by which the heavier elements in a planet sink toward its center while lighter elements rise toward its surface. (pp. 139, 298)

chemical element   A substance that cannot be decomposed by chemical means into simpler substances. (p. 90)

chromatic aberration   An optical defect whereby different colors of light passing through a lens are focused at different locations. (p. 104)

chromosphere   A layer in the solar atmosphere between the photosphere and the corona. (pp. 313–314)

circumstellar shell   Gases immediately surrounding a star from which they have escaped. (p. 393)

close binary   A double star system in which the stars are separated by a distance roughly comparable to their diameters. (pp. 393–394)

cluster of galaxies   A collection of galaxies containing a few to several thousand member galaxies. (pp. 485–487)

CNO cycle   A series of nuclear reactions in which carbon is used as a catalyst to transform hydrogen into helium. (pp. 323–324)

cocoon nebula   The nebulosity surrounding a protostar. (p. 369)

cocreation theory   The hypothesis that the Earth and the Moon formed at the same time from the same material. (p. 173)

cold dark matter   Slowly moving, weakly interacting particles presumed to constitute the bulk of matter in the universe. (p. 562)

collisional ejection theory   The hypothesis that the Moon formed from material ejected from the Earth by the impact of a large asteroid. (p. 173)

color index   The difference between the magnitudes of a star measured in two different spectral regions. (pp. 345–347)

color–magnitude diagram   A plot of the magnitudes of stars in a cluster against their color indices.

coma (of a comet)   The diffuse gaseous component of the head of a comet. (pp. 300–301)

coma (optical)   The distortion of off-axis images formed by a parabolic mirror. (p. 109)

comet   A small body of ice and dust in orbit about the Sun. While passing near the Sun, a comet's vaporized ices give rise to a coma and tail. (pp. 132–133, 300–303)

comet–asteroid hypothesis   The theory that planetary atmospheres resulted from the impacts of comets and asteroids.

Compton wavelength   A characteristic length associated with the size of an elementary particle. (p. 531)

condensation temperature   The temperature at which a particular substance in a low-pressure gas condenses into a solid. (pp. 136–137)

conduction   The transfer of heat by directly passing energy from atom to atom. (p. 325)

configuration (of a planet)   A specific geometrical arrangement of a planet, the Sun, and the Earth. (p. 58)

confinement   The permanent bonding of quarks to form particles like protons and neutrons. (p. 566)

conic section   The curve of intersection between a circular cone and a plane; this curve can be a circle, ellipse, parabola, or hyperbola. (pp. 68–69)

conjunction   The geometric arrangement of a planet in the same part of the sky as the Sun; a planet with an elongation of 0°. (p. 58)

conservation of angular momentum   A law of physics stating that the total amount of angular momentum in an isolated system remains constant. (pp. 71, 424–425)

conservation of energy   A law of physics stating that the total energy in an isolated system remains constant. (p. 70)

conservation of momentum   A law of physics stating that the total momentum in an isolated system remains constant. (pp. 70–71)

constellation   A configuration of stars, often named after an object, person, god, or animal. (p. 19)

contact binary   A double star in which both members fill their Roche lobes. (p. 394)

continental drift   The gradual movement of the continents over the surface of the Earth due to plate tectonics. (p. 150)

continuous spectrum   A spectrum of light over a range of wavelengths without any spectral lines. (p. 91)

continuum   The intensity of a continuous spectrum. (pp. 351)

convection   The transfer of energy by moving currents of fluid or gas containing that energy. (pp. 152, 200, 325)

convection cell   A circulating loop of gas or liquid that transports heat from a warm region to a cool region. (p. 200)

convective zone   The region in a star where convection is the dominant means of energy transport. (p. 326)

convective envelope   The outer layers of a star where energy is transported by convection. (p. 326)

coorbital satellites   Satellites that share the same orbit. (p. 266)

core helium burning   The thermonuclear fusion of helium at the center of a star. (p. 388)

core hydrogen burning   The thermonuclear fusion of hydrogen at the center of a star. (p. 385)

Coriolis effect   The deflection of moving objects on a rotating surface.

corona   The Sun's outer atmosphere, which has a high temperature and a low density. (pp. 314–315, 334–335)

coronagraph   An instrument for photographing the solar corona in which a disk inside the telescope produces an artificial eclipse. (p. 314)

coronal hole   A region in the Sun's corona that is deficient in hot gases. (p. 316)

cosmic microwave background   An isotropic radiation field with a blackbody temperature of about 2.7 K that permeates the entire universe. (p. 532ff)

cosmic particle horizon   An imaginary sphere, centered on the Earth, whose radius equals the distance light has traveled since the Big Bang. (p. 529–530)

cosmic rays   Atomic nuclei (mostly protons) that strike the Earth with extremely high speeds.

cosmic singularity   The Big Bang. (p. 530)

cosmic string   A hypothetical long, thin, wirelike concentration of matter left over from the Big Bang. (p. 566–567)

cosmological constant (Λ)   In the equations of general relativity, a quantity signifying a pressure throughout all space that helps the universe expand. (pp. 527, 536)

cosmological model   A specific theory about the structure and evolution of the universe.

cosmological principle   The assumption that the universe is homogeneous and isotropic on the largest scale. (p. 529)

cosmological redshift   A redshift that is caused by the expansion of the universe. (p. 528)

cosmology   The study of the structure and evolution of the universe.

coudé focus   A reflecting telescope in which a series of mirrors direct light to a remote focus away from the moving parts of the telescope. (p. 106)

crater   A circular depression on a planet or satellite caused by the impact of a meteoroid. (p. 163)

crescent moon   One of the phases of the Moon in which less than one-half of its illuminated surface is visible from the Earth. (p. 37)

critical density   The average density throughout the universe at which space would be flat and galaxies would just barely continue receding from each other infinitely far into the future. (p. 536)

critical surface   The surface of the Roche lobes in a double star system. (p. 394)

crust (of a planet)   The surface layer of a terrestrial planet. (pp. 146, 149)

current sheet   A broad, flat region in Jupiter's magnetosphere that contains an abundance of charged particles. (p. 238)

cyclone   The circulation of winds around a low-pressure system in a planet's atmosphere. (p. 235)

cyclonic motion   Circular wind motion in a planet's atmosphere due to the Coriolis effect.

D ring   One of several faint rings encircling Saturn. (p. 260)

dark matter   Subluminous matter that seems to be quite abundant in galaxies and throughout the universe. (pp. 465, 490)

dark-matter problem   The enigma that most of the matter in the universe is severely underluminous. (pp. 465, 490)

dark nebula   A cloud of interstellar gas and dust that obscures the light of more distant stars. (pp. 365, 368)

daughter isotope   An isotope that results from the radioactive decay of another isotope.

decametric radiation   Radiation whose wavelength is about 10 meters. (p. 237)

decay series   A sequence of isotopes that radioactively decay one into another.

deceleration parameter (q_0)   A number whose value indicates how fast the expansion of the universe is slowing down. (pp. 510, 537–538)

decimetric radiation   Radiation whose wavelength is about a tenth of a meter. (p. 237)

declination   Angular distance of a celestial object north or south of the celestial equator. (p. 25)

deferent   A stationary circle in the Ptolemaic system along which another circle (an epicycle) moves, carrying a planet, the Sun, or the Moon. (p. 56)

deflagration   A sudden, violent burning. (p. 415)

degeneracy   The phenomenon, due to quantum mechanical effects, whereby the pressure exerted by a gas does not depend on its temperature. (p. 388)

degenerate electron (neutron) pressure   The pressure exerted by degenerate electrons (neutrons). (pp. 388, 422)

degenerate gas A gas in which all the allowed states for particles (electrons or neutrons) have been filled, thereby causing the gas to behave differently from ordinary gases.

degenerate pressure The pressure exerted by a degenerate gas. (p. 388)

degree A basic unit of angular measure, designated by the symbol °. (p. 7)

degree (Fahrenheit/Celsius) A basic unit of temperature, designated by the symbol °. (p. 85)

density The ratio of the mass of an object to its volume. (p. 13)

density fluctuation The variation of density from one place to another. (p. 561)

density parameter (Ω_0) The ratio of the average density of the universe to the critical density. (p. 538)

density-wave theory An explanation of spiral arms in galaxies proposed by C. C. Lin and colleagues. (p. 467–468)

descending node A point along an orbit where an object crosses a reference plane (usually the ecliptic or celestial equator) from north to south.

detached binary A binary in which neither star fills its Roche lobe. (p. 394)

detonation An explosion.

deuterium An isotope of hydrogen whose nucleus contains one proton and one neutron; heavy hydrogen. (p. 324)

deuterium bottleneck The situation about 3 minutes after the Big Bang when deuterium inhibited the formation of heavier elements. (p. 559)

differential rotation The rotation of a nonrigid object in which parts adjacent to each other at a given time do not always stay close together. (pp. 229, 317)

differentiation (geological) The separation of different kinds of material in different layers inside a planet. (pp. 139, 298)

diffraction The spreading out of light passing the edge of an opaque object.

diffraction grating A piece of glass containing thousands of closely spaced lines that is used to disperse light into a spectrum. (pp. 113–114)

diffuse nebula A reflection or emission nebula consisting of interstellar gas and dust.

dilation of time The slowing of time due to relativistic motion. (p. 440)

direct motion The apparent eastward movement of a planet seen against the background stars. (p. 55)

disk (of a galaxy) The disk-shaped distribution of popula-tion I stars that dominates the appearance of a spiral galaxy. (pp. 458–459)

dissociative recombination The capture of an electron by a positive molecular ion, which results in breaking the molecule into two neutral atoms.

distance modulus The difference between the apparent and absolute magnitudes of an object. (p. 343)

diurnal Daily.

diurnal motion Motion in one day. (p. 21)

Doppler effect The apparent change in wavelength of radiation due to relative motion between the source and the observer along the line of sight. (p. 96)

double-line spectroscopic binary A spectroscopic binary in which spectral lines of both stars can be seen. (p. 357)

double radio source An extragalactic radio source characterized by two large regions of radio emission, typically located on either side of an active galaxy. (p. 512)

double star A pair of stars in orbit about each other and held together by their mutual gravitational attraction; a binary star. (p. 355)

dredge-up The transporting of matter from deep within a star to its surface by convection. (pp. 402–403)

dust tail The tail of a comet that is composed primarily of dust particles. (p. 301)

dwarf elliptical galaxy A low-mass elliptical galaxy that contains only a few million stars. (p. 480)

dynamo effect The creation of a magnetic field by a moving electric current. (p. 189)

dyne A unit of force; the force needed to accelerate 1 gram by 1 cm/s^2.

E ring A very broad, faint ring encircling Saturn. (p. 260)

earthquake Sudden vibratory motion of the Earth's surface. (pp. 148–149)

eccentricity (of an ellipse) The value $\sqrt{1 - b^2/a^2}$, where a is the length of the semimajor axis of the ellipse and b is the length of the semiminor axis. (pp. 68, 125)

eclipse The cutting off of part or all the light from one celestial object by another. (p. 41)

eclipse path The track of the tip of the Moon's shadow along the Earth's surface during a total or annular solar eclipse. (p. 44)

eclipse season A period of time when an eclipse is possible. (p. 285)

eclipse year The interval between successive passages of the Sun through the same node of the Moon's orbit. (p. 47)

eclipsing binary A binary system in which, as seen from

Earth, the stars periodically pass in front of each other. (pp. 358–359)

ecliptic The apparent annual path of the Sun on the celestial sphere. (p. 24)

Eddington limit A constraint on the rate at which a black hole can accrete matter. (pp. 516–517)

Einstein ring The circular image of a remote light source produced by a gravitational lens in which the source, the observer, and the deflecting mass are nearly perfectly aligned. (p. 447)

Einstein–Rosen bridge A speculative, topological feature of a black hole that connects our universe with another universe. (p. 450)

electromagnetic radiation Radiation consisting of oscillating electric and magnetic fields including gamma rays, X rays, visible light, ultraviolet and infrared radiation, radio waves, and microwaves. (p. 83)

electromagnetic spectrum The entire array or family of electromagnetic radiation. (p. 83)

electromagnetic theory The branch of physics that deals with electricity, magnetism, and electromagnetic radiation. (p. 74)

electromagnetism Electric and magnetic phenomena. (p. 83)

electron A negatively charged subatomic particle usually found in orbits about the nuclei of atoms. (pp. 88, 131)

electron volt The energy acquired by an electron accelerated through an electric potential of one volt. (p. 89)

element A chemical that cannot be broken down into more basic chemicals. (pp. 90, 129–130)

ellipse A conic section obtained by cutting completely through a circular cone with a plane. (p. 62)

elliptical galaxy A galaxy with an elliptical shape and no conspicuous interstellar material. (pp. 478–479)

elongation The angular distance between a planet and the Sun as viewed from Earth. (p. 58)

emission line A bright spectral line. (p. 91)

emission line spectrum A spectrum that contains emission lines. (p. 91)

emission nebula A glowing gaseous nebula whose light comes from fluorescence caused by a nearby star. (p. 372)

energy The ability to do work. (p. 70)

energy flux The rate of energy flow, usually measured in joules per square meter per second. (p. 86)

energy level (in an atom) A particular amount of energy possessed by an atom above the atom's least energetic state. (p. 94)

energy-level diagram A diagram showing the arrangement of an atom's energy levels. (pp. 94–95)

Encke division A narrow gap in Saturn's A ring. (p. 259)

epicenter The location on the Earth's surface directly over the focus of an earthquake. (p. 148)

epicycle A moving circle in the Ptolemaic system about which a planet revolves. (p. 56)

epoch A date and time selected as a fixed reference. (p. 29)

equation of state An equation relating the temperature, pressure, and density of a gas.

equation of time The difference between apparent and mean solar time.

equations of stellar structure The equations used by astrophysicists to calculate the structure, properties, and evolution of a star. (p. 325)

equinox One of the intersections of the ecliptic and the celestial equator. (p. 24)

equipotential contour/surface A surface along which gravitational potential energy is constant. (pp. 294, 394)

equivalent width A measure of the amount of light subtracted from the continuous spectrum by an absorption line. (p. 351)

era of recombination The moment, approximately one million years after the Big Bang, when the universe had cooled sufficiently to permit the formation of hydrogen atoms. (p. 538)

erg A unit of energy; the work done by a force of 1 dyne moving through a distance of 1 centimeter.

ergosphere The region of space immediately outside the event horizon of a rotating black hole where it is impossible to remain at rest. (p. 445)

eruptive variable A star whose brightness changes unpredictably as the result of some energetic event.

escape speed The speed needed by one object to achieve a parabolic orbit away from a second object and thereby permanently move away from the second object. (pp. 70, 135)

event horizon The location around a black hole where the escape velocity equals the speed of light; the surface of a black hole. (pp. 442–443)

evolutionary track The path on an H–R diagram followed by a star as it evolves. (p. 369)

excitation The process of imparting energy to an atom or ion.

exponent A number placed above and after another number to denote the power to which the latter is to be raised, as n in 10^n. (p. 8)

extinction The attenuation of light due to absorption by material between the source and the observer. (p. 367)

extragalactic Beyond our Galaxy.

eyepiece A magnifying lens used to view the image produced at the focus of a telescope. (p. 103)

F ring A thin, faint ring encircling Saturn just beyond the A ring. (p. 259)

false vacuum Empty space with a high-energy density that possibly filled the universe immediately after the Big Bang. (p. 566)

favorable opposition A Martian opposition that affords good Earth–based views of the planet. (p. 210)

filtergram A photograph of the Sun taken through special filters that are transparent only to a narrow range of wavelengths.

first quarter moon The phase of the Moon that occurs when the Moon is 90° east of the Sun. (p. 37)

fission theory The hypothesis that the Moon was pulled out of a rapidly rotating proto-Earth. (pp. 172–173)

Fitzgerald contraction The shrinking of distances due to relativistic motion; Fitzgerald–Lorentz contraction. (p. 441)

flare A sudden, temporary outburst of light from an extended region of the solar surface. (p. 319)

flat space Space that is not curved; space with zero curvature. (p. 539)

flatness problem The dilemma posed by the fact that the average density throughout the universe is very nearly equal to the critical density. (p. 555)

flocculent spiral galaxy A spiral galaxy with fuzzy, poorly defined spiral arms. (p. 467)

flux The number of particles or the amount of energy flowing across a given area per unit time. (p. 327)

focal length The distance from a lens or mirror to the point where converging light rays meet. (pp. 103, 106)

focus (of an earthquake) The location below the Earth's surface where an earthquake occurs.

focus (of an ellipse) One of two points inside an ellipse such that the combined distance from the two foci to any point on the ellipse is a constant. (p. 62)

focus (optical) The point where light rays converged by a lens or mirror meet. (pp. 103, 106)

force That which can change the momentum of an object. (p. 66)

frequency The number of wave crests or troughs that cross a given point per unit time; the number of vibrations per unit time. (p. 84)

full moon A phase of the Moon during which its full daylight hemisphere can be seen from Earth. (p. 37)

fusion crust The coating on a stony meteorite caused by the heating of the meteorite as it descended through the Earth's atmosphere. (pp. 297–298)

G ring A thin, faint ring encircling Saturn. (p. 260)

galactic cannibalism A collision between two galaxies of unequal mass and size in which the smaller galaxy seems to be absorbed by the larger galaxy. (p. 494)

galactic cluster A loose association of young stars in the disk of our Galaxy. (p. 374)

galactic equator The intersection of the principal plane of the Milky Way with the celestial sphere.

galactic nucleus The center of a galaxy. (pp. 458–458, 468–471)

galaxy A large assemblage of stars, nebulae, and interstellar gas and dust. (p. 6)

galaxy merger A collision between two galaxies that causes them to coalesce. (p. 493)

Galilean satellite Any one of the four large moons of Jupiter. (p. 242)

gamma rays The most energetic form of electromagnetic radiation. (p. 84)

gauss A unit of magnetic field strength. (p. 321)

general theory of relativity A description of gravity formulated by Albert Einstein, which explains that gravity affects the geometry of space and the flow of time. (pp. 74–76, 441ff)

geocentric cosmology An Earth-centered theory of the universe. (p. 55)

geomagnetic Referring to the Earth's magnetic field.

giant A star whose diameter is typically 10 to 100 times that of the Sun and whose luminosity is roughly that of 100 Suns. (pp. 350, 352)

giant elliptical galaxy A large, massive, elliptical galaxy containing many billions of stars. (p. 480)

giant molecular cloud A large cloud of interstellar gas and dust. (pp. 375–376)

gibbous moon A phase of the Moon in which more than one-half, but not all, of the Moon's daylight hemisphere is visible from Earth. (p. 37)

glitch A sudden speedup in the period of a pulsar. (p. 428)

globular cluster A large spherical cluster of stars, typically found in the outlying regions of a galaxy. (p. 391)

globule A small, dense, dark nebula. (p. 368)

gluon A particle that is exchanged between quarks. (p. 564)

grand design spiral galaxy A galaxy with well-defined spiral arms. (p. 467)

grand unified field theory (GUT) A theory that describes and explains the four physical forces. (p. 565)

granulation The rice-grain–like structure of the solar photosphere. (pp. 312–313)

granule A convective cell in the solar photosphere. (pp. 312–313)

grating An optical device, consisting of thousands of closely spaced lines etched in glass or metal, that disperses light into a spectrum. (p. 113)

gravitation The tendency of matter to attract matter.

gravitational lens The deflection of light from a remote source caused by the presence of an intervening mass. (pp. 445–447)

gravitational radiation/waves Oscillations of space produced by changes in the distribution of matter. (pp. 418, 444)

graviton The particle that is responsible for the gravitational force. (p. 564)

gravity The force with which matter attracts matter. (p. 67)

Great Attractor A huge concentration of matter toward which many galaxies, including the Milky Way, appear to be moving. (p. 533)

Great Dark Spot A prominent high-pressure system in Neptune's southern hemisphere. (p. 279)

Great Red Spot A prominent high-pressure system in Jupiter's southern hemisphere. (pp. 230, 231, 233)

greatest elongation The largest possible angle between the Sun and an inferior planet. (p. 58)

greenhouse effect The trapping of infrared radiation near a planet's surface by the planet's atmosphere. (pp. 158, 197–198)

Greenwich meridian The meridian of longitude that passes through the old Royal Greenwich Observatory near London; the longitude of 0°.

group (of galaxies) A poor cluster of galaxies. (p. 485)

H I region A region of neutral hydrogen in interstellar space.

H II region A region of ionized hydrogen in interstellar space. (pp. 372–373)

hadron A particle composed of quarks.

half-life The time required for one-half of the radioactive nuclei of an isotope to disintegrate. (p. 172)

halo (of a galaxy) A spherical distribution of globular clusters and population II stars that surround a spiral galaxy. (p. 459)

harmonic law Kepler's third law. (p. 62)

head–tail source A radio source consisting of a bright "head" and a long, dim "tail." (p. 512)

Heisenberg uncertainty principle A principle of quantum mechanics that places limits on the precision of simultaneous measurements. (pp. 544, 556)

helio- A prefix referring to the Sun.

heliocentric cosmology A Sun-centered theory of the universe. (p. 57)

helium burning The thermonuclear fusion of helium to form carbon and oxygen. (p. 388)

helium flash The nearly explosive beginning of helium burning in the dense core of a red giant star. (pp. 388–389)

helium shell flash A brief thermal runaway that occurs in the helium-burning shell of a red supergiant. (p. 404)

Helmholtz contraction The contraction of a gaseous body, such as a star or nebula, during which gravitational energy is transformed into thermal energy; Kelvin–Helmholtz contraction. (p. 137)

Henry Draper Catalogue A famous catalogue of stellar spectra compiled at the Harvard College Observatory between 1911 and 1924. (p. 23)

Herbig–Haro object A small, luminous nebula associated with the end point of a jet emanating from a young star. (pp. 370–371)

Hertzsprung–Russell (H–R) diagram A plot of the absolute magnitude (or luminosity) of stars against their spectral type (or surface temperature). (pp. 349–350)

heterogeneous accretion theory A theory of planetary formation which argues that the composition of planetesimals changed as the planets formed. (p. 139)

high-pressure system A region of high atmospheric pressure. (p. 235)

high-velocity star A star traveling at an exceptionally high speed relative to the Sun. (p. 459)

highlands Ancient, heavily cratered terrain on the Moon. (p. 165)

Hirayama family A group of asteroids that have nearly identical orbits about the Sun. (p. 296)

homogeneous accretion theory A theory of planetary formation which argues that the planets formed from planetesimals of generally the same composition. (p. 139)

horizontal branch A group of stars on the Hertzsprung–Russell diagram of a typical globular cluster near the main sequence and having roughly constant absolute magnitude. (p. 391)

hot dark matter Dark matter consisting of particles moving at high speeds. (p. 562)

hot-spot volcanism Volcanic activity that occurs over a hot region buried deep within a planet. (pp. 152, 201)

Hubble classification scheme A method of classifying galaxies as spirals, barred spirals, ellipticals, or irregulars according to their appearance. (pp. 478–481)

Hubble constant (H_0) The constant of proportionality in the relation between the recessional velocities of remote galaxies and their distances. (p. 483)

Hubble flow The recessional motions of remote galaxies caused by the expansion of the universe. (p. 483)

Hubble law The empirical relationship stating that the redshifts of remote galaxies are directly proportional to their distances from Earth. (pp. 483, 528)

hydrocarbon Any one of a variety of chemical compounds composed of hydrogen and carbon. (p. 265)

hydrogen burning The thermonuclear conversion of hydrogen into helium. (pp. 322–324)

hydrogen envelope A huge, tenuous sphere of gas surrounding the head of a comet. (pp. 300–302)

hydrostatic equilibrium A balance between the weight of a layer in a star and the pressure that supports it. (p. 323)

hyperbola A conic section formed by cutting a circular cone with a plane at an angle steeper than the sides of the cone. (p. 68–69)

hyperbolic space Space with negative curvature. (p. 539)

hypothesis An idea or collection of ideas that seems to explain a specified phenomenon; a conjecture. (p. 3)

igneous rock A rock that formed from the solidification of molten lava or magma. (p. 147)

impact breccia A type of lunar rock consisting of rock fragments cemented together by the impact of a meteoroid.

impact crater A crater formed by the impact of a meteoroid. (p. 296)

inertia The property of matter that requires a force to act on it to change its state of motion. (p. 66)

inferior conjunction The configuration when an inferior planet is between the Sun and Earth. (p. 58)

inferior planet A planet that is closer to the Sun than the Earth is. (p. 59)

inflation A sudden expansion of space. (p. 555)

inflationary epoch A brief period shortly after the Big Bang during which the scale of the universe increased very rapidly. (p. 555)

infrared radiation Electromagnetic radiation of wave-length longer than visible light, yet shorter than radio waves. (p. 83)

inner core The solid portion of the Earth's iron core (p. 149)

inner Lagrangian point The point between the two stars comprising a binary star where their Roche lobes touch; the point across which mass transfer can occur. (p. 394)

instability strip A region of the H–R diagram occupied by pulsating stars. (p. 396)

interferometry A technique of combining the observations of two or more telescopes to produce images better than one telescope alone could make. (p. 115)

intermediate-vector boson The particle that is responsible for the weak nuclear force. (p. 564)

internal rotation period The period with which the core of a Jovian planet rotates. (p. 238)

interplanetary medium The sparse distribution of gas and dust particles in interplanetary space.

interstellar dust Microscopic solid grains of various compounds in interstellar space. (pp. 365–367)

interstellar extinction The dimming of starlight as it passes through the interstellar medium. (pp. 367, 456)

interstellar gas Sparse gas in interstellar space. (pp. 365–367)

interstellar medium Interstellar gas and dust. (pp. 365–367)

interstellar reddening The reddening of starlight passing through the interstellar medium, caused by the fact that blue light is scattered more than red. (p. 367)

inverse-square law The statement that the apparent brightness of a light source varies inversely with the square of the distance from the source. (p. 342)

Io torus A doughnut-shaped ring of gas circling Jupiter at the distance of Io's orbit. (pp. 248–249)

ion An atom that has become electrically charged due to the addition or loss of one or more electrons. (p. 198)

ion tail (of a comet) The relatively straight tail of a comet produced by the solar wind acting on ions. (p. 301)

ionization The process by which an atom loses electrons. (p. 94)

ionization potential The energy required to remove an electron from an atom.

ionopause The boundary around a planet like Venus where the pressure exerted by ions counterbalances the pressure of the solar wind. (p. 198)

ionosphere A layer in the Earth's upper atmosphere in which many of the atoms are ionized.

iron meteorite   A meteorite composed primarily of iron. (pp. 298–299)

irregular cluster (of galaxies)   A sprawling collection of galaxies whose overall distribution in space does not exhibit any noticeable spherical symmetry.   (p. 485)

irregular galaxy   An asymmetrical galaxy having neither spiral arms nor an elliptical shape.   (p. 480)

isochron   A line on a graph indicating the abundance ratios of various isotopes for a rock of a certain age.

isotope   Any of several forms for the same chemical element whose nuclei all have the same number of protons but different numbers of neutrons.   (p. 131)

isotropic   Having the same property in all directions. (p. 480)

isotropy problem   The dilemma posed by the fact that the cosmic microwave background is isotropic.   (p. 555)

Jeans length   The smallest scale over which a density fluctuation in a medium will contract to form a gravitationally bound object.   (p. 561)

joule (J)   A unit of energy.   (p. 84)

Jovian planet   Any of the four largest planets: Jupiter, Saturn, Uranus, or Neptune.   (p. 125)

Kaluza–Klein theory   A theory that describes the four physical forces in terms of the geometry of an 11–dimensional space–time.   (p. 568)

kelvin (K)   A unit of temperature on the Kelvin temperature scale.   (p. 84)

Kelvin–Helmholtz contraction   The contraction of a gaseous body, such as a star or nebula, during which gravitational energy is transformed into thermal energy.   (p. 322)

Kepler's laws   Three statements, formulated by Johannes Kepler, that describe the motions of the planets. (pp. 62–63)

kiloparsec (kpc)   One thousand parsecs; about 3260 light-years.   (p. 10)

kinetic energy   The energy possessed by an object because of its motion.   (pp. 70, 134)

Kirchhoff's laws   Three statements about circumstances that produce absorption lines, emission lines, and continuous spectra.   (p. 91)

Kirkwood gaps   Gaps in the spacing of asteroid orbits, discovered by Daniel Kirkwood.   (pp. 292–293)

Lagrangian points   Five points in the orbital plane of two bodies revolving about each other in circular orbits where a third object of negligible mass can remain in equilibrium. (pp. 293–294)

Lamb shift   A tiny shift in the spectral lines of hydrogen caused by the presence of virtual particles.   (p. 557)

last quarter moon   The phase of the Moon that occurs when the Moon is 90° west of the Sun.   (p. 37)

lava   Molten rock.   (p. 147)

law of cosmic censorship   The hypothesis that all singularities must be surrounded by an event horizon.   (p. 443)

law of equal areas   Kepler's second law.   (p. 62)

law of inertia   Newton's first law.   (p. 66)

laws of physics   A set of physical principles with which we can understand natural phenomena and the nature of the universe.   (p. 3)

leap year   A calendar year with 366 days.   (p. 32)

lenticular galaxy   A galaxy with a central bulge and a disk but no spiral structure; an S0 galaxy.   (pp. 480–481)

lepton   Any member of a class of particles that includes the electron and neutrino.

libration   A slight rocking of the Moon in its orbit whereby an Earth-based observer can, over time, see slightly more than one-half the Moon's surface.   (p. 163)

light   Electromagnetic radiation.   (p. 81ff)

light curve   A graph that displays variations in the brightness of a star or other astronomical object.   (pp. 358–359)

light-gathering power   A measure of the amount of radiation brought to a focus by a telescope.   (p. 107)

light-year (ly)   The distance light travels in a vacuum in one year.   (p. 10)

limb (of Sun or Moon)   The apparent edge of the Sun or Moon as seen in the sky.   (p. 312)

limb darkening   The phenomenon whereby the Sun is darker near its limb than near the center of its disk. (p. 312)

limiting magnitude   The faintest magnitude that can be observed with a certain telescope under certain conditions.

line of apsides   The major axis of an elliptical orbit.   (p. 42)

line of nodes   A line connecting the nodes of an orbit. (p. 42)

line profile   The shape of a spectral line.   (p. 351)

liquid metallic hydrogen   Hydrogen compressed to such a density that it behaves like a liquid metal.   (p. 237)

lithosphere   The solid, upper layer of the Earth; essentially the Earth's crust.   (p. 152)

LMC   The Large Magellanic Cloud.   (pp. 481–482)

Local Bubble   A large cavity in the interstellar medium in which the Sun and nearby stars are located.   (p. 462)

Local Group   The cluster of galaxies of which our Galaxy is a member.   (p. 488)

long-period comet   A comet that takes hundreds of thousands of years to complete one orbit of the Sun. (p. 303)

long-period variable   A variable star with a period longer than about 100 days. (p. 396)

Lorentz transformations   Equations that relate the measurements of different observers who are moving relative to each other at high speeds. (pp. 440–441)

low-pressure system   A region of low atmospheric pressure. (p. 235)

luminosity   The rate at which electromagnetic radiation is emitted from a star or other object. (pp. 86, 323, 343, 346–347)

luminosity class   A classification of a star of a given spectral type according to its luminosity. (pp. 353–354)

luminosity function   The numbers of stars of differing brightness per cubic parsec. (p. 343)

lunar   Referring to the Moon.

lunar eclipse   An eclipse of the Moon by the Earth; a passage of the Moon through the Earth's shadow. (p. 43)

lunar month   The time the Moon takes to complete one cycle of its phases; the synodic month. (p. 39)

lunar phase   The appearance of the illuminated area of the Moon as seen from Earth. (pp. 37–39)

Lyman series   A series of spectral lines of hydrogen produced by electron transitions to and from the lowest energy state of the hydrogen atom. (p. 94)

Magellanic clouds   Two nearby galaxies visible to the naked eye from southern latitudes. (p. 481–482)

magma   Molten rock beneath a planet's surface. (p. 147)

magnetic-dynamo model   A theory that explains the solar cycle as a result of the Sun's differential rotation acting on the Sun's magnetic field. (pp. 320, 322)

magnetic axis   A line connecting the north and south magnetic poles of a planet or star possessing a magnetic field. (p. 280)

magnetic field   A region of space near a magnetized body within which significant magnetic forces can be detected.

magnetogram   An artificial picture of the Sun that shows regions of magnetic polarity in various colors. (p. 319)

magnetometer   A device for measuring magnetic fields. (p. 186)

magnetopause   That region of a planet's magnetosphere where the magnetic field counterbalances the pressure from the solar wind. (pp. 156)

magnetosheath   The region of turbulent flow between a planet's magnetopause and bow shock.

magnetosphere   The region around a planet occupied by its magnetic field. (pp. 155, 189, 238)

magnification   The factor by which the angular size of an object is apparently increased when viewed through a telescope. (p. 104)

magnifying power   The number of times larger in angular diameter an object appears through a telescope than when viewed with the naked eye. (p. 104, 106)

magnitude   A measure of the amount of light received from a star or other luminous object. (p. 341ff)

magnitude scale   A system for denoting the brightnesses of astronomical objects. (p. 48)

main sequence   A grouping of stars on the Hertzsprung–Russell diagram extending diagonally across the graph from the hottest, brightest stars to the dimmest, coolest stars. (pp. 349, 352)

main sequence star   A star whose luminosity and surface temperature place it on the main sequence on an H–R diagram; a star that derives it energy from core hydrogen burning. (p. 349)

major axis (of an ellipse)   The longest diameter of an ellipse. (pp. 62, 68)

mantle (of a planet)   That portion of a terrestrial planet located between its crust and core. (p. 149)

mare   Latin for "sea"; a large, relatively crater-free plain on the Moon; *plural* maria. (p. 163)

mare basalt   A type of lunar rock commonly found in the mare basins. (pp. 170–171)

marginally bounded universe   A universe throughout which the average density equals the critical density. (p. 536)

mascon   A localized concentration of dense material beneath the lunar surface.

maser   A device or naturally occurring situation that amplifies microwaves. (p. 379)

mass   A measure of the total amount of material in an object. (pp. 66–67)

mass density of radiation   The energy possessed by a radiation field per unit volume divided by the square of the speed of light. (p. 534)

mass function   A numerical relationship involving the masses of the stars in a binary system and the angle of inclination of their orbit in the sky. (p. 343)

mass loss   A process by which a star gently loses matter. (p. 392)

mass–luminosity relation   A relationship between the masses and luminosities of main sequence stars. (pp. 356–357)

mass–radius relation A relationship between the masses and radii of white dwarf stars. (pp. 406–407)

mass transfer The flow of gases from one star in a binary to the other. (p. 393)

matter-dominated universe A universe in which the average density of matter exceeds the mass density of radiation. (p. 534)

Maunder butterfly diagram A plot of sunspot latitude versus time. (p. 318)

Maunder minimum An interval of about 70 years, starting around 1645, during which very few sunspots were seen. (p. 321)

maximum eastern elongation The configuration of an inferior planet at its greatest angular distance east of the Sun. (p. 58)

maximum western elongation The configuration of an inferior planet at its greatest angular distance west of the Sun. (p. 58)

mean solar day The interval between successive meridian passages of the mean Sun; the average length of a solar day. (p. 30)

mean solar time Time reckoned by the location of the mean Sun. (p. 30)

mean Sun A fictitious object that moves eastward at a constant speed along the celestial equator, completing one circuit of the sky with respect to the vernal equinox in one tropical year. (p. 30)

mechanics The branch of physics dealing with the behavior and motions of objects acted upon by forces.

megaparsec (Mpc) One million parsecs. (p. 10)

meridian (local) The great circle on the celestial sphere that passes through an observer's zenith and the north and south celestial poles. (p. 29)

meridian transit The crossing of the meridian by any astronomical object. (p. 29)

mesosphere A layer in a planet's atmosphere above the stratosphere. (p. 154)

metal-poor star A star which, compared to the Sun, is underabundant in elements heavier than helium. (p. 392)

metal-rich star A star whose abundance of heavy elements is roughly comparable to that of the Sun. (p. 392)

metamorphic rock A rock whose properties and appearance have been transformed by the action of pressure and heat beneath the Earth's surface. (pp. 147–148)

meteor The luminous phenomenon seen when a meteoroid enters the Earth's atmosphere; a "shooting star." (p. 297)

meteor shower Many meteors that seem to radiate from a common point in the sky. (pp. 304–305)

meteorite A fragment of a meteoroid that has survived passage through the Earth's atmosphere. (pp. 297–299)

meteoritic swarm A collection of meteoroids moving together along an orbit about the Sun. (p. 305)

meteoroid A small rock in interplanetary space. (pp. 296–297)

micrometeorite A very small meteoroid; a grain of interplanetary dust.

microwaves Short wavelength radio waves. (p. 85)

Milky Way Our Galaxy; the band of faint stars seen from the Earth in the plane of our Galaxy's disk. (p. 456ff)

millibar A unit of atmospheric pressure; one-thousandth of a bar. (p. 155)

millisecond pulsar A pulsar with a period of roughly 1 to 10 milliseconds. (p. 429)

mineral A naturally occurring solid composed of a single element or chemical combination of elements, often in the form of crystals. (p. 146)

minor axis (of an ellipse) The smallest diameter of an ellipse. (p. 68)

minor planet An asteroid. (p. 291)

minute of arc One-sixtieth of a degree, designated by the symbol ′. (p. 8)

model A hypothesis that has withstood experimental or observational tests. (p. 3)

model (of the Sun or a star) The results of a theoretical calculation that gives the values of temperature, pressure, density, and so forth throughout the Sun or a star. (pp. 323, 326)

model atmosphere The results of a theoretical calculation that gives the values of temperature, pressure, density, and so forth throughout the outer layers of a star.

molecule A combination of two or more atoms. (p. 130)

momentum A measure of the inertia of an object; an object's mass multiplied by its velocity. (p. 70)

monochromatic Of one wavelength or color.

moonquake Sudden, vibratory motion of the Moon's surface. (p. 168)

mountain range A series of more or less connected mountains.

moving cluster method A technique for determining the distance to a cluster of stars from the motions of the cluster's members. (pp. 340–341)

muon A subatomic particle that behaves like a heavy electron. (p. 559)

nadir The point on the local meridian 180° from the zenith.

nanosecond One-billionth (10^{-9}) second.

neap tide An ocean tide that occurs when the Moon is near first-quarter or third-quarter phase. (pp. 169–170)

nebula A cloud of interstellar gas and dust. (p. 5)

negative curvature A surface or space in which parallel lines diverge and the sum of the angles of a triangle is less than 180°. (p. 539)

neon burning The thermonuclear fusion of neon to produce heavier nuclei. (p. 408)

neutrino A subatomic particle with no electric charge and little or no mass, yet one that is important in many nuclear reactions. (pp. 323, 326–328, 559)

neutrino oscillation The switching of a neutrino from one type to another. (p. 328)

neutrino telescope A device capable of detecting neutrinos from supernovae. (p. 417)

neutron A subatomic particle with no electric charge and with a mass nearly equal to that of the proton. (pp. 92, 131, 559)

neutron capture The buildup of neutrons inside nuclei. (p. 408)

neutron star A very compact, dense star composed almost entirely of neutrons. (pp. 410, 422ff)

New General Catalogue A famous nineteenth-century catalogue of nebulae, galaxies, and star clusters. (p. 23)

new moon The phase of the Moon when the dark hemisphere of the Moon faces the Earth. (p. 37)

new technology telescope A telescope whose design incorporates innovative features, such as adaptive optics. (pp. 109–110)

Newtonian focus An optical arrangement in a reflecting telescope in which a small mirror reflects converging light rays to a focus on one side of the telescope tube. (p. 106–107)

Newtonian mechanics The branch of physics based on Newton's law that deals with gravitation. (pp. 3, 68)

Newtonian reflector A reflecting telescope that uses a small mirror to deflect the image to one side of the telescope tube. (p. 106)

Newton's laws of motion Three statements about the nature of physical reality on which Newtonian mechanics is based. (pp. 66–67)

no-hair theorem A statement of the simplicity of black holes. (p. 445)

node The intersection of an orbit with a reference plane such as the plane of the celestial equator or the ecliptic. (p. 42)

nonradiogenic Referring to an isotope that is not created by the radioactive decay of another isotope.

nonthermal radiation Radiation emitted by charged particles moving through a magnetic field; synchrotron radiation. (pp. 237, 509)

north celestial pole The point directly above the Earth's north pole where the Earth's axis of rotation, if extended, would intersect the celestial sphere. (p. 24)

northern lights Aurorae; aurorae borealis. (p. 156)

nova A star that experiences a sudden outburst of radiant energy, temporarily increasing its luminosity roughly a thousandfold. (p. 434)

nuclear bulge The central region of our Galaxy; the central bulge.

nuclear density The density of matter in an atomic nucleus; about 4×10^{17} kg/m^3. (p. 410)

nuclear fusion The combining of lighter nuclei to make heaver ones. (p. 322)

nucleus (of an atom) The massive part of an atom, composed of protons and neutrons, about which electrons revolve. (pp. 92, 131)

nucleus (of a comet) A collection of ices and dust that constitute the solid part of a comet. (pp. 300–301)

nucleus (of a galaxy) The concentration of stars and dust at the center of a galaxy. (pp. 458–459)

nutation A small periodic wobbling of the Earth's axis superimposed on precession.

OB association A grouping of hot, young, massive stars, predominantly of spectral types O and B. (pp. 373, 378)

OBAFGKM The temperature sequence of spectral types. (p. 348)

objective lens The principal lens of a refracting telescope. (p. 103)

oblate Flattened at the poles. (p. 236)

oblateness A measure of how much a flattened sphere (or spheroid) differs from a perfect sphere. (p. 236)

obliquity (of the ecliptic) The angle between the planes of the celestial equator and the ecliptic (about $23\frac{1}{2}°$). (p. 27)

obscuration (interstellar) The absorption of starlight by interstellar dust.

observable universe That portion of the universe inside the cosmic event horizon. (pp. 529–530)

Occam's razor The notion that a straightforward explanation of a phenomenon is more likely to be correct than a convoluted one. (p. 57)

occultation The eclipsing of an astronomical object by the Moon or a planet. (p. 243)

oceanic rift A crack in the ocean floor that exudes lava. (p. 152)

Olbers's paradox The dilemma associated with the fact that the night sky is dark. (p. 527)

Oort cloud A presumed accumulation of comets and cometary material surrounding the Sun at a distance of roughly 50,000 to 100,000 AU. (p. 304)

opacity The ability of a material to impede the passage of light.

open cluster A loose association of young stars in the disk of our Galaxy; a galactic cluster. (pp. 340, 374)

opposition The configuration of a planet when it is at an elongation of 180° and thus appears opposite the Sun in the sky. (p. 58)

optical window The range of visible wavelengths to which the Earth's atmosphere is transparent. (pp. 116–117)

optics The branch of physics dealing with the behavior and properties of light. (p. 101ff)

orbit The path of an object that is moving about a second object or point.

outer core The molten portion of the Earth's core. (p. 149)

outgassing Volcanic processes by which gases escape from a planet's crust into its atmosphere.

overcontact binary A close binary in which the two stars share a common atmosphere. (pp. 394–395)

oxygen burning The thermonuclear fusion of oxygen to produce heavier nuclei. (p. 408)

P wave One of three kinds of seismic waves produced by an earthquake; a primary wave. (pp. 148–149)

pair production The creation of a particle and its anti-particle from energy. (p. 557–558)

Pallas The second asteroid to be discovered. (p. 291–292)

Pangaea The name of a hypothetical continent that fragmented into several of the continents on Earth today. (p. 150)

parabola A conic section formed by cutting a circular cone at an angle parallel to one of the sides of the cone. (p. 68–69)

parallax The apparent displacement of an object due to the motion of the observer. (pp. 61, 337)

parent isotope An unstable isotope that radioactively decays into another isotope.

parsec (pc) A unit of distance; 3.26 light-years. (pp. 10–11)

partial eclipse A lunar or solar eclipse in which the eclipsed object does not appear completely covered. (pp. 43–44)

partial lunar eclipse A lunar eclipse in which the Moon does not appear completely covered. (p. 43)

partial solar eclipse A solar eclipse in which the Sun does not appear completely covered. (p. 44)

pascal (Pa) A unit of pressure.

Paschen series A series of spectral lines of hydrogen produced by electron transitions to and from the third ($n = 3$) energy level of the hydrogen atom. (p. 94)

Pauli exclusion principle A principle of quantum mechanics stating that two identical particles cannot have the same position and momentum. (p. 388)

penumbra The portion of a shadow in which only part of the light source is covered by an opaque body. (p. 43)

penumbra (of a sunspot) The grayish region that usually surrounds the umbra of a sunspot. (pp. 316–317)

penumbral eclipse A lunar eclipse in which the Moon passes only through the Earth's penumbra. (p. 43)

perfect gas An idealized gas that obeys a very simple equation of state. (p. 388)

perigee The point in its orbit where a satellite or the Moon is nearest the Earth. (p. 45)

perihelion The point in its orbit where a planet is nearest the Sun. (p. 62)

period (of a planet) The interval of time between successive geometric arrangements of a planet and an astronomical object, such as the Sun. (p. 58)

period–luminosity relation A relationship between the period and average density of a pulsating star. (p. 397)

periodic table A listing of the chemical elements according to their properties, invented by Dmitri Mendeleev. (pp. 129–130)

permafrost Frozen subsoil. (p. 215)

perturbation A small disturbing effect.

phases of the Moon The appearances of the Moon at different times as it orbits the Earth. (pp. 37–38)

photodisintegration The breakup of nuclei by high-energy gamma rays. (p. 410)

photoelectric effect The phenomenon whereby certain metals fire off electrons when exposed to short-wavelength light. (p. 88)

photometry The measurement of light intensities. (p. 345)

photon A discrete unit of electromagnetic energy. (pp. 82, 88–89)

photosphere The region in the solar atmosphere from which most of the visible light escapes into space. (p. 311)

pixel A picture element. (p. 112)

plage A bright region in the solar atmosphere as observed in the monochromatic light of a spectral line. (p. 321)

Planck length A fundamental length defined by basic physical constants. (p. 531)

Planck mass A fundamental mass defined by basic physical constants. (p. 531)

Planck time A fundamental interval of time defined by basic physical constants. (pp. 530–531)

Planck's constant (h) The constant of proportionality between a photon's energy and its frequency. (p. 89)

Planck's law The relationship between the energy of a photon and its wavelength or frequency; $E = h\nu$. (p. 89)

planetary nebula A luminous shell of gas ejected from an old, low-mass star. (pp. 401, 404–405)

planetesimal A small body of primordial dust and ice from which the planets formed. (p. 138)

plasma A hot ionized gas. (pp. 238, 318, 535)

plate A large section of the Earth's lithosphere that moves as a single unit. (pp. 150–151)

plate tectonics The motions of large segments (plates) of the Earth's surface over the underlying mantle. (pp. 151–152)

polarized radiation Electromagnetic waves whose electric fields are aligned in a particular direction. (p. 510)

polymer A long molecule consisting of many smaller molecules joined together. (p. 265)

poor cluster (of galaxies) A cluster of galaxies with very few members; a group of galaxies. (p. 485)

population I star A star whose spectrum exhibits spectral lines of many elements heavier than helium; a metal-rich star. (p. 392)

population II star A star whose spectrum exhibits comparatively few spectral lines of elements heavier than helium; a metal-poor star. (p. 392)

population III star A star virtually devoid of elements heavier than helium. (p. 392)

positive curvature A surface or space in which parallel lines converge and the sum of the angles of a triangle is greater than 180°. (p. 539)

positron An electron with a positive rather than negative electric charge; an antielectron. (p. 323)

potential energy The energy possessed by an object because of its elevated position in a gravitational field. (p. 70)

powers of ten A shorthand method of writing numbers, involving 10 followed by an exponent. (pp. 8–9)

precession (of the Earth) A slow, conical motion of the Earth's axis of rotation caused by the gravitational pull of the Moon and Sun on the Earth's equatorial bulge. (p. 29)

precession (of the equinoxes) The slow westward motion of the equinoxes along the ecliptic due to precession of the Earth. (p. 29)

primary wave One of three kinds of seismic waves produced by an earthquake; a P wave. (pp. 148–149)

prime focus The point in a telescope where the objective focuses light. (p. 106)

primordial black hole A hypothetical black hole that may have been created during the Big Bang. (p. 544)

primordial fireball The extremely hot gas that filled the universe immediately following the Big Bang. (p. 535)

principle of equivalence A principle of general relativity stating that, in a small volume, it is impossible to distinguish between the effects of gravitation and acceleration. (pp. 440–442)

principle of totalitarianism The notion in quantum mechanics that if a process is not strictly forbidden, it must occur. (p. 557)

prism A wedge-shaped piece of glass that is used to disperse white light into a spectrum. (p. 81)

prominence Flamelike protrusions seen near the limb of the Sun and extending into the solar corona. (pp. 318–319)

proper distance A length measured by a ruler at rest with respect to an observer. (p. 441)

proper mass The mass of an object measured at rest. (p. 441)

proper motion The angular rate of change in the location of a star on the celestial sphere, usually expressed in seconds of arc per year. (p. 338)

proper time A time interval measured with a clock at rest with respect to an observer. (p. 441)

proto- A prefix referring to the embryonic stage of a young astronomical object (planet, star, etc.) that is still in the process of formation. (pp. 137, 138)

proton A heavy, positively charged subatomic particle that is one of two principal constituents of atomic nuclei. (pp. 92, 131)

proton–proton chain A sequence of thermonuclear reactions by which hydrogen nuclei are built up into helium nuclei. (p. 323–324)

protoplanetary disk A disk of material encircling a protostar or a newborn star. (p. 372)

protostar A star in its earliest stages of formation. (pp. 368–369)

pulsar A pulsating radio source believed to be associated with a rapidly rotating neutron star. (pp. 5, 423–428)

pulsating variable A star that pulsates in size and luminosity.

quadrature The configuration in which a planet has an elongation of 90°.

quantum electrodynamics A theory that describes details of how charged particles interact by exchanging photons. (p. 564)

quantum mechanics The branch of physics dealing with the structure and behavior of atoms and their interaction with light. (pp. 94, 556)

quark One of several hypothetical particles presumed to be the internal constituents of certain heavy subatomic particles such as protons and neutrons. (p. 564)

quarter moon A phase of the Moon when it is located 90° from the Sun in the sky, so that one-half of its daylit hemisphere is visible from the Earth. (p. 37)

quasar A starlike object with a very large redshift; a quasi-stellar object or quasi-stellar radio source. (pp. 6, 507ff)

quasi-stellar object A starlike object with a very large redshift; a quasar. (p. 507)

quasi-stellar radio source A quasar that emits detectable radio radiation. (p. 507)

r process A process of nuclear transformation initiated when certain isotopes rapidly capture large numbers of neutrons. (p. 409)

r-process isotope An isotope that is created only through the r process. (p. 409)

radar A technique of reflecting radio waves from a distant object. (p. 196)

radial velocity That portion of an object's velocity parallel to the line of sight. (pp. 96, 339)

radial-velocity curve A plot showing the variation of radial velocity with time for a binary star or variable star. (p. 358)

radiant (of a meteor shower) The point in the sky from which meteors of a particular shower seem to originate. (pp. 304–305)

radiation Electromagnetic energy; photons.

radiation-dominated universe A universe in which the mass density of radiation exceeds the average density of matter. (p. 535)

radiation pressure The transfer of momentum carried by radiation to an object on which the radiation falls. (p. 301)

radiative diffusion The random-walk migration of photons from a star's center toward its surface. (p. 325)

radiative zone A region within a star where radiative diffusion is the dominant mode of energy transport. (p. 326)

radio astronomy That branch of astronomy dealing with observations at radio wavelengths. (pp. 114–116, 505)

radio galaxy A galaxy that emits an unusually large amount of radio waves. (p. 512)

radio lobe A region near an active galaxy from which significant radio radiation emanates. (p. 512)

radio telescope A telescope designed to detect radio waves. (p. 115)

radio waves The longest-wavelength electromagnetic radiation. (p. 83)

radio window The range of radio wavelengths to which the Earth's atmosphere is transparent. (pp. 116–117)

radioactivity The process whereby certain atomic nuclei naturally decompose by spontaneously emitting particles.

ray (lunar) Any one of a system of bright, elongated streaks on the lunar surface.

recombination The process in which an electron combines with a positively charged ion. (p. 372)

recurrent nova A nova that has erupted more than once.

red giant A large, cool star of high luminosity. (pp. 350, 386)

red supergiant An extremely large, cool star of luminosity class I. (p. 402)

reddening (interstellar) The reddening of starlight as it passes through the interstellar medium. (p. 367)

redshift The shifting to longer wavelengths of the light from remote galaxies and quasars; the Doppler shift of light from a receding source. (pp. 96, 482–483)

reflecting telescope A telescope in which the principal optical component is a concave mirror. (pp. 106–108)

reflection The return of light rays by a surface. (p. 106)

reflection grating A diffraction grating that produces a spectrum when light is reflected off it.

reflection nebula A comparatively dense cloud of dust in interstellar space that is illuminated by a star. (pp. 366–367)

reflector A reflecting telescope. (pp. 106–107)

refracting telescope A telescope in which the principal optical component is a lens. (pp. 103–104)

refraction The bending of light rays passing from one transparent medium to another. (pp. 102–103)

refractor A refracting telescope. (pp. 103–104)

refractory element An element with high melting and boiling points. (p. 173)

regolith The layer of rock fragments covering the lunar surface. (p. 171)

regression of the line of nodes The slow motion of the line of nodes of the Moon due to the gravitational pull of the Sun. (p. 42)

regular cluster (of galaxies) A spherical cluster of galaxies. (p. 485)

regular orbit An orbit in the plane of a planet's equator along which a satellite travels in the same direction that the planet rotates. (p. 265)

relativistic cosmology A cosmology based on the general theory of relativity. (p. 535)

relativistic particle A particle moving at nearly the speed of light.

residual polar cap Ice-covered polar regions on Mars that do not completely evaporate during the Martian summer. (pp. 214–215)

resolution The degree to which fine details in an optical image can be distinguished.

resolving power A measure of the ability of an optical system to distinguish, or resolve, fine details in the image it produces.

rest energy The rest mass of a particle multiplied by the square of the speed of light. (p. 559)

retrograde motion The apparent westward motion of a planet with respect to background stars. (p. 55)

retrograde rotation The rotation of an object in the direction opposite to which it is revolving about another object. (p. 196)

revolution The motion of one body about another.

rich cluster (of galaxies) A cluster of galaxies containing many members. (p. 485)

right ascension A coordinate for measuring the east–west positions of objects on the celestial sphere. (p. 25)

rille A trenchlike depression on the lunar surface.

ringlet One of many narrow bands of particles of which Saturn's ring system is composed. (p. 259)

Roche limit The smallest distance from a planet or other object at which a second object can be held together by purely gravitational forces. (p. 282)

Roche lobe A teardrop-shaped volume surrounding a star in a binary inside which gases are gravitationally bound to that star. (pp. 259, 394)

rock A mineral or combination of minerals. (p. 146)

rotation The turning of a body about an axis passing through the body.

rotation curve A plot of the orbital speeds of stars and nebulae outward from the center of a galaxy. (pp. 465, 490)

rotation of the Moon's orbit The slow motion of the major axis of the Moon's orbit due to the gravitational pull of the Sun. (p. 42)

RR Lyrae variable A class of pulsating stars with periods less than one day. (p. 396, 457–458)

s process A process of nuclear transformation that occurs when certain isotopes capture neutrons at a slow rate. (p. 408)

s-process isotope An isotope that is created only by the s process. (p. 408)

S wave One of three kinds of seismic waves produced by an earthquake; a secondary wave. (pp. 148–149)

saros A particular cycle of similar eclipses that recur about every 18 years. (p. 47)

satellite A body that revolves about a larger one.

scarp A line of cliffs formed by the faulting or fracturing of a planet's surface. (pp. 185, 187)

scattering The deflection of photons or particles by other particles. (p. 260)

Schmidt telescope A reflecting telescope invented by Bernard Schmidt that is used to photograph large areas of the sky. (p. 111)

Schwarzschild radius The distance from the singularity to the event horizon in a nonrotating black hole. (p. 443)

scientific method The basic procedure used by scientists to investigate phenomena. (p. 3)

seafloor spreading The separation of plates under the ocean due to lava emerging in an oceanic rift. (p. 150)

second of arc One-sixtieth of an arc minute, designated by the symbol ". (p. 8)

secondary wave One of three kinds of seismic waves produced by an earthquake; an S wave. (pp. 148–149)

sedimentary rock A rock that is formed from material deposited on land by rain or winds, or on the ocean floor. (p. 147)

seeing disk The angular diameter of a star's image. (p. 109)

seismic waves Vibrations traveling through a terrestrial planet usually associated with earthquakelike phenomena. (pp. 148–149)

seismograph A device used to record and measure seismic waves, such as those produced by earthquakes. (p. 148)

seismology The study of earthquakes and related phenomena.

self-propagating star formation The process by which the formation of stars in one location in a galaxy stimulates the formation of stars in a neighboring location. (p. 467)

semidetached binary A binary in which one star fills its Roche lobe. (p. 394)

semimajor axis One-half of the major axis of an ellipse. (p. 62)

Seyfert galaxy A spiral galaxy with a bright nucleus whose spectrum exhibits emission lines. (p. 512)

shepherd satellite A small satellite that gravitationally confines particles to a ring around a planet. (p. 261)

shell helium burning The thermonuclear fusion of helium in a shell surrounding a star's core. (p. 402)

shell hydrogen burning The thermonuclear fusion of hydrogen in a shell surrounding a star's core. (p. 386)

shell star A star, usually of spectral type A to F, that is surrounded by a shell of gas.

shepherd satellite A satellite whose gravity restricts the motions of particles in a planetary ring, preventing them from dispersing. (p. 261)

shield volcano A volcano with long, gently sloping sides. (p. 201)

shock wave An abrupt, localized region of compressed gas caused by an object traveling through the gas at a speed greater than the speed of sound. (pp. 156, 198)

short-period comet A comet that orbits the Sun with a period of less than about 200 years. (p. 303)

SI The International System of Units, based on the meter (m), the second (s), and the kilogram (kg). (p. 12)

sidereal clock A clock that measures sidereal time. (p. 31)

sidereal day The interval between successive meridian passages of the vernal equinox. (p. 31)

sidereal month The period of the Moon's revolution about the Earth with respect to the stars. (p. 39)

sidereal period The orbital period of one object about another with respect to the stars. (p. 58)

sidereal time Time reckoned by the location of the vernal equinox. (p. 30)

sidereal year The orbital period of the Earth about the Sun with respect to the stars. (p. 32)

silica A mineral whose chemical composition is based on the element silicon. (p. 146)

silicon burning The thermonuclear fusion of silicon to produce heavier elements, especially iron. (p. 408)

single-line spectroscopic binary A spectroscopic binary in which the spectral lines of only one star can be detected. (p. 357)

singularity A place of infinite space-time curvature; the center of a black hole. (pp. 442–443)

small-angle formula A relationship between the angular and linear sizes of a distant object. (p. 8)

SMC The Small Magellanic Cloud. (p. 481–482)

solar activity Phenomena that occur in the solar atmosphere such as sunspots, plages, flares, and so forth. (p. 316ff)

solar atmosphere The outer layers of the Sun, consisting of the photosphere, chromosphere, and corona. (p. 311)

solar constant The average amount of energy received from the Sun per square meter per second, measured just above the Earth's atmosphere. (p. 86)

solar corona Hot, faintly glowing gases seen around the Sun during a total solar eclipse; the uppermost regions of the solar atmosphere. (pp. 44–45, 314–315)

solar cycle The semiregular 22-year interval between successive appearances of sunspots at the same latitude and with the same magnetic polarity. (p. 320)

solar eclipse An eclipse of the Sun by the Moon; a passage of the Earth through the Moon's shadow. (pp. 41, 44–46)

solar flare A violent outburst on the Sun's surface. (pp. 156, 319)

solar interior Everything below the solar atmosphere; the inside of the Sun. (p. 312)

solar nebula The cloud of gas and dust from which the Sun and solar system formed. (pp. 136–137)

solar nebula hypothesis The theory that planetary atmospheres came from gases in the solar nebula.

solar nebula wind The theory that planetary atmospheres came from the solar wind.

solar seismology The study of the Sun's interior deduced from observable vibrations of its surface. (p. 329)

solar system The Sun, planets and their satellites, asteroids, comets, and related objects that orbit the Sun. (pp. 4, 124–132)

solar transient A short-lived eruption that moves rapidly outward through the solar corona. (p. 316)

solar transit The passage of an object in front of the Sun. (pp. 18, 183)

solar wind A radial flow of particles (mostly electrons and protons) from the Sun. (pp. 141, 316)

solstice Either of two points along the ecliptic at which the Sun reaches its maximum distance north or south of the celestial equator. (pp. 24, 26)

south celestial pole The point directly above the Earth's south pole where the Earth's axis of rotation, if extended, would intersect the celestial sphere. (p. 24)

southern lights Aurorae; aurorae australis. (p. 156)

space–time A four-dimensional combination of the three dimensions of space and time. (p. 567)

special theory of relativity A description of mechanics and electromagnetic theory formulated by Albert Einstein, which explains that measurements of distance, time, and mass are affected by the observer's motion. (pp. 74, 439–441)

spectral analysis The identification of chemical substances from the patterns of lines in their spectra. (p. 90)

spectral class/type A classification of stars according to the appearance of their spectra. (pp. 347–350)

spectral line In a spectrum, an absorption or emission feature that is at a particular wavelength. (p. 90)

spectrogram The photograph of a spectrum. (p. 114)

spectrograph An instrument for photographing a spectrum. (pp. 113–114)

spectroheliograph An instrument for photographing the Sun in the monochromatic light of one spectral line.

spectroscope An instrument for directly viewing a spectrum.

spectroscopic binary A binary star whose binary nature is deduced from the periodic Doppler shifting of lines in its spectrum. (p. 357)

spectroscopic parallax The distance to a star derived by comparing its apparent magnitude to an absolute magnitude inferred from the star's spectrum. (pp. 354–355)

spectroscopy The study of spectra. (p. 129)

spectrum The result of dispersing a beam of electromagnetic radiation so that components with different wavelengths are separated in space. (p. 81)

spectrum binary A binary star whose binary nature is deduced from the presence of two sets of incongruous spectral lines. (p. 356)

spherical aberration The distortion of an image formed by a telescope due to differing focal lengths of the optical system. (pp. 109–110)

spherical space Space with positive curvature. (p. 539)

spicule A narrow jet of rising gas in the solar chromosphere. (pp. 313–314)

spin The intrinsic angular momentum possessed by certain particles. (p. 460)

spin–flip transition A transition in the ground state of the hydrogen atom, which occurs when the orientation of the electron's spin changes. (p. 460)

spin–orbit coupling The relationship between the rotation of a body and its orbital period. (pp. 183–184)

spiral arms Lanes of interstellar gas, dust, and young stars that wind outward in a plane from the central regions of a galaxy. (p. 460)

spiral galaxy A flattened, rotating galaxy with pinwheel-like spiral arms winding outward from the galaxy's nucleus. (pp. 478–479)

spontaneous symmetry breaking A process by which certain congruities in the mathematics of particle physics are altered to produce new particles and forces. (p. 565)

spring tide An ocean tide that occurs at new moon and full moon phases. (pp. 169–170)

standard candle An astronomical object of known intrinsic brightness that can be used to determine extragalactic distances. (pp. 483–484)

star A self-luminous sphere of gas.

starburst galaxy A galaxy that is experiencing an exceptionally high rate of star formation. (p. 491)

starquake A sudden shift in the crust of a neutron star. (p. 428)

Stefan–Boltzmann law A relationship between the temperature of a blackbody and the rate at which it radiates energy. (pp. 86–87)

Stefan's law *See* Stefan–Boltzmann law.

stellar association A loose grouping of young stars. (p. 374)

stellar evolution The changes in size, luminosity, temperature, and so forth that occur as a star ages. (p. 369)

stellar model The result of theoretical calculations that give details of physical conditions inside a star. (p. 325)

stony iron meteorite A meteorite composed of both stone and iron. (p. 298)

stony meteorite A meteorite composed of stone. (pp. 297–298)

stratosphere A layer in the atmosphere of a planet directly above the troposphere. (p. 154)

strewnfield An area over which fragments of a meteorite are scattered.

strong nuclear force The force that binds protons and neutrons together in nuclei. (p. 564)

subduction zone A location where colliding tectonic plates cause the Earth's crust to be pulled down into the mantle. (p. 152)

subdwarf A star of lower luminosity than main sequence stars of the same spectral type.

subgiant A star whose luminosity is between that of main sequence stars and normal giants of the same spectral type. (p. 354)

subtend   To extend over an angle. (p. 7)

summer solstice   The point on the ecliptic where the Sun is farthest north of the celestial equator. (p. 24)

Sun   The star about which the Earth and other planets revolve. (p. 310ff)

Sun-grazing comet   A comet that passes quite near the Sun. (p. 305)

sunspot   A temporary cool region in the solar photosphere. (pp. 316–317)

sunspot cycle   The semiregular 11-year period with which the number of sunspots fluctuates. (pp. 317–318)

sunspot maximum/minimum   That time during the sunspot cycle when the number of sunspots is highest/lowest. (pp. 317–318)

supercluster   A collection of clusters of galaxies. (pp. 486, 560)

superconductivity   The phenomenon whereby a flowing electric current does not experience any electrical resistance. (p. 429)

superfluidity   The phenomenon whereby a fluid flows without experiencing any viscosity. (p. 429)

supergiant   A very large, extremely luminous star; stars of luminosity class I. (pp. 350, 352)

supergrand unified theory   A complete description of all forces and particles, as well as the structure of space and time. (p. 565)

supergranule   A large convective feature in the solar atmosphere, usually outlined by spicules. (p. 314)

supergravity   A supergrand unified theory that uses higher dimensions to describe the four forces. (p. 568)

superior conjunction   The configuration of a planet being behind the Sun as viewed from the Earth. (p. 58)

superior planet   A planet that is more distant from the Sun than the Earth is. (p. 59)

superluminal motion   Motion that appears to involve speeds greater than the speed of light. (pp. 515–516)

supermassive black hole   A black hole with a mass in the range of a million to a billion solar masses. (pp. 471, 517–521)

supernova   A stellar outburst during which a star suddenly increases its brightness roughly a millionfold. (pp. 5, 379, 410–415)

supernova explosion   The detonation of a supernova. (pp. 133, 300, 379)

supernova remnant   The gases ejected by a supernova. (p. 379, 415–416)

surface gravity   The weight of a unit mass at the surface of an object such as a planet.

symmetry   Any situation in which certain properties of particles or forces are equivalent. (p. 557)

symmetry breaking   A process by which certain congruities in the mathematics of particle physics are altered to produce new particles and forces. (p. 558, 565)

synchronous rotation   The rotation of a body with a period equal to its orbital period. (pp. 163, 183)

synchrotron radiation   The radiation emitted by charged particles moving through a magnetic field; nonthermal radiation. (pp. 237, 428, 509)

synodic month   The period of revolution of the Moon with respect to the Sun; the length of one cycle of lunar phases. (p. 39)

synodic period   The interval between successive occurrences of the same configuration of a planet. (p. 58)

T Tauri stars   Young variable stars associated with interstellar matter that show erratic changes in luminosity. (p. 371)

T Tauri wind   A flow of particles away from a T Tauri star. (p. 141)

tail (of a comet)   Gas and dust particles from a comet's nucleus that have been swept away from the comet's head by the radiation pressure of sunlight and the solar wind. (p. 300)

tangential velocity   That portion of an object's velocity perpendicular to the line of sight. (p. 339)

tektites   Rounded glassy objects believed to have a meteoritic origin.

telescope   An instrument for viewing remote objects. (*see* Chapter 6)

temperature (Celsius)   Temperature measured on a scale where water freezes at $0°$ and boils at $100°$. (p. 85)

temperature (color)   The temperature of a star determined by comparing the intensity of starlight in two wavelength bands.

temperature (effective)   The temperature of a blackbody that would radiate the same total amount of energy that a particular star does.

temperature (excitation)   The temperature of a star determined from the strengths of various spectral lines that originate in atoms with different stages of excitation.

temperature (Fahrenheit)   Temperature measured on a scale where water freezes at $32°$ and boils at $212°$. (p. 85)

temperature (Kelvin)   Absolute temperature measured in units (kelvin, abbreviated "K") equivalent to the degree Celsius. (p. 85)

temperature (kinetic)   Temperature directly related to

the average speed of atoms or molecules in a substance. (p. 134)

terminator The line dividing day and night on the surface of the Moon or a planet; the line of sunset or sunrise. (p. 187)

terra Cratered lunar highlands. (p. 165)

terrestrial planet Mercury, Venus, Earth, and Mars; the classification sometimes also includes the Galilean satellites and Pluto. (p. 125)

theory A hypothesis that has withstood experimental or observational tests. (p. 3)

theory of everything (TOE) A supergrand unified theory that completely describes all particles and forces as well as the structure of space and time. (p. 565)

thermal energy The energy associated with heat stemming from the motions of atoms or molecules in a substance. (p. 134)

thermal equilibrium A balance between the input and outflow of heat in a system. (p. 323, 558)

thermal pulses Brief bursts in energy output from the helium-burning shell of an aging low-mass star. (p. 404)

thermal radiation The radiation naturally emitted by any object that is not at absolute zero. (pp. 237, 509)

thermodynamics The branch of physics dealing with heat and the transfer of heat between bodies.

thermonuclear fusion The combining of nuclei under conditions of high temperature in a process that releases substantial energy. (p. 323)

thermonuclear reaction A reaction resulting from the high-speed collision of nuclear particles that are moving rapidly because they are at a high temperature. (p. 323)

thermosphere A region in the Earth's atmosphere between the mesosphere and the exosphere. (p. 154)

third quarter moon The phase of the Moon that occurs when the Moon is $90°$ west of the Sun; last quarter moon. (p. 37)

threshold temperature The temperature above which photons spontaneously produce particles and antiparticles of a particular type. (p. 558–559)

tidal force A gravitational force whose strength and/or direction varies over a body and thus tends to deform the body. (pp. 169, 259)

time zone A region on the Earth where, by agreement, all clocks have the same time. (p. 30)

total eclipse A solar eclipse during which the Sun is completely hidden by the Moon, or a lunar eclipse during which the Moon is completely immersed in the Earth's umbra. (pp. 43–44)

total lunar eclipse A lunar eclipse during which the Moon is completely immersed in the Earth's umbra. (pp. 43–44)

total solar eclipse A solar eclipse during which the Sun is completely hidden by the Moon. (p. 44–45)

transit The passage of a celestial body across the meridian; the passage of a small object in front of a larger object. (p. 29, 183)

transmission grating A diffraction grating that produces a spectrum when light is shone through it.

triple alpha process A sequence of two thermonuclear reactions in which three helium nuclei combine to form one carbon nucleus. (p. 388)

triple point The pressure and temperature at which a substance can exist simultaneously as a solid, a liquid, and a gas.

Trojan asteroid One of several asteroids that share Jupiter's orbit about the Sun. (p. 293)

Tropic of Cancer A circle of latitude $23\frac{1}{2}°$ north of the Earth's equator. (p. 27)

Tropic of Capricorn A circle of latitude $23\frac{1}{2}°$ south of the Earth's equator. (p. 27)

tropical year The period of revolution of the Earth about the Sun with respect to the vernal equinox. (p. 32)

troposphere The lowest level in the Earth's atmosphere. (p. 153)

Tully–Fisher relation A correlation between the width of the 21-cm line and the absolute magnitude of a spiral galaxy. (p. 485)

turnoff point The point on an H–R diagram where the stars in a cluster are leaving the main sequence. (p. 391)

21-cm radio emission Radio radiation emitted by neutral hydrogen atoms in interstellar space. (p. 460)

Type I Cepheid A metal-rich Cepheid variable.

Type I Seyfert galaxy A Seyfert galaxy whose Balmer lines are significantly broader than any of its other emission lines.

Type II Cepheid A metal-poor Cepheid variable.

Type II Seyfert galaxy A Seyfert galaxy whose Balmer lines have about the same width as its other emission lines.

UBV filters Three colored filters that are transparent to ultraviolet (U), blue (B), and visible (V) light. (p. 345)

UBV system A system of stellar magnitude involving measurements of starlight intensity in the ultraviolet, blue, and visible spectral regions.

ultraviolet radiation Electromagnetic radiation of wavelengths shorter than those of visible light but longer than those of X rays. (p. 84)

umbra The central, completely dark portion of a shadow. (p. 43)

umbra (of a sunspot) The dark, central region of a sunspot. (pp. 316–317)

unbounded universe A universe throughout which the average density is less than the critical density. (p. 536)

universal constant of gravitation (G) The constant of proportionality in Newton's law of gravitation. (p. 67)

universal law of gravitation A formula derived by Isaac Newton that expresses the strength of the force of gravity that two masses exert on each other. (p. 67)

universal time Local mean time at the prime meridian.

universe All space, along with all the matter and radiation in space.

Van Allen belts Two doughnut-shaped regions around the Earth where many charged particles (protons and electrons) are trapped by the Earth's magnetic field. (pp. 155–156)

variable star A star whose luminosity varies.

velocity The speed and direction with which an object moves. (p. 66)

vernal equinox The point on the ecliptic where the Sun crosses the celestial equator from south to north. (p. 24)

very-long-baseline interferometry (VLBI) A method of connecting widely separated radio telescopes to make observations of very high resolution. (p. 115)

virtual pair A particle and antiparticle that exist for such a brief interval that they cannot be observed. (pp. 544, 557)

visible light Photons detectable by the human eye. (p. 83)

visual binary A binary star in which the two components can be resolved through a telescope. (p. 355)

void A large volume of space, typically 100 to 400 million light-years in diameter, that contains very few galaxies. (p. 489)

volatile element An element with low melting and boiling points. (p. 173)

waning crescent moon The phase of the Moon that occurs between third quarter and new moon. (p. 37)

waning gibbous moon The phase of the Moon that occurs between full moon and third quarter. (p. 37)

watt (W) A unit of power; 1 joule per second. (p. 86)

wave–particle duality The quantum mechanical notion that particles have wavelike properties and radiation has particlelike properties. (p. 531)

wavelength The distance between two successive wave crests. (p. 83)

waxing crescent moon The phase of the Moon that occurs between new moon and first quarter. (p. 37)

waxing gibbous moon The phase of the Moon that occurs between first quarter and full moon. (p. 37)

weak nuclear force The short-range force that is responsible for transforming one particle into another, such as the decay of a neutron into a proton. (p. 564)

weight The force with which an object presses downward due to the action of gravity. (p. 67)

white dwarf A low-mass star that has exhausted all its thermonuclear fuel and contracted to a size roughly equal to the size of the Earth. (pp. 352–353, 405–407)

white hole A black hole from which matter and radiation emerge. (p. 543)

white oval A round, whitish feature usually seen in Jupiter's southern hemisphere. (p. 233)

Widmanstätten patterns Crystalline structure seen in certain types of meteorites. (pp. 298–299)

Wien's law A relationship between the temperature of a blackbody and the wavelength at which it emits the greatest intensity of radiation. (p. 87)

WIMP A hypothetical massive particle proposed to explain the low neutrino flux from the Sun. (p. 328)

winter solstice The point on the ecliptic where the Sun reaches its greatest distance south of the celestial equator. (p. 25)

Wolf–Rayet star A class of very hot stars that eject shells of gas at high velocity. (p. 393)

wormhole A speculative, topological feature of a black hole that connects our universe with another universe. (pp. 450–451)

X rays Electromagnetic radiation whose wavelength is between that of ultraviolet light and gamma rays. (p. 83)

year (yr) The period of revolution of the Earth about the Sun. (p. 32)

ZAMS Zero-age main sequence. (p. 389)

Zeeman effect A splitting or broadening of spectral lines due to a magnetic field. (pp. 318, 320–321)

zenith The point on the celestial sphere opposite to the direction of gravity. (p. 25)

zero curvature A surface or space in which parallel lines remain parallel and the sum of the angles of a triangle is exactly $180°$. (p. 539)

zero-age main sequence The main sequence of young stars that have just begun to burn hydrogen at their cores. (p. 389)

zodiac A band of 12 constellations around the sky centered on the ecliptic. (pp. 22, 26)

zone A light-colored band in Jupiter's atmosphere. (p. 230)

ANSWERS TO SELECTED EXERCISES

Chapter 1
3. 3600″ 9. (a) 10^7 (b) 4×10^5 (c) 6×10^{-2}
(d) 1.7×10^{10} 11. 3.8 arc sec 12. (a) 8.19×10^{13} km
(b) 8.64 years 13. 8.92×10^{56} hydrogen atoms 14. 3350
km 15. (a) 1.43 m (b) 85.9 m (c) 5.16 km 16. 8.3 minutes 17. 5500 kg/m³, which is 5.5 times the density of
water 18. The observable universe is about 3×10^{36} times
larger than a hydrogen atom. 20. 8700 km
21. 29 million Suns 22. 4.3×10^9 km 23. 4 km
24. 4.7×10^{17} s

Chapter 2
18. 56 minutes earlier, at 9:04 PM 19. (a) Both sidereal
and synodic days would decrease. (b) Both sidereal and
synodic days would decrease. (c) The synodic day would
be shorter than the sidereal day. 20. 18:00 21. On
the date of the autumnal equinox (September 22)
27. $0^{hr}\, 00^{min}\, 00^{sec}$ or $12^{hr}\, 00^{min}\, 00^{sec}$ 28. $26\frac{1}{2}°$ above the
southern horizon

Chapter 3
2. (a) waning crescent (b) full moon (c) waxing crescent
(d) at sunrise (e) at noon (f) at sunrise 3. (a) new moon
(b) first quarter (c) full moon (d) third quarter 7. One
20. (a) July 22, 2009, in southeast Asia (b) August 12,
2045 21. 49 arc sec

Chapter 4
7. 5.2 AU² in 1991; 26.0 AU² in five years 8. Average distance is 100 AU; maximum distance is 200 AU 9. 2.8
years 12. 2×10^{20} N, which is about 180 times weaker
than the force between the Sun and the Earth 13. It would
be reduced by a factor of 100. 18. 0.24 year 19. It is
possible only for a superior planet. Mars is close to having
its sidereal period equal its synodic period. 20. (a) 42,365
km 21. (a) 6.11 km/s (b) 3.68 km/s 22. The star is 16
times more massive than the Sun. 23. (a) 2.38 km/s
(b) 59.5 km/s (c) 617 km/s 24. You would weigh one-
quarter of your Earth weight. 25. You would weigh 38%
of your Earth weight.

Chapter 5
1. $7\frac{1}{2}$ trips around the world 5. 3.64×10^{14} Hz
6. 2.91 m 8. 238 nm 9. 6520 K 10. 10 μm 18. 5800 K
= 10,000°F 19. 9.35 μm, which is infrared radiation
20. 10,000 K 21. 2.4×10^{-12} m, which is in the γ-ray region of the electromagnetic spectrum 23. 1.57 times that
from the Sun's surface 24. 250 nm; 16 times brighter than
the Sun 25. The star would be $6\frac{1}{4}$ times brighter than the
Sun. 26. −13.0 km/s 27. 21 km/s

Chapter 6
15. Only 1/25 of the incoming light is obstructed. 16. The
Palomar telescope has a light gathering power one million
times that of the human eye. 18. The smallest visible features on Jupiter's moons are about 300 km across. Only
those features on our Moon larger than about 110 km
across can be distinguished with the unaided human eye.
19. (a) 222× (b) 100× (c) 36× 20. No

Chapter 7:
17. 8.2×10^{20} J, which is about 10 million times more energy than the bomb that destroyed Hiroshima 18. 12
km/s 19. The Sun's escape velocity is 617 km/s. 20. 2.74
km/s from Ganymede, and 15.4 km/s from Jupiter at the
distance of Ganymede's orbit 22. 6.43×10^{23} kg and 3900
kg/m³ 23. −40°C = −40°F and $574\frac{1}{4}$°F = $574\frac{1}{4}$ K 24. The
Earth's mass would have to be about 50 times greater that
it is today. 25. Assuming an average density of 1500
kg/m³, the planet's diameter would be about $1\frac{1}{2}$ times the
Earth's diameter.

Chapter 8
17. 2.2×10^8 years 19. 17% for the core, 82% for the
mantle, and 1% for the crust 20. The density of the core
is about 15,300 kg/m³. 21 At 55 km, the atmospheric
pressure is about 1 millibar.

Chapter 9
16. The Moon cannot have a significant iron core because,
if it did, its average density would be higher than the density of typical crustal rock. 17. 28 pounds on the Moon;
176 pounds on the Earth 18. 1.5×10^5 kg/yr, which
would double the Moon's mass in about 5×10^{17} years
19. 552,000 km

Chapter 10
14. Features larger than about 660 km will be resolved.
16. 4.1 μm 18. 10^{-5} nm 19. μ > 31 20. 684 N = 176
pounds on Earth; 298 N = 69 pounds on Mercury; 130 N
= 29 pounds on the Moon 21. 0.61 AU

Chapter 11
14. 3.9 μm 15. 6.64×10^{-6} nm 19. 0.93 km

Chapter 12
15. 480 km for a 1 arc sec resolution; 48 km for the HST
16. Yes 17. 6.41×10^{23} kg

Chapter 13
14. 7.43×10^5 km from the Sun's center, which is nearly
50,000 km above the solar surface 15. 1.90×10^{27} kg
16. 381 pounds 17. Nearly 600 km/hr

Chapter 14
13. Assuming 1 arc sec seeing, the smallest visible features
are about 3000 km across. 14. 0.05 arc sec 15. About
60 km 16. 310 billion years 17. About 8.1 minutes

Chapter 15
12. Pan 13. 300 km 14. Yes; Tethys, Dione, Rhea,
Titan, and Iapetus (Mimas and Enceladus are quite near
the limit of HST's resolving power). 15. 2.6 km/s; μ > 12
17. 16.7 km/s at the outer edge of the A ring; 20.3 km/s at
the inner edge of the B ring; the maximum wavelength shift
is 0.068 nm

Chapter 16
12. About 7000 km 13. The HST can resolve the Great
Dark Spot on Neptune and shows Pluto as a resolved
disk. 14. 2 hours 6 minutes 15. Neptune exerts about

1.6×10^{-4} as much gravitational force on Uranus as does the Sun.   **18.** The ratio of the diameter of Jupiter to the diameter of Neptune is 2.9 to 1. The ratio of the size of the Great Red Spot to the size of Great Dark Spot is 2.7 to 1.

Chapter 17
14. 678 years; 0.53° per year   **15.** 50 km   **18. (a)** $P = 354$ yr **(b)** $P = 11,200$ yr **(c)** $P = 3.54 \times 10^5$ yr **(d)** $P = 1.12 \times 10^7$ yr   **19.** The diameter of the iron-rich core of a typical parent asteroid is about $\frac{1}{3}$ of the asteroid's overall diameter.

Chapter 18
21. (a) 1.8×10^{-9} J **(b)** 9.0×10^{16} J **(c)** 5.4×10^{41} J
22. (a) 4.6×10^{-36} s **(b)** 2.3×10^{-10} s **(c)** 4.4×10^7 yr
23. A mass of 1.73×10^{11} kg is completely converted into energy each second, which means that 2.5×10^{13} kg of hydrogen is burned each second.   **24.** 4.9%; chemical composition will be 69% hydrogen and 30% helium   **25.** photosphere: 500 nm = visible light; chromosphere: 58 nm = ultraviolet light; corona: 1.93 nm = 'X rays   **26.** The energy flux from a sunspot is about 30% of that from the undisturbed photosphere.

Chapter 19
3. When viewed from Jupiter, the Sun is about 3.7% as it is from the Earth.   **4. (a)** 4.85 pc **(b)** 0.2 arc sec   **5.** 2100 pc; about $\frac{1}{3}c$ **6.** 55 km/s   **15.** 10 $M_\odot$; 0.001 $L_\odot$ **16.** 17th magnitude   **17.** 4.31 pc   **18.** 81 km/s   **19. (a)** 101 km/s **(b)** 486.29 nm   **20.** 37 AU   **22.** 16 kpc   **23.** Procyon's absolute magnitude is +2.64 and its luminosity is 7.6 $L_\odot$.
24. 0.35 pc, or about one-fifth of its present distance from Earth   **26.** 5 magnitudes   **27.** After the outburst, its diameter is twice as large as before and its luminosity has increased by a factor of 64.   **28.** 1.48 $M_\odot$ **29.** $\frac{3}{8}M_\odot$ and $\frac{1}{8}M_\odot$ **30.** Rigel's diameter is about 28 times larger than the Sun's.   **31.** temperature = 4700 K; luminosity = 45.7 $L_\odot$; radius = 10.3 $R_\odot$ **32.** 9.9 pc

Chapter 20
14. 3.36×10^{10} atoms/m³, which is roughly 100 times larger than the densities listed in Box 20-1   **15.** 6.3%
16. Approximately 1000 times the present diameter of the Sun   **18.** 1.3×10^{31} m³

Chapter 21
4. (a) 3 million years, or only about 3×10^{-4} as long as the Sun **(b)** 140 million years, or about 1.4×10^{-2} as long as

the Sun   **15.** 12%   **16.** 1.66 $M_\odot$ **17.** 7.3×10^8 yr
19. 620 pc   **20.** About 100 pc

Chapter 22
6. 15,200 km   **14.** Its radius is 0.106 $R_\odot$.   **16.** Nearly 9000 years ago   **17.** 1.83×10^9 kg/m³; 6450 km/s   **19.** A magnitude of −12, which is about as bright as the full moon   **20.** 6.3 Mpc

Chapter 23
13. 5466 BC   **16.** The maximum correction is 10^{-4} of the pulsar's period.   **17.** density of neutron = 4.1×10^{17} kg/m³; density of neutron star = 1.5×10^{17} kg/m³   **19.** The expansion rate is 0.32 arc sec/yr. A change of 1 arc sec requires a wait of about 3.1 years.   **20.** 2.2 kpc; 1300 years ago

Chapter 24
2. 2.8 m   **12.** 1.8×10^{17} kg/m³   **13.** 1.4×10^8 $M_\odot$
14. (a) 9 mm; 2×10^{20} kg/m³ **(b)** 3 km; 2×10^{19} kg/m³
(c) 2200 AU; 2 g/m³   **15.** 0.98 c **16.** 0.99995 c
17. 0.87 c **18.** 2.8 $M_\odot$

Chapter 25
8. (a) 300 million years **(b)** 7.4×10^{11} $M_\odot$ **9.** 25 times
13. A 10% error in radius results in a 10% error in mass. A 10% error in velocity results in a 20% error in mass.
15. 3.9×10^{11} $M_\odot$ **16.** 18,000 ly   **17.** Once every 90,000 years

Chapter 26
14. 54 km/s/Mpc   **15.** 6.3 Mpc   **16.** 8.2×10^{12} $M_\odot$
17. 470 million light-years, if H_0 is 75 km/s/Mpc   **18.** Orbital period is about 440 million years; mass interior to 20 kpc is about 3.6×10^{11} $M_\odot$

Chapter 27
12. 13.1   **13.** 0.94 c **14.** 0.96 c **15.** 1.4×10^8 $M_\odot$
16. 2×10^{30} kg/m³   **17.** 20 AU   **21.** 3×10^{11} $L_\odot$
22. 0.32 c

Chapter 28
11. 20 billion years, 13 billion years, 10 billion years
12. 1.6×10^8 km/s/Mpc   **13.** 12.6 K   **14.** 1.9×10^{-26} kg/m³   **15.** 2.2×10^{-42} m   **18.** 2.1×10^{70} yr
19. 4.1×10^{22} kg   **20.** 1.35×10^{-43} s   **21.** 2×10^{11} kg

Chapter 29
12. 1.0×10^{-26} J   **13.** 3.5×10^{-25} s   **14.** 5.9×10^9 K
16. Between 4.2×10^{-35} kg and 1.8×10^{-34} kg

ILLUSTRATION CREDITS

Smithsonian Center for Astrophysics; **Fig. 26-22:** S. J. Maddox, W. J. Sutherland, G. P. Efstathiou, and J. Loveday, Oxford Astrophysics; **Fig. 26-23:** V. de Lapparent, M. Geller, and J. Huchra, Harvard–Smithsonian Astrophysical Observatory; **Fig. 26-24:** C. Jones-Forman, Harvard–Smithsonian Center for Astrophysics; **Fig. 26-26:** Ultra-deep CCD image taken by Gary Bernstein and Tony Tyson using the 4-m telescope on Kitt Peak, NOAO; **Fig. 26-27:** Lick Observatory photograph; **Fig. 26-28:** (a) Copyright © 1960 National Geographic Society–Palomar Sky Survey, reproduced by permission of the California Institute of Technology, (b) M. Yun and P. Ho, Harvard–Smithsonian Center for Astrophysics; **Fig. 26-29:** William C. Keel, University of Alabama; **Fig. 26-30:** Joshua Barnes, Canadian Institute for Theoretical Astrophysics; **Fig. 26-31:** Palomar Observatory; **Fig. 26-32:** Lars Hernquist, Institute for Advanced Study, with simulations performed at the Pittsburgh Supercomputing Center; **Figs. 26-33, 26-36:** NASA, ESA; **p. 500:** Copyright © 1960 National Geographic Society–Palomar Sky Survey, reproduced by permission of the California Institute of Technology.

Essay p. 502: Josh Grimes.

Chapter 27 p. 504: Courtesy of *Astronomy* Magazine, Kalmbach Publishing Co.; **Figs. 27-1, 27-2:** Palomar Observatory; **Fig. 27-3:** National Optical Astronomy Observatory; **Fig. 27-5:** Copyright © 1960 National Geographic Society–Palomar Sky Survey, reproduced by permission of the California Institute of Technology; **Fig. 27-8:** Alan Stockton, Institute for Astronomy, University of Hawaii; **Fig. 27-9:** Photograph by David F. Malin, © Anglo-Australian Telescope Board; **Fig. 27-10:** (a) National Optical Astronomy Observatories, (b) J. E. Grindlay, Harvard–Smithsonian Center for Astrophysics; **Fig. 27-11:** (a) Photograph by David F. Malin, © Anglo-Australian Telescope Board, (b) NASA, ESA; **Fig. 27-12:** National Radio Astronomy Observatory, operated by Associated Universities, Inc., under contract with the National Science Foundation (VLA observations by P. A. Scheuer, R. A. Laing, R. A. Perley); **Fig. 27-13:** National Radio Astronomy Observatory, operated by Associated Universities, Inc., under contract with the National

Science Foundation (VLA observations by C. P. O'Dea, F. N. Owen); **Fig. 27-14:** (a) National Optical Astronomy Observatories, (b) David A. Chartee, University of Illinois, and Jack O. Burns, New Mexico State University, and also the National Radio Astronomy Observatory, operated by Associated Universities, Inc., under contract with the National Science Foundation (VLA observations by Jack O. Burns and others); (c) E. J. Schreier, Harvard–Smithsonian Center for Astrophysics; **Fig. 27-15:** T. D. Kinman, Kitt Peak National Observatory, NOAO; **Fig. 27-19:** National Optical Astronomy Observatories; **Fig. 27-21:** (a) Palomar Observatory, copyright © California Institute of Technology, (b) NASA, ESA; **Fig. 27-22:** European Southern Observatory; **Fig. 27-23:** John F. Hawley and Larry L. Smarr; **Fig. 27-26:** NASA, ESA; **p. 524:** (upper left) National Radio Astronomy Observatory, operated by Associated Universities, Inc., under contract with the National Science Foundation (VLA observations by F. N. Owen, C. P. O'Dea, M. Inoue), (lower right) Copyright © 1960 National Geographic Society–Palomar Sky Survey, reproduced by permission of the California Institute of Technology.

Chapter 28 p. 526: Goddard Space Flight Center, NASA; **Fig. 28-3:** Bell Labs; **Fig. 28-4:** John Mather, NASA: **Fig. 28-6:** NASA.

Essay p. 548: Ramono Cagnoni/Black Star; **pp. 549, 550:** Photographs by David F. Malin, © Anglo-Australian Telescope Board.

Chapter 29 p. 553: Lawrence Berkeley Laboratory; **Fig. 29-5:** John P. Huchra and Margaret J. Geller, Smithsonian Astrophysical Observatory; **Fig. 29-6:** Goddard Space Flight Center, NASA; **Fig. 29-8:** NASA; **Fig. 29-9:** James M. Gelb and Edmund Bertschinger, MIT; **Fig. 29-13:** E. L. Turner, J. E. Moody, and J. R. Gott.

Afterword p. 572: Jon Lomberg; **Fig. A-1:** From the collection of Ronald A. Oriti; **Fig. A-4, A-6:** NASA; **Fig. A-5:** Jet Propulsion Laboratory.

Appendixes pp. 579, 580: Stephen P. Meszaros, NASA.

INDEX

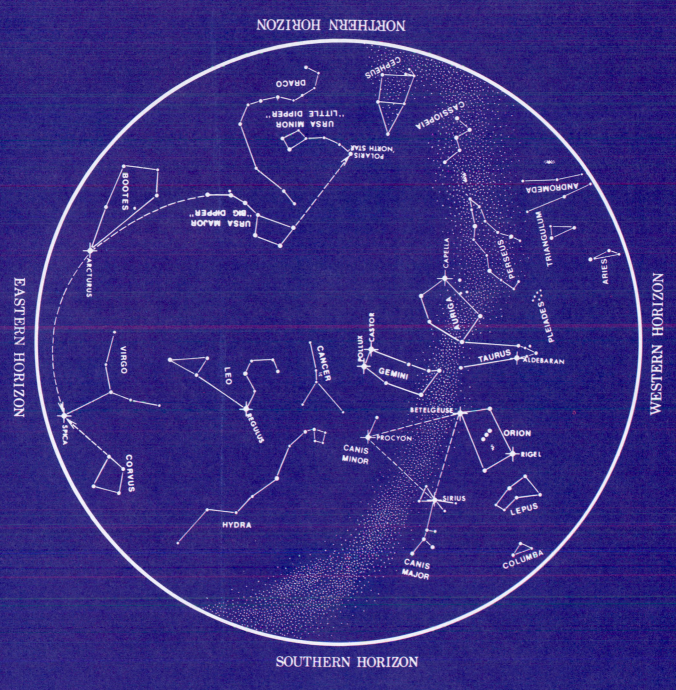

THE NIGHT SKY IN MARCH

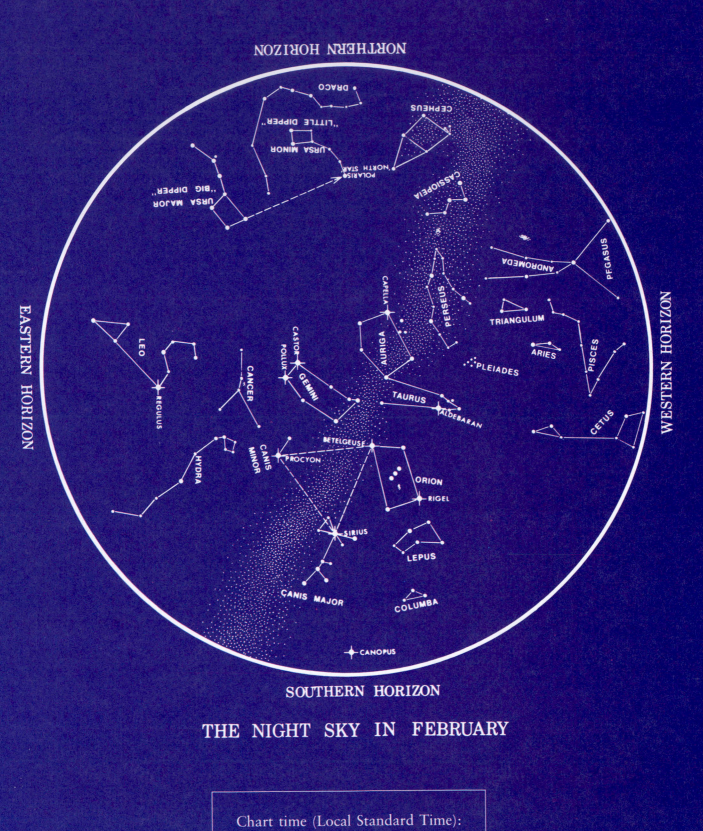

THE NIGHT SKY IN FEBRUARY

Chart time (Local Standard Time):

10 pm...First of February
9 pm...Middle of February
8 pm...Last of February

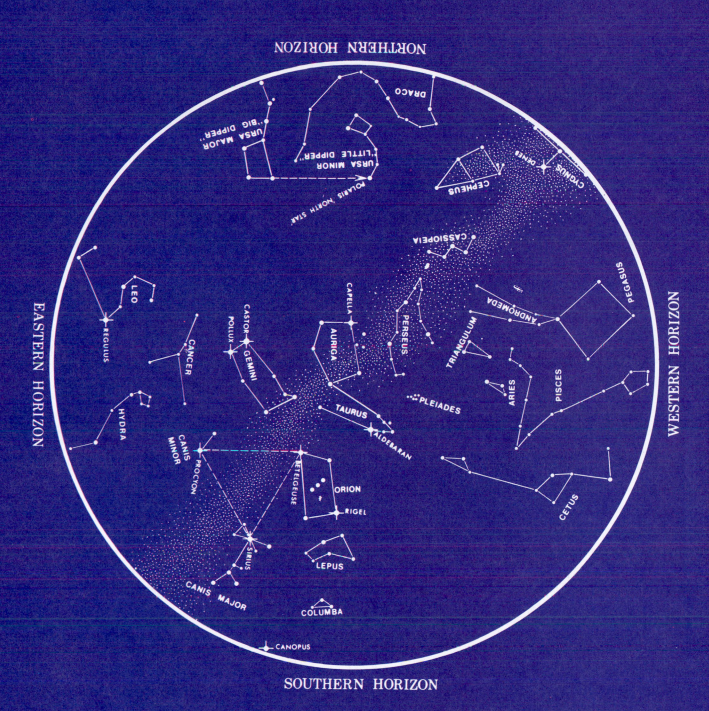

THE NIGHT SKY IN JANUARY

Chart time (Local Standard Time):

10 pm...First of January
9 pm...Middle of January
8 pm...Last of January

STAR CHARTS

The following set of star charts, one for each month of the year, is from *Griffith Observer* magazine. To use these charts, first select the chart that best corresponds to the date and time of your observations. Hold the chart vertically and turn it so that the direction you are facing shows at the bottom.

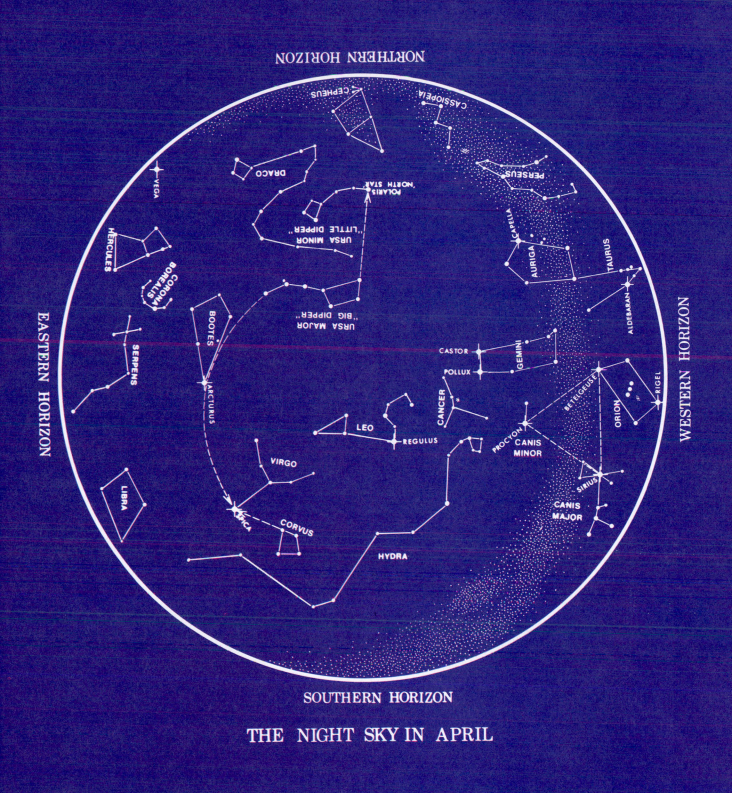

THE NIGHT SKY IN APRIL

Chart time (Daylight Savings Time):

11 pm...First of April
10 pm...Middle of April
9 pm...Last of April

THE NIGHT SKY IN MAY

Chart time (Daylight Savings Time):

11 pm...First of May
10 pm...Middle of May
9 pm...Last of May

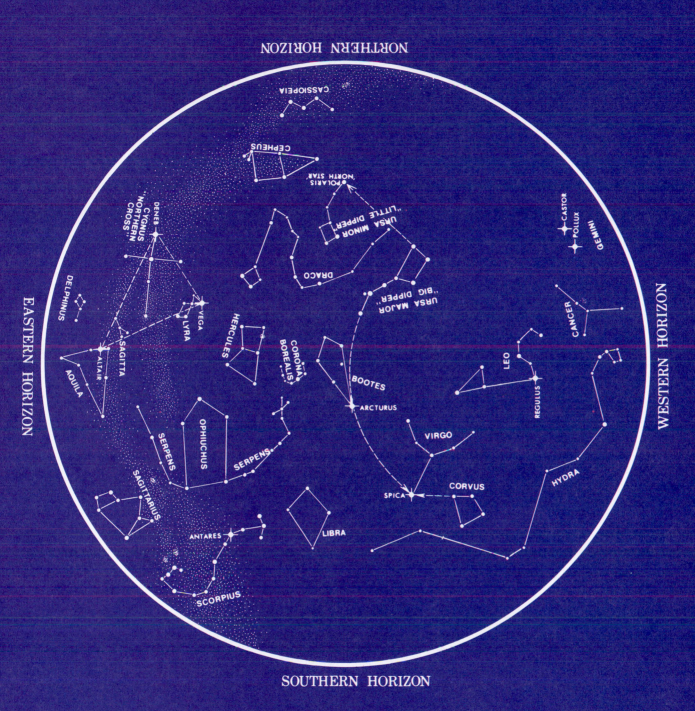

THE NIGHT SKY IN JUNE

Chart time (Daylight Savings Time):

11 pm...First of June
10 pm...Middle of June
9 pm...Last of June

THE NIGHT SKY IN JULY

SOUTHERN HORIZON

Chart time (Daylight Savings Time):

11 pm...First of July
10 pm...Middle of July
9 pm...Last of July

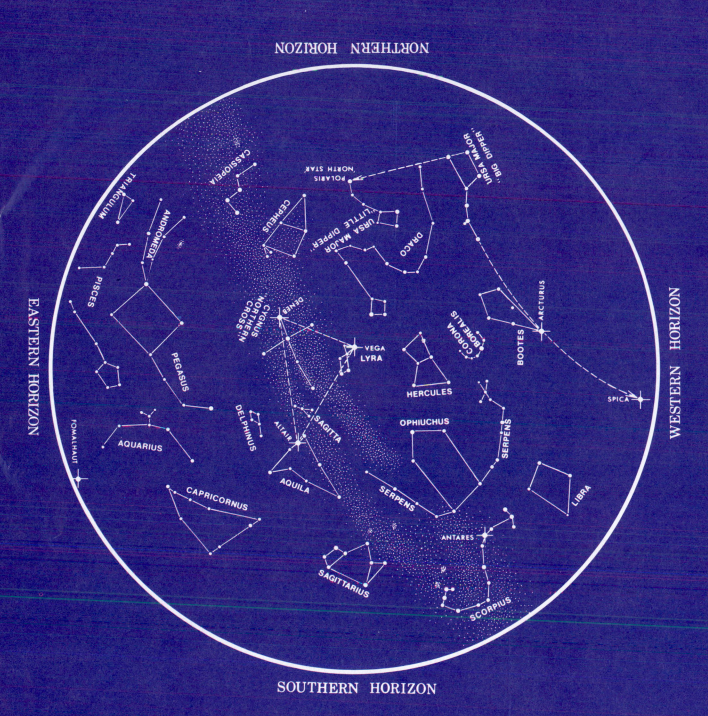

SOUTHERN HORIZON

THE NIGHT SKY IN AUGUST

Chart time (Daylight Savings Time):

11 pm...First of August

10 pm...Middle of August

9 pm...Last of August

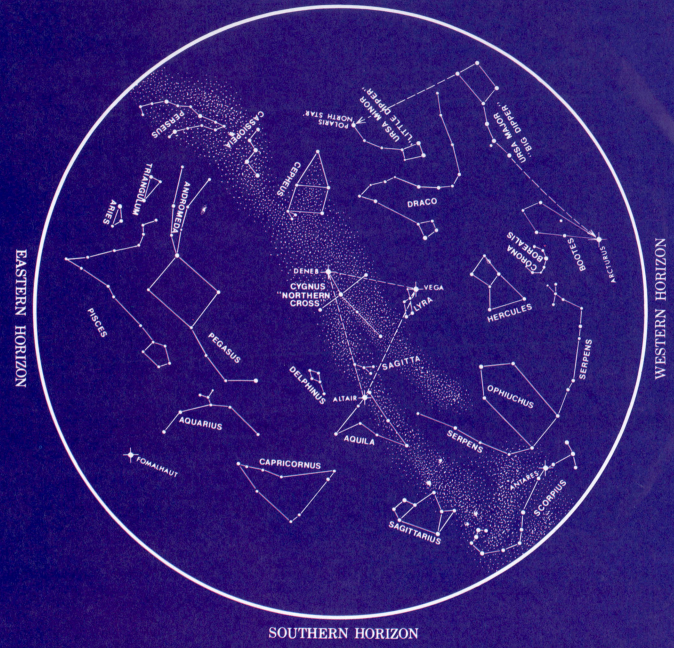

THE NIGHT SKY IN SEPTEMBER

Chart time (Daylight Savings Time):

11 pm...First of September
10 pm...Middle of September
9 pm...Last of September

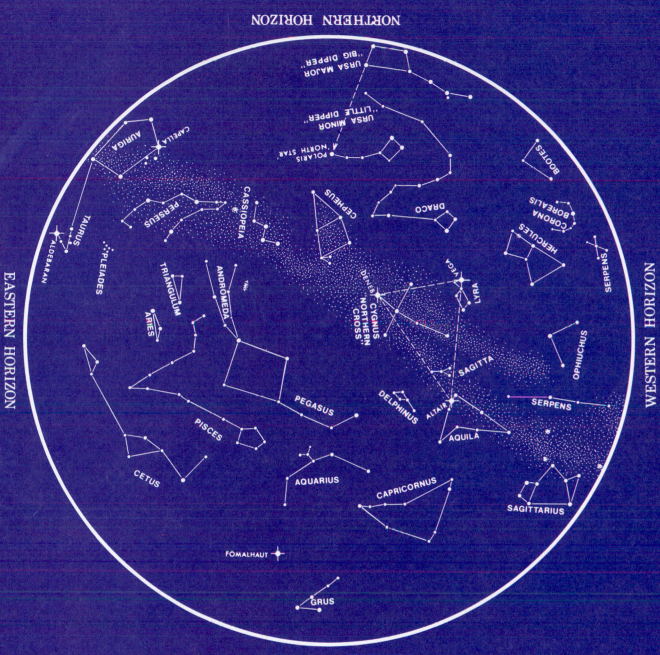

THE NIGHT SKY IN OCTOBER

SOUTHERN HORIZON

THE NIGHT SKY IN NOVEMBER

Chart time (Local Standard Time):

10 pm...First of November
9 pm...Middle of November
8 pm...Last of November

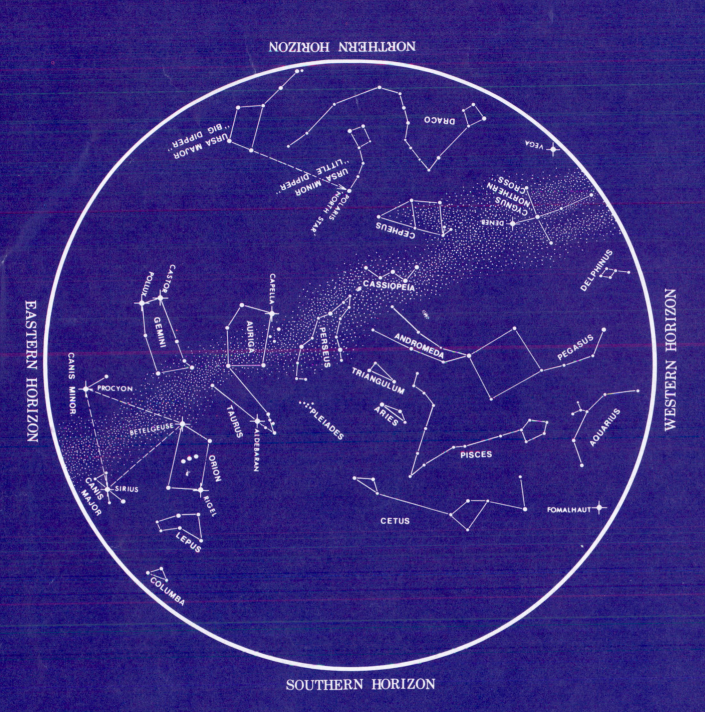

THE NIGHT SKY IN DECEMBER

Chart time (Local Standard Time):

10 pm...First of December
9 pm...Middle of December
8 pm...Last of December